Comparative Biomechanics

Illustrated by Annette deFerrari

STEVEN VOGEL

Comparative Biomechanics

LIFE'S PHYSICAL WORLD

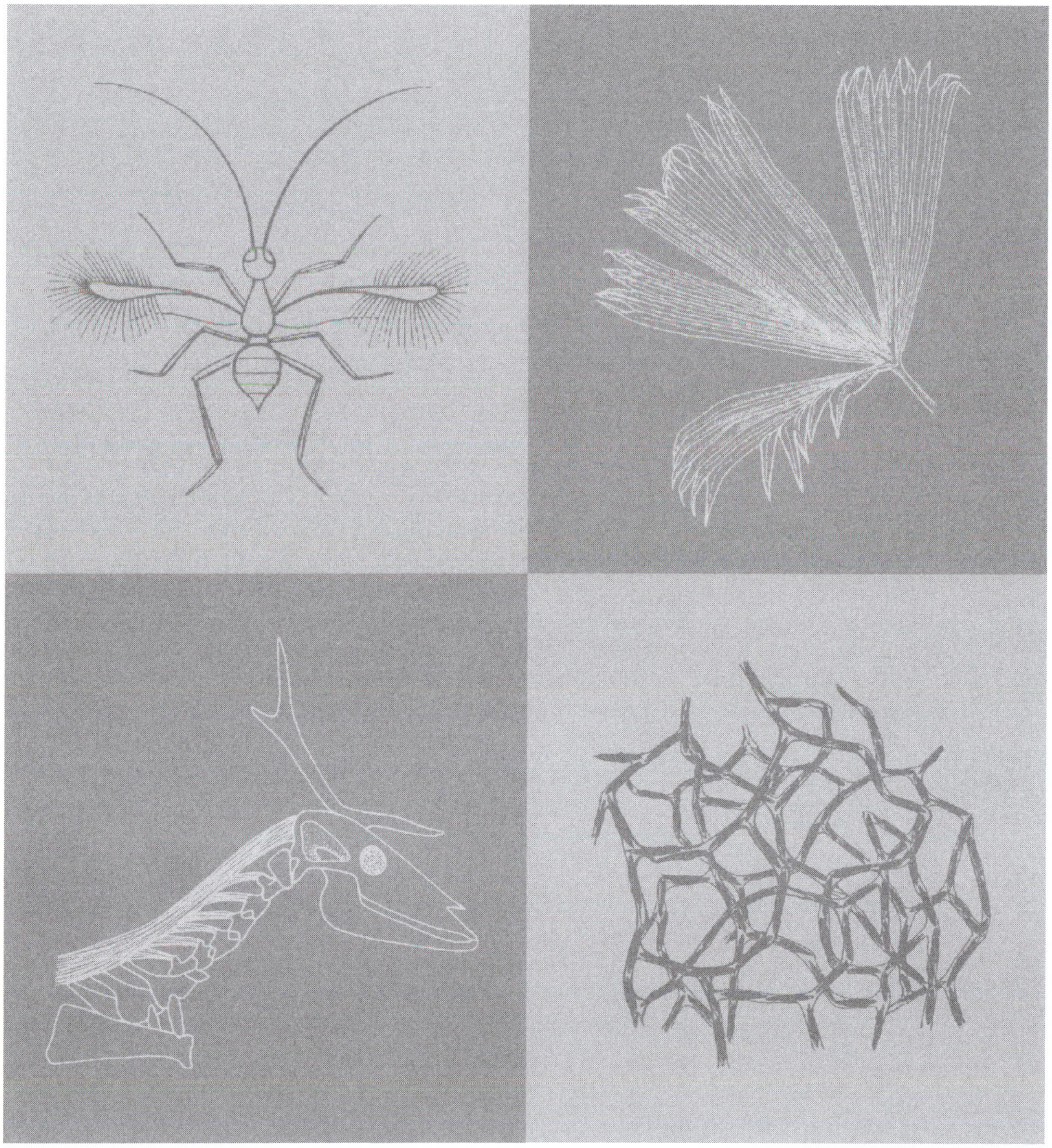

PRINCETON UNIVERSITY PRESS
Princeton and Oxford

Published by Princeton University Press, 41 William Street, Princeton,
New Jersey 08540

In the United Kingdom: Princeton University Press, 3 Market Place, Woodstock,
Oxfordshire OX20 1SY

Library of Congress Cataloging-in-Publication Data

Vogel, Steven, 1940–
Comparative biomechanics : life's physical world / Steven Vogel.
p. cm.
Includes bibliographical references and index.
ISBN 0-691-11297-5 (alk. paper)
1. Biomechanics. 2. Physiology, Comparative. I. Title.

QH513 .V643 2003
571.4′3—dc21 2002042723

British Library Cataloging-in-Publication Data is available

This book has been composed in Sabon

Printed on acid-free paper.∞

www.pupress.princeton.edu

Printed in the United States of America

10 9 8 7 6 5 4 3 2 1

I intend this book as partial repayment to colleagues and students, present and past, who have both supported my professional life and made that life a rich source of personal pleasure

CONTENTS

PREFACE

COMPARATIVE BIOMECHANICS, MEANING A BROADLY BIOLOGICAL RATHER than a narrowly human subject, deals with the cheap, physical stuff—to rehabilitate a phrase from my youth. It worries about why trees so rarely fall over, why legged creatures switch from walking to running at particular speeds, how slugs slither, how flies fly, how samaras spin. In short, it starts with the ordinary activities of ordinary organisms, posing the questions a person might ask while exploring a coastline or tramping through a forest.

What draws the particular questions of comparative biomechanics together is the kind of explanations to which those questions yield. Initially, at least, they draw on a physical and macroscopic world, a world closer to the purview of mechanical engineers than that of most contemporary biologists. The bad news is the unfamiliarity of its context. But that's a minor snag next to its wonderfully commonplace character. Comparative biomechanics invokes notions close to one's everyday experience and intuitive sense of reality. Gravity and elasticity have an immediacy that cells and molecules do not.

While that physical world may be immediate and attractive, we have to face up to the fact that explaining and quantifying it don't draw on well-used items in the biologist's intellectual toolbox. The present book tries to lead the reader into that world as gently as the author knows how; it tries both to make the case that biomechanics matters and that the subject need not be approached with fear, trepidation, and memories of the difficulties of first-year physics. It tries to demystify biomechanics, making it something to bear in mind, something to pursue more specifically when it proves relevant to a particular question, whatever the origin or context of the question. The field shouldn't be the private domain of initiates, isolated and insulated within its jargon and journals.

In particular, the book is intended to be used as a textbook for undergraduates in biology at any level beyond an introductory course—or undergraduates in any other area of science or engineering willing to refer from time to time to a basic textbook of biology, or graduate students or other scientists and engineers who need an entry point that gets them into the content and literature of biomechanics. It grows out of the disparity between a book I wrote about 15 years earlier, *Life's Devices*, and the needs of the undergraduate biology and engineering students I now teach. That earlier book (from which substantial chunks have been appropriated) aims at nonscientists, using the very commonplace character of biomechanical questions as a way of getting them to think in scientific terms. It presumes little—I tiptoed around logarithms and even the simplest calculus—but it doesn't take the subject far enough. More troubling, its contents are just too topical and idiosyncratic for a proper introductory textbook. In a sense, it's all appetizer and dessert rather than proper entrée. Still, the world liked *Life's Device*s well enough, and it remains in print and (judging by my correspondence) remains useful.

When I decided to have a go at a proper textbook, I sent e-mailings (e-missions or e-missives or e-missiles?) to all the people I knew who might be teaching comparative biomechanics to undergraduates. The overwhelmingly positive response came as no surprise—what had anyone else to lose? What mattered more were all the useful suggestions, lists of references, and offers to test-run any preliminary version. A little worrisome, though, were the course outlines. With no standard book or long tradition, courses have become divergent and eclectic; and I envision a bunch of people waiting for this book, each with the expectation that it will follow a specific course description sent in a couple of years earlier.

And that brings up the main problem here. Breaking with some canon is one thing, inadvertently creating a canon in a field blessedly unconstrained by one is quite another. So I must emphasize, more as warning than apology, that the book is almost as idiosyncratic as its predecessor. In particular, the diversity of intended users has led to considerable peculiarity in both content and organization.

Content. Biomechanics covers an awesome diversity of things. The sheer mass of the book (and it has more than 230,000 words) may strike the reader as some assurance of comprehensiveness (never mind comprehensibility). Do not be misled—major areas of perhaps equivalent relevance have been left out. Indeed, I've been driven nearly nuts with the continuous decision making about what to put in and what to leave out. My first rule has been to begin at the beginning, focusing on material that forms the background needed to go further into the subject. My second rule has been to pick material that tells useful stories, cases with general explanatory value. And my third rule has been to focus on topics that could be presented without presuming much background in the particulars of other areas of biology—and the various fields of biology have no end of particulars. And, good intentions aside, one tends to use material with which one is familiar, biasing coverage toward ones own work and that of people with whom one has had association.

As a consequence the relative emphasis on different subjects doesn't reflect the relative attention they've received by investigators. It has also led to an underrepresentation of "hard" subjects such as unsteady flows and viscoelasticity and of the work of some of the most impressive contemporary workers. As a further consequence, subjects whose understanding depends on a lot of morphological information have been given short shrift—however impressed I remain with what functional morphologists have done in recent years. Locomotion gets the lion's share of current effort; it's not an equivalent presence here. Concomitantly, animals get more attention than plants; I've no brief for equal time, but sap and wood matter at least as much as blood and bone. Paleobiomechanics gets ever more interesting, but I felt myself too far from the subject to have any decent perspective. Computational fluid mechanics has begun to make major contributions in such areas as circulatory systems and flight; it's also not represented here. And so on.

Organization. We're dealing with a subject of no obviously sequential character. So I've mostly made it up myself, and I present it with no great conviction of its superiority, with a sense of decisions forced on me, many of which were ultimately arbitrary. I have, though, tried to facilitate the use of the book by courses with different sequences, at least as far as I could with naturally interdependent topics. Part I serves a role that's both necessary and initiatory—although engineers may find most of it laughably primitive. By contrast, parts II and III can be done with either coming first or, for short courses, omitting one or the other. Part IV forms a kind of coda that can be used entirely, in part, or augmented, or else ignored altogether. On a finer level, certain chapters can be omitted without laying traps that will catch people further along—in particular, chapters 10, 13, 14, 17, 21, 24, and, of course, 25. Gaining versatility has cost some redundancy, whether enough, too much, or too little I can't yet say. Once again, I'm painfully conscious of the possibility that I'm defining a canonical course in comparative biomechanics, and I want to impose as few constraints as I can.

Level. The course I give prerequires basic college physics and calculus, which I can afford since we require no less for our biology majors. But the book presupposes little prior exposure to either Newtonian physics or calculus, proving what's minimally needed in chapters 2–4. It does assume familiarity with algebra and with exponents and logarithms.

A few other prefatory notes . . .

- The choice of symbols may strike the nonbiomechanic as inefficient or worse. I've tried to follow the conventions used in more specialized books, especially the ones by engineeers. The awkward consequence is a marriage of various fields, traditions, and histories, in particular because solid and fluid mechanics have been pursued by different people.

- The use of text references to the primary literature may be a bit unusual for books intended for undergraduates, but much of the biomechanical literature ought to be accessible to the better students. In addition, I want to encourage at least recognition of the existence and role of the literature. Similarly, the personal anecdotes intend to point out the immediate relevance of the material to everyday life. They (and the quotes) also provide some equation-offsetting levity.

- The presentation takes a more didactic and expostulatory tone than do most undergraduate textbooks. That and its opinionated character should not be viewed as accidental or matters of editorial negotiation. I carry a lot of scars from reviewing scientific papers that make basic and avoidable mistakes.

- Nonbiologists might find it useful to buy an old (thus cheap) edition of an introductory biology textbook—all have copious indices and will, even for terminological matters, be more useful than a biological dictionary. Alternatively, instructors can accumulate a pile of loaners from the recipients of publisher's samples. My long experience indicates that returns hit one hundred percent when you suggest that return of the loaner is prerequisite for getting a grade.

- A collection of teaching resources may be obtained for the asking, via svogel@duke.edu. These include suggestions for classroom demonstrations, small out-of-class projects, homework problem sets, useful Web sites, and so forth—plus an on-going list of errors that have turned up in the text. Contributions to the collection will be welcomed.

- Physicians take an oath by which they promise at least to do no harm; textbook writers should take an equivalent oath at least to tell no lies. For some subjects, that's no simple matter.

The number of people who have helped me in this endeavor goes beyond what I can conveniently list here. Students, other faculty members, other writers, friends began contributing long before the book became a specific task. And I kept no proper list. I note, with appreciation and good memories, that the book was sketched out while I was a guest at the Raman Research Institute in Bangalore, India—I'm grateful in particular to my host, V. Radhakrishnan ("Rad"). Several chapters were written at the Friday Harbor Laboratory of the University of Washington, where I enjoyed residence at the Whiteley Center—I must especially thank Arthur Whiteley and Dennis Willows. I wrote in a room called "Arthurs' Study"—the position of the apostrophe alludes to Arthurs Martin and Whiteley, both of whom, as it happens, appear in the present bibliography.

The people at Princeton University Press have been as good as ever, and the project (and sometimes the author) has been pressed ever forward by Sam Elworthy, Sarah Harrington, Dimitri Karetnikov, and Linny Schenck. Barbara Clauson checked the prose and my impulse to test repeatedly (biomechantics, of course) the fit of foot into mouth. I had more than what I hope is the usual amount of trouble with computers and my rotten apples were repeatedly resuscitated by John Spady, David Gourley, Susan Gerbeth Jones, and Susan Gillispie.

While the new drawings here are the products of Annette deFerrari, I've reused ones done by Sally Schrohenloher, Rosemary Calvert, Susan Tanner Beety, and Kate Davis. People who do illustrations for me somehow feel compelled to leave town thereafter. All the illustrations have been drawn specifically for this or my previous books, although a few represent only slight modifications of specific originals. Where single published sources have been used, acknowledgment of permission has been made in the legends. In addition, I thank David R. Maddison for permission to use the drawing from which figure 6.12 was derived, and Richard B. Emlet for use of a photograph upon which figure 19.3 was based.

PART ONE

Life's Physical Context

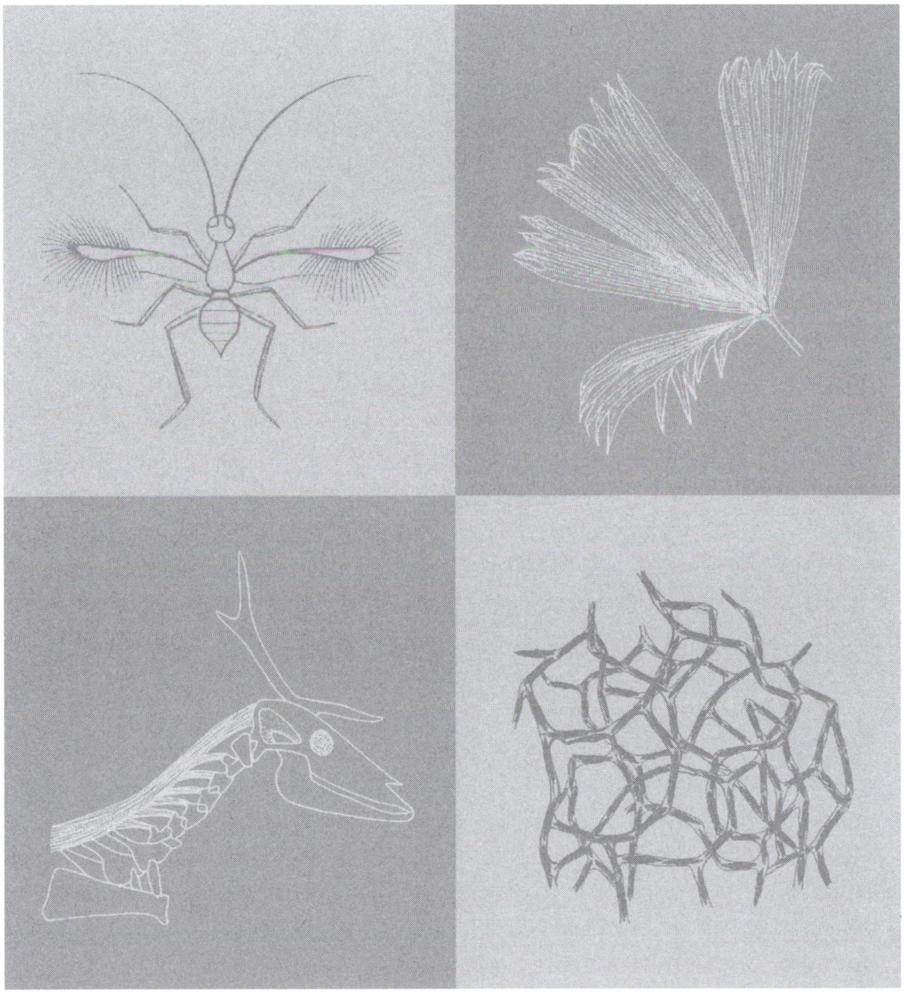

In which we take a quick trip through material that forms the context of biomechanics—physical, mathematical, and biological. After setting the stage with a glance at a few specific biomechanical phenomena, we introduce the relevant physical variables and talk about how they can be manipulated. And we look at the basic issues raised by the extreme variability of both sizes and shapes of organisms.

CHAPTER

Preambulations

> Physics-envy is the curse of biology.
>
> **JOEL COHEN,** **biologist**

WHERE TO START? PERHAPS THE WORLD OF BIOMECHANICS CAN BEST BE introduced by looking at several specific situations, matters part of—or at least close to—your everyday experience. If "mechanics" raises the specter of some formidable physics course, set your mind at rest. No area of contemporary, grown-up science takes as its subject anything closer to home. And no area gives greater explanatory yield with as little investment. So we begin not just with questions posed but with questions answered—at least as part of our open-ended work in progress—questions that might occur to any ordinary person.

How Walking Hits a Speed Limit

Consider an especially literal "preamble." Think for a moment about how you move about on your hindlegs. Ordinarily you walk, swinging your legs alternately back and forth. Pressed a bit, you break into a trot, an easy jog; pressed harder, you run. But note what you've long known but may have never thought about. You don't make a continuous transition between a walking gait and any trotting or running gait. No—the shift is abrupt, and at no point have you the least doubt about which you're doing.

Questions. What's the basic difference between walking and running, and why do you make this abrupt shift? At what speed do you make the shift? Do other legged creatures shift gaits at the same speed or do they at least shift at a speed determined by the same rule we feel compelled to obey?

Walking versus running. Formal recognition of the most obvious distinction between the two goes back to Weber and Weber (1825). When walking, each leg contacts the ground for more than half of each stride, so you always have at least one leg on the ground. When running, each leg stays in contact less than half the time, so you're fully airborne for a time, twice during each full stride. True enough, but it doesn't lead us very far. Less obviously, you go up and down as you move forward. When walking, your body is highest in midstride, when one or the other leg extends almost vertically. By contrast, when running, your body is slightly higher near the extremes of a stride, with one leg extended forward and the other backward. As first recognized by R. McNeill Alexander, this difference in the body's trajectory says more about why we make the transition; his extensive investigations, starting in the late 1970s, have left little doubt about at least the general picture of what's going on.

The underlying problem emerges from a basic feature of animal propulsive systems. Almost all the appendages with which animals push backward on air, water, or the earth's surface reciprocate rather than rotate. Going back and forth (or up and down) means repeatedly starting and stopping legs, fins, wings, and such. And that means accelerating and decelerating masses, which takes force and ultimately energy, force and energy beyond whatever gets usefully invested in propulsion. In short, locomotion with reciprocating appendages suffers a disability that wheeled vehicles never run up against. Consider—you can dramatically reduce the total energy you need to cover a given distance by riding a bicycle, despite the fact that you're now hauling around 20 or 30 percent more mass.

While nature has never achieved a large-scale solution to the problem (as human technology does with rotating propellers and wheels), she has repeatedly come up with at least a partial fix, a way to reduce that accelerative-decelerative cost. The trick consists of storing the braking energy of deceleration for reuse in the subsequent acceleration. Put another way, instead of doing work on the system to stop an appendage's motion, the system does what can be described as "negative work"—it absorbs work for subsequent use.

But how to store the energy? In walking, gravity provides the battery, just as it does when we store energy as a mass of water behind a high dam. You work like a pendulum, which alternately trades off gravitational energy against kinetic energy. In the familiar version of figure 1.1a, gravitational energy is greatest at the high extremes of the swing; kinetic energy is greatest during the fast midswing. A pendulum banks the energy of motion as height and then withdraws it to reaccelerate. But the walking pendulum uses a less

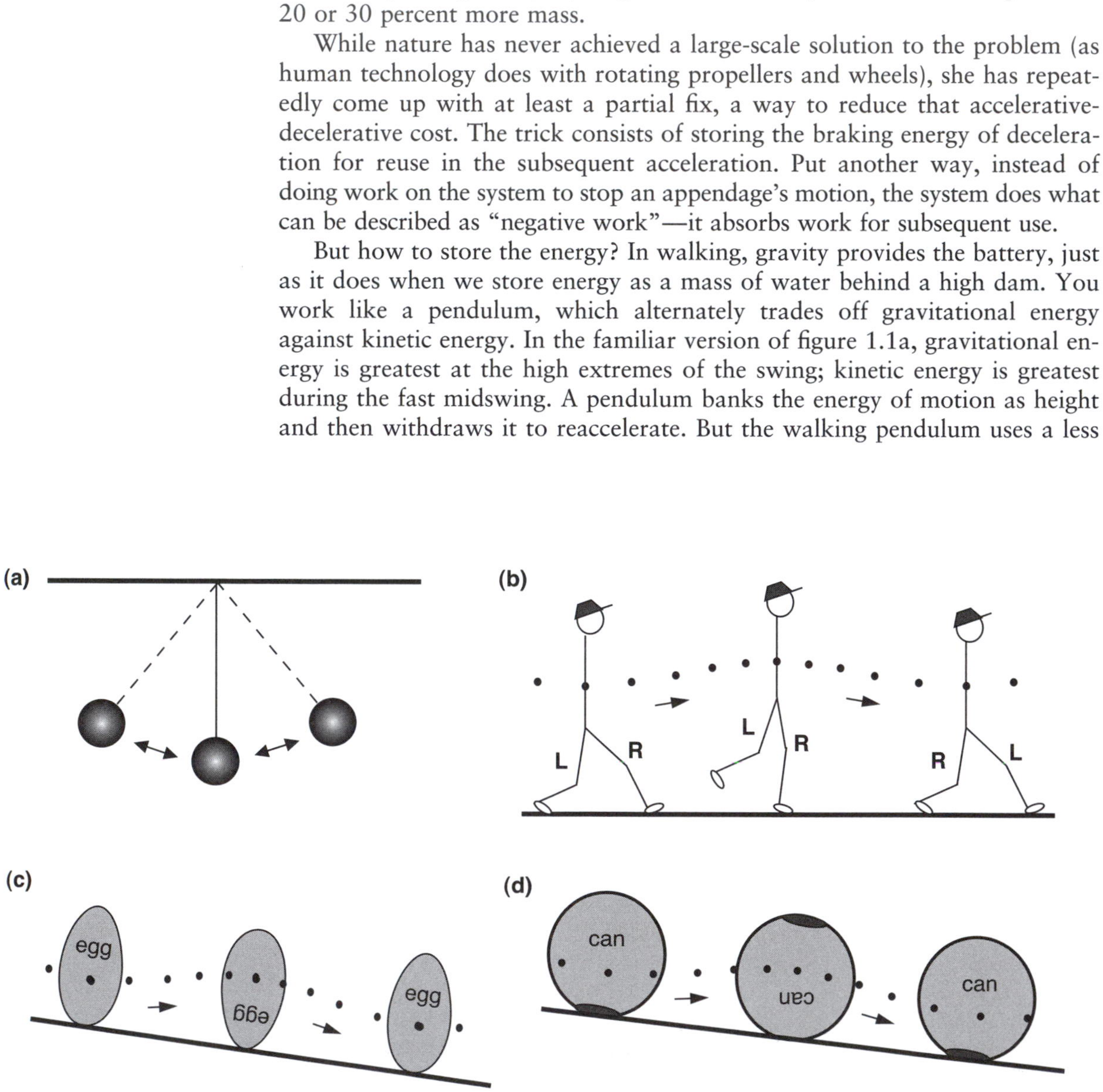

FIGURE 1.1 (a) A conventional pendulum, (b) the inverted pendulum model for walking, (c) an egg rolling down an incline, and (d) a cylindrical can with an off-center weight rolling similarly. In (b), (c), and (d), the centers of gravity are shown as dark spots—their spacings (assume equal time between adjacent spots) give their velocities.

self-evident version. When walking, your body (really your center of gravity) is lowest when your legs are furthest fore and aft, and it's highest in midstride when one or the other leg touches the ground directly beneath your torso. Concomitantly, your center of gravity moves most rapidly at the extremes of the stride and least rapidly at midstride—so gravitational and kinetic energies indeed interchange, pendulum-fashion, as in figure 1.1b.

An egg, rolling end-over-end down an incline (figure 1.1c) provides a partial model. Or, avoiding omelet, tape a weight to the inner side wall of an empty soup can and give it enough of a push so it rolls for a few turns, as in figure 1.1d. It will roll least rapidly when the weight is uppermost—when its center of gravity is highest—and fastest when the weight is at the bottom—when its center of gravity is lowest. So gravitational and kinetic energies rise and fall with opposite time courses, even if these models involve no sharp pendulum-like tuning. In short, we make use of the pendulum principle but, between motion of torso and jointed appendages, not in any simple manner.

Gravity, used pendulum-fashion, has its limits. In particular, it limits the frequency range over which a leg can be swung and a body raised and lowered, much the way a particular pendulum swings at a particular frequency. Try it—to go faster, you swing your legs farther rather than more often. But increasing the amplitude of leg-swinging can't be pushed too far, any more than a practical pendulum can swing through more than a limited arc. What Alexander (see Alexander and Jayes 1983; Alexander 1988) recognized was that this limit sets the maximum speed of a walking gait. To go faster and still store energy from stride to stride, you have to do something different—running gaits store energy elastically, mainly as stretched tendons, rather than gravitationally.

Identifying the nature of the shift leads directly to a prediction about how the shift ought to vary with the size of an animal. One can get a properly quantitative rule in several ways, all more-or-less equivalent in physical terms. All require some assumptions that sound hopelessly crude but are of a type that has been of great value for studies of how animals of different sizes are put together and how they carry out many of their activities. (The parent subject goes by the name of "scaling" and will be taken up in chapter 3). We assume, first, that legs (and bodies) resemble each other in proportions—that they're geometrically similar. Second, we assume that all legs get swung the same way—that strides are dynamically similar. Thus all can achieve about the same maximum swinging angle. Of course we've swept under the rug a staggering amount of biological variability, using the large size range of walking creatures to excuse our sin, and further admitting that we'll get only a crude rule.

Consider the textbook formula for the period of a pendulum:

$$t = 2\pi\sqrt{\frac{l}{g}}, \tag{1.1}$$

where t is the period, l is the pendulum's length, and g is the gravitational acceleration of a freely falling object. Assume that the top walking speed of an animal, v_{max}, will be proportional to its leg length divided by the swinging period of a pendulum of that length. Thus,

$$v_{\max} \propto \frac{l}{t} \propto \sqrt{lg}\,, \tag{1.2}$$

where $\propto$ is the symbol for "is proportional to"—it avoids confusing ourselves with numerical constants of little immediate relevance, and we'll use it often.

Does this work? An answer requires that we specify the length in the formula; the distance from hip joint to ground for a standing animal gets about as close to pendulum length as anything easy to measure and generally applicable. So we're asking whether any single constant of proportionality relates the maximum walking speeds of animals of all sizes to the lengths of their legs. We now have a great deal of data on the speeds at which animals switch from walking to some other gait, initially from the work of R. McNeill Alexander and C. Richard Taylor, each with various collaborators. And examination of the data indicates quite clearly that the formula does work. That body of data, together with the acceleration of gravity on earth (9.8 meters per second squared), produces a practical formula,

$$v_{\max} = 2.2\sqrt{l}, \tag{1.3}$$

with speed given in meters per second and length in meters. The constant can be trusted to about 10 percent each way, not bad when we're talking about animals whose leg lengths span several orders of magnitude. The number of legs per walker—two or four—makes no difference.

As ordinary animals, any one of us can try the formula. For instance, for a leg length of 1 meter (about my datum), maximum speed is 2.2 meters per second, or about 12.1 minutes per mile in the U.S. vernacular. One can walk faster, but the gait gets increasingly awkward and tiring as one does such tricks as rotating the lower back and swinging the hips to increase effective leg length and stride amplitude. The bigger person indeed makes the transition at a higher speed, and an adult walks alongside a trotting child or dog.

Application of the formula isn't limited to mice, elephants, and all extant animals in between. It tells us that a large dinosaur such as a brontosaurus could have walked very fast. Probably no adult brontosaurus did much running. Since walking imposes much lower stresses on the bones of an animal's legs than does running, the brontosaurus's large size might not have caused especially high leg loadings, and we needn't worry whether ordinary bone could support such a large creature. Alexander (1984) also showed that the footprints of some fossil hominids (the Laetoli footprints from Tanzania) were consistent with walking in a fashion similar to the gait of modern humans. But because of their shorter stature, these hominids would have gone more slowly than we do—instead of taking about 20 minutes to stroll a kilometer (the typical speed of a person in a small town), they would have required nearly 30.

We're clearly looking at a constraint that physics imposes on how we move, not at some mere accidental feature of a common vertebrate ancestor. And we have some answers to the questions posed earlier about the difference between walking and running and about the rule for the transition point. The story has a lot more bits and pieces, and we'll return to terrestrial locomotion on legs in chapter 24.

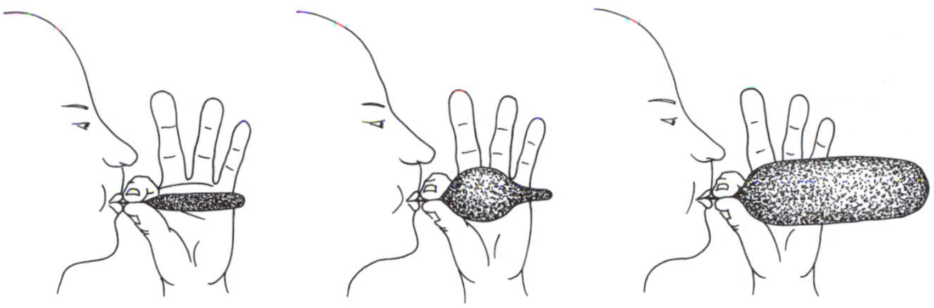

FIGURE 1.2 The way a cylindrical balloon inflates—an initial aneurysm doesn't quite burst but extends with little further expansion toward the balloon's ends.

How Arteries Inflate

Blow up a cylindrical balloon; one of any length will do. Notice the familiar—it doesn't inflate evenly. No, one part expands almost to the point of bursting, as in figure 1.2, and then the expansion extends lengthwise. We put up with this peculiar behavior in balloons, but it's not a nice thing at all when we're concerned with our personal, internal, pressurized cylinders—arteries, especially. Proper arteries need to be stretchy, like rubber balloons, if they're to reduce the pressure fluctuations of our pulsatile hearts. A ventricle, after all, produces no pressure output at all while refilling, and a rigid system of pipes would reflect that directly. But proper arteries need to be stretchy in a way quite distinct from the stretchiness of rubber balloons—physicians have long recognized a dangerous pathology, an aneurysm, when one part expands to near bursting before expansion elsewhere. Simple, rubberlike arteries simply won't do.

Now feel the surface of the balloon when partially inflated. Curiously, it feels as if it has a high-pressure region beneath the expansion and a lower pressure everywhere else. With no internal barriers, though, it's hard to imagine how the internal pressure can be anything but uniform throughout. So what you feel can't be place-to-place variation in the pressure difference across the balloon's walls. No, it's the degree to which the walls have been stretched by that constant pressure difference. Perhaps counterintuitively, the stretch ("tension") caused by a given pressure depends on the curvature of the cylinder's walls.

The rule turns out to be straightforward, dependent on simple geometry rather than on any arcane behavior of materials. Biologists usually refer to it as Laplace's law; we'll say more about it in chapters 3 and 20. Put explicitly, wall tension, T, equals transmural pressure, Δp, times radius of curvature, r:

$$T = \Delta pr. \tag{1.4}$$

Thus, a given transmural pressure difference generates more tension in the wall of a part of the balloon that has already begun to expand (greater radius of curvature) than in a part that has not. In ordinary rubber, extension is roughly proportional to stress applied, and so a given pressure extends the already expanded part more than the more flaccid unexpanded part. Only a slight extra resistance to stretch just before breaking makes cylindrical balloons practical at all.

In effect, Laplace's law rules out the use of ordinary elastic materials for arterial walls, requiring that an appropriate material fight back against stretch, not in direct proportion to how much it's stretched, but disproportionately as stretch increases. Which, again in obedience to the dictates of the real world, our arterial walls do—aneurysms, fortunately, remain rare and pathological. We accomplish the trick first, by incorporating fibers of a nonstretchy material, collagen, in those walls, and second, by arranging those fibers in a particular way. In an unstretched wall, the fibers are folded into kinks. Thus, as the wall expands outward, more and more of these inextensible fibers are stretched out to their full lengths and add their resistance to stretch to that of the wall as a whole—figure 1.3 gives the relevant histology in simplified form.

Arterial walls that resist stretch disproportionately as they extend characterize circulatory systems that have evolved within lineages quite distinct from our own—in cephalopods and arthropods, for instance. Recruitable collagen fibers don't represent the only possible solution to the basic problem, and they're not nature's inevitable choice. Laplace trumps Mendel—both microstructure and molecular composition differ far too much to support the idea of a common artery-endowed ancestor underlying the similarity in mechanical behavior. Yet the force-extension curves (technically stress-strain curves, as will be explained in chapter 4) match remarkably. At least they match when adjusted for the transmural pressure differences—blood pressures—that occur in the arteries of the different animals. And that permits an odd but useful prediction. If one has the force-extension curve for a sample of arterial wall, one can guess the arterial blood pressure of an animal. So from fresh samples of arterial wall, Shadwick (1994) estimated the blood pressure—about the same as our own, incidentally—of a giant squid, an animal that inevitably dies in the process of capture.

We'll return to circulatory systems in chapter 10.

How Weak Leaves Deal with Strong Winds

Next time a storm blows up, watch one or a few leaves on nearby trees; a pair of binoculars will help focus your attention and minimize the distraction of other storm-related activities. Where I live, maple, sweet gum, and tulip poplar leaves flutter and show their undersides with only modest wind; as the wind rises further the various oaks get into motion. Amid the swaying of branches one can easily miss noticing that few leaves flutter much when the wind rises to substantial intensity—any initial excitability dies down as if they're then taking the storm more seriously.

Leaves—trees, really—have a problem, the same one as our solar panels. Their function, trapping solar energy photosynthetically, demands exposure

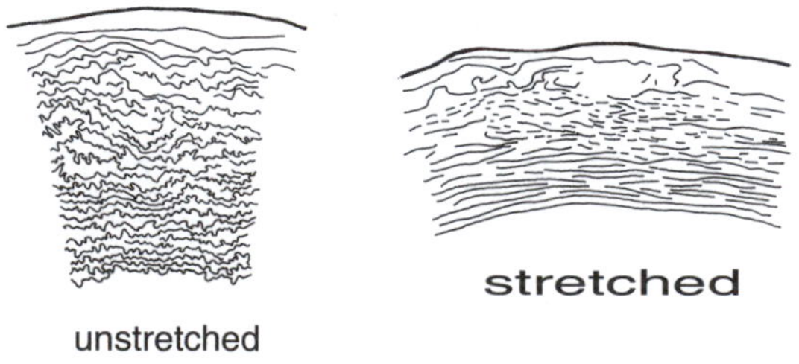

FIGURE 1.3 A bit of arterial wall, showing kinky fibers and the way they stretch out and take an ever-increasing fraction of the load.

of lots of area skyward. But biology compounds the problem. With adequate water and soil, more trees would like to expose more leaves to sunlight than their acreage allows. So trees compete, synthesizing great wooden columns that elevate their leaves above, ideally, those of neighboring trees. Few naturally isolated trees grow tall—competition drives trunks rather than any advantage from getting closer to the sun. Result? Large areas of surface well above ground level. Surface translates into drag; height translates into leverage; high wind then means real trouble, possibly breaking trunks or wrenching roots out of the ground. Figure 1.4 puts the matter diagrammatically.

What to do? Natural selection doesn't lend itself to any mutual trunk-reduction treaty. Streamlining (chapter 7) fits poorly with an area-maximization imperative. Building photosynthetic structures of sufficient rigidity that they stay parallel to the wind and don't flutter, the latter concomitant with high drag, takes lots of material and generates lots of weight. Nature does something relatively uncommon in human technology. She (by which pronoun we assume natural selection) arranges leaves and their attachments so they adjust their configurations and thus reduce their exposure and flutter as the wind increases. Motive force presents no problem, even for these nonmuscular structures, because the wind itself provides more than enough. Photosynthesis? Intermittently strong winds come mostly with reduced sunlight, so temporary reduction in exposure to sky can't entail a great long-term cost.

Here I refer to my own work (a privilege of writing the textbook oneself), in particular to a brief project (Vogel 1989) that took an initial (and thus far unpursued) look at what leaves do in winds that might put trees in (to use the political euphemism) harm's way. Figure 1.5a shows what one kind of leaf, that of a tulip poplar, does at a series of increasing wind speeds. This curling into a tightening cone characterizes quite a few kinds of leaves—

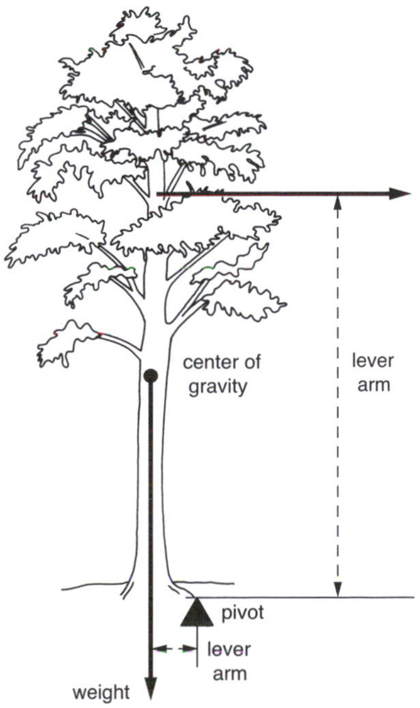

FIGURE 1.4 The problem of a tree in a wind—the sideways force from drag acts far above the tree's base, so its drag causes a large turning force. That must be resisted by the tree's weight, with the small lever arm given by the radius of its base. (Attachment to the ground also helps a little.) The advantage of minimizing drag should be obvious.

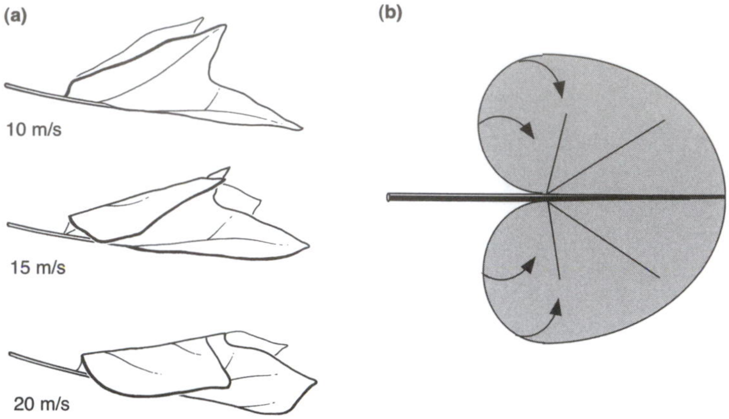

FIGURE 1.5 (a) The reconfiguration of a leaf of the tulip poplar tree (*Liriodendron tulipifera*) in an increasing wind. (b) The way the leaf's basal—hence upwind—lobes catch the wind and facilitate the process.

including maples, sweet gums, sycamores (plane trees), and redbuds among the ones abundant where I live. All are characterized by relatively long petioles (leaf stems) and lobes on their blades that protrude back toward their parent branches from the point of attachment of their petioles. What appears to happen (based on observation and crude models) is that the lobes, upwind because leaves always extend downwind like kites on strings, bend upward (abaxially, technically) and get the curling started, as in figure 1.5b.

This curling into cones dramatically decreases drag—at least if we pick the right item for a comparative baseline. Relative to surface area, the drag of a tulip poplar leaf at 20 meters per second (45 miles per hour) is only a third of that of a square flag that's free to flutter. Still, the drag remains several times higher than that of the gold standard, a rigid, flat plate oriented parallel to the wind. Another wind-dependent reconfiguration, one in which the leaflets of a pinnately compound leaf such as black walnut or black locust roll up around their axial rachis, does a bit better. Nor do these two exhaust the reconfigurational possibilities. The very stiff leaves on a branch of a holly (*Ilex americana*) swing inward toward the branch and lie, one against another, in a common sandwichlike pile. Pine needles cluster instead of being splayed outward. Yet other modes surely occur as well.

A common problem has called forth variations on a common mode of solution. Some, but not all, members of at least fifteen families of broad-leafed trees use the tulip poplar's conical reconfiguration, so it must have arisen on quite a number of separate occasions in the evolutionary history of these trees. The solution, again, is one we don't ordinarily encounter in the devices of our own nonbiological technology. It involves two aspects as common in nature's technology as they are rare in our own. First, shape isn't held constant, but rather shape and the forces of flow interact complexly, each dependent on the other. The local wind forces on a leaf depend (in part) on its shape; its shape, in turn, depends (in part) on the local wind forces. Second, variable circumstances are dealt with by altering functional priorities. Photosynthesis, overall, matters far more than drag minimization; in storms, though, such priorities must reverse.

But that distinction between what we do and what nature does pervades each of the three examples we've discussed. We usually put wheels on our vehicles and so needn't worry about pendulum-based cycle-to-cycle energy storage. We most often use rigid pipes (or at least ones of fixed maximum diameter) and do any buffering of pump pressure fluctuations elsewhere in our systems of plumbing. And we build stiff structures whose shapes vary little with the vagaries of environmental insults. Such comparison of human and natural technologies consumed an entire book I wrote a few years ago (Vogel 1998a). We'll return to the subject in chapter 25; until then human technology need be only a parenthetical presence. Still, it will be our constant companion. To be painfully honest, biomechanics has mainly been the study of how nature does what engineers have shown to be possible. Nature may have gotten there first, but human engineers, not biologists, have provided us with both analytical tools and practical examples.

Joining Physical and Biological Worlds

These stories about gait transitions, inflated arteries, and draggy leaves represent a concert of contexts—the properties of biological materials such as arterial walls, the fluid mechanics of laminar liquid flow through pipes and turbulent atmospheric wind, the evolutionary preference of organisms for doing things with minimal expenditure of energy, and the peculiar dependence of biological design and performance on the size of the organism. These matters, and most of what follows, address a single general issue, that of the ways organisms deal with their immediate physical world. Before going further, we ought to place that issue within biology. Biology, I mean to emphasize, is a particular and exceedingly peculiar domain.

Biology conveys two curiously contrasting messages. In a strictly genetic sense all organisms are unarguably of one family. Our numerous common features, especially at the molecular level, indicate at least a close cousinhood, a common descent from one or a few very similar ancestors. On the other hand, what a gloriously diverse family we are, so rich and varied in size and form! The extreme heterogeneity of life impresses us all—trained biologists or amateur naturalists—with the innovative potency of the evolutionary process. The squirrel cannot be mistaken for the tree it climbs, and neither much resembles its personal ménage of microorganisms. The apposition of this overwhelming diversity with the clear case for universal kinship tempts us to assume that nature can truly make anything—that, given sufficient time, all is possible though evolutionary innovation.

Some factors, though, are beyond adjustment by natural selection. Some organisms fly, others do not, but all face the same acceleration due to gravity at the surface of the earth. Some, but not many, can walk on the surface of a body of water; but all face the same value of that liquid's surface tension if they attempt the trick. No amount of practice will permit you to stand for long in any posture other than one in which your center of gravity, an abstract consequence of your form, is above your feet. If an object, whether sea horse or sawhorse, is enlarged but not changed in shape, the larger version will have less surface area relative to its volume than did the original. In short, there is an underlying world with which life must contend. Put perhaps more pretentiously, the rules of the physical sciences and the basic properties of practical materials provide powerful constraints on the range of designs available for living systems—a point put persuasively by R. McNeill Alexander (1986).

Were these restrictions the extent of the physical world's impact on life, we might be content to work out a set of limits—quantitative fences that mark the extent of the permissible perambulations of natural design. There is, however, a more positive side, at least from our point of view as observers, investigators, and rummagers for rules. The physics and mathematics relevant to the world of organisms are rich in phenomena and interrelationships that we find far from self-evident, and the materials on earth are themselves complex and diverse. Tiny cells with thin walls can withstand pressures that would produce a blowout in any vertebrate artery. Yet the materials of cellular and arterial walls have similar properties. The slime upon which a snail crawls may alternately be solid enough to push against and sufficiently liquid for a localized slide. An ant can lift many times its own weight with muscles not substantially different from our own. (But no Prometheus could exist among ants—as Went [1968] remarked, the minimum sustainable flame in our atmosphere is too large for an ant to come close enough to add fuel.) By capitalizing on such possibilities the evolutionary process appears to our unending fascination as a designer of the greatest subtlety and ingenuity.

This book is about such phenomena—the ways in which the world of organisms bumps up against a nonbiological reality. Its theme is that much of the design of organisms reflects the inescapable properties of the physical world in which life has evolved, a world that at once imposes constraints and affords opportunities. In one sense it will be a long essay defending that single argument against a vague opponent—the traditional disdain or disregard of physical science by biologists. In fact, the theme will function mainly as a compass in a walk through a miscellany of ideas, rules, and phenomena of both physical and biological origin. We'll consider, though, not the entire range of biologically relevant parts of physics, but a limited set of mostly mechanical and largely macroscopic matters, matters more commonly claimed by mechanical engineers than by physicists.

The macroscopic bias should be emphasized. In places, this book deals with some rather bizarre phenomena, but it never gets far from a kind of everyday reality—like shifting from a walk to a run. Explanations, where possible, will deliberately ignore the existence of atoms and molecules, waves and rays, and similar bits of deus ex machina. Not that the latter aren't as real as our grosser selves (or so implies some very strong evidence); rather, as the bases for explanations, they have an unavoidable air of ecclesiastical revelation. More importantly, it would take far more space and complexity than a single book for them to help a person take a more ordered view of the immediate world. After all, can you think of any part of your perceptual reality that demands the odd assumption that matter is ultimately particulate—that if you could slice cheese sufficiently thin it would no longer be cheese? Maybe the "invention" of atoms by Democritus was just a lucky guess, an accident of his inability to imagine anything infinitesimally small! Only when we consider diffusion and a few other phenomena will we need to recognize atoms and a real world in which matter cannot be subdivided ad infinitum.

We may make too much of the distinction between biological and physical science, between living and nonliving devices. It certainly isn't a practice sanctified by antiquity. Galileo, whom we regard as a physical scientist, figured out that jumping animals, from fleas on up, should jump to about the same maximum height irrespective of their body sizes (Haldane 1928). A key element in developing the idea of conservation of energy was established by a German physician, Mayer, in 1841, from observations on the oxidation of blood. The basic law for laminar flow of fluids in pipes was established about the same time by a French physician, Poiseuille, who was concerned with circulatory systems (Pappenheimer 1984).

Physics and biology, with separate histories for the past few centuries, have developed their necessarily specialized terminologies in different and virtually opposite ways. Biology goes in for horrendous words of classical derivation, from *Strongylocentrotus droebbachiensis* (a species of sea urchin, whose roe some consider a delicacy) to anterior zygopophysis (a particular protuberance on a vertebra). Each has been defined in a manner more precise than your workaday household noun in order to reduce misunderstanding and terminological controversy. That the jargon tends to exclude the uninitiates and those without youthfully spongelike memories gets (for better or worse) little consideration.

By contrast, physics and engineering eschew Greco-Latin obfuscation and pretension; in doing so, they create an equivalent difficulty. They take the most ordinary, garden-variety words and give them precise definitions that unavoidably differ from their commonplace meanings. It takes *work* to pull something upward but not to hold it suspended. *Stress* and *strain* are entirely distinct, with the former commonly causing the latter. *Mass* is not *weight*, even if they're functionally equivalent on terra firma. Both physical and biological practices will plague the reader, but the former tends to be more subtly subversive—a bit of biological jargon is jarring when you don't know its meaning, but the special definition of an ordinary word put to technical use easily passes unnoticed.

One term from physics needs special attention at the start—*energy*, which gets the most cavalier treatment by press and politicians. We ought to be able simply to define it with care and proceed from there. While energy has a precise meaning in the physical sciences, the meaning doesn't lend itself to expression in mere words. Basic dictionaries and textbooks help little—they define energy as the capacity for doing work, unblushingly evading or offloading the issue! Feynman (Feynman, Leighton, and Sands 1963), comes right out with an unusually candid admission (no company man was he, whether teaching or serving on the commission probing the space shuttle explosion of 1986): "It is important to realize that in physics today, we have no knowledge of what energy *is*. We do not have a picture that energy comes in little blobs of a definite amount."

In practice the idea of energy explains so much—the law of conservation of energy may be the greatest generalization in physics. Ultimately that's the advantage of energy. For us it is more of a difficulty—it's just too easy to hide behind a word with no easy definition and thereby to avoid some crucial explanations. So word and concept will play only a minor role throughout most of this book.

The next chapters will be largely given to the task of establishing a necessary physical base, with a fair dose of the associated terminology. Biological terminology (and biology itself) will enter piecemeal—for present purposes the physical material does a better job of providing a logical framework.

Adaptation and Evolution—The Biological Context

Nonetheless, biology as well as physics needs a bit of introduction. The words "evolution" and "design" have already surfaced; I find it hard to avoid either in any general discussion. Used together, they represent a subtle contradiction. If the process of evolution is incapable of anticipation, that is, if it's blindly purposeless, the term "design" misleads seriously—design ordinarily implies anticipation and purpose. The problem isn't just terminological. Why do organisms appear to be well designed if they're not designed at all? Perhaps we should to begin by reviewing the logical scheme for which "evolution by natural selection" is our facile encapsulation.

First, some observations (in logic these would be axioms). Every organism of which we have any knowledge can produce more than one offspring, so populations of organisms always are capable of increasing. It takes, though, some minimum quantity of resources for an organism to survive and reproduce; and, in the long run, the resources available to no population are unlimited. Next, three consequences of the observations. A population in an area ought to increase to some maximum. Once that maximum is reached, more individuals will be produced than can find adequate resources. So some individuals will not survive and reproduce. Then two further observations. Individuals in any population vary in ways that affect their success in reproduction, and at least some of this individual variation is passed on to their offspring. And then the final consequence. Characteristics that confer increased relative success in reproduction will appear more often among the individuals of the next generation. We say, in short, that these features will have been "naturally selected." By that we imply only selection among preexisting variations, not design in our usual intentional sense.

At this level, the model is about the least controversial item of modern biology—every aspect has been observed and tested, and competing models for the generation of biological diversity (even if without logical flaw) uniformly fail to correspond to reality. Indeed, given geological time and the variation generated by an imperfect hereditary mechanism, how could evolution have been avoided? Where argument remains, it devolves about details—whether the process is usually steady or episodic, the roles of specific genetic mechanisms (such as sexual recombination), the relative importance of selection and pure accident, and so forth. The model leaves no place for anticipatory design, and it doesn't require (and, indeed, no evidence supports the notion) that environmental challenges determine the character of the variation upon which natural selection acts. But, as a moment's consideration should persuade you, most results of such selection will look as if deliberately designed. How come?

Selection, quite clearly, operates most directly on organisms. An organism's success in engendering progeny defines its "fitness." (Some adjustment has to be made for indirect contributions by way of aiding the reproduction of one's kinfolks, but that's a matter of little present concern.) The selective process knows next to nothing about species; there's little evidence that any organism ever does anything "for the good of the species." Nor does the process care directly about parts of an organism. Legions of cells die on schedule in the development of an individual; in no way can we speak of such cells as more or less "fit" than any others. A tree commonly sheds leaves; those leaves were not less fit than its others—the term fitness is inapplicable here since it refers only to the reproductive potential of a potentially reproductive individual, here the whole tree.

Since this book focuses on organisms, it considers a level of biological organization upon which the invisible hand of the selective process should have fairly immediate consequences. That immediacy of the operation of that unseen hand makes organisms appear well designed—as a colleague of mine put it, "The good designs literally eat the bad designs." But note again the unusual sense in which we use "design," implying only a functionally competent arrangement of parts resulting from natural selection. In its more common sense, implying anticipation, it's a misnomer—it connotes the teleological heresy of goal or purpose. Still, verbal simplicity is obtained by talking teleologically—teeth are for biting and ears for hearing. And the attribution of purpose isn't a bad guide to investigation—biting isn't just an amusing activity incidental to the possession of teeth. If the arrangement of an organism seems functionally inappropriate, the most likely explanation (by the test of experience) is a faulty view of its functioning. As the late nineteenth-century physiologist Ernst von Brücke supposedly said, "Teleology is a great mistress, but no one with whom you'd like to be seen in public" (Gray [1893], quoted by Swanson [1973]).

We functional, organismic biologists are sometimes accused of assuming perfect design in the living world, largely because we find the presumption of a decent fit between organism and habitat a useful working hypothesis. To some "adaptationism" has become the perjorative term (see, for instance, Gould and Lewontin 1979) for the practice. And, indeed, it's all too easy to take literally what often amounts to no more than verbal convenience or baseline presumption. It's not trivial to show that a particular structure not only serves some particular and biologically useful function but that it evolved under selection for that particular function. No direct functional test in the laboratory can prove such a point. One must turn to comparisons among organisms whose evolutionary ("phylogenetic") relationships have been determined. Harvey and Pagel (1991) faced the issue directly; we'll return to it in chapter 25.

Furthermore, the designs of nature must be imperfect. At the very least, perfection would require an infinite number of generations in an unchanging world, and a fixed world requires not only a stable physical environment but a preposterous scenario in which no competing species underwent evolutionary change. Furthermore, we're dealing with an incremental process of trial and error. In such a scheme, major innovation is no simple matter—features that will ultimately prove useful will rarely persist through stages in which they do no good. So-called hopeful monsters are not in good odor. Many good designs must not be available on the evolutionary landscape because they involve unbridgeable functional discontinuities. Obviously jury-rigged arrangements occur instead because they entail milder transitions. The ad hoc character of many features of organisms are recounted with grace and wit in some of the essays of the late Stephen Jay Gould; his collection entitled *The Panda's Thumb* (1980) is particularly good. The more multifunctional the structure, the greater must be the constraints on what evolution can come up with. Finally, a fundamentally poorer but established and thus well-tuned design may win in competition with one that is basically better but still flawed.

I make these points with some sense of urgency since this book is tacitly adaptationist in its outlook and explicitly teleological in its verbiage. The limitations of the viewpoint won't get much repetition, so the requisite grain of salt should be kept in the mind of reader as well as author.

In the final analysis, this book is about organisms rather than physical science—the latter simply provides tools to disentangle some aspects of the organization of life. But, beyond using physics to organize the sequence of things, we'll take an approach more common (historically, at least) in the physical sciences. Biologists love their organisms, whether singly, collectively, sliced, macerated, or homogenized. Abstractions and models are vaguely suspect or reprehensible. As D'Arcy Thompson (1942)—the godfather of our subject—put it, biologists are "deeply reluctant to compare the living with the dead, or to explain by geometry or by mechanics the things which have their part in the mystery of life." But we will repeatedly use the "dead" to explain the "living." Explanation requires simplification, and nothing is so unsimple as an organism. And the most immediate way to simplify capitalizes on nonliving models, whether physical or (even) mathematical. Look how far a few simple models took us toward explaining the upper speed limit for walking and the requisite design of arterial walls.

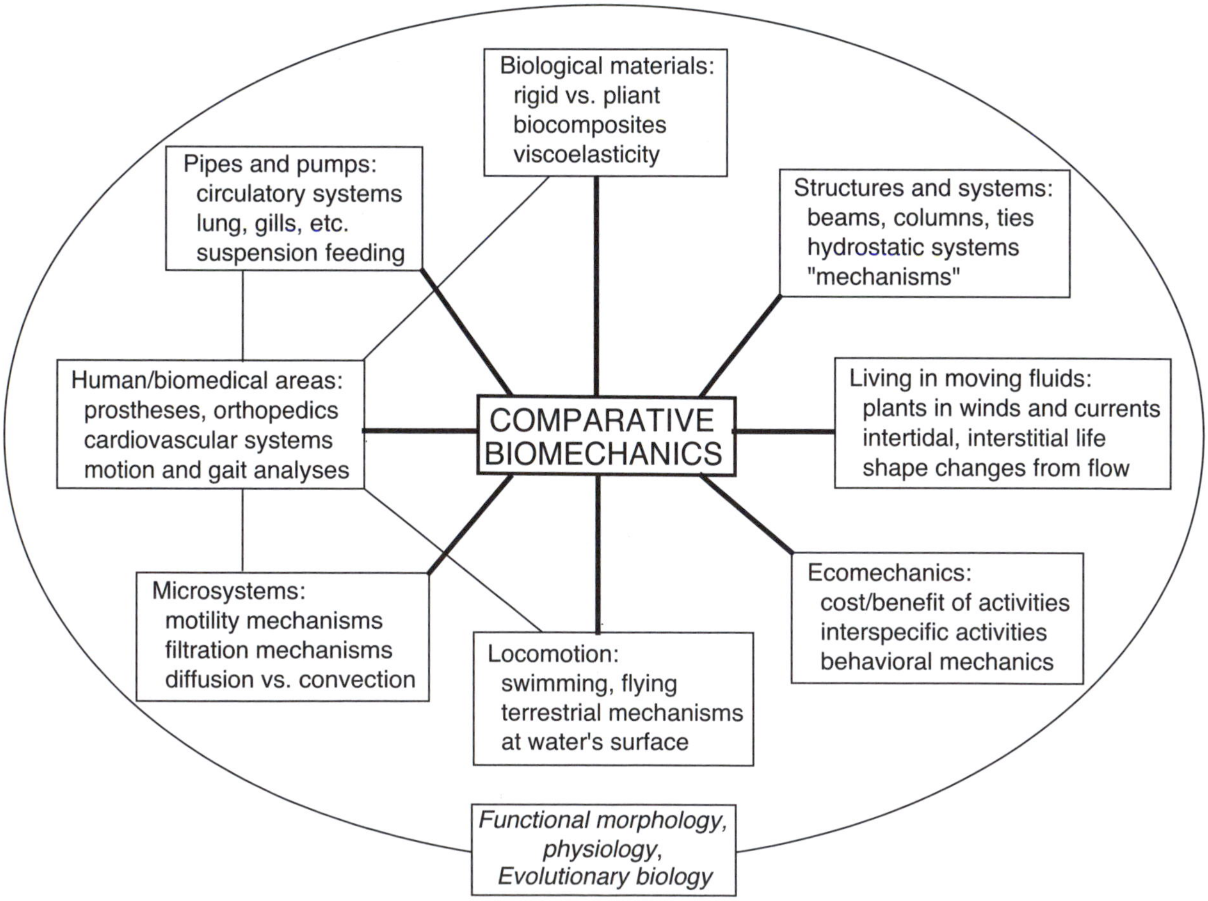

FIGURE 1.6 An opinionated view of the place of comparative biomechanics in the larger world of biology, medical science, engineering, and so forth.

The Two Fields of Biomechanics

Back down to earth, with a few words to show how our subject nestles within its niche among the various areas of biology. Biomechanics at present consists of two distinguishable fields. Current usage, at least in North America, considers "biomechanics" without verbal qualification to be a branch of human functional biology. Its concerns include the efficient design of devices to be used by humans, mechanical prostheses, locomotion as related to rehabilitation or athletics, and similar matters. It has several specific journals and active national and international organizations; its practitioners mostly inhabit schools of engineering and medicine.

The other field takes for its concerns biological systems in their full diversity of size, structure, ancestry, and habitat. It considers biological materials, structural mechanics, and every kind of locomotion. It looks at fluid-mechanical matters from how organisms, both plants and animals, resist the forces of flow to the operation of circulatory and other internal fluid transport systems. It asks questions about both living and extinct organisms and about how environmental factors impinge on biological design. Put another way, it looks at how the design and operation of organisms, including of course ourselves, reflect the values of physical variables such as gravity, viscosity, elastic moduli, and surface tension.

Some of us have begun calling this latter field "comparative biomechanics" to draw a distinction analogous to that between, on the one hand, "comparative anatomy" and "comparative physiology" and, on the other, their human or medical analogs. It's part of a larger subject most succinctly described as the study of function at the level of the organism, or more formally as physiology and functional morphology. Figure 1.6 shows where it sits among its academic parents, sisters, and cousins. Comparative biomechanics can claim no historical novelty—it traces its ancestry at least to Borelli's *De Motu Animalium* of 1680. But it appears to be enjoying something of a renaissance by at least one measure. I've done some (admittedly subjective) counts of the abstracts for the annual meetings of the Society for Integrative and Comparative Biology (formerly the American Society of Zoologists). Identifiably biomechanical contributions increased from about 5 percent to 12 percent between 1985 and 1995; by the 2001 meeting the fraction had increased to 27 percent. A similar picture emerges from the tables of contents of issues of the *Journal of Experimental Biology*.

CHAPTER

Setting the Stage

Nature and Nature's laws
lay hid in night:
God said, *Let Newton be!*
and all was light.

ALEXANDER POPE

It did not last: the Devil
howling 'Ho!
Let Einstein be!' restored
the status quo.

J. C. SQUIRES

THIS WILL BE A CHAPTER ABOUT TOOLS—TERMINOLOGICAL, CONCEPTUAL, and logical—a set of tools from the physical folk for us to use in analyzing biological situations. If terms were all that mattered, a glossary would do; as it happens, we need something more logical. The relevant items are simply too intricately interrelated to make much sense except as components of sequences and hierarchies. It's a big dose of abstractions, but you can easily return to the present pages whenever necessary, so complete mastery isn't crucial. If you recall the contents of a course in physics, much of it will be old hat.

Quantities and Units

Even when we're trying to decide some essentially qualitative question, we almost always use quantitative reasoning and methods (e.g., measurement and calculation). There's nothing really strange about the process—houses come in a range of prices; if you want a house, you use quantities such as price and monthly payment to determine a qualitative issue—should you buy one in particular? Multifaceted quantitative data may be required—which is preferable in some application, a suspension bridge or an arch bridge? The only novelty here is the ubiquity of this sort of reasoning. An outsider may view science as an endless pursuit of data, but for all our addiction to numbers they're just a means, not an end in themselves.

There's a more subtle aspect to measurement and quantification. For some purposes, very good data are mandatory. If you want to predict eclipses, a very accurate measure of gravitational forces is crucial, much more accurate than for worrying about gait transition points. Sufficient accuracy depends on—and only on—the question one is trying to answer. The process is the same as that used by the designer of a product to be made on an assembly line, who has to specify just how inaccurately formed each component can be and still fit satisfactorily into the final product. A good design is one that's tolerant of sloppily made components; similarly, a robust hypothesis often can be tested with crude data. The common notion that scientists are as accurate as possible represents a serious misunderstanding. The trick is to figure out beforehand what level of accuracy is required and then to waste no effort doing substantially better. I think I've never needed to work to closer than a part in a thousand when doing science, but the numbers on my tax return pretend to better than a part in a million.

Indeed, on occasion we'll talk in completely quantitative terms with a *lack* of accuracy quite beyond everyday experience. For some issues, order-of-magnitude accuracy (up to 1,000 percent error!) is entirely sufficient, although most often we do need better than that. Given the variety of shapes in which terrestrial animals come, using hip-to-ground lengths of great precision for gait-transition calculations would be self-deluding.

Measurement implies units. It would be handiest to have only a single set of units, built up from something other than historical accident. The metric system, one of the lasting results of the French revolution, is popularly considered such a system. (I've always viewed the scheme as insufficiently revolutionary—they should have gone further and begun by changing the counting system from a decimal one with a base of ten to a duodecimal one with a base of twelve, rather than merely decimating the units.)

Recently a more rigid version of the metric system has been urged on the scientific community—the Système Internationale d'Unités, "SI" for short (but not "SI system," which is redundant). We'll stick with it pretty consistently here. It involves a few unfamiliar units that will be taken up as needed—kilograms, watts, and meters may be ordinary, but newtons, joules, and pascals are less so. It provides us with a most convenient internal consistency and circumvents all sorts of constants and conversion factors. The main features of SI are the following:

1. The fundamental units are kilograms for mass, meters for distance, and seconds for time. The prefix, "kilo," on "gram," is a minor bit of awkwardness tolerated to keep a preexisting unit of handy size without rejecting the original metric name. It means that a milligram, a thousandth of a gram, is a millionth of the fundamental unit. A biologist, accustomed to small-scale work, would surely have set the gram as the basic unit of mass.

2. Prefixes for larger and smaller units go in steps of three orders of magnitude (steps of 10 times 10 times 10); millimeters are okay but centimeters are officially suppressed (but we'll occasionally use centimeters anyway on account of their handy size and familiarity). For our needs, six prefixes will suffice:

giga-	(G)	a billion (U.S.) times	$10^9 \times$
mega-	(M)	a million times	$10^6 \times$
kilo-	(k)	a thousand times	$10^3 \times$
milli-	(m)	a thousandth of	$10^{-3} \times$
micro-	(μ)	a millionth of	$10^{-6} \times$
nano-	(n)	a billionth (U.S.) of	$10^{-9} \times$

It's easy if you're a proper classicist (which I am not)—the enlarging prefixes come from Greek (giga = giant; mega = large; kilo = thousand), while the diminishing prefixes come mainly from Latin (milli = thousand; micro = small; nano = dwarf).

3. Rules govern the way units can be combined. The main one for us is that prefixes in complex units can attach only to numerators, not denominators. Thus, the kilogram per cubic meter is fine, but a microgram per cubic millimeter (logically the same thing) is unlawful. The results are sometimes

amusing—the breaking strength of a spider's thread is properly given in meganewtons of force (a meganewton is about the weight of a 100 metric ton mass) per *square meter* of cross section (a *very* thick strand, extruded, one supposes, by a terrifyingly large spider).

Dimensions

To begin with, explaining even our limited bit of biology involves lots of things such as force, mass, density, pressure, and acceleration—we'll call them "physical quantities" or "physical variables." What these have in common is that they're quantitative attributes of physical objects or situations. To put them into a systematic scheme, we require some explicit way to express, not just the quantities, but the relationships among them. The simplest approach is to show that some things called the "dimensions" of the quantities are just combinations of a very few basic ones. (Note here the peculiarity of specialized jargon—use of the term "dimension" in quite a different sense from the familiar length, width, and height.)

The actual choice of which dimensions to declare basic is a bit arbitrary. Common practice uses the following three. (Whether temperature is a fourth provokes hot debate, and force sometimes substitutes for mass.)

Mass	given the symbol	M
Length	given the symbol	L
Time	given the symbol	T

Each dimension is a property of the physical quantity with the same name—but not necessarily a property unique to that quantity. Thus, the quantity "mass" refers to an amount of matter; the dimension "mass" is more general and is a property of anything that can be measured on the same scale (or with the same units) as mass. For instance, pressure, stress, and stiffness share the same dimensions. This curious and abstract distinction between dimensions and quantities raises little practical trouble. Nor does definition of the physical quantities corresponding to these basic dimensions—we'll just assume common knowledge of what is meant by mass (quantity of matter), length (the distance between two points on a specified path), and time (the rate of spontaneous conversion of future to past). Quite obviously, each can be measured in a proper quantitative fashion; almost as obviously you'll recognize that each has a real zero-point (not like the arbitrary zero of the ordinary temperature scales) and that none can be explained in terms of the other two.

Note that neither physical quantities nor dimensions are synonymous with units. The dimensions of a quantity constitute the scale on which we measure it; the units are the rather more specific matter of how we mark or subdivide that scale. Notice, though, that the names of units (such as "meters per second") often reveal the underlying dimensions (here "length over time")—the practice is a great help to a fallible memory trying to remember the dimensions of some variable!

The business of figuring out the basic dimensions for a physical quantity is neither mysterious nor difficult. "Speed" is a measure of how long (time) it takes to go a certain distance (length). Or length per unit time. Or length/time. Or in dimensions, L/T. Or, as we'll express the matter henceforth, *speed has dimensions of (length)^1(time)$^{-1}$ or* LT^{-1}.

Volume, to give another example, can be viewed as the product of a length, a width, and a height, three length dimensions, so,

volume has dimensions of length cubed or L^3.

Formulas for volumes may contain other items, but any such item in a formula for volume must be a pure number quite devoid of any dimension. Thus, the formula for the volume of a sphere is the product of a fraction (four-thirds), the ratio of the circumference to the diameter of a circle (pi or 3.1416), and the radius to the third power—"four-thirds" and "pi" are dimensionless, while "radius" has the dimensions of length. Any expression for volume *always* has three length dimensions, whether a single physical variable like radius provides all three or whether, as for the volume of a rectangular solid, we need separate measures of depth, width, and height.

A physical quantity may be dimensionless; it gets that way by being defined as the ratio of a variable having certain dimensions to another variable with the same dimensions. For example, "strain" (yet again a special meaning of an ordinary word) is the distance a material is stretched divided by its original length, the ratio of one length to another. Since length divided by length leaves no resulting dimensions at all (recall pi), strain is quite nakedly dimensionless and, of course, unitless as well. You might say, if it helps your intuition, that strain is measured in meters per meter, but that's a sophomoric sort of sophistry. We'll encounter other dimensionless variables and find them surprisingly ordinary. Their main oddity is really a nice convenience: they come out the same no matter what units are used to obtain them! Thus, for strain, one gets the same numerical result whether original length is in inches and extension is also in inches or whether the two are both in cubits or fathoms. A strain of 0.01 just means that something has been stretched 1 percent beyond its original length.

It's worth being careful to keep the dimensions distinct from the physical quantities, and it's even more important to maintain a careful distinction between the quantities and their units. Don't *ever* indicate units within a formula; when units must be specified, mention them in a separate bit of text. In an equation such as $F = mg$ there shouldn't be even the merest hint of the abbreviation for a milligram—the formula talks about quantities, not units. The biological literature has too many instances of violations of this rule, which not only produce confusion but subject us to well-deserved ridicule. And don't mix systems of units in using a formula—first convert each datum to a consistent system, preferably SI.

Physical Variables

I once wrote a polemical paper arguing that the physical characteristics of an animal's world are relevant to its behavior (Vogel 1981a); in the process I got up a list of physical quantities that seemed to be at least occasionally important ("bioportentous parameters" in the paper). The list was compiled from contributions solicited as I took lunch at a different table each day at a marine laboratory—after collecting about thirty items I began to imagine that people were avoiding me; with no real end in prospect, I arbitrarily ended the exercise. We'll not worry about quite so many quantities in this book, and we certainly don't want to begin with all of those that we will eventually encounter. But it will be useful to step through a set of explanations or evasions, moving from the simpler and the more intuitively real to the more derived and abstract. Table 2.1 summarizes the prose and gives the combination of basic dimensions for each physical quantity; you might find it useful to dog-ear its page (unless, of course, you don't own the book).

Mass, length, and time

These three have already appeared and we'll not attempt further definition. Note, though, that lowercase italic "*m*," "*l*," and "*t*" will designate these quantities, with the uppercase letters reserved for the corresponding dimensions.

Area

An area is a length in one direction multiplied by a length at a right angle to the first and often multiplied by some dimensionless factor (perhaps pi) as well. So it's always proportional to length times length, or length squared. We'll use the symbol, *S* (think of "surface"), whether we're talking about surface area, cross-sectional area, projected or shadow area, or any other area. (Occasionally it's useful to invent nominal areas—quantities that have the dimensions of an area and thus behave like areas in formulas—for example,

TABLE 2.1 A Variety of Variables

Quantity	Symbol	Origin	Dimensions	SI Units
Mass	m	(fundamental)	M	kilogram
Length	l	(fundamental)	L	meter
Time	t	(fundamental)	T	second
Area	S	$\propto$* to length squared	L^2	square meter
Volume	V	$\propto$ to length cubed	L^3	cubic meter
Velocity	v	length/time	LT^{-1}	meter/second
Acceleration	a	speed change/time	LT^{-2}	meter/second^2
Force	F	mass $\times$ acceleration	MLT^{-2}	newton
Momentum	mv	mass $\times$ velocity	MLT^{-1}	kilogram-meter/second
Stress	σ	force/area	$ML^{-1}T^{-2}$	newton/meter^2 or pascal
Pressure	p	force/area	$ML^{-1}T^{-2}$	
Work	W	force $\times$ length	ML^2T^{-2}	joule
Power	P	work/time	ML^2T^{-3}	watt
Density	ρ	mass/volume	ML^{-3}	kilogram/meter^3

*Recall that $\propto$ is the symbol for "is proportional to."

volume raised to the two-thirds power is a nominal area of some use for comparing biological items of diverse shapes.)

Volume

As mentioned earlier, volume is the product of three lengths (and some constant that depends on the shape at hand). Each length with which a volume is specified is at right angles to the other two—this puts us clearly in a three-dimensional world and solid geometry. Thus, volume is proportional to length cubed (third power); we'll use the symbol V.

Speed *and* velocity

Speed is the ratio of a distance to the time it takes to go that distance; to it we'll assign the symbol v, using the lowercase to distinguish it from volume. The v really stands for "velocity," a special sort of speed that properly includes some specification of the direction as well as the rapidity of travel, and about which more will be said shortly.

Acceleration

If speed is viewed as the rate at which your location changes, then acceleration can be visualized as the rate at which your speed changes. In short, it's the ratio of a difference in speeds to the time that elapses as the speed alters from one to the other. It includes the notion of deceleration—when the speed happens to be decreasing and acceleration is negative. If you run into a wall, the magnitude of your acceleration (negative) may be inconveniently great—your speed changes a lot over a short period of time. Increasing the time period through controlled collapse of your vehicle can usefully reduce the absolute value of the acceleration and the corporeal consequences. We use the symbol a, and recognize its dimensions as length divided by time (as in speed) and divided again by time, that is, length divided by the square of time.

One particular value of acceleration is especially relevant here, and allusions to it appeared in the first chapter—it's the downward acceleration of an object under the sole influence of the gravitational force at the surface of the earth. This value is about 9.8 meters per second squared (or 32 feet per second squared); it's usually referred to as g. In our perceptual world that's an unpleasantly high value, corresponding (to use the automotive benchmark) to a mere 2.7 seconds for steady acceleration from 0 to 60 miles per hour.

Force

Here things begin to get more complex. It's more difficult to get a large mass into motion than a small one or to speed up a given mass in a shorter period of time—to give something a large acceleration. In these situations "difficulty" is essentially synonymous with "force." (Similarly, a large mass slowed down by impact can give something a substantial push, a clear disadvantage of being kicked or punched.) Two factors are involved, mass and acceleration, and force, with symbol F, is defined as the product of the two: perhaps the most useful of all the equations we'll encounter is this one, simply put as $F = ma$. A special name has been coined for the SI unit of force, the "newton"

(N) (after Sir Isaac). It's the same as a kilogram-meter per second squared (from mass times acceleration).

But force can be exerted without causing any noticeable motion, as when you stand on the floor or press on an incompressible object such as a brick wall. So, although force is defined in terms of acceleration, the former can exist without the latter—the squeeze on the brick is measured as the force that would, if applied to one side of a movable mass, give it a certain acceleration. Acceleration, it turns out, happens only as a result of a *net* or *unbalanced* force. Objects stay at rest or in steady motion (constant velocity) unless acted upon by an unbalanced force. When you press the brick wall it responds by pushing outward against you with equal force because it can't head off anywhere. The reality of this countervailing force will be no small matter when we talk about the behavior of nonrigid materials.

What about the force on the floor? Here gravity enters—an object is attracted to the earth with a force proportional to its mass times the mass of the earth and inversely proportional to the square of its distance from the earth's center. Because the earth's mass is constant and the distance to the earth's center (about 4,000 miles) is practically the same for all objects, the proportionality of the force to the mass of the object is all that normally matters. Then $F = ma$ becomes $F = mg$, where g is the acceleration due to gravity. As a result, objects left to themselves (dropped, for instance) accelerate downwards. Even if the object isn't free to fall, though, the force is still there, and it's still proportional to the mass of the object.

A bathroom scale measures the unhappy consequences—the force, whether measured in pounds, stones, or newtons, is what we normally call "weight." Again, $F = mg$: an object weighs an amount equal to its mass times the acceleration it would experience if freely falling. On the earth, as mentioned, that acceleration is about 9.8 meters per second squared. Since 1 kilogram times 9.8 meters per second squared is 9.8 newtons, a mass of 1 kilogram has a weight (here on earth) of 9.8 newtons. Note an important distinction—mass and weight have different dimensions and units. (Strictly speaking, a pound is a unit of force, not mass; but practice introduces ambiguity. We'll avoid trouble the easy way, by avoiding pounds. If you need to convert, a 1-pound force equals 4.45 newtons; a 1-pound mass equals 454 grams.)

In all of our personal experience, weight is proportional to mass with the same constant of proportionality (the gravitational acceleration on earth). So it's difficult to get a personal sense of the fundamental difference between the two quantities. On the moon a 1-kilogram mass will weigh, not 9.8 newtons, but a mere 1.6 newtons (or 0.37 pounds) because lunar gravity is about six times lower. The consequences range from obvious to subtle. Consider the gait of an astronaut on the moon. Hopping upward goes quite well—the same muscular force makes body mass go further upwards when the downward acceleration is less. Hopping forward works as nicely, but we notice that it takes quite a forward tilt to get going—the same mass must be accelerated, but the concomitant push backward on the ground requires weight, and the latter is in short supply. Even odder is the business of cornering—to change the direction of motion requires as much force as on earth—again it's a matter of accelerating an unchanged mass. But the force must be exerted against the lunar ground, and the scarcity of weight means that a moon-walker must lean farther over to come around a turn than an earth-walker. What fascinates me is just how splendidly the human neuromuscular machine copes with this personal and evolutionary novelty! To repeat, on the moon an object has less weight than on earth but isn't 1 gram less massive.

In everyday practice we "weigh" something and specify the outcome in either pounds or kilograms depending on local custom. Conversion is trivial—there are 2.2 pounds in a kilogram. The latter conversion, though, masks a simultaneous shift from weight to mass using a conversion factor applicable only on the surface of the earth. As long as we only patronize terrestrial emporia there's no problem—weight equals the product of mass and a dependably constant gravitational acceleration.

And we use beam balances and spring scales more-or-less interchangeably even though one indicates mass and the other weight. It's amusing to contemplate the consequences of a small change in the earth's gravity. One class of device would require recalibration; the other would work quite as before. The reader should figure out which is which.

Stress

If you exert 10 newtons of force pushing the broad side of a knife against a carrot, the carrot will be none the worse; if you use the same force and the sharp edge, the carrot will be cleaved. Clearly the area over which the force is applied makes a difference—more often than not what matters isn't force per se but force divided by the area against which it's applied, what we mean by "stress" in the physical as opposed to the psychological sense. The symbol used is the Greek lowercase sigma (σ); the SI unit is designated the newton per square meter or alternatively the pascal (Pa), after Blaise Pascal (1623–62). A stress inevitably causes a strain (mentioned earlier); the two are most emphatically not the same thing—the former is how much you aggress upon something, the latter is how much it is bent out of shape as a result.

We need to distinguish among several sorts of stresses, as done in figure 2.1.

1. Tensile stress. Tensile stress is force per unit cross-sectional area, as usual, but specifically where the two ends of something are being pulled away from each other by the stresses. Pull on a rope or on taffy or even on a beam of wood or metal—the stress is applied whether or not the material stretches appreciably. The area for converting force to stress is, of course, at a right angle to the force—the cross section of rope, taffy, or beam.

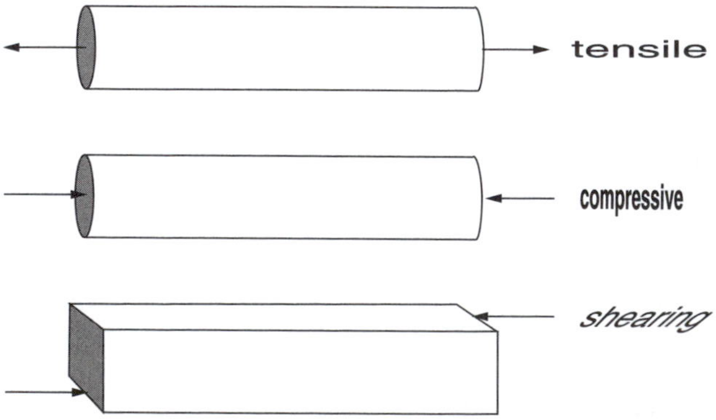

FIGURE 2.1 Three kinds of material stresses—less formally, these are pull, push, and slide. (Bend and twist will come up as structural properties in chapter 18.)

2. Compressive stress. Compressive stress is just the same except that it pushes the material together. Push on a wall, step on the top of a full can of beer, push on the handle of a bicycle pump with a finger over the orifice—you've applied a compressive stress to a solid, a liquid, a gas. Again, the area is at a right angle to the force.

3. Shear stress. This stress differs more. Apply a pair of forces to an object, forces from opposite directions but along lines of action that don't coincide. Sometimes if we apply such a pair of forces we observe rotation as the result—here assume instead that the object will deform rather than rotate. It makes it easier to picture if one of the forces (the resisting force) keeps a face of the object stationary. With a soft material the deformation will be noticeable—try a kitchen sponge, bread dough, a lump of jello. A cube will be distorted so some of its angles are no longer right. Whether the material is initially cubical or whether we merely view it as an amalgam of coalesced cubes, a given shear stress will give a certain angular deformation, a certain distance of movement in the direction of the force on the unanchored face for each unit distance between the lines of action. Here the relevant area differs from that for tensile and compressive stresses. If the applied force is exactly above the resisting force, the area is any horizontal plane between them—pull on a sheet and you put a shear stress on the mattress.

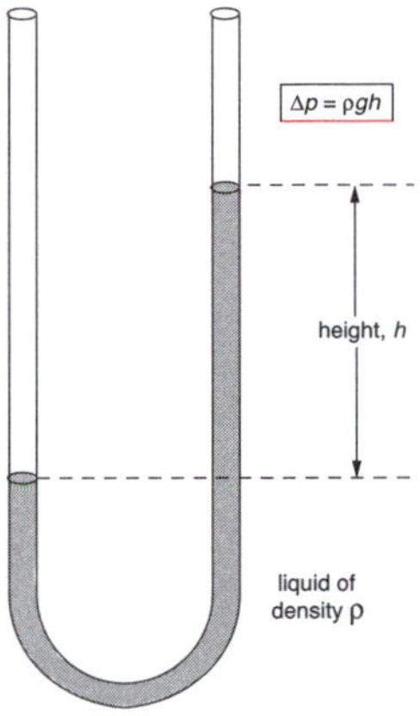

FIGURE 2.2 A simple U-tube manometer. If there's another liquid above the manometer liquid (shaded) as when working with pressures in water, then one uses the difference in their densities in the formula. If (as usually) the device is used in air, the slight density of the latter can be ignored.

Two further kinds of stress will appear when we impose forces on objects of specific shape—flexural (bending) and torsional (twisting) stresses. These "structural" stresses, though, can be reduced to combinations of the three basic "material" stresses just described. When you bend something, you stretch one side, compress the other, and shear what's in between. When you twist something (such as a wet towel), you stretch the outside, compress the core, and shear concentric cylindrical layers.

Pressure

Related in a slightly confusing way to stress is pressure, because both are forces divided by areas. Stress, though, is a unidirectional phenomenon—you press (should I say stress?) downward with the knife. Pressure is omnidirectional—you press air into a tire and the tire inflates in any direction it can get away with. We'll use a lowercase p for pressure; the SI unit is again the newton per square meter or (more often for pressure) the pascal. It's a very small unit—atmospheric pressure is about 101,000 pascals (Pa), and my (nonhypertensive) arterial blood pressure ranges from about 14,000 (systolic) to 11,000 pascals (diastolic).

In addition to pascals, pressure comes in lots of other units. Some raise no difficulties—pounds per square inch (14.7 of them equal 1 atmosphere), for instance. Others are distinctly odd—inches of mercury or millimeters of water are units of height with some peculiar specification of a material. They refer to the results of applying a pressure to a column of liquid (fig. 2.2). The more pressure, the higher the column that can be supported. Neither the area of the exposed surface of liquid nor the diameter of the column nor the empty volume above the column make any difference to the height that a given pressure will support—only height matters. To get a height into a respectable force per area, one multiplies the height by the gravitational acceleration on earth and then by the density (see below) of the indicated material. Consider a figure of 29.9 inches of mercury given by the (U.S.) weather reporter for barometric (atmospheric) pressure. That datum times 0.0254 meters per inch (converting inches to meters) times 9.8 meters per second squared (gravitational acceleration) times 13,600 kilograms per cubic meter (density of mercury) gives 101,000 pascals—1 sea-level atmosphere in honest SI units.

One other detail about pressure. We live with a background pressure of about an atmosphere. Its value must be considered only in rare devices such as the earliest steam engines—in these a stream of water condensed steam in a cylinder, thus creating a partial vacuum, and the outside pressure of an atmosphere then pushed the piston inward. Or in the inspiratory movements of the breathing scheme of animals such as us—you force your rib cage outward and upward and your diaphragm downward, and air enters your lungs, pushed inward by the atmosphere of pressure outside (101,000 pascals) working against a slightly lower pressure (minimally about 90,000 pascals) inside (fig. 2.3). For most purposes, though, the background pressure is presumed and ignored, and what's cited is pressure above atmospheric, called "gauge pressure"—that's what a pressure gauge for tires or blood indicates.

Momentum

Like weight, momentum is another composite variable, but here we're up front about it. Momentum is the product of mass and speed (or, really, velocity). It turns out to be an especially useful variable, as first recognized by Descartes, who referred to it as the "quantity of motion"—a 10-kilogram object going 1 meter per second has as much motion as a 1-kilogram object going 10 meters per second. No special symbol or SI unit attaches to momentum—we merely use the symbol *mv* and the kilogram-meter per second. We'll find momentum to be an especially useful variable.

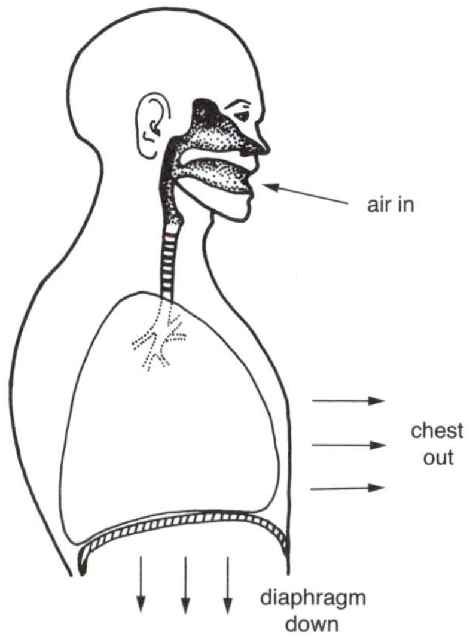

FIGURE 2.3 A person inspiring (having an aspiration, one might say). Air enters the mouth because the diaphragm flattens, moving downward, and the rib cage moves forward and upward.

Work and energy

Applying a force to something needn't accomplish much at all in a strictly physical sense. We accomplish something when work is done. The work done is defined as the product of the force applied times the resulting distance an object moves while the force is acting and in the direction of application of the force. Thus, a chain on a chandelier may exert an upward force on the chandelier without pause for a century or two; but it does no work, requires no motor, and consumes no food. By contrast, the carriage of an ascending elevator must be worked on by a motor or descending counterweight. For work we'll use the symbol W; the SI unit is the joule (after James Joule, 1818–89), sometimes called the newton-meter.

We tend to confuse force and work, probably because our muscles have to degrade food to produce force even without doing work—holding a weight or even standing erect is fatiguing. Why, once we've lifted something, we can't just lock up the relevant muscles isn't at all clear. After all, you do work to raise a flag, but you tie its rope to a cleat rather than holding the rope all day. Odder still, the problem isn't a peculiarity of all muscles. Those that hold the halves of a clamshell together for long periods have largely solved the problem with a so-called "catch" mechanism (Muneoka and Twarog 1983). One wonders about Atlas, but the ancient myth makers say nothing of any affinity with clams.

A textbook of high school physical science to which my son was subjected showed a man pushing a desk sideways across the floor. The desk weighed 100 pounds (force) and was pushed 10 feet (distance); the legend, a more literal term than usual, declared the work to be 1,000 foot-pounds. Don't make the same mistake—weight is always a downward force but proper floors are horizontal. The man does no work directly against the weight of the desk and, if the floor is slick or wheels are provided, he may do almost no work even against the frictional resistance to sliding. To put it another way, the weight of the desk gives us little idea of its resistance to sideways motion, the resistance the force of the man "works" against.

It always takes energy to do work; it needn't take energy to apply a force, contrary to our experience with those flawed motors we call muscles. As noted in chapter 1, we define energy as the capacity to do work. Energy comes in a variety of manifestations—work itself plus any measure of the capacity to do work, what we might call "latent work." So energy can appear as force times distance-in-the-direction-of-the-force:

$$E = W = Fd. \qquad (2.1)$$

Or it can appear as so-called kinetic energy, energy by virtue of motion, $\frac{1}{2}$ times moving mass times the square of velocity:

$$E = \frac{1}{2}mv^2. \qquad (2.2)$$

Or it can appear as gravitational potential, energy by virtue of being above (height; h) some reference surface, energy that could instigate and maintain a fall:

$$E = mgh. \tag{2.3}$$

All of these, of course, have the same dimensions and units. But no more about that queerest of concepts for quite a while—seventeenth- and eighteenth-century mechanics did quite well without it and so will we.

Power

Power is very simply the rate at which work gets done, or work done divided by time taken. A more powerful motor can either lift a given weight faster or else lift a heavier weight at the same rate—the physical quantity corresponds reasonably well with our everyday usage. We'll use the symbol P; the SI unit is the watt (after James Watt, 1736–1819) and, as an electrical unit, is familiar to all (except those who keep light bulbs in all the sockets to prevent electricity from leaking into the house). A watt is, of course, a joule per second. If you multiply power by time you get energy or work, so a watt-second is a joule—a trivial point except for the predilection of electric companies to bill us for kilowatt-hours instead of megajoules (1 kilowatt hour = 3.6 megajoules) for our purchases of energy. "Metabolic rate," often given in anachronistic kilocalories per some unit of time is, of course, just power. Because 1 calorie equals 4.2 joules, a kilocalorie (or Calorie) per hour is 1.16 watts. The other common household (or, better, barnyard) unit of power is the horsepower—746 watts.

Density

Density is mass divided by volume—the ratio of how much substance is present to how much of the world it takes up. It's a useful "material property," that is, all samples of the same substance have the same density under the same conditions. Material properties, others of which will appear in later chapters, characterize substances rather than particular objects. Water, for instance, has a density of about 1,000 kilograms per cubic meter. Most organisms have about this value; the minerals of which organisms are built are denser; most metals are denser yet (table 2.2). Gases are all much less dense (air at sea level and room temperature has a density of about 1.2 kilograms per cubic meter). When we say something is very heavy we (ambiguously) either mean that it's massive or that it's dense. Fortunately, the distinction can usually be made from the context of the remark. For a symbol, we'll use the Greek "rho" (ρ).

Motion

With the introduction of speed and acceleration, we encounter motion, something about as ordinary as mass, but in practice a little more complicated. To begin with, there's that distinction between *speed* and *acceleration*—the former is a measure of the amount of motion, the latter of its unsteadiness. The difference involves the general issue of rates and gradients, which will occupy a chunk of chapter 4. For now, the main point is a reminder that the two are fundamentally different; however, we confuse them in everyday affairs. Then

TABLE 2.2 Densities of a Variety of Materials

Material	Density (kg/m^3)
Dry timber, white pine	370
Fresh wood, red oak	600
Gasoline	680
Lipids—fats and oils	915–945
Freshwater	1,000
Seawater	1,026
Femur, cow (typical bone)	2,060
Tympanic bulla, fin whale (densest bone)	2,470
Aluminum	2,700
Shell, mollusk	2,700
Calcite (mineral of shell)	2,800
Tooth enamel, human	2,900
Apatite (mineral of bone, tooth)	3,200
Iron	7,870
Mercury	13,550

comes another important division of the types of motion, yet again using terms that deviate from their everyday meanings.

Translational motion

In translating, an object changes its *position* in space with respect to some frame of reference. Its orientation remains as before. The frame of reference is something assumed to be stationary—for our purposes it usually will be the surface of the earth. The arbitrary nature of the frame of reference needs emphasis; if you're uncomfortable with it, so were the ancients and so, apparently, are some contemporary biologists. I once worked on the flight of small insects and made a tediously large number of measurements on fruit flies "flying" in a (small) wind tunnel—the air moved while the flies stayed put. I recall having to stammer out in several talks ad hoc justifications of the shift of frame of reference—I never found a generally satisfactory explanation of something I found intuitively obvious, and I usually resorted to analogies. Try observing a dandelion seed (fruit, really) moving in a gentle wind—one imagines the fluff on top, having more drag, will be tilted downwind relative to the seed beneath. But the flier is fully vertical! In reality it knows nothing about the wind—it's traveling at wind speed, so the wind doesn't exist in its frame of reference, quite detached from the ground. The stillness of the air around a drifting hot-air balloon is remarked upon by almost every first-time passenger. Yet one still encounters suggestions that birds flying downwind may be troubled by ruffled feathers.

Note that by this definition, translational motion may be along a curve or even in a circle. The pedals and your feet *translate in a circle* when you ride a bicycle. Similarly, the seats of a Ferris wheel translate in a circle around its axis of rotation, because they're always horizontal. More peculiarly, a small element of bath water translates in a circle as it moves around the micromaelstrom toward the drain. Both these situations—circular translation of a solid element and circulatory motion of a fluid in a vortex—turn out to be of biological relevance, the first because of the scarcity of truly rotational wheel-and-axle arrangements in nature and the second in the explanation of fluid-dynamic lift.

Rotational motion

But the rotor of the Ferris wheel and both the wheels and cranks of the bicycle do rotate—that is, they change their *orientation* (angle) with respect to a line of fixed direction, the essential element in rotational motion.

In practice, these two kinds of motion may be combined—the wheel of a moving bicycle both rotates and advances. A particularly interesting case is the motion of the moon around the earth. The same side of the moon always faces us, so it must rotate once around an internal axis every time it translates in a full circuit around the earth. Incidentally, our sensory machinery makes the proper distinction between rotation and circular translation—the former makes you dizzy whereas the latter does not. The cultural evolution of dances may reflect this fundamental distinction—the circular translation in social as opposed to exhibitionistic dancing must keep many of us from falling to the floor.

Forces as Vectors

Force, velocity, acceleration, and momentum are what are called "vector" quantities as opposed to a mere "scalar" quantity such as mass. The difference is that a vector has a direction as well as a magnitude. (For a force, a third specification, called "line of action" often is important.)

In a one-dimensional world, a force can be exerted in only two directions, both along the same line. We arbitrarily designate one direction of force as positive and the other as negative, we add the forces algebraically, and we thereby get the resultant (net) force. If the net force *is not* zero, the object on which the force acts accelerates ($F = ma$, again)—the simplest example is an object falling at the urging of gravity just after release, before it's going fast enough to experience much drag from the air. If the net force *is* zero, the object upon which the forces act does not accelerate. An example is an object falling in air at the urging of gravity when the object has already fallen a considerable distance. The object has now had time enough to accelerate until the upward force of drag (drag is always opposite to the direction of motion with respect to the air) just reaches the downward force of gravity. With no unbalanced force, the object falls at a steady speed. (If there's a wind, the object may simultaneously move laterally with respect to the ground, but nothing else alters.) For smallish items of low density, that is, for biologically interesting bodies, such steady descent constitutes most of a fall.

In a two-dimensional world, combining forces gets a little more elaborate. Two or more forces acting in the same or opposite directions add or subtract in the ordinary way. Two forces at an angle to each other will act on an object in a direction between those of the forces, and they will do so with a resultant force less than the ordinary sum of the two forces. We can make a useful diagram of the situation by representing the directions of the forces by arrows and making the length of each arrow proportional to the magnitude of the force (fig. 2.4a).

A simple graphical technique (vector summation) can then combine the forces to determine the resultant force. Leave one arrow attached to the original point and then move (translate) the others without stretching or rotating them. Fix the tail of one of the loose forces onto the head of the attached one, put the tail of another loose one on the head of the newly fixed one, and continue in this way until all forces are represented. If the head of the final force gets just back to the original point, there is no net force, and we say that the forces are "balanced." If the final force has its head somewhere else, one merely draws a line from the original point to the head of the final force, with an arrow pointing toward the head. The line represents the magnitude and direction of the net, or resultant, force (fig. 2.4b). This graphical process permits an end run around a bit of trigonometry. If you prefer something more concrete than force vectors, think of two people pulling a heavy object, each pulling with a separate rope and in slightly different direction, and ask about the net effect on the object. Or think of a migrating bird heading in one direction with a wind blowing in another and ask just where and how fast the bird is going—velocity vectors work just the same way as force vectors. And subtraction of vectors works the same way as summation—one merely takes the vector to be subtracted, points it in the opposite direction, and then adds this reversed vector instead.

Just as two plus three make five, five can be dismembered into two and three. Similarly, a vector can be dismembered ("resolved") into two components. One particular resolution of a vector will be useful when we consider the aerodynamic forces, lift and drag. A vector in any direction can be resolved into a pair of components at right angles to each other. Thus, any velocity can be viewed as the combination of a velocity on a north-south axis and a velocity on an east-west axis. Or a force directed obliquely upwards can be resolved into a vertical component and a horizontal component. Resolution takes only a pair of dotted lines on a graph, as shown in figure 2.4c. Alternatively, one just applies the Pythagorean theorem for triangles with a right angle—the square of the length of the side opposite the right angle (the hypotenuse) equals the sum of the squares of the lengths of the other two

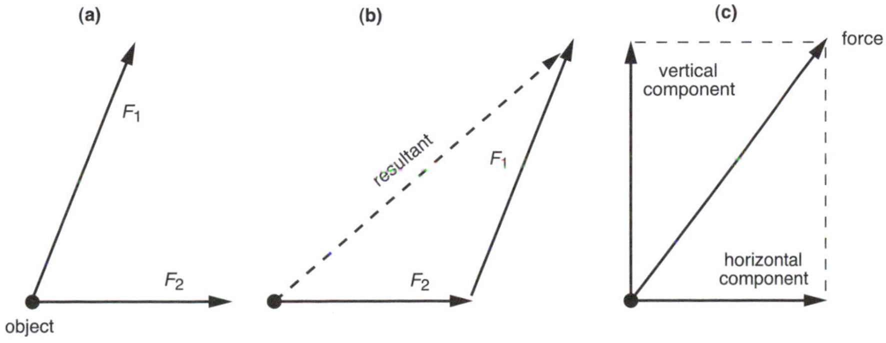

FIGURE 2.4 In (a) and (b) a pair of force vectors acting on an object are combined (summed) to give a resultant (dashed line). In (c), a vector is resolved into two components at right angles; these would restore the original vector if combined.

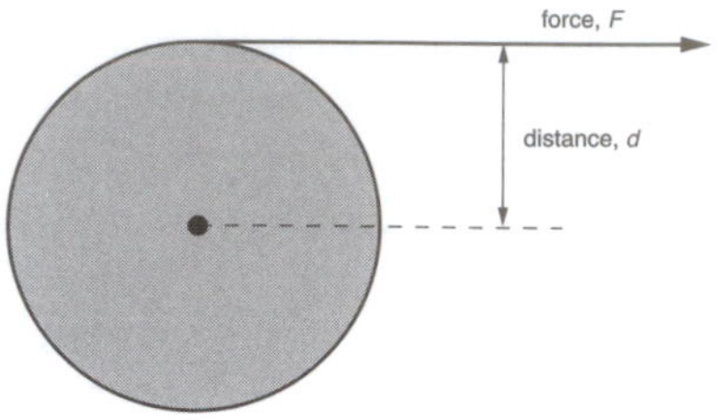

FIGURE 2.5 A wheel may be set into motion by a force on its rim; the effectiveness of the force is given by its "moment," the product of the force (*F*) and the perpendicular (shortest) distance (*d*) from its line of action to the axis of rotation.

sides. Put symbolically, $c^2 = a^2 + b^2$, where c is the overall force and a and b are its components.

Dealing with vectors in three dimensions, in principle at least, doesn't differ. It's awkward, though, with two-dimensional paper, so we'll stick to a two-dimensional slice of reality and leave the third to your imagination.

Moments of Force

So far, we've used forces only to get objects into translational motion. What makes things rotate? If you push on the axis of a wheel you merely make it translate; if you apply a force to the rim it begins to spin. (The decently skeptical will argue that pushing on the axle of bicycle or rolling pin will cause rotation. It does so because the rim touches the ground. Taking bicycle or rolling pin as a translationally fixed frame of reference, the ground or dough is moving and applying a force by contact and friction with the rim.) *Just how effective is a force in inducing rotation depends on how far from the axis of rotation is its line of action.* So the effectiveness of a force, what we call the *moment of the force* (and which has nothing whatsoever to do with time), is the product of the magnitude of the force and the perpendicular distance from the line of action of the force to the axis of rotation, F times d in figure 2.5. The action of a force, then, depends on three factors—its magnitude (as a quantity), its direction (as a vector), and, finally, its line of action. Note that a moment has the same dimensions as does work, but the two differ fundamentally—for work, force and distance run in the same direction, whereas for a moment the two are perpendicular.

Equipped with the notion of a moment of force, you can make sense of lots of practical devices. Levers, for instance, are practical applications of nothing more than these moments. To extract a nail with a pry-bar you have to exert a moment in one rotational direction that just exceeds (at the start) and then equals (for steady motion) the moment in the opposite direction caused by the resistance of the nail to extraction (fig. 2.6a). A pair of pliers, wire snips, scissors, or shears operates in a similar way—you can understand why some have long handles and short business ends and others the opposite. And that's why you can turn a nut more effectively with pliers if you put the nut as near the hinge point as you can and hold the handles out near their far ends.

Another large class of devices, from crankshafts to bicycles and including just about all of our muscles, tendons, and joints, can be illustrated by a so-called windlass (fig. 2.6b). With such a device, a person applying a small force can generate a very large force. It isn't a matter of something for nothing—to

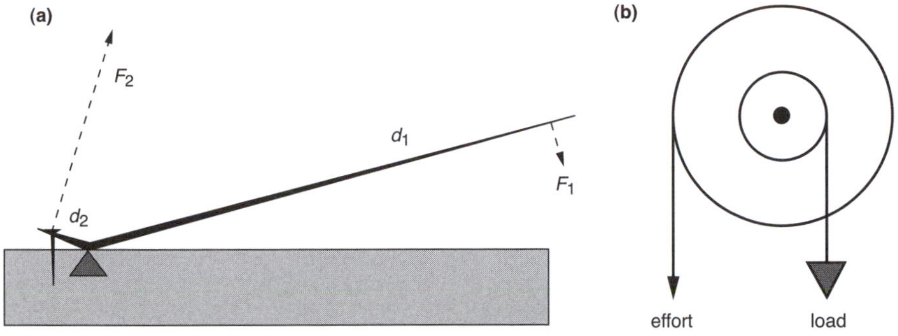

FIGURE 2.6 (a) A pry-bar extracts a nail—a small force (F_1) at the end of a long handle (d_1) produces a large force (F_2) on a short lever arm (d_2). (b) A windlass gets a large upward force by doing the same thing—the same moments (as in fig. 2.5) but with different distances and hence different forces.

make the large force move something a short distance the small force must move a long distance. An extreme version of the windlass was used (with teams of horses) to pull the thousands of stumps involved in excavation of the Erie Canal between the drainages of the Hudson River and the Great Lakes in the 1820s (Tarkov 1986). In the arrangements of our muscles, the load and effort are ordinarily reversed—muscle generates lots of force but doesn't shorten all that much, so, as in figure 2.7, the load is at the end of the long arm (which may be literally an arm). Nature takes muscle, an engine that produces lots of force but shortens only slightly, and reduces force and increases distance by the way she hooks it onto our bones. We then devise tools (such as that pry-bar) that reverse the process, increasing force and decreasing distance.

Force: The Newtonian Summary

Newton's three great laws of motion have been introduced without specific notice—they bear repetition and emphasis. The first law is that matter of unbalanced forces . . .

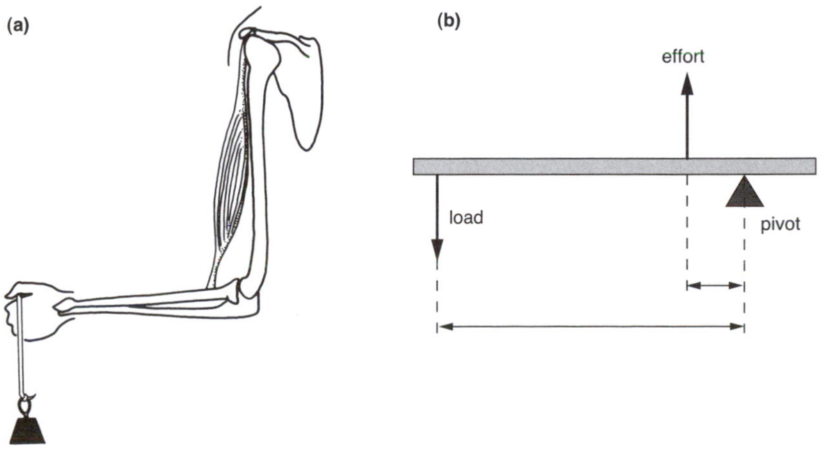

FIGURE 2.7 (a) A force exerted by the biceps muscle of the upper arm supports a handheld weight. (b) The lever arrangement involved, showing how a little shortening of the biceps can give a wide sweep of the forearm and weight.

A body stays at rest or in uniform motion in a straight line unless a force is applied to it.

The second law is equivalent to our definition of force, using the equation, $F = ma$ . . .

Acceleration is proportional to the applied force and is in the same direction as the force.

The third is the assertion that the brick fights the squeeze and the floor pushes up against your weight . . .

When one body exerts a force on another, the second always exerts a force on the first; the two forces are equal in magnitude, opposite in direction, and act along the same line.

Conservation Laws

Perhaps the most elegant and powerful statements in physics are the conservation laws. Each states that in any process that takes place in a properly isolated system something remains quite dependably unchanged. That is, amidst all of the confusing alterations of states of matter, chemical arrangements, velocities, shapes, and so forth, some item will add up to the same total amount at any time as at any other time. Several of such laws will be of especial use to us.

Conservation of mass

It may seem obvious to us, but it was far from clear to the ancients that mass does not change in any normal process. Most of the difficulty concerned chemical changes, and the confusion came mostly from ignorance about the nature and mass of gases.

Conservation of mass may not make you sit up and take notice, but the notion has an important application in fluid mechanics, where it's called the "principle of continuity." Consider a functionally incompressible fluid—water, for instance—flowing through a rigid pipe. Because the fluid is incompressible, mass and volume remain proportional, so we can talk about either just as well. Clearly, the rate at which mass flows past any one point along the pipe must be the same as past any other point—nothing can accumulate anywhere, because the water is incompressible and the pipe is rigid. (Neither is ever *perfectly* true, but both may be *very* good approximations.) In particular, the volumetric rate at which water enters one end of the pipe must be the same as that at which water leaves the other. That's the basis for the use of a nozzle—constrict the exit, and, if the volume leaving per unit time is unchanged, then the water has to leave at a higher speed. In some cases the pipe needn't have walls. Water in a column falling downward from a vertical spigot accelerates as it falls; because its speed increases, the cross-sectional area of the column decreases. The mass of water passing each horizontal plane in a unit of time must be the same or the conservation of mass would be violated.

It turns out that the principle works for gases in quite as simple a form—unless the speed of flow is higher than ever happens in living systems, gases are not compressed by flowing and the relationship between mass and volume doesn't depend on speed. (For cases where compression is appreciable, density is no longer constant—the conservation law still holds, but its application gets more complicated.) This principle of continuity is basic to the understanding of all circulatory, respiratory, and similar fluid transport systems in organisms. Because flow in the tiny capillaries of the body or in the small tracheoles of the lungs is vastly slower than flow in the aorta or bronchi, the total cross-sectional area of capillaries or tracheoles *must* exceed that of aorta or bronchi—the details will get attention in chapters 9 and 10.

A numerical example (put diagrammatically in fig. 2.8) will illustrate the way the principle can be used. Let's estimate roughly how many individual capillaries are open and functional in a person at rest, certainly a tough thing to get by a direct count. But by the principle of continuity, all the blood leaving the left side of the heart in a given time must pass through the capillaries of the body in the same time. And all the capillaries have about the same diameters. So cardiac output (volume per time) must equal the product of capillary flow speed (distance per time), the cross section of a capillary (distance squared), and the number of capillaries. We know the output of the heart from its stroke volume and rate of beating—1×10^{-4} cubic meters per second. (I've put everything in SI units—the numbers may look strangely small, but it makes the calculations easy.) It's easy to measure the area of a capillary and the speed of blood flow through it—functioning capillaries and moving blood cells are visible with a good microscope in thin tissues such as the extended tongue or interdigital webbing of a frog, or the mesenteries of a rabbit. The results are 3×10^{-11} square meters and 5×10^{-4} meters per second. So the number of capillaries must be about seven billion (U.S.).

The principle of continuity is a handy tool in all sorts of situations. After a storm, a creek rises. Does the speed of flow in the creek increase in direct proportion to the total water it carries? Certainly not: if this were true the creek wouldn't rise. Incidentally, the piers of the old London Bridge (built in 1209, it lasted until 1832) were protected with "cutwaters" (essentially tiny islands) so wide that almost half of the cross-sectional area of the Thames was blocked. As a result, the speed of flow nearly doubled at the bridge, giving rise to a substantial hazard and to the expression, "shooting the bridge," for a boat passing under.

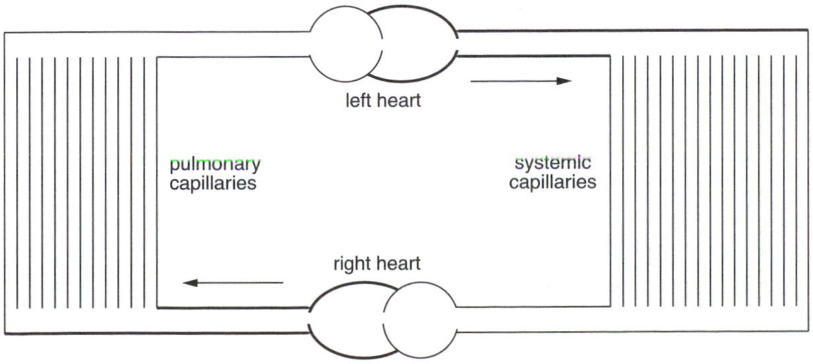

FIGURE 2.8 The principle of continuity demands that average flow speed in a circulatory system be inversely proportional to the aggregate cross section of the vessels at any stage in the system. That lots of vessels run in parallel makes no difference—aggregate cross section is what matters. Thus flow in the capillaries will be slower than flow in either side of the heart or in the large vessels.

Conservation of momentum

The main reason we bother with the composite quantity, momentum (mass times velocity), is that recognition of the concept of momentum allows us to state another and even more powerful conservation law. Mass is a scalar quantity, whereas velocity is a vector quantity. So their product has a direction, that of the velocity. We thus can recognize that an object traveling in one direction will have positive momentum whereas one traveling in the opposite direction will have negative momentum. (Which we call positive and which negative is entirely arbitrary.) If two blobs of clay of equal mass come together at equal and opposite speeds and stay stuck, the resulting mass will be at rest. The sum of their momenta before the collision was zero; it remains so afterward. Nothing momentous yet.

But consider the hose and nozzle again. Constrict the nozzle and the hose pushes back on you as the water exits. What has happened is that the speed of the water leaving the hose has increased from its speed when it was traveling along the hose, but the mass of water leaving in any period of time has not decreased (the principle of continuity forbids it). So the mv of water leaving the nozzle is greater than that of water flowing within the hose, and the mv difference pushes the hose and its holder backward. If the movement of the hose is not damped (by holding it), things can get rather well dampened. If you stand and hold the hose, we assert that the counteracting mv appears as a change in the rotation of the earth transmitted through your feet. Because the earth has a generous mass it doesn't complain. In fact, after leaving the nozzle, the water pushes against air and earth and thus returns the borrowed momentum.

Other cases are less frivolous. The physics books make much of rocketry (expel enough gas backward and you'll move forward), and the recoil of cannon; in each case the conservation law permits nice quantitative predictions about the performance of the devices. A fish swims forward by imparting rearward momentum to the water—a liquid or gas has mass and thus when moving has momentum. Conversely, water passing an attached object slows down—its momentum decreases. The object experiences a force, drag, that pushes it, increasing the momentum of the object plus the substratum. If you can measure the rate of decrease of the momentum of the water flowing over the object, it turns out that you can calculate its drag without even touching the object. The measurement isn't especially difficult—because the water is incompressible, the principle of continuity permits you to figure out the momenta upstream and downstream solely from measurements (quite a few, though) of velocity. The trick is useful when dealing with shy or uncooperative organisms.

Conservation of energy

This one receives more public notice than any other conservation law, and it rejoices under the name of the "first law of thermodynamics." But it's less useful for biomechanical problems than the notion of momentum conservation, mainly coming into play when we step out of a purely mechanical domain—when heat or metabolism or the various ways of storing potential work come into play. What needs emphasis here is the distinction between these two conservation laws, so deceptively similar in the way they preserve

mv and $mv^2/2$, respectively. As just noted, momentum preserves the direction of its velocity—it's a vector quantity. By contrast, energy is not a vector quantity—squaring velocity gets rid of its directionality. Thus, in any collision, whether or not the bodies rebound, momentum stays the same, as with a pair of blobs of clay (or automobiles). In a collision, energy is conserved as well, but in quite a different sense. If two bodies of equal mass collide and stick, then kinetic energy, energy of motion, disappears. The conservation principle tells us that the energy must appear in some other guise—mainly, as it happens, as heat. If two bodies both having a resilience (chapter 15) of 100 percent were to collide, then kinetic energy would be unchanged and no other form of energy would increase.

Nonconserved quantities

Mass, momentum, and energy take their places in conservation laws—one can make a good case that we'd otherwise not have bothered defining the second and third of these. Force, the subject of all three of Newton's laws, is not conserved. You can exert a small force and make it into a large force with nothing more than lever, windlass, gear train, or pulleys. Archimedes blamed his inability to move the earth on his lack of a fulcrum, not on any muscular insufficiency. Or you can make a large force moving a short distance move a smaller force for a longer distance. That's how we typically use our muscles, as noted earlier.

Nor is power conserved, at least in the short term. Store up a quantity of energy and you can put out immense power, albeit briefly. That's the basis of almost all muscle-powered weaponry, from huge catapults to the various forms of one-person archery. Some small animals do mighty powerful tasks—consider the jump of a flea. It can jump a good fraction of a meter upward, which demands a substantial take-off velocity. But, because the creature has such short legs, that take-off velocity has to be reached in less than a millisecond. In fact, a flea accelerates at about 2,000 meters per second squared, roughly two hundred times gravitational acceleration. That takes a power output about ten times what a muscle ordinarily does even without applying a discount for internal energy losses. Put another way, a flea couldn't do its jump on direct muscle output even if it were 100 percent muscle, much less its 20 percent jumping muscle. Instead, a flea crouches down for a fraction of a second before a jump, putting energy into highly resilient elastic pads at a (relatively) slow rate, and it then triggers release of that energy at a high rate—multiplying power (Bennet-Clark and Lucey 1967; Bennet-Clark 1977) in a kind of leveraged fly-out.

States of Matter

Gases and liquids have provided some of the examples used up to this point; solids have provided others. In our everyday world, matter comes in these three states (although sometimes other states such as "dissolved" are recognized), and we need to point out the key differences among them. We're familiar enough with these to need little in the way of description, what matters is how the states of matter differ in behavior. One can base a proper explanation of the differences between states on molecular behavior; but that's not usefully intuitive. Alternatively, one can distinguish among the states of matter by the way each resists the application of forces or, more realistically, stresses. Table 2.3 summarizes the distinctions. Table 2.3. Which stresses each state of matter resists.

Solid

A solid resists compression, tension, and shearing. No matter what you do to a solid, it fights back. As a rule, the more vigorously you aggress upon it, the more it deforms; but the relationship isn't necessarily linear—doubling the stress needn't exactly double the strain, whatever you may have heard previously.

Liquid

A liquid resists compression and tension—squeezing and stretching. It's very hard to squeeze a liquid—squeeze it in a chamber and it will be forced out of any opening, a phenomenon underlying the design of quite a few pumps, including our own hearts. It's also difficult to stretch a liquid, although we rarely have occasion to try. Trees, as we'll discuss in chapter 5, have long columns of water under very high tensile stress running up their trunks, so the issue is no mere curiosity. But liquids are magnificently disdainful of shear—the coffee fits equally agreeably into a cup of any shape.

Gas

A gas resists only compression. Squeezing a gas may be easier than squeezing a liquid, but it still takes work. On the other hand, a gas doesn't mind being stretched in the slightest—indeed it's always trying to stretch (expand) itself and will fill the volume of any container rather than lying limply low like a liquid. When you think you're stretching a gas—pulling on a piston in a cylinder, for instance—you're really only working against the external atmosphere. This less than self-evident point may seem more real if you consider that the *less* gas in the cylinder, the *harder* it is to pull the piston outward (fig. 2.9)—any gas inside is working with you, not against you. And the gas cares as little about shear as did the liquid. Tabor (1979) provided a more elaborate account of the states of matter from this particular viewpoint.

The relationships among these three states hold some distinct peculiarities. Most of us will recognize that under everyday conditions there exist materials between solids and liquids. We have noncrystalline solids such as glass that behave like liquids if our time scale is sufficiently long. And we have viscoelas-

TABLE 2.3 Which Stresses Each State of Matter Resists

	Compression	Tension	Shear
Solid	X	X	X
Liquid	X	X	
Gas	X		

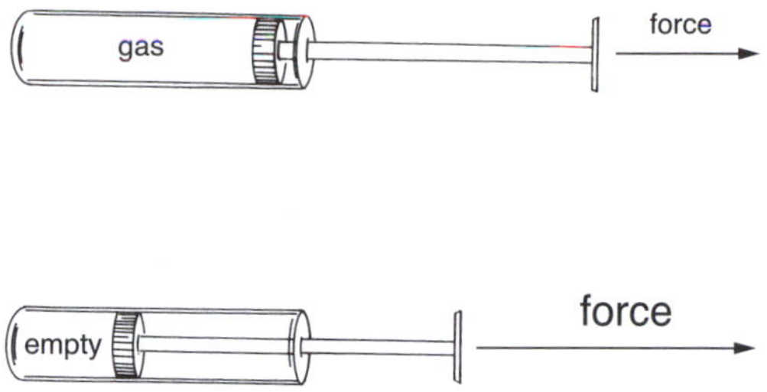

FIGURE 2.9 The presence of some gas makes pulling the piston outward easier than if a cylinder is completely evacuated. You're really pushing on the atmosphere outside, not pulling on the internal gas, so any internal gas at all makes the task easier—it's obligingly eager to expand.

tic stuff like paints, bread dough, and "Silly Putty" that have uncertain convictions about their shapes. Yet we make a useful distinction between, on the one hand, solids, and on the other, liquids and gases—as we will between parts II and III of this book. Conversely, we run into nothing in between liquids and gases. Yet we find it equally useful to lump the two as "fluids" because, as states of matter that flow, they share many behavioral features.

Maybe the best distinction among the three states is the simplest of all. A solid has size and shape; a liquid has size but not shape; a gas has neither size nor shape.

C H A P T E R

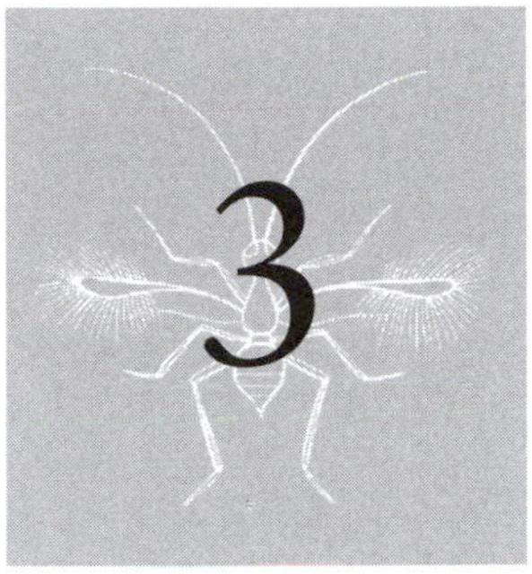

Size and Scale

Birdie, birdie,
in the sky—
Must your poop fall
in my eye?
I'm a big kid; I don't cry;
But I'm sure glad that
cows don't fly!

THE EASE WITH WHICH WE CAN AVOID WORRYING ABOUT ATOMS REflects the vast gap in scale between them and us, between the size of atoms and small molecules and even small organisms. Cells (or unicellular creatures) may be small, but inhabitants of the atomic realm are *much* smaller. There are, roughly, as many molecules in a cell as there are cells in the cat observing me write. (The point is crucial in Schrodinger's 1944 classic essay *What is Life*. One of his arguments is that well-ordered structures can be built of individually ill-behaved atoms only if enough atoms are used so that their actions are statistically dependable.)

But from smallest to largest, we organisms ourselves occupy an extensive size range—from the tiniest bacterium about 0.3 micrometers long (about a hundred-thousandth of an inch) to a whale about 30 meters (100 feet) long. (Some trees are 100 meters high, but they're no more massive than the whale.) The range is about a hundred millionfold; eight orders of magnitude we call it, counting the zeroes or factors of ten. An excellent introduction to the truly cosmic subject of size is *Powers of Ten*, by Morrison and Morrison (1982). And size looms large for comparative biomechanics, treating, as it does, organisms spanning the full size scale.

Among organisms we're near an extreme—our meter or two in length makes us relatively big creatures. Only a little over an order of magnitude separates us from the largest living things, but six to seven orders lie between us and the smallest. On a scale of orders of magnitude, a "typical" organism would be between a millimeter and a centimeter in length—roughly an eighth of an inch. The point about size isn't trivial—the nature of the physical environment of an organism, the phenomena of immediate relevance to its life, depend most strongly on how big it is. You may not need to imagine the world of an atom, but you'll find challenge enough in trying to get some intuitive sense of the physical world of small creatures. Incidentally, for all of our fixation on microscopes, biologists have not usually had much of that intuitive sense to which we'll aspire here.

The relationship between size and reality can be best put with a half-serious example. Consider all animals that live in air; that is, neither in water nor some solid material. These creatures are much denser than the medium around them and therefore can fall if released from a height. But size enters in any examination of this business of falling. We can divide organisms according on the consequences of a fall into four categories that depend mainly on size.

In the first category, made up of creatures above roughly 100 kilograms (220 pounds) in mass, injury is possible if the animal falls from a distance as short as its own height—tripping is a potential danger to cows, horses, and the like. The fall of an elephant is a matter of the utmost gravity. (We, especially as we get older, run a similar risk even at a lower mass; the upright posture of humans gives us an unusually great height relative to our mass.)

In the second category, comprising animals with masses between about 100 kilograms and 100 grams (4 ounces), falling may be injurious, but the fall must involve a distance greater than the height of the animal. Dogs should avoid cliffs, cats must climb down trees with deliberation, but squirrels, near the lower limit, can take riskier-looking leaps of faith. Hedgehogs (about 500–1,000 grams in mass), are also just above the lower limit, but, according to Vincent and Owers (1986), cope with falls by using a special device—spines that can act as shock absorbers.

In the third category, from 100 grams down to perhaps 100 milligrams (give or take an order of magnitude), no height is great enough to cause substantial injury from a fall—the hazard, if any, is the predator at ground level. Falls may all too often befall nestling birds, but do we ever notice one injured by impact? Some years ago, at the instigation of my skeptical colleague, Knut Schmidt-Nielsen, I dropped two adult mice onto pavement from the roof of a five-story building. Not only were they uninjured (very briefly stunned, though), but they adopted a spread-eagle, parachute-like posture and fell stably. It certainly looked as if the neural circuitry of these small rodents was arranged to deal with the circumstance. (The extent to which this posture reduces falling speed might bear looking into.)

The fourth category includes the smallest airborne organisms, for whom falling itself takes on a peculiar meaning. Upon release, the creature (by which I mean either plant or animal—the word "organism" is awkwardly deficient in commonplace synonyms) goes downward only in an uncertain, statistical sense. Air is never still, and at falling speeds no greater than the speeds of upward and downward movement of air a fall no longer goes dependably earthward. In fact, air is host to quite a diversity of seeds, pollen, spores, and tiny animals, to the great discomfort of those of us with allergies.

On the surface of the earth gravity (gravitational acceleration, strictly) is everywhere the same. Yet its practical effects are widely divergent, depending mainly upon the size of the organism in question. As Haldane (1928) put what took me far more words, "you can drop a mouse down a thousand-yard mine shaft and, on arriving at the bottom, it gets a slight shock and walks away. A rat is killed, a man is broken, a horse splashes."

So size is no small consideration in the design of organisms. We living things come in a wide range of sizes—one can emphasize the point by comparing a bacterium and a whale. Since our intuitive scale of size seems more closely associated with length than with volume, we get little sense of the situation from the bland assertion that 10^8 cubed or 10^{24} of the smallest bacteria would fill the volume of a whale. So let's lay those 10^{24} bacteria end-to-end and note the length of the line. It turns out that the line would encircle the earth 7.6 billion times—still too large to imagine! Alternatively, it would go to the sun and back fully a million times—still hard to envision. Or it would do the round trip to the nearest star (3.3 light years away) about five times, quite an astronomical distance, and not a lot of help either.

But a whale and a bacterium, whatever their similarities in overall shape or biochemical processes, are profoundly different sorts of creatures. Let's look at the range of lengths of some more biologically homogeneous sets of organisms (table 3.1). Not only are the ranges less than 10^8-fold, they are *much* less—the commonest factor is 1,000, a *hundred thousandfold* less than 10^8. And the sets that have larger factors, the algae and vascular plants, are structurally less uniform than the others. Clearly organisms are not like stretch socks—nature seems to have found that one size does not fit all, in both Darwinian and ordinary senses of "fit." That physically peoplelike organisms are roughly a meter rather than a millimeter in length can be no mere accidental historical accident.

Lengths, Surface Areas, and Volumes

Max Kleiber (author of the essay *An Old Professor of Animal Husbandry Ruminates*) calculated that if a mouse consumed food at the same rate, relative to its body mass, as does a cow, the mouse would need fur 20 centimeters thick to stay warm. To accommodate the fur coat, it would, (un)naturally, need either stilts or very long legs. By contrast, if a cow used fuel at the same rate, again relative to its body mass, as does a mouse, its body temperature would exceed the boiling point of water, and we'd be supplied with precooked steaks (Kleiber 1961). The difference boils down to the fact that the mouse has vastly more heat-losing surface area relative to its heat-producing volume than the cow.

Without a doubt, nothing is more important in determining how size affects biological design than the relationship between surface area and volume. Contact between an organism and its surroundings is a function of its surface, while its internal processes and structure depend mainly on its volume. And the two do not maintain a simple proportionality—unless you change the shape of a body you cannot simultaneously double both its surface area and its volume.

While perhaps not intuitively obvious, the situation can be easily illustrated with a few quick and facile calculations. We use rules relating changes in length to changes in area and volume all the time, but we rarely recognize the underlying generalities. Consider cubes of different sizes (fig. 3.1). Surface

TABLE 3.1 The Range of Sizes Within General Categories of Organisms

Group	Length Range (m)	Factor
Insects	10^{-4} to 10^{-1}	1,000 ×
Fish	10^{-2} to 10^{+1}	1,000 ×
Mammals	10^{+1} to 10^{+2}	1,000 ×
Vascular plants	10^{-2} to 10^{+2}	10,000 ×
Algae	10^{-5} to 10^{0}	100,000 ×

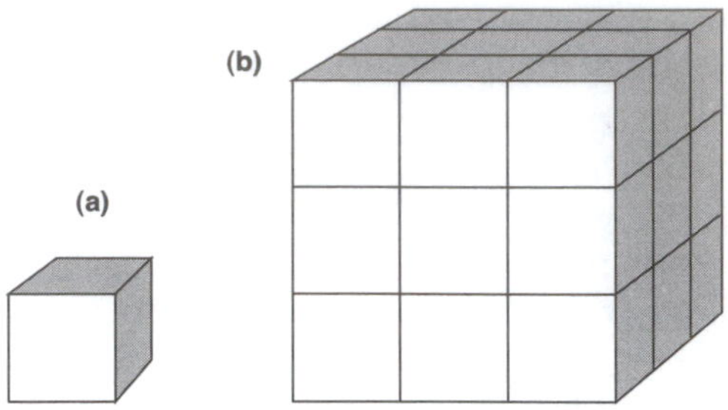

FIGURE 3.1 A pair of cubes. Each edge of cube (b) is three times as long as an edge of cube (a). Cube (b) has nine times the surface and no less than twenty-seven times the volume of cube (a).

area (S) and volume (V) are related to the length (l) of a side by the following formulas:

$$S = 6l^2 \text{ and } V = l^3. \tag{3.1}$$

Now consider a set of spheres and the analogous formulas for surface and volume in relation to radius (r):

$$S = 4\pi r^2 \quad \text{and} \quad V = \frac{4}{3}\pi r^3. \tag{3.2}$$

And then imagine a set of cylinders in which radius and height are equal (so $r = h$)—tuna fish cans, roughly. Their surface areas and volumes are

$$S = 4\pi r^2 \text{ and } V = \pi r^3. \tag{3.3}$$

In each case, the formula for area specifies the length of something squared and the formula for volume requires some length cubed. The situation is general, since areas always have dimensions of L^2 and volumes of L^3. At least it's general for any set of "isometric" objects, that is, objects of varying size but the same shape. It is, of course, not usually true for a collection of nonisometric (= "allometric") objects, objects differing in shape. But shape can be a secondary matter when sizes extend over orders of magnitude.

So what happens if you double *any* length measure of a solid body without changing its shape? You thereby increase both surface and volume, not twofold but fourfold (two squared) in surface and eightfold (two cubed) in volume. If you triple any length, you increase surface by three squared or ninefold, and volume by three cubed or twenty-sevenfold. The constants—6, 4π, etc.—in the formulas, being constant, make no difference to the argument.

Put as formal proportionalities then, the rules for sets of isometric (= geometrically similar) objects are the following:

Surfaces are proportional to the squares of lengths.
Volumes are proportional to the cubes of lengths.

We began by asking about the relationship between surface and volume as size (length) changes; we can now state that relationship merely by dividing each element of the surface proportionality by the corresponding one in the volume proportionality:

Surface/volume is proportional to one/length, or
Surface/volume is inversely proportional to length, or

$$\frac{S}{V} \propto l^{-1}. \tag{3.4}$$

The latter implies that a large object will have *less* surface relative to its volume than will a small object of the same shape. From this viewpoint, the main difference between bacterium and whale is that the bacterium has 100,000,000—or 10^8—times as much surface, *relative to its volume*, as does the whale. The long and the short of it is that the whale has a lot of inside with little outside while the bacterium has a lot of outside with little inside.

In case after case, this reduction in relative surface as a body gets larger has elicited a specific response from the evolutionary process. The essence of the response isn't a defiance of geometry, for that's simply not possible. Instead it involves evasions of the premise of isometry—great proliferations of extra, often hidden, surface in large organisms. A whale and a bacterium are vastly further from geometrical isometry than they appear, one is tempted to say, on the surface. A few "gee whiz" examples will emphasize the point:

1. A tree looks far less rotund than either a whale or a bacterium—that is, it has a lot of very obvious leaf surface. In fact, gas exchange between the photosynthetic cells of leaves and the atmosphere occurs at the walls of tortuous internal passages, so the functional surface area is from ten to more than thirty times greater than meets the eye. For an orange tree with two thousand leaves, the outer surface was measured as 200 square meters; but the internal surface for gas exchange is thirty times greater—6,000 square meters or 0.6 hectares (1.5 acres).

2. Plants absorb water and minerals from the soil; again an exchange process depends on having adequate surface. A figure is available for the surface area of the root hairs of a square meter of lawn planted with Kentucky bluegrass—the ten billion (U.S.) hairs have a surface of 350 square meters, double that of the floor of my whole house.

3. Our lungs are the functional interface between us and the atmosphere. The capacity of a pair of lungs is about 6 liters, but this modest volume, divided among three hundred million alveoli, is bounded by a surface of 50–100 square meters, about the floor area of a large classroom.

4. Topologically and physiologically (but not ordinarily) speaking, the cavity of one's entire digestive system is outside the body. Absorption of food into bloodstream and cells takes place mainly across the wall of the small intestine. Each of us has about 7 meters of the latter, looped around within the abdomen. This 7-meter pipe, though, has projections (villi) on it, which have smaller projections (millivilli?), and the surface cells have a border of microvilli. All the villification results in a surface of 2,000 square meters, or 0.5 acres.

5. Even within an organism, sites of exchange processes are sites of surface proliferation. Our muscles exchange various substances, oxygen and carbon dioxide in particular, with the bloodstream. They do so across the

> walls of capillaries, which have a combined surface area of about 6,000 square meters, the same as the total internal surface of the leaves of the orange tree.

What happens, then, are increasingly large deviations from functional sphericity in increasingly large biological systems, whether organisms or the internal components of organisms. The arrangement is recognizable even at the cellular level—many crucial bits of metabolic machinery operate on surfaces, and cells are full of internal membranous surface. If a stack of surfaces maintains its absolute (*not relative*) internal spacing whatever the size of cell or organism (as with the spacing between sheets in a pile of paper), then this particular allometry will keep the system's surface area proportional to its volume. Increase in size, whether at the level of cells or of organisms, and whether during the growth of an organism or the evolution of some lineage, is typically allometric rather than isometric—the parts increase out of proportion to one another. And thus shape turns out to be a function of size.

Pressure, Tension, and Radius of Curvature

Another biologically important relationship that depends on size—touched on in chapter 1—is that between the pressure difference across the curved walls of cylinders and spheres and the resulting tension in the walls of those cylinders and spheres. What we're talking about are biological balloons—thin-walled structures in which a tensile wall surrounds a pressurized volume. We'd like a formula describing how the relationship between the pressure within a cylinder and the tension in its walls depends on size, a rule analogous to that of equation (3.4). This isn't a trivial item—it's tension that splits cylinders, not pressure inside per se. And on that difference between pressure and tension hang some odd and severely size-dependent goings on.

Consider a spherical balloon that one slowly inflates or deflates while monitoring its internal pressure, as in figure 3.2a. The larger the balloon, the tenser its surface membrane and, one might imagine, the greater its internal pressure. But the manometer tells a different story, with pressure remaining more nearly constant than proportional to volume (fig. 3.2b). A little higher pressure happens near zero inflation, and (if tested destructively) a little higher pressure occurs just before the balloon explodes. (And, once pointed out, you'll recognize that you experience just this behavior of pressure every time you blow up a balloon—hard to start, but little harder to expand as it gets bigger.) The same pressure-versus-volume behavior happens in urinary bladders and in other biological bags. A significant medical literature once invoked such notions as "reflex softening" to rationalize this constancy of pressure, a literature presumably authored by people beyond such trivia as balloons.

Why should pressure and tension not remain in simple proportion? Just as their dimensions—length squared and length cubed—provided the key to the relationship between area and volume, the dimensions of pressure and tension provide the key here. Pressure is expressed as a force per unit area, tension as a force per unit length. We haven't yet talked about the latter, but it's not all that strange. Stretch a thin sheet of, say, rubber—the wider the sheet, the more force you have to exert to stretch it a given distance. Press on a resilient surface—the more the area of surface, the more force you have to exert to

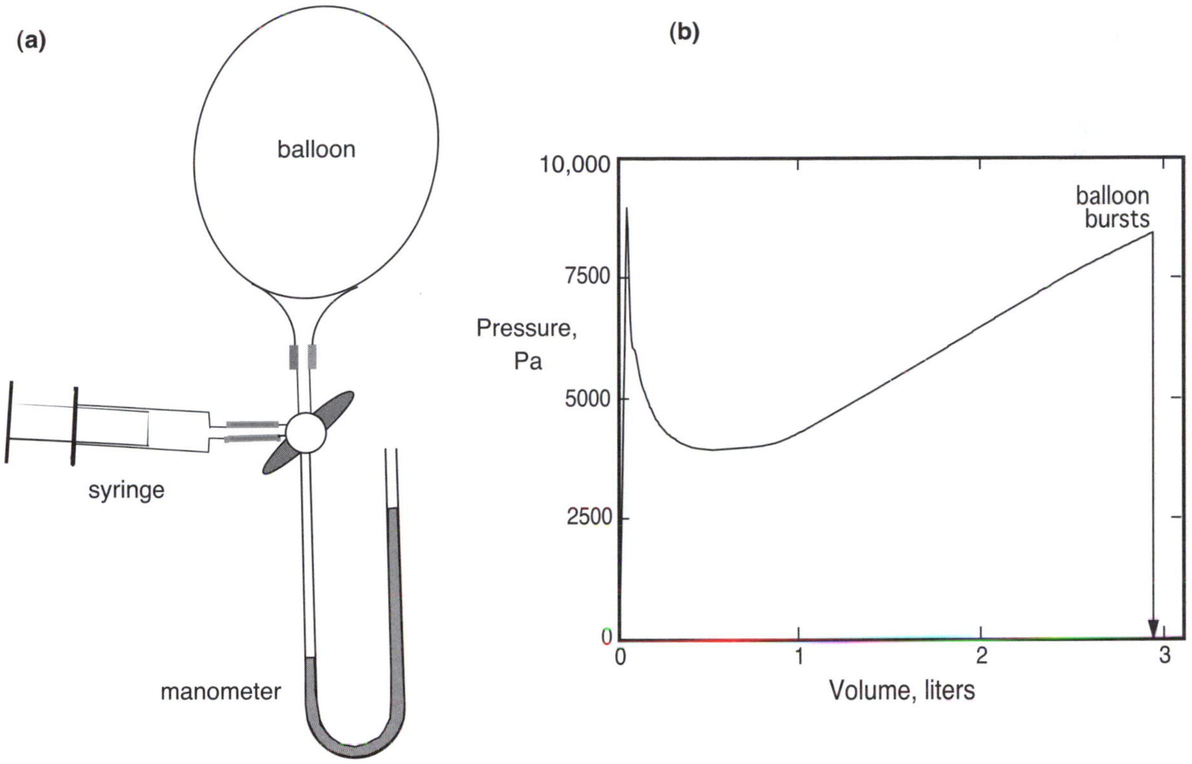

FIGURE 3.2 (a) A balloon with a manometer monitoring its pressure as it's inflated in standard increments of volume. (b) The resulting graph of pressure versus volume for such a balloon.

press it a given distance. So to convert a tension to a pressure, that is, to get the internal pressure a wall tension generates (or the wall tension caused by some internal pressure), we have to divide tension by some length. The specific formula depends on whether we're dealing with cylinders or spheres—cylinders curve only way, while spheres curve in two directions. Using Δp for the pressure difference between inside and outside and T for the wall tension, for a cylinder

$$\Delta p = \frac{T}{r} \quad \text{or} \quad T = \Delta pr \tag{3.5}$$

$$\Delta p = \frac{2T}{r} \quad \text{or} \quad T = \frac{\Delta pr}{2}. \tag{3.6}$$

The underlying proportionality between pressure, tension, and radius for curved vessels has come to be called, at least by biologists, "Laplace's Law," after the French mathematician and astronomer, Pierre Simon Laplace (1749–1827). In fact, Thomas Young (1773–1829) may have gotten it first, and "Laplace's law" gets confused with the "Laplace transform" and "Laplace's equation" in mathematics.

The rule, call it what you may, has a simple origin. Consider the balance of forces of a cylinder that has been split lengthwise into halves (fig. 3.3a).

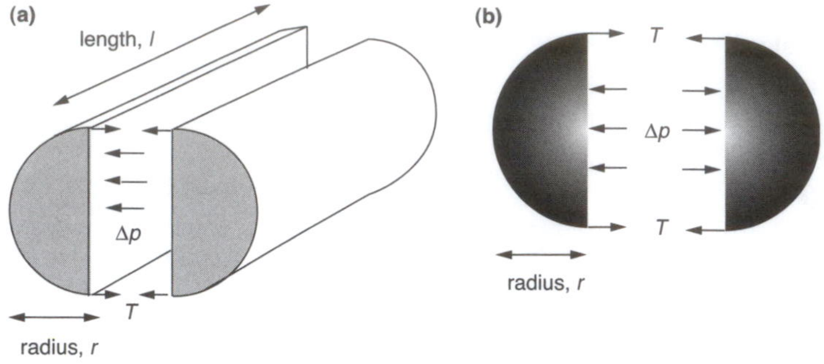

FIGURE 3.3 (a) A piece of a hollow, thin-walled cylinder of length l and radius r, split lengthwise. Internal pressure Δp pushes radially outward; for the split shown, it acts as if pushing on the interior rectangular planes—thus over an area of $2rl$. Wall tension T opposes such a split over the length $2l$ of the cut wall. (b) The analogous case of a sphere. The outward pressure force is $\pi r^2 \Delta p$, while the opposing tensile force is $2\pi rT$.

The halves are being pushed apart by pressure pushing each way on an effective area equal to that of the split surface—the length (l) times twice the radius ($2r$)—so the outward force will be $2\Delta prl$. But the cylinder is held together by the tensile force at the cut top and bottom edges; the latter will be $2Tl$. Equate the two forces, rearrange, cancel, and you have equation (3.5). For a split sphere (as in fig. 3.2b), $\pi r^2 \Delta p$ will push outward while $2\pi rT$ will hold the halves together—pressure and the sphere's cross section versus tension and the sphere's circumference. Equating the forces and rearranging gives equation (3.6).

Consequences—cylinders

The implications and applications are at once profound and ordinary. The skinny tires of a racing bicycle are normally inflated to pressures around 700 kilopascals (7 atmospheres or 100 pounds per square inch); the tires of an automobile take only about 200 kilopascals (2 atmospheres or 30 pounds per square inch). Yet the latter must withstand greater tensions in their walls, which is why their walls must be considerably thicker.

Some years ago I built a flow tank so I could subject organisms and models to water currents. Water pumped through the top channel had to return through a similar channel a few feet beneath—so the water in the bottom had a hydrostatic pressure of about 10 kilopascals (1.5 pounds per square inch). I was all too aware that the cost of pipe goes up with the *square* of the diameter. Larger piping has not only a greater circumference, the latter increasing in proportion to the diameter, but it also needs thicker walls to withstand the same pressure; Laplace's law demands that thickness also increase in proportion to the diameter. This matter of economics (or economics of matter) led me to build my first big flow tank with a return channel of cheap plywood instead of expensive plastic pipe, and using plywood meant a square cross section for that channel. Laplace then hit from a different angle, and I learned the hard way why pipes should be round, not square. A square pipe has flat walls; these have an infinitely large radius of curvature (or, strictly, a very large radius, since the walls have some curve-concealing thickness). According to equation (3.5), an infinite radius implies that *any* internal pressure (such as my 10 kilopascals) will generate an infinite tension in the walls. In fact, the walls bulged a bit, greatly stressing the corner joints and generating recurrent and persistent leaks. Since then, I've purchased cylindrical pipe for return channels.

It should come as no surprise, then, that very high pressures are recorded in small nematode worms, cylindrical animals less than a tenth of a millimeter in diameter. Nor should it be surprising that the relatively thin walls (around 20 micrometers) of the tubular xylem elements (with radii of about 0.1 millimeters) that carry water up the trunks of trees commonly withstand inward pressures of the order of 2 megapascals or 20 atmospheres. Here both pressure and wall tension will be negative, with failure a matter of buckling collapse rather than explosion, but the same relationship between pressure and tension must apply. Our sense of what constitutes a high pressure needs adjusting for the size of the system.

Blood vessels are cylindrical, withstand internal pressure, and (as we'll explore further in chapter 10) they vary greatly in diameter. So the operation of Laplace's law ought to be evident in their design. Our capillaries have an internal pressure averaging about a third of that of the aorta—4 kilopascals instead of 13 kilopascals—but (fortunately for material diffusing in and out of them) they take the pressure with walls only a single cell layer thick. The tension in the walls drops not threefold, but ten thousandfold, as you can readily check by plugging the radii of aorta (13 millimeters) and capillaries (4 micrometers) into equation (3.5). Wall thickness drops from 2 millimeters to 1 micrometer—two thousandfold (Burton 1954), so the tensile stress (tension divided by wall thickness) actually drops by 10,000/2,000—a mere factor of five. Thus the stresses on the material of the aorta and on a capillary wall are relatively similar.

One item about tension in the walls of cylinders will take on particular significance in chapter 20. We've been talking about circumferential stresses, so-called hoop stresses. Internal pressure in a cylinder generates tensile stress in a lengthwise direction as well. The lengthwise ("longitudinal") stress comes from a balance of forces equivalent to that used to derive Laplace's law for a sphere (fig, 3.3b)—pressure tries to blow the ends, circular in outline, off the cylinder. Tensile stress in any wall is equal to tension divided by wall thickness. So we can rewrite equations (3.5) and (3.6) in terms of tensile stress, σ, using Δr for wall thickness. For the hoop stress in a thin-walled cylinder, we get

$$\sigma_h = \frac{\Delta pr}{\Delta r}. \tag{3.7}$$

For the wall stress in a sphere *or the longitudinal stress in a cylinder*, we get

$$\sigma_l = \frac{\Delta pr}{2\Delta r}. \tag{3.8}$$

Notice that the hoop stress is exactly twice the longitudinal stress. Blow up a cylindrical balloon until it bursts—the slit at which failure occurs will run lengthwise. A lengthwise slit means that failure was circumferential—the balloon was blown apart radially rather than having an end blown off. A blood vessel must be (and is) more strongly braced against circumferential than against longitudinal failure. If a cylinder is helically (diagonally) wound with reinforcing fibers, these should run a bit more circumferentially than longitudinally—the stresses will be equally accommodated if the fiber angle is somewhat over 45 degrees, something upon which we'll elaborate in chapter 20. More specifically, the best angle is a little less than 55 degrees. Gordon (1978) commented that a cooked sausage fails with a lengthwise slit. That certainly agrees with my experience, at least with sausage made with artificial and presumably isotropic casing. But I wonder about the character of failure old-fashioned sausage made with a natural casing of intestinal wall.

Consequences—spheres

With its two-directional curvature, a sphere is automatically stiffer and stronger than a cylinder of the same radius of curvature. Or, to put it another way, a given pressure is half as effective at generating tension in the walls of a spherical container as it is in the case of a cylindrical container. That's, by the way, why there's no particular problem in putting hemispherical ends on a cylindrical tank and why the ends may have a larger radius of curvature (be flatter) than the side walls of the cylinder.

For spheres, Laplace's law has an additional consequence. The situation is best introduced by inflating balloons again. With the aid of a Y-tube and a few rubber bands, there's no great difficulty arranging so two spherical balloons with their insides interconnected can be pressurized at once (fig. 3.4a). But when you blow, inevitably one balloon inflates fully (or even explodes, sometimes) before the other begins to fill—the set-up is dramatically counterintuitive, with a soft balloon connected to a turgid one and yet with no possibility of different pressures inside the two. If the turgid balloon is squeezed sufficiently, it will begin to fill the limp one; at a certain point the squeeze becomes unnecessary and the (formerly) turgid balloon spontaneously completes the process, becoming fully flaccid and nearly empty. As with the cylinder, pressure generates more tension in the wall of a big curved surface than in the wall of a small one, but here we're looking at quite separate vessels.

An immediate biological application seems to have escaped notice by the people who design laboratory material for introductory biology courses. Each of us is equipped with two lungs—subdivided balloons inflated through a Y-tube. Inhale, and *both* lungs inflate—something that should now seem less than automatic. The fact that the lungs are inflated by reducing the pressure outside them (dropping the diaphragm and expanding the rib cage as in fig. 2.3) rather than by increasing the pressure inside (throat pumping, as in frogs) is inconsequential. What makes both lungs fill is that they are inflated separately—each one is in its own compartment, separated from the other by a

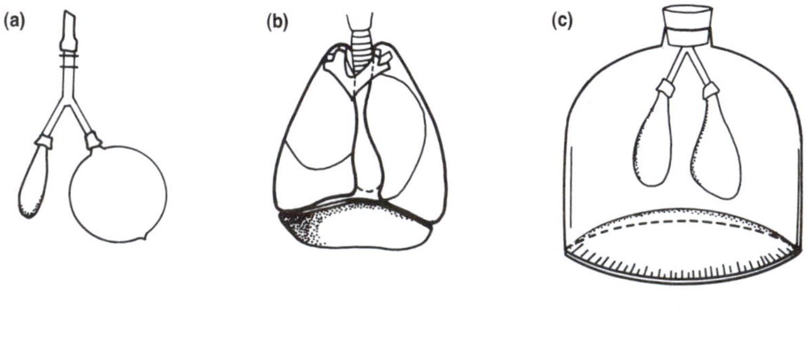

FIGURE 3.4 (a) A pair of balloons on Y-tube—only one inflates. (b) A pair of lungs in a chest—both inflate (in part) because they're in separate compartments. (c) The common classroom bell-jar model of a lung with balloons as in (a) and a rubber diaphragm on the bottom—pulling downward on the diaphragm won't make both balloons inflate by stretching, whatever the accompanying text might say.

sturdy wall, the mediastinum (fig. 3.4b). Separation is a good thing in another way as well—puncture of one lung isn't uncommon but the condition, pneumothorax, also isn't life-threatening since only one lung collapses. The common (and commercial) model, with two balloons in a bell jar (fig. 3.4c), is really a fraud perpetrated on generations of innocent biology students. Teachers routinely rationalize its failure as a matter of nonuniformity of balloons. I leveled scorn at the situation to little effect more than 30 years ago (Vogel and Wainwright 1969)—another impetus to write the present book!

The individual alveoli have somewhat the same problem as the pair of lungs—why doesn't one alveolus expand to the point of explosion (pneumothorax, again) before the others begin to inflate? Here the balloons quite clearly lie in the same compartment. Lungs filled with air take more force to inflate than do lungs deliberately filled with a salt solution. With air inside, the outward pressure difference across the alveolar walls must work against two resisting components of about the same magnitude, the elasticity of the tissue and the surface tension of the layer of water inside the alveoli. The latter opposes the formation of additional air-water interface as the alveoli expand. The surface tension, though, is drastically reduced by a wetting agent secreted by cells in the alveolar walls. But, and here's the trick, the effectiveness of the wetting agent depends on its concentration, which falls as the alveoli expand. Thus the force of surface tension rises sharply as an alveolus inflates, opposing further inflation. As a result of this wetting agent (or surfactant or detergent), the alveolar wall has a functionally curved stress-strain plot (chapter 5) and the requisite nonlinear elasticity. Underlying hyaline membrane disease of infants is a failure to produce sufficient wetting agent (Clements 1962). (As noted in chapter 1, blood vessels use an analogous trick to expand uniformly.)

And Laplace's law explains why very small spheres with only thin walls can withstand enormous internal pressures—plant cells often have internal pressures of over a million pascals (ten times atmospheric pressure) with walls only a few micrometers thick. (Discussion of the origin of such high pressures from osmotically generated forces will wait for chapter 22.) Nothing very special need happen, although cellulose fibers are certainly good at resisting tensile stresses—the cells are just small. Carpita (1985) tested the pressures a variety of thick-walled cells could withstand by subjecting them to sudden decompressions. He got values ranging from 3 megapascals for cultured carrot cells to around 10 megapascals for a bacterium and a green alga. Using equations (3.7) and (3.8), he showed that these pressures corresponded to a narrow range of tensile strengths—450 to 840 megapascals (less than twofold)—although the cell radii varied from 0.25 to 30 micrometers (120-fold) and their wall thicknesses varied from 0.003 to 0.1 micrometers (thirty-threefold).

The basic idea behind Laplace's law also applies to large domes. Here the analog of pressure is gravity. The loading forces are inward instead of outward, so the dome must take negative tension, that is, a compressive force per unit distance. And the load is no longer exactly radial except at the peak, so the analysis gets less like that of a sphere as one moves down from the peak of the dome. These differences notwithstanding, we can easily see that the dome with the larger radius of curvature (the bigger or flatter dome) is the weaker—a given weight produces a greater compression in its skin. Thus one has to make it stronger; if the latter means that it gets proportionately heavier then nothing is gained, and one needs a different design or better materials. Domes do occur in nature—a human skull is one, and we'll talk further about them in chapter 19—but even after some heavy thinking, I haven't come up with an example of a gravitationally loaded one.

The Scaling of Various Phenomena

The original example arguing the importance of scale involved falling under the influence of the earth's gravity. By this time it should be clear that many other biologically relevant phenomena are strongly size-dependent. To bludgeon the point, consider another example of our difficulty imagining the world of organisms differing in size from ourselves.

Animals attack each other in various ways. Chimps and humans throw projectiles, mainly nicely dense rocks, among their other aggressive tactics. Smaller animals, even manually dexterous ones, do not. Are we simply more clever? Probably not—our size merely preadapts us for throwing things. The momentum of a projectile, a reasonable measure of its destructive capability, is its mass times its velocity. Mass is proportional to length cubed, but projectile velocity is a little harder to guess. It's probably about constant, by the following reasoning. Muscle force varies with muscle cross section, and thus length squared. Mass will vary with length cubed. So (by $F = ma$) acceleration will vary inversely with length; the longer appendage has a proportionately longer path for acceleration but a proportionately lower value of acceleration—it will cast its projectile at roughly the same launch speed.

Thus if an animal can throw a projectile of mass proportionate to its own, then the momentum of the projectile (mass times velocity) will be proportional to the animal's length to the third power—quite a drastic relationship. Probably no small animal can impart enough momentum to a projectile to make it a significant weapon! There are, someone will counter, the nematocysts produced by the stinging cells of jellyfish and other coelenterates. But they aren't projectiles in quite the familiar sense—they require direct contact for their operation, so the prey, not the predator, pulls the trigger.

General rules can be suggested through such arguments, rules that ought to be relevant to students of animal behavior, natural history, and ecology. Kicking and hitting, for the same reason as throwing, are the prerogatives of large creatures—the projectile is simply part of the aggressor. Biting, crushing, and squeezing will work for somewhat smaller as well as for very large organisms, and these forms of violence are extremely common in nature. But for still smaller systems, the advantage gradually shifts to the defense—with more sharply curved surfaces, relative resistance to deformation improves (Laplace's law once again). A predator will find it more practical to pierce or burrow into the prey if a mechanical sort of aggression is to be used—mosquitoes and chiggers really get under one's skin. Small prey, though, have a very serious nonmechanical disability. Their surface-to-volume ratios are large, and the distances from their surfaces to their innermost recesses are short, so they are exceedingly vulnerable to chemical assault. The typical scheme for disposing of small prey, whatever the size of the predator, involves enticing them into some enclosed place and then pouring some digestive concoction on the intact creature. A predator may be unable to chew tiny food, but it need not do so anyway!

Gathering a lot of diverse items, most of which will reappear later, we can make a useful classification based on how things scale; that is, how relevant items change size as a system as a whole changes size. Note at the start that, for isometric objects, any length is directly proportional to any other, any area (surface, projected or shadow, etc.) is proportional to any other, and likewise for any measure of volume.

Some items vary more or less in proportion to the length of the system or organism—we say that they scale linearly with length. Height may seem a trivial case, but it does represent the distance an object will fall if it trips (to use an example developed earlier). For small organisms traveling slowly with respect to air or water the drag they experience is proportional to their length. The perimeter of an organism or of one of its feet quite obviously varies with its length (still assuming isometry). As a result, the peculiar upward force of surface tension (chapter 5) on a creature standing (not floating) on the surface of water rises with the creature's length. The distance from the edge to the center of a solid body also varies in direct proportion to its length—this distance is proportional to the distance material must move in order to be exchanged with the surroundings or between parts of the body.

Other things vary more nearly with the area, that is, the square of the length of the system. External surface is again the obvious case; its importance in exchange of material or heat (recall Kleiber's furry mouse) needs no additional emphasis. Organisms of moderate size and speed encounter a drag that varies nearly in proportion to their areas. The maximum force with which bones, muscles, and tendons can resist being stretched varies in proportion to their cross-sectional area, as does the force an actively contracting muscle exerts. And the capacity of a wing to produce lift or a tail to generate thrust varies similarly.

Still other items vary with volume. Mass, for instance—density varies little among organisms, so mass follows volume very closely. Indeed mass is the common measure of volume since it's easier to use a balance than to immerse something in a full bucket of water and measure the overflow. The water a submerged organism displaces, and thus the upward buoyant force on it, is just proportional to its volume. And so is its momentum at a given speed.

Some things scale both more and less drastically than any of these geometric quantities. The frequency of sound produced by many animals varies inversely with their lengths. Big dinosaurs probably had deep voices (Alexander 1989), and elephants communicate by using infrasound—sound of frequencies below those audible to humans (Payne 1998). A bigger fruit fly or mosquito has a lower frequency of wingbeat and thus a deeper-toned buzz—as is most common among animals, the males are smaller (higher-toned buzz), and sex recognition takes advantage of the difference. Similarly, a female toad prefers to mate with the largest possible male and approaches the one with the bassest voice. But the males "know" what's going on and, since frequency also tracks temperature, a male picks the coldest place, whether air or water, to sit while calling (Fairchild 1981).

At the other extreme, the speed at which a large organism hits the ground when it trips is proportional to the square root of its height (drag in air being minimal for short falls by large creatures). Since its mass is proportional to the cube of its length, the momentum (mv) with which it hits the ground is proportional to its length raised to the power 3.5, and the kinetic energy

($mv^2/2$) that has to be off-loaded is proportional to its length to the fourth power (Went 1968). A doubling of height increases momentum on impact more than elevenfold and increases energy sixteen times. Literally and drastically, the bigger they are, the harder they fall. Infants learning to walk are far more likely to be frustrated than fractured in falls.

What if the moderate-to-large organism falls far enough to approach terminal velocity, the speed at which the downward force of gravity (mg) equals the upward force of drag? Gravitational force is proportional to mass and thus to the cube of length. Drag is proportional to area (square of length) times the square of velocity, so, equating the force of gravity with that of drag,

$$l^3 \propto l^2 v^2. \tag{3.9}$$

Dividing each side by the square of length and taking the square roots we get

$$v \propto \sqrt{l} \text{ or } l^{0.5}. \tag{3.10}$$

Thus terminal velocity is proportional to the square root of length. The square root of length is, by the way, proportional to the *sixth* root of volume or mass. Double the mass of an animal and you increase its terminal velocity by two to the one-sixth power, a factor of 1.1225 or an increase of 12.25 percent—certainly less than one would guess. The increase in injurious momentum on impact reflects the increase in mass far more than it does the increase in speed.

Scaling Empirically

Up to this point we've talked about size in fairly loose terms, using some slightly vague length as a measure. And we've predicted relationships—with considerable success—from the underlying physical and geometrical rules. But we can reverse the process, going from measurements on organisms of different sizes to descriptions of the scale-dependence of their designs. A well-developed subject does just that. It considers scaling, or allometry—how scaling involves deviations from simple geometric isometry. One shouldn't forget, though, the empirical nature of the subject, which describes how organisms are built, even how we might expect a newly encountered organism to be built, but certainly not how they ought to be built based on any underlying theory.

What's the appropriate measure of size for such purpose? The choice of units is trivial; the real issue is the choice among quantities with different fundamental dimensions. We might measure length, or width, or diameter, or circumference; or we might use total surface or cross-sectional area; or we could use volume or even mass as an index of size. Which we choose matters less than does appreciating the consequences of the choice. In the physiological literature, body mass is the variable of choice, mainly because it's the easiest quantity to measure. It also avoids (at least initially) worrying about shape. (A few problems crop up involving an organism's inert portions, such as thick shells or internal water, but they're uncommon enough not to loom large.) And, as noted, body mass tracks body volume quite well in organisms.

So we'll stick with it, even if a length may be easier on the intuition. If you prefer to refer the scaling factors that follow to length, just multiply them by three.

We now have our measure of size—our independent variable that we can put on the x-axis of graphs. There's yet another problem. It's an easy thing to plot a graph for, say, the way the surface area of a cube varies with its length, putting length on the x-axis and area on the y-axis, as in figure 3.5a. The result is a set of points that define a curve (in this case a parabola). But the graph is awkward. First, we'd like to read from the graph a single number that somehow specifies the way area varies with length. If we had plotted the length of the diagonal of a face against the length of an edge, the graph would have been a straight line, and the slope of the line would give the relationship between the two variables—the constant of proportionality (1.414 in this case, incidentally). But length against area gives a nonlinear graph, so its slope has different values at different points on the curve. Second, the scales don't cover much range—if one wanted to consider cubes of lengths from 0.000001 to 10 meters (the size range of organisms) most of the points would be indistinguishably clustered right near zero. It turns out that we can get around both problems at once by a cleverly nonlinear set of labels for the axes.

Figure 3.5b has the axes relabeled as so-called geometric scales. These scales have no honest zero, and the progression of numbers is what's commonly called "geometric"—the distance between 1 and 2 is the same as from 2 to 4 or from 4 to 8 or from 10 to 20. The distance between 1 and 10 is the same as from 10 to 100 or 100 to 1,000 (or from 0.001 to 0.01 if an axis goes down that low). Let's consider what we've done in making this transformation.

The original graph, with linear axes, was convenient for plotting a function of the form

$$y = bx + a. \tag{3.11}$$

The function yields a straight line; one can get the value of a from where the line intersects the y-axis and the value of b from the slope of the line. The slope, to be tediously explicit, is the number by which an interval on the x-axis must be multiplied to get the corresponding interval on the y-axis. Or,

$$y\text{-interval} = (b)(x\text{-interval}). \tag{3.12}$$

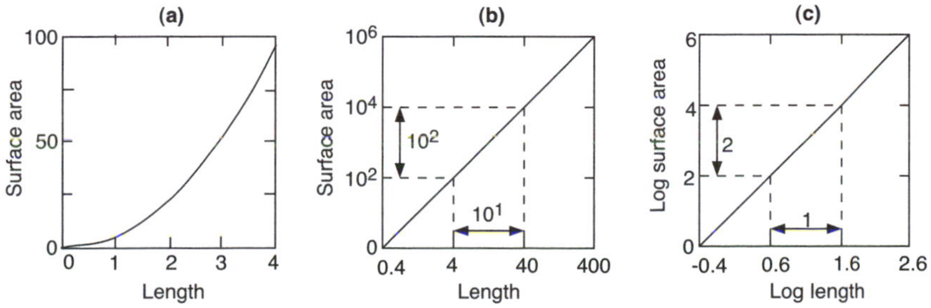

FIGURE 3.5 (a) How the surface of a cube varies with its length, using a linear scale. (b) How the same surface varies with length, using a geometric scale. (c) The same again, with the axes scales in logarithmic units. From (b) or (c) one can get the useful information that a tenfold change in length results in a hundredfold change in surface area.

The graph with geometrically scaled axes is convenient for plotting functions of the form of equations (3.1) to (3.4),

$$y = ax^b. \tag{3.13}$$

With these axes, one gets straight lines for such functions, and the value of b is the slope of the line. Except that the slope must be read in a way appropriate for these axes. Intervals on the axes are read, not as arithmetic differences, but as factors of increase. For instance, an interval between 4.1 and 41 is a factor of 10; between 100 and 10,000 there is an interval factor of 100. For these geometrically scaled axes, the slope is the *exponent* to which a given interval *factor* on the x-axis must be raised to get the corresponding interval *factor* on the y-axis. Thus, the rule for obtaining b is

$$y\text{-interval factor} = (x\text{-interval factor})^b. \tag{3.14}$$

The exponent b may be obtained by trial and error with a calculator if it has a key for exponents (labeled "y^x"). Or, if the calculator has a key for logarithms (any sort of logs will do), one can avoid trial and error and instead use the formula

$$b = \log(y\text{-interval factor})/\log(x\text{-interval factor}) \tag{3.15}$$

That, though, brings up logarithms, further use of which simplifies both the graphical presentation and extraction of exponents. Having first viewed the process in terms of exponents ought to minimize their magic and mystery. They provide an end run around the nonlinear axes—and any need for special paper—as well as permitting slopes to be read off as ratios of differences rather than factors of increase. One simply converts the data to be plotted into the corresponding logarithms and plots them as in Figure 3.5c. In effect, we've gone back to equation (3.11), so the slope is now b and the y-intercept is now log a:

$$\log y = \log a + b \log x. \tag{3.16}$$

We won't be concerned with the a in equations (3.13) and (3.16)—it's largely an accident of the particular units employed. If you ever need it, it's the value of y at an x of unity. (When $x = 1$, then $x^z = 1$ *for any* z; thus $y = a$.)

The whole thing was well summed up by D'Arcy Thompson; the second chapter of his *On Growth and Form* (1942) is the great classic of the literature on scaling, and Thompson's splendid language tempts one to quote again and again.*

> *It is a remarkable thing, worth pausing to reflect on, that we can pass so easily and in a dozen lines from molecular magnitudes to the dimensions of a sequoia or a whale. Addition and subtraction, the old arithmetic of the Egyptians, are not powerful enough for such an operation.*

*Sir Peter Medawar called the writing in *On Growth and Form* the best English prose in the scientific literature; J. T. Bonner provided a nice abridgment in 1961.

Our main use of a graph with geometrically scaled axes will be for finding the exponent of proportionality, the scaling factor, b, from empirical data. To be biological about it, consider the following example. You want to buy fish, paying in proportion to their weight, but you have only a tape measure for routine use. To get the scaling factor relating length to weight, you get a set of fish that vary in size. You measure the length of each fish with the tape measure and determine its volume by dipping it into a full pail of water and catching the overflow. Then you plot volume against length on a geometric scale or log volume against log length on a linear scale. Either way, you get a persuasively noncurved line with a slope of very nearly 3.0. What do you conclude? That the fish are for practical purposes isometric—shape is the same whatever the size—and that weight thus is proportional to the cube of length. If a 10-inch fish costs \$1.00 you should pay no less than \$8.00 for a 20-inch fish. (Incidentally, I tested the scheme, using it to pay my son for scaled fish to feed the two of us during the summer of 1981—it's surprisingly dependable—a little longer means a *lot* heavier.)

Scaling factors such as that exponent of 3.0 are becoming matters of great interest to biologists, and a spate of books—McMahon and Bonner (1983), Schmidt-Nielsen (1984), Peters (1983), Calder (1984), Pennycuick (1992), Niklas (1994), and Brown and West (2000)—deal quite extensively with them. The first two seem to be of the greatest general interest and accessibility. Harvey and Pagel (1991) give useful advice about techniques and subtle pitfalls in fitting allometric equations to empirical data.

Some Specific Scaling Factors

The literature on scaling is particularly extensive for mammals, giving quite a complete picture of the size-dependence of the construction and operation of the animals most like ourselves. Various aspects of the topic will reappear later; however, discussing a group of scaling factors (including nonmechanical ones) ought to help put the occasional additional factors into a proper context. The figures for b in table 3.2 have been extracted mainly from Schmidt-

TABLE 3.2 Relative Magnitude in Mammalian Design: Scaling Factors for the Exponent b in the Equation $y = ax^b$

Item	Scaling Factor
Surface area	0.65
Skeletal mass (terrestrial)	1.08
Skeletal mass (cetaceans)	1.02
Muscle mass	1.00
Metabolic rate, minimal	0.75
Effective lung volume	1.03
Frequency of breathing	–0.26
Kidney mass	0.85
Liver mass	0.87
Brain mass	0.70

Nielsen (1984). Bear in mind that $b = 0$ implies size independence; $b = 0.33$ represents scaling in proportion to length; $b = 0.67$ means scaling with area; and $b = 1.00$ indicates scaling with volume. And bear in mind also that the scaling factors have variable but usually substantial uncertainty—the original data are often fairly untidy. So on to the *b*s.

Surface

Are mammals isometric with respect to external surface area? That is, does our surface-to-volume ratio vary as it would for, say, a set of cubes, with a scaling factor of 0.67? Surface area proves to be an especially awkward quantity to measure, but some data do exist for animals ranging from mice to cattle; the factor seems to be between 0.63 and 0.67, or very near but perhaps slightly below the isometric value. The lower value implies that surface increases a little more slowly with size than the isometric expectation—that larger animals are a little more rotund and smaller ones a little more elongate or spindly overall. But not much, and data obtained within individual species stay very close to 0.67. (The size range within a species is, naturally, more limited, reducing confidence in the accuracy of the factor; but the data represent consistent technique in making area measurements, which rather improves matters in compensation.)

Skeleton

The realization that large mammals have disproportionately thick bones goes back at least to Galileo. The problem faced by the large creature is that a structure stressed by twisting, bending, crushing, or anything other than simple pulling (tension) is relatively weaker as it gets larger. In compensation, the scaling factor for skeletal mass is above 1.0—it's 1.08. Over the range of mammalian size, the difference is appreciable. An 8-gram shrew is about 4 percent skeleton; a 5-kilogram cat about 7 percent; a 60-kilogram person about 8.5 percent; a 600-kilogram horse 10 percent; a 7,000-kilogram elephant nearly 13 percent. You can tell from a drawing or photograph of a skeleton about how large was the animal merely from the proportions of the bones (fig. 3.6).

But the scaling only partially compensates for increased weight in large animals. We might crudely assume that the force or weight a bone can bear is proportional to its cross-sectional area and that the animals, overall, are an

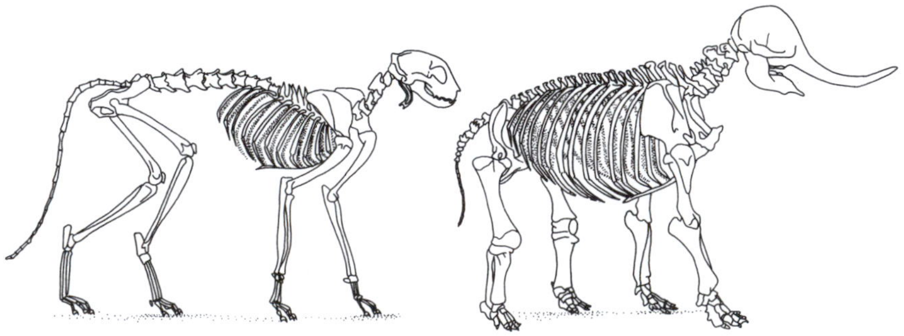

FIGURE 3.6 The skeletons of a cat (left) and an elephant (right) drawn to about the same size. Telling which is which presents no problem, even ignoring the elephant's tusks. Notice, in particular, the differences in both shape and position of the bones of the legs.

isometric set. Then the weight that must be borne increases with the cube of length—double the length and you increase the weight eightfold, but the force a bone can withstand goes up only fourfold. Meanwhile, the length of each kind of bone must increase in proportion to the length of the animal. So the bones of big animals must be disproportionately thick—their cross sections must increase in proportion to the mass of the animals—and their own masses must increase in proportion to their cross sections times their lengths. Thus, bones, overall, should increase in mass or volume in proportion to the *fourth* power of length—a b of 1.33, not 1.08. Built to the exacting standards of cats, a horse would therefore have to be no less than 34 percent skeleton and an elephant an impossible 78 percent. (As we'll see in chapter 18, this analysis tacitly assumes that bones cannot bend, that they fail or hit their practical limit in what's called "crushing failure." But long, thin columns more often fail by bending than by crushing. The difference, though, has little impact on this scaling argument—longer may be weaker, but wider is stronger than we've assumed.)

Big mammals are indeed fragile fellows, as the number implies, and they make do with limited margins of safety. A structure must be designed or evolved with some kind of worse-case criterion, and the stresses associated with locomotion greatly exceed those for standing. The larger the runner, the less time, proportionately, it spends airborne, so the peak stresses in its leg bones don't go up as fast as does body mass. But that single scaling factor of 1.08 may overgeneralize misleadingly, as emphasized by Christiansen (1999). Big mammals comport themselves differently. Elephants don't gallop at all, and their running gait differs from ordinary trotting and incurs less stress. What offsets the problem, at least to some extent, are the increased speeds at the gait transition points. Elephants can achieve respectable speeds while withstanding only (only!) the stresses of their trotlike gait. And (as suggested in the first chapter) really big dinosaurs may have been walkers.

All these figures are for terrestrial creatures. Because of the long interest in whaling, data exist for marine cetaceans, for whom impact and gravitational loads have little immediacy. Not only do they have less bone than terrestrial mammals relative to their body masses, but the amount of bone turns out to be more nearly proportional to length cubed or mass—b for mature whales is 1.02. The value for fish, incidentally, is a statistically indistinguishable 1.03.

Muscle

A muscle functions by contracting, which ordinarily pulls toward each other the bones to which its ends are attached. The force with which the muscle pulls is proportional to its cross-sectional area. By contrast, the work it does in pulling varies with that area times the distance it shortens, and that distance is proportional to the muscle's length. So the work a muscle can do ought to be proportional to its volume or mass. And, as far as data are available, muscle mass is proportional to length cubed or to body mass—mammals of all sizes are 40–45 percent muscle. Thus, the work a body's muscles are called upon to do ought to be proportional to the body's mass or weight—in short, to the load on the body. So things are nicely balanced.

This constancy has an everyday implication—the edible fraction of a mammal ought to be independent of its size. Size-related considerations of growth rate, relative fat content, and so forth may influence our choice of cuisine, but edible fraction is not a factor.

Metabolic rate

This is the power input to an animal, the rate at which it consumes fuel to carry out its activities. It's usually measured (and commonly quoted) as a rate of oxygen consumption, but with a few assumptions oxygen consumption rate can be converted to watts of power—1 liter of oxygen represents 20.1 kilojoules of work, or 1 milliliter per second represents about 20 watts of power. Here the data, for mammals at least, are voluminous.

First, though, the expectation. Virtually all of the power input should appear as heat, and heat loss depends on surface area. Unless heat production is proportional to surface rather than volume, big animals will get excessively hot—as with Kleiber's mouse and cow, mentioned earlier. If heat production and hence metabolic rate is proportional to surface (and the animals are isometric), b should be 0.67. In fact, the commonly quoted value of b, the slope of the classic mouse-to-elephant graph, is around 0.75, slightly higher.

The difference has been the subject of much debate, conjecture, and calculation—especially quite recently, when various nonphysiologists have taken that slope of 0.75 as a fundamental law of biological design and who have then been widely cited. So a note, really several notes, of sobriety ought to be offered. First, the specific value may be none too dependable a generalization, either statistically or biologically (see, for instance, Dawson [2001] and Dodds, Rothman, and Weitz [2001]). The measurements must be made under equivalent resting conditions, and animals vary in the degree to which they relax in metabolic chambers. If any of that variation happens to be size-dependent, then the slope of the graph will shift. For instance, if small animals remain more alert and jumpy, then the slope will be diminished. If large herbivores have a lot of residual digestive activity, then the slope will be increased. Second, the values refer to minimal metabolic rate. That may be the easiest standard to enforce, but it also may be the least relevant to biological design—it's equivalent to fuel consumption of an idling automobile. The more relevant maximum metabolic rate, much trickier to obtain, almost certainly scales with a higher exponent. Taylor and his collaborators (Taylor et al. 1981) got an exponent of 0.81, with a lot of scatter. Recently, Bishop (1999) reanalyzed all the data on both mammals and birds, cleverly standardizing for heart mass and blood hemoglobin concentration. That corrects for different styles of disportation—cats (small-heart lungers) versus dogs (big-heart chasers), for instance. The result, 0.88, is even higher and has a lot less scatter—I think that's as reliable a number as we have, and it simply won't support the arguments based on 0.75.

The ratio of maximum metabolic rate to minimal (or basal) metabolic rate—it's called "metabolic scope" or "factorial increase"—averages somewhere around 10 for mammals. It's a bit lower for small ones, higher for large ones since we can't equate slopes of 0.75 and 0.88. That is, a typical mammal can increase its resting metabolic rate about ten times. Not only size but lifestyle matter—cats and dogs again. Well-conditioned humans get well into the 20s; dogs can do 30; the scopes of horse and pronghorn antelope may exceed 50.

Whatever the precise value or explanation of the metabolic scaling factor, the fact that it's below 1.0—that metabolic rate increases with size more slowly than does mass—is profoundly important in determining differences in the ways of life of the small and the large. The oddity is baldly exposed by converting the basal value of 0.75 to one for metabolic rate *relative to body mass* by subtracting 1.0—the scaling factor relating mass to volume for isometric objects of uniform density. We get not zero but $b = -0.25$—*relative* food consumption is less for the larger animal. It can tolerate longer periods without food, it incurs a smaller relative price for maintaining a body temperature above that of its surroundings, and so forth. A bear sleeps through the winter with a body temperature near normal; it's sensibilities are easily offended, as one physiologist did not bear in mind when attempting to measure core (rectal) temperature. By contrast, many small mammals hibernate with a temperature that may drop nearly to freezing—they could never store enough fuel to stay warm all winter without feeding. Some hummingbirds, tiny bats, and a few other of the smallest mammals undergo short periods of "torpor," permitting body temperature to drop for a few hours when they're not engaged in specific tasks.

Metabolic scaling matters. Some psychologists once tested the effects of the hallucinogenic drug LSD on an elephant. They took a dose known to give a minimal effect on a cat and multiplied the feline dose by the ratio of the weight of the elephant to that of a cat. Upon receiving the injection, the elephant had violent convulsions and promptly died. It was concluded (West, Pierce, and Thomas 1962) that elephants are remarkably sensitive to LSD. But (as was soon pointed out) drug dosages scale more nearly with surface area (or metabolic rate) than with weight, and one shudders at the magnitude of the miscalculation.

A few internal items

The big mammal needs more skeleton than does the small one, but because of its relatively lower metabolic rate it should place lower demands on its respiratory system. Where might these lower demands be manifested? Effective lung volume ("vital capacity") is about proportional to body mass ($b = 1.03$) but the frequency of breathing is less in larger animals ($b = -0.26$); the algebraic sum of the two, in essence the oxygen supply, is 0.77. Similarly, the circulatory system needn't do as much in a big animal. Heart mass scales with body mass ($b = 0.98$) (in mammals but not birds—see Dawson [2001]) but the rate of beating goes down in large animals ($b = -0.25$); the sum is 0.73. That both come close to the 0.75 of minimal metabolic rate suggests that it might approximate maximum as well as minimum rate, or (as seems more likely) that other constraints operate. Organs that function in altering the composition of the blood seem to scale between the exponents of 0.75 for metabolic rate and 1.00 for mass. For kidneys, $b = 0.85$, and for liver, $b = 0.87$. The smaller animal has, relatively, the larger liver and kidneys, something perhaps familiar to butchers.

In addition, the larger animal has a relatively smaller brain ($b = 0.70$), but one wonders why (nerve cells being roughly the same size in all) the scaling factor is even this high. The only sense organ that lends itself to straightforward measurement is the eyeball; I collected some data from the literature and calculated a b of 0.60. Again, one wonders why eyes of the same size (equivalent to $b = 0$) will not suffice for all. Perhaps bigger animals commonly look at more distant objects—if rods and cones don't vary in size, then a large eye gives the same benefit (and without the extra cropping) as a telephoto lens.

If we divide one quantity by another where the two scale in about the same way, we get something that's size-independent. (Remember that these factors are exponents, so to get the scale factor for a ratio you subtract the denominator's b from the b of the numerator.) Heartbeats per breath is a fairly understandable ratio—it's about 4.5 whatever the animal or level of exercise. Some are more curious. For example, the life span of mammals has $b = 0.20$—the bigger live longer, on the average. Since frequency of breathing has a factor of −0.26, the duration of a breath will give one of +0.26 (time per cycle is the reciprocal of, or one divided by, a frequency). The ratio, breaths per lifetime, will scale with a b of 0.26 minus 0.20 or 0.06—not far from size independence! Heartbeats per lifetime ($b = 0.05$) is even more nearly independent of size. Perhaps the near constancy of these quantities is coincidental or incidental to some other factors—I've never heard a direct, functional rationalization.

Some more obviously important things do not vary much with the size of an animal; preeminent among the latter is cell size. To a first approximation, a cell is a cell—its size varies far more (and even that isn't too extreme) with cell type than with host organism. Any attempt at a functional explanation will have to wait for a discussion of diffusion (chapter 8). In fact, the constancy of cell size is quite a lot stranger than you might guess. Recall how different are the relative metabolic rates of, say, mice and elephants. If they have cells of the same size, then a typical mouse cell will have a metabolic rate (*not* just relative rate) more than twenty times higher than that of an elephant. It will need (and has) quite a lot more internal machinery—just as you can guess the size of the animal from a picture of its skeleton, you can guess the size of an animal from an electron micrograph of one cell of its liver! But why, with such a profound difference in metabolic rates, are the cells still the same size? A cell must be a pretty versatile bit of metabolic machinery. Thus, the cells of a human fetus have metabolic rates appropriate for the maternal size; very shortly after birth they shift to the higher rates normal for cells within an organism of infant size.

CHAPTER

More Tools

> I never used a logarithm in my life, and could not undertake to extract the square root of four without misgivings.
>
> **GEORGE BERNARD SHAW**

IN TALKING ABOUT SCALING, WE KEPT ORIENTED BY BEARING IN MIND the dimensions of lengths, areas, and volumes. Still, that use of the notion of dimensions was no more than a minor convenience, and you might well wonder why they got so much notice in chapter 2. As it happens, dimensions hold further utility. Perhaps most of us missed something important by learning mathematics from mathematicians who unconsciously abhorred any intrusion of physical reality into their subject. And dimensions form part of a nice scheme to help apply mathematics to practical matters. The scheme can be a real godsend even (possibly especially) to our crude and cursory view of things.

At some point you were undoubtedly informed that an equation is "true" or "correct" if the value of whatever is mentioned on one side exactly equals whatever is mentioned on the other. True—each equation we'll use states that the expressions on opposite sides of the "equals" sign have the same numerical value for at least one set of values of the variables. In addition to numerical equality, though, for the two sides to be identical, *they must be dimensionally the same*. That means that if the left side has dimensions of, say, length over time, the right side must also. In fact, each term of each side must have the same dimensions. (Terms are the parts of mathematical expressions completely separated by either pluses or minuses.) Otherwise one might be adding chocolate chips to pork chops or lengths to forces, both frowned upon. In the jargon, then, all terms must be "dimensionally homogeneous" in any equation that correctly describes an aspect of the real world.

Dimensional Analysis

So what of it? Dimensional homogeneity sounds like just another admonition to worry about, another place to go wrong. In fact, it's a fine thing, a blessing informally discovered by countless students trying to remember formulas for physics exams—whether there's a v or a v^2 in an equation can be determined if one knows the rest of the equation and the dimensions of the variables. The essence of its merit is a powerful (if a mite inelegant) bit of applied mathematics called "dimensional analysis." Langhaar (1951), in a book on dimensional analysis, describes it as "a method by which we deduce information about a phenomenon from the single premise that the phenomenon can be described by a dimensionally correct equation among certain variables." He continues, "The generality of the method is both its strength and weakness. With a little effort, a partial solution to nearly any problem is obtained. On the other hand, a complete solution is not obtained nor is the inner mechanism of a phenomenon revealed by dimensional reasoning alone."

How, then, to do it? For complex problems, formal theorems provide rules and guidance; here we'll get along with a semi-intuitive approach. The first step is to make an astute guess (yes, a guess, reprehensible as it sounds) as to what variables are relevant to a physical phenomenon. Then one represents each variable by its basic dimensions (M, L, and T). Finally one arranges the variables in the form of a dimensionally correct equation or proportionality. If the original guess was reasonable, then the resulting equation might just be useful.

A particularly good illustrative example was given by P. W. Bridgman (1961) in his article on dimensional analysis. He asked for a formula for the period of a simple pendulum—the time it takes to swing back and forth, something that we invoked when talking about the walk-trot transition in the first few pages of this book. One notices that the period is independent of the amplitude (width) of the swing, at least for small swings. And one guesses that the period (t) might depend on three variables—the mass of the bob on the end of the pendulum (m), the length of the string from bob to attachment (l), and the acceleration (g) of gravity pulling the pendulum towards its lowest position. We can obtain the dimensions of each of these variables from table 2.1:

t has dimensions of T
m has dimensions of M
l has dimensions of L
g has dimensions of LT^{-2}

What combination of the last three variables will equal the first? Somehow we have to get

$$t = m^a l^b g^c,$$

where a, b, and c are unknown exponents (that may be positive, negative, or even zero). Or, writing out the dimensions of the variables,

$$T = [M]^a [L]^b [LT^{-2}]^c.$$

Simple cases like this can be done by inspection—dead reckoning or trial by error:

1. To get the right side to come out to T just like the left, c must be $-1/2$. It's the only way possible since only gravity, on the right, has any time dimension.
2. If so, the last L has an exponent of $-1/2$.
3. Since there's no L on the left, the Ls on the right must cancel, which means that the first L must have an exponent of $1/2$.
4. Which is to say that b must be $1/2$.
5. And a can only be 0.

One then goes back and reconstructs the original equation with the exponents ($a = 0$; $b = 1/2$; $c = -1/2$) inserted. A dimensionless constant factor (call it k) must be assumed as well—remember Langhaar's comment about the limitations of dimensional analysis. Thus,

$$t = km^0 l^{-1/2} g^{-1/2}, \quad \text{or} \quad t = k\sqrt{\frac{l}{g}}. \tag{4.1}$$

(In fact $k = 2\pi$, as introduced earlier as equation (1.1) but the value of that constant had to be found in some way other than by dimensional analysis.) Look how much could be derived just from the assumption that the dimensions had to come out right! Not only did the proper square root relationship emerge, but, amazingly, the zero exponent for a told us that the mass of the bob cannot matter in determining the period. The original inclusion of mass was revealed as a red herring.

It's worth checking all equations you use to see if they are dimensionally correct; conversely, one can use familiar equations to help one figure out dimensions. (I always think back to $f = ma$ and $w = mad$ to get the dimensions of force and work.) In addition, one can use the definitions or dimensions of physical quantities to jog one's memory about how different units are related. Thus, by knowing either that power is the rate of doing work or the dimensions of the two, we know that the watt must be a joule per second; by knowing that work is force times distance, we know that the joule is a newton-meter.

So all terms of an equation must have the same dimensions; in use, their values must be expressed in the same units. But the correctness of the equation does not depend on the actual choice among systems of units as long as an internally consistent set of units is used—even the constants are unaffected by the choice among systems. There is, though, a partial exception to the business about units. In the allometric equation, $y = ax^b$, used in the last chapter, the value of a (but not b) depends on the specific units used for x and y. The alternative, making these equations truly general by specifying the approximate (and noninteger) dimensions of every a, is usually judged to be more trouble than it's worth.

There's another way of viewing Bridgman's example. We can ask of the pendulum whether some dimensionless (pure) number characterizes this sort of device. The question is procedurally trivial—just rearrange equation (4.1) to put k alone on the left. Then k becomes the pure number we sought. Similarly, the equations relating wall tension, pressure difference, and size (equation (3.5) through (3.8))—using either tension or tensile stress—can be obtained (less, of course, their constants) by dimensional analysis. In short, the general procedure turns out to be loaded with both analytical and intuitive utility.

Dimensionless Numbers

Pure numbers, such as the 2π in the pendulum formula, can have quite a direct bearing on the real world. One might reasonably expect that decent physical variables would require some sort of dimension and (thus) unit—else why the designation "physical"? Oddly, an especially useful class of variables have no units at all—their values are pure numbers. These get their dimensionlessness by being ratios of two variables with the same dimensions. Indeed, we met one earlier—strain, the ratio of the distance a specimen is deformed to its original, undeformed length (chapter 2), thus a length divided by a length. And a good way to derive dimensionless numbers is through dimensional analysis, even if it sounds like a contradiction.

Some dimensionless numbers, named after the person who first either mentioned or used them extensively, are given a two-letter symbol from the first letters of the perpetrator's name: Reynolds number, *Re*, and Womersley number, *Wo* (no apostrophe on the name—it's Reynolds number but Stokes' law—and no subscript in the symbol). Others are just called indices or coefficients—with some more-or-less descriptive designation—ponderal index, drag coefficient. And there's an ongoing tradition of concocting home-brew dimensionless numbers as smitten by need or inclination. (McMahon and Bonner [1983] and Pennycuick [1992] give biologically contextualized introductions to the general subject.) We can illustrate the uses of dimensionless numbers by considering several that vary in complexity, applicability, and frivolity.

Mechanical advantage

The efficacy of a lever, a system of pulleys, a windlass, or any similar device (chapters 2 and 23) often depends on the relationship between the force the device exerts against the load (output force, F_{out}) and the force you or some motor has to exert to operate it (input force, F_{in}). The ratio of these two forces is termed its "mechanical advantage." We'll call it "force advantage," *FA*, to avoid ambiguity as to what's being increased in the process. That's because muscles most often attach to bones in ways that reduce their force and increase the distance over which they pull. Thus, their force advantages run below one, and we find it more convenient to use the reciprocal of force advantage, "distance advantage," *DA*, when talking about their operation. So:

$$FA = \frac{F_{out}}{F_{in}} \tag{4.2}$$

and

$$DA = \frac{F_{in}}{F_{out}}. \tag{4.3}$$

As ratios of a pair of forces, neither force nor distance advantage can possibly have dimensions or units. They're not merely constants but depend on the construction of the device at issue. And they have obvious and immediate physical applicability. Perhaps an item mentioned two chapters ago bears repetition: since a dimensionless number is automatically unitless as well, its numerical value does not depend in the slightest on what units were used in its calculation. Pounds or newtons—the value of the ratio doesn't change.

Efficiency

Most instances where efficiency is mentioned boil down to the same thing—a ratio of the work gotten out of some contrivance (W_{out}) to the work (or energy) put into the thing (W_{in}). Thus, efficiency (e) is a measure of the energetic leakiness of mechanical arrangements. That the character of the input and output may be quite different makes no difference—electrical input and mechanical output for a motor, an input of the product of power and time versus an output of force times distance for some locomotory system. Efficiencies are always less than 1.0 (in the much lamented absence of perpetual motion machines), although they're usually multiplied by 100 to express their values as percentages:

$$e = 100 \frac{W_{out}}{W_{in}}. \tag{4.4}$$

In practice, applying the term "efficiency" can prove awkward, especially for living systems. Any number of inputs might be appropriate for, say, a jog around a track—total amount of food needed, food needed beyond what's required for ordinary body maintenance, extra work done by heart and other muscles to keep you jogging, or just the work of the locomotor muscles? Output is even harder to define. If, ideally it takes no work to keep an object moving at constant speed on a level surface, does car or runner have zero output and thus zero efficiency unless going uphill? Does a fish swimming steadily do useful work? Is the power output of a fish's muscles a good "output" or is the rate of momentum transfer to the water a better one? The bottom line is that all figures cited for efficiency have to be viewed cautiously, giving particular attention to what items are being used for input and output.

Ponderal index

Here's one that may be too close to home for comfort. How corpulent are you? Weight is easy to measure, and it tells the sad story of one's past indulgences. But it fails badly in providing a comparison among different individuals—a tall person ordinarily has a greater mass than a short one even if they are of similar "builds." That one can do better was long ago (1835) recognized by the Belgian statistician, Adolphe Quetelet, to whose work D'Arcy Thompson (1942) gave substantial attention. For geometrically similar bodies, volume is proportional to the cube of *any* consistent length; the handiest length for humans is our height. For two humans of the same shape, volume divided by the cube of height ought to be constant and is, of course, dimensionless. A higher value would indicate a more corpulent corpus. Volume, though, as mentioned, is harder to measure than mass, so Quetelet used the ratio of mass to the cube of length.

But that ratio isn't dimensionless, so one needs to specify the units for its variables; to make matters worse, the medical profession hasn't adopted SI. One can restore dimensionlessness by recognizing that all humans have nearly the same density, close to 1,000 kilograms per cubic meter—a little lower for the very fat, a little higher for the very bony (recall table 2.2). So we might use mass over density (ρ) for volume and write a revisionist index, which we'll call the "corporal index," *CI*:

$$CI = \frac{m}{\rho h^3}. \tag{4.5}$$

Assuming constant density solely to make the dimensions come out right may seem a sophistry, but measuring density with some scheme based on submergence will dampen the cooperativity of potential sources of data. You might at this point figure out your corporal index, using the density above and consistent SI units (1 kilogram = 2.2 pounds; 1 meter = 39.37 inches) to get a figurative figure. In my experience, females have slightly lower *CI*s than males, either with respect to the lowest values one encounters or relative to one's visual and cultural perceptions. The lowest value (for a nonanorexic individual) of the fifty I've collected is 0.0100; there's a long tail on the high (obese) end of the data. Values from 0.0105 to 0.0140 seem to be about the range for people of ordinary appearance.

Versions of this index have been used by people in different fields. It is, among other things, exactly a hundredth of what is known as the "Rohrer body-build index" (Rohrer 1921) among anthropologists and other people interested in the human form—the Rohrer index (not dimensionless) puts mass in grams over the cube of height in centimeters. Sheehan (1986) described its use in a fascinating book about the identification of the remains of victims of an old airplane crash.

The contemporary medical literature most often uses something called the "ponderal index," or the "body mass index" (*BMI*). Sometimes (always, I think, for *BMI*) mass is divided by the square rather than the cube of height (and thus isn't dimensionless)—if you're small, you're expected to be less skinny. The usual formula,

$$BMI = \frac{m}{h^2}, \tag{4.6}$$

uses mass in kilograms and height in meters. Values of 20–25 are considered normal; 27 and above is obese, with body weight more than 20 percent above what's desirable (Beers and Berkow [1999] and Abdel-Malek, Mukherjee, and Roche [1985] provide a general discussion). Put another way, isometry doesn't underly contemporary tables of desirable weights—we expect short people to be somewhat stockier. The commonest medical use of such indices seems to be in exploring correlations between obesity and longevity or obesity and cardiovascular disease—after all, one must have some sort of quantitative data to put on the x-axis for the fatness of the subjects, and the data must be obtained from quick and nondestructive measurements.

A similar index, introduced by Fulton (1904), has found extensive use in ichthyology and fisheries management—it's called "condition factor" or "Fulton's condition factor," with no apparent connection to the old Fulton Fish

Market in New York. Fish, as noted in the last chapter, are (at least within a species) close to isometry. The index,

$$CF = 100\frac{m}{l^3}, \tag{4.7}$$

(with mass in grams and length in centimeters) corrects for size and gives values conveniently near unity (Bagenal 1978; Lambert and Dutil 1997). Here higher is better, with no nonsense about the evils of obesity.

Drag coefficient

A quantity called "C_d" or "drag coefficient" looms large in advertisements for slick and speedy cars whenever the price of gasoline rises; it will reappear here in chapter 7. The coefficient amounts to a dimensionless measure of the retarding force of gas or liquid flowing past an object. But what's wrong with the direct use of force for drag? The limitation is that drag then refers only to the conditions operative for a particular measurement—it's handier to have an index that characterizes the "dragginess" of an object in a more general context. Thus, drag coefficient indicates the drag of an object relative to a standard force; in this case the standard is the force that would be exerted on the object if the fluid directly upstream from it was brought to a halt upon impact. The standard is distinctly queer, since the halted fluid would have to disappear in order to keep out of the way of the principle of continuity (chapter 2), but reality needs no mollification since it's only a mathematical trick.

So what is this peculiar standard force? Again we can begin with $F = ma$ and again note that neither mass nor acceleration are easy items for steady flows of a fluid that extends far out in several directions. So let's replace mass by volume times density, volume by the product of length (in the direction of flow) and the frontal area, S, (perpendicular to flow) of the body, and acceleration by velocity over time:

$$F = \frac{\rho S l v}{t}. \tag{4.8}$$

(Density, of course, refers to that of the gas or liquid, not, as with corporal index, to that of the solid body.) However, l/t is also v, because l was specified as being in the direction of flow, and thus l is the distance the fluid travels in time t, so:

$$F = \rho S v^2. \tag{4.9}$$

That gets rid of time. The drag coefficient, then, is drag (D) divided by this standard force (F), with a factor of two included for reasons beyond the present discussion:

$$C_d = \frac{2D}{\rho S v^2}. \tag{4.10}$$

Two comments must be made about the casual use of drag coefficients. First, at least for cars, aerodynamic drag is not a terribly crucial matter except at highway speeds. More importantly, the necessary inclusion of S in the coefficient can be misleading. If (and there are other conventions as well) it is taken as frontal area, a perfectly useful decrease in drag accomplished by reducing that area will not decrease the drag coefficient. Conversely, a well-designed bulgy body may have a low coefficient but a relatively high absolute drag.

Walking on water

Which water striders and some other creatures can do (chapter 5). There's no trick at all to concocting a dimensionless index to the practicality of standing or walking (not floating or swimming) on the surface of water, held up by the force of surface tension acting on some hydrophobic (nonwettable) foot. All that's needed is the ratio of the force holding the creature up to that pulling it down. The upward force is surface tension, γ, which, like the tension in the wall of a balloon (chapter 3) has dimensions of force over distance, times the length of the air-water-foot contact line (usually the combined perimeter of the feet). The downward force is gravitational, the usual $F = mg$ or approximately the density (of the organism once again) times length cubed times gravity. We get, then, as an index (call it *Je*, the Jesus number)

$$Je = \frac{\gamma l}{\rho l^3 g} \text{ or } \frac{\gamma}{\rho l^2 g}. \tag{4.11}$$

Since the surface tension of water, the density of organisms, and gravity are all essentially constant, the index provides a simple scaling rule. The main variable is the square of length in the denominator, which shows that the practicality of walking on water becomes dramatically less as animals grow larger. (The choice of an appropriate length isn't especially critical for such order-of-magnitude judgments.) On the other hand, there should be nothing startling about the phenomenon in very tiny creatures, and it's uncommonness perhaps comes mainly from the comparative scarcity of smooth water in nature. Still, the index gives only a distant view of the matter, to which we'll return later. Incidentally, $1/Je$, the reciprocal of the Jesus number, is a recognized quantity called the Bond number, which finds use in chemical engineering.

With that mild bit of sacrilege (apologies to the offended), we have probably plumbed enough of these indices to persuade any skeptic that discrete dimensions are no prerequisite for applicability to the physical world. More importantly, it ought to be apparent that dimensionless numbers can prove useful in our attempts to uncover biological generalizations, a point I attempt to make elsewhere with further examples (Vogel 1998b). An imposing list of named dimensionless numbers can be found near the back of the *Handbook of Chemistry and Physics*.

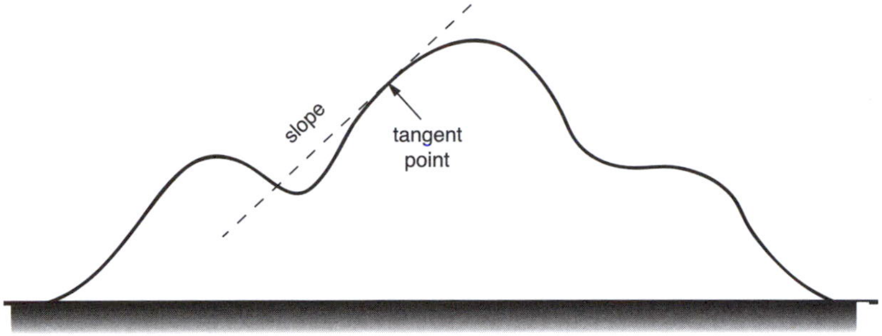

FIGURE 4.1 Side view of a mountain. The slope of a line tangent to the mountain is a measure of the steepness of ascent at the point of tangency.

Rates and Gradients

Earlier we defined velocity as the rate of change of position—the rate at which the location of something changed as time progressed. Acceleration was then defined as the rate of change of velocity as time progressed. Thus, acceleration is related to velocity in just the same way that velocity is related to position. We need this idea of *rate* to take an orderly view of the physical world. It applies to much more than speeds, accelerations, and other time-related quantities—rates may refer equally well to how something changes with distance. How the elevation of your position on a mountain changes with distance from your base camp is such a rate, a rate of vertical ascent relative to a horizontal progress. Such a rate is more commonly referred to as a "gradient," but gradients in ordinary usage are simply rates referred to distance instead of to time. Concentration or intensity gradients do a lot for us—how the concentration of a pollutant changes with distance from its site of discharge, how the intensity of a sound drops with distance from a caller. And, as we'll see, rates and gradients may refer to still other variables besides time and distance.

Rates and gradients really jump out at you from graphs. Consider a vertical cross section of a mountain (fig. 4.1). The rate of ascent with distance (gradient, or steepness) at any point is the slope of a straight line tangent to the mountain at that point—a line that just grazes the curve of the mountain without crossing it. The idea of slope applies as well to temporal gradients. On a graph of some variable plotted against time you merely take a ruler and draw a tangent line at any point you want. Just as with mountains, it's how far one progresses upward relative to how far one progresses horizontally.

All of this seems quite simple; in reality there's a subtle but profound problem. Previously, for the average slope of a curve, as in Figure 4.2a, we've used

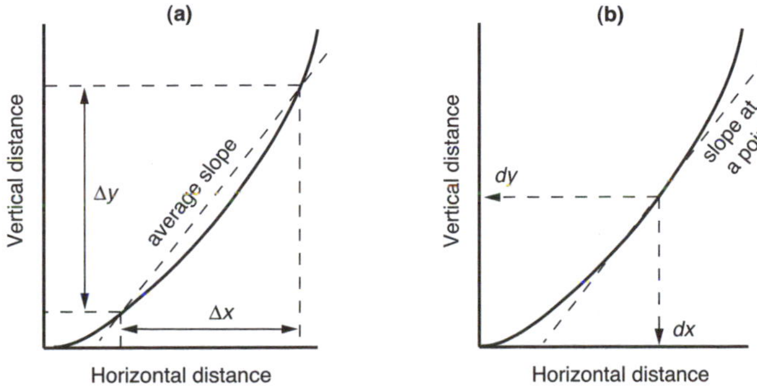

FIGURE 4.2 The average slope of a portion of a curve may be obtained as in (a) by dividing an interval Δy on the y-axis (distance, here) by the corresponding interval Δx on the x-axis (time, here). If Δx is made infinitesimally small, one has the slope at a point on the curve, as in (b).

some vertical distance (call it Δy) divided by the corresponding horizontal distance (Δx). Where the "curve" was a straight line (as when we figured out scaling factors) we ran into no trouble. But can we apply the same approach to get the slope of a curved line *at a point*, a trivially simple operation on a graph (fig. 4.2b)? At a point, Δy is zero, which is awkward enough. Worse, though, Δx is zero; so our usual technique involves dividing by zero, something flagrantly lawless. Does this mean that the slope at a point is mathematically meaningless? One has to do a bit of mental sleight of hand and reduce both vertical and horizontal distances until they're infinitesimally small (but, please, not zero!). On a graph, the resulting slope at a point becomes nothing other than the tangent line to the curve. Instead of $\Delta y/\Delta x$, we most often call this slope at a point dy/dx.

We owe the odd concept of slope at a point to Newton and Leibniz (independently) in the seventeenth century. The operation, though, remained a theoretical embarrassment (rather like sailing upwind) until Cauchy clarified things almost 200 years later. What we have here is nothing less than the basic idea behind the differential calculus—something we can't do without, even if we use graphical methods and other devices to avoid the formal complexity of the subject. As Leibniz, more generous than Newton, put it, "Taking mathematics from the beginning of the world to the time of Newton, what he has done is much the better half."

To reiterate, then. If you plot distance (y-axis) versus time (x-axis), the slope of the graph at any point is the "instantaneous" velocity; if you plot velocity versus time, the slopes are values of acceleration; if you plot height versus distance, the slopes are steepnesses. Velocity, acceleration, and steepness—all are rates or gradients, whether temporal or spatial. As we'll see from a few examples, this idea of a *gradient* is as powerful as any notion in all of science.

Slipping down a gradient is as easy as rolling off a log. The direction of a gradient commonly determines the direction of processes that go by themselves—"proceed spontaneously" is the usual phrase. The stone rolls downhill, requiring Sisyphus only to restore its uphill position. Strictly speaking, it rolls down a gradient of what's called gravitational potential. If a fluid flows, you can usually presume the existence of a pressure gradient. If heat flows, there's ordinarily a temperature gradient. If a substance appears to diffuse, then there's likely to be a concentration gradient. If electrical current flows, there's a voltage or electrical potential gradient. If a chemical reaction proceeds, it reflects a gradient in chemical potential. And the magnitude of the gradient, its steepness, looms large in determining the speed of the spontaneous process.

(Complex biological systems, though, needn't always work according to such a simple rule. Who eats whom, that is to say how energy gets passed up from one trophic level to the next, depends in no obvious way on a gradient or potential difference. Thus, no rule demands that the biomass or its concentration at one level exceed the one above it, elementary textbooks notwithstanding.)

Temperature gradients

What determines the rate at which heat passes out through the walls of my heated house is not simply the difference in temperature between inside and out. The rate, in fact, depends on the magnitude of the thermal gradient. If a given kind of thermal barrier is made twice as thick, the temperature difference is unchanged but the rate of heat loss is halved. What has happened is that *the temperature change per unit distance* through the wall has been halved, the slope of the temperature versus distance curve is half as steep, the gradient, in short, has been halved.

The relationship between heat loss and temperature gradient is an enormously important constraint in the design of animals that maintain body temperatures substantially different from those of their surroundings. Fur serves as an air trap, preventing movement of air en masse from augmenting conductive heat transfer. Fur thus effectively lengthens the path heat must traverse in dropping from body temperature to the temperature of the local atmosphere. A layer of fat plays a similar role, quite a critical one for a creature moving through cold water, where fur functions poorly. For a large animal in anything but a very cold climate, heat retention is not a serious problem, as mentioned earlier when we talked about surface areas and volumes. For a small animal, preventing heat loss may be crucial. And that, too, has odd but ordinary implications. Beef is well marbled with fat, skinned chicken is naturally lean. The difference is less a contrast between mammals and birds than between a large animal that can afford to distribute its fat and the small one that makes any fat do double duty—as metabolic reserve and as thermal insulation beneath the skin.

Concentration gradients

These may be slightly harder to envision, but they work in exactly the same manner as do thermal gradients. If you put a bit of dye to a beaker of really still water (water stabilized by adding gelatin or agar), the dye will dissolve and over a few weeks spread through the beaker (fig. 4.3). In this latter diffusive process, there has been a net movement of dye from a region of high concentration, near the crystal, to areas of lower concentration. The operative rule, Fick's law (chapter 8), has precisely the same mathematical form as does Fourier's law for heat transfer by conduction. Movement of oxygen from lung alveoli to capillaries and from capillaries to active tissues of necessity involves diffusion down concentration gradients. These systems have a nicely self-regulating character—if blood enters the lungs already well charged with

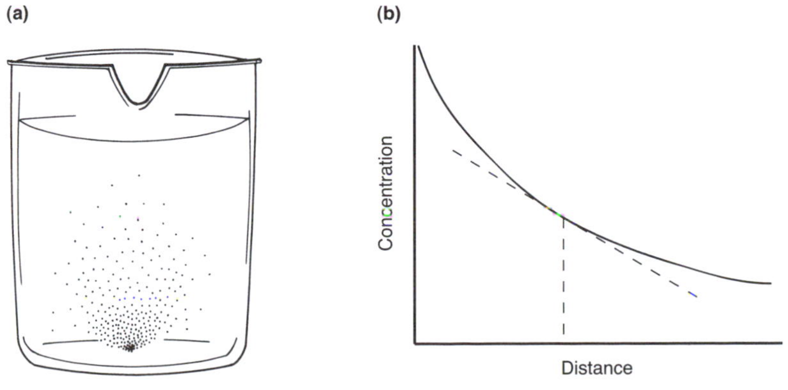

FIGURE 4.3 Dye slowly diffuses away from a speck of dye on the bottom of a beaker. There's no edge to the cloud of dye but just a continuously changing concentration as shown in (b), a graph of how dye concentration varies with distance from the speck. At each point on the curve, the concentration gradient (slope) has a different value.

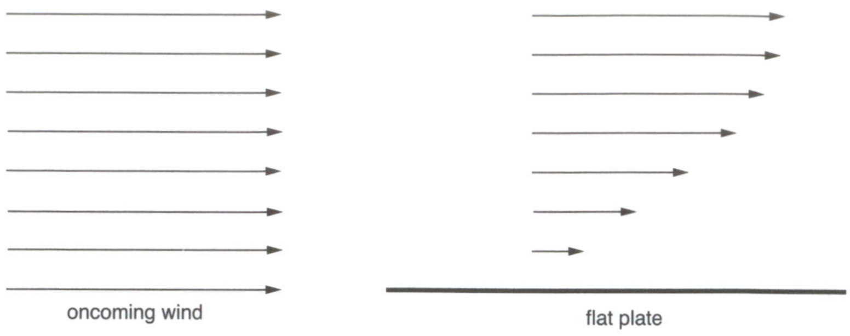

FIGURE 4.4 The velocity gradient adjacent to the surface of a flat plate exposed to a wind along its surface. As one moves vertically away from the plate, the local wind increases, at first steadily and then more slowly as it approaches the free-stream, or ambient, wind speed. The thickness of the gradient region has been much exaggerated here.

oxygen, then the gradient of oxygen concentration is reduced and less oxygen diffuses from air into blood.

Velocity gradients

When one solid moves across another solid, only people interested in lubrication pay much attention to any velocity gradient between the two. With moving gases and liquids, velocity gradients are an unavoidable part of practically every situation. The reason for their ubiquity is an oddly counterintuitive phenomenon called the "no-slip condition." It turns out that whenever a fluid flows across a solid surface the fluid immediately adjacent to the surface does not move—it has the same speed as the surface. The fluid slightly more remote from the surface has a speed slightly different from that of the surface, closer to the speed of what we call the "free stream" some distance away. And fluid still further from the surface has a speed still closer to that of the free stream (fig. 4.4).*

As a consequence of the no-slip condition, a velocity gradient with respect to distance extends outward near the surface of every solid over which a gas or liquid flows. Have you ever asked yourself how dust and grime can accumulate on fast-moving fan blades? In reality, flow right near the surface where such tiny bits of debris are located goes far slower than the general airstream. Or have you ever wondered why a dishcloth works so much better than running water in removing even water-soluble stuff like jam from a plate? An electric, nonscrubbing dishwasher has to use repetitive jets of strong detergent solution, and it still works best with prescraped plates. The faster the flow, the thinner the gradient region and thus the steeper the gradient—due both to the greater speed difference across the region and to that shorter distance. If you sit in a sauna bath, air moves almost imperceptibly, and the gentle velocity gradient provides some insulation, slowing heat gain from the hot air. If you fan yourself you can get a burn—you've made the velocity gradient and hence the thermal gradient much steeper, and heat gain rises in proportion.

Temporal velocity gradients are more obvious than spatial ones. To return yet again to falling, we note that it's not the fall that hurts but the stop at the end—the velocity gradient associated with impact. Thus, we make padded

*The reality of the no-slip condition was a matter of hot dispute in the midnineteenth century. Goldstein (1938) has a special appendix that reviews the arguments.

dashboards and collapsible bumpers on cars; and Vincent and Owers's (1986) hedgehogs, as mentioned earlier, have their crushable spines. Cats have no direct skeletal connection through a collarbone between the bones of their forelimbs (pectoral girdle) and those of their vertebral columns. In effect, they have shock-mounted forelimbs, which cushion a landing after a jump. None of these schemes reduce the extremes of velocity one bit; what they reduce are the velocity gradients.

Stimulus intensity gradients

Even your perceptual world is a bunch of gradients, both temporal and spatial; the gradients you perceive are, in fact, grossly exaggerated beyond mere physical reality. Admittedly, in some instances, organisms find it useful to monitor the intensity or direction of some environmental factor over a long period of time—gravity detection in plants, measurement of day-length (photoperiod) in all sorts of organisms, and so forth. But what is usually of immediate behavioral relevance is a change in the outside world; and, typically, the more rapid the change the more important it is that a response be initiated.

So we're equipped with a bunch of gradient detectors. Some are tactile and temporal—when you sit down you feel the chair but you're then unaware of it until you get up and feel the lack of chair. Others are tactile and spatial—you're more aware of the edge of something pressing against you than of the rest of its surface. Some are visual and spatial. Thus, you interpret each of a series of gray stripes (fig. 4.5) not as uniform in brightness but each appearing brightest adjacent to the less bright next stripe. And others are visual and temporal—a sudden shadow causes you (or just about any other animal) to take notice, and a small item in motion "catches the eye" amid a jumble of stationary detail. It's an edge that matters, whether in space or time, and the prudent potential prey makes no sudden motion lest it be detected by edgy predators. Even with sound, a sudden change in intensity stimulates with especial effectiveness. Music written before electronic amplification got most of its dramatic effect from changes in intensity, rather than from loudness itself, since even the maximum amplitude of a full orchestra isn't all that loud. The auditory sense, though, has an interesting asymmetry—for good behavioral reasons the sudden onset of a sound catches one's attention better than any

FIGURE 4.5 The gray stripe illusion. The computer contends that each stripe is of uniform intensity. But our visual system accentuates edges or boundaries and thus misinterprets the series of stripes when put together, as in (a). Even a small contrasting separation puts our judgment back on track, as in (b).

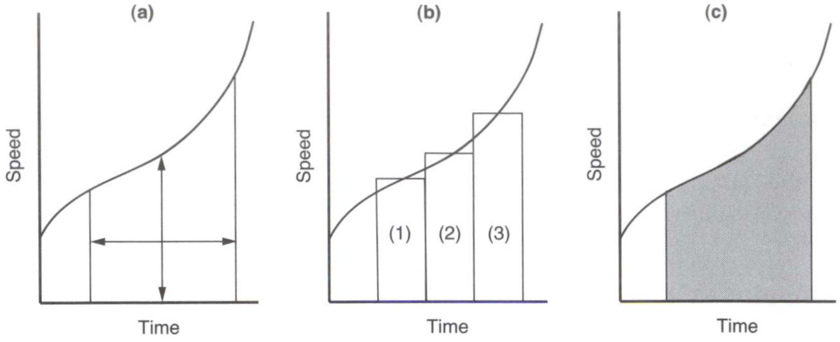

FIGURE 4.6 Distance traveled when one's speed isn't constant (a) can be figured by dividing the trip into intervals (b), taking the average speed for each, and adding up the distance-time products for the intervals. Ideally, one takes infinitesimally short intervals, obtaining, in effect, the area under the curve of speed versus time, as in (c).

sharp termination. It's disconcerting to hear a recording of any percussive instrument, such as a piano, played backwards—the music seems both quieter, which it isn't, and quite drab. We've designed such musical instruments around the gradient-exaggerating properties of our auditory equipment! (But with the advent of cheap amplification, we now subject ourselves to truly loud music, with some auditory hazard.)

Summation of Continuously Changing Items

The first bit of math we learn is counting. After that we learn to add, first combining sets of discrete items (two apples and three apples) and then numbers with fractional parts (three and a third plus two and a half plus five and a quarter pages of text). Even in the latter, though, we do a discrete number of summing operations (three, here). We need to extend the notion of adding or summation a bit farther. What if you travel at a varying speed, recorded on a speedometer, and want to know how far you've gone during a particular period of time? Or what if you have a crop growing, not by a fixed amount each day but at a rate that's always proportional to the existing amount (a lawn or young forest, approximately), and you want to know a week's accumulation? You need to be able to do a summation, not of a set of items, but of a so-called continuous function.

Again, let's view the matter in terms of manipulating an ordinary graph. Say we plot speed versus time (fig. 4.6a) and wish to ask how far we've gone over a given interval of time. Distance is, of course, the product of speed and time, but simple multiplication won't suffice—speed isn't constant. But just as length times width (two variables at right angles) gives area, area on our graph represents the product of speed and time, hence distance. What we're really looking for is the area under the curve for the elapsed time of interest—that area represents the distance traveled.

In practice we run into a problem similar to that that bedeviled figuring slopes at points. For a trip of an hour we might figure the average speed for each 20-minute period and from these get the distance for each period; we could then add the three distances (fig. 4.6b). Better, we might take 1-minute periods and add sixty distances. Best, we take infinitesimally short periods and add a large number of tiny distances (fig. 4.6c). Conceptually the matter is simple (at least armed with graphs); mathematically it's tricky—again Newton and Leibniz got people thinking correctly about the matter.

Either of two schemes can ease the task. The ancient and honorable fallback consists of weighing a sample of graph paper of known area, cutting out the offending area under the curve, and weighing the latter. The ratio of the mass of paper under the curve to the mass of the known area equals the ratio of the unknown to the known areas. With decent paper and a steady hand, achieving 1 percent accuracy, usually better than the original data, presents no problem. The other easy way uses a computer. It has both speed and patience, so it can add the areas for *very* small time increments and thus get quite an accurate summation. Even better, one can instruct it to try successively shorter increments and to stop when it finds that making the increments still briefer doesn't change the result by more than some tolerable uncertainty.

You'll recognize that the summation process is nothing more than the opposite of the instantaneous value or slope-at-a-point problem. These are, in fact, the essences of the two branches of calculus—differential (for slopes) and integral (for summations). Change in position, for instance, is the summation (the "integral") of tiny increments—instantaneous velocities times infinitesimal intervals of time. Change in velocity, similarly, is the summation of instantaneous accelerations times intervals of time. And, as with gradients, the independent variable (on the x-axis) may be something other than time—often length, area, or volume. Several examples, anticipating material further along, ought to add sufficient emphasis to the latter point.

Work of extension

Take a sample of some material—a string, a tendon, a piece of arterial wall—and slowly stretch it until it breaks, while keeping track of both its increasing length and the force (weight, usually) you apply. Data of this sort can be displayed as a force versus distance graph (fig. 4.7a). Alternatively, the data can be massaged into a more general form so they apply to the kind of material rather than to a certain sample. Force can be converted to stress (force per unit cross-sectional area; recall chapter 2), length can be converted to strain (fractional extension), and the results can be plotted as stress versus strain (fig. 4.7b). For most materials, neither graph takes the form of a straight line. The form of the second graph, including the particulars of its curvature, is of especial interest—from it one can read (as we will in chapter 15) strength

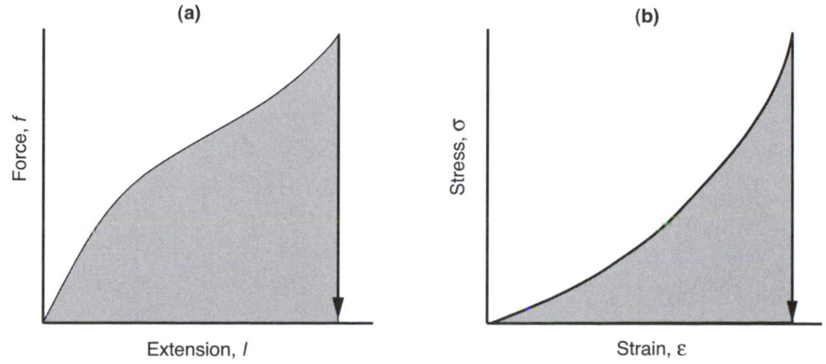

FIGURE 4.7 We stretch something by an increasing force until it breaks. The basic data (a) may be generalized (b) by converting force to stress and extension to strain. The area under curve (a) is the work done on the sample; the area under curve (b) is the work per unit volume, a property of the material rather than the particular sample.

(maximum height), extensibility (maximum horizontal distance), and stiffness (slope of a curve).

But those aren't all. Work, remember, is force times distance. The area under the force-distance curve from one distance to another is the work it took to stretch the sample from the first to the second length. The area under the stress-strain curve—what is that? Stress has dimensions of force per area while strain is dimensionless, so the area has dimensions of force per area also. Force per area, though, is dimensionally equivalent to work per volume—we now have a form of work independent of the particular specimen, a property of the material just like stress and strain. This property, work of extension (for the full test sometimes called work to fracture), can be an important one. For instance, if you want to store up work for later use by stretching a material, you pick one that has a high work of extension for loading together with a work of extension for unloading nearly as high (some loss being unavoidable).

As mentioned in chapter 1, running animals such as ourselves or horses use tendon (the protein collagen, mainly) for just that brief storage. Similarly, insects beating wings up and down use pads of a material called "resilin" to use the work absorbed in decelerating one stroke to start the next (Weis-Fogh 1960), as do fleas preparing to jump. Both collagen (fig. 4.8) and resilin meet our requirements well. By contrast, arterial wall has quite a different functional role (as we noted in chapter 1) and a stress-strain curve of a different shape, one where stiffness increases with strain and one with a minimum of potentially dangerous stored work. So we've done two integrations, one with respect to a length, the other with respect to a (dimensionless) strain—one gave work as a result, the other work per unit volume.

Total flow through a pipe

Consider water flowing smoothly ("laminarly," not "turbulently") through a circular pipe, well downstream from where it entered. The no-slip condition must hold at the walls of the pipe, so the speed there is zero. Automatically, then, the speed of flow must be fastest in the middle (fig. 4.9). In fact, a graph of speed versus distance across the pipe takes the form of a parabola; a three-

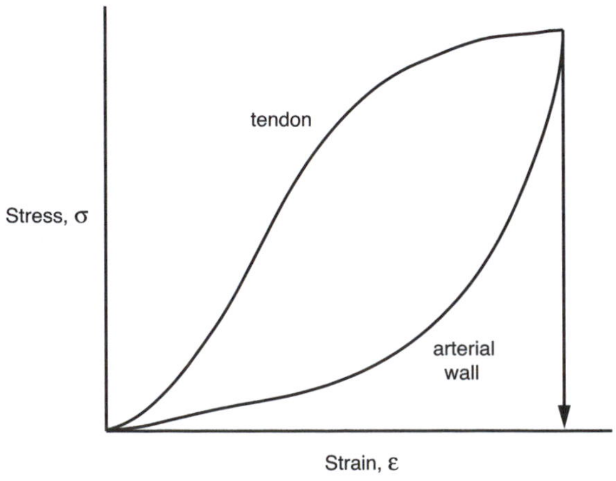

FIGURE 4.8 Two materials (here tendon and arterial wall, with some adjustments in the axes to match their endpoints) can have very different stress-strain curves. Such differences hold great functional significance.

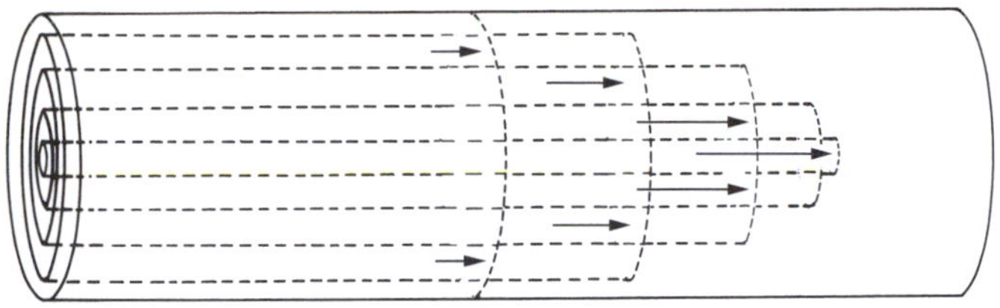

FIGURE 4.9 Flow through a pipe can be approximated as if through a set of concentric channels. Speed will be greatest in the center channel and least in the most peripheral one.

dimensional graph of speed across the area of the pipe would be a "paraboloid of revolution"—the parabola spun around the long axis of the pipe.

Question—if you know the speeds of flow, can you figure from them the total volume of water that passes (recall the principle of continuity) any cross section along the pipe's length? You can do it by integrating across the paraboloid in the appropriate manner. But here it isn't simply a matter of getting the area under the curve of speed against distance—since we're integrating with respect to area rather than distance, we have to use concentric elements of cross-sectional area (these are annuli like the pieces of a telescoping radio antenna). We could make the right paraboloid of revolution out of clay and do a three-dimensional version of the paper-weighing game. Alternatively, we can take the average speed for each concentric annular area, multiply each speed by the area of the corresponding annulus, and add up the products. The dimensions come out right—speed times area is the same as volume per unit time.

Dealing with Shape

Of the characteristics of organisms, only shape approaches size in its variability. And shape proves far harder to reduce to one or a few numerical specifications. It has been a subliminal presence in everything said so far—but we've mainly managed to sweep it under the rug. For instance, most of the slop in the data behind the scaling exponents probably came from shape variability. In our search for general rules of biological organization and operation, how might we deal with shape? Perhaps the best I can offer is an old but especially apt aphorism, that one size doesn't fit all. How one specifies shape depends on the task at hand. With that in mind we might look at some shape descriptors that have proven useful in the past or that at least clutter up the literature.

Center of gravity

(And *center of mass*, for our purposes the same thing.) A beam balance or a seesaw is a device whose operation is based on the equality or near-equality of a pair of moments, one clockwise and the other counterclockwise. A small weight may balance a large one if their respective distances from the pivot point are large and small, respectively. We can view such arrangements from the opposite perspective as well. Consider a beam with fixed weights in which only the balance point is adjustable—one can move the beam around to find that point; it's called the "center of gravity" of the beam. That's the point (or axis) at which the entire downward force of beam and weights acts as if concentrated—we can express the weight of the beam by a single downward vector through that point.

If the beam is weightless, aside from some weights that hang from it, then the location of the center of gravity is easy to calculate—it's the unique place that makes the moments due to those weights equal and opposite. If the beam, of some nonuniform thickness, *is* the weight in question, then finding the center of gravity gets a little messier. One has to treat it as lots of little elements, each with its tiny weight and a distance from an unknown balance point. Or one can let gravity do the test. Merely hang the object so it's free to rotate; it will reorient so the point from which it's hung is at or above the center of gravity (fig. 4.10). Doing the test twice, with different hanging points, will usually locate the center of gravity in two dimensions; doing it three times is necessary for an object irregular in all three dimensions.

This odd abstraction, center of gravity, is crucial to what's often called gravitational stability. A person just can't stand not having the body's center of mass above an area that includes the feet and any floor between them. Note that in the real world the center of gravity must be above an area, not a point or a line between the feet—two peg legs won't suffice. We get away with having only two feet by virtue of their area and with a set of fine postural reflexes that permit us to handle some instability. Thus, you can walk (with a little practice) on stilts; you can't stand on them without moving one or the other (or shifting posture) from time to time.

We snuck around the center of gravity when talking about the pendulum model for gait transition; after all, the formula for the period of a pendulum (equation (1.1)) includes the pendulum's length. What we did was to tacitly assume proportionality among the variables—the overall leg length, the distance between hip joint and a leg's center of gravity, and the height above the ground of the body's center of gravity. Candor requires admission that some problems were blithely ignored—see, for instance, Lee and Farley (1998).

Moments of area

We need to look at this shape-based issue, the distribution of the bits of a system around some center, in a larger context. Consider a two-dimensional body—one whose shape can be described on a flat page, however far it projects at a right angle in or out of the plane of the page. Take each little bit of the area of that body in turn and multiply it by its distance from a pivot point

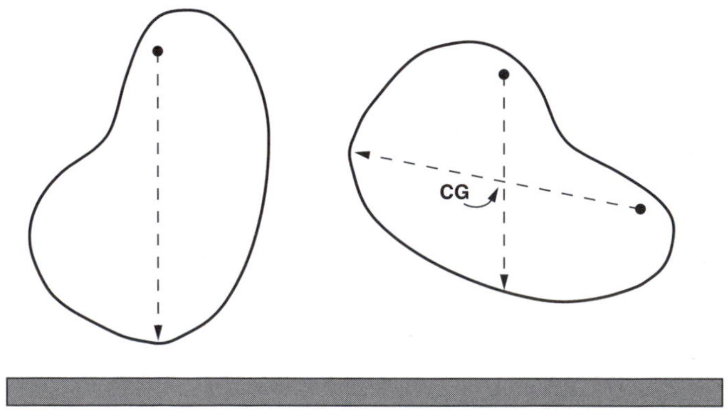

FIGURE 4.10 Locating the center of gravity (CG) of a two-dimensional irregular object. The object is first hung from an arbitrarily located point and a vertical line is drawn through that point. The object is then hung from another point and a second vertical line is drawn. The two lines intersect at the center of gravity.

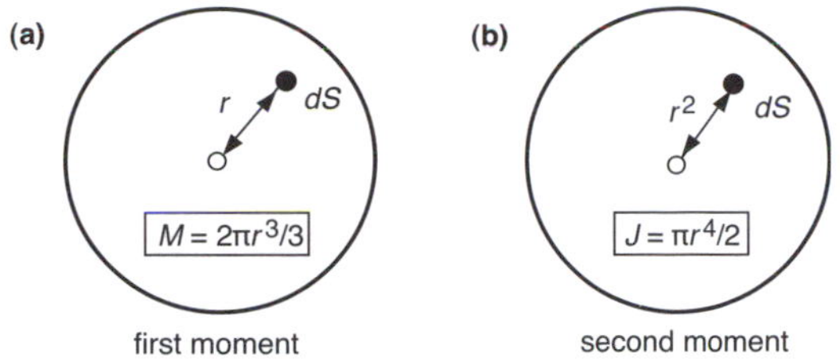

FIGURE 4.11 (a) The first moment, M, of an element of area, dS, about an axis. (b) The second moment, J, of such a bit of area. Note that the first moment has units of length cubed, the second of length to the fourth power.

or an axis of rotation, as in figure 4.11a. The sum (really the integral) of all these products is called the *first moment of area* of the body. First moments are important for determining leverage, balance points, gravitational stability, and pendulum periods—just as above.

Now take each little bit of area of the body in turn and multiply it by the square of the distance from a pivot point or axis of rotation, as in figure 4.11b. The sum of the products is called the *second moment of area* (sometimes, a bit misleadingly, the *moment of inertia*) of the body. Second moments come into play in at least two kinds of situations. One comes up when you want to "balance" the tires of your car or (in general) keep a rotating mass from exerting any sideways force on its shaft. The process usually goes by the name "dynamic balancing" to distinguish it from the less appropriate "static balancing" that deals with first moments instead. Similarly, the frequency at which insects beat their wings (and that of similar vibrating systems) varies (inversely) with their second moments of area. The other involves the resistance of beams and columns to bending and twisting (chapter 18)—their stiffness depends on that square of the distance of their bits of material from a central plane (for bending) or a central axis (for twisting). We alluded to the matter in the last chapter when talking about how bones scaled.

These moments describe shape. To get a shape factor uncontaminated by size takes nothing more than dividing the first moment by a length cubed and the second by a length to the fourth power. For either size-corrected moment, a circular section will give the lowest value. But beyond that, the relationship between sectional shape and relative value will usually differ between the two measures. Which is better? Again, it depends on the application.

Center of buoyancy

Closely related to centers of gravity and mass is center of volume—it's most easily viewed as the center of gravity of a body of uniform density. Center of volume plays a major role for bodies submerged in fluids of densities similar to their own. A submerged body will ordinarily reorient so its center of gravity lies exactly beneath this more abstract center of gravity—that of the water that would have been present had the object not taken up the space, its center of buoyancy. Why do dead fish commonly go belly up? A bit of ventrally located gas, from decomposition in stomach or intestine, is all it takes to ensure that the center of volume or buoyancy drops beneath the center of gravity. Even the concentration of denser material in the dorsal backbone may be enough to instigate inversion once the neuromuscular machine turns off.

We considered situations in which the square of a distance mattered, but we can also recognize ones in which the square root of a distance ought to have functional relevance. Consider a leaf that's heated up by exposure to direct sunlight with minimal wind. Heat loss from any part of its blade should depend on the temperature gradient at its surface, and that will depend, in turn on the velocity gradient—it's a problem I looked at many years ago (Vogel 1970). But the steepness of that velocity gradient depends on the square root of the distance from the upwind edge. So the average temperatures of leaves under equivalent thermal loading ought to follow some shape index that reflects the sums of the square roots of the distances of each bit of leaf surface from an edge.

One can contrive other shape factors; indeed, a large, if scattered, biological literature deals with them. They've been of particular interest to people concerned with the way shape changes during the course of an organism's development and the way it changes as a lineage evolves—Roth and Mercer (2000) provided a recent review of the subject.

Fractal dimension

One intriguing approach to shape has come in for a great deal of attention in the last decade or so, and it forms a logical extension of the way biologists have dealt with scaling. Recall that for geometrically similar bodies, surface area increased with the square of length and volume increased with length cubed. These exponents of increase, two and three, happen to be integers. Can we imagine objects of increasing size in which the exponents depart from integer values? While they can't retain geometric similarity, perhaps they can have other regular features. Mathematicians have explored such systems; Mandelbrot (1983) called them "fractal" for "fractional dimension."

Before telling you what they look like, let me describe how to specify a fractal dimension. Consider again how area varies with length for a body of unchanging shape. Increase length by any factor and area goes up by the square of that factor—by that factor to the exponent two. We'll now say that the area of such a body has a fractal dimension of 2.00. In effect, we've taken the logarithm of an increase factor of our variable, area, and divided it by the logarithm of the corresponding increase factor of length needed to produce it. In formal terms, then, increase factor = length change factor$^{\text{fractal dimension}}$, or

$$\frac{\text{log increase factor}}{\text{log length change factor}} = \text{fractal dimension.} \tag{4.12}$$

If, for instance, we increase length of a body tenfold, we increase area a hundredfold; the ratio of the logs of a hundred and of ten is two. For volume referred to length, we get three rather than two.

Now imagine a body that changes shape in a strange fashion. Every time we triple its length we increase its perimeter, not threefold, but fourfold. By the formula above, perimeter has a fractal dimension of, not 1.0000, but 1.2619. Figure 4.12 shows a pair of nicely regular objects (called Koch snowflakes) that work that way (adapted from Peterson [1988]). In figure 4.12b, a threefold overall diameter increase has been accompanied by a fourfold perimeter increase—shape has subtly changed.

The process generates a strange regularity, evident in the figure. A bit of the object looks just the same at any magnification! The length of a coastline

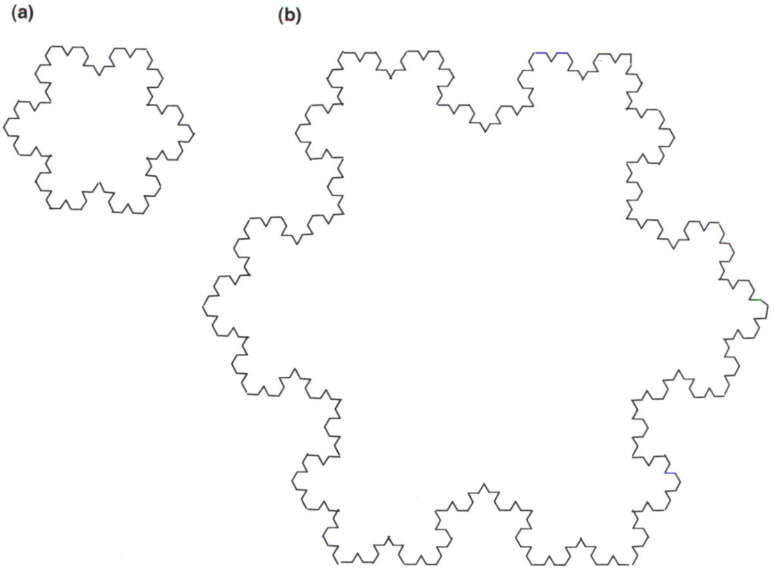

FIGURE 4.12 Two Koch snowflakes, with (b) three times the diameter of (a).

provides the classic illustration—a coastline has about the same jaggedness at any scale, and the finer the resolution with which it's measured, the greater its length. Fractals are often spoken of as "self-similar" to emphasize the way their appearance very often doesn't depend on magnification; the coastline case points out that the self-similarity may be of a purely statistical kind. Computers generate truly self-similar fractals that look like fancier snowflakes and statistically self-similar ones that look like mountainous terrain—one isn't limited to planar figures. Computers also generate fractals that look like fern fronds, branching trees, intestinal linings, and vascular systems—biological entities.

Such a tidy mathematical approach to complex shapes holds great appeal, and its utility for biological systems has been extensively explored. We're all too aware of the intricacy of biological organization at every size scale. Querying colleagues in a variety of fields, though, has failed to turn up any example of a biological revelation from applying fractal geometry. Part of the problem is that so few features of organisms can be described as fractal, even in a rough way. And that's not surprising in light of all that has been said here (and much more that will yet be said) about how the operative physical factors depend on the size of the system. Nor is it surprising when one considers the developmental processes that make organisms. Nor is it surprising from an evolutionary viewpoint—what might be the selective advantage of a fractal geometry? The other part of the problem is that fractal mathematics mainly provides a descriptive tool, much like the scaling exponents discussed earlier. As LaBarbera (1989) pointed out, for these exponents to do more than summarize data, the living world would have to be organized around the underlying power-law functions, something we can't assert with any confidence. I must admit, though, that others (see, for instance, Pennycuick [1992]) take a more optimistic view of what fractal mathematics can do for biology.

Still, discovery that a biological system is dependably fractal over a reasonable size range would say something about how it comes about or operates—that's the positive consequence of a world in which fractal organization cannot be the default. I think, for instance, that sponges may be fractal, with large ones organized mainly as foldings and refoldings of proliferated small ones, thereby making transitions between the organizational grades we learned as ascons, sycons, and leucons. Fractal geometry allows a systematic change of shape with size in which surface area can increase in proportion to volume, something that might direct attention to a system whose volume is limited by diffusion inward or outward across its surface.

Flatness index

The ratio of surface area to volume loomed large in the last chapter—its magnitude is certainly the most important single consequence of the size of an organism, and size (as length) varies over at least eight orders of magnitude. Everybody talks about surface-to-volume ratios, but few specific values are ever mentioned. The trouble is that the ratio, S/V, varies with size as well as shape, so it's not a very useful measure of shape per se. Can we do better, that is, can we concoct some version of a surface-to-volume ratio that's independent of size and reflects only variation in shape? What we need is a ratio that doesn't have any dimension of L, and preferably no dimension of M or T either while we're at it, instead of the $1/L$ of the ratio S/V.

There's no unique solution. Perhaps the simplest (you might think of others) consists of dividing the square root of surface area by the cube root of volume, so both numerator and denominator have dimensions of length to the first power. Since area is in the numerator, the further from sphericity the object, the higher the value—we might call the ratio the "flatness index" (FI):

$$FI = \frac{S^{1/2}}{V^{1/3}}. \tag{4.13}$$

You ought to check that this index is, indeed, size-independent by trying calculations for a few shapes over a range of sizes. I did a few myself—spheres give a value of 2.20, the minimum possible value since spheres are the least flat solids by the present criterion. Cubes give a value of 2.45—they're a little flatter.

The main use for such an index ought to be as an axis alternative to surface-to-volume ratio itself, a way to look at shape untainted by size. Thus, a paper on sinking rates for phytoplankton presented a graph showing that these rates vary inversely with S/V, and concluded that more irregularly shaped phytoplankters sank more slowly (Smayda 1970). Replotting the data with a dimensionless surface-to-volume ratio on the x-axis, though, eliminated the inverse relationship—the data scattered about a horizontal line. So the original inverse effect must have been a matter of size rather than shape: bigger plankton sank faster, and big ones have lower surface-to-volume ratios on account of their size alone.

That's the way O'Dor and Hoar (2000) have recently used the index, which they've generously designated the "Vogel number." They showed a striking difference between squid and fish—in squid, shape depends strongly on size; in fish it scarcely does at all. (As we noted earlier, fish are reasonably isometric.) Juvenile squid a few millimeters long have slightly higher values, around 4, than ones a few centimeters long, around 3.5, which are slightly rounder. But above that size, squid get steadily less rotund, with an increase rate similar to that of an elongating tube, and 10-meter-long giants reach about 7. They examined the consequences of this unusual trajectory on propulsive efficiency (which we'll take up in chapter 13), respiration, and the results of fishing.

Organism-to-organism variation in the index indicates deviation from isometry—squid are extreme deviants; fish are a pretty isometric bunch. What about mammals? Calder (1984) and Schmidt-Nielsen (1984) cited data for weight and surface area that encompass a size range from 0.3-kilogram white rats to 550-kilogram dairy cattle. Assuming the density of water (to convert weight to volume), their data sets confirm the reported near-isometry of mammals—index values average 3.31, with no noticeably regular variation with size. Noncorpulent humans, around 3.3, are certainly not aberrant.

One general note on dimensionless numbers. As emphasized earlier, all these are ratios of quantities (variable or constant) that have the same dimensions. So they tell you how something changes with respect to something else. An important use of such comparisons is just what we've seen with the flatness index (or Vogel number; *Vo*, I suppose)—how shape (uncontaminated by size) changes with change in size. Drag coefficient showed how the drag of something varied relative to an idealized impact pressure. What both of these (and many others) can do is to direct attention to the unexpected or extraordinary; they do this by normalizing for the expected or ordinary. A constant flatness index indicates isometry; a constant drag coefficient (usually) indicates that flow patterns remain unchanged.

Shape—Geometric and Morphogenetic Constraints

Almost everything so far said about size and scaling hinges on a single geometric constraint—as size changes, the ratio of any aspect of surface to any measure of volume will vary unless shape simultaneously changes. The beauty of scaling arguments is that they transcend the host of details that we normally think of as constituting the shape of any complex object. Still, among those details lurk generalizations, and these (yet more promises!) will make sporadic appearances further along. To set the stage, though, we can look at a few matters pertaining to the geometry of life, inquiring both about why certain shapes occur and why certain others either do not or are notably rare.

The pentamerous echinoderms

Starfish have five arms; sand dollars have five radial food grooves on their undersides—this arrangement of five elements radiating from a center point ("pentaradial symmetry") is widespread among the echinoderms but unknown elsewhere in nature. Why so and why not? Accident of ancestry seems unlikely—the fossil record is good, and the ancestral symmetry seems to have been trimerous, not pentamerous. More likely is an explanation based on the construction of early (and many contemporary) echinoderms and the geometric limitations of the scheme.

Early echinoderms were covered with a skeleton made up of discrete plates of calcium carbonate. Now one can pave a floor with triangles, squares, or hexagons, but using pentagons alone inevitably leaves gaps. One can't make an array of squares close on itself to form a hollow solid unless at eight special locations the apices of three rather than four squares touch, a distinct complication. And one can't make any array solely of hexagons close on itself at all. Conversely one can get a closed, space-enclosing structure from triangles (tetrahedrons are the simplest, but others such as twenty-sided icosahedrons are possible) and pentagons (the simplest being the twelve-sided dodecahedron). Among the pentagons (fig. 4.13) hexagons can be intercalated practically without limit, but twelve basic pentagons must remain. In the most symmetrical arrangement, these pentagons are in six pairs with members of a pair at the opposite extremities of the solid. If we run an axis between members of one pair, the ten other pentagons then arrange themselves in two nearly equatorial rings. If enough hexagons are intercalated, these can form the key elements of five arms. And a look at any book treating the paleontology of echinoderms reveals a host of hexagonal plates. Perhaps pentaradial symmetry is, in fact, a "natural" or easy way to organize a radially symmetrical creature built of a shell of little solid elements!

And the conical mollusks

A profound difference between an organism and any piece of human technology is that the organism is capable of growth without loss of function—the possibilities for taking a time-out to enlarge oneself are exceedingly limited. And nowhere does this problem pose worse difficulties than in supportive systems. Mollusks, a widespread and clearly successful group, have accepted a most peculiar constraint undoubtedly related to the need for functional continuity during growth.

Consider shapes that satisfy the following set of conditions. To provide both support and protection for the organism, the shape must be a hollow one, but an opening must exist somewhere. Growth can occur only by addition to the inner surface or the free edge. And the shape should change only minimally as it grows. A cubic shell with an open face won't work: addition to walls will give more shell relative to its contained volume, and addition to the edge will make the rectangular shape increasingly elongate. Similarly, a cylinder doesn't meet the conditions—addition to the edge will move it from short and fat to long and (relatively) thin. What will work are cones, whether circular or elliptical. Add to the edge and thicken the walls and one gets a bigger cone, isometric with the original.

With only slight violations of the condition of isometry, all sorts of wild derivatives of cones are possible—and these latter are the shapes in which

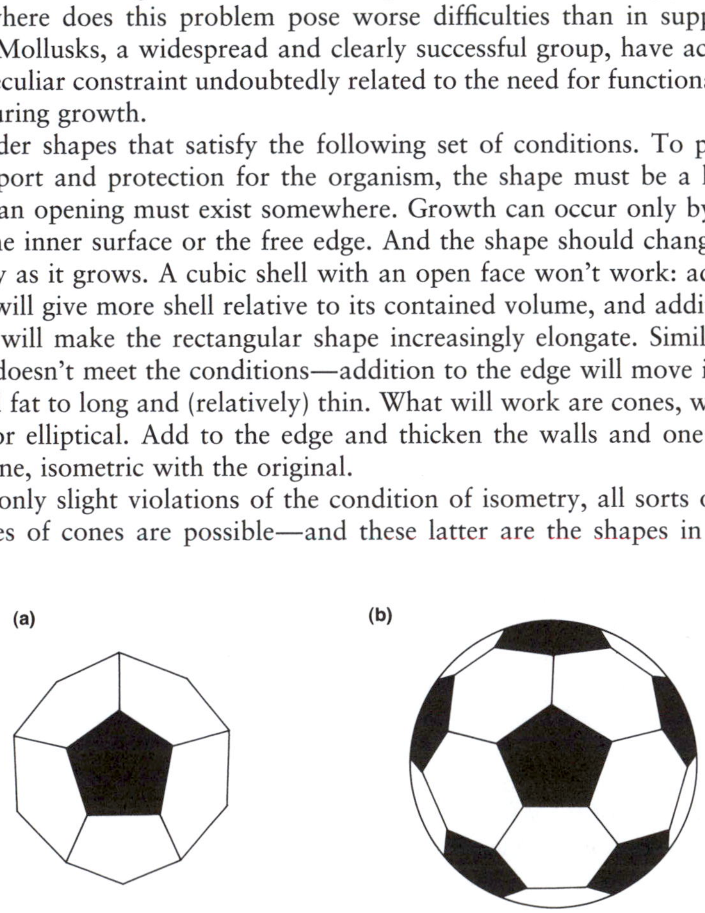

FIGURE 4.13 (a) A twelve-sided solid made up of regular pentagons—the edges of the peripheral pentagons of this view radiate in an echinoderm manner. (b) Hexagons now separate the twelve pentagons without changing the underlying pentaradial symmetry of the resulting thirty-two-sided solid.

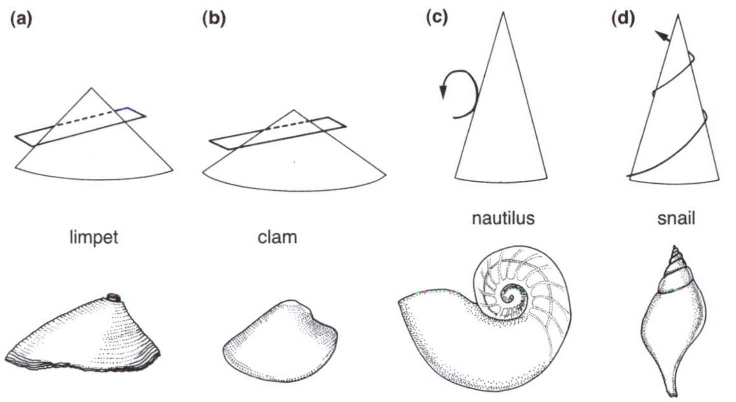

FIGURE 4.14 Deriving mollusks from cones. A limpet (a) is approximated by the surface above a slightly oblique slice of a low cone. A clam half-shell (b) is roughly a set of similar slices in which the tilt of successively lower slices increases a little. A nautilus (c) is a cone rolled up by stretching one side and squeezing the opposite. A snail (d) is similar, but with the apex shifted to one side of the center of the coil.

shelled mollusks occur. As shown in figure 4.14, a broad cone, with an opening almost at a right angle to the cone's axis, is the shape of a limpet. An even broader cone, with a very oblique opening, is the shape of each valve (half-shell) of a clam. Elongate the cone and wind it up into a flat spiral, and one has a chambered nautilus or one of a great diversity of extinct shelled cephalopods. Wind the spiral with some amount of helical offset, and there's a snail. The notion of molluscan shape as reflecting such isometric shell growth isn't new—Raup (1966) traced it back via D'Arcy Thompson to a paper in 1938. Raup, though, has initiated what is now a fine, biologically interesting game. A computer can be programmed to describe and draw examples of a universe of hypothetical mollusks, some of which exist or have existed (shells fossilize very well), and others that don't appear to have evolved—one might look at pretty books by Prusinkiewicz (1990) or Meinhardt (1995) for a glimpse. Why are some but not others real? Are the limitations matters of evolutionary accident or of relative functional disability? If our biology is *really* competent it ought to be able to rationalize not only what has occurred but what might have but has not.

But we're touching yet another subject—morphogenetic and functional constraints on shape—that goes far beyond the scope of this book. Arthropods must molt as they enlarge, necessarily constraining possible forms. Most trees can grow radially as well as apically, but the large group (monocots) that includes palms and bamboos lacks the cell layer requisite for radial increase. So their girth gets set once and for all, with obvious (and perhaps nonobvious) constraints. On the other hand, ordinary trees have no way of reabsorbing their core wood as they enlarge, so they can't (as do bamboos) take advantage of the mechanical efficiency of a hollow cylinder. Vertebrate bones can grow without loss of supportive function, so we need neither molt nor limit ourselves to molluscan geometries—perhaps that's the basis of the success of the vertebrates as relatively large-sized organisms.

PART TWO

Fluids

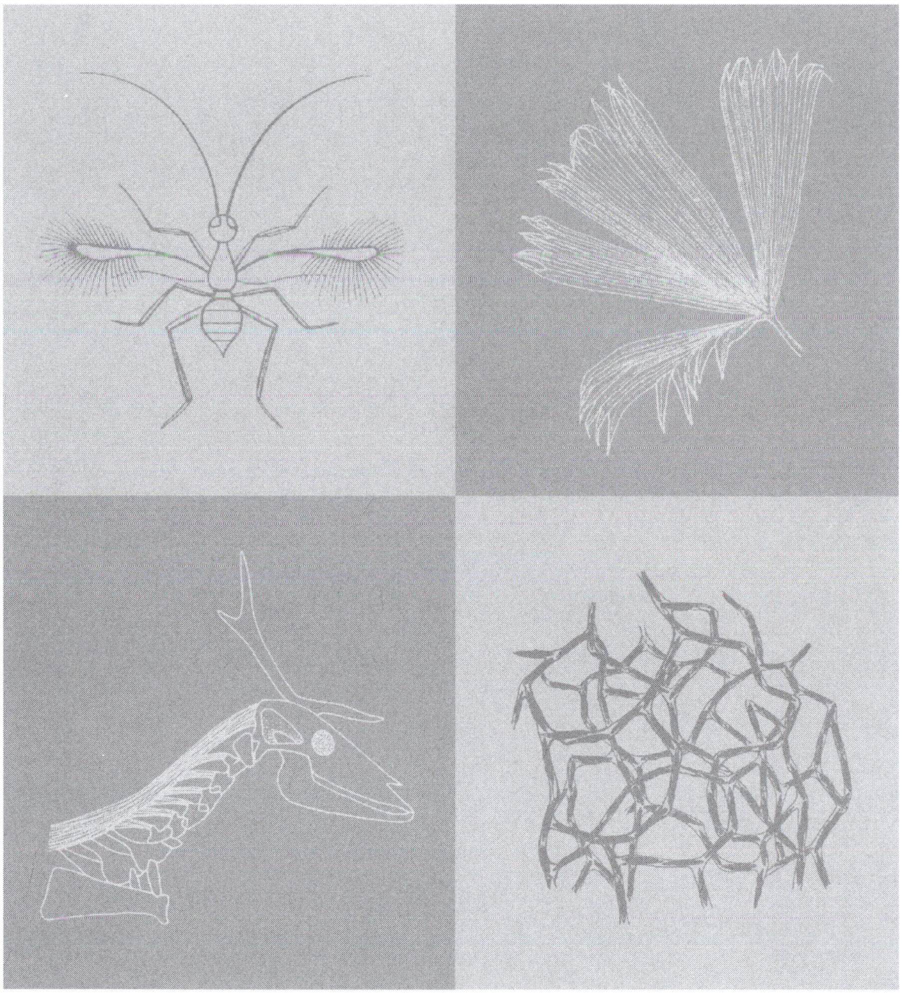

In which we begin with a look at how gases differ from liquids. We end up recognizing how the two nonetheless turn out to behave in remarkably similar ways, especially when they're in motion. Along the way, we're exposed to a host of physical phenomena whose counterintuitive character is surpassed only by the remarkable ways organisms contend with them.

CHAPTER

Gases and Liquids: Fluids at Rest

> You may have inner tranquillity, but you can't escape surface tension.
>
> **V. LOUISE ROTH**

WE START WITH SOMETHING CURIOUS. GASES AND LIQUIDS CONSTITUTE two distinct states of matter, with no intermediates. In motion, though, they're remarkably alike—for dynamics, we lump them as fluids and cite rules that apply equally well to either. But they don't behave at all alike when at rest—statically, they differ qualitatively. So an initial chapter on fluid statics has to look at them one at a time even if subsequent chapters devoted to fluid dynamics don't need the distinction. Except, that is, for a pervasive difference in biological contexts—water, our common liquid, has a density close to that of organisms; air, our common gas, has a density nearly three orders of magnitude less. This and lots of other contrasts come in for thorough examination in Mark Denny's (1993) magnificent book, *Air and Water: The Biology and Physics of Life's Media.*

About Gases—The Basic Rules

For a surprisingly large number of situations, a gas is a gas is a gas. Properties that vary widely among liquids or between one solid and another are either invariant among gases or change according to simple and regular rules. The gas of main interest biologically is, of course, that mixture we breathe and call air; but neither the fact that it isn't a pure substance nor the specifics of its composition are of fundamental importance to its physical behavior. So it's fairly easy to state some general rules for dealing with gases:

1. For a given gas at a fixed temperature, either a particular pure gas or a mixture, the pressure it exerts on the walls of its container is proportional to its density. In short, if the amount of gas in the container and thus the density is doubled, then the pressure within the container (with rigid walls, we'll assume) is also doubled. But the relationship between pressure and density does vary from gas to gas—when one speaks of a "light" gas such as hydrogen or helium, one means that the gas exerts a rather high pressure for a given density or has a low density for a given pressure. What matters for pressure turns out to be the number of molecules of gas relative to the volume of the container, not the mass of each molecule, which sets the density, nor anything else about the character of the gas collectively or molecularly. If you want to tie things to molecules, the relevant formula is

$$pV = nRT, \tag{5.1}$$

where p is pressure; V is the volume of the container; n is the number of molecules or moles (depending on which R you use) of gas in the container, R is another constant, this one universal, 8.314 joules per mole-degrees Kelvin; and T is the temperature in degrees Kelvin (degrees Celsius plus 273).

Notice that for a given volume, pressure rises in precise proportion to temperature—and that at a given pressure, volume likewise varies with temperature. The skeptical reader will gasp—the formula claims that at absolute zero, the gas disappears altogether. So the formula has a residual problem; let someone else worry. For our mundane uses, gas equals "ideal gas" or "perfect gas," to which the formula applies strictly.

2. That direct proportionality of the pressure-volume product to temperature serves us well. For a given mass of gas (and again at a fixed temperature), the pressure and volume vary inversely; that is, the product of pressure and volume is a constant. Double the pressure and you halve the volume. The relationship was shown back in 1660 by Robert Boyle; it's sometimes called "Boyle's law." The skeptical reader will again protest. The rule implies that matter can be squeezed out of existence by a sufficiently high pressure—there ought to be some irreducible volume that is much harder to compress. Yes, but the gaseous state is so rarefied that we can ignore the actual material volume of the underlying matter in all problems of present interest. Incidentally, Boyle's law implies that you can measure atmospheric pressure by taking a volume of air and noting the pressure needed to compress it by half—you will have thereby doubled the pressure or added an additional 1 atmosphere (which turns out to be 101,000 pascals).

The handiest formula to bear in mind takes account of temperature as well, asserting simply that for a given mass of gas

$$\frac{p_1 V_1}{T_1} = \frac{p_2 V_2}{T_2}. \tag{5.2}$$

In effect, it's a conservation law, conserving this combination of variables for a given body of gas with no concern for molecules or universal constants.

3. In the absence of specific chemical reactions (as when hydrogen and oxygen combine to make water vapor), the individual gases in a mixture act independently, and properties such as the pressure or density of the mixture are just the sums of the properties of the components. Thus, the air at sea level has a pressure of 101,000 pascals. Of its components, oxygen makes up 20.95 percent (by volume, not mass). Oxygen thus contributes 21,160 pascals to the overall pressure—its "partial pressure." Were all the other gases removed from a rigid container of air, the pressure inside would then be that of oxygen alone—21,160 pascals. In fact, in an atmosphere of pure oxygen at about a fifth of normal atmospheric pressure, your respiratory system would behave almost normally. The situation is equivalent to breathing pure oxygen at ambient pressure at an altitude of 12,000 meters (39,000 feet).

4. Gravity can be ignored as an immediate influence on the density and pressure of a gas in the atmosphere around any terrestrial organism. It's thus

not dangerous to hold your breath while ascending in an elevator (although it may hurt your eardrums a bit). From the weight per volume of air of about 12 newtons per cubic meter (which is the same as the air's density times gravitational acceleration) and the sea-level pressure of 101,000 pascals (newtons per square meter), it's easy to calculate that a column of air (of constant density) 101,000/12 or 8,400 meters high would be needed to exert 1 atmosphere. Ascending 84 meters would therefore expand the air in your lungs by only 1 percent. Thus, the volume of a flexible container of gas will not vary much as an organism ascends or descends biologically reasonable heights in air. A soaring bird might be the exception—it ought to (and does) vent internal air spaces to the outside.

5. But the situation is quite different for a container of gas surrounded by water rather than by air—the analogous calculation yields another result altogether. Water weighs, not 12, but about 9,800 newtons per cubic meter. A column only 101,000/9,800 or 10.3 meters (33.8 feet) instead of 8,400 meters high exerts 1 atmosphere of pressure. If you dive down about 10 meters in water the pressure on your body increases by an atmosphere—from 1 to 2 atmospheres. The effect on the liquids and solids of your body is trivial, but the air in the lungs reacts by decreasing its volume by half. Note, though, that the fractional volume change of the air in the lungs per unit distance up or down depends on how deeply one is submerged. Going 1 meter downward from the surface will decrease lung volume by about 7 percent. A 7 percent decrease at a depth of 40 meters takes an additional descent of almost 4 meters. As divers (and fish) know, buoyancy compensation is most demanding near the surface.

Incidentally, it's possible to escape without any special machinery from a sunken ship or submarine at depths of as much as 100 meters—you just have to get out fast and start to ascend with your lungs filled with air at the local hydrostatic pressure. The expansion of air in your lungs will then maintain the ascent; you exhale continuously until you reach the surface. (When you need further instructions, consult Jensen [1986]).

For essentially the same reason one can't use a snorkel tube to breathe at any great depth—air is drawn in at atmospheric pressure, but one must displace water at an additional pressure determined by the depth. A meter down, the extra pressure is a tenth of an atmosphere, or 10 kilopascals, or 1.5 pounds per square *inch* of chest wall. That's about the limit for the muscular system with which we inhale. Schmidt-Nielsen (1972) punctured a prevalent presumption that long-necked brontosaurs lived in water using their necks as snorkel tubes, pointing out the unreasonable pressure differences involved. But mosquito larvae happily hang from the water's surface with their internal air pipes (tracheae) connected to the atmosphere—the problem is tiny if you are also.

6. Gravity may enter the picture in another guise. Pressure depends, again, on the number of molecules; density depends on their molecular weights. So only if gas composition remains constant can one count on a tidy proportionality between the pressure and density. Increase the proportion of a light gas in a mixture and the density of the mixture will decrease. And even a small

fractional decrease can have practical consequences—a lighter bit of gas will rise by what's called "free convection."

We normally think of temperature differences as the cause of such convective updrafts, whether over a radiator in a house or over a plowed field surrounded by forest. But consider—water vapor (molecular weight 18 grams per mole) is lighter than dry air (a nitrogen-oxygen mix with a weight of 28.6 grams per mole). As a result (vernacular speech notwithstanding), at 25 degrees, air of 100 percent humidity has a density 7.8 percent less than that of dry air. That's equivalent to the density drop caused by a 23-degree temperature increase. In short, even with no temperature difference, moister air will rise by convection. I don't know if anyone has looked for biological consequences—certainly both the breathing of animals and the transpiration of plants produce local increases in humidity.

Buoyancy

A body submerged in water that weighs less than the water it displaces will be subjected to an upward force. It "displaces," of course, a volume of water equal to its own volume, so if its density is less than that of water it feels a net upward force (it's "positively buoyant"). Conversely, if its density is more than that of water, the body is forced downward (it's "negatively buoyant"). The force, upward or downward, is just that of gravity—equal to the density difference times the volume of the body times gravitational acceleration:

$$F = mg = (\rho - \rho_o)Vg. \qquad (5.3)$$

(Density with the subscript "o" is that of the medium, here water; the formula assumes that a downward force is positive.)

Here's the problem of the escaping submariner—with air-filled lungs, body volume, and thus density, depends on depth; thus, as ascent proceeds the upward force gets greater and greater because the air in the lungs expands and thereby reduces the submariner's density. But the cure, exhalation, is simple, as mentioned a few paragraphs ago. Not so simple is the situation of the submarine diving too deep—the pressure on the hull increases by an atmosphere for each 10.3 meters in freshwater or 10 meters in the ocean. For moderate depths, the stiffness of the hull means that additional inward pressure makes little difference to the volume of the craft, and any small increase in its density can be offset by pumping water out of its ballast tanks. But if it goes too deep, the hull begins to compress too much, the craft gets too dense, and *the downward force increases, propelling it still deeper*. To the extent that either submarine (or submariner) has a nonrigid hull and a gas (air) inside, it (or he or she) is unstable with respect to depth. If it ascends, it wants to ascend faster; if it descends, it wants to descend faster.

The material of which organisms are made is almost always slightly denser than either freshwater (1,000 kilograms per cubic meter) or sea water (1,024 kilograms per cubic meter). Fat may be a little less dense—about 94 percent of water, but bone is much more so—170 to 250 percent of water, and other mineralized tissues are similarly dense. So organisms would sink were they not otherwise provided with some buoyant ballast. If the ballast is as incompressible as water itself, they face no problem of instability. The less dense the material, though, the more space-efficient the buoyant ballast it provides. Thus, best is a gas, something unavoidably compressible.

How then do pelagic (non-bottom-living) animals maintain buoyancy? Schmidt-Nielsen (1997) gave a short but comprehensive account of the various mechanisms and their relative advantages and disadvantages. Some creatures assiduously avoid any dense materials even if used by their sessile relatives (jellyfish and nudibranchs or shell-less snails), some use body fluids enriched in ionic solutions of low density (a few large marine algal cells), some retain large reserves of fats and oils (sharks and some other deep-sea fish), and some use gas-filled organs.

Two sorts of animals manage to store gas in structures of sufficient structural rigidity that the compressibility of the gas mixture is inconsequential (Alexander 1983); they thus maintain neutral buoyancy almost independently of depth. Both are cephalopod mollusks, but I'd guess that that evolutionary relationship doesn't reflect any common ancestor equipped with a rigid gas tank. Cuttlefish (which look like bulgy squid) have a foamlike or corrugated cuttlebone containing a gas mixture at nearly constant pressure (fig. 5.1). And the chambered nautilus (fig. 4.14c) has air pockets in the unoccupied chambers of its shell.

Many of the bony fish, though, keep gas in a flexible swimbladder whose internal pressure cannot differ much from that outside. The details of their arrangements vary. Some retain a connection between swimbladder and esophagus (the former develops as an outpocketing of the latter) and can gulp air from the surface or disgorge bubbles if need be. If, for instance, you put a goldfish in a large jar and reduce the pressure by attaching a vacuum source, the fish will give a mighty belch of gas. If the pressure is then restored by detaching the vacuum source, the fish sinks rapidly to the bottom. It may not be able to swim immediately to the surface again, but it can replenish its swimbladder from the dissolved gases in its blood. Other fishes, especially ones that live at considerable depths in the ocean, lose any connection between swimbladder and esophagus and depend entirely on secretion from dissolved gas in their blood—being hauled to the surface typically causes a fatal expansion of the swimbladder. But one way or another, fishes actively control the volumes of their swim bladders, even if the control mechanisms don't operate instantaneously. Bottom-living fishes commonly lack swim-

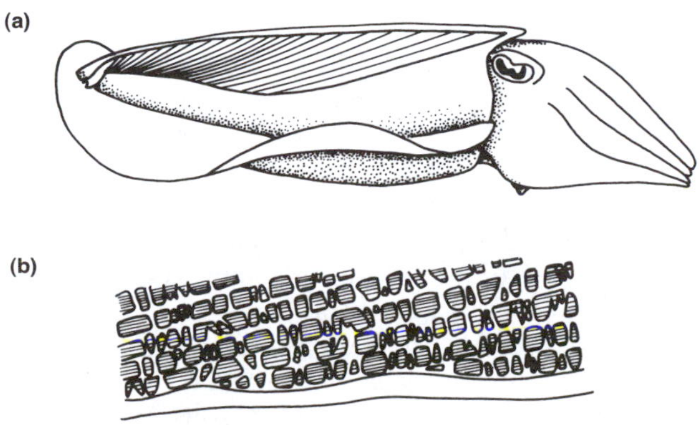

FIGURE 5.1 (a) A cuttlefish; the striped material near its upper surface is the nearly incompressible cuttlebone. (b) A slice of cuttlebone, with the air spaces shaded.

bladders, as do a few (mackerel, for instance) that simply swim continuously. Swimbladders have the odor of an awkward necessity, the price most pelagic bony fish pay for being bony. We air-breathing vertebrates have converted them to respiratory lungs, but we still use them, when necessary, to maintain buoyancy.

Sea snakes, air-breathers like us, dive deep and stay under for prolonged periods while swimming for considerable distances. They deal with this problem of variable buoyancy in what must be the simplest possible way. One such snake, *Pelamis platurus*, typically descends about 30 meters with enough air in its lung (it has only one) to be neutrally buoyant at that depth. But it loses gas, mainly carbon dioxide and nitrogen, from its skin, whose permeability lets it act as a kind of gill and supplement its lungs. So as it swims along it gradually ascends, thus maintaining its neutral buoyancy as part of its requisite return to the surface (Graham et al. 1987).

Taking air under water can raise problems of excessive as well as variable buoyancy. Consider a duck that dives beneath the water's surface with air trapped in its plumage. That air makes it buoyant, and it's a lot of air—typically enough to displace half a kilogram of water per kilogram of body mass (Wilson et al. 1992). So the duck has to swim harder to stay submerged. By one account, 95 percent of the work done during a duck's dive goes into overcoming buoyancy (Stephenson et al. 1989). A partial solution consists of trapping less air and using more fat (which is good insulation as well.) And waterfowl that routinely pursue prey underwater do just that, even though the device raises the cost of flight. But the problem diminishes as a bird goes deeper and the air compresses; thus the deeper the dive, the longer the bird can remain submerged—it should cost half as much to stay down 15 meters as to maintain position just beneath the surface (Wilson et al. 1992). Some birds even dive deep enough to become negatively buoyant—that is, to become sinkers (Lovvorn and Jones 1991). The problems afflict small birds more severely than they do larger ones, since they're matters of surface area, not volume—although larger birds may have somewhat thicker layers of trapped air. One bird, at least, entirely evades what would be a particularly severe version of the problem. Anhingas (or snakebirds), which swim with only their heads above water, simply get wet to the skin. But they stick to warm water, look quite bedraggled when they get out, and spread their wings to dry before doing much flying.

Pressure and Solubility

Carbonated beverages are familiar to us all, as is the observation that the bubbles don't appear until the bottle top is removed. Clearly the gas, carbon dioxide, is soluble in water and just as transparently obvious (for bottles if not cans) is the fact that the solubility varies with the ambient pressure. In quantitative form, the latter is known as "Henry's law" for William Henry (1775–1836), who pointed out that a given volume of water would take up a fixed volume of a gas whatever the pressure on the system. The fixed volume in solution, of course, contains a mass proportional to the pressure (Boyle's law), so the higher the pressure, the more gas goes into solution—in a direct proportionality. Again, only the partial pressure of the gas matters—the same amount of pure oxygen at 21 kilopascals will dissolve in a container of water as will oxygen from atmospheric air whose total pressure is 101 kilopascals and in which about one molecule in five is oxygen.

Here chemistry does come in—different gases have different solubilities. On a volumetric basis, nitrogen is half as soluble in water as is oxygen, while carbon dioxide is about thirty times more soluble than oxygen.* The range is very wide—helium is a fifth as soluble as oxygen while ammonia is thousands of times more so.

Henry's law has major implications for organisms. A deep dive means not only that an air store will be compressed but that its components will much more readily go into solution in either body fluids or ambient water. A diver breathing compressed air gradually absorbs a lot more nitrogen than usual in the bloodstream and tissues. A rapid ascent is like taking the top off a bottle of beer—gas bubbles (here nitrogen, not carbon dioxide) form. The effect ranges from unpleasant through crippling to fatal—circulatory systems don't like bubbles much better than they like blood clots. One partial remedy is to mix the oxygen for the dive with helium—which is much less soluble in water—rather than nitrogen. The commonest remedy (the solution, literally) consists of a very slow ascent, guided by tables that take into account the depth and duration of the dive. The alternative (the "bends") was first noted in workers preparing bridge footings from caissons of compressed air. Well over a century ago, Eads recognized the necessity for slow ascents by workers leaving the caissons for his bridge over the Mississippi at St. Louis. His contemporary, Roebling (himself a victim), took no such measures when building the Brooklyn Bridge in New York, and he lost quite a few workers.

Natural waters normally contain dissolved gas—indeed rapidly flowing streams may be slightly overloaded ("supersaturated") with atmospheric gases. But if a pond or ocean absorbs gases mainly from the atmosphere above it, concentrations of dissolved gases will reflect the their solubilities at atmospheric pressure. Due to biological activity and other causes, concentrations of some gases at substantial depths may exceed those that would result from direct dissolution from the local atmosphere. But the concentration of oxygen at appreciable depths is rarely even as high as atmospheric—the abysmal darkness of the abyss precludes photosynthetic oxygen production.

As a result, a fish with a swimbladder faces a real problem—two things are awkwardly unequal. On the one hand, the partial pressures of the gases in the swimbladder of a deep-sea fish must add up to the local hydrostatic pressure, which is equal to 1 atmosphere (surface) plus an additional atmosphere for each 10 meters of depth. On the other hand, the dissolved gases in the water (and in fish blood, via the gills) will be equilibrated with air at pressures more like those of the atmosphere above than with the gas in the swimbladder. But the swimbladder must be filled from dissolved gases in the blood, and it must not lose gas through redissolution into the blood. So secretion of gas from blood into swimbladder faces a big barrier, and swimbladder gas will all too readily go into solution in the fish's blood and thence out into the ocean. Two devices stand in the way. First, a layer in the swimbladder wall provides a very effective barrier to the passage of oxygen (Lapennas and

---

*The word "tension," here in a nonmechanical sense, is often used for the concentration of a dissolved gas in a liquid—it specifies the partial pressure of that gas in an atmosphere with which the liquid containing the dissolved gas is equilibrated.

Schmidt-Nielsen 1977). Second, blood leaving the so-called gas gland in the wall of the swimbladder passes through an exchanger (fig. 5.2) in which blood leaving the swimbladder loses excess dissolved gas specifically to blood moving toward the swimbladder (Scholander 1954).

In short, air under water is unstable stuff. Not only does it like to bubble upward, impelled by buoyancy, but it compresses into smaller volumes with depth and increasingly tends to dissolve in the surrounding water. Still, some organisms routinely carry unconfined air downward—in an evolutionary sense, air-breathing seems a hard habit to break. A number of beetles, for instance, breath from submerged gas stores. But the scheme gets rapidly more difficult with depth—not only does the insect have to swim further downward with its buoyant cargo, but any bubble will arrive diminished in size and then have a shorter life as the gases more rapidly dissolve in the water. One beetle, *Potamodytes tuberosus*, contrives to get the pressure in the bubble below atmospheric, taking advantage of the pressure drop attending fluid flow. That makes gas (at atmospheric partial pressure) move into the bubble, which can thus act as a permanent lung (Stride 1955). But it works only in very shallow water (about 4 centimeters) and very rapid flows (over 1 meter per second). We'll return to *Potamodytes* in chapter 7 when we consider Bernoulli's principle.

Perhaps the most impressive use of gaseous air under water is that of the European water spider, *Argyronetes aquatica* (Foelix 1996). It is the only spider that lives constantly underwater—walking, swimming, feeding, mating, and raising young. It makes a bell-shaped web in the submerged vegetation of ponds and then travels repeatedly up to the surface and down again, in each trip carrying down a coating of air on its abdomen (the allusion to silver in the name of its genus reflects this shiny layer). Thus it fills the bell with air; the fine mesh and a water-repellent coating on the strands prevent upward leakage, as will be explained shortly. From time to time the spider adds air to its diving bell to offset use and dissolution. Incidentally, it has one other, perhaps unrelated, oddity—uniquely among spiders (and unusual among animals in general), the males are larger than the females.

The Liquid State

A liquid may be a familiar matter, but the liquid state boils down to quite a queer business. It's something between solid and gas, although "between"

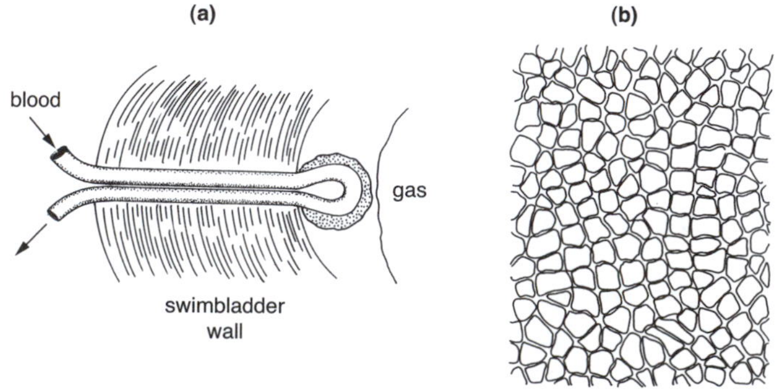

FIGURE 5.2 (a) Diagrammatic view of the exchanger in the wall of a swimbladder. Blood vessels going inward through the wall are in close contact with those coming out, as seen in (b), a cross section drawn from a photomicrograph. Half of the vessels carry blood inward and the other half carry blood outward.

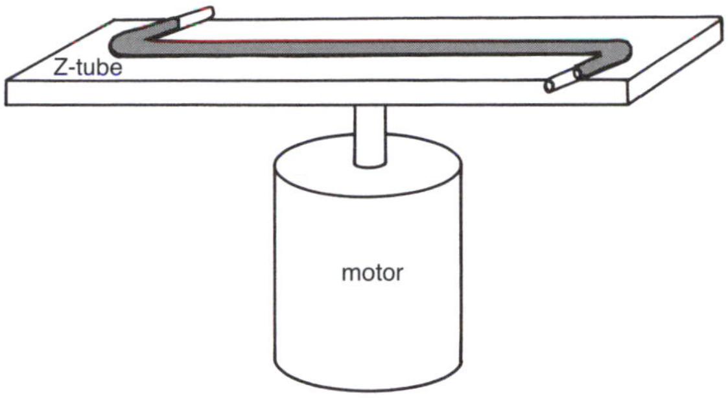

FIGURE 5.3 A diagram of the device of Briggs (1950) for measuring the cohesion of liquids—a motor spins a plate to which is glued a glass Z-tube containing liquid. The cohesion can be calculated from the speed at which the column of liquid ruptures in the middle. The inwardly bent ends of the tube ensure that the liquid is self-centering about the axis of rotation.

implies nothing like a continuous quantitative scale. Rather, a liquid shares certain properties with each of the other states—for some problems a liquid behaves like a gas of exceptional density and coherence; for others it can be thought of as a broken-up solid. (Conversely, a fine powder is often better treated by the rules for liquids than by those for solids.)

Internal cohesion

Internal cohesion is the basic attribute that distinguishes liquids from gases—a gas withstands no tension; a liquid does. Under conditions of weightlessness and windlessness, a glob of liquid will assume a spherical shape, spontaneously minimizing its surface area as its molecules huddle together. But how strong is that internal cohesion? Measuring it requires that we pull on a liquid, a task as ostensibly impractical as pushing on a rope. Ingenuity, though, has not been lacking, and the thing has been done by several techniques; the one devised by Briggs (1950) and shown in figure 5.3 is perhaps the most straightforward. Water (that's the liquid we'll worry about) is loaded into a glass tube of fine bore that has been bent into a Z-shape and that's open at both ends.* The tube is then spun about its midpoint at increasing speeds like a helicopter rotor. By Newton's first law, the water prefers to go straight rather than turn; that is, its mass requires that a force be applied along the axis of the tube to keep the water rotating. But the only source of force is the internal cohesion of the water column, pulling both halves of the column toward the midpoint. By Newton's third law (action and reaction) half the column is pulling on the other half, connected together at the axis of rotation. When the tensile forces can no longer maintain the water's continuity, the column ruptures at the axis. From the rotation rate and the dimensions of the column, the tensile strength (force per unit of cross-sectional area) comes out to as much as 28 megapascals—that is, to a stress equivalent to the pressure exerted by *280 atmospheres*. Put another way, a rope 1 centimeter in diameter made of water, if you can imagine something so preposterous, would support a mass of about 225 kilograms or a weight of almost 500 pounds.

This phenomenon of tension in liquids bothers people used to thinking about gases. Recall that you can't really pull on a gas since it has negligible

*If the tube were I-shaped rather than Z-shaped, the system would be unstable, and the water would be flung out of one end or the other as soon as the tube began rotating.

internal cohesion. Negative pressure, for that's what tension is, is illegal in a gas (at least a perfect one). It does look as if the stunt can be done—a pump at the top of a pipe can draw out air and raise water in the pipe. But by pumping out air the pump can't raise the water to a height beyond that set by the local atmospheric pressure—at sea level 10.3 meters. The limitation arises because the pump, by removing air, is merely allowing the atmosphere to *push* the water upward (fig. 5.4), and no *pull* is actually involved. In theory, a pump dealing with a column of water only, with no air inside, can avoid this height limit. The ultimate limit, calculated from intermolecular attraction (see Hayward 1971), far exceeds the 280 atmospheres mentioned above. But even 280 atmospheres corresponds to a column of water approaching 3 kilometers (2 miles) high. In practice, though, water pumps cannot do much better than 10 meters—either the water fails in its adhesion to the walls of the pipe or dissolved gas in the water boils out at the low pressure, and the column of water is ruptured.

Are trees permissible?

The difficulties of adequately containing a liquid under tension are so great that the phenomenon of liquid tension is ordinarily dismissed as a curiosity without technological ramifications. It has received most of its attention from plant physiologists, for whom it is at once crucial and commonplace. Running upward in trees from roots to leaves are a set of pipes (xylem vessels and tracheids, technically) through which water ascends. While roots provide some upward push ("root pressure"), the main agency impelling ascent clearly consists of pull from the top, as first suggested to a skeptical world by Dixon and Joly, a botanist and a physicist, in 1894. If vessels are suddenly cut, the hissing sound of air entering can be detected. Water evaporates ("transpires") at the leaves, and the vacant space can only be refilled, as things are arranged, by raising water from the roots. An early bit of support for the existence of internal tension was the observation of Macdougal (1925) that when the transpiration rate increased, tree trunks shrank. Trees are just a little skinnier at midday!

And trees exceeding 10 meters in height have been around for the past hundred or so million years—trees get up to about 100 meters, implying an

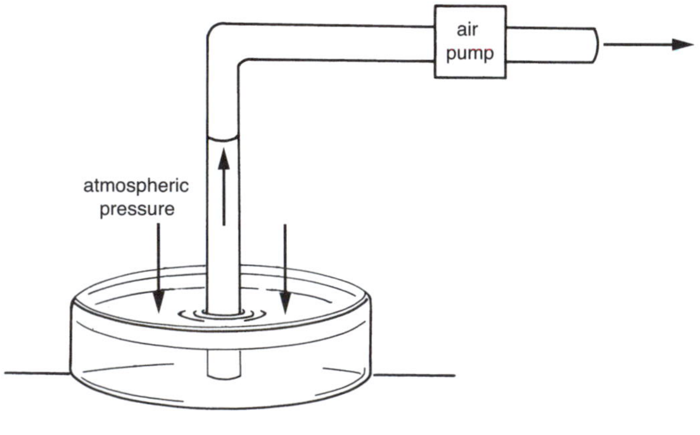

FIGURE 5.4 An air pump can raise water only to a height set by the local atmospheric pressure—in removing air from the pipe it merely permits the atmosphere to push the water column upward.

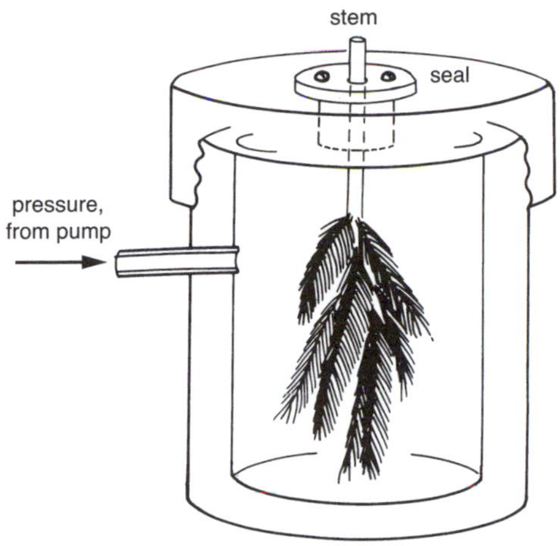

FIGURE 5.5 A Scholander bomb, with a stem protruding. Cutting the stem from the plant relieves the internal pressure, so the air-sap interface retreats inward as the vessels expand. Sealing the stem in the bomb and applying positive pressure through the side port mimics the effect of the original negative internal pressure. The applied pressure that just returns the interfaces to their original locations at the cut surface is thus equal (if opposite in sign) to the original negative pressure in the sap.

internal tensile stress of 1 megapascal, which corresponds to a pressure of –10 atmospheres. With a clever device called a "Scholander bomb" (Scholander et al. 1965; fig. 5.5), it's possible to check these expectations directly—in fact, the pressures are often still more extreme, with –20 atmospheres or more not at all uncommon high up in a tree. A negative pressure of –10 atmospheres might account for water staying up, but more pull is needed to move water through the narrow vessels as well as to detach it from the soil in the first place. Interestingly, the cells making up the vessels are not alive at what amounts to functional maturity. Perhaps the release of any dissolved respiratory carbon dioxide into a vessel with a large negative pressure would cause a microcatastrophic bubble! While virtually no doubt remains that sap is being pulled up by strongly negative pressures, it must be admitted that not all the details of the scheme are clear yet—questions remain about the properties of the internal walls of the vessels, how the continuous columns are initially established, how ruptures such as those caused by freezing can be repaired, and so forth. Good references to these matters are Tyree and Sperry (1988) and Tyree and Ewers (1991) or the textbook by Nobel (1999).

Do animals make analogous use of negative pressures; that is, do they ever intentionally put liquids in tension? No one has suggested anything approaching the Herculean pulls of trees, but a few cases of negative pressure have turned up. Bennet-Clark (1963) calculated that a blood-sucking bug, *Rhodnius prolixus* (once the *Drosophila* of insect developmental biologists), has to pull at several atmospheres in order to get blood through the canal of its stylet (10 micrometers in diameter and 200 micrometers long) at the observed flow rate of 0.33 cubic millimeters per second. And Smith (1996) found that the suckers of cephalopods—octopus, cuttlefish, and squid—could develop up to an atmosphere of tension when just submerged. The limit appeared not to be the muscular strength of the suckers, since they could pull far harder in deeper water, where higher ambient pressures had to be offset before true negative pressures occurred. Instead, the limit traced to cavitation of ordinary sea water on typical marine surfaces. The plants have two things going for them that the cephalopods do not—they make all exposed surfaces (the vessel walls) themselves, and they process the water in their roots before it enters the vessels of xylem.

That must matter—only with carefully distilled and degassed water could Lawrence Briggs reach the high tensions of his Z-tube measurements.

The Air-Water Interface—Surface Tension

That internal cohesion of liquids, something especially strong in water, has another consequence. Liquids often form sharp interfaces with other liquids (vinegar and olive oil in the kitchen, for instance), and they always form sharp interfaces between themselves and gases—as when air meets water. Some peculiar things happen at the boundary between air and water, phenomena rich with biological significance. So we have to worry not just about life's gaseous and liquid media, but about the special features of their interface, which we might view as a weirdly two-dimensional state of matter. Two features have particular biomechanical relevance. For us, the more familiar one is surface waviness, which we'll take up in chapter 14. Less familiar is a more peculiar property, surface tension.

For water deep within a large body of the stuff, internal cohesion gives rise to forces in all directions that, taken together, balance out and thus generate few immediate consequences. For a bit of water at the surface of a liquid, the situation can be quite different. Water below it pulls down, but the air above doesn't provide any balancing upward force. Since the bit of water certainly doesn't accelerate downward, it must encounter some counterbalancing force—in practice the resistance of the water to being compressed. To some extent, then, the presence of a surface must increase the hydrostatic pressure everywhere within a body of water, but (except as we'll discuss shortly) that's normally no big deal. At the surface, though, the asymmetry of forces gives the liquid its surface tension.

Another way to look at the matter is to recognize that the internal coherence of a liquid makes a surface something that the system doesn't like and that it will rearrange itself to minimize. Just as a droplet will spontaneously round up into a sphere, it takes work to force it into any other shape—it takes work to create surface. To put the phenomenon into quantitative form, either we need to measure the force necessary to maintain a surface against the surface's urge to submerge, or we need to measure the work necessary to create some unit area of surface. It's easy to construct a semiquantitative apparatus to view the force or work; it's a little harder, though, to make precise measurements.

A film of soapy water amounts to two air-water interfaces with a thin layer of water between. If you stretch such a film, then you're stretching two air-water interfaces. And that requires no more than one wire sliding down two others (fig. 5.6a). The results contrast strikingly with what would happen if you pulled on a sheet of rubber or other elastic. The "material" is an elastic in one sense—it does pull back against the applied force. But "stress" is apparently independent of the "strain"—the force needed to hold the slider against the surfaces' desire for oblivion or to move the slider and create more surface doesn't at all depend on the particular position of the slider (fig. 5.6b). It takes work to make surface, but the work is independent of the size of the preexisting surface. What happens is that stretching the surfaces *creates new surface area*! Water from between the surfaces simply moves out to the interfaces to increase their areas, and the distance between the parallel interfaces decreases.

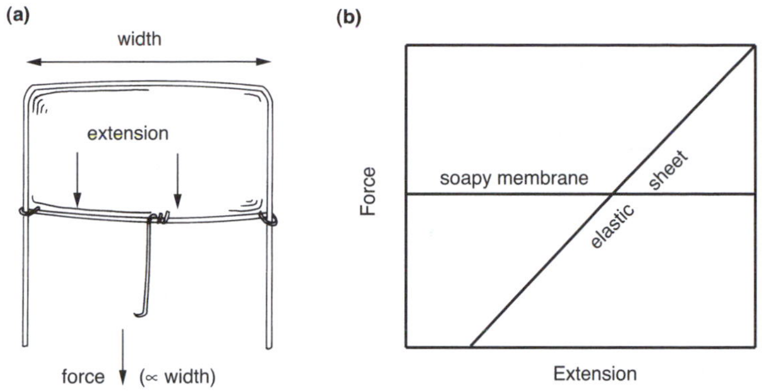

FIGURE 5.6 (a) Expanding the pair of air-water interfaces that enclose a film of soapy water by pulling down on a slide wire. It's really as simple as it looks. (b) The force needed to stretch a sheet of a simple elastic material increases with the distance it's stretched; by contrast, the force needed to stretch a pair of air-water interfaces shows no such dependence—stretching just generates more interface.

The force may not depend on the thickness of the water between the interfaces or on the distance the interfaces have been stretched, but it does depend on the width of the two interfaces. Double the length of the slider and you double the requisite force. The appropriate dimensions in which to express surface tension as a material property are therefore force divided by distance. Similarly, the work depends on that width and on how far the constant force has moved the bar—in short on the area of surface created. It also does not depend on thickness—the interfaces have no thickness, and the water between them plays no role. Both force-to-make-distance and work-to-make-surface reduce to basic dimensions of MT^{-2}; they're equivalent in dimensions, units, and numerical values. Thus, surface tension constitutes a tension in the same dimensional sense as the tension we used for thin-walled spheres and cylinders in chapter 3. We usually think of force per unit distance and most commonly use a Greek lowercase gamma (γ) for its symbol.

For present purposes we need only one or two values for this physical quantity—the interface between water and air is the main one relevant to organisms. For pure water the value is 0.073 newtons per meter; for seawater it's 0.078 newtons per meter. For liquids against air these are unusually high values, another of many peculiarities of water. Temperature affects the values only a little, but they're drastically lowered by soaps and detergents, the main reason we use the latter.

You've undoubtedly noticed that a small quantity of water seems to know something about the surface on which it resides. On a block of paraffin, a Teflon-coated pan, or a freshly waxed car, it forms little beads of the form of flattened spheres. By contrast, on scrupulously clean glass or on any surface with a trace of soap, it spreads out. It isn't enough to worry just about the air-water interface—somehow any solid contacting that interface influences the behavior of the line along which the three phases (gas, liquid, and solid) meet. The phenomenon goes under the heading of *wettability*—we say that water "wets" some solid surfaces but not others; the wettable surfaces are those on which water spreads rather than beads. Wetting occurs when the *adhesion* of liquid to solid exceeds the internal *cohesion* of the liquid; nonwetting indicates that the internal cohesion is greater. There are degrees of wettability—the phenomenon is a continuous and quantitative one—but we'll mainly consider the two extremes.

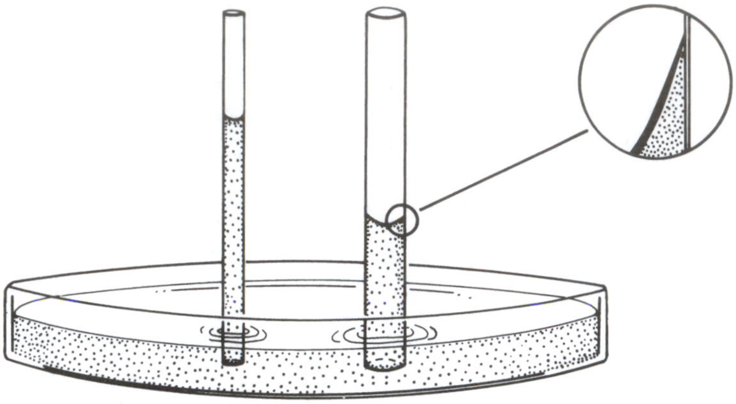

FIGURE 5.7 Water spontaneously rises in a clean glass tube; the thinner the bore of the tube, the higher the column goes. The inset shows the "contact angle" between glass and water.

Capillarity

The classic demonstration of the effect of wettability consists of immersing a clean, narrow-bore, glass tube in water. The air-water interface in the tube rises to a level above that of the surrounding water (fig. 5.7)—it's being pulled up by some force, and the fact that the periphery of the interface is highest points a clear finger at the location of the force. Attraction of water to glass draws water at the edge up; the resistance of the interface to expansion of its area (as we've just discussed) raises the rest of the water to nearly the same height. Wetting is nearly complete, as shown by a low contact angle between glass and water—at the point of contact, the interface is almost vertical. The low contact angle also means that water gets pulled almost purely upward and minimally outward. Pressure in the water is subatmospheric since it's being pulled from above, not pushed from below. So you can't put a side tube on the apparatus and use the elevation due to surface tension to pump a continuous stream of water up into a reservoir higher than the original body of water and then let the water fall back through a generating turbine. Too bad—another perpetual motion machine grinds to a halt.

How high (h) will the water rise in the tube? It will rise until the upward force due to surface tension is just equal in magnitude (and opposite in direction) to the downward, gravitational force. The upward force is surface tension times some length, the appropriate one being that of the line of contact of glass, water, and air—the circumference of the tube. The downward force is the volume of water in the tube above the external water level times its density times gravitational acceleration. For a cylindrical tube of radius, r, setting forces equal, we have

$$(\gamma)(2\pi r) = (\pi r^2 h)(\rho g). \tag{5.4a}$$

Canceling and solving for height gives

$$h = \frac{2\gamma}{\rho g r}. \tag{5.4}$$

Clearly, water will rise higher in smaller tubes. The upward force varies with the first power of length; the downward force varies with the second;

thus, the rise varies inversely with radius or diameter of the tube. So we're looking at a small-scale phenomenon, one that ought to be of more immediate relevance to small organisms than to ourselves. But we do run into them. A bath towel works because it has lots of wettable surface, lots more than a wet person. Rub the two together, and capillarity among the fibers of the towel assures that the bulk of the water will end up transferred to the towel.

Physical scientists occasionally claim that water rises in trees due to capillarity. In fact, the vessels of the xylem are just too large. Simply plug the radius of a typical xylem vessel of about 50 micrometers—some come much larger—into equation (5.4). Capillary rise comes out to 0.3 meters, or about a foot, rather shorter than anything outside the bonsai business.

Capillarity—if in an odd guise—does matter, at least to a few plankton-feeding birds. According to Rubega and Obst (1993) and Rubega (1997), it's used by several phalaropes (*Phalaropus lobatus* and *P. tricolor*) and sandpipers (*Calidris mauri* and *C. minutilla*). Instead of filtering, as do most planktivores, these small birds "tweezer" prey. But that puts prey at the tip of the bill, not the pharynx. Gaping the bill slightly, though, creates an interface, as in fig. 5.8a). The water then does its part—surface tension reduces the area of interface by making the droplet of water move up and back from the bill's tip.

What if the solid isn't "wetted," if the cohesion of the liquid exceeds the adhesion between it and the solid? The usual demonstration (before we took proper cognizance of mercury's toxicity) consisted of inserting a thin glass tube into liquid mercury; the interface was depressed beneath the level of the outside liquid. Mercury may not be the stuff of which life is made, but deliberate nonwettability may be quite important to organisms—it's as ordinary as

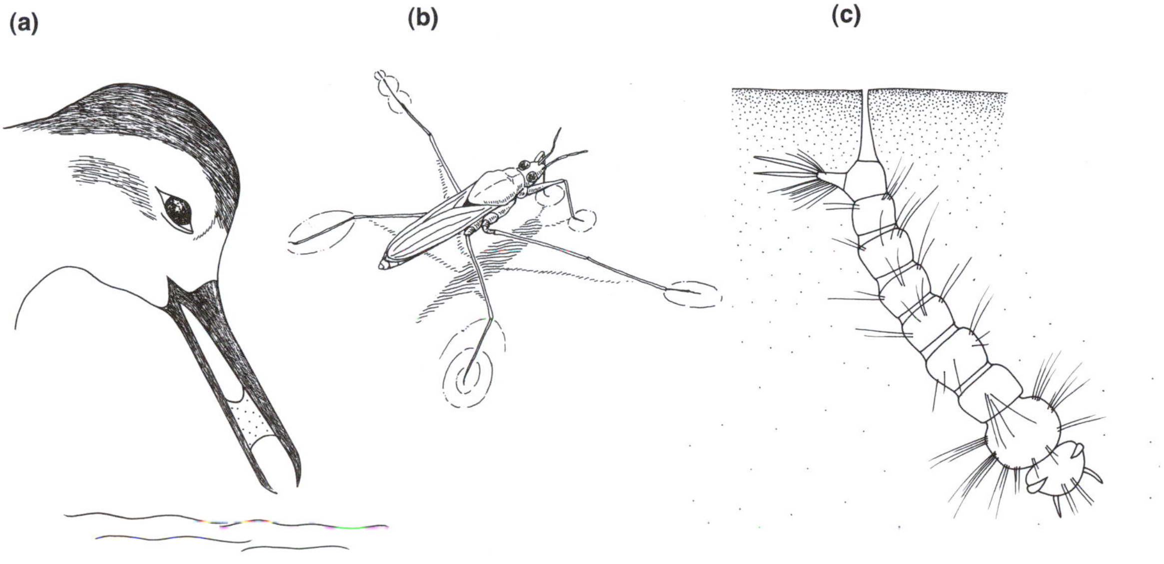

FIGURE 5.8 (a) The moving interface in the bill of a phalarope that draws food upward. (b) A water strider with its legs dimpling the surface of a pond; fisher spiders, less common, do the same thing. (c) A mosquito larva and its snorkel. Part (a) redrawn from Rubega, M. A., 1996, *J. Morphol.* 228: 56 with permission of the copyright holder, Wiley-Liss, a subsidiary of John Wiley & Sons, Inc.

water off a duck's (nonwettable) back. If you're small, as living things usually are, then being covered by a layer of water of substantial thickness isn't exactly a convenience. Contact with the surface of a body of water can be a perilous entanglement, best (but not inevitably) avoided by arranging to be as repulsive as possible to this clingy liquid. Water may be crucial to all life, but the air-water interface can hold a fatal attraction. Unsurprisingly, insects typically have strongly hydrophobic outer cuticles.

Walking on water

The surfaces of bodies of water aren't the commonest of habitats, but a decent number of creatures (some bugs, beetles, springtails, and an occasional spider) get around on it by using surface tension for support. The possibility was raised in the last chapter, and an index was developed (equation (4.6)) relating size and the practicality of such support. We can now take a more specific look at the problem, one similar to that of capillarity. Absolutely prerequisite is a surface resistant to wetting, but, again, arthropods commonly have waxy coatings. So an insect pushes down with four to six feet against the surface, usually with the feet splayed out and thus not interacting—each foot makes a separate dimple in the surface. How much insect can be supported? What matters is the combined perimeter of the feet, the total length of the air-water-insect contact line, although exactly how to measure that length isn't quite as clear as it was for the capillary tube.

A marine water strider, *Halobates*, has a mass of about 10 milligrams; using the surface tension of seawater for support would require a minimum of about 1.3 millimeters of contact line, about 0.2 millimeter per leg. *Halobates*, though, can jump several centimeters upwards from the surface, so it must be able to exert a downward force about ten times its weight (Andersen and Polhemus 1976) and so should require 13 millimeters of contact at takeoff to avoid wet feet; it has, in fact, specialized long hairs on its legs. *Halobates* and the freshwater striders can propel themselves around on the surface by pushing on the local indentations beneath the legs. Andersen and Pohlhemus (1976) reported top speeds of 0.8–1.3 meters per second, most impressive for such small animals—that's faster than a duck can paddle.

Going somewhere by pushing on the surface turns out to be a subtle business. Suter et al. (1997) recently looked at how a fisher spider, *Dolomedes triton*, does it—we have no reason to doubt the wider applicability of their results. The guesses of neither Denny (1993), who invoked waves, or me (Vogel 1994a), who imagined pushing directly against the surface, hold up. What happens is that the leg and dimple (the latter from the downward weight transferred by the leg) act as a unit. Both move rearward as the animal pushes, and the rearward drag of the unit generates the forward thrust (fig. 5.8b). (Moving a dimple backwards forces water to move forwards, hence this nonobject has perfectly ordinary drag.) Once surface tension can be handled, such a surface must be quite a nice environment—it has neither the hilliness of land nor the high drag encountered under water.

There's an even more peculiar way to get around on a liquid's surface. A small beetle, *Stenus*, normally walks slowly on the surface. When speed is desired, though, it secretes a substance from its last abdominal segment that locally reduces the surface tension. The result is an asymmetrical force on the beetle, which moves forward at up to 0.7 meters per second (Chapman 1982).

The trick, though, is probably an expensive one inasmuch as the surfactant chemical cannot be reused. You can easily demonstrate *Stenus*'s stunt in either of two ways. A loop of string can be loosely placed on the surface of a bowl of water and a piece of soap touched briefly to the surface within the loop—the loop will quickly be pulled outward and become circular. Or a small surface boat can be made of a piece of shiny paper and a tiny piece of soap stuck on its stern—the boat will be propelled forward.

What would be the maximum weight of a human who could walk on water? My size 9 sandals have perimeters of 0.62 meters each; that figure times the surface tension of water gives 0.045 newtons of force or 4.6 grams (a sixth of an ounce) of terrestrial mass—so 9.2 grams to stand (two feet in contact) or half that to walk. That's an explicit view of what was presented as just a scaling rule (recall *Je*) in the last chapter. The theological implications are beyond the scope of the present book.

Mosquito larvae don't walk on water but instead hang downward from the surface, breathing, as already mentioned, with a snorkel system (fig. 5.8c). Adding a substance to the water that reduces surface tension is highly detrimental to their well-being; it may be done deliberately or may be just a not-so-undesirable consequence of detergent pollution. Alternatively, some light and nonaqueous liquid can be poured on the water (kerosene, for instance), which will "wet" the waxes of the larval cuticle and plug the snorkel. If you live by surface tension you can die without it.

Laplace's Law and Surface Tension

Two chapters ago, we encountered a scaling rule that described how size set the relationship between the pressure in a closed vessel and the tension in its curved walls—Laplace's law, as it's called in the biological literature. Now we're again talking about tension, as surface tension, and curved walls, or at least curved interfaces. No surprise—Laplace's law again plays a role, for bubbles of air in water, for droplets of water in air, and for partial bubbles, or curved interfaces. Recall the equation relating pressure difference, tension, and radius for a sphere:

$$\Delta p = \frac{2T}{r} \quad \text{or} \quad T = \frac{\Delta p r}{2}. \tag{3.6}$$

If surface tension is the tension at issue, we just substitute γ for T:

$$\Delta p = \frac{2\gamma}{r} \quad \text{or} \quad \gamma = \frac{\Delta p r}{2}. \tag{5.5}$$

The only unusual aspects are two. First, the value of tension is generally a given—the surface tension of the particular interface. Second, if one has something such as a soap bubble with air both inside and out, there are two interfaces and the pressure inside is therefore twice what equation (5.5) predicts. (Boys [1902] gave the classic account of the physics associated with soap bubbles; it's still informative, enjoyable, and available.)

Bubbles

As an immediate result of Laplace's law, the pressure inside a bubble or droplet is greater than otherwise expected—surface tension compresses whatever is on the concave side of the interface, whether the inside is liquid or gaseous. As you'd expect from equation (5.5), the smaller the bubble or droplet, the greater is this extra internal pressure. As a quick illustration of how surface tension becomes a big thing for small spheres, we can calculate the extra pressure (above atmospheric pressure) inside some air bubbles in water. For bubbles 1 millimeter, 10 micrometers, and 0.1 micrometers in diameter, the pressures come out (assuming a surface tension of 0.073 newtons per meter) to 290, 29,000, and 2,900,000 pascals. The first may be trivial, but the second is over a quarter of atmospheric pressure, and the third is nearly 30 atmospheres.

The consequence of an extra pressure of 30 atmospheres in a sphere of water may not amount to much, but for a sphere of air, Henry's law predicts drastic effects. Gases become increasingly soluble in liquids as the hydrostatic pressure is increased, and we're talking about a huge hydrostatic pressure, equivalent to what you'd encounter beneath about 300 meters, or 1,000 feet, of water. And the gas in the tiny bubble can't be far from the interface, so the times need for diffusive transport to the inner surface are very short. Result—the gases of the air go into solution in the water. Whether the water already contains dissolved gas is immaterial—since the water is at a pressure 30 atmospheres lower it can't be saturated with gas relative to the pressure in the bubble. The bubble gets still smaller, the pressure rises further, more gas goes into solution, and quick as a wink the bubble ceases to exist.

This probably explains why most tiny organisms achieve buoyancy with lipid droplets rather than air bubbles or bladders. Lipid is costly—it's the most concentrated form in which organisms can store chemical energy, and it can't be secreted as a low-density gas and reabsorbed into solution. But it does avoid the peculiar volumetric compression suffered by small gas bubbles as well as their instability due to surface tension.

There's even a worse aspect to the high pressures in small gas bubbles. Imagine a liquid supersaturated with gas—a state in which the gas would "prefer" to bubble out of the dissolved state. A bubble, though, must start small—it's hard to imagine enough gas molecules coming together at the same time to create a big bubble at one stroke. But a small bubble of gas will collapse in accord with the laws of Laplace and Henry. Small, spherical bubbles should be unlawful! So how can a bubble ever form, even in a liquid supersaturated with gas—a freshly poured or slowly warming glass of beer, for example? The problem isn't trivial, and certain aspects are still argued. Presumably this is what a physical scientist ponders while staring at a glass, deep in thought.

Still, bubbles do form, and *careful* observation of the effervescence of the beer in the glass reveals the basis of the best explanation. The latter, incidentally, was provided by some biologists (Harvey et al. 1944)—gas bubbles let loose within organisms are not nice things at all. Notice (if you haven't) that bubbles seem always to form on the surface of the glass and that certain places on the surface are particularly effective in generating bubbles—sometimes they come off such sites in steady streams. A bubble-generating place has two important characteristics. First, it's hydrophobic—that is, gas spreads

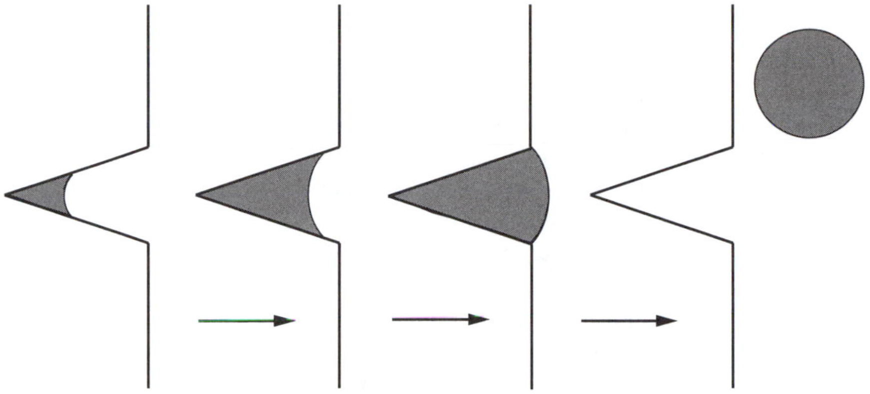

FIGURE 5.9 The formation of an air bubble (shaded) in a scratch on the wall of a water-filled glass container. At no point can a highly concave water surface squeeze the bubble into oblivion.

on the surface while water shrinks away from it. Second, it has a scratch or other irregularity. The combination permits formation of a local gas pocket whose boundary (the gas-liquid interface) curves inward (or is at worst flat) toward the gas phase (fig. 5.9). There a bubble with a concave surface can grow until it reaches a critical size—large enough so that when it bulges out and rounds up into a properly convex sphere it will have a sufficiently large radius of curvature so the excess inward pressure will not cause collapse.

If no hydrophobic surface is available, enormous degrees of supersaturation of gases in liquids can be produced by a variety of simple schemes; such a lack of appropriate surface seems to explain the great (and fortunate) reluctance of bubbles to form within cells and blood vessels. The water columns in the xylem vessels running up the trunk of a tree provide a dramatic example of what's possible when all surfaces are hydrophilic. With megapascals of negative pressures virtually *any* dissolved gas ensures supersaturation, yet bubbles rarely form. It's a good thing, too—a tiny bubble would rupture a water column since any bubble is itself an appropriate surface for gas formation; and, once formed, bubbles grow almost explosively in a supersaturated liquid. But trees persist despite the seeming fragility of this most basic element of their physical existence. They can even refill vessels emptied by bubble formation, that is, by an air or vacuum embolism—perhaps by producing locally positive pressures and taking advantage of Laplace's law and the geometry of the "bordered pits" associated with their xylem (see Holbrook and Zwieniecki [1999]).

Nor is the question of initiation of bubble formation a purely biological one. Sudden and self-driven carbon dioxide outgassing from the supersaturated depths of the lake seems to have been the basis of the disaster at Lake Nyos (or Nios) in central Africa in 1986. The lake (and some others) sometimes becomes strongly stratified, with colder and thus denser bottom water stabilizing it against mixing. If that water contains a substantial amount of dissolved gas, a potential instability then lurks—a little rise of some of that bottom water will produce bubbles, whose buoyant rising draws up more water until the whole lake turns over and releases massive quantities of gas (Kling et al. 1987). If the lake is in a valley, and the gas (as does carbon dioxide) exceeds air in density then everyone in the vicinity can be killed by poisoning or asphyxiation.

Critical matters in our daily lives depend on the peculiarities of bubble formation and growth. Some truth lies in the rumor that champagne tastes better when imbibed from the best quality glassware—fewer scratches means less bubble formation, more persistent supersaturation, and release of more effervescence on the palate. Treatment with dishwasher detergent may help matters—the stuff leaves a little residue to make glass less hydrophobic, so glasses will air dry without local deposits from streaks and bubbles of impure tap water. Plastic glasses are bad for champagne—most plastics are terribly hydrophobic. Drinking beer from an unglazed mug appreciably changes the taste—the irregular surface precipitates a great flurry of bubble formation, so the head is high while the beer is flat. Of course you may prefer beer this way, especially if you like to drink it very cold but would rather avoid the concomitant gassiness. Water boils more easily in an old kettle with scale inside its bottom—in a new, stainless steel kettle water can get a bit above the normal boiling temperature and then suddenly puff out into vapor. Vaporization then lowers the temperature and the process may repeat as a set of pulsations. In chemistry labs, rough-surfaced "boiling chips" are often introduced into reaction vessels to provide enough irregular surface so the bubble formation of boiling occurs steadily and smoothly.

If a large number of small bubbles coalesce into one large one, the total gas volume increases—for a thousand bubbles, each 1 micrometer in diameter, the final volume will be almost triple that of the original, individual bubbles. Still, I haven't heard of any bit of biology being hung on the phenomenon. It may be important in some circumstance where ice melts—freezing commonly causes bubble formation because less air can dissolve in ice than in cold water—during melting, bubbles might coalesce instead of their contents going back into solution.

Pores

Surface tension is the manifestation of a liquid's abhorrence of surface. Water rises against gravity in a capillary tube rather than make a greater area of air-water interface at the top. Put another way, the liquid does its best to avoid interfaces—and under proper provocation it does other things besides squeezing tiny bubbles of air out of existence. With a sieve of sufficiently fine mesh you can bail water—the interfaces get caught in a meshwork through which either air or water can freely pass. If the phenomenon seems preposterous, consider something less hypothetical. About to drown, you take off your pants, knot the end of each leg, fill the pants with air, and hold the waist closed and downward. If the knots are tight enough you have a good float. But note—either air or water can pass freely through the pants; it's only the interface that seems to get caught. If you don't wet your pants, the float won't last! Again, for an interface to get pressed through a tiny pore, its radius of curvature has to get very small, and so a lot of pressure is needed to push the liquid through against surface tension. Lowering the surface tension of the water reduces the required pressure, one of the advantages of laundering with detergent.

Further practical matters are germane here. Synthetic fabrics are said not to "breathe." In the damp southern summer that I annually endure, cotton is clearly preferable to polyester in any circumstance in which a person produces liquid perspiration. The trouble with most synthetics is that they're very hydrophobic, so when they get a liquid film across their pores they're about as vapor-proof as a raincoat. The opposite side of the coin is the treatment of tents and raincoats to make them hydrophobic, so they pass air freely but don't pass water even when the fabric feels wet as a result of stuck interfaces. (A truly waterproof and closed tent is not pleasant—the water vapor you lose from lungs and skin condenses on the cold walls and runs down as a liquid.) Leave a wet piece of absorbent cotton to dry, and it spontaneously contracts into a tight wad. The fibers get pulled inward by the receding interfaces.

Plants again. Even in a desert the soil a little ways below the surface contains liquid water. It's called "capillary water" and is often thought of as firmly stuck to soil particles. The binding, though, is as much physical as chemical—the water in the soil interstices lies in tiny recesses between soil crumbs where it has minimized its exposed interface with air (Rose 1966). For the roots of a plant to extract the water requires making more surface, and thus it takes a very great pull, one that appears as an additional (negative) component of the pressure in the vessels running up a stem or trunk. The lowest (most negative) pressures known in plants occur in desert shrubs, which must suck really hard on the ground to get any water out. The most extreme value on record is, I think, minus 120 atmospheres (Schlessinger et al. 1982)—that would hold up a column water over 1,200 meters (4,000 feet) high. So the pull needed to get water free of the soil can exceed both the pull that keeps water moving in the vessels and the pull that counteracts gravity.

Of the several places where the resistance of an interface to passage through a pore is biologically relevant, one overwhelmingly matters most. Earlier we used the internal coherence of water to explain how sap can be held up a tall tree by tensile pull from the top rather than by compressive push from the bottom. But these columns of sap in the xylem are in some real way open at their tops—very large quantities of water evaporate (transpire) from leaves. Why, then, isn't air sucked into the leaves, filling the entire system? Clearly, the columns of sap are in some equally real way closed! The explanation isn't especially mysterious (more mysterious is why the situation is so rarely mentioned in textbooks), and a good treatment is given by Nobel (1999). The functional interfaces are in the interstices of the cell walls within the leaves, in their feltwork of cellulose fibers. The gaps between the fibers are about 10 nanometers across; making the rough-and-ready assumption that these gaps amount to pores, it's no trouble at all to calculate (equation (5.5) again) that it would take an internal negative pressure of 29 megapascals or 290 atmospheres to suck air into the plant's vascular system. That's far greater than the record negative pressure noted in the last paragraph.

Some aquatic arthropods make good use of the resistance of an interface to passage through a pore. They make a surface that's virtually nothing but pore and acts as a stable underwater lung, as first pointed out by Thorp and Crisp (1947). Short but dense coats of hydrophobic hairs or cuticular protuberances—from tens of thousands to millions per square millimeter—cover these creatures. When they submerge, air remains within those coats, forming what looks like a tent roof supported by innumerable vertical poles (fig. 5.10). Pressure in the air stays well below the local hydrostatic pressure because water can't penetrate the interstices of the fur coat without creating lots of additional surface, to which water doesn't take kindly. What set the decrease in pressure below the local hydrostatic value are the radii of curvature of the interfaces, and a high density of hairs permits very small radii of curvature. Going deeper may increase the hydrostatic pressure, but that's offset by increasing inward curvature of the interfaces. Thus, gas doesn't get forced into

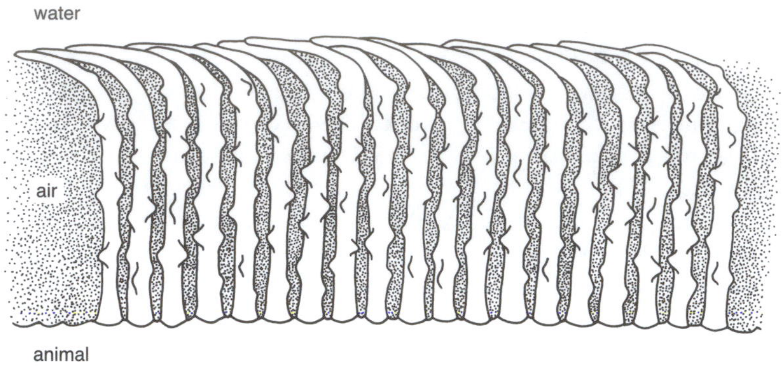

FIGURE 5.10 The "plastron," or surface air layer, on a submerged insect.

solution, and the insects can breath air from this so-called plastron. As they deplete the oxygen in the plastron, it's restored by net inward diffusion from the surrounding water, and carbon dioxide moves in the opposite direction. Diffusion of dissolved gas across an air-water interface turns out to be magnificently indifferent to surface forces.

The trick has evolved repeatedly, mainly in insects, but also in millipedes, mites or acarids, and whip scorpions or amblypygids (Hebets and Chapman 2000). As pointed out by Hinton (1976), the hairs are apparently just barely strong enough as columns to take the compression (they're but a hair's breadth away from collapse), based either on observations where pressures are artificially increased or on calculations based on their dimensions and the properties of chitin (chapter 15).

Staying at the Air-Water Interface

To float, a body need only be less dense than the surrounding water. For flotation in a stable orientation—an orientation that will persist without an external force and that will restore itself following rotational displacement—the conditions are slightly more complicated. Consider, first, not a floating but a submerged body of the same density as the medium, a fish with a swimbladder of air, for instance (fig. 5.11), as mentioned in the last chapter. It

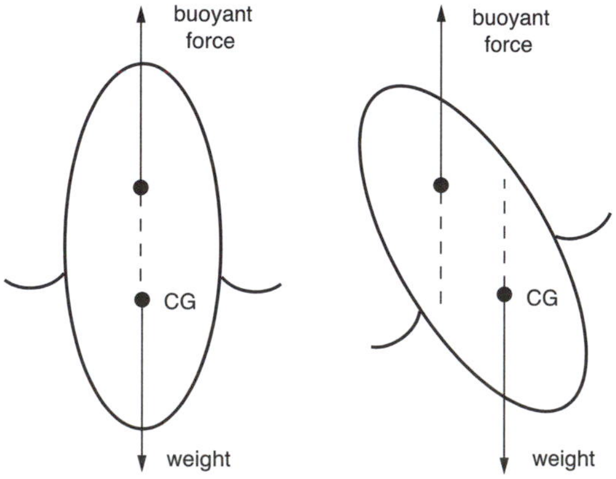

FIGURE 5.11 The orientation of a fully submerged body such as a fish is stable if the center of gravity (CG) is below the center of buoyancy—a deflection either way produces a pair of turning moments that act to restore the body's original orientation. It's unstable if the center of gravity lies below the center of buoyancy.

neither rises nor sinks since the upward force of buoyancy equals its weight. In what orientation will it be stable? The upward force equals the product of gravitational acceleration and the mass of the water the fish displaces; its line of action runs through the "center of buoyancy," the center of gravity of the displaced water. The weight of the fish acts as if concentrated at the actual center of gravity of the fish. In only two orientations do the lines of action of these two forces coincide and thus generate no turning moments—the two are those in which the centers of buoyancy and gravity are on the same vertical line. But one of these, that in which the center of gravity is above the center of buoyancy, is unstable in that any slight roll will create a moment that will increase the roll. Stability requires that the center of gravity lie below the center of buoyancy, for only in this latter case does a roll generate a moment that opposes the roll.

What does a typical fish do about the matter? If it put its swimbladder above the center of buoyancy then its weight would be concentrated below, and it would be stable. It doesn't, though. Fish, as least an eel, rock bass, perch, and bluegill, stay near neutrality, with the perch just a trifle unstable—its center of gravity is slightly above its center of buoyancy. They're also stable longitudinally; thus they have no strong tendency either to pitch up and down or roll side to side (Webb and Weihs 1994). That dead fish float belly up turns out to be a postmortem artifact.

What does a fish gain from neutral stability or slight instability? A good argument can be made that dispensing with strong intrinsic stability has an advantage, at least if an animal has a competent active control system. Stability and maneuverability tend to be antithetical qualities. We stand on two legs, and some birds routinely stand on only one—with good postural reflexes, stability matters less. And that holds for flying and swimming as well, which both we and nature have discovered—a stuffed bird can't be thrown as a stable glider and the most maneuverable military aircraft depend on active computerized controls to deal with their lack of stability. And instability, imperfectly managed, should exact a lesser penalty among swimmers than among fliers, since loss of orientation can't be as bad as plunging earthward. Most fish do keep their fins moving even when their bodies remain in place, as they deal with local disturbances—and they're impressively maneuverable.

Surface floating presents another kind of stability problem. First, a body sinks down until it displaces a weight of water equal to its own weight. The upward force of the displaced water acts ("action" of displaced water being a convenient linguistic fiction) as if concentrated at a center of buoyancy, defined as above. With much of the hull exposed, this center of buoyancy almost inevitably lies below the center of gravity of the object, which sounds like trouble. But if the object rolls in the water (fig. 5.12), the water line shifts and so (most often) does the location of the center of buoyancy. Because the center of gravity remains fixed, the lines of action of the two forces no longer coincide, which generates a turning moment. If the turning moment tends to offset the roll ("right" the boat), the body is stable; otherwise it "turns turtle," an expression certainly offensive to even the thickest shelled chelonian. So what matters is how the center of buoyancy shifts in a roll, and that depends solely on the shape of the hull. Sides that flare outward as they extend upward give a stable craft; sides that flare inward generate instability. It's just the opposite of what our land-based intuition suggests ought to happen, but a

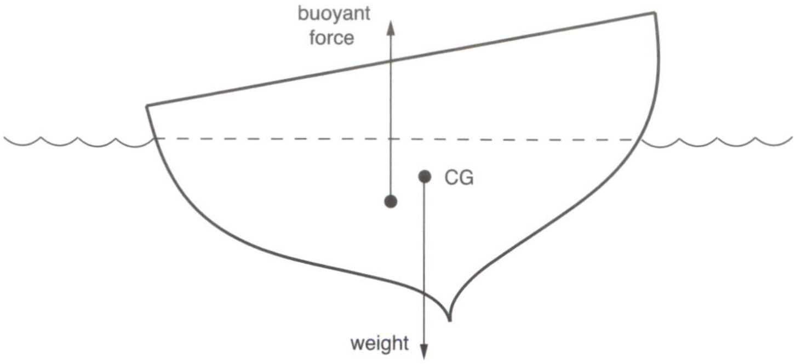

FIGURE 5.12 Cross section of a stable boat hull. A roll shifts the center of buoyancy in the direction of the roll; thus (if the cargo doesn't shift) the center of buoyancy and the center of gravity (CG) generate a pair of turning moments that tend to right the boat.

few minutes of sketching and counting the squares on graph paper or a little crude carving of softwood or buoyant soap ought to be persuasive.

Surface swimming isn't especially common among organisms; we'll explore why that might be so in chapter 14. But ducks do it, and they have appropriately flared hulls, at least if they don't swim too low in the water. Sea turtles sometimes swim on the surface; Jeanette Wyneken tells me that the flared, V-shaped bottom is characteristic of buoyant baby sea turtles, which are obligatory surface swimmers. With maturity and the shift to submerged swimming, the hull shape changes to one more characteristic of submarines. The change, she says, is most noticeable in greens and least in loggerheads; the differences in hull shapes are evident in the figures published by Davenport, Munks, and Oxford (1984).

CHAPTER

Viscosity and the Patterns of Flow

> Blood is thicker than water.
>
> **SIR WALTER SCOTT, on viscosity**

IN EVERYDAY PARLANCE, "FLUID" MEANS "LIQUID"—THE STUFF WE drink or use to fuel our cars. Properly, though, a fluid is any substance that responds to unbalanced forces by flowing, making the term a generic for both gaseous and liquid states. Indeed, for problems involving flow, liquids and gases are remarkably similar, more so than one has any a priori reason to suspect.

Because of their lower densities, gases are easier to push around—but that's a quantitative rather than a qualitative difference. One might worry about the vast differences in compressibility between gases and liquids—double the pressure on a gas and you almost exactly halve its volume; do the same to, say, water, and the reduction in volume is less than one part in 20,000. But it turns out that in the sorts of flows encountered by organisms, gases are functionally incompressible as well. Consider a wind of 20 meters per second (45 miles per hour) brought locally to a halt by hitting the beak of a flying bird—the increase in pressure will be 240 pascals (chapter 7) at the upstream tip. This increase of a trivial 0.24 percent relative to ambient atmospheric pressure will cause only a comparably trivial compression of the air. Compressibility becomes significant only when a gas encounters something at an appreciable fraction of the speed of sound (about 320 meters per second or 720 miles per hour).

Or one might worry about the different effects on liquids and gases of gravity and surface tension. These agents are certainly significant—they make waves, for instance—but their effects are almost entirely limited to fluid-fluid interfaces. So such effects can be ignored for natural flows that don't involve interfaces between fluids, which is to assert that they don't much matter for most situations of biological interest—again, gases and liquids behave quite similarly.

Very young children really appreciate fluid flow—just watch one intensely absorbed by the transfer of a liquid from one container to another. But we (most of us, anyway) grow into a world dominated by the more staid and apparently practical domain of solid mechanics, losing sight of both the beauty of fluids in motion and the variety of their counterintuitive behavior. I must admit, though, to a persisting infantile prejudice in favor of fluids, a partiality that has for many years supplied me with notions to investigate and that abetted the creation of my first book (Vogel 1981b). The beauty of flow, no mere intellectual aesthetic, is nowhere better presented than in a book of photographs, *An Album of Fluid Motion* by Van Dyke (1982).

Fluid flow has been a bit player in every chapter so far. Chapter 1 worried about the drag of leaves on trees. Chapter 2 talked about the very important principle of continuity and made the point that one could derive forces just by measuring speeds of flow. Chapter 3 casually mentioned that drag varied either with the first or the second power of length, depending on the sizes and speeds involved. Chapter 4 introduced the drag coefficient and the crucial notion of velocity gradients. And Chapter 5 alluded to water motion in connection with water striding and slapping locomotion.

Solids Versus Fluids—Viscosity

Gases and liquids share a property, *viscosity*, that's absent in solids, and lack another, *stiffness* or *elasticity*, that characterizes solids. Note that the distinction we make isn't just one of degree but a matter of fundamental applicability of these analogous but clearly different properties. Solids resist being deformed in shear—fluids simply don't. Fluids, though, do still care about shear—what's different is *how* they respond to shear stresses.

Consider (assume weightlessness for convenience) rectangular blocks of solid and fluid, each distorted in shear (fig. 6.1). For the solid, the greater the force, the further it will be distorted. One can write a rough formula for what the shear stress does—more stress means more strain (distortion), and the ratio of stress to strain (what a biologist might call stimulus to response) is the "shear modulus," a measure of stiffness under shearing loads:

$$\text{shear stress} = \text{shear modulus} \times \text{shear strain}. \tag{6.1}$$

For the fluid, by contrast, a greater force or stress doesn't lead to more distortion or strain—for all practical purposes the stuff is capable of infinite distortion since it takes on any shape equally happily. But what does depend on the force is the *rate of strain* of the fluid cube—the resistance of the fluid depends only on how *fast* it's distorted. This resistance to *rate* of shear is what we mean by viscosity; in short,

$$\text{shear stress} = \text{viscosity} \times \text{shear strain } \textit{rate}. \tag{6.2}$$

Since the flow of a fluid inevitably involves shearing or distortion (I'll defend that bald statement below), viscosity comes down to a measure of the resistance of fluids to flowing across surfaces or through conduits. Thus, the more

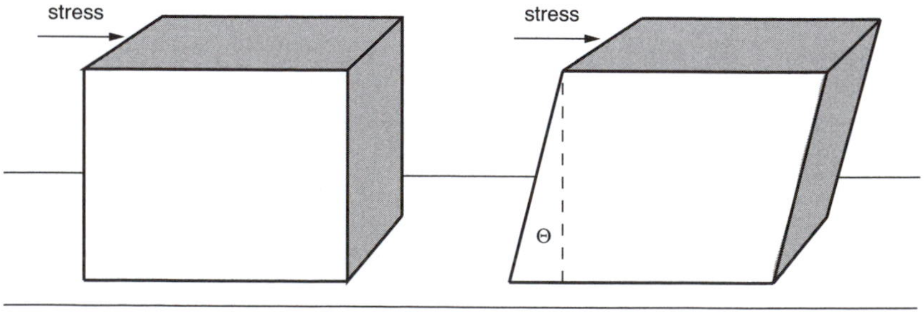

FIGURE 6.1 Pushing across the top of a block whose bottom is fixed distorts the block—two faces of a rectangular solid become nonrectangular parallelograms as they're tilted to an angle θ.

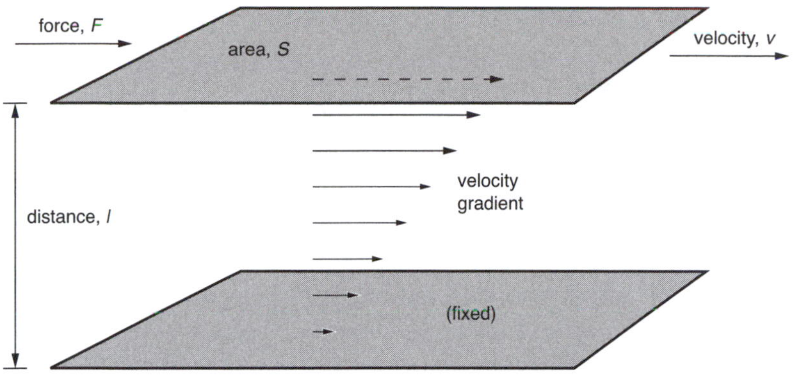

FIGURE 6.2 A pair of flat plates with fluid between; the top one moves with respect to the bottom one. Since the no-slip condition applies to both plates and the fluid's properties are the same throughout, a constant velocity gradient develops between them; the viscosity of the fluid determines how much force it takes to give the top plate a given velocity relative to the bottom one—for a given plate area and distance between them.

the viscosity, the less "fluid" the fluid (which may seem backward, but it would be even worse if put the other way around).

This notion of shear rate needs a few more words. It amounts to a measure of how rapidly "layers" of fluid are sliding with respect to each other—as if you push across the top sheet of a pile of paper and each sheet moves a little faster than the one beneath it. Shear rate is thus a rate of change of velocity with distance—a velocity gradient, but one in which velocity varies, not in the direction of travel (as with acceleration), but instead at a right angle to flow. It's the gradient you'd experience when wading out into a creek through the quiet waters near shore toward the torrent in midstream. As mentioned in chapter 4, the lack of "slip" of a fluid at a solid surface implies the presence of such a shearing region with its velocity gradient wherever a fluid flows past a solid.

But we need a more quantitative view of viscosity. Imagine our previous cube now transformed into two thin, flat plates with fluid between (fig. 6.2). The lower plate is fixed, a force moves the upper one horizontally, and the no-slip condition applies to both inner faces. As a result, the fluid between them shears, producing a smooth variation in flow rate, that is, a constant velocity gradient. With this diagram, we can put equation (6.2) in more formal terms, using the Greek mu (μ) for viscosity (sometimes called the "coefficient of viscosity" or the "dynamic viscosity"):

$$\frac{F}{S} = \frac{\mu v}{l}. \qquad (6.3)$$

The speed at which the upper plate moves follows the magnitude of the force directly. And the force needed to get a given speed varies directly with the area of the plates and inversely with the distance between them.

From this equation we get the basic dimensions of viscosity—they come out to $ML^{-1}T^{-1}$, a combination that has no obvious intuitive significance. The *SI* unit is the kilogram per meter-second or, shorter, the pascal-second (Pa s).

Looking up viscosities isn't especially difficult, but it's rarely necessary since two materials, air and water, dominate the biological scene. Table 6.1 gives their viscosities at a temperature of 20 degrees Celsius, along with the

TABLE 6.1 Viscosities and Densities at 20 Degrees Celsius

	Viscosity, μ (Pa s)	Density, ρ (kg/m³)	Ratio (m²/s)
Air	18.1×10^{-6}	1.20	15.00×10^{-6}
Water	1.00×10^{-3}	1.00×10^{3}	1.00×10^{-6}
Seawater	1.07×10^{-3}	1.02×10^{3}	1.05×10^{-6}

densities and the ratio of viscosity to density. (The latter is called the "kinematic viscosity" to distinguish it from "dynamic viscosity.") Notice that while the substances have very different viscosities and densities, the ratio of the two properties—upon which in practice a lot depends—differs only fifteenfold. It's surprising, and will prove a considerable convenience, to find such similarity between two such different materials.

Since the no-slip condition is universal, so are velocity gradients adjacent to surfaces—that's the main reason why viscosity matters. Pipes and arteries develop deposits on their walls, it takes a strong wind to remove a spore resting on a leaf, and water without suspended grit lacks erosive action on rock. If a fluid is very resistant to *rate* of shearing, then the velocity gradients in it will tend to be gentle, and relatively great distances will separate the surface and the region where the fluid moves at full speed (called the "free-stream velocity"). If, conversely, the effects of viscosity are slight, then the gradients will be steeper and the distances shorter. Figure 6.3 (which repeats fig. 4.4, so important is the matter) gives a graphical look at how speed changes within the gradient region for flow parallel to a flat plate at moderate speeds.

People sometimes ignore the picture given by figure 6.3 and retain a fuzzy notion of a discrete region of nonmoving fluid rather than the discrete notion of a fuzzy region with a strong velocity gradient. The physical fuzziness must be kept in mind—local speed hits zero precisely at the surface, but it asymptotically approaches that of the free stream as one moves outward from the surface. Part of the trouble comes from misleading terminology—the gradient regions usually carry the label "boundary layer," and "layer" sounds ever so discrete. I think the best practice is to think of (and in writing allude to) velocity gradients, retaining "boundary layer" for the quantitatively defined region that will be introduced in chapter 8. Whatever the designation, though,

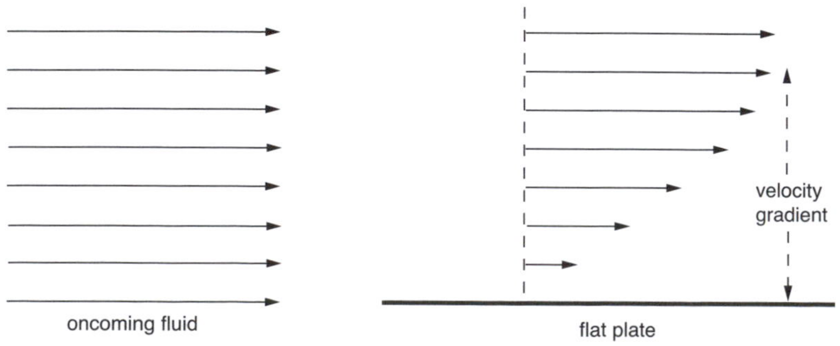

FIGURE 6.3 The velocity gradient on a flat plate oriented parallel to flow; the length of each arrow above the plate is proportional to the velocity at its base. The thickness of the gradient region has been grossly exaggerated for the purpose of illustration.

what we're concerned with are the places near surfaces where viscosity exerts its pernicious effect on flows and forces.

How Much Does Viscosity Matter—The Reynolds Number

Only in some specialized instances can viscosity be wholly ignored for problems of fluid flow. But, curiously, just how much difference it makes varies widely. One might guess that flows of gases approached some "inviscid" limit more closely than those of liquids, but that turns out not to provide a dependable rule. One gets the sense that viscosity dominates flows of treacle or molasses (as in the expression "slow as molasses") and corn syrup more than it does flows of water or acetone—especially when one notes the relatively small range encompassed by their densities. One's puzzlement can be excused when faced with authoritative statements that airplane wings would give no lift without viscosity. A strange business is viscosity.

More, yet. Even where fluid moves across or through solid elements of fixed geometry, flows vary widely in character. Sometimes fluid moves smoothly, with each bit following a course nearly parallel to adjacent ones. Under other circumstances, fluid moves less regularly, with the overall direction of flow mainly a statistical matter and the individual bits tracing all sorts of erratic courses. We speak of these situations as "regimes" of flow—"laminar" and "turbulent" in these particular cases. You surely have experience with both—syrup pours in a laminar fashion, water emerges turbulently from a nozzle, and deliberately induced turbulence disperses the cream in the coffee.

The consequences of the existence of such distinct regimes underlies most of the apparent complexity of the phenomena attending flow—the definition of viscosity and the Newtonian laws may not change, but the manner in which they apply to particular situations obviously varies. Fortunately, we have a single, simple variable that, although no everyday item, goes far toward putting all these disparate matters in decent order. That famous dimensionless expression goes by the name of Reynolds number (with no apostrophe on either side of the *s*).

Osborne Reynolds (1883) knew that flow through pipes was sometimes laminar but could abruptly become turbulent, that is, sometimes a streak of dye introduced upstream stayed discrete whereas at other times it spread rapidly throughout the flowing fluid. By using a variety of fluids, flow speeds, and pipe diameters, he was able to show that transition from laminar to turbulent flow happened at about a certain value, 2,000, of a dimensionless, composite variable to which we now give his name—the Reynolds number—in cylindrical pipes. That variable proves to be the best single index to the character of flow, not just in pipes, but in streams, lungs, ocean currents, blood vessels, or industrial vats; in fact, just about everywhere that real fluids flow steadily at substantially subsonic speeds reasonably far from fluid-fluid interfaces. Which includes most biologically interesting flows—without mention of Reynolds number, no sensible treatment of flow in living systems is possible.

What the Reynolds number provides is an indication of the importance of viscosity—actually the unimportance, since higher Reynolds numbers are associated with less viscous situations. One might expect that the value of viscosity would suffice, but other factors have equal influence in determining its relative role. We've noted that viscosity opposes the existence of steep velocity gradients—equation (6.3) says that the higher the viscosity the more force it takes to work against a velocity gradient and move one plate across another. So viscosity might be thought of as a measure of the "groupiness" of a fluid, the tendency of bits of fluid to flow coherently, to flow in unison. What has the opposite effect, promoting gradients? Recall the notion of inertia, the tendency of a bit of matter to keep going as it has been in the face of any retarding or accelerating influences. Thus, inertial resistance to acceleration, the force needed to change its speed, opposes viscous forces and reflects the "individuality" of a bit of fluid.

We thus have two opposing sorts of forces, loosely called "inertial" and "viscous." The Reynolds number, *Re*, is simply the ratio (dimensionless, of course) of these two. Inertial force, F_i, is represented by the familiar $F = ma$, although disguised a bit into a form useful for steady flow of fluids—as we noted two chapters back. Mass is replaced by density times area (across the flow) times length (with the flow), and acceleration is replaced by velocity over time. Since length/time is velocity,

$$F_i = \frac{\rho S l v}{t} = \rho S v^2. \tag{4.8, 4.9}$$

For viscous force, F_v, we rearrange equation (6.3), which was introduced earlier to define viscosity:

$$F_v = \frac{\mu S v}{l}. \tag{6.3}$$

The result is surprisingly ordinary, without even an exponent to challenge the calculator:

$$Re = \frac{F_i}{F_v} = \frac{\rho l v}{\mu}. \tag{6.4}$$

None of the factors in the Reynolds number presents much problem of evaluation for practical situations. Density and viscosity refer to the fluid and (if not found in table 6.1) are usually matters of public record. Notice that the ratio of the two appears in the Reynolds number—that ratio matters more than the individual values, which is why it appeared in the table. Velocity is that of the mainstream fluid relative to the object or substratum or the wall of a pipe. Length is a "characteristic length" of an object chosen largely by convention and convenience—maximum length in the direction of flow for flows across solids, or the diameter of pipe or channel for internal flows.

For a given size and speed, then, flows of water have Reynolds numbers fifteen times greater than those of air—flows of water are, oddly, less viscous in character. But in our everyday world, water flows are slower than air flows by about the same factor, so the same Reynolds numbers characterize, for instance, flight by birds and swimming by fish. It must be emphasized that, as with other scaling rules, one is dealing with wide-ranging situations and a fairly crude index. In fact, one almost never cites a Reynolds number with more than two significant figures, and one can't compare the Reynolds numbers for different shapes to even that rough level of precision. It's a ballpark business (to use a bit of American vernacular that puzzled the students in a course Mark Denny and I once taught in Sweden).

And organisms operate over an especially wide range of values (table 6.2). The extreme range results from the fact that small creatures typically move slowly or encounter slow flows deep within the velocity gradients near surfaces; large creatures experience more rapid flows; and the formula includes the product of length and speed.

One complication needs to be mentioned right up front, lest errors in the biological literature go unnoticed. Reynolds found a laminar-to-turbulent transition at $Re = 2{,}000$, based on equation (6.4) for a straight, circular pipe with a gently tapered inlet. (In fact, using radius rather than diameter for a characteristic length, he gave his actual value as 1,000, which better emphasizes its approximate character.) The value 2,000 can't be taken as the transition value for any other case. For instance, transition for flow through a circular aperture occurs at a Reynolds number of about 3 (Happel and Brenner 1983), and transition for flow around a sphere or cylinder happens at around 100,000. Furthermore, as we'll see shortly, transition itself isn't always such a tidy singularity.

The Reynolds number, again, provides an index to the importance of viscosity in determining the character of flow. As such, it tells us a great deal about what patterns of flow we ought to expect in a given situation. The lower the value, the more do viscous phenomena dominate; that is, the more the fluid dislikes turbulence and steep velocity gradients. One consequence is that what's a good shape for one organism may be quite inappropriate for another if the two operate at substantially different Reynolds numbers—even if they are engaged in seemingly similar activities. Another is that similarity in form may not reflect an underlying similarity in physical situation—spermatozoon and eel may look superficially alike, but they operate in quite different hydrodynamic regimes.

At low Reynolds numbers (roughly 10 and lower), flows across solids are laminar and orderly; if vortices occur at all, they are relatively large and distinct. Velocity gradients are gentle and, in general, drag is proportional to the

TABLE 6.2 Reynolds Numbers—Examples

Organism	Reynolds Number
Bacterium, swimming	0.000001
Pollen grain, falling, or sperm, swimming	0.01
Fruit fly in flight	100.
Small bird, flying	100,000.
Squid, fast jetting	1,000,000.
Large whale, swimming	200,000,000.

first power of velocity and doesn't depend very strongly on shape. As we'll explain shortly, streamlining is almost without value as a scheme for reducing drag at low Reynolds numbers.

As Reynolds numbers rise (from roughly 100), flows become increasingly disordered, with turbulent chaos setting in, first in one situation or location and then in others. Velocity gradients are increasingly steep, and drag may be proportional to the square of velocity (but the relationship is rarely tidy) and becomes strongly shape-dependent. For instance, total drag may be reduced as much as fiftyfold by shifting from a sphere to a well-streamlined form where both have the same frontal area perpendicular to flow. But no Reynolds number can be so high that viscosity ceases to matter; that's a consequence of the no-slip condition. The higher the Reynolds number, the thinner (relative to other dimensions) are the velocity gradients near surfaces—but the more extreme are the gradients, since they always extend from zero to free-stream velocity.

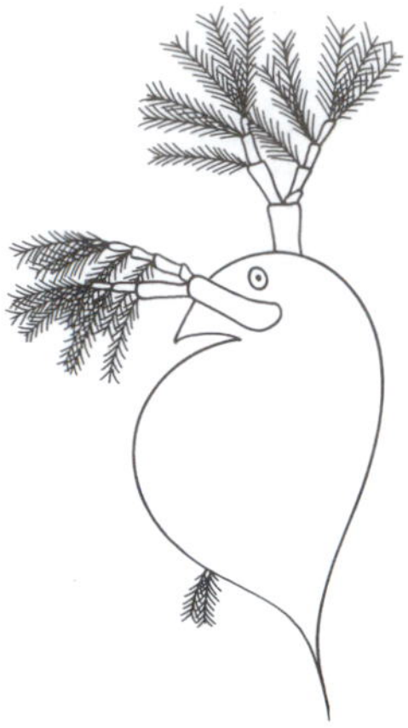

FIGURE 6.4 The bristly appendages of a tiny crustacean, *Daphnia* by both common and scientific name.

Reynolds Number and Consequences of the No-Slip Condition

We've now introduced a set of concepts that, even if not especially abstruse, aren't exactly household items—the no-slip condition at surfaces, the resulting ubiquity of velocity gradients adjacent to surfaces, and the Reynolds number as (among other things) an indicator of the steepness of such gradients. One implication is that in flows at low Reynolds numbers, objects ought to be surrounded by quite a lot of semistagnant fluid. No discrete layer of zero-flow fluid may exist, but the slow stuff has serious consequences, both good and bad. It can provide useful insulation; it impedes penetration of diffusible material.

In a series of studies, Mimi Koehl and her collaborators have looked at (among other things) the bristle-equipped appendages of tiny crustaceans, as in figure 6.4. These creatures swim with such appendages, using them as paddles—with a lot of semistagnant water around each bristle, an appendage can serve as a paddle. The creatures also use such appendages as rakes, filtering edible particles from the water around them. That requires the passage of water between the bristles. Paddle or rake? The appendages look pretty much alike. What determines how one can be used is the Reynolds number at which it operates. Viscosity and density are givens; size and speed of motion provide the operative variables. Large and fast lets an appendage rake; small and slow promotes paddling. (See, for instance, Koehl [1993] and Loudon, Best, and Koehl [1994].) More on the issue will be given in chapter 11.

Aquatic organisms are not alone in facing this peculiar Reynolds number-dependent design problem. Consider the problem of male saturniid moths (sometimes called giant silkworm moths). Females attract males for mating by releasing a volatile attractant chemical, but these moths typically live at very low population densities. Males have equipment to smell out females at distances of miles—large, feather-shaped antennae (fig. 6.5) on which some 70 percent of the sensilla are sensitive to nothing other than the females' perfume—truly an olfactory sensation. A nineteenth-century French entomologist, Jean Henri Fabre, did a few experiments on the phenomenon but still could not believe that any sense of smell could work so well (Teale 1949)!

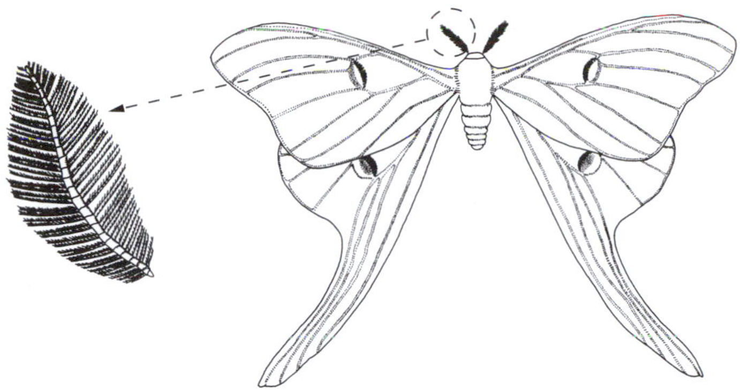

FIGURE 6.5 A male luna moth and one of its antennae.

In order for an odorant to be picked up from the airstream, air must be made to pass through an antenna. People interested in olfaction often presume that air directly upstream from an antenna would pass through it. But gases as well as liquids have viscosity, and passing through a dense array of sensilla involves shearing and velocity gradients. So the air might well prefer to go around an antenna as if it were a solid plate. The moth thus faces the problem of designing a rake that doesn't end up as a paddle—more sensilla would give more sensitivity to odorant but at the same time would reduce the passage of odorant-laced air. For another investigation, I had built a very tiny device to measure airspeeds, so I looked at the speeds immediately behind antennae, where air had just passed through (Vogel 1983). Comparing these speeds to that of the free stream gave figures of from 8 to 18 percent for the fraction of the air approaching an antenna that went through and not around. I called the fraction the "aerodynamic transmissivity," but "leakiness" seems to be catching on as the general designation. The exact figure depended on the particular free stream speed. Lower speeds gave lower transmissivities, as one would expect—the lower Reynolds numbers of lower speeds mean gentler velocity gradients and more semistagnant fluid around each sensillum. The figures are far below the generally assumed transmissivity of 100 percent and well below even the antenna's optical transmissivity of 43 percent. With less air passing immediately by them, the sensilla must be even more sensitive than previously imagined—such are the vicissitudes of viscosity.

Velocity gradients take on great importance wherever organisms live on surfaces. Free-stream velocity lacks any immediate relevance to an organism that spends its life deep within such a gradient region. As with filtering and paddling appendages, the gradients can be either trouble or assistance. Getting down near a surface provides protection from the forces of fast flows of air or water (although nothing is so flat as to have no drag at all—at the surface itself the velocity may be zero, but the all-important velocity gradient is not). Conversely, an organism feeding from the passing water must somehow get its feeding equipment or intake pipe up high enough to encounter adequate flow speeds. In short, there's lots of biology here, and we'll take a more extensive and quantitative look at these gradients in chapter 8.

Modeling

The Reynolds number provides an especially useful tool for making models. For two geometrically similar situations, equality of Reynolds numbers assures equality of the patterns of flow, whatever the individual values of length, speed, density, and viscosity. We biologists seem always to be dealing with extremes of size, so a good scaling rule for models is a big boon, although it must be admitted that biologists are often skeptical of its legitimacy—those μs and ρs look like Greeks bearing gifts.

I've had recourse to the use of the Reynolds number for scaling experiments on quite a few occasions. At one time I was interested in induced air flow through the burrow systems of prairie dogs at low ambient winds (Vogel, Ellington, and Kilgore 1973). A burrow in nature is 3 meters deep, 15 meters long, and about 0.1 meter in diameter—quite an awkward item to bring into (or reproduce) in a laboratory. But reduction of lengths by tenfold permitted use of a simple model made of copper water pipe beneath a plywood "ground" that would fit in the local wind tunnel. In compensation for the reduction in size (l) all I had to do was increase speed (v) by the same factor, making, say, 1 meter per second into 10 meters per second. As a bonus, the internal speeds were raised into a more easily measured range.

A few years later, I felt I needed to know how pressures varied along the length of a squid swimming at top speed (Vogel 1987). My best pressure measuring device works only in air, so I had to factor in a shift in media (changing μ and ρ) as well size and speed. The local engineers kindly loaned me a wind tunnel that went up to 75 meters per second (170 miles per hour), and I made my model half again as large as life. As a result I could simulate a squid jetting at nearly 8 meters per second in water, as the reader can easily verify from equation (6.4) and table 6.1.

Such modeling requires only that the conditions for applying the Reynolds number be met—no fluid-fluid interfaces, decently subsonic flow—and that any shift from, say, air to water not cause abnormalities (due to the change in forces) in the shape of the organism or model. The shifts can be extreme—using corn syrup instead of water permits over a thousandfold increase in size and thus macroscopic modeling of microorganismic phenomena. By contrast, modeling surface ships is far messier—ships make surface waves (chapter 14), so gravity can't be ignored; Froude number (inertial versus gravitational force) as well as Reynolds number matter. As a result, no alteration in size or speed can be made without changing either one dimensionless index or the other. So compromises are inevitable, the results always err a bit, and extreme size or speed shifting can make trouble.

Note that one can blithely change frame of reference for such problems. That a model of something that swims or flies now remains at rest while the surrounding fluid (air or water) moves makes no difference whatsoever. Were this not the case, wind tunnels would be far less useful! At least one needn't worry for motions (of object or fluid) that aren't either accelerating or changing direction. Thinking in terms of frames of reference ought to be familiar—when looking out of an airplane, you take it on faith that the airplane moves and the ground is stationary rather than the other way around. After all, the furnishings of the plane provide you with a nicely fixed reference frame, one that your sensory equipment does nothing to contradict.

Streamlines and the Appearance of Flows

The principle of continuity was presented (chapter 2) in terms of flow through pipes—the principle states that, if the fluid's density remains unchanged, the rate at which volumes of fluid enter a rigid system of pipes must exactly equal the rate at which volumes of fluid leave the system or pass any cross section of it. For an incompressible fluid, that requires that the product of velocity and cross-sectional area must be the same for all cross sections of the system:

$$S_1 v_1 = S_2 v_2. \tag{6.5}$$

The principle turns out to be quite as useful for flows across objects and surfaces—open fields of flow—where no rigid pipes provide boundaries. Fluid (for our purposes) remains incompressible, and mass remains conserved. But applying the principle takes some analog of the walls of pipes; in practice the scheme is a simple one. Consider a two-dimensional (to fit on paper) field of flow (fig. 6.6), where the speed of flow varies from place to place but does not, at a given place, vary from time to time. (We're talking about what's defined as "steady flow," which, somewhat confusingly, may involve local accelerations and decelerations.)

Assume that you have some device to tell the direction of flow at any point (a thread on a stick, a stream of dye, a time-exposed photograph of particles passing through the field, and so on). The directions of flow can be mapped by drawing short lines that follow the local flow directions—lines bits of fluid will pass along as they flow. With enough points mapped, the short lines can be connected into longer ones running completely across the field in the directions of flow. Mapped this way, fluid is automatically prohibited from crossing the lines, so the lines amount to nonmaterial partitions and play exactly the same role as the walls of rigid pipes! Between any pair of lines, the principle of continuity must hold. If a pair of lines diverges, the fluid must be slowing down; if a pair converges, the fluid must be accelerating. The lines are called "streamlines" since they follow the stream of fluid at all points. In three dimensions, the set of lines just takes the form of a set of tubes, and the principle applies as well. "Streamlining" in the popular sense means shaping something so fluid flows smoothly around it and causes a minimum of drag; it connects only indirectly with these streamlines.

Armed with streamlines, we can take a more physical view of the way the Reynolds number delineates flow regimes. Figure 6.7 gives the streamlines for flow around a cross section of a long circular cylinder. The real messiness of the world with which fluid mechanics must deal becomes immediately obvious, but so also is the way the Reynolds number at least orders the mess. At

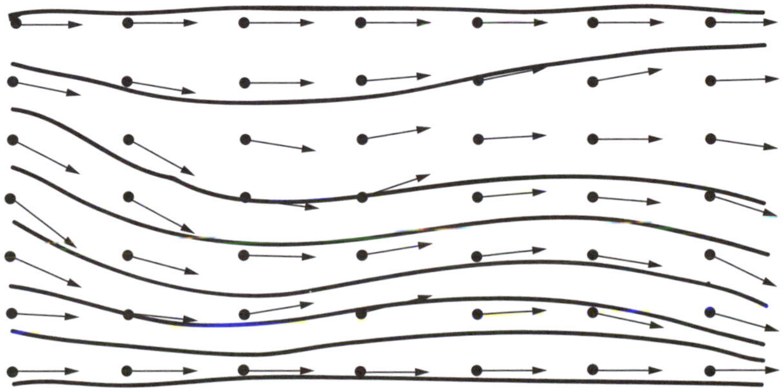

FIGURE 6.6 A field of flow, with arrows indicating flow directions at their basal points. Streamlines (dashed) connect upstream points with others successively further downstream, everywhere following the local direction of flow.

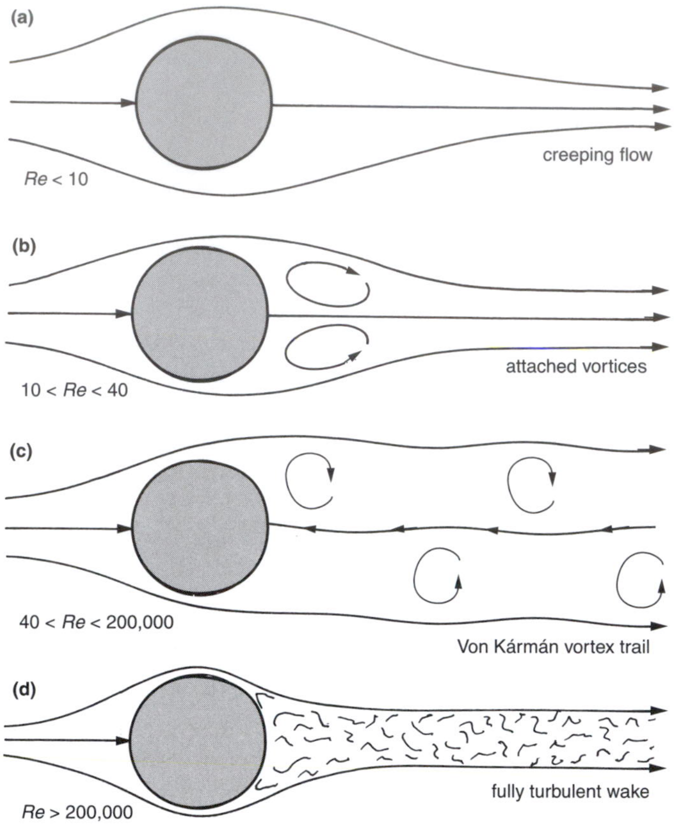

FIGURE 6.7 The character of flow around a circular cylinder (shown in cross section) depends strongly on the Reynolds number. Orderly flows (a) break into attached vortices (b), which are then periodically shed (c), and finally become disorganized in a wake that's narrower, overall (d).

the lowest values, flow is smooth—in this situation at least, vortices aren't generated. But, a little higher, a pair of vortices forms behind the cylinder; they remain in position. With flow from left to right, the top one rotates clockwise and the bottom one counterclockwise, kept going by the flow lateral to them and the resistance of the fluid to being sheared rapidly. That implies something that, strange to say, is no mirage—just downstream from where the main flow leaves the surface of the cylinder, flow near that surface goes upstream, here from right to left. So there's a point of convergence, where the two meet and proceed away from the surface of the cylinder—it's called the "separation point."

At still higher Reynolds numbers, the pattern becomes unstable, and the vortices alternately detach and get carried downstream, forming the familiar zigzag pattern of trailing vortices known as a Von Kármán trail. A separation point (really a line perpendicular to the page) still exists on either side, but it oscillates slightly rearward and forward as vortices detach. Higher yet, the tidy pattern becomes disorganized by turbulence (but will once again reappear at superhigh Reynolds numbers.) At the same time as the pattern breaks up, the wake becomes narrower, with the separation point moving further back. What has happened is that the increasing momentum of the oncoming fluid (an inertial effect) can carry it further around the curved surface (against an increasing pressure, about which more later) before it breaks loose to flow straight downstream. We speak of this narrowing point as the "turbulent transition" and mean that turbulence in the flow close to the surface now keeps that fluid sufficiently momentum-charged to continue further around the cylinder. Again, for a given shape, only the Reynolds number matters—transition depends only on the particular combined value of size, speed, viscosity, and fluid density.

Note that for such flows around objects, regimes don't neatly divide into laminar and turbulent. We can recognize these as extremes—fully laminar below $Re = 10$, fully turbulent above 200,000. But four orders of magnitude intervene, representing a vast region of shifting flow patterns, a region where vortices are ubiquitous, even without definitive turbulence. As it encompasses a vast diversity of biologically relevant flows, the region provides nature with manifold opportunities to surprise us.

Vortices

Talking about flow behind a cylinder brought up the subject of vortices, objects at once stranger than they look (just concentric streamlines), biologically important (remarkably so), and peculiar to fluid mechanics (unless we talk about galaxies). Much of the strangeness turns (the word chosen deliberately) on the distinction we made in chapter 2 between translational and rotational motion.

Making a body of liquid—or even gas—rotate presents no great problem. One just spins a container of the stuff and, after a little lag for viscosity to make the stuff in the middle catch up, the body of fluid rotates. But we've put the fluid into rotational motion by stepping outside the system, rotating the container and (choose your metaphor) cheating or playing God. Imagine, by contrast, some fluid flowing through a broad pipe or channel or a wind blowing steadily. Can we put some obstacle in the flow that redirects it in such a way that either the whole fluid or a portion of it rotates? That turns out to be far trickier, no easy thing at all.

Consider again that moderate Reynolds number flow behind the cylinder in figure 6.7. Vortices formed and persisted without requiring heroic measures. At first glance, we've made fluid rotate. Oddly enough, we have not. The presence of vortices need not (and typically does not) indicate rotation in the fluid.

We need to invoke the conservation of angular momentum, a conservation law as yet unmentioned here. The usual illustration involves a skater turning about the body axis while pulling arms closer in toward torso. The skater turns ever faster as the arms move inward. A springboard diver provides another example, turning several times head over heels while tucked into a ball but almost ceasing to rotate when arms and legs are reextended. In such instances the product, $mr^2\omega$, of the body's moment of inertia, mr^2, and the angular velocity of the turn about its axis, ω, remains constant. A fluid vortex follows such a rule, but with quite a different outcome. As a fluid, it can shear indefinitely—it may not be altogether happy with the speed of shear, but it has no preferred shape. So the middle can turn faster than the periphery *at the same time*. A bit of fluid that moves inward would go around the axis more often; a compensating bit of fluid (continuity!) that moves outward would go around less often.

Vortices normally turn faster in their middles than farther outward. They do so in a way that gives them a strangely irrotational character—like the pedals of a bicycle, the seats of a Ferris wheel, or the body movements in most social dances, their bits *translate around circular paths*. It's easy to show that irrotational motion follows from the assumption that each bit of fluid mass has the same angular momentum. (We'll ignore the question of why angular momentum should be uniform.) For each bit of fluid, angular velocity, ω (as radians per second), equals tangential velocity, v_t, over radius, so for the conserved quantity,

$$mr^2\omega = mv_t r. \tag{6.6}$$

Conservation of the product of v_t and r for any bit of m implies that tangential velocity must vary inversely with radius. The further the bit from the axis, the slower will be the motion—as in figure 6.8. Scale that bit up to, say, a small boat in a large whirlpool—it will translate in a circle rather than rotating around the axis, as you can convince yourself by drawing some objects and vectors on a piece of paper.

Rotation, again, can't easily be started within a system that's either stationary or simply translating. Perhaps one shouldn't be all that surprised. Skater and diver did rotate, but both had to go outside the system to get going, pushing against ice or springboard in the first place. That's not so easy for a moving fluid, so the irrotational vortex is the norm and not some oxymoron.

But, again, achieving irrotational motion demands that concentric layers of fluid shear against each other. Worse, the closer to the core of the vortex, the more intense that shear—since the slope of the line in figure 6.8 gives the steepness of the velocity gradient, it's proportional to the shear stress. Far worse is the implication of what happens at the axis. Where $r = 0$, ω, v_t, and

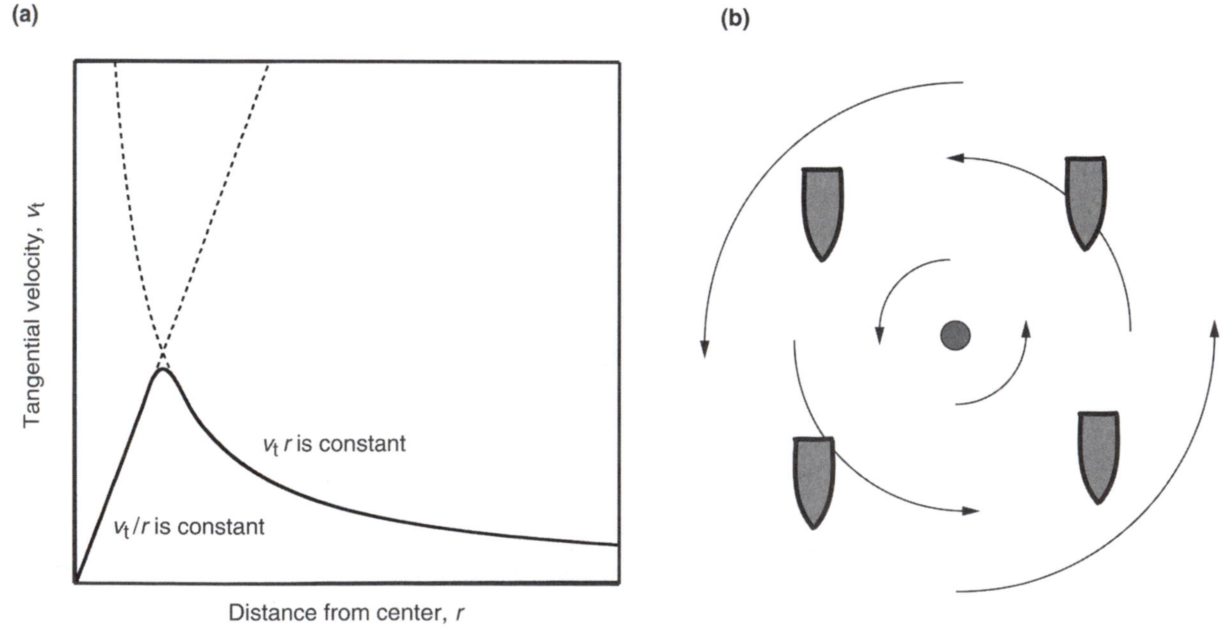

FIGURE 6.8 (a) How tangential velocity varies with the distance from the center of a vortex. The core, with low values of radius, may be rotational, but farther outward the vortex becomes irrotational. (b) Successive orientations of an object going around in the irrotational part of the vortex.

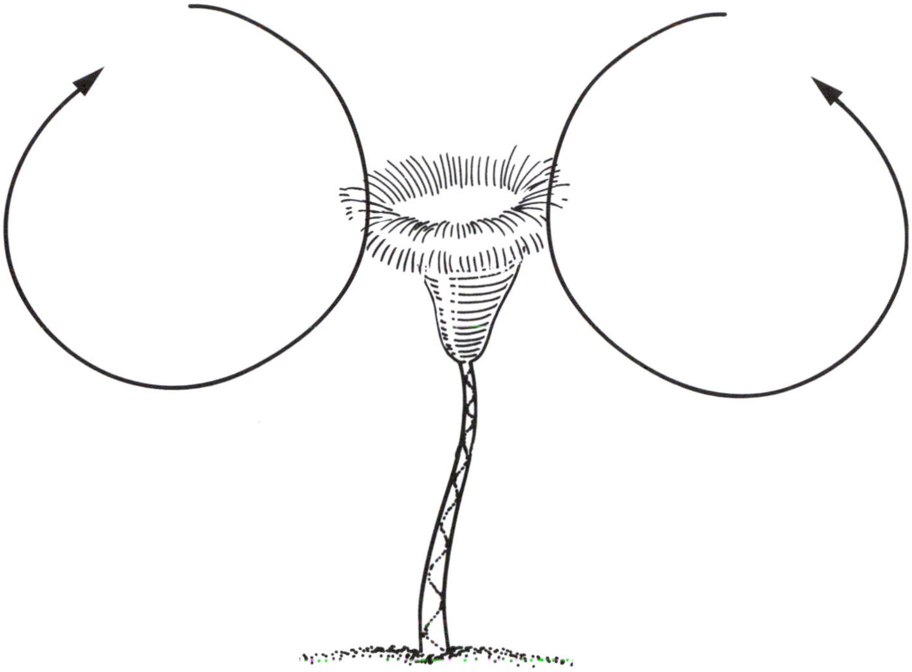

FIGURE 6.9 The toroidal feeding vortex formed by the cilia of the stalked protozoan, *Vorticella.*

the velocity gradient all must be infinite! No surprise—reality differs. In practice, every irrotational vortex contains at its center a rotational core, a region near the axis where angular velocity varies little, and tangential velocity is proportional to radius. The dashed line in the figure describes speeds in this core, which behaves much like a rotating solid.

How big, relatively speaking, is the core? That's a matter of the resistance of the flow to shear, which once again depends on the Reynolds number. The lower the Reynolds number, the bigger the core relative to the overall size of the vortex. So the assumption of complete irrotationality may mislead the biologist, who works so often with tiny systems. Vortices do occur at fairly low Reynolds numbers (the record is $Re = 0.01$)—they're tidy-looking things, they can be mostly core, and they demand a continuous input of energy lest their leakage to viscosity cause them to grind to a halt. The protozoan, *Vorticella*, has not been misnamed (fig. 6.9), and it has considerable kin and company.

Yet other problems travel in these circles. What happens out of the plane of the paper or screen? We keep talking about an axis of rotation, implying something more cylindrical than, say, spherical. A vortex, in fact, does not normally have an exposed end. Instead it wraps around in the hermaphroditic unity of a toroid. The torus may be circular or highly elongate, it may look like an almost holeless bagel or a long loop of rope. Flow in it may just go around radially, or it may have lengthwise flow as well. The combination of the toroidal shape and the conservation of angular momentum for any element leads to yet another odd feature. If you slice across the torus anywhere, you encounter fluid going around clockwise in one place and counterclockwise in another. We noted that the product of tangential velocity and radius was conserved; for such a slice, the overall product (ignoring core effects)

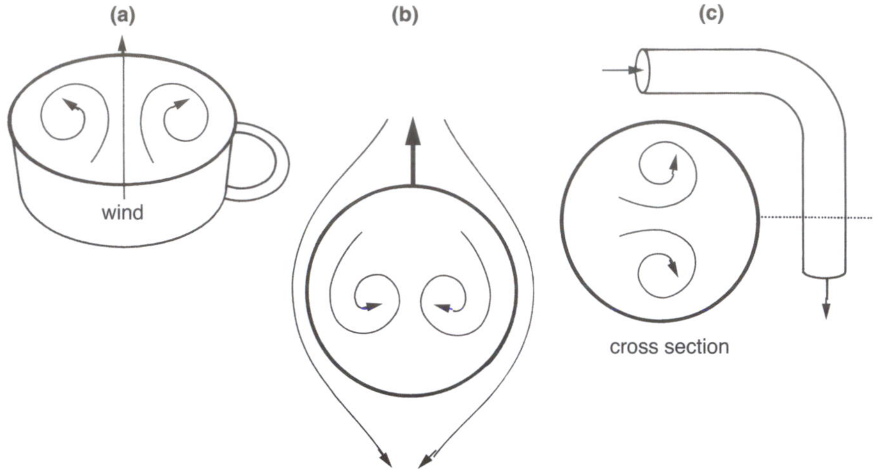

FIGURE 6.10 Various vortices—(a) toroid in a teacup, the same as what happens in a rising bubble of warm air; (b) a vortex in ascending gas bubble; (c) the paired vortices downstream from a bend in a pipe.

adds up to zero. If each skater (to return to the analogy) starts the other spinning, the ice need do nothing—but there's no net spin in the sense of angular momentum. (Of course the one with the greater moment of inertia, ordinarily the heavier, will rotate a bit more slowly.)

Vortices include a huge diversity of things—if the following paragraphs don't prove persuasive, the reader need just peruse the book by Lugt (1983). They include the tempest in the stirred teacup (fig. 6.10a), where flow goes not just around but outward near the top, downward along the sides, inward across the bottom, and upward in the middle as a whole-cup torus. (The velocity gradient at the solid bottom means that fluid turns more slowly than at the top; thus the top tea leaves are flung more strongly outward, and flow ascends in the center—the possibility of lengthwise flow was noted just above.) Vortices include, as well, violent storms such as hurricanes, cyclones, and tornadoes. In all, things get worse toward the center, but some have calm zones at their precise cores—what we call the "eye of the storm"—where, in the reference frame of an observer on the ground, the wind reverses direction as the storm passes.

And they occur as thermal vortices—if a patch of ground heats up more than its surroundings, a bubble of buoyant warm air will periodically break loose and ascend. Passage upward, by viscous interaction with the surrounding air, slows the upward flow on the bubble's sides relative to upward flow in the middle; that soon generates a toroidal vortex, as in figure 6.10a. Birds such as vultures can soar upward in these thermal vortices by circling near the middle of the torus. Since the air in the middle ascends faster than the torus as a whole, a bird can glide downward relative to the air immediately around it without dropping below the ascending torus as a whole.

On a vastly smaller scale, a bubble of gas ascending through a liquid (fig. 6.10b) develops an internal toroidal vortex by just the same mechanism. As it ascends, the gas at its sides is slowed by passage through the liquid; that, by continuity, makes the gas in the middle rise faster than the overall bubble. The relative downward motion at its sides reduces shear with the surrounding liquid, so a bubble of gas ascends faster than if it were solid or had a weightless membrane around itself—its internal motion amounts to a partial evasion of the no-slip condition. A liquid drop descending through a gas does less in the way of toroidization—its greater viscosity minimizes the phenomenon.

Vortices even form with only slight provocation in pipes. Consider what happens when flow encounters a bend in a pipe (or blood vessel). Since flow in the middle goes faster than flow near the walls (recall chapter 4), midpipe fluid has more momentum, won't swerve as readily, and will be carried outward toward the far wall as it goes around. As in figure 6.10c, fluid near the walls will move toward the inside of the bend, and flow will leave the bend as a pair of vortices resembling side-by-side ropes with oppositely twisted strands.

We've already noted that a Von Kármán vortex trail (fig. 6.11a) forms downstream from a cylinder. The same pattern occurs behind other elongate shapes as well. With flow from the left, the top vortices rotate clockwise and the bottom ones counterclockwise, as they're rolled by shear with the passing flow. A similar-looking set of vortices develops behind a swimming fish. They differ, though, in one crucial respect, reflecting the fact that the fish puts energy into the flow while cylinders and such withdraw it. The fish's vortices rotate the opposite way, counterclockwise above and clockwise below, as in figure 6.11b. I've heard that Theodore Von Kármán himself first pointed out the contrast. Just as the fish's oscillations generate the vortices, vortex shedding by a passive body makes it oscillate from side to side. While such shaking can cause mischief for human-made structures, it hasn't been tied to much that's biologically consequential.

Drawing attention to a lot of peculiar physical phenomena increases the chance that someone will recognize some clever way some organism uses one or another. That's the main reason I describe phenomena for which I know of no biological relevance. But there's another reason, one that may matter more in this instance—recognizing what dangers are either too remote to be of concern or else long since accommodated by biological designs. For instance, repeated shaking can be hazardous—gusty winds uproot trees at speeds well below those of steady winds and forces lower than those predicted by

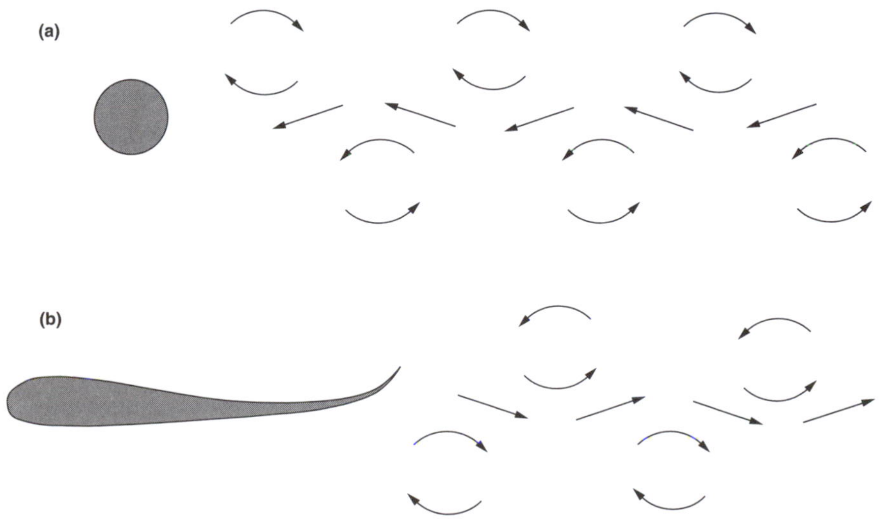

FIGURE 6.11 (a) A Von Kármán trail. (b) Vortices behind a swimming fish. Note how the vortices of the two systems rotate in opposite directions—one system removes momentum from the passing fluid, the other injects it.

experimental winchings. Steady winds would undoubtedly present greater dangers if trees shed vortices at frequencies in the same range as their natural vibrational frequencies. But the frequencies of their Von Kármán vortex shedding run much higher than those at which they naturally sway. A simple, dimensionless index, the Strouhal number, gives the frequency of vortex shedding:

$$St = \frac{nl}{v} \tag{6.7}$$

(where n is the frequency and l is a characteristic length, usually cross-flow diameter).

For circular cylinders across a flow, the value of the Strouhal number remains at about 0.2 over a Reynolds number range from about 150 to 300,000. That gives a frequency of around 20 hertz for a substantial tree trunk in a substantial wind—you might try some values in the equation. But the natural frequencies of trees run around 0.5 hertz or less, and trees are well damped (Baker 1997). Furthermore, the Reynolds numbers of the trunks of really big trees in storms will be well above 300,000; so vortex shedding won't be so tidy, and the effective Strouhal number (and thus frequency) will be still higher. So we need to look elsewhere to explain the troublesomeness of gusts.

Ground-Level Vortices

Near substrata, Von Kármán trails take on an additional characteristic that sessile animals sometimes capitalize upon. Flows in the paired vortices typically move upward. Sometimes smoke reveals such ascending flows behind tall chimneys; one can (at the price of littering) discover them by sending off paper gliders from the downwind side of the top of a tall, thin building. What's at work is this. Near the top of the protrusion the local speed of flow is greater than it is near the bottom. Fluid, as repeatedly mentioned, doesn't like to shear rapidly, and the fluid behind the top of the cylinder is subjected to more shear from the faster-moving fluid passing by it than is the fluid near the bottom. So fluid behind the top is drawn, essentially by its viscosity, into the passing flow; and that fluid is replaced by fluid moving up behind the cylinder from the bottom. In short, a protruding vertical cylinder causes a net upward flow—the effect is occasionally referred to as "viscous entrainment."* It's sometimes misattributed to Bernoulli's principle (chapter 7); the latter, though, doesn't apply to velocity differences solely due to position within the velocity gradient caused by viscosity near a surface.

Protruding vertical cylinders are extremely common in nature—all sorts of plants grow that way. They occur, as well, as attached animals sticking up into moving water. It was once assumed that arrays of cylinders inevitably had a stabilizing effect on any erodible substrate, that they caused what's termed "skimming flow" in which the cylinders shelter the water between

*"Viscous entertainment"—which sounds like mud wrestling—after the spell checker does its work.

themselves from the current, which then "skims" over the top. But it's now clear that skimming requires a very dense array of cylinders, enough to cover at least about a twelfth of the surface's area (Jumars and Nowell 1984), and that lower densities have the opposite effect, eroding and lifting sediment in ascending vortices by the mechanism just described (Eckman, Nowell, and Jumars 1981). The protruding tubes of several kinds of marine worms effectively resuspend detritus—edible dirt, loosely speaking (Carey 1983)—one worm switches from feeding on detritus to suspension feeding with its pair of tentacles when the water velocity increases (Taghon, Nowell, and Jumars 1980).

Some black-fly larvae that live in shallow, rapid streams have an interesting variation on the device. They are attached at their posteriors and have a pair of food-trapping fans on their anterior ends. A larva twists lengthwise so one fan is uppermost and feeds from material suspended well out in the velocity gradient (fig. 6.12a). The other fan is then lower and a bit downstream; it feeds on material resuspended in the vortices rising behind the body. Upward flow of water in the vortices is further encouraged by a downstream tilt of the body as a whole. Groups of larvae, positioned as they are in nature, don't compete for food but instead enhance each other's feeding efficiency (Chance and Craig 1986). Using vortices near surfaces for feeding provides not just a little lift for particles but also a neat way of sorting them by density. Material only a little denser than water will suspend and rise more readily than stony particles of sediment. And lower-density material is much more likely to be properly nutritive stuff. There's more than sex to being social; sometimes it's just a way to stir up some dirt!

Another way of using vortices to put edibles where they can be eaten underlies (literally) a peculiar behavior of phalaropes (noted as surface-tension transporters in the last chapter). A phalarope spends much of its time in a single place on the surface of shallow water, spinning rapidly around its own body, as in figure 6.12b. According to Obst et al. (1996), the spinning represents the recoil of the peculiar paddling pattern with which it creates an upward vortex like a tiny, underwater tornado or water spout. That vortex

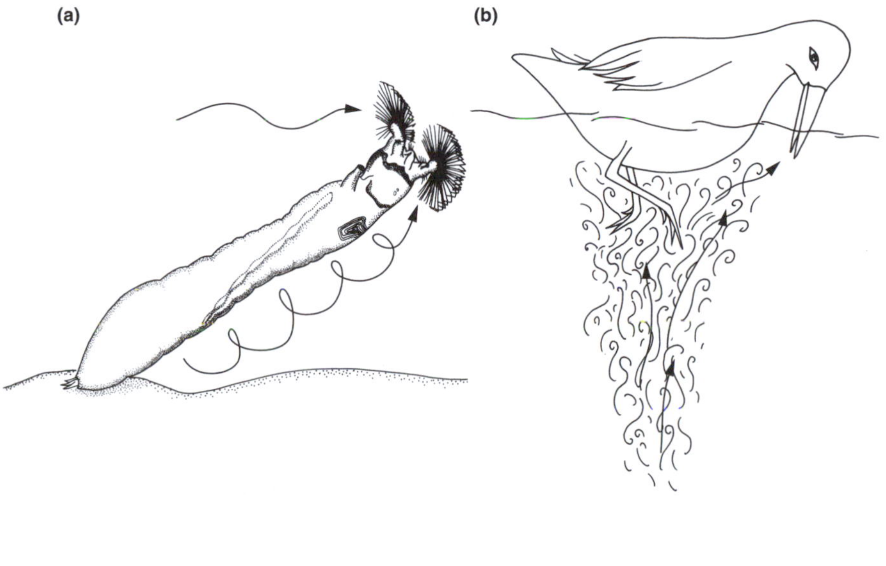

FIGURE 6.12 (a) The larva of a black fly on a rock in a stream; one of its fans feeds on mainstream plankton, the other from resuspended detritus. (b) The upward flow induced by leg motion in a spinning phalarope. Part (b) redrawn from Obst et al. (1996) with permission from *Nature*, copyright 1996 Macmillan Magazines, Ltd.

brings up edible items, which, with barely a pause in the spinning the bird snags in its bill.

Perhaps the most biologically important locus of vortex generation is the solid-fluid interface—where earth meets moving air or water. That ubiquitous velocity gradient represents what's called in more formal fluid mechanics "vorticity." At our more intuitive level it represents incipient vortices—vortices looking for an excuse to materialize. Imagine a solid object moving parallel to a surface, in the velocity gradient caused by the presence of that surface (fig. 6.13). It will translate along at a speed reflecting the average speed of flow to which it's exposed. Now shift your frame of reference from substratum to solid object—think of yourself as an observer on the object. On the side of the object farthest from the substratum you'll encounter a flow that pulls you ahead, since local flow speed exceeds the average flow along your path. On the side of the object nearest the substratum, you'll see a flow pulling you back. Such forces in opposite directions, each with a moment arm (the radius of the object) must make the object rotate. A tumbleweed would tumble even if it never quite touched the substratum, and off-axis blood cells rotate as they move along in our vessels. Or, if the "object" is a chunk of fluid, it just might become a vortex, moving along the surface at a speed below that of the free stream flow. In short, velocity gradients have vorticity, a physical variable we'll not deal with specifically; they may (but don't always) generate a train of moving, cross-flow vortices.

The puffy breeze you may notice on nearly calm days comes from periodic passage of such cross-flow vortices; all lake sailors where I live are familiar with the phenomenon, although transplants from the coast find it initially upsetting. These vortices commonly "catch" around protuberances from surfaces. At some point you may have stood on a beach, watching the ocean, with your feet exposed to the swash that ascends and falls back across the sand. Why do your heels usually sink into the sand? The down-beach backwash of the wave surge generates a vortex that persists behind your feet, and it erodes the sand just up-beach from your heels. And you may have noticed an odd, cuplike hole in the snow on the upwind sides of trees, with a compensating build-up on the downwind side. You might expect the opposite, with more snow upwind, but another velocity-gradient vortex has been at work. The full vortex is best described as a horseshoe, bent around the protuberance and trailing off downstream on both sides. Animals can use these as deliberate excavation devices for feeding in erodable substrates—as do some insect larvae in streams (Soluk and Craig 1990) that feed from the upstream pits created by the presence of their own bodies.

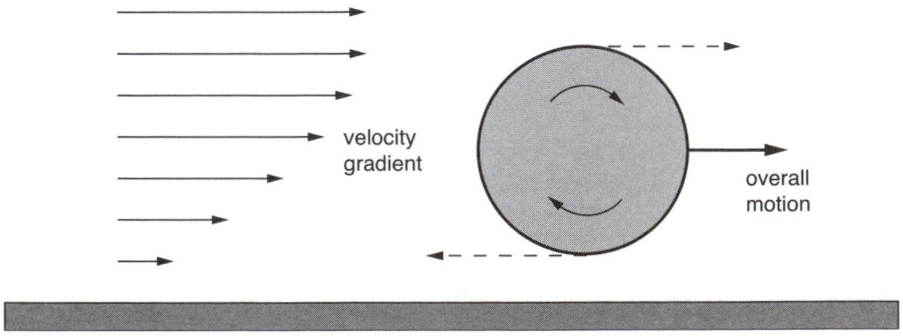

FIGURE 6.13 How the shearing motion within a velocity gradient near a surface can induce rotation in either a solid body or a cylinder of the fluid itself.

Yet a Few More Vortices

Similar horizontally rolling vortices can break out further up in the atmosphere. They need only a velocity gradient of sufficient strength relative to the local density gradient (another dimensionless index, the Richardson number, provides a criterion). Nighttime inversions may routinely generate them. Such inversions represent particularly stable atmospheres—radiation to the sky cools the ground and the air near it, so colder air lies beneath warmer air, with a consequent vertical density gradient. At the same time, wind provides a velocity gradient.

What of them? Several decades of tracking the migrations of small insects (no minor agricultural consideration) with short-wavelength radar has shown discrete layers of migrants at altitudes between about 100 and 500 meters. These layers seem to correspond to the tops of inversions (Drake and Gatehouse 1995; Gatehouse 1997). One can speculate that the insects seek out the same thing as does a bird circling in the middle of a thermal vortex—a location where they can descend with respect to the local (rising) air without descending with respect to the earth itself. Night inversion-layer vortex transport ought to matter. Rice leafhoppers, for instance, travel for many hours on two or three successive nights, probably consuming more fuel than they could supply for that long a level flight. But direct evidence remains lacking.

For that matter, with rising air, powered flight may be avoidable entirely. Spiderlings pay out a long thread, let loose, and travel great distances. The late Lloyd Trefethen suggested (and made a computer model indicating) that they might use the drag of the thread to keep themselves in a vortex for long periods. Analogous transport happens in tiny mussels, who pay out long threads as they're released into ocean currents (Lane, Beaumont, and Hunter 1985). Vortices routinely occur at salinity gradients in estuaries, where freshwater flows seaward above a so-called "salt wedge"; those gradients matter physiologically to the local fauna and most likely matter biomechanically as well. For neither aerial nor aquatic transport have we much direct evidence for such specific roles of vortices. What we do have are observations that lack more persuasive explanations, some of proper antiquity and authority. Charles Darwin observed the following when off the coast of Argentina, as reported in *The Voyage of H.M.S. Beagle* (1845):

> *On several occasions, the rigging has been coated with the web of the Gossamer Spider. One day I paid particular attention to this subject. The weather had been fine and clear, and in the morning the air was full of patches of the flocculent web, as on an autumnal day in England. The ship was sixty miles distant from the land, in the direction of a steady though light breeze. Vast numbers of a small spider, about one-tenth of an inch in length, and of a dusky red colour, were attached to the webs. There must have been, I should suppose, some thousands on the ship.*

Inversion layers happen when high-density fluid lies beneath low-density fluid. Vortices can be produced as a consequence of the opposite situation as well—where the lower density of fluid beneath that above makes the fluid want to turn over. If no lateral bulk movement or substantial turbulence interposes itself, then a pattern of local turnovers appears—Bénard cells, they're called—at least if still another dimensionless criterion, the Rayleigh number,

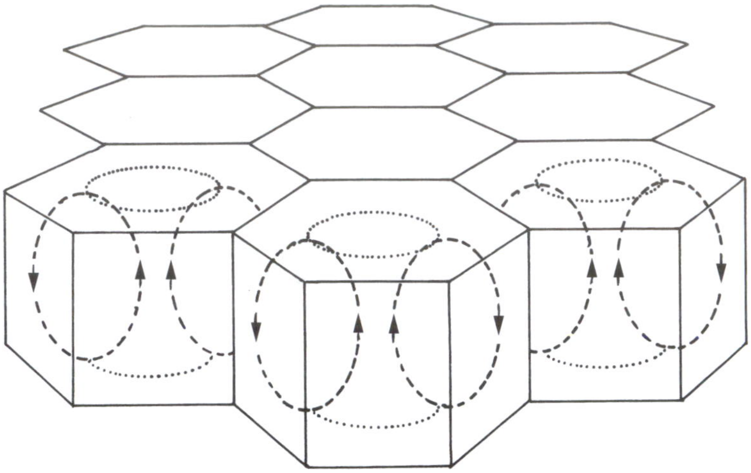

FIGURE 6.14 Bénard cells can occur when fluid of lower density lies under fluid of higher density, and nothing else impels much fluid motion. They take the form of a more or less hexagonal array.

is satisfied (fig. 6.14). You can observe a set yourself if you mix a small amount of pearlescent liquid soap with water in a dark frying pan, let things settle down, and then heat the pan very gently and uniformly from below. (The phenomenon illustrates a very general point of thermodynamics. Order appears where none was present; the order comes ultimately from energy flow through the system. That's an analog of the energy transfer from hot solar energy hitting the earth to reradiation out to a cold sky, the flow that makes life thermodynamically permissible.)

Direct biological relevance of physically generated Bénard cells remains, I think, undemonstrated. Microorganisms in dense culture, though, can make their own, in a process that goes by the name (useful for searching) "bioconvection." They just have to do a bit of synchronized swimming, either upward or downward, and the familiar pattern appears. The phenomenon has been given quite a bit of attention, both theoretically and experimentally—Lewin (1995) and Czirók, Jánosi, and Kessler (2000) provide entry points to the relevant literature.

Vortices will get further attention when we consider the origin and uses of lift in chapter 12 and the mechanisms of swimming and flying in chapter 13. What's important here that one bear in mind their multifarious origins, their ubiquity, and their actual and potential biological relevance.

CHAPTER 7

The Forces of Flow

> Wind is caused by the trees waving their branches.
>
> **OGDEN NASH**

THE LAST CHAPTER FOCUSED ON THE WAYS FLUIDS FLOWED—IN PARTICular, on patterns of laminar and turbulent flow, flow near surfaces, and vortices. A good qualitative sense of how fluids move around can go far toward anticipating what happens in and around organisms, toward generating rational and even testable hypotheses. But ultimately most problems require more than recognizing patterns and regimes of flow. So now, armed with some mental images, we turn quantitative, looking at momentum, power, pressure, drag, and so forth.

Force as Momentum Change

The main forces of concern are drag, lift, and thrust; and three separate stories explain their origins. But, before worrying how these forces come about and how they depend on shape, speed, and such, we should view them in terms of momentum. While momentum hasn't gotten much notice since chapter 2, it now becomes a major player.

At least dimensionally, our old, familiar $F = ma$ is equivalent to $F = mv/t$—as we've seen several times already.* Of course, mv/t is momentum over time, or the rate of change of momentum. In other words, force can be viewed as a rate of change of momentum. Apply a force and something gains or loses momentum at a rate determined by the size of the force. As noted in the last chapter when inertial force was made the numerator of the Reynolds number (equation (4.9)), putting mv/t as $\rho S v^2$ means we don't have to worry about specific masses, a lot handier when dealing with unbounded, flowing fluids. Putting the business more specifically,

$$F = \rho S_2 v_2^{\,2} - \rho S_1 v_1^{\,2}. \qquad (7.1)$$

As noted earlier, density can be regarded as constant for biologically relevant flow speeds.

What does equation (7.1) say in physical terms? Consider a propeller or fan through which a stream of flow flows (fig. 7.1a) and assume, for simplic-

---

*Properly, force equals mass times the slope of the curve of velocity versus time—the acceleration at a particular moment of time, as explained in chapter 4; formally, $F = m\, dv/dt$.

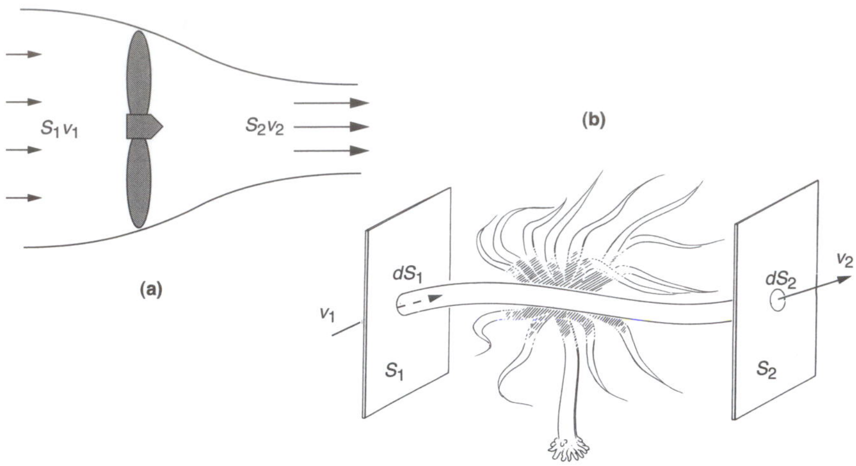

FIGURE 7.1 (a) Flow through a fan, with streamlines to show how flow constricts as it speeds up. (b) A stream tube crossing an object. The rate at which momentum passes an upstream plane is its cross section times the density of the fluid times the square of the local velocity; the analogous rate for a downstream plane uses its cross section and velocity in the same way.

ity, that the speed of that flow stream doesn't vary across the stream. If passage through the fan speeds up the flow, then, v_2 is greater than v_1 and, by the principle of continuity (chapters 2 and 6), S_2 is less than S_1—the stream must constrict:

$$S_2v_2 = S_1v_1. \tag{6.5}$$

The amount of constriction or the change in flow speed—at constant density we can use either—tells us the force the fan exerts on its mounting. Or, if we're dealing with the propeller on a craft, the thrust with which it pulls the craft.

If, instead of propeller or fan (or the idealization called an "actuator disk") in the flow, we consider a passive body that just resists the force, equation (7.1) works as well, except that F now indicates drag instead of thrust—v_1 is greater than v_2, and the force goes with rather than opposite the flow. If, to go a step farther, we're interested in a body that generates lift with a fixed wing, we need just look at the rate at which it creates a downward component of momentum. The value of v_1 becomes zero, presuming a horizontal oncoming flow, and we have

$$L = \rho S_v v_v^2 \tag{7.2}$$

(with the subscript v indicating the downward component). If we're considering a craft that hovers by blowing fluid downward (helicopter or hummingbird), then we simply revert to equation (7.1), with subscript 1 indicating a plane above the craft and subscript 2 below the craft.

So far we've assumed uniform flow speeds across our various planes fore and aft of bodies. That may be reasonable upstream of fan or fly, but it's rarely a safe assumption downstream. Fortunately, we don't really need it. Imagine that a body sticks up into a uniform oncoming flow of speed v_1. We want to measure its drag, but it doesn't take kindly to contact with an instrument. All we need to do (easier said than done, of course) is to divide the plane behind into a grid of squares or rectangles anywhere that v_2 differs significantly from v_1—as in figure 7.1b. Equation (7.1) now gets applied to each stream tube defined by our grid, and adding up the results gives the drag. That takes a set of speed measurements, one for each stream tube over which flow is assumed uniform, plus a single upstream datum. Oh, yes—one needs to use the principle of continuity (equation (6.5)) to eliminate S_1. After all, if the upstream flow is uniform, the areas of individual stream tubes don't matter. So momentum change, continuity, and streamtubes combine quite usefully.

Using this approach and a tiny (1/8-millimeter-diameter) flow sensor, I once managed to measure the drag of the bodies of fruit flies mounted in a wind tunnel. The procedure was nothing if not laborious—especially with data read from analog meter and chart paper and then reduced on a mechanical calculator. Eventually I figured out how to make direct force measurements on the flies' mounts; at my 5 percent precision the two procedures gave indistinguishable results.

As an alternative to moving flow sensors around, one can get speed data from photographs of passing particles taken with stroboscopic illumination. Several people have done fancier versions of what we've been considering, persuading birds to fly through paper confetti or neutrally buoyant bubbles (for instance, Spedding [1987]) and getting aquatic animals to swim in flow tanks in which tiny particles are suspended. Things do get more complex when flows are directionally nonuniform, especially when vortices are formed. But the general approach still proves useful, and modern digitizing techniques and software for particle image velocimetry ("DPIV" or "PIV") makes the approach quite practical—see, for instance, Raffel, Willert, and Kompenhans (1998) or Drucker and Lauder (2002).

Recognize, though, that assertions about momentum insertion, removal, or deflection rates say nothing whatsoever about the physical mechanisms by which thrust, drag, or lift happen. Useful though they may be, they merely declare the applicability of the three Newtonian laws. As we'll see shortly, the origins of each of these three forces involves a lot that's devious or at least counterintuitive.

Power and Propulsion Efficiency

Consider a propulsive device (such as a propeller but not a jet engine or rocket) that simply speeds up air that passes through it. Reverting, for simplicity, to mass and time rather than density, speed, and cross section, we can express its thrust, T, the rate at which it inserts momentum into the flow, as

$$T = \frac{mv_2}{t} - \frac{mv_1}{t} = \frac{m}{t}(v_2 - v_1), \tag{7.3}$$

where v_2 represents the speed of fluid coming out of the propulsive unit and v_1 is the speed at which fluid enters the propulsive unit, effectively the speed of craft or creature through the fluid. Thrust happens when v_2 exceeds v_1. Since work is force (here thrust) times distance and power is work per time, power becomes force times velocity, in this case the speed of the craft. So

$$P_{out} = \frac{mv_1}{t}(v_2 - v_1). \tag{7.4}$$

(That pulling produces no power if it doesn't move something, as happens before an aircraft releases its brakes for its takeoff run, may seem peculiar, but we have to remind ourselves that that's just the faulty intuition of creatures that must expend power to produce muscular force even if no load gets moved. Odder still, but equally important, lift generation by an aircraft in level flight may cost power, but it involves no useful work or power output.)

Power output divided by power input gives a measure of efficiency. Efficiency, though, can be a slippery business for the biologist, who needs to define with care exactly what input is being considered. No right or wrong is at issue, just explicitness and utility. Metabolic rate in excess of basal level provides one useful input—with this input efficiencies above 20 percent for swimming or flying must be considered exemplary. For many purposes, we don't want to go that far upstream for our input. Kinetic energy per unit time gets us outside the animal, restricting our purview to propulsive hydrodynamics. It's simply mass per time times speed squared over two; in present terms,

$$P_{in} = \frac{m}{2t}(v_2^2 - v_1^2). \tag{7.5}$$

A little algebra give a simple result for efficiency, dimensionless as always:

$$\frac{P_{out}}{P_{in}} = \eta_f = \frac{2v_1}{v_2 + v_1}. \tag{7.6}$$

This particular ratio, η_f, is called the "Froude propulsion efficiency," after the nineteenth-century British naval engineer William Froude. Its origin may not be biological, but it's fraught with implications for organisms. It says that for highest efficiency, the velocity of the fluid issuing from the propulsive unit—paddle, propeller, jet, or whatever—should be as close as possible to the velocity of the craft. Any increase in v_2 beyond v_1 drops efficiency below the ideal 100 percent. But we noted earlier (equation (7.3)) that thrust production requires v_2 to exceed v_1. Clearly the way to maximize Froude propulsion efficiency consists of moving the largest possible mass-per-time (m/t) of fluid and giving it the least possible increase in speed ($v_2 - v_1$). In practical terms that means maximizing S, the cross section of the propulsive flow stream.

What are the Froude efficiencies of organisms? Putting aside fliers, since the necessity for generating lift complicates the picture, I've gathered some data in table 7.1. While these vary in quality and involve differing underlying assumptions and simplifications, the picture that emerges is satisfyingly consistent with our expectations.

- Moving water with undulating body, beating wing, or swinging tail beats squeezing water out of a jet, as anticipated. A squid may jet fast, but when it wants to go far, it's more likely to use its fins.

TABLE 7.1 Froude Propulsion Efficiencies for a Variety of Swimmers

Organism	Froude Propulsion Efficiency
Bacterial flagella (rotating helix, model)	0.34–0.78
Eucaryotic flagella (undulating lengthwise)	0.2–0.4
Hydromedusae (thin, jet propelled)	0.5
Hydromedusae (broad, jet propelled)	0.9
Copepods (tiny crustaceans, paddling)	0.34
Mosquito larva (gliding with "brushes")	0.23
Scallop (jet propelled)	0.3–0.5
Squid (jet propelled)	0.42, 0.56
Tadpole tail (undulating, modeling)	0.45
Tadpole tail (undulating, experimental)	up to 0.82
Penguin (wing beating)	0.4
Siren (eel-like salamander, undulation)	0.83
Fish (pectoral paddling locomotion)	0.60–0.65
Fish (body and tail undulation)	0.75–0.90
Muskrat (underwater paddling)	0.30
Fully marine mammals	0.8–0.9

Sources: Data from Webb 1974; Videler and Hess 1984; Morris, Gust, and Torres 1985; Wassersug and Hoff 1985; Webber and O'Dor 1986; Bschorr 1988; Webb 1988; Culick and Wilson 1991; McHenry, Pell, and Long 1995; Cheng and DeMont 1996; Fish 1996; Gillis 1997; Liu, Wassersug, and Kawachi 1997; Purcell 1997; Anderson and DeMont 2000; Ford and Costello 2000; Brackenbury 2001.

- The same undulating devices do better than systems that move water backwards with a paddling system, with its alternating power and recovery strokes. We'll return to this comparison between "lift-based" and "drag-based" propulsion in chapter 13.
- Bigger (or at least moderate size) is better than smaller. Except for one questionable datum for a bacterial flagellum, no creature below about a centimeter in length does better than $\eta_f = 0.5$. The pernicious effects of low Reynolds number (chapter 11) cannot be denied.
- The broad hydrozoan medusae (essentially small jellyfish) may use jet propulsion, but they do it by pushing out an especially large volume (relative to their own) through a wide aperture. So they have a much higher m and lower v_2 than the other jetters, and thus evade most of the difficulty inherent in equations (7.5) and (7.6).

The Great Principle of Daniel Bernoulli

For fluids in motion, pressure matches momentum in importance. We last met pressure in its hydrostatic manifestation—exerted equally in all directions and increasing with the height of the column of liquid above our point of reference. Fluid dynamics treats hydrostatic pressure as some background baseline, much the way a tire pressure gauge starts at zero, whatever your altitude. It divides the pressure in a flowing fluid into two components. One is *static pressure*, the ambient pressure in the flowing fluid, which can be above or below the zero set by the local hydrostatic pressure. The other is *dynamic pressure*, the additional pressure that would be generated by abruptly reducing the flow speed to zero. In other words, dynamic pressure is the pressure that would appear if the kinetic energy of motion got entirely transmuted into pressure. Two consequences right off. First, fluid motion provides a source of pressure that can affect organisms—they might put it to use or might have to deal with its awkward consequences. Second, converting the kinetic energy of flow into pressure permits measurement of energy (really energy per unit volume) with a pressure-measuring device—a manometer. Since fluid density remains constant (in our fluid-dynamic domain, at least), energy per volume converts to flow speeds. Thus, a manometer can indicate speed.

But we need a rule. Fortunately, we have one, a simple and straightforward way to link speed of flow and dynamic pressure—at least if we're willing to live with a peculiar assumption. Up to this point, we've considered viscosity the defining property of fluids. Now viscosity will be assumed negligible, and we'll look at the behavior of that grotesque misnomer, an "ideal fluid." Still, simplification underlies the success of science—we follow the philosopher William of Ockham, trying the simplest things first and adding complications only as absolutely necessary. Ignoring viscosity doesn't differ from treating the interactions among solids as frictionless—useful abstractions even if the world doesn't quite work that way. Still, ideal fluids remain slippery, yielding both quantitatively reliable results and terrible paradoxes. Presupposing an inviscid fluid demands caution and vigilance. Zero viscosity implies perfectly free slippage between a flowing fluid and the walls of pipe or channel—neither velocity gradients at surfaces nor transfer of momentum from moving fluid to walls of pipe or channel.

That rule, Bernoulli's principle, amounts to a statement that energy is conserved, analogous to mass conservation (the principle of continuity) and momentum conservation. Energy can take the form of kinetic energy ($mv^2/2$) or pressure-volume energy ($V\Delta p$), and one trades off against the other. Not that conservation of energy depends on neglecting viscosity—we just find that it permits a simplified formulation that handles a lot of our problems. Bernoulli's principle can be easily and directly derived from the notion of conservation of energy, although that sweeps a few relevant matters under the rug. Alternatively, and more honestly, it can be derived from forces via Newton's second law, but that approach is a bit awkward with our present mathematical tools.

What matter here are the assumptions involved—that flow is steady, that the fluid of interest is incompressible, and that gravity can be ignored, either as a sink or source of energy or as something that changes the density of the fluid. In effect, we're presuming (1) that at no point in space is velocity time-dependent, (2) that velocities in air remain below about 100 meters per second, and (3) that fluid doesn't move upwards or downwards with respect to the earth to any appreciable extent. Pressure does vary with altitude, as explained in chapter 5—the change is 9,806 pascals per meter in water and

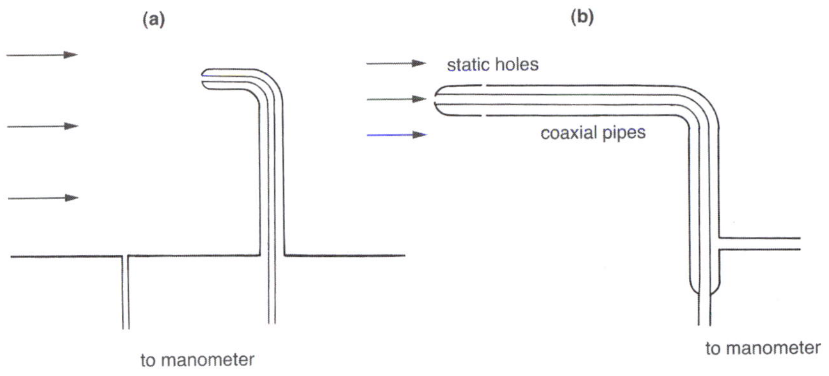

FIGURE 7.2 Two forms of the manometric device for measuring flow speed at a point. The device in (a) is a so-called Pitot tube, after Henri Pitot, who used it to measure flow speeds in the Seine in 1732; (b) is a "Pitot-static" tube that uses a coaxial pair of channels to put the whole thing in a single probe.

about 12 pascals per meter in air,* but it's easiest to treat these variations as items that at most require minor post hoc adjustments.

Pitot tubes and an aquatic insect larva

Consider a rounded probe with a small opening that faces upstream (fig. 7.2) connected to a manometric device. Since nothing actually flows through a manometer or other pressure gauge once equilibrated (and we're still talking about steady flows), that upstream-facing opening is exposed to flow that has suddenly stopped—the condition for converting kinetic energy to pressure energy—as well as to ambient pressure. So it sees dynamic pressure, the other item in Bernoulli's principle, as well as static pressure. How does kinetic energy translate into dynamic pressure? Conversion of kinetic energy per unit volume of fluid into dynamic pressure ($\rho v^2/2$) is just a formality, since they're the same thing:

$$\frac{mv^2}{2V} = p_{\mathrm{d}} = \frac{\rho v^2}{2}. \tag{7.7}$$

We can then assert a simple form of Bernoulli's principle, constancy of the sum of dynamic and static pressures:

$$\frac{\rho v^2}{2} + p_{\mathrm{s}} = H. \tag{7.8}$$

(The constant sum, H, is sometimes called the "total head.") Now the typical pressure-measuring device looks not at some absolute pressure but at the difference in pressure between one place and another, Δp. In the case of either device of figure 7.2, one end sees dynamic plus static pressure, H, while the

*The difference, 9,794 pascals per meter, is, in fact, the factor used to convert height to pressure in an ordinary U-tube manometer filled with water and working in air.

other sees static pressure, p_s, only. The manometer thus does the subtraction for us, eliminating H, and reading dynamic pressure in the form of its Δh:

$$\Delta p = \frac{\rho_a v^2}{2} = (\rho_m - \rho_a) g \Delta h. \tag{7.9}$$

The last term puts matters in shape for using an ordinary U-tube manometer filled with a liquid whose density, ρ_m, differs from that, ρ_a, of the ambient fluid. Admittedly, it may seem odd that the device ends up measuring the speed of flow at a point where its presence reduces that speed to zero! But at least it's intuitively comfortable to get an increase in pressure on something (even a hole) that faces into a flow.

The biologist, though, must give equation (7.9) skeptical scrutiny before rushing out to use it in lab or field. The problem lies less in its assumptions, limiting as they sound, and more in that square of velocity and what it implies as far as sensitivity for biologically interesting flows. Air flowing at a meter per second gives a pressure difference of 0.6 pascals, merely six millionths of an atmosphere or, with manometers, 4.5 micrometers of mercury or 61 micrometers of water. That's well beyond the sensitivity of the devices used to measure such variables as blood pressure (around 13,000 pascals). A water flow of a tenth of a meter per second gives 5.0 pascals—better, but still no manometric pushover. Pitot tubes are great for airplanes, though, as the reader can verify with some simple calculations. So as an air-speed indicator (anemometer), a Pitot tube is standard gear on all small airplanes—a tube protrudes, facing forward, and the body has a little hole with the cryptic legend next to it—"static orifice, do not block."

Nature makes occasional use of the device. One case consists of the silk-bound bent tube spun by the aquatic larva of a caddisfly, *Macronema* (fig. 7.3a). It lives in streams and with this arrangement capitalizes on the current

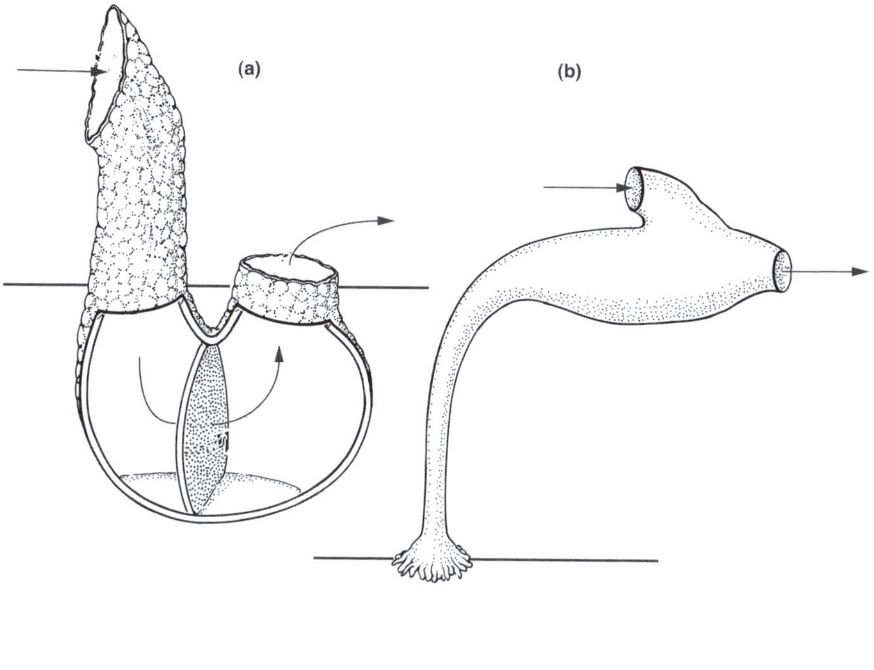

FIGURE 7.3 Two natural Pitot tubes: (a) the case and catch-net of the larval caddisfly, *Macronema*; and (b) the incurrent and excurrent siphons of the ascidian, *Styela montereyensis*.

to drive flow (and food) through its catch-net, spun across the middle of the overall tube. As Wallace and Merritt (1980) put it, *Macronema* "perfected the Pitot tube long before Henri himself." (But, in fairness, Monsieur Pitot probably knew nothing of caddisfly larvae.)

An equivalent scheme serves the same function in at least one tunicate or ascidian (fig. 7.3b), a sessile marine animal about as distantly related to a caddisfly as are we. These "sea squirts" have the relevant plumbing inside themselves rather than in some external domicile, but the fundamental problem of suspension feeding remains—how to move water through a separation device in such a way that the yield from food exceeds the cost of moving the water. The particular ascidian, *Styela*, takes advantage of local water motion to pass water through itself, but to do so it must contend with the rapidly changing direction of coastal wave surge. It has, though, quite a flexible stalk, and it manages to reorient like a weathervane so the input opening always confronts the flow (Young and Braithwaite 1980).

Venturi tubes and prairie dogs

There's another way to view equation (7.8), one that also generates both a measuring device and some biological adaptations. If the sum of a pair of terms is conserved, then that sum can supply both sides of the equation—which gets rid of the awkward constant, H, another way:

$$\frac{\rho v_2^2}{2} + p_{s2} = \frac{\rho v_1^2}{2} + p_{s1}$$

or

$$\Delta p = \frac{\rho}{2}(v_2^2 - v_1^2). \qquad (7.10)$$

This recognizes two variable velocities, one of which the Pitot tube dropped to zero. As a practical matter, it corresponds to the device of figure 7.4, in which a pressure-measuring device connects portions of a pipe that

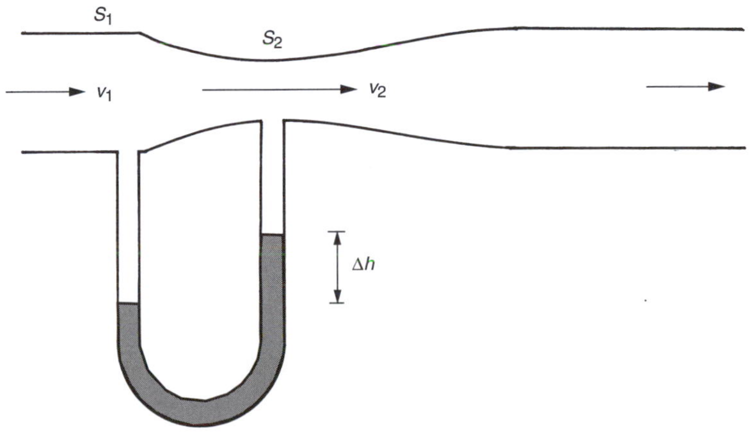

FIGURE 7.4 A Venturi meter—a constriction of known size makes the fluid flow faster and thus lowers the pressure locally. The difference in pressures can be used to find the speed of flow in the unconstricted pipe.

differ in diameter. That device exposes the counterintuitive implication of Bernoulli's principle, which I've avoided up to this point—higher speed corresponds to lower pressure. You might want to recage your intuition by reversing the logic, viewing pressure as a cause rather than as a result of flow. Thus, lower pressure ought to be needed to make flow speed up as it enters a constriction in a pipe. Continuity says that the volume per unit time will be the same for wide and narrow portions, but with faster flow in the narrower portion, kinetic energy per unit volume and time must increase. (Continuity contains the first power of velocity; kinetic energy requires the second.) That demands a decrease in pressure.

Such a Venturi meter turns out to be quite useful, in particular because we can get rid of one of the velocities with another trick, by applying the principle of continuity. Imagine that you have a pipe through which you want to measure the flow. You just need to insert in the pipe a slight constriction of known character—one for which you know the reduced cross section—and arrange pressure taps to connect to a manometer. Applying the principle of continuity (equation (6.5)) gets rid of v_2 (which you don't really care about anyway), and the expression in parentheses remains constant for your setup:

$$\Delta p = \frac{\rho v_1^2}{2}\left(1 - \frac{S_1^2}{S_2^2}\right). \tag{7.11}$$

Since Bernoulli's principle gives the relationship between speed and pressure difference, and since S_1 and S_2 are known (you made or bought the constriction), then a measured pressure difference can be used to figure the speed of flow in the pipe outside the constriction. In short, this deliberate but slight and local constriction is the basis for a fine meter for flow through a pipe. You merely have to pay a little attention to the shape of the constriction, you have to keep the two pressure-measuring holes reasonably close to each other to minimize the effects of viscosity, and you can't trust the thing at really low Reynolds numbers. In practice, the problem of low speed sensitivity isn't as bad as for the Pitot tube.

The basic principle underlying the Venturi tube works as well on a surface beneath an open field of flow—it's just not immediately handy as a measuring tool. Let's remove the indicating fluid from the pressure-measuring manometer so pressure differences will drive fluid through the manometer's pipe, just as it did through the caddisfly's catch-net. And let's simply erase the line in figure 7.4 that represents the upper surface of the pipe. We now have an arrangement that occurs in nature in several places, most notably and straightforwardly in the burrows of prairie dogs (fig. 7.5a).

These rodents that live on the relatively windy plains of North America are consummate burrowers even by the high standards of their order—a typical burrow is two-ended, about 15 meters long, and has a maximum depth of about 3 meters. It's not difficult to calculate (using Fick's law, chapter 8) that if oxygen is supplied by diffusion alone, a prairie dog in the bottom of a burrow will asphyxiate, so some system of forced ventilation is needed. But work with models in a wind tunnel showed that an animal just has to maintain a slight difference between the geometries of the two openings of its burrow for even a modest breeze to do the rest. Tests in the field with a smoke-generator showed that any noticeable wind above caused air to move through burrows, and various observers have noted that the animals assiduously maintain the difference in the form of their two openings. While the scheme is direct and effective, its physics has a few complications—in practice Bernoulli's principle acts in concert with viscous entrainment, and flow through the burrow to some extent relieves the predicted pressure difference (Vogel, Ellington, and Kilgore 1973; Vogel 1978a). In addition, the system turns out to be wildly overbuilt—a minimal wind of a mile per hour will give a full change of air about every 10 minutes, while an animal has about a 6-hour oxygen reserve. So maybe the scheme serves as a giant olfactory accessory, telling what's up to animals below.

Nor are prairie dogs the only creatures that use the scheme. Some combine the Venturi-like aspiration of an elevated opening with the Pitot-like supercharging of upstream-directed inputs, as in figure 7.5b. While omnidirectionality may be lost (but can be regained with some ancillary devices), they gain access to a much greater pressure difference. African termites living in open country use it to ventilate their mounds (Weir 1973; Turner 2000). It allows sponges to use ambient flows to reduce the cost of pumping water through their filtering equipment (Vogel 1977). It helps force air through the tracheal pipes of large insects in flight (Miller 1966). It even allows reeds to use wind across broken stalks to aid diffusion in moving air around inside (Armstrong, Armstrong, and Beckett 1992). And it helps ventilate a wide variety of buildings, from half-buried domiciles to tepees and hay lofts. Among other things.

Pressures Over Objects in Flows

Pressure measurements hold more interest than just as data on the road to velocity. Consider what happens if we set up a remote static aperture such as

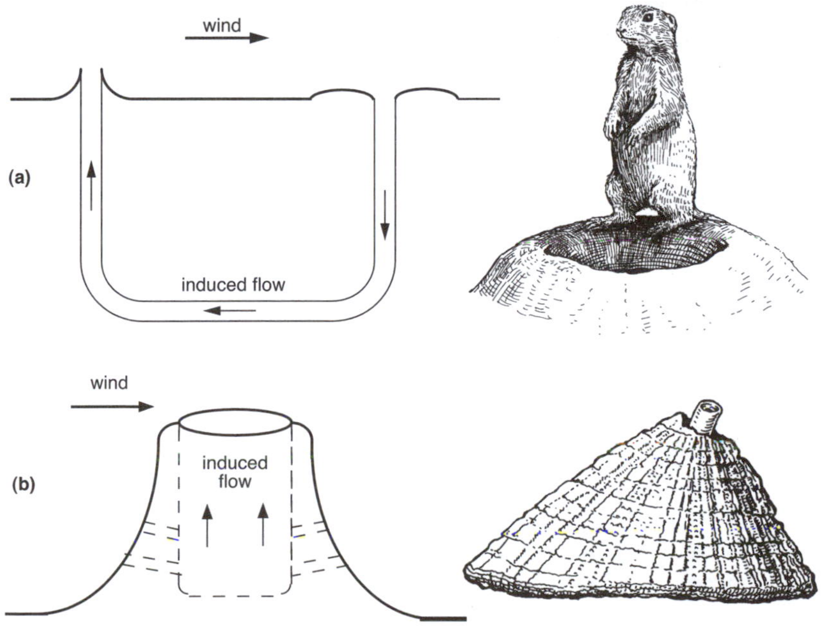

FIGURE 7.5 (a) The below-the-substratum ventilation system of a prairie dog burrow; and (b) an above-ground version used by sponges, termite mounds, keyhole limpets (shown), and other creatures.

that in figure 7.2a and connect the other aperture first to one and then to another and then to still another orifice along the length of some object in the flow. We get a graph or map of the way pressure varies along the object. The latter tells us several things. First, this pressure distribution is in the most immediate sense a set of forces trying to distort the object. If, as organisms often are, the object is nonrigid, the pressure distribution tells us a lot about how shape will change as a result of exposure to flow. Second, for real (as opposed to ideal) fluids, the magnitudes and directions of these pressures add up to some overall force on the object—if the resultant force is in the direction of flow it's a major part of the drag on the object; if it has a component perpendicular to flow, it contributes a lifting force as well.

How might such a plot of pressure versus location best be presented? We mainly care about the influence of the shape of an object on the pressures it feels. So we'd like to compare pressures around objects of a variety of sizes. That can be done by using a dimensionless x-axis, dividing the distance from the upstream end to a given point by the overall distance from upstream to downstream ends. Like Procrustes' bed, such an axis fits all comers. There's a good universal y-axis as well. For any object in any flow, net pressure at the extreme upstream end will be the full dynamic pressure of the flow, ($\rho v^2/2$), as at the hole of a Pitot tube. So dividing each datum for pressure by this dynamic pressure makes the plots for all objects and flow speeds begin at the left at 1.0. To this new dimensionless variable, measured pressure divided by dynamic pressure, the name "pressure coefficient" is given; in formal terms, it's

$$C_p = \frac{2\Delta p}{\rho v^2}. \qquad (7.12)$$

Velocity, of course, is the free-stream speed of air or water. Objects operating in different media or at different speeds or of different sizes all yield data that fit comfortably on the same axes with the same scales. Restoring a properly dimensioned pressure requires nothing more than the reuse of equation (7.12). The overall plot now displays one dimensionless variable against another, but with that one soon becomes contemptuously familiar.

Figure 7.6 gives such a graph for pressures around a sphere at a Reynolds number of 28,000 (roughly a 5-centimeter object going through water at 0.5 meters per second). The line has negative values back where (by continuity) flow has had to speed up beyond its free-stream value to get around the obstacle. Bernoulli's principle is clearly operating—pressure drops as flow speeds up and vice versa. Two points are especially noteworthy from our viewpoint. First, although one result of flow is an inward pressure near the very front, the overall effect is an outward pressure over the surface of the object. Second, the net force on the object pushes it rearward. There's a region of high pressure near the front that will push it backward, but no equivalent region of high pressure near the back to push it forward. Which is to say that a sphere suffers a relatively high drag.

As a contrast, figure 7.6 gives the analogous curve for an object with a slightly elongated but still rounded front along with a much elongated and pointed rear. We loosely refer to this teardrop shape as "streamlined"—this one was a toy water-rocket with its fins filed off. (Again, "streamlined" has

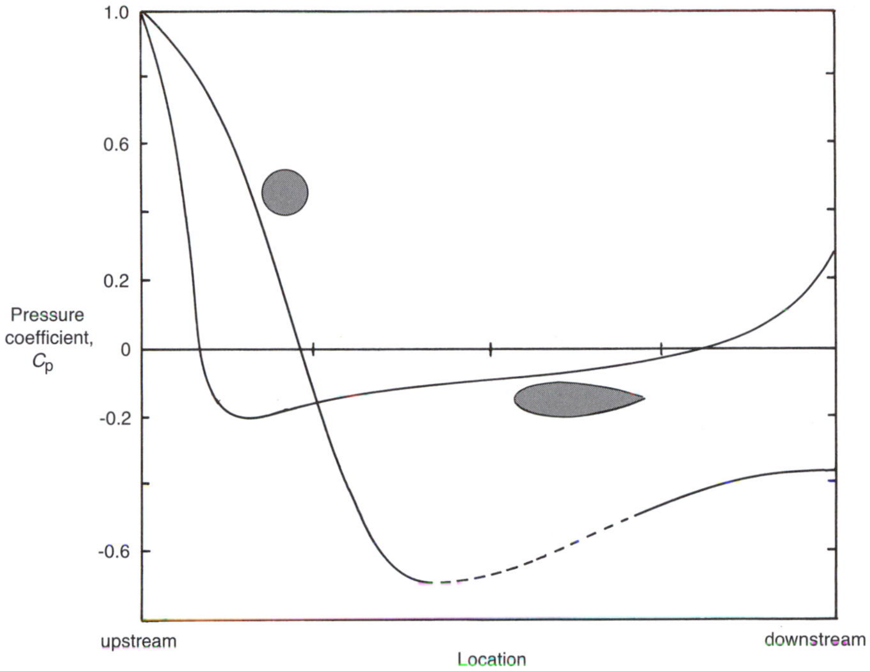

FIGURE 7.6 The variation of pressure coefficient with location for a sphere and a streamlined object. The dashed portion of the line for the sphere marks the region behind the separation point where pressure varies erratically.

only a loose relationship with "streamlines" as introduced earlier.) Not only are the pressures on the sides less negative (it's, after all, relatively more slender), but now a respectably positive pressure pushes on the rear to offset in part the rearward push on the front. Which is to say that this streamlined object has a lot less drag than a sphere. We'll return to that shove from behind shortly; at this point we need note only that the curve differs only trivially from published data for very large airships—once one exceeds a few thousand, streamlining turns out to be nearly scale- (really Reynolds number-) insensitive.

Blimps, fast fish, and jetting squid

The engineer mainly uses these plots of pressure coefficient versus location for calculating overall drag and noting the effects on drag of details of shape. I mentioned earlier that for flexible objects the pressure distribution itself takes on immediate importance. Blimps, in fact, have reinforcing battens near their front ends to offset the potentially distorting forces—it wouldn't do to have the front flattened in flow or for shape to change with flying speed. But among the products of our technology blimps are unusual in their lack of rigidity. What effects might these variations of pressure with location have on the design and operation of organisms?

Some years ago, DuBois, Cavagna, and Fox (1974) managed with some difficulty to make a set of pressure measurements on a bluefish in a flow of about 2 meters per second—it's a lot tougher to run pressure-conveying conduits through a fresh fish than through a model! The results (fig. 7.7a) were clearly worth the trouble even though they proved gratifyingly similar, overall, to expectations from data such as that in figure 7.6. Strictly speaking, one

FIGURE 7.7 Inward and outward pressures (as pressure coefficients) for (a) a swimming bluefish, and (b) a model of a jetting squid.

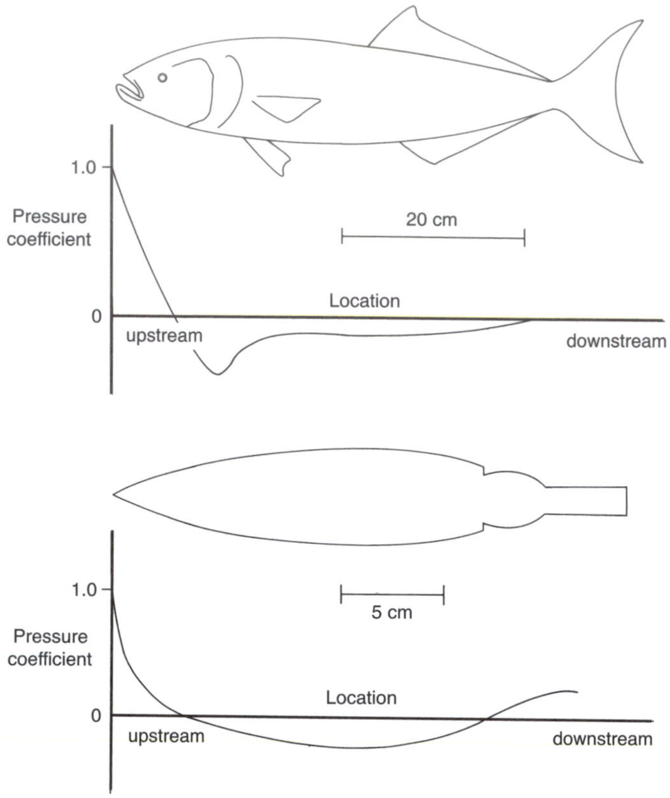

can't prove that such pressures were decisive in the determining the design of fish; but they're at least consistent with the usual functional rationality of evolutionary products.

Fish respire by taking water into their mouths, passing it across their gills, and ejecting it behind a lateral shield, the operculum. The mouth is obviously at the point of most positive (inward) pressure while the operculum is located at the site of most negative (outward) pressure; no greater pressure difference to augment a fish's pumping could be easily contrived. Indeed, fish such as mackerel do no pumping but swim continuously and depend on this so-called "ram ventilation" to keep from suffocating (Randall and Daxboeck 1984), and lots of fish such as trout switch to pure ram ventilation at high swimming speeds (Steffensen 1985).

The eyes of bluefish (and, almost certainly, of many other fish) are located at the point where the pressure coefficient passes through zero; at this unique location pressure doesn't vary with swimming speed. Focus of an eye depends in large part on the curvature of the cornea; variable pressure might awkwardly affect that curvature. The heart is located beneath the area of minimum pressure; hearts are made of muscle, which can actively contract but which needs some special device to ensure reexpansion. The low pressure might aid reexpansion and thus increase the heart's stroke volume during rapid swimming, which, after all, is the only heavily aerobic exercise of fish. As a possible corollary, fish commonly increase stroke volume more than the frequency of beating during activity, the opposite of our own compensation

(Jones and Randall 1978). Finally, the skull of a fish reinforces it against inward pressure throughout the anterior area where such pressures can occur; further aft its largely inextensible skin ought to be adequate bracing against forces that are dependably outward. Nothing fishy about the design of a fish.

When startled or when pursued by predators, a squid squirts water from a tube just beneath its head and moves rapidly rearward. The water comes from a cavity between the viscera and the muscular body wall (the "mantle"); this mantle cavity is refilled through openings at each side of the head. To refill, though, the mantle must be expanded. In part, expansion involves very short muscle fibers running radially in the mantle—when not in motion a squid has little to use other than these muscles and some residual elasticity of the mantle. During motion, though, flow-induced pressures can be brought to bear. Refilling occurs back where pressure begins to turn inward again; the greatest area of the mantle covers the region of greatest outward pressure, as in figure 7.7b. As a result, a difference in pressure coefficient of about 0.25 is available for refilling. At a swimming speed of 3 meters per second this difference can provide about half the total refilling pressure, while at 9 meters per second (about 20 miles per hour—a generous estimate of top speed), flow-induced pressure would account for somewhere around 90 percent of the total (Vogel 1987; determining these numbers was why, as mentioned in the last chapter, I fussed with a model squid in a wind tunnel).

In chapter 5, the problem of depletion of underwater bubble-lungs was mentioned, along with a beetle (*Potamodytes*) that uses moving water as an evasion. Its scheme depends on the way the pressure, averaged across the surface, of a submerged body in a flow is outward—as evident in figure 7.6, the net pressure coefficient is negative whether or not the body is streamlined. So flow reduces the pressure in its bubble (Stride 1955); with sufficiently fast flow and low enough hydrostatic pressure (shallow water), the internal pressure can get subatmospheric. If, in addition, the water is well saturated with oxygen (loaded to atmospheric partial pressure), then the dissolved oxygen will spontaneously move into the bubble to resupply what the beetle consumes. The conditions are demanding, though—speeds over 1 meter per second and depths of less than 4 centimeters—which limits the trick to fairly rare habitats. Once again, that square of velocity in Bernoulli's principle is daunting.

Pressure and Drag

Recall that one of the corollaries of the pressure distribution across the surface of an object is drag. Earlier (chapter 3), a rough relationship was asserted between drag and the square of velocity at moderate and high Reynolds numbers. The origin of that v^2 is clearly related to Bernoulli's principle—where drag depends more on pressure than viscosity, drag should vary with the second rather than the first power of velocity. (We'll tiptoe around any implicit contradiction in using a notion derived for ideal fluids to say something about drag, which ideal fluids lack.) But the business of drag has further complications.

Two kinds of drag

Figure 7.6 presented a severely asymmetrical graph for pressure against location around a sphere—pressure was maximal at the upstream extremity and minimal near the widest portion, but it didn't rise again in the rear by very much. In an ideal fluid, a sphere would give a symmetrical plot, rising up to 1.0 in the rear. In a real fluid, that certainly doesn't happen. In physical terms, flow follows around to somewhere near the widest part—easily, because pressures are dropping, and flow goes naturally from high to low pressure. But the flow fails to follow the surface farther back—here pressure begins to rise again, so to follow the surface, flow would have to go toward higher pressure. Which it doesn't do—instead it "separates" from the surface near the widest point (figs. 6.7 and 7.8), and we notice a wide "wake" of disordered flow behind the object. The result is that front-to-rear asymmetry of figure 7.6 and the high drag of a sphere.

Just where separation occurs determines, as much as anything, the relative magnitude of the drag. And where separation occurs depends largely on the momentum of the fluid near the surface. At higher speeds (really, higher Reynolds numbers), velocity gradients are steeper and momentum near the surface is higher—so separation happens further aft, the wake is narrower, and the relative drag is less. For a sphere or cylinder this occurs abruptly at a Reynolds number somewhere between 100,000 and 250,000 (fig. 6.7), the value at which turbulence invades the velocity gradient near the surface. While we can blame all of this drag on pressure, specifically on the fore-to-aft pressure difference, viscosity plays an indirect but crucial role. Separation turns out to be a consequence of viscosity since, without viscosity, moving fluid near the surface would not have its momentum taxed by a velocity gradient and would continue around the surface without separation.

We therefore have to recognize two elements of this force we call drag. What we've just been talking about is *pressure drag*—it may come ultimately from viscosity but it can be calculated from the pressure distribution and varies (assuming the separation point doesn't move) with the square of velocity. In addition, another sort, something called *skin friction*, comes directly from viscosity through the fluid's resistance to the unavoidable shearing motion near any surface. It varies with the first power of velocity. At low Reynolds numbers (around one or less), skin friction is mainly what matters, generating a world that may be unfamiliar (and will get a chapter of its own) but is at least tidy. At moderate and high Reynolds numbers—that is, at values above 10 or so where separation can happen, pressure drag may be far greater than skin friction. But skin friction remains a rub—nature cannot evade the no-slip condition.

How, then, might drag be minimized at moderate and high Reynolds numbers? The best approach is to delay separation—to make flow follow the

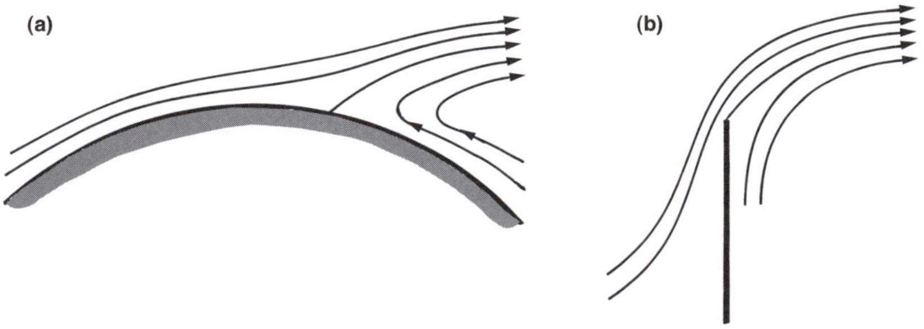

FIGURE 7.8 Separation of flow from a surface for (a) flow around a cylinder, and (b) flow past a broadside flat plate. Notice the paradoxically upstream flow just behind the separation point on the cylinder.

surface further rearward. A gentler curve near and behind the widest point is very effective, and we can see in figure 7.6 how the pressures behind rise further with such a streamlined shape. As with so much in life, it's how things end that matters most. Or, as Shakespeare put it, "All's well that ends well." Conversely, how might drag be maximized? One makes the wake as wide as possible. A flat plate broadside to flow or a hollow hemisphere concave upstream (a parachute, for instance) has a wake substantially wider than the object itself—flow heads outward and its momentum keeps it going beyond the edge of the object. A rounded front, although far from fully streamlined, gives less drag than a flat front, as truck and car designers have recently begun to appreciate.

Drag is drag—a force on an object carrying it back in the direction of flow, a rate of transfer of momentum from fluid to object. But one sort, skin friction, depends mainly on total exposed surface. The origin of the other, pressure drag, would be simpler if it depended just on frontal area, but in fact it's a complex function of both shape and Reynolds number. Thus, drag as a whole is still more complexly dependent on Reynolds number. At low Reynolds numbers streamlining does almost no good—any reduction in residual pressure drag is offset by a greater total exposed surface. Microscopic swimming creatures don't often come with the familiar rounded front and tapered rear. Fish, squid, whales, the larger swimming insects and crustaceans—these by contrast are inevitably streamlined. And streamlining can make a world of difference at moderate and high Reynolds numbers. Just how much turns on the particular comparison—the length-to-diameter ratio of the streamlined body, the unstreamlined body used for reference, and the particular Reynolds number. It also depends on how we compare bodies—whether we compare bodies of equivalent frontal area, overall surface, or total volume. But rather than perhaps 10 or 20 percent, streamlining decreases drag around a full order of magnitude, whatever the basis of comparison.

The drag coefficient

All the complications mean that for almost no shape do we have a simple and general formula for drag as a function of speed at moderate and high Reynolds numbers—we have no handy analogs of Stokes' law (chapter 11) for Reynolds numbers below about one. So we're driven to the use of graphs and tables, and our task comes down to figuring out the most efficient presentation. It's obviously impractical to plot drag separately against all the factors that might affect its value—speed, size, density, and viscosity—for each shape of interest. The Reynolds number collects these quantities into an appropriate composite variable, so we might plot drag against Reynolds number. But it turns out that, even for a given shape, a particular Reynolds number doesn't give a unique value for drag.

Still, something related to drag does vary uniquely with the Reynolds number. Not to be coy about the variable, it's essentially a version of the pressure coefficient, introduced earlier (equation (7.12)). Recall that the pressure coefficient, dividing measured pressure by dynamic pressure, provided a dimensionless pressure adjusted for the density and speed of the medium. Drag, of course, is just pressure times area; the drag on an object is the average pressure difference across its projected area multiplied by that area. ("Projected area" is the area of the shadow that a collimated light shining in the direction of flow would cast.) So if we divide drag per area by dynamic pressure we get something dimensionless; this not-so-new dimensionless variable is the "drag coefficient," touched on in chapter 4. Thus,

$$C_d = \frac{D}{S}\frac{2}{\rho v^2}. \tag{7.13a}$$

To put the formula in a form analogous to equations (7.1) and (7.2),

$$D = \frac{C_d \rho S v^2}{2}. \tag{7.13b}$$

This drag coefficient turns out (for a given shape) to vary only with the Reynolds number. Thus, a graph with the drag coefficient on the *y*-axis and Reynolds number on the *x*-axis "says it all" with regard to the drag of an object of a particular shape and orientation. Note that equation (7.13) is no "equation for drag"; it's just a formula for converting drag to drag coefficient and vice versa.

A plot of drag coefficient versus Reynolds number for a sphere or cylinder (fig. 7.9) has bumps and discontinuities that correspond satisfyingly to the

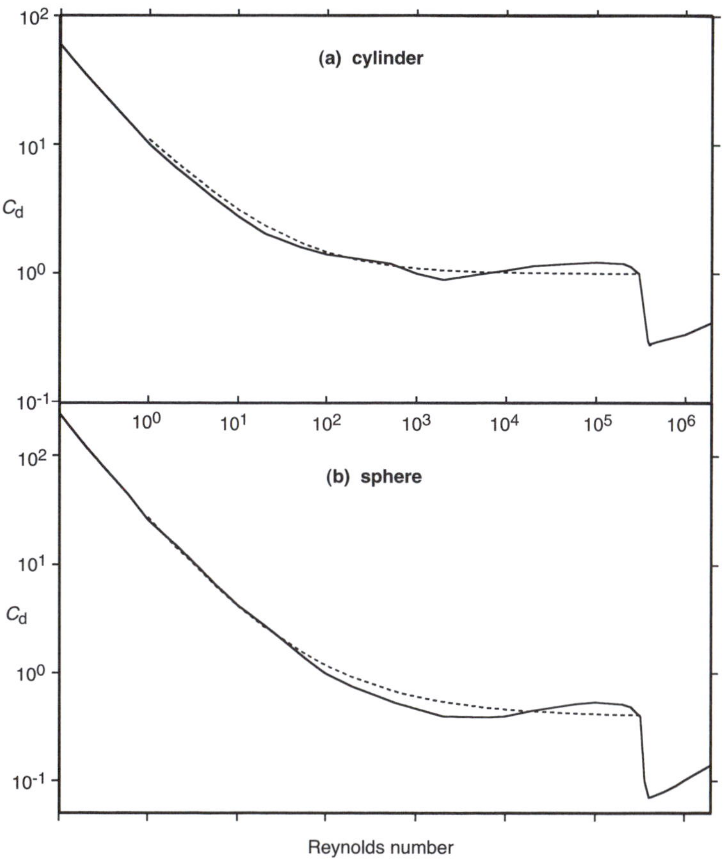

FIGURE 7.9 Drag coefficient (based on frontal area) versus Reynolds number for flow (a) across a long circular cylinder, and (b) over a sphere. The dotted lines provide adequate estimates for the somewhat irregular shapes the biologist encounters.

changes among the regimes of flow illustrated in figure 6.7 and described in the last chapter. The graph itself, though, takes some getting used to—one dimensionless variable has been plotted against another. Viewing it as drag per speed squared versus speed may help the intuition. The generality obtained excuses its obscurity. From the drag coefficient and the Reynolds number for a given shape you can quickly figure the drag for any combination of the variables in these expressions—any size, any speed, any fluid medium—just plug 'n chug with equations 7.13 and 6.5. And these drag coefficients (at least for regular shapes) are matters of public record in handbooks and compendia such as that of Hoerner (1965).

One final element of complication must be admitted. The explanation of drag coefficient presumed that the area involved was the projected area. Any area, though, preserves dimensional propriety; and sometimes other reference areas are used. Projected, or "frontal," area in the direction of flow finds most use for high-drag bodies (such as automobiles)—which is why low-drag sport cars have coefficients only a little lower than high-drag minivans. For streamlined bodies the commonest is total surface ("wetted area"), ignoring any fractal character of the surface. For lift-producing bodies such as wings, we encounter the area at right angles to flow ("plan form area"). For blimps and other airships, and increasingly for organisms where the payload consists of guts and gonads, a strange but especially relevant area comes into use. It's the nominal area created by taking the two-thirds power of the volume of blimp or bluefish. So any use of equation 7.13 has to specify the area of reference—the same drag will translate into a different coefficient if a different reference area is used. Don't forget—reality is drag, not its coefficient.

Drag and flexibility

For some objects, as figure 7.9 shows for spheres and cylinders, the relationship between drag and speed or between the drag coefficient and the Reynolds number is a bit volatile. For others, such as a round disk perpendicular to flow, the drag is almost precisely proportional to the square of speed. For such a disk the relationship proves to be regular enough ($C_d = 1.2$ using frontal area, for *Re*s between 10^3 and 10^6) so its drag can be a reliable and regular indicator of speed. Why the disk gives such a tidy result presents no great mystery—with a sharp outer edge, the separation point can't creep back and forth to complicate matters. The drag coefficients of many objects of biological interest, such as a leafless tree in a winter storm or a hard coral, behave in yet a third manner. They vary, but without abrupt changes, since what odd transitions may happen no longer happen at the same speed (even if at the same Reynolds number) for all parts of the object.

As noted earlier, streamlining lowers drag coefficients fivefold to fiftyfold, depending on the Reynolds number and the area used as reference. Moreover, the drag increases less sharply with speed. At least for certain Reynolds number ranges, the drag of unstreamlined objects ("bluff bodies") varies with roughly the square of speed. By contrast, at fairly high Reynolds numbers, the drag of streamlined objects increases with speed to the power 1.5. Thus, the drag coefficient drops with increasing Reynolds number ($\propto Re^{-1/2}$) rather than remaining nearly constant. As mentioned, streamlined shapes are very much the rule among macroscopic swimmers and all but the smallest fliers. Indeed, the first deliberately streamlined shape, contrived by George Cayley (1773–

1857), was based on girth measurements of trout and dolphin (Von Kármán 1954; Vogel 1998a).

What's at least initially puzzling (like Sherlock Holmes's dog that didn't bark) is the comparative *un*commonness of streamlined shapes among sessile, attached organisms, even those exposed to potentially hazardous forces from winds and water currents. One obvious difficulty is that a streamlined shape achieves its low drag only in a very specific orientation with respect to flow, and an attached organism must commonly contend with erratic flows. And the flows are not only directionally erratic but variable in intensity—a low-drag shape may be crucial in a storm or surge but needlessly compromising of other functions otherwise. Obviously, drag minimization cannot be an end in itself—it isn't exactly the common keystone of reproductive success.

I know of only one streamlined, attached organism whose drag has been investigated—a marine alga, *Halosaccion*, of the rocky intertidal areas of sheltered shorelines in the Pacific northwest. The body ("thallus") of the alga is about the shape and size of a cigar; it's known (at least at the Friday Harbor Laboratory in Washington State) as the "sea condom." The creature manages the problem of orientation in variable flows with a short but flexible kite-string (the "stipe"), so it repositions continuously like a windsock. Its drag coefficient (using frontal area) is 0.084 at a Reynolds number of 40,000, fully fourteen times lower than that of a perpendicular disk of the same frontal area (Vogel and Loudon 1985). Put to it, nature clearly can do a good job with conventional streamlining in a sessile organism.

But the usual approach for an attached organism involves a clever combination of solid with fluid mechanics. A tree responds to the wind with what one is wont to call a "deformation." Since the deformation results in a dramatic reduction in the drag of the tree, perhaps we ought to stop using a term with unavoidably pathological overtones—"reconfiguration" would be better. The key is flexibility. An organism occasionally subjected to strong forces from flow may respond by accepting some temporary loss of other activities (photosynthesis, feeding, and reproduction come immediately to mind) as its shape briefly changes. This seems to be the basic tactic of trees, as noted back in chapter 1.

Two points about drag reduction by reconfiguration ought to be borne in mind. First, conventional drag reduction, as by streamlining, works over a wide speed range. By contrast, reconfiguration depends strongly on speed—the organism employs the wind to force (quite literally) the critical shape change. One has to look both at drag coefficients per se and how those coefficients (or drag itself) change with speed. And second, these drag coefficients are easiest to calculate and their consequences most readily envisioned if the areas used are the original ones for the still air or still water form. Adjusting area as reconfigurations proceed complicates everything; moreover the biologically relevant area ought to be that available for photosynthesis, filtration, or whatever the basic fair-weather function.

What range of exponents characterize the way drag changes with speed for flexible organisms in general? Bear in mind that such exponents are a kind of scaling factor, emerging from log-log plots in the same manner as the factors cited back in chapter 3. Ignoring data for very low speeds on the grounds that drag would present no problem, the exponents are far below the value of 2.0 found in rigid, unstreamlined bodies—they're on the order of 1.0 for plants and animals, both terrestrial and marine—softwood and hardwood trees, a large sea pen and a tiny colonial hydroid (both coelenterates), a diversity of marine algae, a host of freshwater algae, insect larvae in streams, and others (see Vogel [1984, 1994a]).

Further flourishes. Telewski and Jaffe (1986) showed that high-altitude fir trees, which normally experience substantial winds, are stiffer than low-altitude trees of the same species. The result is puzzling unless one bears in mind the possible loss of other functions due to reconfiguration. Perhaps the tree afflicted with chronic wind, like those of us with chronic illnesses, must manage to be adequately functional (photosynthetic, for them) in all but the most extreme episodes. The work of Telewski and Jaffe also draws attention to a general phenomenon among terrestrial plants, something called "thigmomorphogenesis" that gets fuller treatment in chapter 25. Just as bones or muscles enlarge when repeatedly or chronically used, plants alter their structure in response to mechanical loads, of which drag is as important as any.

Shape and Drag—Motile Animals

Overstating only a little, one can declare that while drag matters for some organisms that remain attached to a substratum, drag matters for all organisms that swim or fly. Economics makes the difference. Work need not be expended to counteract one force with another, and that's all the attached organism must do. Structure must be synthesized, but (maintenance aside) that's a one-time expenditure. By contrast, the motile creature has to expend power, force times velocity, which can rarely be done without cranking up the metabolic machine.

One can further recognize that flexibility plays a smaller role in minimizing the problem, while conventional streamlining plays a larger part. Locomotion up in the air or down in the water establishes the direction of flow. The organism always faces upstream relative to its egocentric reference frame, occasional queries about feather-ruffling problems of downwind-flying birds notwithstanding. And streamlining works far better when a body maintains a specific in-your-face orientation to flow. Still, flexibility does matter; indeed, it greatly complicates determinations of the drag of swimmers and fliers.

Streamlining reduces pressure drag, ultimately to insignificance. That permits us to establish a gold standard for streamlining—one, though, that sweeps aside a lot of relatively relevant matters. The ultimate streamlined object consists of a rigid plate of zero thickness (easy) oriented parallel to flow (also easy)—it experiences only skin friction. For simplicity, we consider a very long plate extending indefinitely across the flow, that is, one with only a flowwise dimension, and look at drag per unit cross-flow length. From drag per length one can get drag coefficients, based on wetted area (meaning both sides of a plate). Good formulas exist for these coefficients as functions of the Reynolds number for two conditions; they set the lines shown in figure 7.10. The distinction between laminar and turbulent flow depends both on the smoothness of the plate and on that of the oncoming flow. At low Reynolds numbers, flows are laminar unless deliberately stirred up, while at high Reynolds numbers, increasingly heroic measures need be taken to maintain laminarity.

Curiously, a crossover occurs at a Reynolds number of about 17,000. Below this value, turbulent flow gives lower drag; above this value, you're better off flow is laminar. Turbulent flow below 17,000 probably represents

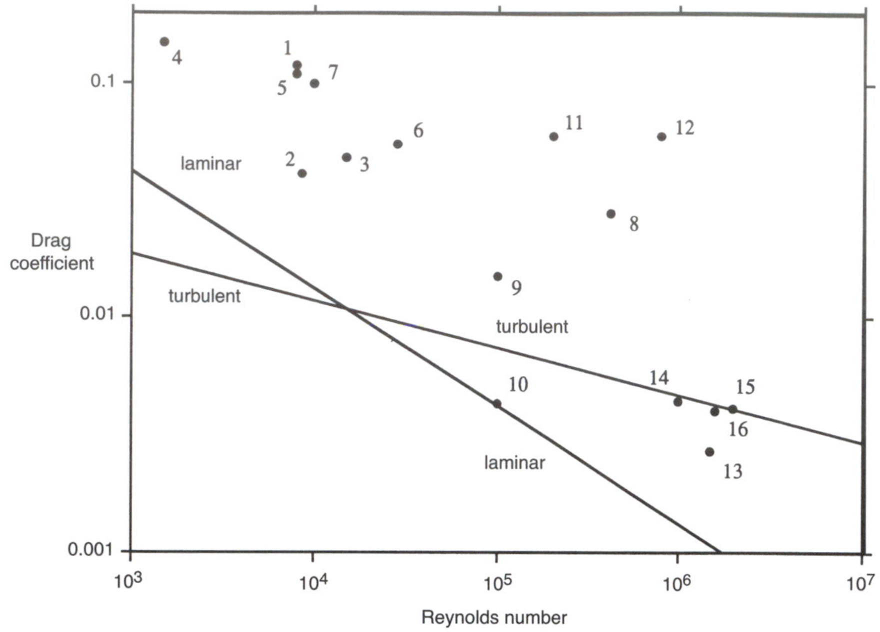

FIGURE 7.10 Drag coefficient (based on wetted area) versus Reynolds number for long flat plates whose widths (chords) are oriented parallel to flow. The equation for the less steep, laminar line is $C_d = 1.33Re^{-0.5}$; that for the steeper, turbulent line is $C_d = 0.072Re^{-0.2}$ (sources in Vogel 1994a). The animals are: (1) desert locust body; (2) large beetle body; (3) larger beetle body; (4) bullfrog tadpole; (5) small frog; (6) medium frog; (7) crab; (8) duck (submerged); (9) trout; (10) mackerel; (11) pigeon body; (12) vulture body; (13) large penguin; (14) smaller penguin; (15) sea lion; and (16) harbor seal.

no practical option for organisms—flow normally shifts from laminar to turbulent at around 500,000. But staying on the laminar curve above 500,000 does seem to happen; still, one has to say "seem" in light of a lot of bedeviling complications. The graph of figure 7.10 includes some biological data points, and they run high—bearing in mind that a little goes a long way on a logarithmic scale! Part of the problem, a small part as it happens, has an innocent explanation. A flat plate has no thickness, as the biologist might say, for guts and gonads, and as the engineer might note, to give flexural stiffness. As the body fattens, the drag coefficient (again, based on surface area) increases. Consider a streamlined strut—our long, flat plate now with rounded front and sharpened rear. If its thickness is 6 percent of its flow-wise length ("chord"), its drag goes up 20 or 30 percent; if its thickness is 12 percent, drag increases about 50 percent.

Beyond that, though, lies a slough of uncertainties. What's the drag of a swimming fish? If the fish swims at constant speed, forces must balance, and thus drag equals thrust. The problem, though, is that for most fish the same structures contribute to both. The simplest solution consists of towing, say, a pickled herring through a tank of water. That might be reasonable for a stiff-bodied mackerel or tuna, which produce most of their thrust with their tails, and such fish yield especially low drag data in towing tests. But we have reason to believe that the drag for towed flexible fish exceeds their drag when swimming. A somewhat better but similarly problematic approach involves calculation from speed changes during gliding—the greater the drag, the greater the deceleration. Here the problem lies in equating drag while gliding with drag while actively swimming. Yet another approach combines recent advances in both particle imaging and computing to look at the strength of the vortices in the wake that represent thrust, which, again, must equal drag. Still, one has to bear in mind that both drag and thrust affect the wake of a fish, albeit in different places. And a fish with a Froude propulsion efficiency of 100 percent would leave no wake at all since the momentum change due to drag would exactly offset that due to thrust. One cannot, as an alternative, use data from fish pushing against or pulling on force transducers—that's unreal in yet another respect.

The problem proves less severe for flying animals. The wings produce thrust, the body incurs drag. So one can remove the wings and measure the drag of bodies in wind tunnels; those drag data times flying speed give the thrust that the wings need to generate. But fliers, too, give trouble. In all cases, wings and body interact in ways that are hard to evaluate. For birds, the feathers present a special problem. The data one gets for the drag of bird bodies in wind tunnels seem too high for their shapes or for the argument that they should be well streamlined—as George Cayley assumed two centuries ago. Coefficients up around 0.4 (based on frontal area) look scarcely better than those for spheres, as in figure 7.9. Examination of data from stuffed bird bodies suggests that feather control must be important in keeping down the drag of living birds. Pennycuick et al. (1996) found that bird bodies did better with a bit of hair spray—the drag coefficient of a snow goose body dropped from 0.33 to 0.28. The lowest reliable value seems to be that of Tucker (2000), who measured 0.11 on a featherless model of a peregrine falcon's body. These, of course, are high-speed specialists. So perhaps values ranging between 0.17 and 0.77 obtained from field measurements on diving passerine birds (Hedenström and Liechti 2001) aren't unreasonable.

Clouding the issue further is a peculiar general issue. Should a flying bird be terribly concerned about the drag of its body? On the face of it, of course. Maybe, though, body drag isn't as important as it sounds. The wings incur drag, lots of it, in the process of producing lift and thrust, as will be explained further in chapter 13. The main difference from body drag is that the drag of beating wings, as a set of vectors following the local flow, has a complex and time-dependent direction. But it's still drag, and the cost of keeping the wings beating may dwarf that of moving the rest of the body forward.

Tricks for Dropping One's Drag

Fast swimmers, again, might be expected to take all reasonable steps to minimize their drag—drag should be a selective force whether fitness depends on maximizing peak speeds or minimizing the cost of sustained locomotion. And we have no reason to assume that conventional streamlining exhausts the devices for drag minimization. Perhaps no aspect of aquatic animal locomotion or biological fluid dynamics as a whole has received the attention of such a diverse community of investigators. The great stimulus, of course, comes from possible applications to human technology. At this point, though, the hypothetical outweighs the actual, the apocryphal outweighs the experimental—too much hope and hype, in monosyllabic terms. Still, we now know of a wide variety of devices that may be both useful and used; we just don't have a good sense of exactly where and how much.

Dimpling

One device that we make some use of has not been found in nature. Golf balls with dimples have lower drag and thus go further than smooth balls. But either way, they're high-drag spheres rather than decently (or even crudely) streamlined shapes. Moreover, the mechanism behind the dimples operates under severe restrictions. Dimples shift the transition from laminar to turbulent flow right near the ball's surface and thus permit fluid to flow further around the ball before separating, so the ball's wake is narrower. In effect, they reduce the Reynolds number of the onset of that near-surface turbulence from around 200,000 to slightly under 100,000. They thus reduce the speed of the great drop in drag coefficient shown on the right side of figure 7.9b in what happens to be the Reynolds number range of a well-struck drive. Might tree trunks use the scheme? It might make a slight difference for a whole, leafless tree. While trunks over about 10 centimeters in diameter hit $Re =$ 200,000 at too low a speed for the roughening to be of much use, some branches might get some help some time. If looking for cases, one ought to focus on situations in which flow direction cannot be anticipated and on stiff and unstreamlined structures. Dimpling, after all, deals with pressure drag, of which streamlined bodies have little to begin with.

Flow-wise ribbing

Fine ridges running in the direction of flow occur as parts of the scales of fast-swimming sharks and the shells of scallops. Tests of artificially produced microgrooved surfaces indicate a drag reduction of 7 or 8 percent beyond conventional streamlining, and a commercial product ("Riblets") was briefly available for use on racing boats. Ridges and grooves appear to work, at least in part, by breaking up cross-flow vortices near the surface and thus reducing skin friction at high Reynolds numbers. Just how much benefit sharks realize remains uncertain, for reasons already discussed.

Mucus

Adding some long-chain polymeric molecules to a flow can also reduce the skin friction of a surface. The effect traces, so it appears, to gentler velocity gradients and thus less surface shear. On the one hand, the effect has been well documented in pipes and for flow over test bodies; much is known about the kinds of molecules that are most effective; and fish as well as other animals produce surface mucuses of the appropriate kinds. On the other hand, demonstration of drag reduction by mucus secretion in nature has remained problematic. Quite possibly the cost of making mucus in sufficient quantities may preclude its routine use, so the trick may be reserved for times when imminent and involuntary transfer to the next trophic level offsets any issue of cost. One complication is that surface mucuses have other functions as well, stopping in its tracks any tidy correlations between locomotory habit and their use, such as we noted for ribbing in sharks.

Compliant surfaces

No mechanism has had more publicity or generated more controversy—it could provide a fine case study for a social scientist. Part of the noise must reflect its origin—James Gray (1936), for long the most notable name in biomechanics, calculated that dolphins (and some other small cetaceans) could not produce sufficient power with their muscles to swim as fast as they did unless flow along their surfaces remained laminar. At the operative Reynolds numbers, above 10,000,000, that demands some special mechanism. Worse, the problem got named—Gray's Paradox—apparently by Parry (1949). A solution was proposed by Kramer (1960)—that dolphin skin, known to change texture locally during swimming, damped (not "dampened") incipient vortices and thus maintained laminar flow. Worse still, Kramer had a product to sell, his "Lamiflo" coating. Subsequent tests cast severe doubt on the efficacy of the coating but (of course) provided additional publicity for the problem. And even worse, Cold War paranoia afflicted both Pentagon and Kremlin in the form of wildly exaggerated estimates of the speeds of each other's submarines and the results of each other's classified work on compliant coatings.

Recent work should have relegated Gray's Paradox to its proper status as historical footnote. The practical power output of a muscle-powered organism, unlike that of most of the engines of our technology, depends strongly on how long the organism must sustain that output—as we'll explore more fully in chapter 22. The ability to deliver oxygen rather than intrinsic contractile power becomes ever more limiting after a few tens of seconds, which is why record speeds for human running depend so much on the length of the race. Not only do cetacean muscles work well, but they may have an even stronger duration-dependency than we. And top cetacean speeds, notoriously hard to measure anyway, vary a lot with how long the animals sustain them. In other words, cetaceans don't swim as fast as commonly believed except for very short bursts. Calculations based on realistic speeds and durations (see, for instance, Lang and Pryor [1966], Hui [1987a], and Fish and Hui [1991]) tell us that the muscular performance of small cetaceans should be adequate to offset the drag of turbulent flow.

Still, one has to add, recent work doesn't deny that dolphins swim with especial hydrodynamic efficiency but merely that they needn't. Dolphin skin remains pretty special stuff, the drag of a swimming (as opposed to coasting) animal remains uncertain, and efficient locomotion ought to matter to the fitness of these animals. Work continues, if in a less supercharged context (see, for instance, Gad-el-Hak [1996]).

Other mechanisms

Of which no shortage of candidates exists. *Surface heating* ought to reduce the viscosity of water near an animal's skin, which would decrease skin friction, and fast-moving animals often need to dump heat. But calculations suggest that the effect should be negligible—at least barring clever local heat release at some hypothetical critical location. *Injection of fast water along the surface* is done by fish, which push water out their gill coverings (opercula), with a drag reduction for a trout near top speed of about 10 percent (Steffensen 1985). *Surface roughening* in critical regions remains a possibility for a variety of fast fish. Lighthill (1995) found evidence of *active control*, in which pressure detectors along a fish fine-tune its head turning in order to reduce drag. Good reviews of the range of possibilities have been provided by Bushnell and Moore (1991) and Fish (1998).

One must bear in mind the different quantitative perspectives of biologist and biomimeticist. We're wondering about the major factors that determine drag, about variations of 50 percent or more. For someone interested in swimsuits or racing yachts, a 5 or 10 percent improvement is monumental, and even 2 percent is serious business. But for me the bottom line remains as I put it earlier (Vogel 1994a): "A claim of drag reduction should be viewed with skepticism until (1) it has been tied to a plausible physical mechanism, (2) has been shown to work on physical models under biologically relevant conditions, and (3) has been shown to work by some direct test on real organisms under controlled and reproducible conditions."

CHAPTER 8

Fluid Events near Surfaces

> But somewhere, beyond
> Space and Time
> Is wetter water,
> slimier slime.
>
> **RUPERT BROOKE**

AS PROMISED IN CHAPTER 6, I'VE DELIBERATELY AVOIDED USING THE term "boundary layer" when talking about flow near surfaces, opting instead just to allude to velocity gradients. Velocity gradients must be ubiquitous; that's demanded by the unavoidable no-slip condition. Although closely related, a boundary layer isn't quite the same thing. While definitionally specific, it's considerably slipperier as a matter of practical reality. Properly construed, though, the boundary layer proves an important and useful concept.

The Laminar Boundary Layer

What we most often want to know about a gradient is its steepness, the rate of change of one variable, here speed, with another, here distance, outward from a solid-fluid interface. That can be viewed as the slope of the speed versus distance plot or, in the formal terms of first algebra and then calculus,

$$\frac{\Delta v_x}{\Delta z} = \frac{dv_x}{dz}. \tag{8.1}$$

Note that subscript x. It makes explicit that we're talking about flow in the normal downstream or x-direction, even if we're at the same time considering the gradient–how flow changes–in a cross-flow or z-direction.

So how does flow change as one moves outward, but never very far, from a surface? Consider a flat surface oriented with the flow. The speed of flow begins (nearest the surface) by increasing linearly from zero—that is, in proportion to distance, with a constant slope, as we saw back in figure 6.3 and reiterate here (with a little more detail) in figure 8.1. After a while the rate of increase gradually begins to diminish, shifting from linear to asymptotic. So it asymptotically approaches the speed of the fluid that's unaffected by the presence of the plate—the "free-speed" flow. But bear in mind what "asymptotic" implies: flow never truly reaches free-stream speed. That implies, of course, that the presence of a surface anywhere denies the reality of free-stream speed everywhere—a bit of sophistry we choose to ignore. The implication that does matter, though, is that any definition of the thickness of the gradient region must be a bit arbitrary. Since speed in the gradient region

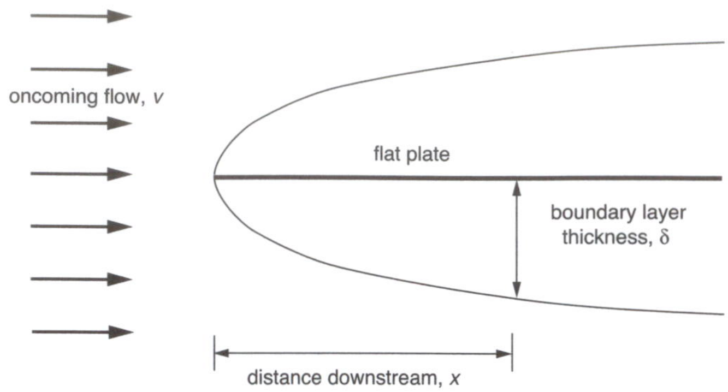

FIGURE 8.1 The variation of the thickness of the boundary layer on a flat plate with distance from the upstream edge. The velocity gradient gets less steep as one moves downstream. The shape of the curve may be real, but the line as drawn is arbitrary in that it represents no physical discontinuity in the flow.

never quite equals free-stream speed, we have to decide how close to free-stream speed amounts to having reached it.

The classical definition was established by Ludwig Prandtl early in this century (Von Kármán [1954] gives an engaging historical account), who set the outer limit as where the speed achieved 99 percent of the final free-stream value, so we can speak of that convention as the "Prandtl boundary layer" (Prandtl and Tietjens 1934). Using that definition, the following formula gives its thickness:

$$\delta_{0.99} = 5\sqrt{\frac{x\mu}{v\rho}}. \tag{8.2}$$

Here x is the distance downstream from the leading (upstream) edge of the flat plate, and v is free-stream speed; μ and ρ are, as previously, viscosity and density. So the boundary layer grows in thickness as one moves further downstream (x); it's thinner at higher free-stream speeds (v), and it varies with the ratio of viscosity to density, the so-called kinematic viscosity. All nicely tidy and intuitive.

Better yet—these variables ought to be pretty familiar from the last two chapters. So we can rewrite the formula to put boundary layer thickness in dimensionless terms, expressing its thickness relative to the distance downstream, or δ/x:

$$\frac{\delta}{x} = 5Re_x^{-0.5} \tag{8.3}$$

We thus see that this dimensionless boundary layer thickness varies only with our familiar dimensionless Reynolds number. We just add the subscript x and call the variant the "local Reynolds number," since the distance used is that downstream from the leading edge rather than some overall size of the plate. The formula declares that, *relatively speaking* boundary layers are thicker at lower Reynolds numbers—in slower, smaller, gooier realms.

But the ointment has several flies; indeed one can so rarely make explicit use of the formula that it provides more of an intuitive guide than something to reckon with literally. One has to shed any notion picked up in elementary physics courses that physical science (by contrast with biology) provides simple and universally applicable equations for reality. Perhaps physics courses select those aspects of reality for which such equations can be proudly presented. And perhaps that underlies the omission of much talk about fluids in such courses. In any case:

1. We've assumed a flat plate parallel to flow, which means that all the resistance it imposes on the flow comes from skin friction and none from pressure drag. In particular, separation never rears its ugly head. (But a few degrees of tilt don't make a lot of difference, since flow will tend to follow the surface.)
2. We've assumed that the plate has a discrete, specific, and sharp upstream edge. That puts the velocity gradient upward from the earth's surface off limits—it's messier and has to be handled differently.
3. We've assumed that flow is everywhere downstream, that is, in the x-direction. But, as you can now figure out, that amounts to a violation of the principle of continuity. If flow in the boundary layer slows down as it goes downstream, its streamlines must diverge, which is to say that flow must have a z-component. The lower the Reynolds number, the larger the error, so the formula can't be taken too literally below about 500.
4. We've assumed that flow in the boundary layer is laminar, making a specific rather than simply statistical statement that bits of fluid move downstream. That gets us into trouble if the Reynolds number is over about 500,000. For higher values, boundary layers are turbulent, and another formula must be invoked.

Finally, the arbitrary element in the definition should be emphasized. The figure of 99 percent has biological, not physical, roots—the exact value is an accident of our bipedal stance and pentadactylic forelimbs. Prandtl wanted a generous definition. To facilitate calculating drag, he was making a distinction between the main part of a flow, where viscosity could be ignored and any region where viscosity might be doing anything significant. If, instead, one wants to know whether the spore-bearing stalk of some fungus extends out into substantially full wind, one might pick a value of 90 percent rather than 99 percent of free-stream speed for the outer limit (Silvester and Sleigh 1985). The resulting boundary layer will be only three-fifths as thick—a constant of 3 will replace the 5 in equations (8.2) and (8.3).

Perhaps more useful for the biologist interested in what flow a small, attached organism might face, is a formula for the speed distribution within the boundary layer. One for the entire layer is a mess—there's nothing algebraically tidy about a function that starts linear and then goes asymptotic. By contrast, a simple one describes how the horizontal velocity, v_x, varies as one goes outward (z-ward) from the surface in the inner, linear portion (roughly half) of the whole 99 percent layer:

$$v_x = 0.33zv\sqrt{\frac{v\rho}{x\mu}}, \qquad (8.4)$$

(v, again, is free-stream velocity). The formula says that the speed increases at a given location with free-stream speed to the power 1.5. So boundary layer speeds may be lower, but they increase faster than free-stream speeds—the layer itself is thinning, so as the free stream increases, a given place is (relatively) less deep within the layer. That has implications for the drag experienced by attached organisms. And, as one might have guessed from equations (8.2) and (8.3), speed decreases as one moves further downstream (x-ward) at a given z.

Incidentally, equation (8.4) provides an alternative way to define the thickness of a boundary layer, one that closely corresponds to the 90 percent layer mentioned just above. Imagine if the inner, linear portion extended outward, with its speed increasing steadily until it equaled that of the free stream. If the distance outward (z) at that point defines the boundary layer thickness, a little algebra (setting $v_x = v$ and $z = \delta_{0.90}$) gives us:

$$\delta_{0.90} = 3\sqrt{\frac{x\mu}{v\rho}}. \qquad (8.5)$$

Drag and the Boundary Layer

Quite obviously, a boundary layer can provide local shelter from flow. Whether that's for better or worse depends on whether drag detaches a seed or spore and flow disperses it, or whether drag damages by dislodging and flow carries something away from its proper habitat. But either way, drag and boundary layers can interact in complex and biologically significant ways.

Perhaps we can best introduce the matter with some idealized examples. Let's look at the three related cases shown in figure 8.2—first, the drag of a nonstreamlined body out in the free stream; second, the drag of the same body well down within the boundary layer; and third, an effective minimum for the body's drag. The exercise ought to put some reality around some of the equations that have been bandied about.

First, consider the drag of a sphere, a millimeter in diameter, exposed to a free-stream water flow of 0.5 meters per second. That corresponds to a Reynolds number of 500 (equation (6.4)). From figure 7.9a or an equation fitted to it (Vogel 1994a), we get a drag coefficient (based on frontal area) of 0.70. And that, by equation (7.13), translates into a drag of 69 micronewtons. (Or

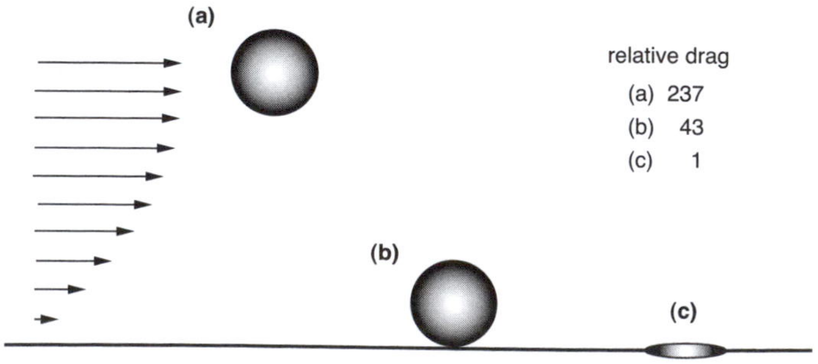

FIGURE 8.2 The drag of three bodies. (a) A sphere in the mainstream; (b) the same sphere in the lower portion of a laminar boundary layer; and (c) a circular flat plate right on a surface, essentially the area formed by the shadow of the sphere.

the weight of a 7-milligram mass—we're working at a biologist's rather than an engineer's scale.)

Second, consider the drag of the same sphere, now resting on the surface of a flat plate parallel to flow, with the same free-stream speed. We'll assume that it's 10 centimeters downstream from the edge of the plate. At that point, the boundary layer, by equation (8.2) is 2.2 millimeters thick; a local Reynolds number (using 0.1 meter as length) of 50,000 means that the equation ought to be applicable. With a boundary layer of that thickness, the ball sits safely in the lower half, and equation (8.4) should describe the velocity gradient. To get the average speed on the ball (good enough for present purposes), we use the ball's radius and that equation—it comes out to 0.18 meters per second. Now using the average speed on the ball and the ball's diameter, the Reynolds number is 185; the corresponding drag coefficient is 0.94. And that gives a drag of 12.6 micronewtons, five and a half times lower than the drag of the sphere when in the free stream.

Third, consider the drag of a nonprotruding patch on the surface of a flat plate parallel to flow, a patch of the same area as the sphere's cross section. At the surface, the velocity is zero, in accord with the no-slip condition. That suggests zero drag. But, in fact, the minimum drag on something in a flow cannot be zero—fluids have viscosity, and, while the speed is zero, the velocity gradient isn't. Equation (8.4) says as much; and converted to the slope, v_x/z, it gives just that velocity gradient. Multiplying that velocity gradient by the fluid's viscosity gives the shear stress on any bit of surface at a distance x from the upstream edge of the plate. That shear stress times the area of the patch gives its drag, called the "local drag" to go along with local Reynolds number. Of course skin friction bears full responsibility, with pressure drag, the main item for the spheres, now out of the picture.

Just as in the last chapter we defined a drag coefficient to provide a dimensionless drag for a whole object, we can use equation (8.4) to define a "local drag coefficient" for a small patch of a surface of area S:

$$C_{dx} = \frac{2D}{\rho S v^2} = 0.66\sqrt{\frac{\mu}{v x \rho}} = 0.66 Re_x^{-0.5}. \tag{8.6}$$

(Remember that drag coefficient is still drag divided by dynamic pressure, even if skin friction is the culprit; the doubling of the coefficient from equation (8.4) comes from the ½ in the dynamic pressure, $\rho v^2/2$.)

If our patch, assumed to be a circle a millimeter across, is the same 10 centimeters downstream, the local Reynolds number remains 50,000. So the local drag coefficient is 0.003, and the drag of the circular patch is a mere 0.29 micronewtons. While not zero, the drag on a patch well downstream from a leading edge doesn't amount to much—it's over two hundred times less than that of an unstreamlined body of the same diameter in the free stream.

(Parenthetical item. Integrating equation (8.6) from leading edge downstream—thus with respect to x—gives the equation for the drag coefficient of one side of a flat plate in laminar flow, as in the legend for figure 7.10. Good science provides a satisfying web of interrelationships, one too easily obscured by a necessarily sequential account.)

Silvester and Sleigh (1985) gave a lot more guidance and formulas for calculating the drag on tiny things near and on surfaces, including cylinders with long axes both normal and parallel to flow as well as flat plates.

Living in Boundary Layers

One can too easily dismiss all the talk about what happens on flat plates oriented parallel to flow as a bit far from common reality. Certainly we can't casually assume strict applicability of the various equations that apply to that situation. But they give at least a semiquantitative view of life near the surface of an unbounded substratum—the ground beneath a wind, the bottom beneath a current. They give a useful approximation of what happens upstream from separation points—the situation of attached organisms that live as epifauna on swimming animals or on rocks that stick up into the flow of streams. And they may apply with greater specificity to microscopic life on the leaves of trees and the fronds of macroalgae. Flexibility of form on such a macroscopic level provides surfaces that maintain orientation parallel to flow even as the flow itself changes direction. In all such situations, free stream velocity bears only indirectly on an object. A few examples ought to illustrate the range of biological relevance.

Hiding on a surface

We saw that while something that didn't protrude at all from a substratum still suffered drag, drag on the surface in the form of skin friction amounted to no great shakes. That must provide effective shelter for flat flora and fauna, reducing their need for sturdy attachment. We're talking about a diverse batch of little, but not microscopic, things, some of which appear in figure 8.3—

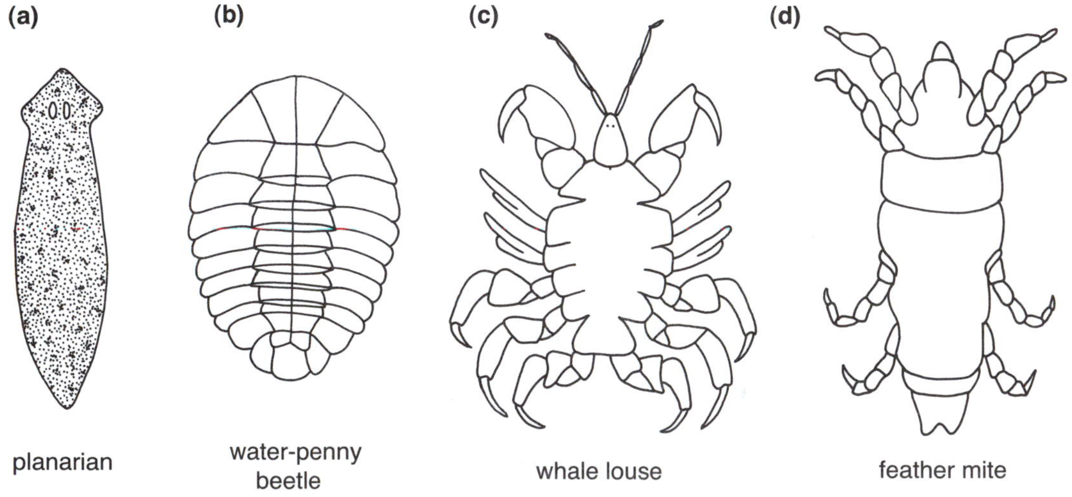

FIGURE 8.3 Some of the flat creatures that hide on surfaces. (a) A flatworm or planarian, (b) a water-penny beetle, (c) a cyamid or whale louse, and (d) the feather mite of a large seabird.

flatworms (planarians) and water-penny beetles on the rocks of streams, barnacles and cyamids ("whale lice") on cetaceans, and feather mites on large seabirds. What must be mentioned (and will reappear in chapter 12) is that protruding just a little may raise a problem of lift that's worse than a bit of drag. A given force can be more easily resisted if, as does drag, it's directed downstream and loads the attachment in shear than if, like lift, it pulls upward, and peels with tension chapter 21). Enough protrusion to cause separation makes lots more drag, but it eases the stress of lift—so racing cars have unstreamlined "spoilers" atop their hind ends to help keep the rear wheels on the ground. That seems to be the ploy of mayfly nymphs in streams and limpets of both freshwater and marine persuasions (Statzner and Holm 1982; Denny 1989).

The business may be far more important further down on the scale of size. Recall how much time, turbulence, and strong detergent it takes to remove a bit of residue from a plate in a dishwasher, doing no more than what a quick swipe with a dishcloth can accomplish. Blame no-slip, velocity gradients, boundary layers. It's a good thing if you're a bacterium with only minimal attachment to a smooth surface. It's equally good if you're a spore newly settled on a leaf or a barnacle larva that has landed on a rock. Fouling organisms don't need crevices to get an initial foothold—a wonderful community often establishes itself on the inner walls of the glass piping sometimes used to distribute seawater at marine laboratories. Plumbing scales up in places where the water is "hard" with dissolved minerals. Worse, we all too often scale up with atherosclerotic plaques in our arteries.

Looking for gradients

Attached organisms commonly depend on flow across them for food, gas exchange, heat exchange, and other interactive functions—even ones that drive fluid movement themselves. For a small, sessile creature, then, a boundary layer can be a mixed blessing. One tactic consists of picking a location, not by the strength of the adjacent free stream, but by the steepness of the velocity gradient at the surface. Indeed, it may be the rule rather than the exception among both freshwater and marine organisms in which the larvae are motile and the adults sessile.

The way barnacle larvae (cyprids) chose attachment sites may have provided the first clear demonstration. Crisp (1955) forced water containing such larvae through circular glass pipes; by manipulating both flow rate and pipe diameter, the effects of velocity and velocity gradient could be unambiguously unscrambled. What determines whether they settle down is what happens within a half millimeter of the wall of a pipe. And what the larvae like best is a gradient or shear rate ($\Delta v/\Delta z$) of 100 inverse seconds, although they will accept rates from 50 to 400. (Inverse seconds may be unfamiliar units, but that's what dividing speed by distance gives. As an example, a head-high wind of 20 meters per second or 44 miles per hour corresponds roughly to a gradient of 13 inverse seconds.)

Fifty years later we have a large literature on larval settlement (or "recruitment"), enough to assert with confidence that Crisp was looking at something widespread. Unfortunately, though, we have few comparable numbers to provide a context for his data. Black-fly larvae like strong gradients, and they will shift their positions after initial establishment if the flow conditions change

(Lacoursière 1992). What has become clear are complexities both of such hydrodynamic responses and of the other factors—surface roughness, chemical cues, proximity to conspecifics, and such—that determine who lives where when given the chance.

Moving fluid across the gradient

Gradients matter in yet another way. Both adult barnacles and attached larval black flies snag edible particles from passing fluid with appendages as far from the substratum as either can reach. But a variety of other animals pump fluid through internal filtration equipment and out again. These face the problem of wastefully recirculating water that already has been processed. One solution consists of putting up a chimney to gain access to fresh input, but that's rarely done except for cases (recall the caddisfly larva) that take advantage of dynamic pressure with an upstream opening. More common are omnidirectional excurrent chimneys, which (again recalling the last chapter) take advantage of Bernoulli-effect pressure reduction and viscous entrainment.

Still, a chimney must be stiff enough to stand the flow, and it might be bitten off. So we also find excurrent, hydrodynamic chimneys—nozzles. Recall yet again the last chapter, in particular our discussion about how a flow stream constricts as it passes through a fan or other pumping device. You can expel a jet of greater speed and coherence than any one you can inhale. So what we find are excurrent openings arranged as jets located maximally distant from substrata—in sponges, bivalve mollusks, and ascidians as parts of the organisms, and in burrows of some infaunal worms and crustaceans as low chimneys on burrows (fig. 8.4).

Jets, though, don't work well at small scales. Partly that's a matter of velocity, since smaller means slower and the slower squirt has less momentum to carry it away. And partly it comes down to Reynolds number, since the edge of a jet has to have a steep velocity gradient, at which viscous forces

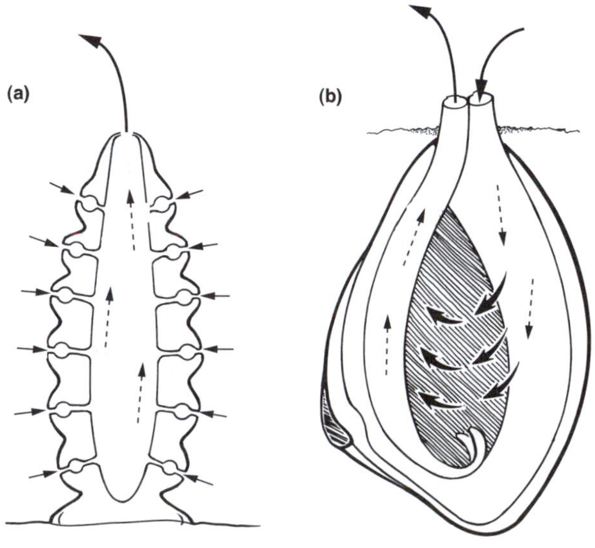

FIGURE 8.4 Chimney-nozzles: output openings arranged as jets maximally distant from substrata. (a) A sponge with its excurrent osculum, and (b) the excurrent and incurrent siphons of a clam.

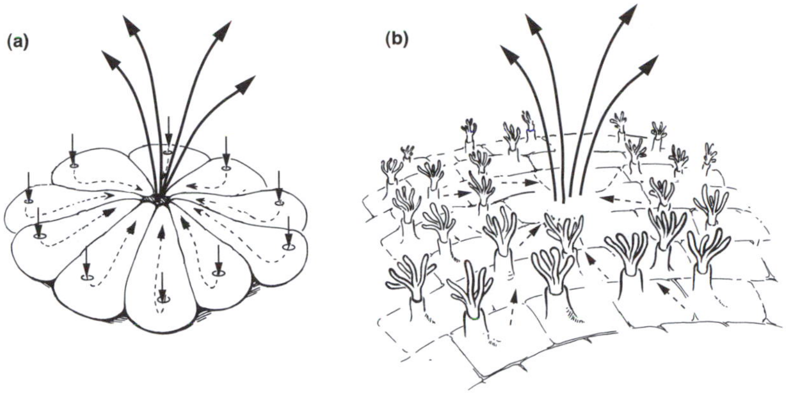

FIGURE 8.5 The communal jets of (a) an ascidian (*Botryllus*), and (b) a bryozoan (*Membranipora*).

look askance. Several small, encrusting organisms make communal excurrent jets—at least one ascidian (or tunicate or sea squirt) and a variety of encrusting (as opposed to erect) bryozoans (or ectoprocts or moss-animals), as in figure 8.5. Living off some kind of filtration process, the common feature here, demands that water be processed at a high rate—a sponge, for instance, which we think of as pretty passive, passes a volume of water equal to its own volume about every 5 seconds; and clams do even better. So these jets are no minor features.

Getting away from the surface

Wind and water currents provide mighty agents for dispersal of seeds, spores, pollen, sperm, eggs, larvae, and even some adult organisms. Getting on board, though, requires more than release from rock, leaf, or mother—that velocity gradient can provide a substantial barrier at a wide range of scales. Thus, the spores of a fungus need a high wind to be blown off a leaf's surface if they're to infect another plant. More generally, sessile organisms commonly bear their wind-borne propagules furthest from the substrata, often on special stalks. That applies from the scale of fungal stalks a tenth of a millimeter high, through moss sporophytes extending perhaps a centimeter above the leafy gametophytes, to the structures that produce wind-dispersed pollen at the tops of trees. Stalks several times the height of the rest of the plant bear the seeds of dandelions, plantains, and other afflictions of lawn lovers. Nor is the practice limited to terrestrial plants. Both wolf spiders and gypsy moth caterpillars, dispersed with the aid of silk threads in their first instars (molts), climb as high as possible before letting go. An alternative, perhaps less effective than gaining altitude, is simply getting some distance from the parental surface. Thus, many mushrooms bear their gills beneath a cap, so that upon release spores will drop free.

A boundary layer can play a peculiarly positive role in dispersal. What if propagules located down within a boundary layer require that flow reach a critical speed before they break loose? For local flow to reach that critical value, free-stream flow will of necessity exceed it. So something that's loosely attached and comes off in a slight flow will end up in a more substantial free stream of air or water. Better, an organism might either gradually weaken the attachment or extend a stalk still further from the surface. In that way it would release into a strong flow if one comes along but would gradually become less discriminating and so avoid missing all chance of wind-borne dispersal. My casual observations of the winged samaras (fruits) of maples point to such a device, but I can't at this point cite specific data.

Many terrestrial plants have evolved schemes to shoot off seeds and spore clusters at high initial speeds. The larger ones achieve their dispersal entirely (or almost so) with the momentum of their projectiles. Many must use a mix of momentum and wind dispersal; the smallest can do no more than use projectile momentum to get through the velocity gradient that surrounds the parent plant. Reasonably reliable measurements put the initial speed of the seeds of a euphorb native to West Africa, *Hura crepitans*, at 50 meters per second. With a mass of nearly a gram, such a seed will travel up to 45 meters if released 7 meters above ground (Swaine and Beer 1977). At the other extreme, the ascomycete fungus, *Sordaria*, shoots spores or clusters of up to eight spores, 40 micrometers in overall diameter (Ingold and Hadland 1959). From the distance they go and estimates of the drag of the largest cluster I calculate that the initial ("muzzle") velocity is an impressive 20 meters per second. So bad is drag that they travel no more than about 60 millimeters, which, in the absence of wind, hardly qualifies as long-distance dispersal. It does, though, get them well away from the maternal surface and her boundary layer.

Turbulent Boundary Layers

What if the Reynolds number for flow across a flat plate exceeds 500,000? That means an airflow of 20 meters per second and a position 40 centimeters downstream on the plate—or any combination of these variables whose product equals 8 square meters per second. In water it means a flow of 2 meters per second and a position 25 centimeters downstream—or a product of 0.5 square meters per second. While large, these aren't biologically bizarre. A cyamid ought to experience larger values more than once in a whale.

Above this transitional Reynolds number, the character of a boundary layer changes in several ways. For one thing, it grows thicker at a slower rate as flow moves downstream or the free-stream Reynolds number increases. For another, it has a schism, with a thin, laminar, "sublayer" developing beneath the main turbulent gradient. The transition differs a bit from what we met when considering separation around a sphere or cylinder. One has to bear in mind the notion of local Reynolds number, recognizing that a laminar boundary layer begins at the upstream edge of a plate and then becomes turbulent at a distance downstream where Re_x first exceeds 500,000. Formulas for the thickness of the turbulent layer (including the sublayer) are analogous to those cited earlier as equations (8.2) and (8.3):

$$\delta = 0.376x\sqrt[5]{\frac{\mu}{\rho x v}} \tag{8.7}$$

and

$$\frac{\delta}{x} = 0.376Re_x^{-0.2}. \tag{8.8}$$

If drag is what matters (and the line for turbulent drag coefficients in figure 7.10 can be derived from velocity gradients just like that for laminar coefficients), then little more need be said. But if (as we'll turn to shortly) heat or mass transfer across a boundary layer is of interest, then one has to bear in mind that a turbulent boundary layer differs fundamentally from a laminar one. In a laminar layer, all bits of fluid move (at least approximately) downstream. So some other agency such as diffusion or conduction has to be invoked to provide any cross-stream transport. By contrast, in a turbulent layer, only the average flow can be counted on to go downstream, and cross-flow movement happens continuously if chaotically. So heat and mass get moved toward and away from the surface by the flow itself—whether for better or worse depends on the situation.

Another form of turbulent boundary layer may have more biological relevance than the one on a flat plate. That's the layer in the air just above the ground or in a body of moving water just above the bottom. Here, of course, no discrete leading edge exists—the point Columbus made to Isabella back in 1492. But this layer proves to be quite a lot messier than either of the ones we've just considered. With no leading edge, we can't assign to it a discrete thickness. No significant linear region occurs near the interface, and the whole thing is better fitted with a logarithmic relationship. How and where the layer extrapolates to zero speed depends on the character of the surface. And that character has a substantial effect on the shape of the curve higher up. So one has to insert some semiempirical numbers into a general formula, numbers that one gets from books on micrometeorology such as Arya (1988) or Geiger, Aron, and Todhunter (1995). That general formula for how horizontal velocity increases with height, the analog of equation (8.4), is

$$v_x = \frac{v_*}{\kappa} ln\left(\frac{z-d}{z_0}\right), \qquad (8.9)$$

where z, again, is height above the ground, and κ is an empirical, dimensionless constant of 0.40—neither make problems. The "zero-plane displacement," d, accounts for an effectively zero velocity (by extrapolation) a little above the actual ground. The "roughness length," z_0, accounts for changes in the steepness of the gradient caused by eddies of different sizes tossed upward. A nominal velocity, the "shear velocity," v_*, accounts for turbulence level in the overall flow. Whatever values get inserted for the three situational variables, the equation generates a linear graph if the logarithm (or *ln*) of height, z, is plotted against velocity, v_x.

Some sense of the range of the three variables might help bring reality to equation (8.9). Zero plane displacement, d, varies from zero for sand or short grass to 20–30 meters for forest. Roughness length, z_0, ranges from about 0.0003 meters over calm water or desert sand to between 1 and 6 meters for forest. Shear velocity varies less (at least terrestrially), from around 0.16 to about 0.6 meters per second, and again is least on the smoothest surfaces.

I cite formula (8.9) and the values of the variables mainly to discourage anyone from using the earlier formulas for situations where no leading edge can be identified. For anything beyond the most casual or preliminary calculation, more specific sources of information should be consulted.

Diffusion

Biological systems invariably involve ingress or egress of material across surfaces, so when considering flow near surfaces, we ought to develop tools to handle such exchange processes. As often as not, two entirely and fundamentally different physical processes participate—the kind of bulk fluid transport we've been talking about up to now and the process of diffusion, not yet mentioned. So we need a brief introduction to yet another peculiar phenomenon.

Diffusion is one of those rare processes in which the behavior of molecules as individuals has practical consequences in our macroscopic, physical world. (In biology, of course, molecules have much more direct impact, a major point of Erwin Schrodinger's [1944] famous essay *What is Life?*) In gases and liquids, individual molecules wander around aimlessly at rates determined by their temperatures. (Or, from another viewpoint, their temperatures are determined by the rates at which they wander.) Collisions are frequent, with the changes in direction we would expect from our earlier discussion of momentum. The only peculiarity about the collisions is that they are fully "elastic"—not only is momentum conserved, but the molecules go, on the average, as fast after colliding as before. The wandering, of course, results entirely from the collisions since, by Newton's first law, a molecule would not otherwise change direction.

The average distance a molecule goes between collisions takes on great importance in this random motion—it's called the "mean free path" and depends on the size of the molecule and the nature of the medium in which it's moving. The more densely the molecules are packed, the more frequent the collisions, so a molecule has a longer mean free path in a gas than in a liquid. Thus, an oxygen molecule in water goes about 0.3 nanometers between collisions, while one in air goes about 90 nanometers, a factor of 300. Molecules in crystalline solids vibrate rather than wander. But some loose molecules do wander around even in solids, with important consequences.

These collisions of wandering particles provide an agency through which momentum is shared around and therefore a molecular explanation of the viscosity of gases. Gas molecules bang into the walls of their containers, and the interaction of molecules and walls provides a good quantitative explanation for pressures in gases. And the collisions have an even more powerful consequence, at least for our understanding of physical science. Small particles in a liquid under a microscope are seen to be in a sort of continuous jiggling motion—"Brownian motion"—first noticed in pollen particles in water by Robert Brown in 1827. If the motion is interpreted as being due to hits by individual molecules, then it is possible, as shown by Einstein early in this century, to calculate the size of a molecule. Pais (1982), in his splendid account of Einstein's science, reminds us that even that recently there remained the suspicion that molecules might be just the polite fiction of a formal chemistry. But arriving at the same size for a species of molecule from quite different physical starting points and by different lines of reasoning, as Einstein did, provided exceedingly persuasive evidence for their reality.

Diffusion represents the macroscopic consequence of this aimless wandering; it's a subject of daunting mathematical elaboration. You may be surprised that equations can describe with enviable precision the results of a process in which we can say nothing about what a particular molecule is doing at a particular time. But making usefully precise statements about statistical aggregates certainly pervades social science—perhaps you just thought physical scientists were above such crudeness.

In reality, very few areas of science are uncontaminated by the pseudocertainty of statistical conclusions describing individually uncertain events. What saves the physical sciences from obvious sloppiness is merely the smallness of molecules relative to us; that's essential to the argument of Schrodinger's essay, mentioned just above. One's relative uncertainty decreases with the number of entities under consideration, following what's called the "square root of *n* rule." Flip a coin once—the uncertainty is complete. Flip a hundred coins; the outcome will more likely than not fall between forty-five and fifty-five heads. The range, 10, equals the square root of the number of trials, 100, and it's fully 10 percent of the number of trials or data. Flip a million coins; between 499,500 and 500,500 will most likely be heads. The range, 1000, still equals the square root of the number of trials, but now the range is only one-tenth of 1 percent of that number. Worry about 10^{20} or more molecules, and you deal with an awesomely lawful aggregate.

The random walk

Fortunately there's a intuitive model from which we can develop a decently quantitative notion of diffusion. Consider an individual at a point in a state of ideal inebriation—a drunk at a light pole. He or she takes a step away from that comfortable support, stops, gets thoroughly disoriented, takes another step in a direction chosen randomly (that is to say, not chosen at all!), stops, goes another step in another random direction, and so on. (We assume the steps occur at uniform intervals of time and that they are of equal lengths.) What happens over a longer time? Does the individual move away from the light? Nothing says he or she must. If an individual does move away does this happen at a uniform rate or does the rate of movement depend on the distance from the light already achieved?

While it might be easy to find volunteers for such experimentation, one does anticipate difficulty obtaining funds for the requisite chemicals. But since the drunk is only a model we might as well use a more socially acceptable model. Handiest is a random-number generator caged within a computer. With it drunks can be as ideal and as numerous as necessary, they can walk very rapidly, the random walks can be in one, two, or three dimensions, and the Volstead Act (U.S.) is not violated. Figure 8.6a gives a typical path. From such simulations two interesting results emerge. First, as in figure 8.6b, almost every drunk does indeed wander away from the light. One may occasionally meander back, but with time these returns become less and less frequent. The average position in space of the whole befuddlement is still the light post, but the average distance between drunks and post increases with time. Second, while that average distance increases, it does not do so in proportion to the time that has elapsed. Queerly, the drunks achieve a distance from the light (*on the average*) that's proportional not to the time but to the *square root of the time* they've spent, as in figure 8.6c. With the passage of time, their "progress" gets slower and slower. The same result occurs whether they keep their feet on the ground (a two-dimensional simulation) or whether (in a more surrealistic three–dimensional wandering) they move away from a point in space.

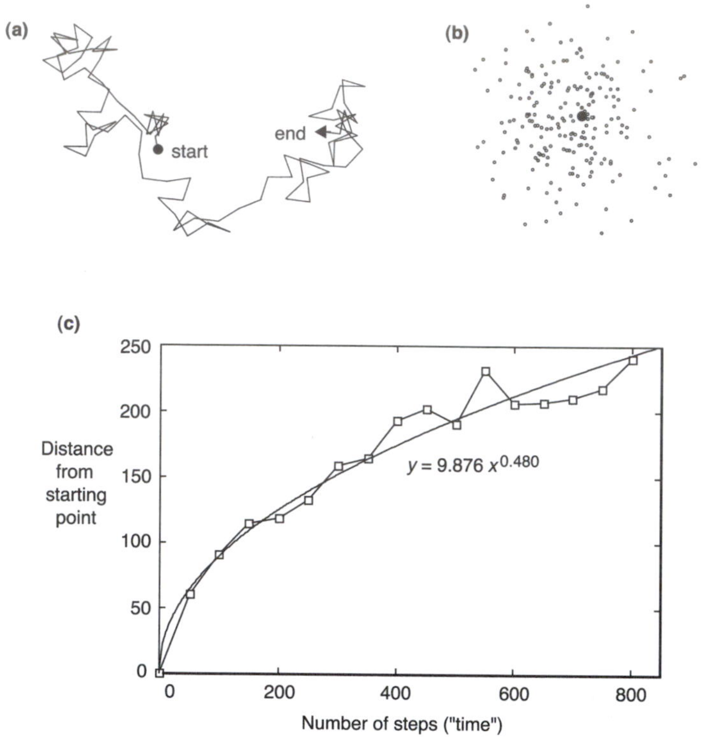

FIGURE 8.6 Diffusion simulations. (a) A computer-generated random walk, with steps of equal length. Almost no random walk ever really looks random! (b) The endpoints of a large number of such random walks of the same number of steps (although not to the same scale). (c) The average distance from the start versus the number of steps taken in these random walks.

If distance is proportional to the square root of time, then the square of distance is proportional to time. For time we can substitute distance over speed, so now we see (with a little algebraic juggling) that speed is inversely proportional to distance. If the verbiage seems too quick and slick, try the sequence on paper. Put back in terms of molecules or particles, *the rate of movement of particles on the average away from a point as a result of a random walk will be inversely proportional to the distance from that point.* So if it takes 1 second for a given amount of a substance to diffuse a given distance then it will take 100 seconds to go only ten times as far.

Dealing with diffusion

We have here a very strange kind of rate—to use yet another analogy, it's as if a trucking company, upon doubling the distance traveled with each load, needs not twice the number of trucks to carry the same total cargo in the same time, but now needs four times the number. It's a world in which the speed limit is inversely proportional to the length of the journey. So we shouldn't be surprised to require similarly strange dimensions and units. Consider an ordinary rate, the kind measured by a speedometer, for which distance equals the product of rate and time. Rate or speed is thus distance divided by time, with units of miles per hour, meters per second, and so on. For diffusion, though, distance *squared* equals the product of rate and time, so rate must have dimensions of distance squared per time, or L^2T^{-1}. Now this kind of rate is not our old, familiar, and comfortable distance per time or speed. Howard Berg, in fact, recommends that we not even use the term diffusion *rate* lest we be confused with normal rates; the term "diffusion coefficient" is, in fact, the normal designation. (His book, *Random Walks in Biol-*

ogy [1993] provides a fine introduction to the subject.) Incidentally, the dimensions are the same as those of kinematic viscosity (viscosity divided by density); kinematic viscosity amounts to a diffusion coefficient for momentum rather than, as here, mass.

The matter of units is comparatively ordinary. The SI unit for the diffusion coefficient is the square meter per second. But more commonly encountered is the square centimeter per second. To get the SI unit, one just divides the later by 10,000.

Diffusion can be viewed in an entirely different way—with a more formal if less explanatory approach. This latter leads more easily to an equation from which equations that apply directly to practical situations can be derived. Consider a slab of material through which something is to diffuse (fig. 8.7). (This conceptually simple arrangement proves even less practical than the two-plate model for viscosity in fig. 6.2.) How much will diffuse through per unit time? According to what's known as Fick's first law, an intuitively reasonable relationship, the mass per time diffusing will be proportional to the product of (1) the diffusion coefficient, D, of the material diffusing through the material of the slab, (2) the area, S, of the slab, and (3) the concentration gradient across the slab, that is, proportional to the concentration difference across the slab divided by its thickness, x:

$$\frac{m}{t} = DS\left(\frac{m_1/V_1 - m_2/V_2}{x}\right). \tag{8.10a}$$

Using C for concentration (mass per unit volume, m/V) and recognizing that a minus sign ought to appear since material is moving down the gradient, Fick's law can be more tersely written as:

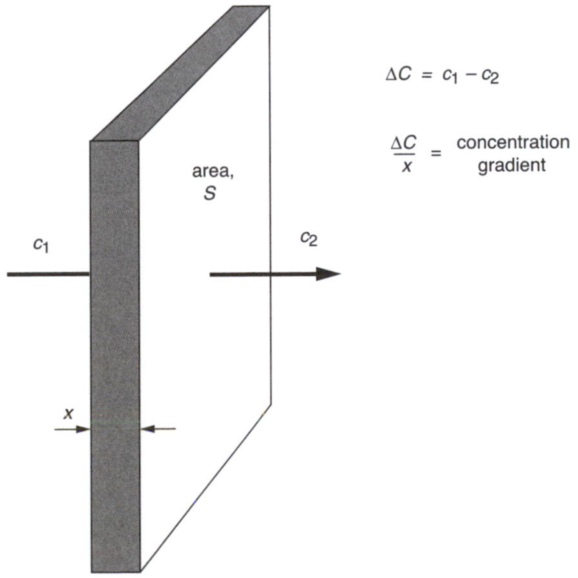

FIGURE 8.7 Steady diffusion down a constant concentration gradient, showing the relevant variables for Fick's first law.

$$\frac{m}{t} = -DS\left(\frac{\Delta C}{x}\right). \tag{8.10b}$$

The skeptical reader will raise the objection that the equation above predicts a decrease in transport proportional to distance, not distance squared—doubling x only halves m/t. True—but the equation presupposes a rather different situation than the one considered earlier. Instead of decreasing in steepness with increasing distance from a point source, this gradient is constant in steepness across the slab—any locally greater gradient would cause a more rapid movement of whatever is diffusing. The latter, then, would restore constancy of the gradient. The equation also presumes that the gradient does not change over time—in some way material is continuously supplied to the side having the higher concentration and removed from the side with the lower concentration, so neither concentration changes with time. Finally, it presumes one-dimensional rather than two- or three-dimensional movement. So it's not inconsistent with the wandering drunks described earlier but just applies to a simpler set of circumstances.

The same skeptic may also ask how the molecules "know" in which direction Mr. Fick requires that they diffuse—the drunks wandered randomly, but here, suddenly, we assert that the process has directionality, that substances diffuse from regions of higher to ones of lower concentration. There's no chicanery or paradox—diffusion really occurs in both (really, all) directions. The initial random walk led to an overall dispersion—on the average, the distance between drunks and post increased with time. Dispersion resulted because the drunks were initially more concentrated in a particular place, so more were available to move away from that place than back toward it. Thus there was a *net* movement from higher to lower concentration, just as with diffusive transport across the slab.

The definition of the diffusion coefficient, using figure 8.7 and equation (8.10), is quite analogous to the definition of quite a different physical property, "thermal conductivity." It turns out that heat transfer by conduction (Fourier's law) gets just the same formal treatment as mass transfer by diffusion (see, for instance, Bejan [1993]). Heat transfer rate equals the product of thermal conductivity (replacing diffusion coefficient), cross-sectional area, and temperature gradient (replacing concentration gradient). The equivalence, incidentally, is more than a mathematical convenience. Situations involving diffusion can be replaced for experimental purposes by models involving heat conduction, a handy thing since we can usually measure temperature more easily than concentration of one substance in another. I once took part in a study (Hunter and Vogel 1986) that used this devious approach. We asked whether the spinning of snail embryos in their jelly mass might have the effect of reducing the distance over which diffusion had to occur and thus of improving their supply of oxygen. The results, by the way, were positive, but not strongly so.

And some real values

Even if you have no occasion to do specific calculations involving diffusive transport, it's worth examining the relative values of diffusion coefficients, a few of which are given in table 8.1. "Molecular weight" refers to the mass of

TABLE 8.1 Various Diffusion Coefficients, *D*

Substance	Medium	Molecular Weight	*D** (cm²/s)
Tobacco mosaic virus	water	40,000,000	0.03×10^{-6}
Human serum albumin	water	69,000	0.61×10^{-6}
Inulin	water	5,500	1.5×10^{-6}
Sucrose	water	342	5.2×10^{-6}
Glycine	water	75	10.0×10^{-6}
Oxygen	water	32	18.0×10^{-6}
Oxygen	air	32	0.2

*These are the usual units; to convert to proper SI units of m^2s, divide by 10,000.

a molecule relative to the mass of a standard unit, approximately a hydrogen atom (really a sixteenth of an oxygen atom).

Two things are immediately clear, and both have important implications for the operation of organisms. First, large molecules diffuse more slowly than do small ones—the diffusion coefficient varies inversely with something close to the square root of molecular weight. Second, diffusion rates for gases or volatile liquids in air are *far* faster than those in liquids—the roughly ten thousandfold difference for oxygen between air and water is typical. It's one of several physical advantages of a terrestrial rather than an aquatic life.

Purely diffusive transport in living biological material is still slower than transport in water. Schmidt-Nielsen (1997) cited data for oxygen in muscle and connective tissue (steak and gristle); the diffusion coefficients are roughly one third to one half of those for water. And for diffusion within individual cells, examination of the data collected by Mastro and Keith (1984) shows that large molecules have disproportionately low diffusion coefficients relative to their values in water—for protein molecules the coefficients may be more than a hundredfold lower in cells than in water. The internal architecture of cells seems to slow macromolecular movement disproportionately. Indeed the loss of much of this structure on the death of a cell is marked by an increase in Brownian motion that's quite conspicuous in time-lapse videos.

Diffusion Versus Convection

Any discussion of diffusion must emphasize how strongly its effectiveness depends on the size of the system in question. Purcell (Purcell [1977], also in Berg [1993]) gives an amusing but important illustration of the issue by asking whether a bacterium ought to swim as it feeds on dissolved small molecules in the broth in which it lives. His answer is that swimming isn't worth the bother—food diffuses to the bacterium faster than it can possibly move. It lives in a world analogous to that of a cow in a bovine nirvana—a pasture of grass growing so fast that the cow need not walk as it grazes. Bacteria do swim, though, and Purcell notes some profit to be gained in explorations that sample more concentrated or salubrious broth—new and perhaps greener pastures.

Closer to home, the delay in transmission of a nerve impulse from nerve cell to nerve cell is only about a millisecond—transmission depends on diffusion of a so-called transmitter substance across the synaptic gap, but the gap is only about 20 nanometers across. But for long distances, the times involved in diffusive processes get devastatingly long. Schmidt-Nielsen (1997) gave the following sobering calculation. If it takes a hundredth of a second for diffusion of a substance such as oxygen between a typical cell and a capillary, then diffusion over a distance of 1 millimeter will take 100 seconds, and diffusion of the same substance between a lung and an extremity in a person will take no less than *3 years*.

Diffusion proceeds much faster in air than in water or tissue, which is to say that it gives decent service over substantially longer distances. But even in air, diffusion remains glacially slow for what we regard as ordinary distances. A demonstration commonly done in high school physics classes purportedly shows the operation of diffusion—a jar of perfume is opened, and over time people increasingly distant from the jar report detection of the odor. The demonstration, however, is a fraud. As a practical matter one can't get sufficiently low wind in a room to observe diffusive transport—the mere presence of warm bodies makes air temperature vary from place to place, and the resulting differences in density produce imperceptible but quite significant and measurable wind. Most of the phenomena we casually attribute to molecular diffusion, such as the spread of the smell of coffee from kitchen bedroomward or of cooking fish toward the family felid, are in fact cases of bulk flow of fluid, what's usually called "convection."

Failure to distinguish properly between convective bulk flow and diffusion is no recent wrongheadedness. As Tabor (1979) pointed out, Lucretius (ca. 95–55 BCE) thought he had discovered what we know as Brownian motion—dancing of particles due to hits from atoms, he said—but the movement of dust particles in a sunbeam are really caused by air currents. It's practically impossible to give a macroscopic demonstration of diffusion in a gas, and it demands some unusual precautions to do so even in a liquid, so ubiquitous are convective motions such as those generated by minor thermal irregularities (Vogel 1994b). To do the latter, you have to create a stable density gradient through deliberate control of temperature or concentration gradients in order to prevent convection. Alternatively, you can add to your water some substance such as agar or household gelatin that will prevent convection—a percent or two of gelling agent should make little difference to the diffusion coefficients. Visually complete equilibration of colorant in even a small (covered!) beaker of gelled water will take weeks or months.

The biologist's world, though, turns on structure, process, and motion at short spatial scales—our classic icon is a light microscope. Diffusion looms large in that world, even if it's macroscopically imperceptible. One can make a reasonable case that, at least in the animal kingdom, the boundary between cellular and supercellular organization marks a shift in the predominant mode of material transport. Above this distance, about 10 micrometers, diffusion becomes too slow for moving material around at the time scale typical of organismic processes. So the free transport system provided by the thermal motion of molecules must be augmented by metabolically powered pumps and their associated plumbing, as we'll elaborate upon in chapter 10.

Diffusion in solid systems is just that—diffusion. By contrast, in fluid systems, diffusion and convection most often coexist. They may work synergistically, when convective motion prevents local diffusional depletion of some source of material or equilibration of diffusing material on two sides of a barrier. Or convection may limit the efficacy of diffusion by prematurely moving material beyond the site of diffusional exchange. In any case, the relative roles of diffusion and convection shift with distance from a surface, with diffusion diminishing and convection increasing. Two factors drive that shift—first, the no-slip condition and velocity gradient, slowing convective motion near the surface; and, second, the concentration gradient that drives diffusion, typically greatest at the surface, where exchange takes place. With only slight oversimplification, one can view what happens as diffusion of material outward (z-ward) from a surface until the bulk fluid flow of the free stream sweeps it away (x-ward). Or vice versa for material coming toward a site of exchange.

Physiologists, both plant and animal, deal with regions near surfaces in which diffusion predominates by invoking a polite fiction much like our velocity boundary layer—the "diffusive boundary layer"—in much of the relevant literature. In the literature on the innards of animals, it's most often called the "unstirred layer" or "USL." No implication of turbulent mixing in the free stream should attach to the term, just a shift from convective x-transport to diffusive z-transport. (Indeed, for the small-scale systems most often considered, diffusion stirs things up as well or better than convection.) Plant physiologists most often refer simply to the region as a boundary layer (especially when talking about the surfaces of leaves), but one defined by concentration gradient rather than velocity gradient. However designated, it's a fiction-of-convenience that divides the world between a region of zero flow and a distinct concentration gradient and one of full flow and zero concentration gradient.

One gets some feeling for the importance of this region—the fiction lies mainly in its formal definition—by looking at what limits water loss from a typical, transpiring leaf in a negligible wind. The diffusive boundary layer imposes as much resistance to water loss as the diffusive resistance of the openings on the underside of the leaf, when these openings, the stomata, are at their widest. The stomatal area, though, amounts to only about 1 percent of the leaf surface. No surprise, then, that water loss varies strongly with wind speed and that stomatal area goes down as wind speed increases. With such high boundary layer resistance, still greater stomatal area would give little increase in access to carbon dioxide$_2$ or control over gas exchange in general.

The practical definition of the thickness of this concentration boundary layer corresponds to the way we defined the velocity boundary layer in equation (8.5)—we set its limit as where local speed reached 90 percent of free-stream speed. One takes the concentration gradient at the surface and extrapolates its line out to the concentration of whatever molecule happens to be of interest in the free stream. The distance from the surface at which free-stream concentration is reached is then the thickness of the layer, as in figure 8.8. Put as a formula, the thickness of the concentration boundary layer or unstirred layer, δ_u, is

$$\delta_u = \frac{c_f - c_m}{(dc/dz)_m}, \qquad (8.11)$$

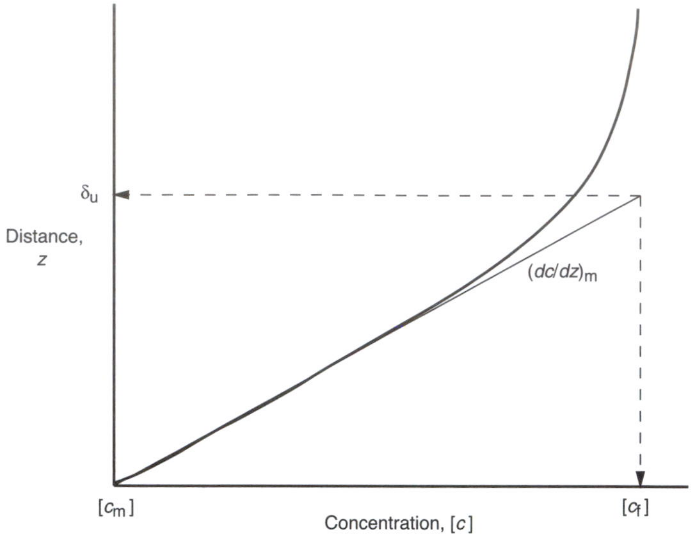

FIGURE 8.8 How we define the thickness of the "unstirred layer" or "concentration boundary layer" for diffusion near a surface across which fluid is flowing. The concentration gradient at the membrane is linearly extrapolated out to the average concentration of the solution.

where c_f and c_m are the concentrations in the bulk fluid and at the surface membrane, respectively, and $(dc/dz)_m$ is the concentration gradient at the membrane.

Two additional points should be noted. First, while the thickness of the unstirred layer has thus been defined in a way that corresponds to our 90 percent velocity boundary layer, the thicknesses of the two will not, in general, coincide. And second, the thickness of the unstirred layer varies slightly with the diffusion coefficient of whatever solute is at issue—the higher the diffusion coefficient, the thicker the layer (Pohl, Saparov, and Antonenko 1998). Pedley (1983) and Barry and Diamond (1994) provide useful introductions to unstirred layers within animals; Slatyer (1967) and Nobel (1999) do the same for plants.

The Péclet Number

The relative importance of two phenomena often can be viewed through the value of their ratio. We've seen several instances in the preceding pages—in particular, the Reynolds number for inertial force relative to viscous force. Another named number, the Péclet number, gives the ratio of mass movement by convection to mass movement by diffusion. This especially simple ratio,

$$P\acute{e} = \frac{lv}{D}, \qquad (8.12)$$

consists only of a characteristic length, l, the free-stream speed, v, and the diffusion coefficient, D. In effect, as Denny (1993) pointed out, access to material by convection depends on velocity, v; access to material by diffusion depends on D/l, assuming the one-dimensional surface penetration of figure 8.7. The Péclet number is just the ratio of the two.

(Once again, a variable gets used in analogous but incompatible ways in different fields. One encounters a thermal Péclet number for convective versus conductive heat transfer and an oceanographic version in which eddy, or turbulent, diffusivity replaces the present molecular diffusivity.)

The Péclet number may be a crude index, but it gives a useful view of some phenomena rarely seen in a common context. Consider again Purcell's contention that a bacterium can't improve its access to dissolved nutrients by swimming out of diffusionally depleted regions. If the bacterium is a micrometer across, swims at a micrometer per second, and feeds on glucose ($D = 6.7 \times 10^{-6}$ square centimeters per second), its maximum Péclet number is 0.0015. In short, it moves far too slowly for convection to be significant. By contrast, imagine a ciliated protozoan faced with the same swim-or-stay-put decision. It's 40 micrometers across, swims at 450 micrometers per second, and seeks oxygen ($D = 18 \times 10^{-6}$ square centimeters per second). Its Péclet number is 10, so in its world convection (by swimming) might well augment diffusion.

Another such problem. Diatoms are photosynthetic algal cells of great importance in both freshwater and the oceans. Under most circumstances, they sink, albeit very slowly, through the water that surrounds them. It has been claimed from time to time that sinking, not an obviously appropriate activity if you're photosynthetic, compensates for local depletion of carbon dioxide. But back in 1952, Munk and Riley showed that that could not be so. Put in our present terms, we multiply a cell diameter of 10 micrometers by a sinking rate of a mere 4 micrometers per second (a foot per day) and divide by the diffusion coefficient of carbon dioxide$_2$ of 14×10^{-6} square centimeters per second. That gives $Pé = 0.03$, a value too low for sinking to augment diffusion. More likely, slow sinking simply assures that they won't slowly rise and perhaps contact the air-water interface and be trapped by its surface tension.

One can press the Péclet number into more general and speculative service. The value for small molecules in our small intestines runs around 10,000, which says that diffusion, always critical, needs a lot of help—which we provide with villi, microvilli, and the mixing associated with peristaltic pumping. By contrast, the values for kidney tubules run in the range of one to ten, so diffusion needs less in the way of special anatomical augmentation. Intestinal physiologists worry about the unstirred layer; renal physiologists do not. Or one can use the number to get some idea of how much wider the pipes can be and how much faster the flows can be in exchange systems that process gas rather than liquid—tubular bird lungs versus fish gills, for instance. The product of l and v ought to run around 10,000 times higher in the former, reflecting the greater values of D in gases. If residence time for a passage of a bit of fluid through comparable systems stays constant, then we can look at the product of l and v as l^2/t or, with t constant, as l^2. If $l^2 = 10{,}000$, then $l = 100$—the gas-processing system can have conduits a hundred times the diameter of those of the liquid-processing system. We'll return to this prediction in chapter 10.

Finally, a variant of the Péclet number may give some insight into the development of animals. Much depends on the diffusion of substances, morphogens, whose concentration gradients establish embryonic fields. Establishing larger fields not only means lower gradients but takes more time, a non-negligible resource in a competitive world. Breaking up velocity into length over time (as above) we get what's often referred to as the Sherwood number, *Sh*:

$$Sh = \frac{l^2}{Dt}. \tag{8.13}$$

To get a situation in which diffusion is not relied on excessively, we might assume $Sh = 1$. For molecules of reasonable size moving through cells, D ought to be around 1×10^{-6} square centimeters per second. A reasonable time for embryonic processes should be a few hours, say 10^4 seconds. The numbers and the equation imply embryonic fields of around 1 millimeter, about what one does indeed find. The case (put somewhat differently) was first made by Francis Crick in 1970.

CHAPTER

Where Flows Are Inside

He who pisseth against the wind wetteth his shirt.

old proverb, from *Oxford English Dictionary*

FOR THE PAST THREE CHAPTERS, WE'VE FOCUSED ON EXTERNAL FLOWS, flows in which gas or liquid passed over, under, and around solid objects. Not since we left continuity and Bernoulli have flows gone through or between solids, despite the obvious importance of these latter. We return to them now, starting with the reassuring comments that the same underlying rules apply, if with slightly different formulations, and that we'll need little further recourse to odd inventions such as boundary layers.

Laminar Flow through Circular Pipes

Nonetheless, as in the last chapter, all will turn on velocity gradients. Consider situations in which the only force resisting flow is that due to viscosity. We've previously looked at two such instances, both cases where flow ran parallel to flat surfaces. In one, the velocity gradient didn't vary—that was where flow between two plates gave us a way to define viscosity (fig. 6.2 and equation (6.3)). In the other, flow over a single surface, the gradient began in the same linear fashion (equation (8.4)), but it asymptotically approached the free stream, giving something quite messy mathematically. We turn now to a third.

For laminar flow through circular pipes, circumstances are nicer than the second, if not quite as simple as the first. At least once fluid gets well downstream from the entrance to a pipe, from a bend, or from any abrupt change in bore, the gradient of velocity across a pipe turns out to be both mathematically tidy and independent of the distance downstream. That, a moment's thought should convince you, must follow from the principle of continuity. An ever-thickening boundary layer forces flow to have some component away from a surface, that is, z-ward. But that's something a pipe of constant bore just won't permit—so the velocity gradient has to come to some steady state with no further downstream change. As a consequence, each bit of fluid then proceeds along at an unchanging distance from the walls.

Resistance comes entirely from the no-slip condition and viscosity. Moreover, a constant velocity gradient makes it just as hard to force fluid through one part of the pipe as another. That means that the pressure loss per unit length remains unchanged down the pipe. And that pressure drop per unit length reflects the same energy dissipation through solid-fluid interaction as did drag—even if the units differ, it measures the same lossiness as flow moves from upstream to downstream.

What would we like to know about flow down the lumen of a pipe? A major matter for pipes in which heat or material gets exchanged across the walls must be that velocity gradient, here most often called the "velocity profile." A pressure difference from an upstream to a downstream location pushes the flow; it acts across the pipe. Viscosity opposes the flow; it acts most emphatically at and near the pipe's walls, with the no-slip condition taking full charge at the walls, analogous to what happens deep within a boundary layer. Quite naturally, flow speed goes from zero at the walls to a maximum in the middle. Less obviously, the profile takes the form of the particular parabola of figure 9.1 and can be described by a straightforward equation with no funny fudge factors,

$$v_r = \frac{\Delta p}{l}\frac{(a^2 - r^2)}{4\mu}, \tag{9.1}$$

where $\Delta p/l$ is the pressure drop per unit length of pipe, a is the radius of the pipe, r is some position on the radius from zero at the center to a at the wall, and v_r is the speed at that position.

A second nice piece of intelligence would be the way the overall flow, as volume per unit time (designated Q), varies with the pressure push ($\Delta p/l$) and viscosity (μ). That comes from an integration of equation (9.1) across the pipe, one used earlier (chapter 4) to illustrate the process of integration. Again, the result turns out to be well-behaved,

$$Q = \frac{\Delta p}{l}\frac{\pi a^4}{8\mu}. \tag{9.2}$$

The only difficulty is in pronouncing its name, "Poiseuille's equation" or the "Hagen-Poiseuille" equation. (The name of the old unit of viscosity, the poise, was a merciful truncation.)

This equation asserts that total flow and pressure drop per unit length are mutually proportional—it takes more pressure to push a flow along faster. It also asserts that total flow varies inversely with viscosity—stickier fluids flow more slowly. All this satisfies the intuition. But it says something more startling as well. Total flow varies with the *fourth* power of the radius of the pipe, which is pretty extreme stuff. Part of the latter should be obvious—if the

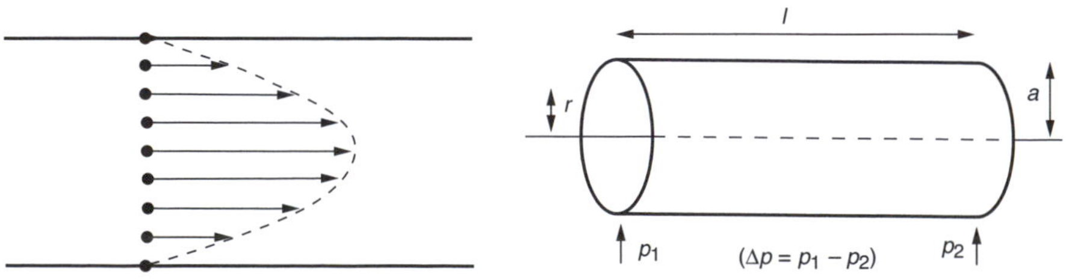

FIGURE 9.1 The parabolic velocity distribution for flow down a pipe, and the symbols used to describe flow through such a cylindrical pipe.

radius is doubled, area is quadrupled, so even without viscosity, total volume flow should be proportional to the square of the radius. In addition, though, a bigger pipe has less wall relative to its area (recall the reasoning in chapter 3), so it has (relatively) much less resistance to flow on that account as well.

Thus, if a big pipe is divided into a pair of small ones without change in total cross-sectional area or pressure drop, the overall volume flow is halved. No surprise, really—you've doubled the area of internal pipe wall per unit length. Or if, to save material, you halve the diameter of a pipe in a plumbing system, you need sixteen times as much pressure to keep the total flow the same. These matters of scale will prove decisive in explaining many features of the plumbing within organisms in the next chapter. It might be noted, at least parenthetically, that this scaling rule differs sharply from that for electrical conduction in wires, where current passing (equivalent to total flow) for a given voltage difference (equivalent to pressure) varies with cross section, or the *second* power of the radius.

Incidentally, if conduits aren't circular, errors can be quite large—again, that fourth power of radius or diameter comes home to haunt us. Fortunately, data for what's called hydraulic diameter can help. But don't use tabulated information from engineering handbooks that apply to turbulent flow. For laminar flow through rectangular and elliptical conduits, Lewis (1992) gave useful correction factors. For noncircular conduits, incidentally, one may have to use a lower estimate of the Reynolds number at which turbulent flow begins—more about that transition in a few paragraphs.

A third and fourth useful relationships drop into our laps from these equations. From the first, equation (9.1), comes the maximum speed of flow—just look at the center by setting $r = 0$:

$$v_{max} = \frac{\Delta p}{l}\frac{a^2}{4\mu}. \tag{9.3}$$

From the second, equation (9.2), comes the average speed of flow—just divide the total flow by the cross-sectional area of the pipe, πa^2:

$$v_{av} = \frac{\Delta p}{l}\frac{a^2}{8\mu}. \tag{9.4}$$

Maximum flow is just twice as fast as average flow, quite independent of the values of any of the specific variables. How nice! Sometimes it's handy to measure the maximum, using the time of first arrival of some tracer substance such as a dye. At other times, it's easier to measure the average, monitoring the discharge with, say, beaker and stopwatch. No matter—one can always convert one into the other.

Still, you may ask, just how useful is all of this talk about laminar flow in pipes? After all, laminar flow around spheres as described by Stokes' law (chapter 11) happens only up to $Re = 1$ or perhaps, at least semiquantitatively, 10. That corresponds to pretty small-scale situations—to an object a millimeter in diameter going a millimeter per second in water or a little over a centimeter per second in air. But recall Reynolds's original observation. Laminar

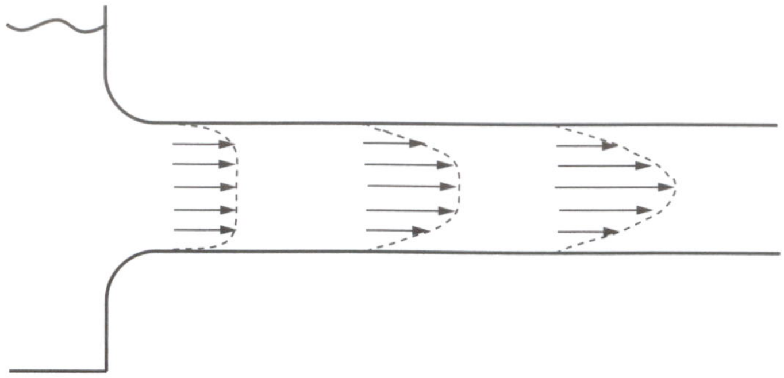

FIGURE 9.2 How the parabolic velocity distribution develops as fluid moves through the entrance region of a cylindrical pipe.

flow in pipes extended up to $Re = 2{,}000$, where the length in the expression for Reynolds number is taken as the diameter of the pipe. That corresponds to a pipe of 1-centimeter bore carrying water at 0.2 meters per second. That may exclude almost all commercial plumbing, but it includes virtually all internal fluid transport systems in organisms. We approach the limit of laminarity in our aortas, but our careful design normally ensures an absence of turbulence and its resulting higher pressure loss. Turbulence makes noise, and the character of this pathological noise, audible with a stethoscope, provides a useful diagnostic tool.

Just as with boundary layers, a few cautions need mentioning—a few complications can easily be overlooked in applying these attractively specific equations. For one thing, that limit of 2,000 shouldn't be taken as having the precision of pi or Avogadro's number. As noted in chapter 6, Reynolds gave a figure of 1,000, based on radius rather than diameter—it comes down to the same thing, but 1,000 has a more realistically approximate odor. A little roughness, some bending or noncircularity or change in diameter, an abrupt intake from a reservoir, or a bifurcation just upstream, and turbulence will set in at half or less of that value. With extraordinary care, transition can be postponed by up to an order of magnitude.

For another thing, flow takes a certain distance (or a little time) to settle down to its final parabolic profile after intake, bore change, or junction. In the so-called entrance region, matters look more like a combination of boundary layer and free stream—the parabola amounts to an amalgamation of the boundary layer all around the pipe, as in figure 9.2. A rule of thumb gives the effective length of this entrance region, expressing the "entrance length" (l') in terms of pipe diameters (d):

$$\frac{l'}{d} = 0.058Re = 0.058\frac{\rho v d}{\mu}. \tag{9.5}$$

Strictly speaking, flow approaches the final parabolic profile asymptotically, so the formula really gives the distance to get within 99 percent of the final form. If no great accuracy is needed, halving the distance may be good enough for use of equation (9.2)—even if not for equations (9.1), (9.3), or (9.4). Notice that the actual length (l') varies with the *square* of the diameter, other things (such as v) being equal. Also, while the formula implies that achieving fully developed flow should take less than one pipe diameter below $Re = 20$, in practice one should always assume an entrance length of at least a diameter.

Many organisms use short pipes—insects that suck sap, blood, or other sources of nourishment are only the most familiar. So entrance length might matter if one wants to calculate the cost of pumping. Loudon and McCulloh (1999) offered equation (9.6) as a correction for the Hagen-Poiseuille equation (9.2):

$$\frac{Q_{\text{real}}}{Q_{\text{calc}}} = \frac{l/l'}{(l/l') + 0.52}, \tag{9.6}$$

where l is the length from the entrance at issue, and l' is the entrance length as calculated from equation (9.5).

As they pointed out, the bottom line says that the correction doesn't amount to much for the usual suspects—aphids feeding on phloem, leafhoppers on xylem, assassin bugs on blood, and so forth. Some larger, extinct insects had mouthparts large enough so the entrance effect may have reduced flow. Still, while insects provided the focus of the journal in which they published, they don't exhaust the world of short, fat, intake pipes.

Laminar Flow Between Parallel Plates

The tidiness of equations (9.1) through (9.4) reflects less the circularity of the pipe than the strictly laminar regime of flow. Analogous equations apply to other geometries. Of especial biological relevance is flow between parallel flat plates—the gills of fish for instance (Stevens and Lightfoot 1986), which generates the similar (but not identical) parabolic profile shown in figure 9.3. Assuming that the plates are much closer together ($2a$, where a is half the interplate spacing) than they are wide (w), we have the following equations. For the velocity distribution between the plates, analogous to equation (9.1):

$$v_h = \frac{\Delta p}{l} \frac{(a^2 - h^2)}{2\mu}. \tag{9.7}$$

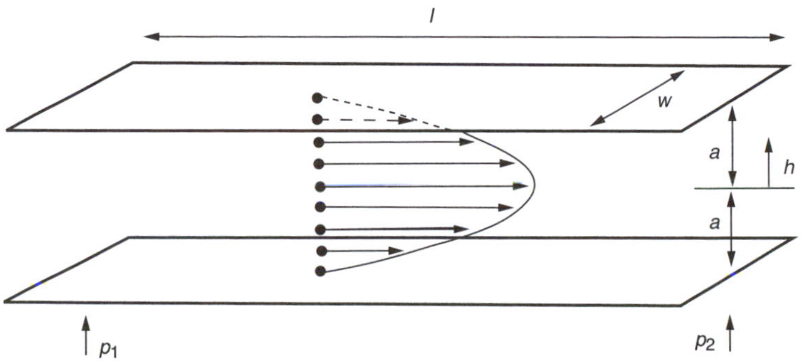

FIGURE 9.3 The parabolic velocity distribution (and symbols) for flow between a closely spaced pair of parallel plates.

For total flow between the plates, analogous to equation (9.2):

$$Q = \frac{\Delta p}{l} \frac{2wa^3}{3\mu}. \tag{9.8}$$

For maximum flow speed, midway between the plates, analogous to equation (9.3):

$$v_{max} = \frac{\Delta p}{l} \frac{a^2}{2\mu}. \tag{9.9}$$

For average flow speed, analogous to equation (9.4):

$$v_{av} = \frac{\Delta p}{l} \frac{a^2}{3\mu}. \tag{9.10}$$

Notice that total flow (equation (9.8)) still depends on the fourth power of the linear dimensions of the system but that it's far more sensitive to the distance between the plates (which determines shear rates) than to their cross-flow width. And note that here maximum flow (equation (9.9)) is 1.5 times average flow (equation (9.10)) rather than, as for circular pipes, twice average flow—different, but still simple.

Manipulating the Parabolic Profile

For many purposes a pipe of circular cross section transmitting a laminar flow works just fine. Its construction uses less wall material relative to its inside volume than does any other shape, and it withstands internal pressure with less bulging or cracking as well. (We examine the mechanics of such cylinders in chapter 20.) No other shape gives as low a pressure drop per unit length for a given flow. The pressure losses in laminar flow are usually less than those associated with any amount of turbulence. And flow is minimally sensitive to the texture of the inner pipe walls—small bumps, scratches, and minor bends make little difference.

But for functions that depend on having a lot of surface relative to volume, no shape may be worse. In particular, organisms commonly combine flow through pipes and channels within organisms with exchange processes. Heat, dissolved respiratory gases, ions, and organic molecules pass back and forth between internal fluid and external material, going across the walls of the pipes—that's the main reason organisms bother to pump liquids and gases around. Obviously, intimate contact between fluid and walls facilitates exchange, and a combination of geometry and fluid mechanics that minimizes this contact sounds bad. A particularly effective way to improve contact is through proliferation of small pipes; the point was made in connection with surface areas in chapter 3. But an organism can play various games with the profile of flow itself.

If flow goes between parallel plates instead of in circular channels it sees more wall and less channel; on the average a bit of fluid flows closer to the wall relative to the overall thickness of the channel. Gills of many aquatic organisms, both vertebrate and invertebrate, take the form of parallel flat plates, as do the so-called book lungs in spiders. What matters, of course, isn't the flatness of the plates but the relatively uniform distance between them. Thus, the intestine of an earthworm has a U-shaped passage, as in figure 9.4a. And the nasal passages of mammals are extensive and flattened in cross section, as in figure 9.4b. Contact between air and wall serves a variety of functions even beyond the exchange involved in olfaction. In small mammals and birds, heat from an exhalation may be used to warm the walls; the same heat then warms the subsequent cold inhalation instead of being lost to the atmosphere. And in the same process, water vapor from the lungs may condense out on the cold walls where it can be used to moisten the next inhalation as another conservation device (Schmidt-Nielsen 1972). Interestingly, adult humans exhale air only slightly below body temperature, but an infant exhales air closer to ambient temperature (at least my son and first grandson did so)—the surface-to-volume ratio within the passages probably makes the difference.

But deviation from circularity costs power, and the cost of respiratory pumping isn't trivial. The power expended in pushing a fluid through a channel equals the product of volume flow and pressure drop, so equations (9.2) and (9.8) generate the following equations. For flow down a circular pipe:

$$P = Q\Delta p = \frac{\pi \Delta p^2 a^4}{8\mu l}. \tag{9.11}$$

For flow between parallel walls:

$$P = Q\Delta p = \frac{2\Delta p^2 w a^3}{3\mu l}. \tag{9.12}$$

Using equations (9.11) and (9.12), we might consider the increase in cost associated with pushing a given flow through a channel with a 20:1 cross

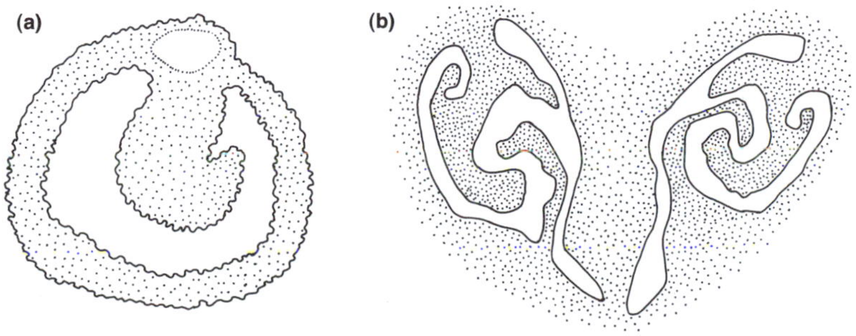

FIGURE 9.4 Examples of pipes in which flow travels between nearly equidistant walls—in effect, flow between parallel plates. (a) A cross section of the intestine of an earthworm, and (b) passageways within the turbinate bones of a pair of mammalian nasal passages.

section ($w = 40a'$) rather than a circular one—setting $\pi a^2 = 2wa'$ and $Q_{pipe} = Q_{channel}$ and fixing l. The cost isn't small—I calculate an increase factor of 2.67. (Checking my figure provides a good exercise in using some of the preceding equations!)

Recall that the parabolic profile comes about solely because the material walls of a pipe cause it to resist flow. What if the walls provide propulsion instead of resistance? A ciliated surface does just that, greatly increasing the steepness of the local velocity gradient. We're a bit big to make much use of the device (or so one guesses)—cilia in our respiratory passages move surface mucus, not the air itself. But lots of invertebrate gills have ciliated surfaces as the basic pump, and the digestive tubes of many small creatures are lined with cilia, so the arrangement is really quite widespread. Such ciliary propulsion ought to facilitate exchange between fluid and walls of conduits by causing flow, on average, to come closer to the conduit walls, but I haven't found specific figures to cite.

One scheme mentioned from time to time for promoting intimacy between flowing fluid and channel wall (for instance, Burton [1972]) may be less effective than claimed. Large vessels with nearly parabolic velocity profiles transport blood from heart to capillaries. (Some distortions arise from the pulsatile flow, from branching, and from the peculiar irregularities of blood viscosity.) In the capillaries the flow is far from parabolic—the red blood cells just squeeze through, generating an odd pattern of flow in the plasma between them. Since the no-slip condition hasn't been repealed, flow slows at the walls. But it can't achieve full parabolic flow since the next red blood cell comes along so soon—red blood cells make up almost half of blood volume. With the space between each pair of cells amounting to a fixed and isolated volume, slowing near the walls must be accompanied by speeding along the axis. That results in a toroidal flow, shown in figure 9.5. Which sounds like a terrific way to improve exchange between the plasma and the endothelial cells lining the capillaries, a purely passive effect of the cells.

But think back to the discussion of the Péclet number in the last chapter. If we plug into equation (8.12) values for capillary length, capillary speed, and the diffusion coefficient of oxygen, we get a value above one. That suggests that the system already does about as well at convection as it can, and augmentation of diffusion by toroidal flow may buy little improvement—see, for instance, Middleman (1972).

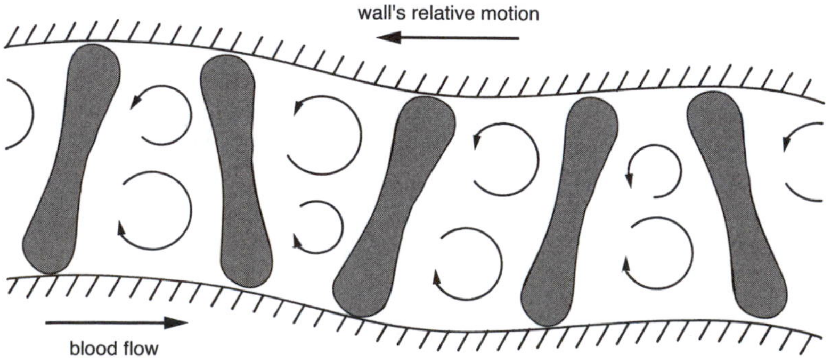

FIGURE 9.5 Red blood cells moving through a capillary. A cell in motion would see the walls as moving backward and the vortices as stationary.

Turbulent Internal Flows

While most flows within organisms remain laminar, a few do get into the turbulent range—blood flow in large mammals during activity and air flow in the largest of the airways. In some cases, turbulence provides a way to improve heat exchange between the flow and surrounding tissue. Not only does turbulent pipe flow develop a flatter velocity profile, with flow on the average closer to the walls, but turbulence assures cross-flow exchange, as we noted in the last chapter in connection with boundary layers. For either heat or mass transfer, that has to be important in nasal passages and the like—stirring takes place almost to the wall itself.

Beyond these instances within organisms, one contends with turbulent flow in equipment designed to look at the interactions between organisms and flowing fluids, even where the organisms are small enough so they see essentially laminar flow. For instance, a flow tank made of pipe 0.1 meters in diameter within which water flows at 0.5 meters per second will be operating at a Reynolds number of 50,000. That's far above the limit of laminarity even without factoring in corners, propeller, and the like.

Engineering handbooks provide excellent guidance for turbulent flow, since that's what practical plumbing must contend with. One can look up the pressure losses caused by corners, contractions, joints, and the like. The main operative formula has a rather different form than those we've seen so far. So it's worth a brief introduction, at least to flesh out the present story. In particular, it lacks the simplicity and tidiness of the formulas that describe laminar flow. The flow profile isn't a straightforward parabola (or any other easy function), the internal roughness of the pipe matters quite a lot (as with the earth's boundary layer), Reynolds number becomes more than a threshold, and pressure drop follows density more than it does viscosity.

In practice, all the eccentricities of turbulent pipe flow get piled onto a universal variable dimensionless constant analogous to the drag coefficient. This one is most often (but not always) called the friction factor, f, and comes down to a ratio of shear stress to dynamic pressure:

$$f = \frac{\Delta p}{l} \frac{4a}{\rho v_{av}^{2}}. \tag{9.13}$$

(Some sources omit the factor of four, which alters all tabulated data.) Friction factor can be used for any flow, laminar or turbulent—equation (9.13) amounts to nothing more than a formal definition. For laminar flow, a little algebra (try it) shows that $f = 64/Re$, but that has little practical significance beyond permitting us to extend the line of figure 9.6 liberally leftward.

Figure 9.6 points up a peculiar source of fluid-mechanical instability, analogous to the periodic vortex shedding behind unstreamlined bodies that we noted several chapters ago. Say flow comes out of a short pipe that's connected (by a wide, low-resistance tube) to a reservoir at a fixed height (or driving pressure), and the Reynolds number is around 2,000. Laminar flow speeds up, exceeding 2,000; turbulence starts abruptly; the higher resistance then slows the flow; laminarity is restored; the flow speeds up again, and so forth. In effect, turbulent plugs appear periodically, and the flow comes out in spurts—with

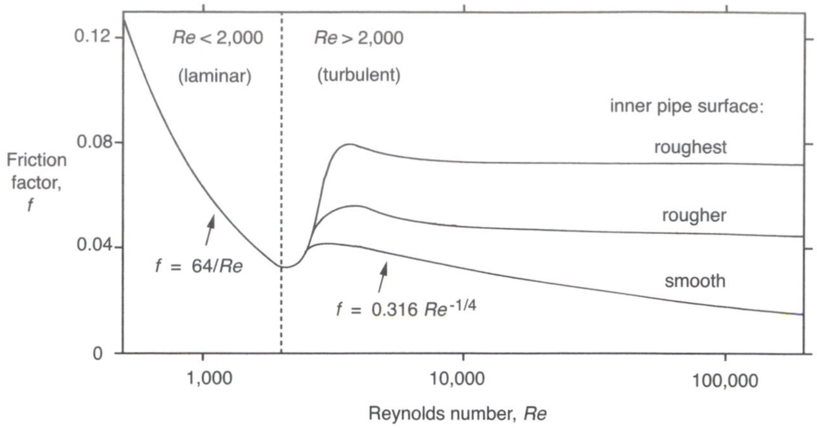

FIGURE 9.6 Friction factor versus Reynolds number for flow through cylindrical pipes of varying degrees of surface roughness. Notice the transition at $Re = 2{,}000$ and how roughness matters only above that transition point.

either higher or lower driving pressures flow remains steady. Demonstrating the phenomenon presents no great difficulty—I do it in class with about 12 centimeters of glass tubing and a reservoir on a ring stand or elevated lab jack. In the process, one gets a good view of the difference between exit flows from laminar and turbulent streams. I'm not aware of any biological example, but it might be involved in some form of pulsed vocalization.

Flow Through Circular Apertures

What if a pipe has negligible length? That means an aperture or a pore, for which analogous formulas tend to be somewhat obscure—one more often encounters treatments of flow through porous media, something quite different. Diffusion through pores came up in the last chapter; here we're talking about bulk flow driven by hydrostatic pressure. Happel and Brenner (1983) gave an equation for laminar flow through single apertures reasonably far from other ones:

$$Q = \frac{a^3 \Delta p}{3\mu}, \tag{9.14}$$

where a, again, is radius. Note that total flow varies with the cube of radius or diameter—not the square as might something like light rays or the fourth power, as in the Hagen-Poiseuille equation for proper pipes. The formula works reliably only up to a Reynolds number of about 3, above which edge effects, vortices, and turbulence raise their ugly heads.

For turbulent flow, a slightly less obscure equation finds use. But this one has the usual fudge factor, analogous to drag coefficient and friction factor, in this case called, a bit salaciously, the orifice coefficient, C_o,

$$Q = C_o \pi a^2 \sqrt{\frac{2\Delta p}{\rho}}, \tag{9.15}$$

so it's little more than a definition. The value of C_o varies a bit, although never straying too far from unity; above $Re = 30{,}000$ it settles at about 0.61.

Flow through circular apertures has biological relevance. In particular, in many trees the upward-conducting pipes, the xylem elements, are periodically interrupted by perforated plates. For instance, the vessels of a red maple, about 50 micrometers in diameter, extend about only 12 millimeters vertically plate to plate; those of a holly get up to 1.3 meters long (Niklas 1992). These plates provide some protection against upward and downward penetration of an air embolism (as noted in chapter 5, an air-water interface doesn't go through a small hole without great urging). But they extract a cost in pressure, one that's usually less but still comparable in magnitude to those imposed by gravity and the resistance (Hagen-Poiseuille) to flow through pipes. Equation (9.14) (solved for $\Delta p/Q$) gives a reasonable approximation of the resistances found by Ellerby and Ennos (1998) for models of simple perforated plates in xylem.

More problematic is application of equation (9.14) to flow through the glomerular filter that serves as the first stage of our kidneys. We filter about 120 milliliters per minute through the glomeruli, driven by a hydrostatic pressure difference of about one third of average arterial pressure. But the pores are truly tiny, excluding anything larger than 3–5 nanometers in diameter—a properly working kidney passes no blood protein. Getting enough stiffness to take the pressure takes some length relative to their diameters, and a little length goes a long ways toward making an aperture into a pipe at their low Reynolds numbers. The same argument should hold for filtration through pores in the capillary walls of the rest of the body.

Bernoulli Versus Hagen and Poiseuille

Recall the pressure curves for flows around bodies, both streamlined and unstreamlined (fig. 7.6). Pressure, following the mandate of Bernoulli, dropped below its ambient value on the sides of the bodies as flow sped up. Faster flow was associated with lower pressure (again, cause and effect can be taken either way). Now we've introduced the Hagen-Poiseuille (equation (9.2)), and variations on it, which implies an association of faster flow with higher pressure. Fluid flows through a pipe—will the walls of the pipe tend to expand as a result of a pressure increase (Hagen-Poiseuille prediction) or constrict as a result of a pressure decrease (Bernoulli prediction)? How can we resolve this seemingly paradoxical question?

Physical models prove both instructive and amusing. A thin-walled rubber cylinder serves as a convenient indicator of what pipe walls will do, and we can get one by doing no more than cutting off the end of a cylindrical balloon. Say, as in figure 9.7, we attach plastic tubing to both ends, a short piece on one end and a long piece on the other. Blow through the short piece, and the balloon will inflate. Blow through the long piece, and the balloon will collapse and oscillate irregularly (perhaps making flatulent noises). The first blow seems, in some way, Hagen-Poiseuille dominated; the second seems Bernoulli dominated. (The irregular oscillation must come from secondary pressure rise every time the thing collapses.) How can we put the result in quantitative, predictive terms?

Our now familiar device, the dimensionless ratio, comes (in part) to the rescue. We can take dynamic pressure, $\rho v^2/2$, as the manifestation of Bernoul-

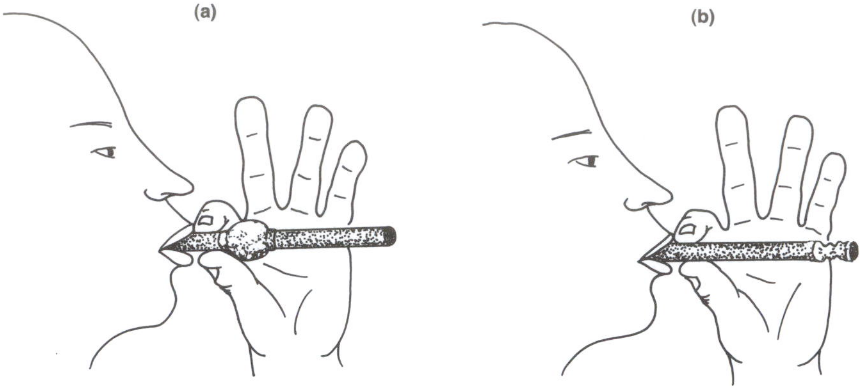

FIGURE 9.7 Blowing through a long cylindrical pipe. The behavior of a section of balloon tells whether flow inside comes with an increase or decrease in internal pressure. In (a) the balloon near the input expands, while in (b) the bit of balloon near the output collapses and oscillates.

li's principle (*B*) and equation (9.4), $8v_{av}l\mu/a^2$, for average Hagen-Poiseuille pressure (*HP*). Putting one over the other, we get

$$\frac{B}{HP} = \frac{\rho v a^2}{16\mu l}. \tag{9.16}$$

This says, assuming the same speed, viscosity, and density, that the fatter tube will tend to be Bernoulli-driven into collapse, and the skinnier tube will tend to be Hagen-Poiseuille-driven into aneurystic expansion. Conversely, the longer tube will be Hagen-Poiseuille dominated and the shorter one will be Bernoulli dominated. The latter explains the behavior of the model—with a longer pipe (more resistance) downstream of the pressure indicator, the balloon expanded; with a shorter pipe it collapsed. (You may also be able to show that speeding up flow may make a previously expanding balloon wall start to collapse.)

What about circulatory systems? For a pipe 100 millimeters long and 1 millimeter in diameter carrying blood at 0.1 meters per second, the ratio has a value of about 0.01, telling us that Hagen and Poiseuille, not Bernoulli, are in charge. Circulatory systems, of course, have much longer pipes, even if they're arranged in complex branching and coalescing arrays. So when the heart squeezes, your arteries expand—as you know from feeling your pulse. Physiology textbooks may begin their sections on circulation by citing Bernoulli's principle, but few ever mention it again. It matters only in a few instances—around valves, where vessels suddenly constrict (coarctations), and other places where viscosity doesn't have time (lots of speed relative to length) to exert much effect.

Where else might pipes be short but wide? The intake chambers of the carburetors with which we used to draw gasoline into the air entering our automobile engines and the aspirators on laboratory faucets have dimensions adequate to suck in gas and gases, respectively. A more biological possibility is the terminal length of a urethra. At some point as urine approaches its end, remaining length (as in the balloon) ought to be short enough so net pressure becomes inward. So one might anticipate, the more so in large mammals, urethras braced near their termini against collapse rather than expansion.

One final note about equation (9.16). If one tosses out the constant in its denominator and casually merges radius and length into some generic length, the Reynolds number emerges. Maybe we ought not be surprised. Bernoulli's principle ignores viscosity and thus becomes increasingly applicable as the Reynolds number increases. The Hagen-Poiseuille equation applies only at Reynolds numbers below the critical value of 2,000, and it depends on viscosity instead. Such things give the subject a satisfying symmetry.

A Pump Primer

So far, the talk about pipes has focused on the loss of pressure as a fluid moves along. We turn now to the opposite process, elevating the pressure to make and keep the fluid flowing—from pipes to pumps. Nature makes a great diversity of pumps, including ciliated surfaces, valved hearts, peristaltic tubes, osmotic and evaporative devices, and paddles. For that matter, we can't draw any sharp line between thrust-producing devices used in locomotion and pumps used for pushing fluid. Thus, bees beat their wings both to fly and, standing still, to ventilate their hives.

Human technology has generated a similar diversity of pumps, which drive fluid with impellers, pistons, jets, and so forth. These are most often divided into two general classes, termed positive-displacement or fluid-static pumps, and fluid-dynamic or rotodynamic pumps (Karassik et al. [1976] or other standard engineering sources). The distinction applies as well to biological pumps, even where the specific devices bear little overt similarity to technological analogs.

Most positive-displacement pumps work by luring successive bits of fluid into a confined space and by then reducing the volume of that space, forcing the fluid to leave by a specific outlet. The everyday example is a tire pump for a bicycle or car. It's equipped with a piston to reduce internal volume and a pair of one-way valves. One valve opens when the piston moves upward, the other when it moves downward, producing an pulsatile output. Hearts of our kind work just the same way, adding only multiple chambers and additional valves. Peristaltic pumping, done by some invertebrate hearts and by our intestines, works on the same principle, although with traveling occlusions instead of discrete valves. The evaporative sap lifter of trees must be considered a positive-displacement pump—removing liquid at the top draws more liquid up from below; it just happens to use reduced rather than increased pressure. Similarly, the osmotic pumps (chapter 22) of roots (and elsewhere) are positive-displacement devices, since they make liquid move by creating hydrostatic pressure.

Most fluid-dynamic pumps, by contrast, force fluid into motion with rotating impellers or propellers. Fluid may move outward centrifugally (as with squirrel-cage fans and centrifugal pumps) or it may move directly forward (as with ordinary axial fans). Nature's versions of these differ more from their human analogs than do her positive-displacement pumps, most likely a consequence of the dearth of rotational devices in her technological armamentarium. But, with lots of cilia and flagella, with countless paddles, fins, and wings, fluid-dynamic pumps abound in organisms.

Any pump does work on its fluid, and its pressure-volume work has the same dimensions as the more familiar force-distance work. A pump applies a pressure (or, strictly, creates a pressure difference) and moves a volume. Put per unit time, it inserts power, offsetting the power extraction described by equations (9.11) and (9.12), and expressed in the same terms of volume flow times pressure difference, $Q\Delta p$. Of no small importance, though, are the relative magnitudes of the two variables, pressure and volume per time. Every pump deals with both, and every pump achieves its greatest effectiveness at a particular combination of the two.

That trade-off can be best seen in graphical form, as in figure 9.8. (Riisgård and Larsen [1995] give less idealized examples.) A curved line describes the pressure a pump can produce against the volume it can pump and, doing so, gives the outer bound of the possible combinations of pressure and flow the pump can manage. Since power is pressure times flow, an alternative y-axis (usually on the right, as here) and second curve indicate the power output of the pump for those maximal combinations of pressure and flow. (In a sense it adds nothing new, power being pressure times flow.) At maximum pressure, the pump moves no fluid, while at maximum movement the pump can't boost the pressure—therefore, neither corresponds to a positive power output, and maximum power happens somewhere in between. A third curve, extending outward from the origin, represents the resistance imposed by whatever system the pump is being called on to work against—it's not, strictly speaking, a characteristic of the pump itself. The point at which that third curve intersects the curve of pressure versus volume (our first curve) gives the maximum the pump can do when working against that particular load.

Clearly, the pump does its best if the intersection of the system resistance line and the pump's capability limit lies directly above or below its peak power output. And there's the rub. To work with reasonable effectiveness and efficiency, a pump must be appropriate for its load. Adequate maximum power output isn't sufficient; the relative magnitude of volume flow and pressure change demanded by the load must match that put out by the pump. As a general rule, positive-displacement pumps produce high pressures but relatively low flow rates. Even though both push air around, a scaled-up bicycle pump won't give decent service as a ventilating fan—room ventilation de-

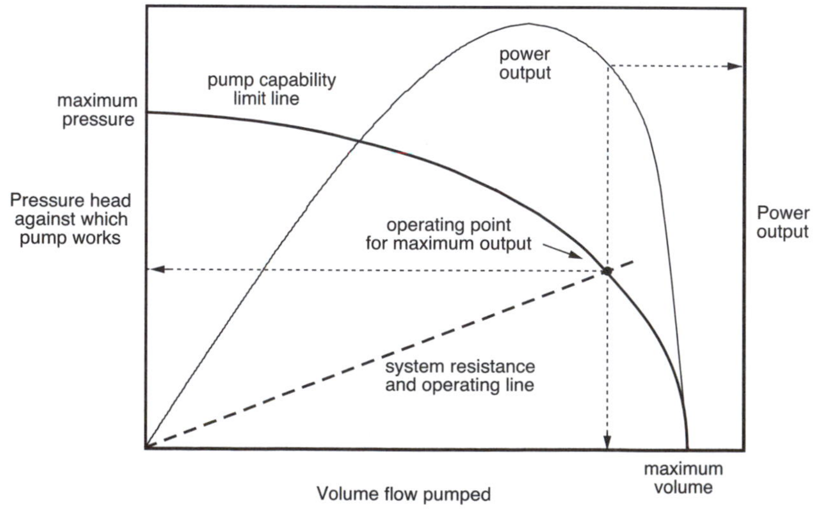

FIGURE 9.8 Pressure increase and power output versus volume flow for a pump. The pump can operate anywhere within the circular curve. The resistance of the particular load sets the system's operating line—the higher the resistance the greater the slope. Maximum output, then, occurs where the operating line hits the limiting curve. Best match of pump with load occurs when that maximum output corresponds to maximum power.

mands lots of flow but entails little increase in pressure. By the same token, one encounters no circulatory systems in which ciliated capillaries replace valve-and-chamber hearts—to achieve a sufficiently low system resistance, a circulatory system would need a mass of capillaries, other vessels, and blood approximating the volume of the organism (chapter 10).

What we're looking at is what's commonly referred to as impedance (or resistance) matching. It's just the same as the trade-off between voltage (the pressure analog) and current (the flow analog) in electrical systems. A 60-watt bulb on household 120-volt electricity takes 0.5 amperes of current; an equivalent bulb on an automobile's 12-volt system draws 5 amperes. The bulbs, although of equal power consumption, can't be interchanged. Nor can pumps of equal maximum power output. Biological pumps encompass not only a wide variety of designs but of choices of relative values of pressure boost and volume flow rate. Ancestry may constrain what a particular organism might employ—plants don't go in for moving parts, and arthropods don't make motile cilia—but nonetheless we can recognize good matches between functional role and pump performance. In short, nature's pumps (is anyone surprised?) seem well matched to the resistances of their loads. Table 9.1 should give at least a semiquantitative view of the diversity of both factors, while figure 9.9 depicts some of the organisms and devices.

Unfortunately, the measure of resistance used for electrical work doesn't adapt to our present problems without obfuscating as much as it enlightens. One can define an equivalent variable, resistance (R), as pressure increase (for pumps) or pressure drop (for the load imposed by the system) divided by volume flow. Thus, for laminar flow through circular pipes,

$$R = \frac{\Delta p}{Q} = \frac{8\mu l}{\pi a^4}. \tag{9.17}$$

The variable, though, has awkward dimensions. Even limiting our purview to a given fluid (to fix viscosity), we still run into systematic variation with size—l/a^4 means that, other things being equal, load resistance will vary inversely with the cube of length. The smaller system will have the greater resistance under what we'd like to think equivalent circumstances, whether ontogenetic, interspecific, or ecological. We'd prefer something less size-contaminated, preferably fully dimensionless. Pressure coefficient (C_p; chapter 7) is such a variable, but it makes a poor fit with systems in which viscosity provides most or all of the resistance, and it has some other problems as well.

An alternative possibility (Vogel 1995a) uses viscosity rather than dynamic pressure as a denominator; that dimensionless pressure-volume flow index (PVI) comes out to be

$$PVI = \frac{\Delta p t}{\mu}. \tag{9.18}$$

The tricky item here is the interpretation of time, t. It amounts to a lengthless velocity in the denominator, and it's probably most usefully interpreted as the time the fluid needs to pass through whatever part of the system imposes the greatest part of its resistance to flow.

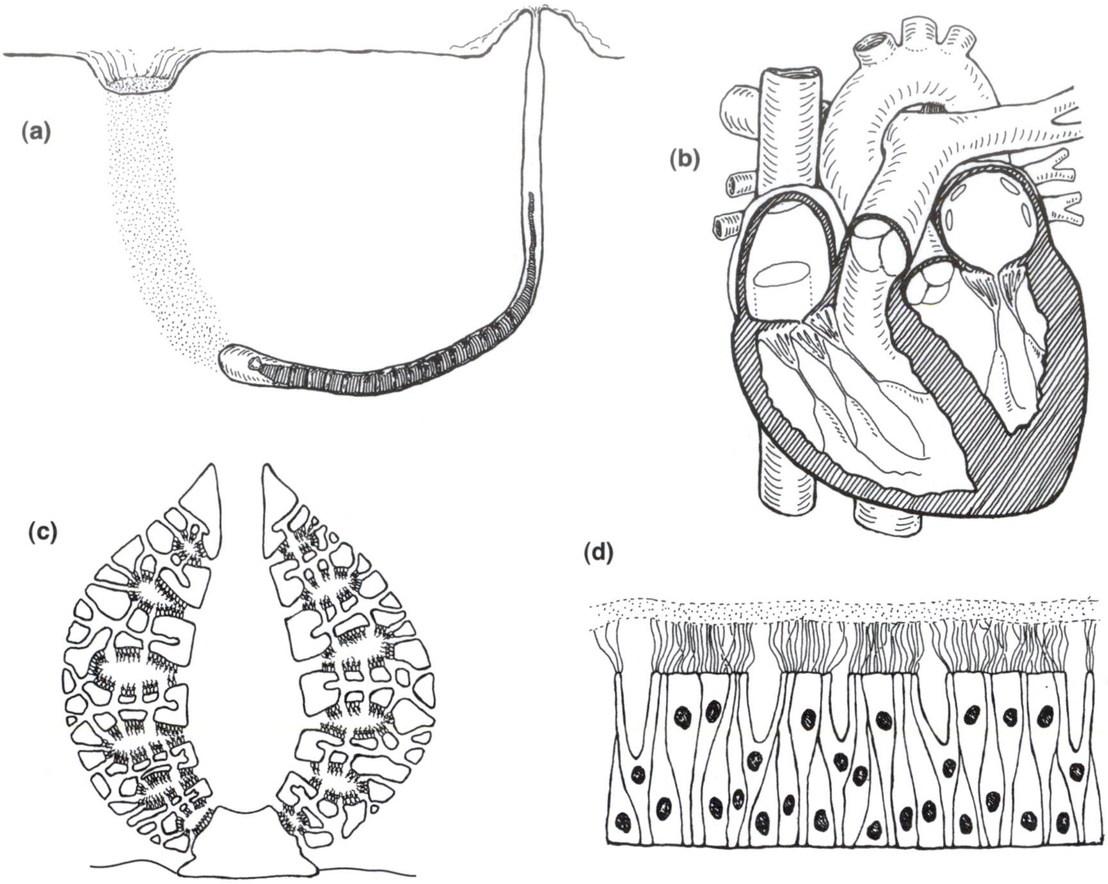

FIGURE 9.9 Pumps in nature. The infaunal marine worm, *Arenicola* (a), pumps water through its burrow by passing waves of contraction down its body, and our hearts (b) pump by squeezing chambers—both are positive-displacement pumps. A sponge (c) uses chambers lined with flagellated cells (choanocytes) to pump, and we line much of our respiratory tract with ciliated epithelia (d) that push mucus along—both are fluid-dynamic pumps.

Using this index for flows of aqueous fluids (including bloods), how do the various biological pumps stack up or sort out? Table 9.2 gives a few data.

While the fluid-dynamic pumps come in at lower indices than most of the positive-displacement ones, the distinction isn't complete. But we do see relatively low values for jets, as necessary to keep their Froude propulsive efficiencies from being intolerably low. Hearts have high values, which allows circulatory systems of relatively high resistance and thus relatively low blood volume, as noted earlier. The filter feeders have to deal with huge Qs—keeping the cost in line with potential yield must mandate very low values. (For these, Riisgård and Larsen [1995] summarized a large amount of data.) Still, one has no reason to expect a perfect dichotomy—the pumps of our technology don't divide perfectly on this criterion either. Size does seem to factor out fairly well, and wide variations in pressure and transit time collapse nicely into the index. But we need more data, especially for fluid-dynamic pumps.

TABLE 9.1 Various Kinds of Biological Pumps

Type	Category	Impedance	Examples
Evaporative	positive displacement	highest	leaf sap sucker
Osmotic	positive displacement	very high	root sap pusher
Valve/chamber	positive displacement	high	heart, bird lungs, squid jet
Peristaltic	positive displacement	high	intestine, some hearts
Piston	positive displacement	medium	some tubicolous worms
Valveless/chamber	positive displacement	medium/low	our lungs, jellyfish jet
Drag-based paddles*	fluid dynamic	medium	crustaceans in burrows
Lift-based propellers*	fluid dynamic	low	hive-ventilating honeybees
Ciliary layer	fluid dynamic	low	bivalve gills
Independent flagella	fluid dynamic	very low	sponge choanocytes

*For the distinction between these, see chapter 13.

TABLE 9.2 The Performance of Various Biological Pumps

Pump	Pressure (Pa)	Transit Time (s)	*PVI*
Positive Displacement			
Sap-lifter, red oak	500,000	3,000	1.5×10^{12}
Heart, human systemic	13,000	50	1.9×10^{8}
Heart, octopus systemic	2,500	120	3.0×10^{8}
Worm as piston in burrow	90	10	9.0×10^{5}
Pumped fish gills	30	20	6.0×10^{5}
Urethral pump, human male	4,000	0.10	5.6×10^{5}
Squid, jetting	30,000	0.0025	7.5×10^{4}
Jellyfish, jetting	30	0.5	1.5×10^{4}
Fluid Dynamic			
Ram-ventilated tuna gills	300	0.25	8.5×10^{4}
Mussel, ciliary pumping	20	2.5	5.0×10^{4}
Sponge, flagellar pumping	15	2.5	4.0×10^{4}

Sources: Data from Zimmermann 1971; Vogel 1978b; Wells 1983; Lauder 1984; Meyhöfer 1985; Stevens and Lightfoot 1986; DeMont and Gosline 1988b; Riisgård 1989; Glemain et al. 1990; Riisgård and Larsen 1995.

CHAPTER 10

More about Circulatory Systems

Nothing venture,
nothing win—
Blood is thick, but
water's thin—
In for a penny,
in for a pound—
It's love that makes
the world go round!

W. S. GILBERT'S (*Iolanthe*)
Viscous Cosmology

NO AREA OF BIOLOGICAL FLUID MECHANICS HAS COME IN FOR MORE INtense or sustained attention than circulatory systems—at every level of science and culture. Heart and blood get more literary allusions than any other aspects of the human corpus, as a look at any dictionary of quotations will quickly verify. And disabilities of our circulatory systems dispatch more of us mature inhabitants of well-fed and well-medicated lands than any other class of pathologies. We'll turn now to an explicit rather than parenthetical look at these systems—after two (yet again!) prefatory comments.

First, up to this point in the book (unless one has read part III before part II), physical science has provided our organizational framework. Biology may be our subject, but taking our sequence of topics from physical science turns out to have major pedagogical advantages. At least it seems easier (perhaps because the author is a biologist) to jump around among phyla and biological functions than to intermingle surface waves, material moduli, and dimensional analyses. So the present chapter reflects a different kind of viewpoint than the previous ones. In the process it uses some solid as well as a lot of fluid biomechanics. We may separate the two, but our distinction-of-convenience carries the perils of oversimplification. Certainly nature makes no such differentiation beyond her obviously greater ability to tamper with the properties of solid materials. If nothing else, the present chapter points up that practical interdigitation of the two states of matter. Shifting organizational mode will entail a little redundancy with earlier material—so you may recall having heard some items before.

Second, the chapter gets more explicitly engaged with a general point that's been little more than a background presumption since chapter 1. Natural selection has done wonders in generating the diversity of biological designs, but the process can't circumvent the physical context in which it takes place—such things as the acceleration of gravity on the earth's surface and the densities of air and water. The larger question behind most of the present chapter is the extent to which the design of systems that move fluids through organisms reflect that context. Or, put the other way around, we seek limits that the underlying physical context places on the range of biologically practical designs. That question will focus our attention on common features of systems of overtly disparate appearance and independent evolutionary origin. Put less pretentiously, we'll look for lessons to be learned from points of commonality amid the rich diversity of multicellular life.

Why Circulate, and Who Does It

Organisms move fluids through themselves and their domiciles in dauntingly diverse ways. Flows may be unidirectional or oscillatory, single passage or repetitively circulatory; fluids include gases and both simple liquids and non-Newtonian mixtures and slurries; pumps work by evaporative suction, ciliary or flagellar propulsion, or muscle-driven peristaltic and valve-and-chamber contraction. Moreover, the purposes for which fluids move range widely. Organisms move water from soil to elevated photosynthetic organs and bring the products of those organs to points of growth or storage; they move air and blood as transporters of oxygen and other substances; they provide currents of ambient water for suspension-feeding organs. In short, forced, bulk (convective) flows appear diverse as well as ubiquitous.

Nonetheless, the basic device used both for moving material within organisms and for bringing needed dissolved substances to them isn't forced convective flow at all—rather it's the energetically undemanding process of diffusion that we talked about in chapter 8. Nor does bulk flow ever fully supplant diffusion—at every surface across which material moves, the final step in transport remains diffusive. What matter here are the limitations of diffusion and, in the main, the severe dependence of its efficacy on the distance over which it must carry material.

When diffusion was considered, back in chapter 8, Fick's first law (we'll not need the second) was defined, by using equation (8.10). It describes a net transport rate (mass per time, *m*/*t* through a slab of material of area *S* and thickness *x*, driven by a concentration difference ΔC and a diffusion coefficient, *D*:

$$\frac{m}{t} = -DS\left(\frac{\Delta C}{x}\right). \tag{8.10}$$

At a glance, diffusive transport appears to vary inversely with distance, much as bulk transport by, say, a fleet of trucks. Double the distance requires that you use twice as many trucks to move the same amount of material in a given time, since each truck must bear each load for twice as long.

But that's not quite the right analogy since, as distance increases, the concentration gradient itself decreases. One does better by thinking of a fleet of vehicles that roll down a hill while carrying their loads. The steeper the hill, the greater will be their steady-state speed. Doubling the distance increases the carrying time both by lengthening the path *and* by reducing its steepness. We can move back toward biological reality by considering how far a particle or molecule might diffuse in a given time. Of course we know nothing about any particular one, so we use instead a kind of average distance; assuming diffusion from a region of high to one of zero concentration, the mean-square distance, which was given by Morowitz (1970), in an equation similar to equation (8.13):

$$(x^2)_{\text{av}} = 2Dt. \tag{10.1}$$

Thus, distance does impose a disproportionate penalty—for most geometries, diffusive transport varies inversely with the square of the distance. Berg (1993) gave a good account in a biological context. So bigger is worse, much worse. As noted earlier, transport that would take 10 milliseconds to cover a micrometer will take no less than 3 *years* to manage a meter. A biological system that relies on diffusive transport must have no active part far distant from the source of what that active part needs, whether oxygen from the medium or transfer RNA from its nucleus.

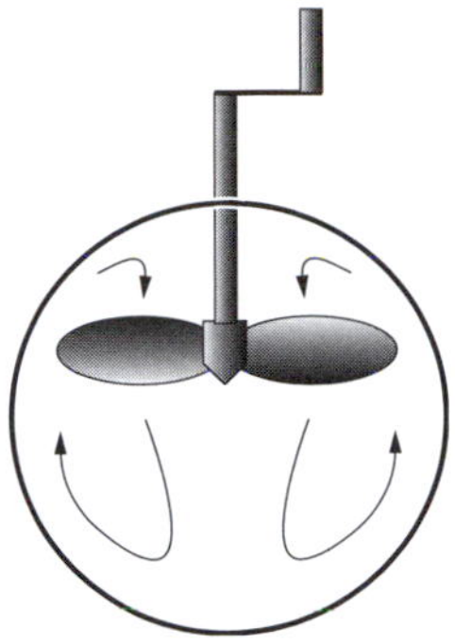

FIGURE 10.1 The minimalist's circulatory system—turning the crank turns a propeller, causing convective transport within the "cell."

That scaling rule for diffusion presents a severe challenge to the design of all but very tiny life forms—the basic physical problem most likely impeded the evolution of macroscopic life forms in many lineages. An organism can get larger if willing to accept a *much* more leisurely life, with its attendant competitive disadvantage. That's mandatory if diffusive transport limits life's processes—for a set of geometrically similar organisms, maximum metabolic rate must follow, not body mass, but body mass raised to the one-third power. (Or the mass-specific rate will vary inversely with length squared or with mass to the two-thirds.) The alternative consists of an equally daunting geometric constraint, adopting shapes in which no active element lies far from a surface for material exchange. The latter clearly does happen, since we know of lots of threadlike algae in several lineages, and we're familiar with even more flat algae and worms. In a host of large anemones and polyps, among the coelenterates, almost all active cells coat their external surfaces or line their internal (gastrovascular) cavities, the last only minimally separate from the external medium.

The general solution, as pointed out long ago by the great physiologist August Krogh (1941), consists of augmenting diffusion with bulk flow—convection—when distance becomes substantial. In practice, "substantial" proves to be 5–10 micrometers, the size of a typical animal cell. Among animals, the cellular world mainly transports material by diffusion; to this the multicellular world adds bulk flow. Despite the wide size range of animals (around 10^8-fold in length), cells vary little in size. We might recognize a general principle that the scaling problem of diffusion limits cell size and that the boundary between cellular and multicellular corresponds to purely diffusive versus convectively augmented domains. A few cell types far exceed 10 micrometers, but for each of these we recognize some special consideration. A nerve cell is very thin, an avian egg does its business on its surface, a muscle cell is multinucleate and penetrated by blood vessels. And some bulk flow does happen within animal cells, although not as the main mode of transport.

Augmentation by bulk flow need not involve an array of pipes and pumps—even minimal, low-Reynolds number stirring of the inside of a compartment will help a lot, as in figure 10.1. Plant cells are commonly an order of magnitude longer than animal cells, thus about a thousandfold greater in mass, suggesting mass-specific metabolic rates a hundred times less. But they commonly stir their contents in the process termed cyclosis, something that will get more attention when we consider motility (chapter 22). Thus, in the plant kingdom, that shift in transport mode doesn't set cell size, and we can't recognize any tidy correspondence of cellular with diffusive transport and multicellular with convective transport. Besides using cyclosis, a large plant cell typically contains a central vacuole of material that's only minimally active metabolically, so most of its chemistry gets done around the edges, an intracellular version of the coelenterates' dodge. That ought to further facilitate large cell size.

So we have, at best, a broad-brush picture and a semiquantitative view, one arguing that even something as distant as cyclosis serves the same basic function as a definitive circulatory system. Indeed, many familiar convective schemes (such as the tracheal systems of insects and our own respiratory systems) don't fit comfortably under a heading of "circulatory systems." And "internal bulk fluid transport system" doesn't even generate a decent acronym. But all accomplish the same thing and so ought to be included in the discussion.

And that function, in essence, consists of reducing the dependence of transport on the square of distance to a dependence on the first power of distance—as with a conventional fleet of trucks. Transport still takes time (if less), but now doubling the distance merely doubles the time for material to make the journey. Matters get worse as size increases, but things deteriorate only in proportion to linear size, not to its square. A logarithmic plot of transport rate of material (or even heat) relative to body mass versus body length has a slope of minus one, not minus two—as in figure 10.2. That's about what we do—the metabolic rates of animals (relative to body mass) typically decrease nearly in proportion to their lengths. Or, as put in chapter 3, basal metabolic rates vary with something close to the two-thirds power, not the first power, of body mass.

Matching Diffusion and Convection

Excessively enthusiastic bulk flow as a way to augment diffusive exchange runs into the twin problems of excessive cost and diminishing returns. Pushing fluid through tiny pipes and channels demands lots of power, and diffusion works quite well, thank you, when distances are small. How might internal bulk fluid transport systems be arranged to best deal with cost—that is, to do their jobs most economically?

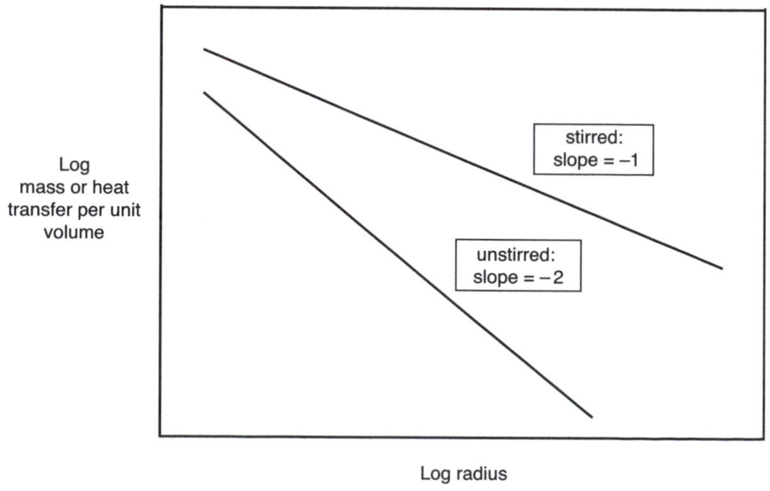

FIGURE 10.2 How the rate of transport of heat or mass relative to the volume of the container varies with size for two situations—with and without stirring the contents of the container. Containers (100-, 400-, and 1000-milliliter laboratory beakers when we tried it) vary in size but not shape.

First, consider cost in a circulatory system of the basic vertebrate type, one in which material is exchanged between the tissues and the smallest pipes, the capillaries. Obviously, an exchanger ought to employ small pipes to ensure intimate contact between circulating fluid and tissue, but just how small ought those tiny pipes be? If excessively small, the circulatory system will be doing work that diffusion might do without metabolic cost. If excessively large, flow will have to be slow in order to allow time to approach diffusive equilibrium; that is, for ample exchange to occur before fluid leaves the capillary bed.

Making fluid flow through unnecessarily thin pipes entails a dramatic increase in cost. The relevant rule comes from the Hagen-Poiseuille equation, noted in chapter 9; total flow (Q) or volume per time translates directly into power (P):

$$Q = \frac{\Delta p}{l}\frac{\pi a^4}{8\mu} \tag{9.2}$$

and

$$P = Q\Delta p = \frac{\pi \Delta p^2 a^4}{8\mu l}. \tag{9.11}$$

To remind ourselves, $\Delta p/l$ is pressure drop per unit length of pipe, μ is viscosity, and a is pipe radius. Note again that inverse *fourth* power of the radius. Making pipes narrower costs disproportionately. And powering our circulatory systems entails significant expense—about 10.7 percent of our resting metabolic rate.

Making pipes wider would save doubly, an attractive prospect. First, there's that radius-to-the-fourth. In addition, for decently efficient exchange, flow would necessarily have to be much slower—adequate cross-flow diffusion would take a lot more time—which is also cheaper. But making pipes wider raises a pair of difficulties of at least equivalent magnitude. Filling wider pipes takes more blood. And slower flow means that each bit of blood must cycle less often, requiring a still greater blood volume. An increase in volume would of necessity come at the expense of other systems, again a significant business. Both mammals and birds invest about 6.5 percent of body volume in blood, and that fraction varies insignificantly with body size (Calder 1984). A similar figure of 5.8 percent in octopus (Martin et al. 1958) argues that we're once again not talking about some accident of ancestry. Bony fishes, incidentally, are a slightly less bloody lot. As expected, animals with open circulatory systems (about which more in a few pages) such as arthropods and noncephalopod mollusks have much larger blood volumes. The cartilaginous fishes—sharks, skates, and rays—with venous sinuses and thus incompletely closed systems have blood volumes in the same range as we, about double that of bony fishes (Prosser and Brown 1961).

So do capillary size and flow speed give a reasonable match between flow and diffusion? A crude device for assessing that match was introduced earlier—the Péclet number, the ratio of convective to diffusive transport:

$$Pé = \frac{lv}{D} = \frac{av}{D}. \tag{8.12}$$

Length here ought to be interpreted as the distance across which material must pass, thus the radius, a, for a circular pipe, rather than the more literal length used in the previous equations. (We're thus guilty of a little fudging, since the length implicit in velocity no longer has the same direction as the length in the formula.) Good balance, a situation neither confusive nor defective, implies a value somewhere around unity for sites of material exchange.

Consider our capillaries. Flow speed is typically about 0.7 millimeters per second and radius about 3 micrometers. Oxygen transport should be the rate-limiting process, and it has a diffusion coefficient (table 8.1) of 18×10^{-6} square centimeters per second. That gives a Péclet number of 1.2, remarkably (and to some extent coincidentally) close to the ideal 1.0. Of course the value reflects the product of capillary radius and flow speed, not one or the other by itself. But decreasing capillary size and increasing flow speed would raise the cost of pushing blood around from its value of around 10.7 percent of metabolic rate, while increasing capillary size and decreasing flow speed would raise circulating volume from its value of about 6.5 percent of body volume.

We might note that not only does the octopus have a similar blood volume to our own, but it has capillaries and blood cells only a little larger than ours (Smith 1963)—although organizational differences cloud the comparison. Squid and octopus pump blood that carries far less oxygen per unit volume than the blood of almost every vertebrate, and neither can do as much revving up of its aerobic metabolism in bouts of activity. But common physical context largely trumps physiological diversity. Nor does capillary size vary in any systematic way with body size—capillaries of elephant and whale differ little from those of bat and shrew. Nor does capillary size vary with whether red blood cells are nucleated, as in birds, or just disposable corpuscles, as in mammals.

Or take an internal fluid transport system as different from our circulatory system as can be. For cyclosis in large plant cells, material flows at about 5 micrometers per second. If we assume a Péclet number of 1 and transport of carbon dioxide (supplying photosynthesis), with $D = 14 \times 10^{-6}$ square centimeters per second, transport distance comes to 280 micrometers, about a quarter of a millimeter. That's large enough to imply that cyclosis accomplishes nothing much—a cell of the typical 100-micrometer radius would have a Péclet number of 0.36 and be amply served by diffusion. If, on the other hand (and not unreasonably), what matters is moving very small proteins (molecular weight = 12,000) around, then (for $D = 1 \times 10^{-6}$ square centimeters per second), transport distance becomes a more appropriate 20 micrometers, and cyclosis appears much more worthwhile.

The Sizes of the Larger Pipes

The Hagen-Poiseuille equation, equation (9.2), implies that the small pipes needed at sites of exchange ought to be kept short in the interest of economy. The corollary is that pipes for long-distance transport, where exchange isn't involved, should be of greater size. Thus, an efficient system should use both large and small pipes, and without exception, fluid transport systems do so. Table 10.1 gives the diameters of largest and smallest pipes of a variety of such systems. So different are these sizes that one can't easily draw to scale a biological fluid transport system—for the small pipes to be visible, the large ones dwarf the page.

But using both large and small pipes raises a serious difficulty for serially arranged circuits. The principle of continuity (chapter 2) requires that flow speed be inversely proportional to cross-sectional area. Tapering a pipe increases the mean speed of flow within it, of which we take advantage when designing nozzles and spouts. Simply reducing the size of a vessel as it approaches the site of exchange simply will not suffice. Not only does rapid flow in the small pipes consume power profligately, it also minimizes the time available for exchange or filtration. Nor will a branching system that maintains constant aggregate cross-sectional area work either—if a pipe bifurcates into two daughter pipes, each of half its cross section, the average speed of flow won't change. If built this way, blood in your capillaries would, on average, flow as fast as blood in your aorta.

Nature, though, has a universal solution. All systems in which fluid moves in discrete vessels employ branching manifolds, manifolds in which aggregate cross-sectional area actually increases as the cross-sectional areas of the individual pipes decrease, as in figure 10.3. Thus, your aorta has an internal cross section of about 6 square centimeters, while your capillaries have a combined cross-sectional area of around 2,000 square centimeters. The final columns of table 10.1 give examples of such aggregate areas with the factors by which aggregate areas increase. By the principle of continuity, these are precisely the factors by which the average flow speeds decrease.

In fact, a more precise prescription proves possible. Given the sizes of the smallest pipes, an economic argument predicts the sizes of almost all the rest of the pipes of almost all the fluid transport systems of organisms. The argument was first made by Hess in 1914 (as noted by Weibel [2000]), apparently independently by Murray in 1926, and with a different derivation by Sherman in 1980, who termed the resulting rule (probably not knowing about Hess's work) Murray's law. It was derived yet again by Gafiychuk and Lubashevsky (2001), who traced the rule back to Thomas Young, in 1809. The basic argument (of all but the last authors) is that fluid transport systems incur two

TABLE 10.1 Diameters of Individual Conduits and Their Range, and Aggregate Cross-Sectional Areas of Conduits of Different Size Classes—for Three Quite Different Internal Fluid Transport Systems

	Diameters			Aggregate Cross Sections		
	Largest (mm)	Smallest (μm)	Ratio	Largest (cm^2)	Smallest (cm^2)	Ratio
Dog vascular	16	6	2,600	2.0	600	300
Human lung	18	460	40	2.5	1,700	700
Small sponge	2	20	100	0.035	200	6,000

Source: Data from LaBarbera and Vogel 1982.

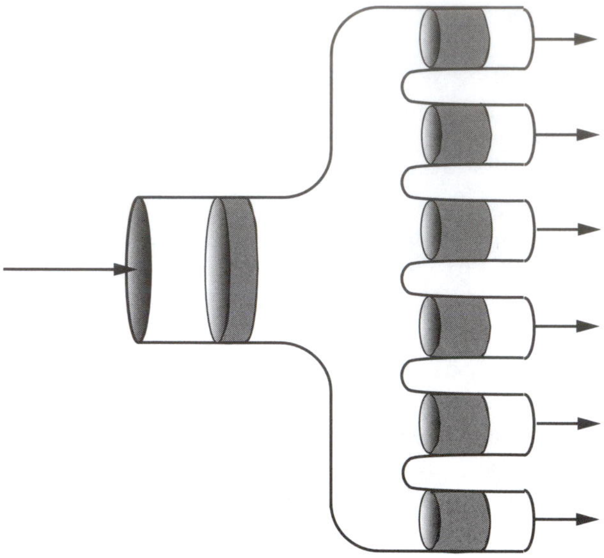

FIGURE 10.3 In a branching array of pipes the speed of flow, here proportional to the length of the arrows, is inversely proportional to the total cross-sectional area at any stage of the array. Thus, individual narrow pipes can have slower flow than individual wide pipes.

kinds of cost. The first, that of making pipes and blood, rises with contained volume and so favors small pipes. The second, that of pumping blood, runs inversely with pipe size, and so favors large pipes. Equating the two generates a formula for cost minimization with only a single constant of proportionality.

If one assumes steady, laminar flow, well downstream from a branch or entry point—the conditions upon which the Hagen-Poiseuille equation is predicated—the rule is a simple one. Volume flow rate through a vessel should be proportional to the cube of the radius of the vessel. To maintain this proportionality, the principle of continuity then tells us that the sum of the cubes of the radii of the vessels downstream from a branch should equal the cube of the original vessel. More generally, at any level in the branching array that makes up a fluid distribution system, the sum of the cubes of vessel radii ought to be the same as that at any other level. Or, in equations,

$$Q = ka^3 \tag{10.2}$$

and

$$a_o^3 = a_1^3 + a_2^3 + \ldots a_n^3, \tag{10.3}$$

where a_o is the radius of the parental vessel, and a_1, a_2, and so forth are the radii of all n vessels that receive the output of that parental vessel after one or more generations of branching.

For a simple bifurcation (as in figure 10.4), each of the daughter vessels will have a radius or diameter 79.4 percent of that of the parental vessel (since 0.794^3 times 2 equals 1). Each daughter will therefore have a cross section 63.0 percent of that of the parent (since 0.794^2 equals 0.63). And, thus, flow in the daughters will be 26 percent slower than in the parent (since 0.63 times 2 equals 1.26). More generally, Murray's law, as given in equation (10.2), gives us a rule that relates vessel diameter and relative flow through the vessel—either total flow, Q, or (dividing by πa^2) average velocity, v_{av}.

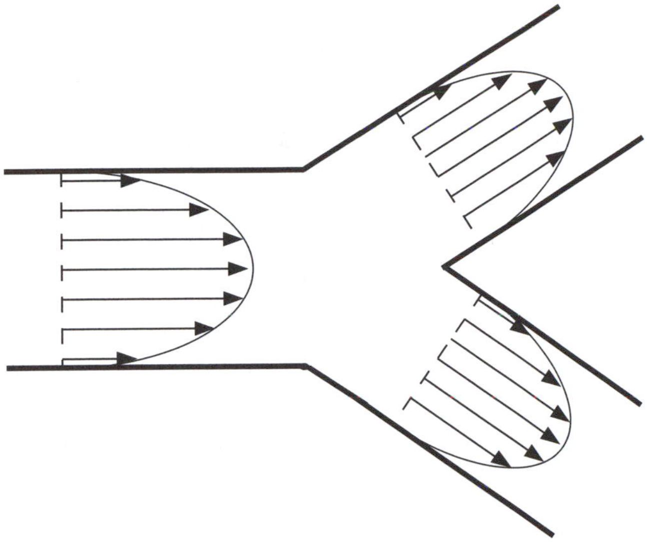

FIGURE 10.4 A bifurcation that obeys Murray's law. Note that the flow arrows decrease in the same proportion as the diameter, so the velocity gradient at the wall remains unchanged.

Does Murray's law actually describe vessel sizes in the fluid transport systems of organisms? In fact, it applies surprisingly well for, among others, the blood-transporting systems that have been examined, both vertebrate and invertebrate, for the channels supplying and draining the choanocytic filter-pumps of sponges, and for the small vessels of sunflower leaves (LaBarbera 1990; Canny 1993; Marcinek and LaBarbera 1994). It also applies to the branching bronchial trees of human (and presumably other mammalian) lungs (Weibel and Gomez 1962). Consider, as in table 10.2, the vessels of our own circulatory system. Veins run a little larger than arteries, but with thinner walls, they must cost less to build. The smallest vessels deviate somewhat, but in them the parabolic profile of flow assumed in the underlying Hagen-Poiseuille equation is substantially distorted by the red blood cells. Certainly an exponent of three fits far better than an exponent of two or four.

Beyond the deviations just noted, Murray's law fails where branching is especially frequent, as in some of the vessels supplying hearts—again the para-

TABLE 10.2 The Radii, Numbers of Vessels, and Cubes of the Radii of Vessels in the Human Circulatory System

	Average radius, r (mm)	Number, n	nr^3 (mm^3)
Aorta	12.5	1	1,953
Arteries	2.0	159	1,270
Arterioles	0.03	14,000,000	380
Capillaries	0.003	3,900,000,000	860
Venules	0.02	320,000,000	2,550
Veins	2.5	200	3,180
Vena cava	15.0	1	3,380

Source: Data from LaBarbera 1990.

bolic profile no longer applies. (But the coronary arterioles of both normal and ventricularly hypertrophic pigs do follow the law, according to Kassab and Fung [1995].) Murray's law also fails where some other condition supervenes. Thus, the larger vessels of tree trunks are much smaller than predicted, thereby almost certainly gaining safety in numbers where embolisms commonly occur. And Murray's law fails where transport is diffusive—long ago, Krogh (1920) showed that the branches of the tracheal system of a goat moth larva preserved the squares of their radii rather than the cubes. This last result, of course, is just what one expects for a diffusive rather than a flowing system. Preservation of the squares of radii means preservation of aggregate cross section as the system branches. Any local expansion gains such a system very little, while any constriction loses significantly.

Murray's law has a subtle but marvelous consequence. Recall several things from chapter 9. The average velocity of flow in a pipe equals volume flow divided by cross section. So we can rewrite equation (10.2) as

$$v_{av} = \frac{ka^3}{\pi a^2} = k'a. \tag{10.4}$$

Maximum velocity, for laminar flow, is just twice the average, so it shares the direct proportionality. Wider pipe, faster flow—exactly in proportion. As a result, the parabolic profile of speeds across the wider pipe shares geometric similarity with that across the smaller pipe. As a result of that result, (and as shown in fig. 10.4) the velocity gradient will be the same at the wall of both pipes. Now recall the definition of viscosity from chapter 6. Shear stress equals viscosity times shear strain rate (equation (6.2)), or viscosity times velocity gradient. Put in terms of what happens at a pipe's wall, using τ for shear stress,

$$\tau = \mu \frac{v}{r}, \tag{10.5a}$$

or, putting the slope properly,

$$\tau = \mu \frac{dv}{dr}. \tag{10.5b}$$

Since dv/dr, the velocity gradient, is constant at the wall, and since viscosity shouldn't do anything strange until we get down to the smallest vessels, then the shear stress should be the same in all vessels, large and small, in an energetically most efficient system. To repeat, the same (time averaged) shear stress will be exerted on the walls of all vessels.

Why might that matter? Vertebrate circulatory systems, and perhaps other fluid transport systems as well, take advantage of the rule during their initial development and in any subsequent readjustments. The fortuitous constancy of the shear stress on the internal walls of properly sized vessels permits the endothelial cells lining the vessels to act as size monitors in an especially simple manner. Such a cell needs no information about the particular size of the vessel in which it's located—it needs to do no more than trigger enlargement (by dividing) if the average shear stress exceeds some standard value. Or, it should trigger diminution if the shear stress drops below that value. If each endothelial cell follows the rule, using the same set point, and if other cells in the walls follow the lead of the endothelials, the system will be self-optimizing with respect to economy of operation.

A variety of manipulations shows with no uncertainty that changes in shear stress induce such responses. Increased stirring of the culture medium of isolated endothelial cells stimulates cell division. Partially occluding a blood vessel induces narrowing of vessels downstream from it, while splicing in an additional blood supply initiates enlargement of downstream vessels. Less definitively (meaning fraught with possible alternative explanations), a sustained increase in use of some oxygen-demanding organ—such as a group of muscles—increases the size of vessels supplying it. The whole scheme has an odor of practical reality. One can't build an organism simply by reading out a genetic program, given the severely limited amount of genetic information available. Much of the construction and operation of an organism must depend on self-assembly and demand-sensitive control. We only map and name the large blood vessels, since the small ones differ even between monozygotic twins. A developing organ (or, unfortunately, a tumor) need only secrete a substance that stimulates proliferation of blood vessels—an angiogenesis factor, well known if imperfectly understood—and Murray's law will fix all else.

Of course, the cells must detect shear stress on their lumenal surfaces. Exactly how that critical transduction happens isn't yet clear, although the question is of intense interest, both scientific and commercial. Getting a foot in that particular feedback loop might be a handy thing in a society commonly afflicted with pathological narrowing of blood vessels in our commonest cause of death and pathological proliferation of vessels in our next most common fatal affliction.

Walls for the Pipes—Thickness

Pressures within the pipes of biological fluid transport systems range from about half an atmosphere above the exterior, in the lower arteries of a giraffe, to as much as 100 atmospheres below ambient, in the xylem of a desert shrub. These hydrostatic pressures generate tensions or stresses in vessel walls. So we now have to invoke some solid mechanics—old hat if you've chosen to do part III before II—and something familiar from chapters 1, 3, and 5. What's at issue is the way transmural (across-the-wall) pressure translates into tension in a cylindrical wall. And what's important here is the way that transformation depends on the size of the pipe. By a purely dimensional and geometrical argument, one can quite simply show that, in terms of tension,

$$T = \Delta p r, \tag{3.5}$$

or in terms of stress,

$$\sigma_h = \frac{\Delta p r}{t}, \tag{10.6}$$

TABLE 10.3 Overall Vessel Size Versus the Thickness of the Vessel's Wall—and the Stress on the Material of the Wall Given the Pressure at That Level—for the Human Circulatory System

	Aorta	Capillaries	Ratio
Vessel radius	13 mm	3 μm	4,000
Wall thickness	2 mm	1 μm	2,000
Stress on wall material	85 kPa	13 kPa	6.5

Source: Data from Caro et al. 1978.

where T is wall tension (force per unit length, assuming negligible thickness), σ_h is hoop or circumferential tensile stress (force per unit area), r is cylinder radius, and t is wall thickness. We remind ourselves that for a given pressure, as the radius decreases, the tension in the wall must also decrease (equation (3.5)) and that the wall can be thinner without its material experiencing an increase in tensile stress (equation (10.6)). The rule, to repeat, is called by physiologists, Laplace's law.

While vessel radius and wall thickness vary widely, as in table 10.3, the tensile stress on the material of the wall changes little—by a small fraction of 1 percent of the variation in the other two parameters. Again, the system seems well balanced, since stress is ultimately what stretches, causes aneurysms, and ultimately ruptures pipes. Capillaries experience lower pressures than the aorta, which contributes as well to a lowering of stress. But even so, a very thin capillary need withstand stresses only about six times lower than a robust aorta. Even the slight difference in stress looks reasonable, since capillaries must be permeable for exchange of material and can't, therefore, be as single-minded about dealing with pure mechanics. The bottom line is that, thanks to Laplace's law, thin pipes can have thin walls, certainly a convenience if exchange takes place in the smallest pipes of a transport system.

Walls for the Pipes—Mechanical Behavior

The basic problem appeared back in the first chapter as part of your initiation into biomechanics. Laplace's law may help the scaling of wall thickness, but at the same time it presents fluid transport systems with a peculiar and subtle problem. A cylindrical balloon inevitably inflates irregularly—one part expands almost to bursting before adjacent portions begin to enlarge. The transmural pressure difference doesn't vary, but it generates more tension in the wall of a part of the balloon that has already begun to expand than in a part that has not. In an ordinary rubber balloon, extension is nearly proportional to stress applied, and a given pressure extends the already expanded part more than the more flaccid unexpanded part. Only a slight increase in stiffness (properly Young's modulus of elasticity), in the slope of the stress-strain curve, just before breaking makes cylindrical balloons practical at all.

Were arteries made of such ordinary elastic, each beat of the heart would most likely cause an aneurysm—a local expansion—somewhere in the system. But aneurysms remain rare and pathological. In effect, Laplace's law rules out the use of ordinary elastic materials and requires that the stress-strain curve for arterial wall be concave upward—that it increase in stiffness as it is stretched. The alternative of having rigid pipes in pulsatile systems such as our circulation is not attractive—wall elasticity provides considerable smoothing of pressure (and thus flow) fluctuations even in persons with severe atherosclerosis (vessel wall hardening).

How, then, do actual vessel walls behave? It's no great trick to apply pressure to a length of artery and look at the fractional increase in its diameter or circumference, as in figure 10.5. For low pressures, it takes a steadily increasing pressure to expand the artery, unlike the constant pressure that inflates a balloon. That makes the system resistant to the aneurysms a balloon suffers (recall figure 1.2)—in mechanical terms (chapter 15), the wall gets stiffer as it stretches. In fact, arterial wall doesn't stop there. Pressures above the resting range, those normal for people working vigorously, push vessels rightward onto ever steeper parts of the curve.

The basis for the complex nonlinearity has long been known; figure 1.3 showed its structural basis. Two different elastic materials, both proteins, play crucial roles; together they make up about 50 percent of the dry mass of a large artery. One, elastin, stretches easily and far—to at least twice its unstretched length. (Tables 15.1 and 15.4 give data in terms of elastic modulus and extensibility.) The other, collagen, resists stretch about a thousand times more strongly and then tolerates stretches of only about 10 percent beyond its unstretched length. In the unstretched arterial wall, discontinuous but more-or-less circumferential fibers of collagen have a zigzag, kinky configuration. So the initial stretching doesn't do more than unfold them—as one might unfold a piece of fan-folded paper. Resistance to expansion in the lower portion of figure 10.5 depends mainly on that of the stretchy elastin. But as more and more of the collagen fibers go straight, the greater stiffness of collagen gradually comes to bear, and the artery becomes increasingly resistant to further stretch—as when the piece of paper has been fully unfolded and one pulls on the material itself. (Smooth muscle contributes as well, but that's another and more complexly physiological story.)

Such upwardly concave stress-strain curves characterize circulatory systems that have evolved in lineages quite separate from our own—as they must

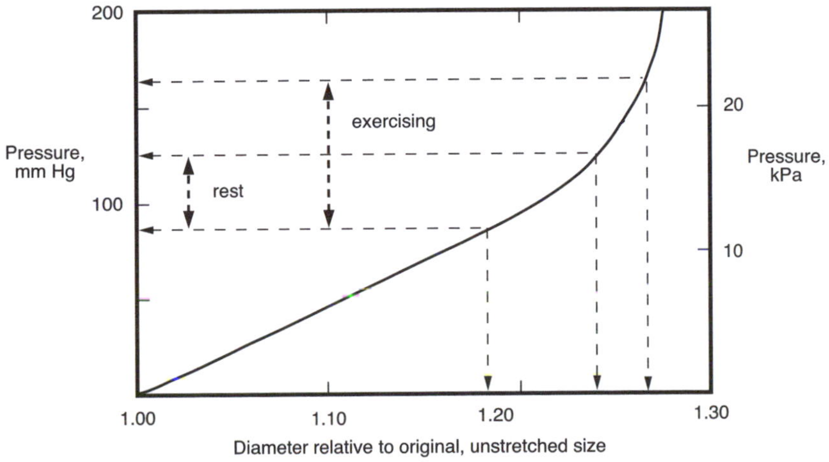

FIGURE 10.5 Pressure (*y*-axes) inside versus diameter (*x*-axis) of an artery that's being inflated. The stiffness rises especially steeply when pressure gets into the upper part of the physiological range.

for other animals to take advantage of the pressure-smoothing of stretchy vessels and yet avoid aneurysms. Cephalopod mollusks provide the most interesting of these alternative realities. These recurrent alter egos use the same (or nearly so) collagen as we do, but they know nothing about elastin, which is strictly vertebrate stuff. Instead, they make fibers of a chemically distinct but mechanically equivalent protein (called, by Shadwick and Gosline [1985], "octopus aortic elastomer"). Large crustaceans, likewise, have well-developed arterial systems that generate substantial pressures (disparaging remarks in elementary textbooks notwithstanding). And, like vertebrates and cephalopods, they build arterial walls of stiff and initially kinked collagen fibers together with yet another much less stiff elastic protein (Shadwick, Pollock, and Stricker 1990). The blood pressures of individual members of all three groups differ considerably—mammals and birds run higher than amphibians and reptiles; squid run higher than octopi, and so forth. Yet each has arterial walls with about the same stiffness in its normal physiological range of pressures—to anticipate material given proper treatment in chapter 15, about 0.5–1.0 megapascals (Shadwick, Pollock, and Stricker 1990). That's the basis of Shadwick's (1994) estimate of the blood pressure of a giant squid, about the same as ours and much higher than in small squid, from a sample of arterial wall off a freshly caught but dead animal.

Pumping Fluids

Pumps and their diversity came up in the last chapter. We saw that pump performance—in particular its operating trade-off between pressure boost and volume pushed—had to match its application. And that applied both to the pumps produced by human technology and by nature (Vogel 1995a). Oxygen-distributing circulatory systems invariably use positive displacement pumps since they operate at relatively high pressures, probably as much as anything to keep their overall volumes low. By contrast, applications demanding really high flow rates inevitably keep pressures low and use fluid dynamic pumps, most often cilia and flagella. Our hearts pump a body mass of fluid about every 10 minutes; a scallop or mussel does that every 3 or 4 seconds—over a hundred times our flow rate (Meyhöfer 1985). Were the lowly mollusk to face the same pressure drop as that imposed by our pipes, it would need a metabolic rate (per unit mass) around five times ours. Conversely, a circulatory system with ciliated capillaries instead of a heart, whatever the overt attractiveness of such decentralization, would suffer a most inefficient mismatch of pump and load.

The design of hearts themselves, in particular the way they vary with body size, brings up Laplace's law once again. The thickness of the ventricular walls of mammals should reflect the tension they can develop. As Calder (1984) pointed out, that thickness varies with body size exactly as would be expected if blood pressure were constant—which (except for especially large mammals) it is. And a rationale for that constant blood pressure can be derived from other scaling relationships, that is, applying physical arguments and biological requirements (Khanin and Bukharov 1991).

We humans have about the normal blood pressure for mammals—16,000 pascals (120 millimeters of mercury) systolic and 10,000 (80 millimeters of mercury) diastolic, or an average of about 13,000 pascals. It probably takes a minimal average of about half that to supply a brain. If so, one can easily calculate that a brain should be no more than 0.7 meters (or 27 inches) above the heart. So we (even hypotensive people) are safe enough. But we're borderline. If you have normal blood pressure (you're not hypertensive), you may feel a momentary dizziness if you stand up suddenly when aroused from a sound sleep. Still, quite a few mammals are quite a lot bigger than we, and one wonders what they do. Nothing all that startling—they operate at higher blood pressures than mammals of up to about human stature (or heart-head distance). Above about our size, normal blood pressure increases in rough proportion to body height.

Giraffes, the extreme case, drive blood brainward at systolic pressures of the order of 37,000 pascals (280 millimeters of mercury) rather than our 16,000 pascals (120 millimeters of mercury). The blood vessels in their legs need (and have) thicker-than-typical walls to withstand resting pressures around 66,000 pascals, or about two thirds of an atmosphere—that produced by the heart plus the added hydrostatic pressure resulting from the heart's height. Similarly, they reinforce the vessels in their necks against aneurysm and rupture, since serious risk of brain blowout when bending down to drink must be avoided.

One obvious solution (or at least amelioration) of the special problem of the giraffe consists of siphoning. Normal mammalian blood pressure would suffice were the venous blood descending from the head at subambient pressure—in effect pulling the blood upward. No subzero pressures need be invoked, just subambient levels, so blood needn't be truly in tension. Dropping a liquid a few meters in a pipe does the trick; does the giraffe use the trick? Siphoning has been suggested on several occasion (see, for instance, Hicks and Badeer [1992]) and argued against (see Pedley, Brook, and Seymour [1996]). Circumstantially, giraffes do pump hard enough to get blood up without necessarily resorting to siphoning. Moreover, siphoning requires that the descending venous pipes remain open (patent, if you're into medical jargon) despite the inward transmural pressure difference attending their subambient internal pressures. The vessels may be well braced against outward pressure, but their flexibility will cause collapse. Still, unless collapse were total, siphoning could occur. Perhaps giraffes do some siphoning but, in a sense, know they can't rely on it to supply blood under all circumstances to the organ least tolerant of any interruption.

An oddly similar problem afflicts reptiles. Their blood pressures run about half of ours, so they ought to get into hydrostatic trouble at half our height—or half our heart-to-head distance. Most contemporary reptiles go around in pretty horizontal postures; fortunately, big crocodilians don't climb steep hills or trees. Some snakes, though, spend most of their lives in trees and must deal with bodies that extend substantial distances vertically. Members of these arboreal species have especially high blood pressures—6,500–12,000 pascals (50–90 millimeters of mercury), snaking into the mammalian range. They do something more startling as well. Arboreal snakes put their hearts closer to their heads, positioned at 15–25 percent of head-to-tail distance rather than nearer 50 percent, as in other snakes—they climb with, almost literally, their hearts in their throats. (See, for the snake story, Lillywhite [1993] and Seymour, Hargens, and Pedley [1993]).

One wonders about big, and especially long-necked, dinosaurs. The basic arrangement of the reptilian circulatory system raises a particular difficulty for dealing with high pressures. Reptiles have a connection between systemic and pulmonary circulations that we lack. For most, it's a hole through the interventricular septum of the heart. For crocodilians, it's a connection at the level of the aortas. That connection has the virtue of permitting the lungs to be shunted—the system can vary the fraction of cardiac output that's directed lungward, even shutting lungs down entirely during a dive. We can't do that. But the connection makes it hard to operate lungs at a lower pressure than the rest of the body, as we do. Thus, our pulmonary artery, from the right side of the heart, operates at only around 2,000–5,500 pascals (15–40 millimeters of mercury), depending on our level of activity. So the giraffe's solution looks especially unattractive for a reptile. At least it does unless big dinosaurs had the fully serial avian and mammalian kind of circulatory system. So they probably did; one can even surmise that evolution of great body height in the dinosaurs set the stage in that lineage for the appearance of the intense activity of avian flight.

Open and Closed Systems

Elementary biology textbooks have long emphasized a division of circulatory systems into so-called closed systems and open systems, illustrated diagrammatically in figure 10.6. A closed system puts all its blood in pipes, with pipes leading both to and from its pump or pumps. Vertebrate circulations fit that pattern, for the most part, as do those of cephalopod mollusks and quite a few annelid worms. An open system forgoes return pipes, pumping blood peripherally in vessels but getting it back through less discrete blood sinuses. So it has arteries but lacks at least some of the corresponding veins. (A "hemocoel" is the usual designation for the vesselless part of the system.) Noncephalopod mollusks, arthropods, and a host of other invertebrates have open systems. Even within the vertebrates, blood sinuses can be found, near the front ends of sharks and within the muscular walls of some hearts.

In general, open systems operate at lower pressures than do closed systems. The common story is that open systems should be regarded as primitive and inferior arrangements, suitable only for leisurely living. Insects do fly and in the process consume oxygen at rates second to none, but they distribute oxy-

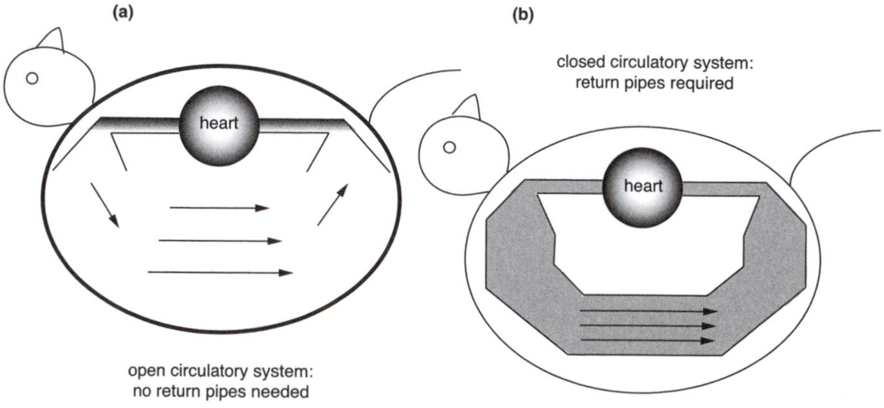

FIGURE 10.6 Open (a) and closed (b) circulatory systems.

gen with a tracheal system, not in their blood. Open systems, though, viewed more closely, turn out to be considerably fancier than the simpler accounts imply, with complex arrays of tiny vessels in tissues and provision for guidance of returning blood—one might look, for instance, at Ruppert and Carle (1983) or Voltzow (1985). And some develop substantial pressures—sharks and spiders in particular, about which more will be said in chapter 20.

We might view the distinction from a somewhat different vantage point. Chapter 2 introduced the principle of continuity for fluid flow, something once again relevant. If an inextensible outer integument enwraps an animal, then pumping blood in one direction ought to be sufficient. Being incompressible, the blood simply has to go back again if the animal doesn't leak. Arthropods have outer cuticles of stiff and inextensible chitin (tables 15.1 and 15.4), so they meet the basic requirement. Shark skin has sufficient strength to be used to make leather; it's not something you might mindlessly ingest the way you can the skin of most bony fishes.

Carrying Oxygen

Mention of two more items will round out this survey of the ubiquity of simple physical determinants of circulatory variables; both concern oxygen transport. For the circulatory systems of vertebrates and cephalopods at least, the critical function is transport of dissolved gas, especially oxygen. Other functions would be amply served by a less substantial system, perhaps like that of insects, where oxygen transport gets handled separately.

First, oxygen passes from our respiratory to our circulatory systems as it moves from the alveoli of the lungs to the pulmonary capillaries that enwrap them. Our nearly four hundred million alveoli are blind, thin-walled pouches at the ends of about twenty-three generations of branching airways. Their diameters vary between about 0.2 and 0.3 millimeters, and they have, incidentally, a combined exchange surface of the order of 50–100 square meters. Pulmonary capillaries are flattened channels (somewhat different than other capillaries) that enwrap the alveoli, and flow through them is better described as flow between parallel flat plates than through cylindrical vessels (Caro et al. 1978). Their diameters, though, are about normal for capillaries, about 6 micrometers, perhaps since the same blood cells must pass through. We shouldn't forget the special circulatory role of lungs and their special position in the system—your lungs receive the entire output of your right ventricle and, in this serially arranged circuit, must pass blood at a rate equal to that of the entire remainder of the body combined.

How should these gas-containing chambers and blood-containing channels be scaled so neither seriously constrains the operation of the other, that is, so the system is efficiently balanced? Transport from alveoli to pulmonary capillaries occurs entirely by diffusion, which simplifies matters—a point made by August Krogh against considerable skepticism early in the twentieth century. Diffusive transport, as noted earlier, diminishes with the square of distance. At the same time, diffusion coefficients in gases greatly exceed those in liquids. For oxygen in air, the coefficient is 20×10^{-2} square centimeters per second; for oxygen in water, it's 18×10^{-6}, about ten thousand times less. The square root of ten thousand is a hundred; one thus might expect an alveolus to be about a hundred times the diameter of a pulmonary capillary. In fact, the

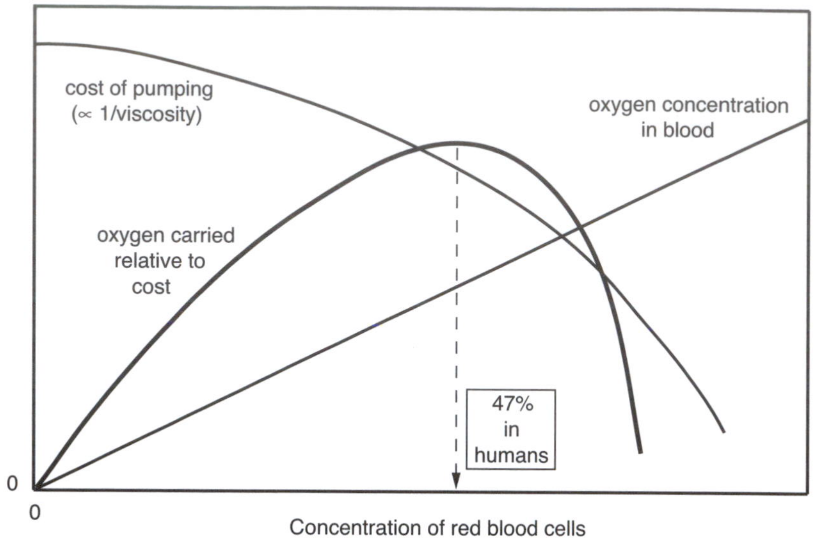

FIGURE 10.7 The cost of moving blood (varying inversely with blood viscosity), oxygen—transport capacity, and their combination oxygen transport relative to pumping cost—all plotted against the fractional content of red blood cells, or hematocrit.

alveoli run just a bit smaller, around forty times capillary diameter. Still, our guess isn't too far off, and we might rationalize the situation by the further supposition that if our alveoli were fully 0.6 millimeters in diameter, our pleural cavities could contain only about a quarter as much of that critical exchange surface.

Second, blood carries nearly all its oxygen bound to hemoglobin in the red blood cells. How concentrated ought these cells be? Too few cells will carry too little oxygen; too many cells will increase the viscosity of blood excessively and thus either increase the cost of pumping or slow down its passage. Were both relationships linear, no optimum would emerge; but, in fact, a curve that's concave upward describes the increase in viscosity with cell concentration (appearing downwardly concave in fig. 10.7 since it puts viscosity in a denominator). Optimum oxygen transport relative to pumping cost occurs when the red cells make up 47 percent of the blood. Appropriately, the cells normally make up about that fraction of our blood. A few animals such as seals and whales have higher concentrations, but they use blood as an oxygen store for diving, which provides a fine rationale for their bias. A few mammals such as goats and camels have lower concentrations, but their red blood cells differ in size, shape, and flexibility, so the optimum again turns out to match reality (Burton 1972; Weibel 2000).

Summing Up the Case

As suggested at the start, the chapter has provided examples of aspects of circulatory systems whose arrangements could best be explained by the constraints of mechanics, situations where evolution must tread relatively well-defined pathways. More of these undoubtedly can be recognized; indeed, quite a lot of the most recent literature on scaling (see, for instance, Brown and West [2000]) does just that. But moving further brings up things more complex than appropriate here or than the present author finds intuitively comfortable. More useful at this point might be summarization. Here then (using presentational bullets as is the fashion) are the main contentions:

- Number of capillaries:
  Fick's first law for diffusion
- Length of capillaries (in part):
  Hagen-Poiseuille equation for pressure drop per unit length
- Diameter of capillaries:
  Péclet number for balancing diffusion and convection
- Relative diameters of other vessels:
  Murray's law for minimizing circulatory work
- Relative thickness of vessel walls:
  Laplace's law for how size sets pressure versus wall tension
- Nonlinear elastic behavior of vessel walls:
  Laplace's law, as above
- Choice of category of pump:
  Required flow rate and pressure boost
- Relative sizes of capillaries and alveoli:
  Diffusion coefficients of oxygen in air and water
- Blood pressure in different places, and for large mammals and snakes:
  Hydrostatics
- Presence or absence of return pipes:
  Tensile stiffness of an animal's outer integument
- Concentration of red blood cells:
  Viscosity versus cell density and the Hagen-Poiseuille equation

Pulsations

Little has been said so far about the nonsteady character of flow in much of most circulatory systems. We noted a few maximum and average pressures and commented that the elasticity of arteries smoothed out the pressure fluctuations of the pulsatile pumps that drive most such systems. Once one gets into explanations of the causes and consequences of fluctuations in pressure and flow, matters rapidly get dauntingly complex, well beyond the scope of the present book. My favorite reference on these matters is Caro et al. (1978)—I find little indication that its age proves much of a drawback; where that matters, Fung (1997) is an alternative. But one general notion can help at least tell where nonsteady flow might matter.

A good index to the importance of unsteadiness has been available for many years—the Womersley criterion, or Womersley number (*Wo*):

$$Wo = \frac{d}{2}\sqrt{\frac{2\pi n\rho}{\mu}}, \tag{10.7}$$

where d is pipe diameter, n is the frequency (assumed sinusoidal) of pressure oscillation, and ρ and μ are, as usual, density and viscosity. The Womersley number indicates how far from the normal, parabolic distribution of speeds across a tube a pulsating flow will be. In effect, it's a kind of Reynolds number for unsteady flow—if below 1, viscosity dominates and the flow can be assumed parabolic, while if above 1, inertia dominates and risk attends application of the Hagen-Poiseuille equation. So the smaller the vessel and the slower the heartbeat the more ordinary the flow. In the human aorta $Wo = 20$; in that of a rat, $Wo = 3$. In us, the value of one corresponds to a vessel diameter of about a millimeter. So pulsations can be ignored in all the very small arteries, arterioles, and capillaries. The Womersley number has applications beyond circulatory systems, as discussed by Loudon and Tordesillas (1998). They considered geometries such as flat plates and fluctuating flows around sensory structures—gills, antennae, and such things.

Although the pressure-buffering role of elasticity in arterial walls isn't especially mysterious, putting it in quantitative terms gets into complications. Were the pulse of a heart to expand all arteries simultaneously, things wouldn't be so bad. That's the assumption behind the famous "windkessel" (literally "air chamber") model proposed by Otto Frank back in 1899—the air provides a compressible, elastic element that smoothes out fluctuations. But the arteries don't expand simultaneously. A wave of pressure runs outward from the heart, a wave fundamentally similar to surface waves on a body of water (as will come up in chapter 14.) What determines the speed of either kind of wave is the interplay of two factors, the inertia of the fluid and the restoring force. For the surface wave (like a pendulum), gravity serves as the restoring force; for the arterial pressure wave, the elastic walls of the vessels serve the same role. The formula for propagation speed isn't complex:

$$c = \sqrt{\frac{Et}{\rho d}}, \tag{10.8}$$

where c designates wave propagation speed, called "celerity" to distinguish it from actual flow velocity; E is the wall's stiffness, properly Young's modulus of elasticity, or the slope of the material's stress-strain graph (chapter 15); t is wall thickness; d is vessel diameter, and ρ is the fluid's density. For mammalian aortas, t/d values run around 0.05, and E values about 0.5 megapascals. Typical speeds run around 5 meters per second.

What makes a mess of the real situation is the branching and tapering character of an array of vessels. That brings up wave reflections, impedance matching, resonance, and such. In concert, the phenomena seem to minimize the load on the heart and increase the efficiency of the circulatory system, at least in mammals. Paradoxically, in amphibia, reptiles, and cephalopods (yes, octopus again), Frank's windkessel model, which was developed for mammals, works fairly well. It also works, interestingly, in whales and, presumably, other diving mammals. These have expanded and highly distensible aortic arches, which effectively localize the pressure buffer. The benefit appears to be a greater flexibility in rerouting blood—especially to the brain—during dives (Shadwick and Gosline 1994; Gosline and Shadwick 1996). So the windkessel has the prints of whales so it might at least be an all-porpoise mechanism.

Yet More

No topic covered in this book approaches the complexity of circulatory systems, particularly mammalian ones. That tempts one to be chauvinistic about our particular sophistication. More likely, though, we're just seeing the result of more investigative effort—no topic in all of biomechanics has come in for more work or has a more extensive primary literature. Should we be surprised? The departure of more of us from this world results from circulatory malfunction than from any other cause. By contrast, tearing tendons and falling trees loom less large in the actuarial listings. We have, in fact, left out major items of biomechanical interest and effort. Nothing has been said about the non-Newtonian characteristics of blood and how its apparent viscosity depends on vessel size. Nor about axial accumulation, the way red blood cells move toward the centers of medium-sized vessels and thus travel faster than the average flow speed. Nor about skimming flow, the scheme by which the plasma-processing kidneys take advantage of axial accumulation to draw off more plasma and fewer cells. Nor about the way red blood cells deform and align as they pass through capillaries, the volume-buffering role of veins, the operation and role of lymphatic systems, and so forth.

CHAPTER

Flows in Small Worlds

> it's like swimming
> in asphalt on a
> summer afternoon.
>
> **HOWARD BERG**

WE LIVE IN A WORLD FULL OF SWIRLS AND TURBULENCE—FLUID MEchanical as well as political and social. The winds we feel, the water in our pipes and baths have the features we now recognize as those of high Reynolds numbers. And we intend that the devices we build, from pumps to aircraft, will cope with that world. But a lot of biology happens in a different world. When exploring flow near the surface of a flat plate we met a special region of laminar predominance, a small one that persisted up to Reynolds numbers of around 500,000. We saw how laminarity mattered in the flow near a sphere or cylinder up to the great drag crisis at 100,000. And we saw how laminar flow persisted up to a Reynolds number of about 2,000 for flow through cylindrical pipes but only up to about 3 for flow through apertures.

Here we turn to the final, most important, and least intuitive of these laminar worlds—external flows in which the whole field, not just some local region, stays turbulence-free. It's a small, or slow, or gooey domain—whatever combination gives Reynolds numbers below about 10. It may be a moving glacier, large but slow and (in a certain practical sense) extremely viscous. Or it may be a settling particle of silt or a droplet in a fog. Or, and of major significance to the biologist, it may be the ordinary events in the domain of cells and small organisms. Bear in mind that such small systems typify life more than we do.

Below $Re = 10$, flows make no further awkward transitions, not just around spheres, but around practically any shape at all. That avoids lots of the complications we've run into earlier. But the simplicity doesn't imply familiarity—a new domain of intuition must be cultivated to deal with this novel regime of flow. A few general words ought to point up the strangeness before we start serious flinging of formulas and numbing with numbers.

Flow without Much Inertia

Think again about what the Reynolds number tells us. Low numbers indicate situations dominated by viscous forces, which means, by the definition of viscosity, that fluids abhor cross-flow velocity gradients. Bits of fluid prefer to follow the crowd. At the same time, inertial forces are relatively low, which means that bits of fluid don't feel any particularly strong urge to keep flowing in a given direction. So flow easily curves around from upstream to downstream portions of sphere or cylinder. Flow may be slowed by the no-slip condition and the resultant velocity gradient, but without enough inertia to give it a strong downstream compulsion, it never separates from the surface, and no reverse flow occurs behind an object. That's, of course, what we saw in figure 6.7a. But that's only the start.

Pressure drag reflects inertia. Fluid encountering an object of appreciable cross-flow dimensions has to turn, which it doesn't like to do if inertial forces bother it. So, one might say, it tries not to turn and, in the process, pushes the object downstream. Of course, some of that downstream push may be recovered by clever design, streamlining, which produces a counteracting upstream push from behind. Skin friction, by contrast, comes from the shear stresses of viscosity. So, at low Reynolds numbers, skin friction matters more than does pressure drag, and that has consequences. In particular, drag no longer depends so strongly on shape. Exposing surface generates drag, which even the shrewdest orientation does little to minimize. Streamlining now offsets something that's no longer of great importance, and it does so at the expense of exposing extra surface. It's a fool's errand; and, as the shapes of swimming microorganisms suggest, nature is nobody's fool.

And in this world turbulence is simply unimaginable—turbulence belongs to a world of high rates of shear and steep velocity gradients. Sustaining the velocity gradients of a statistical mess of proper irrotational vortices goes beyond the poor power of the fluid's inertia when faced with such militant viscosity. While vortices have been made at $Re = 0.01$, creating them is difficult and sustaining them demands constant feeding of their voracious appetites for energy. Moreover, what vortices do occur at lowish Reynolds numbers haven't heard about any ban on rotation—they go around almost as if frozen into solidity, like the rotational cores of vortices at higher Reynolds numbers. Turbulence might augment exchange between flowing fluids and tissues, but it's as unavailable to help mixing at low Reynolds numbers as is lower gravity to help lift an airplane.

We normally and uncritically regard flow as a disordering process—even linguistic conventions reflect the view, as someone "stirs up trouble." At Reynolds numbers below unity, flow is not necessarily disordering—it's very hard to mix two viscous fluids such as corn syrup and molasses or the two components of epoxy glue, both in terms of the work it takes to stir and the amount of stirring needed to have much effect. With a little care one can demonstrate the business of stirring without mixing—it just takes a container of viscous liquid with some colored marker (a solid or, more impressively, a similar liquid) somewhere in it. The effect of three turns clockwise with a spoon can be pretty well undone by three turns counterclockwise.

This reversibility of flow extends as well to the forces involved. At higher Reynolds numbers, one can get a net force (and thus propulsion) by moving something with a rapid stroke in one direction and a slow one in the other—not that in practice the scheme has much to recommend it. The total "impulse" of the force (the product of force and time or integral of force over a period of time, dimensionally equivalent to momentum) differs for the two directions. That permits "power" and "recovery" strokes. The difference in impulse (and momentum transfer) traces to the fact that the force of drag for each stroke varies (roughly) with the square of its speed, whereas the time for each stroke varies inversely with speed to the first power—the product of time and force therefore differs for the two strokes.

By contrast, at low Reynolds numbers, drag varies with the first power of speed, and time remains (as always) inversely proportional. While a faster stroke may still mean more drag, the magnitude of the impulse remains the same for two strokes that cover the same distance. So each stroke, whatever its speed, always cancels the effect of any other. Williams (1991) looked at the paddling locomotion of some small crustaceans as their operating Reynolds numbers increased, molt by molt. With each transition, the loss on the recovery stroke became proportionately less and the motion less jerky and more consistently forward. Purcell (1977) provides a short and engaging look at this odd world; Dusenbery (1996) views it in a wider perspective.

It may be an orderly world, but the dominance of viscous forces makes it a gooey one. Inertia has little practical meaning—Berg (1993) calculated that if a bacterium could instantaneously stop spinning its flagellum, it would almost as instantaneously stop moving through water—it would coast only a distance equal to the diameter of a hydrogen atom, less than a thousandth of its own length! To move is to distort the fluid around you, not just immediately around you but to a remarkable distance away. Among other consequences, the difficulty of moving is affected by the presence of walls far from the immediate vicinity of the mover. As an extreme example, the drag of a long cylinder dropping in a viscous fluid at $Re = 10^{-4}$ is *doubled* by the presence of walls *five hundred cylinder-diameters away* (White 1946). Again, the biologists have not been au courant with these small-scale, slow flows even though we ordinarily work with small items. We describe the motion of microorganisms abnormally crowded between glass slide and cover slip; we have tables of falling speeds for spores and phytoplankton obtained by watching them descend within unreasonably tiny capillary tubes.

Just as walls at even considerable distances affect moving objects, the motion of an object affects other things a long distance from it—opposite sides of the same coin, so to speak. That wide field of affected flow must have ecological, behavioral, and physiological consequences, as pointed out by Rudi Strickler with various coworkers (see, for instance, Strickler [1975]). For a small creature such as a microcrustacean to swing its appendages and swim is, in effect, to announce "here I am." A nearby predator can easily sense its presence. Conversely, at least in theory, such an organism might detect walls, food, or even predators simply by monitoring the forces needed to propel itself. Even an object settling under gravity announces its passage. An analysis by Wu (1977) suggested a partial evasion of the problem. An organism swimming by means of a coating of cilia (a paramecium, for instance), must make far less "noise" than even an inert object falling at the same speed. To do their work of propulsion, cilia must extend from the surface out into fluid that's not moving along with the organism. So the velocity gradient on the surface of a ciliated creature cannot be thicker than the length of its cilia. In fact, cilia must move fluid *faster* than the speed of the free stream if they're to propel an organism. So the gradients are severe, as shown in figure 11.1. The downside is a high cost in terms of force and energy. The upside is a region of flow distortion much less than that of swimming with big paddles (like most microcrustaceans) or even with sinking passively. So perhaps some indirect advantage lies in what is otherwise a necessarily inefficient system of locomotion.

In general, though, movement of an organism or an appendage carries with it a lot of the adjacent fluid and anything in the fluid as well. A tiny

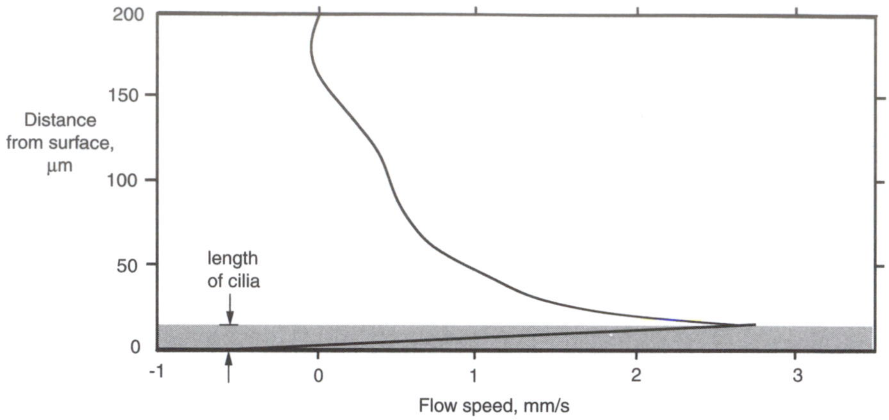

FIGURE 11.1 The steep velocity gradient on a ciliated surface, in this case, the gill cilia of a mussel, *Mytilus edulis*. The cilia, 15 micrometers long, push water through a channel 200 micrometers wide.

attached protozoan on a contractile stalk can contract the stalk when a small but voracious predator is about to bite off its head, but contraction might just draw the predator forward at the same time—the predator's inertia is inconsequential. If only the Reynolds number were higher! Faster contraction has just this effect, and, as it happens, the most rapid contraction of any bit of animal (in lengths per time relative to original length) is that of the so-called spasmoneme of the stalked, colonial protozoan *Zoothamnium* (figure 11.2; Weis-Fogh and Amos 1972). On the other hand, a predator has to get to its prey, not push the prey forward while still out of reach. Again a burst of speed does the deed—one predatory microcrustacean, a millimeter long, can briefly get up to $Re = 500$; to do so, though, it must move at a remarkable speed of 200 body lengths per second (Strickler 1977).

FIGURE 11.2 Contraction of the spasmoneme of *Zoothamnium*—the straight stalk through which the spasmoneme passes hunkers down into a tight helix.

Stokes' Law

If, as at low Reynolds numbers, drag depends on viscous and not on inertial forces, a simple dimensional analysis can suggest a formula for drag. Assuming that the relevant variables are the size and relative speed of an object and the viscosity of the fluid, it predicts that drag ought to be proportional to the product of the three, with each raised only to the first power. (Ignoring inertial forces has the effect of removing density from consideration—if density is included, the analysis is no longer so simple, as you can easily verify.) As usual, the analysis remains silent about any constant; the constant turns out to depend on shape and orientation. For a sphere, George Stokes (1819–1903) long ago showed that the full formula can be given simply as

$$D = 6\pi\mu a v, \tag{11.1}$$

where a is the radius of the sphere; we now call it "Stokes' law" (and note the position of the apostrophe—his name was Stokes, not Stoke).

The formula is tidy and useful, at least up to $Re = 1$; it gives the drag on a fog droplet or a spore sinking in air or on an algal cell sinking in the ocean. Since drag doesn't much depend on shape, it gives a decent approximation for shapes other than spheres (although formulas for other ordinary shapes are available). But several related matters need mention, in part to offset prevalent misconceptions.

- The formula assumes no slip at the surface of the object. For a droplet of gas rising in a liquid, the liquid induces motion in the gas, and their joint motion has the effect of a slipping surface—the factor of 6 in the formula should be replaced by a 4, according to Happel and Brenner (1983).

- Inertia still does matter—viscous forces may cause the larger part of the force on an object, but inertial forces still account for fully a third. That two-to-one ratio doesn't depend on the Reynolds number, as long as one is below the threshold for Stokes' law to apply.

- A sphere may have the least volume for its surface area of any shape, but it doesn't have the least drag relative to volume. A prolate spheroid (a jellybean or egg) twice as long flowwise as crosswise does a little better. So does an oblate spheroid (a pill), again with a two-to-one ratio of axial lengths, oriented with its faces sideways to local flow. For a prolate spheroid traveling lengthwise, a length-to-width ratio of about 2:1 gives least drag relative to volume and hence fastest sinking.

- As should be clear by this point, the formula doesn't work when a wall is nearby, and it's unlikely to be useful for locomoting organisms, for whom the production of thrust complicates the local flow and for whom at steady speed thrust must equal drag.

- Just as a wall nearby increases apparent drag and slows an object, other objects moving in close proximity speed things up. Sperm take advantage of the phenomena, going faster *en masse* than individually. Sinking particles tend to cluster—ones that get ahead face more drag and drop back; ones behind get help from the fluid dragged along by the leaders.

Stokes' law describes the smoothly ascending left end of the curve for the drag of a sphere given in figure 7.6. One needs only to plug equation (11.1) into the equation (4.10) or equation (7.13) defining the drag coefficient, C_d, and then extract the Reynolds number (equation (6.4)) from the result. This comes out as $C_d = 24/Re$, for a drag coefficient based on frontal area and a Reynolds number based on diameter. Of course recourse to a rough-and-ready coefficient isn't necessary with an equation as explicit as equation (11.1) in hand—the point is simply the absence of practical discontinuity.

Stokes' law finds its most common use in deriving a formula for the terminal velocities of spheres. At terminal velocity, you'll recall, the upward forces (here drag and sometimes buoyancy) equal the downward force of gravity. So one just sets the Stokes formula equal to mg and solves for v. Or, where buoyancy is significant or it's handier to use densities than masses, one makes a minor emendation—for mass one substitutes the product of the volume of the sphere and the difference in densities between sphere (ρ_o) and fluid medium (ρ_f):

$$v = \frac{mg}{6\pi\mu a} = \frac{2a^2 g(\rho_o - \rho_f)}{9\mu}. \qquad (11.2)$$

Bear in mind that equation (11.2) gives no universal scheme for calculating terminal velocities, even for spheres. All the limitations on the use of Stokes' law apply to it, in particular, the limitation to Reynolds numbers not substantially above 1.0. This last limitation needs emphasis. Maybe one can get a decent estimate (depending on one's purpose) at $Re = 10$, but beyond that great peril lurks. Consider, for instance, something of the density of water and a diameter of 2.2 millimeters (twelve to the inch) falling in air. A little messing around with equations (6.4) and (7.13), using 0.5 for C_d, gives a sinking rate a bit under 7 meters per second (15 miles per hour) and a Reynolds number of 1,000 (the last why I picked the example). Yielding to the seductive simplicity of equation (11.2), essentially Stokes' law, one predicts a sinking rate of no less than 175 meters per second (almost 400 miles per hour)—no small error!

But, to put things more encouragingly, bear in mind that the rule refers to sinking or rising with respect to the local fluid and applies whatever the larger-scale downdrafts or upwellings. After all, if you're denser than the medium, you sink; if you're less dense than the medium, you rise—with respect to your immediate surroundings. Only violent accelerations alter the picture.

The first part of equation (11.2) provides a handy measure of the effective size of some irregularly shaped object sinking (or, for that matter, rising). What we're talking about are a host of phytoplankton and wind-dispersed seeds, such as those shown in figure 11.3. Given the sinking speed and the mass, one can calculate a value for a—it has been called the Stokes radius or nominal radius. One might take matters a step farther and go dimensionless, dividing the Stokes radius by actual radius to get an indication of the effectiveness of some protruding bristles or fluff.

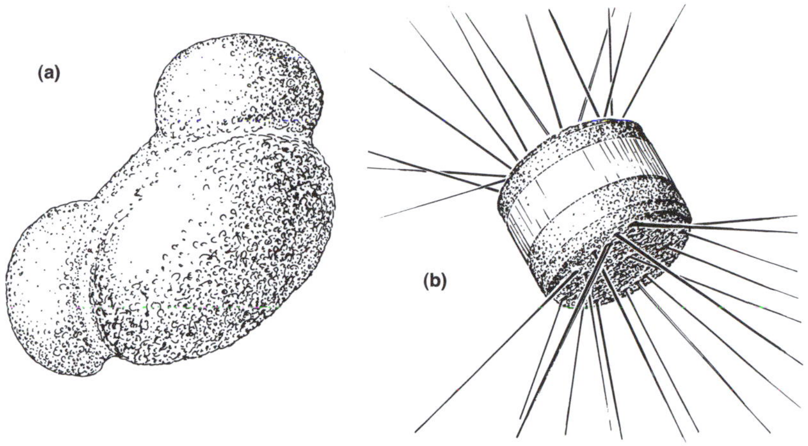

FIGURE 11.3 Examples of items that sink passively at very low Reynolds numbers relative to the fluid around them. (a) A grain of the wind-dispersed pollen of a pine tree, and (b) a diatom, *Thalassiosira*.

The formula for terminal velocity also provides a scheme for measuring viscosity—it isn't always applicable, but it's far more practical than the imaginary apparatus of figure 6.2. One just releases a ball that's sufficiently small or has a density near enough to that of the fluid to get the Reynolds number low enough. The density and size of the ball and the rate of ascent or descent go directly into equation (11.2). I once got a rather surprising result with the technique. I had filled a small recirculating tank with corn syrup and wanted to know its viscosity. So I timed the fall of a small sphere—a bearing ball. My calculated viscosity came out fully two orders of magnitude higher than the standard value for a saturated glucose solution. Finding a hundredfold error hard to explain away as inaccuracy of balance, ruler, or stopwatch I poked around a bit. An answer emerged from measuring the density of the corn syrup and looking at the calorie content on the label. Commercial corn syrup consists mainly of *supersaturated* glucose that just doesn't feel compelled to crystallize. In the process I found out why the stuff gets used in cooking. Table sugar (sucrose) crystallizes readily—if you want sticky frosting or fudge use lots of corn syrup; if you want solid stuff use mainly dissolved table sugar. So from the recipe you can anticipate the mechanical behavior of the product.

How fast do biological entities descend or ascend in free fall, and what are their effective Reynolds numbers? Table 11.1 gathers some representative data.

While Stokes' law for the drag of a sphere gets most of the attention, analogous equations for the drag of other shapes can be found in standard sources such as Happel and Brenner (1983). Some, such as those for the drag of disks broadside and parallel to flow, turn out to be just as tidy. For these latter, the relevant formulas are, first broadside to flow:

$$D = 16\mu a v, \tag{11.3}$$

and, second, parallel to flow:

$$D = 10.67\mu a v. \tag{11.4}$$

TABLE 11.1 Size, Rates of Motion, and Reynolds Numbers for Some Passive Biological Particles in Air (*a*) and Water (*w*)

Object	Cross-Flow Diameter (μm)	Speed (mm/s)	Reynolds Number
Bacterium, generic (*a*)	2.0	0.1	0.000013
Spore, *Lycoperdon* (*a*)	4.2	0.5	0.00014
Pollen, pine (*Pinus*) (*a*)	34.	24.	0.054
Pollen, larch (*Larix*) (*a*)	75.	30.	0.15
Scale insect, *Pulvinariella* (*a*)	340.	260.	5.9
Diatom, *Thalassiosira* (*w*)	18.	0.0038	0.000068
Rotifer, generic (*w*)	200.	4.0	0.80

Sources: Data from Ingold 1953; Walsby and Xypolyta 1977; Niklas 1984; Washburn and Washburn 1984; Dusenbery 1996.

Note that the low ratio of these two formulas, only 1.5, represents the extremes—the drags of the orientations giving maximum and minimum drag. Shape (what orientation effectively changes) does matter, but remarkably little. Also, compare the coefficients in equation (11.1) for a sphere with equation (11.3) for a broadside disk with the same frontal area. At moderate Reynolds numbers, the disk has the greater drag because of its broader wake—drag coefficients of 1.1 versus 0.47 for disk and sphere, respectively. At these very low Reynolds numbers, the situation reverses, with the sphere having $6\pi/16$ or 1.18 times as much drag. Skin friction has become dominant, and the sphere has more skin. But pressure drag still plays a role, as you can see from the ratio of 1.5 between equations (11.3) and (11.4) and from the fact that the sphere has twice the surface area of the disk but only 1.18 times as much drag.

The analogous formulas for cylinders with their long axes parallel and perpendicular to flow come out somewhat messier. Important to the present story, they'll reappear a few paragraphs further along.

Locomotion at Very Low Reynolds Numbers

At moderate and high Reynolds numbers, one scheme for moving through air or water surpasses all others in efficiency. First, an airfoil (or, the same thing, a hydrofoil) produces a force component at a right angle to its motion relative to the fluid—we call this lift, whatever its direction with respect to gravity or the earth's surface. Factoring in its drag, the airfoil produces an overall resultant force, whose direction depends on the ratio of lift to drag. As chapter 13 will elaborate, the greater the lift relative to the drag, the greater will be the angle between the resultant force and the airfoil's motion. Second, we persuade the airfoil to move in some direction other than that of a craft, using it as a rotating propeller blade, a flapping wing, or a swinging caudal fin. As a result, the airfoil faces a flow in a direction other than that of the craft as a whole. And as a further result, the resultant force on the airfoil needn't tilt away from the direction of motion of the craft—it can have a component that's directed forward, not, like drag, rearward. That forward component we call thrust. It will get its moment of fame shortly.

This familiar (if widely misunderstood) scheme works badly at very low Reynolds numbers, and nature uses it for neither flying nor swimming below about $Re = 10$. Viscosity derails it with two pernicious effects. First, it generates all that skin friction alluded to already, so drag gets relatively high for any shape at all. That has a devastating effect on the maximum achievable lift-to-drag ratio—the critical variable. And second, all that semistagnant fluid hanging around a moving airfoil obscures its shape, a shape crucial for producing lift. Less relative lift does its dirty work to the numerator of the ratio just as more relative drag compromises the denominator. In the fitness sweepstakes, bet on no propeller-propelled protista.

Still, microorganisms do swim around, even if none actively fly. But they have to use some other mechanism to push or pull themselves forward. Another possibility that finds use at higher Reynolds numbers takes advantage of the dependence of drag on orientation. Broadside to flow for a power stroke, a paddle or similar object has a lot of drag; parallel to flow ("feathered") for a recovery stroke, it incurs much less. A mechanism based on drag sounds more attractive than one based on lift, and, in fact, it turns out to be much more practical at low Reynolds numbers. Still, it can't be used in quite the same fashion as at high and moderate Reynolds numbers. Unfortunately, a paddle broadside to flow has only about 50 percent more drag than the same one parallel to flow—equations (11.3) and (11.4) can't be denied.

Oddly enough, about the best arrangement possible along these lines consists of using a cylinder—the drag of a cylinder whose long axis moves perpendicular to the flow is nearly twice that of one with the axis moving parallel to flow. Thus, a craft can make some headway by erecting a cylindrical paddle and moving it crosswise for a power stroke and then folding it down and moving it lengthwise for a recovery stroke. And folding it down has the additional advantage of getting the cylinder further from the mainstream so it recovers in fluid that's moving less fast than the oncoming free stream. This arrangement is just what is used by the cilia of many tiny organisms (fig. 11.4). Some organisms that use it (*Paramecium* is the classic case) have cilia over their entire surfaces; others (such as many invertebrate larvae) have ciliary bands and various other localizations. Still, as noted early in the chapter, the scheme demands that the entire velocity gradient between free stream and surface of organism occur in less than the length of the protruding cylinder, which extracts a severe price in a viscous regime. So while not wonderfully efficient, as we demonstrate in every introductory biology class, it does work.

Microorganisms have an alternative to such a ciliary beat, with its separate power stroke and recovery stroke. This other version uses the underlying fluid mechanics in a more devious manner. Recall that a cylinder has nearly twice

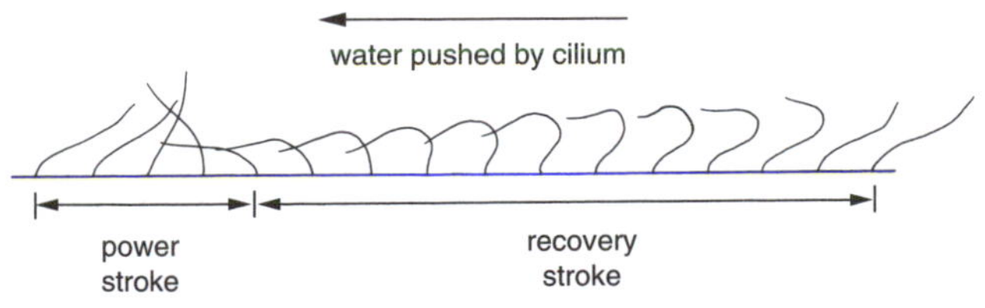

FIGURE 11.4 The motion of a cilium, as a left-to-right time sequence, during which water gets pushed from right to left. Thus, the cilium swings leftward during the power stroke and returns rightward during the recovery stroke.

as much drag when moving crosswise as when moving lengthwise. If, then, it falls in an oblique orientation, it won't fall straight downward but will glide off a bit sideways—in the direction of its downward end. In a sense, it takes the easy way out. You can demonstrate the effect for yourself, since it's not a purely low–Reynolds number business. Just release a length of cylindrical wire from an oblique orientation beneath the water's surface in a large vessel. It won't fall straight downward but will slew off in one direction. The slewing has to indicate a lateral force, since the only other force on the cylinder is that of gravity. The lateral force comes from nothing more than the asymmetry in the cylinder's drag.

(Incidentally, we put to macroscopic use this business of pulling or pushing one way and getting motion with a component off the direction of the pull or push. Even an old-fashioned square-rigged ship can go in a direction other than that of the wind by virtue of having a hull that's far from circular and is oriented obliquely to the wind. And one can go forward on ice skates by pushing outward on properly angled skates, not just by pushing backward with a skate oriented crosswise. So we're not talking about something totally obscure.)

Now imagine an obliquely oriented cylinder that's pulled downward or sideways, as in figure 11.5a. It wants to go laterally as well, so a body that forces it downward or sideways will find itself pushed at a right angle to that downward or sideways motion. Not to mince words, it feels thrust and moves through the fluid. Of course flows are still reversible—if it forces the cylinder back again, it loses everything it gained. BUT . . . What if it reverses the pitch of the cylinder before it forces it back again? Now it feels thrust in the same direction as before, impelling it the same way.

That repeated lateral motion can be achieved in several ways. The organelles that do it mostly go by the name flagella, but they're not either much different in structure or unequivocally distinct in function from cilia. Sometimes they propagate waves of bending in a plane, like sine waves moving down a rope or across the face of an oscilloscope; sometimes the motion

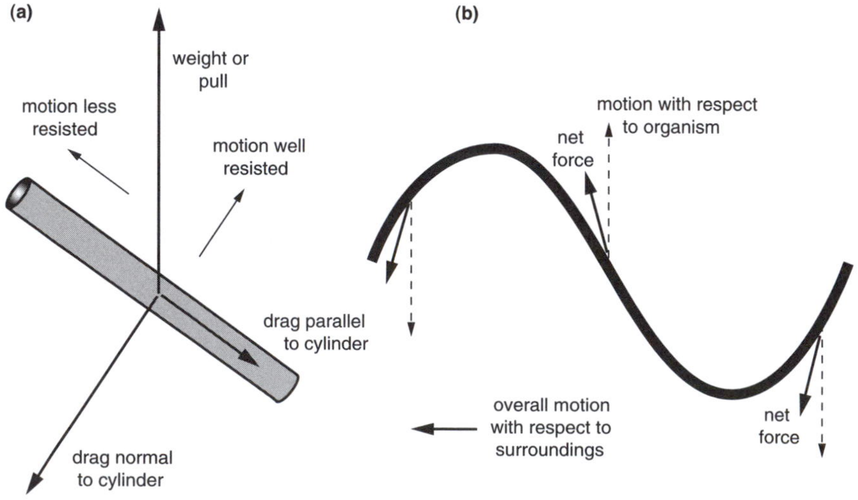

FIGURE 11.5 (a) What happens if a cylinder is pulled obliquely (here downward) to its long axis in a viscous fluid. Forces balance only if the thing slews forward (here to the left). (b) How a two-dimensionally beating flagellum takes advantage of that slewing force to achieve propulsion from right to left.

TABLE 11.2 The Body Lengths, Speeds, Relative Speeds, and Reynolds Numbers for Some Swimming Microorganisms

Organism	Body Length (μm)	Speed (mm/s)	Lengths/Time s^{-1}	Reynolds Number
Bacterium, *Escherichia coli*	3.0	0.03	10	0.00009
Sperm, *Lytechinus* (sea urchin)	5.1	0.16	31	0.0008
Flagellate, *Chlamydomonas*	13.	0.06	4.6	0.0008
Flagellate, *Euglena*	50.	0.08	1.6	0.004
Ciliate, *Tetrahymena*	70.	0.48	6.9	0.03
Ciliate, *Paramecium*	210	1.0	4.8	0.21
Ciliated flatworm, *Convoluta*	2,000	0.60	0.3	1.2

Sources: Data from Holwill 1975, 1977; Brennen and Winet 1977; Sleigh and Blake 1977; Berg 1993.

involves circular (but not truly rotational) motion of a helical structure. The first, two dimensional and thus the easier to envision and illustrate, is shown in figure 11.5b. Parts of the flagellum will be moving uselessly broadside to flow, but a reasonable fraction of its length will consist of a cylindrical profile moving sideways while oriented obliquely. Whether the flagellum sticks out rearward or forward doesn't much matter, something at first disconcerting when looking at how an organism such as *Euglena* or *Chlamydomonas* swims. When a helical flagellum moves around and around like a corkscrew, all parts of the cylinder can be usefully employed all the time. Of course it doesn't advance anywhere near as well as a corkscrew, which doesn't slip a step backwards for each two forward. Or worse—a bacterium reportedly goes only 7 percent as far forward per turn as it would if screwing its way through a solid (Dusenbery 1996). Efficient—no; effective—most definitely, as we'll see below.

But, I hasten to point out, the corkscrew analogy misleads in a serious way. Most flagella behave like nonrotational corkscrews, quite distinct from those in kitchen or on pocket knife. They're active organelles, using energy to bend forcefully along their length, rather than being twisted by a motor at one end. Only bacterial flagella, about ten times smaller and quite distinct in structure and chemical composition, move like the corkscrews of our technology under the impetus of a motor at one end. We'll have more to say about such mechanisms of motility in chapter 22; for now it's sufficient to note the indistinguishable net effect of rotating bacterial (procaryotic) flagella and all other (eucaryotic) circularly translating flagella.

How fast do these microorganisms swim? One gets an exaggerated impression when watching them through a microscope. It may magnify distance, making micrometers into millimeters, but it can't compensate by making microseconds into seconds—one watches a misleadingly real time. Table 11.2 gives a quick look at some speed data. Most of these stand as respectable, even remarkable, speeds by animal standards when normalized to body lengths per time. Few if any fish or cetaceans can manage ten body lengths per second; few humans can sustain even one.

Suspension Feeding

An impressively diverse and abundant array of animals obtain their nutrients by separating edible particles from the medium in which they're suspended. Almost all do the operation in water, either fresh or salt. Edible particles mainly mean living plankton. As Denny (1993) pointed out, water supports (in the most literal sense) a planktonic mode of life far more effectively than does air. The term plankton comes from the Greek for drifting, and, by definition, then, plankton have only local (if any) locomotory capacity. Quite aside from life's internally watery character, the slower sinking rates possible in water (as in table 11.1 above) make planktonic life in water much more practical than in air. For a long time, living off plankton went by the name "filter feeding," but "suspension feeding" has now become the preferred term—as we'll see, filtering tacitly ignores several important mechanisms for separating meal from medium.

Who are the players? Suspension feeders include some or all of the sponges, anemones, jellyfish, bryozoans, marine and freshwater annelids, bivalve mollusks, barnacles, microcrustaceans, aquatic insect larvae, brittle stars, fish, amphibian tadpoles, a few birds, and baleen whales—for starters. They consume particulates ranging from bacteria less than a micrometer across, by sponges, to krill (crustaceans) up to about 10 centimeters long, by the baleen whales—a hundred thousandfold length range. Almost all do their separations at Reynolds numbers below about 10, but a few operate in a higher range—whales in the thousands and tens of thousands.

With all this diversity of eaters and eaten, a comparable diversity of machinery doesn't surprise. Some animals swim around, intercepting edibles and often seeking out particularly rich patches. Many of the larvae of marine invertebrates go through a stage in which they live as plankton while feeding on smaller plankton. A few animals (such as goose barnacles) attach to floating objects. But most live as sessile creatures at the interface between solid substratum and stream, lake, river, or ocean, and somehow manage the inevitable velocity gradients. It's a world as unfamiliar to us as any widespread lifestyle among animals, we being too large, too aerial, and too used to getting large chunks of food by guile rather than tiny bits of food by dumb patience. Wildish and Kristmanson (1997) provide a very good general view of the business.

Even in her schemes for reaching up from the bottom, nature employs quite a range of physical tricks and biological devices, some of which we encountered when talking about boundary layers in chapter 8 and pumps in chapter 9. Some creatures put up a passive screen to intercept particulates, some reorient such a screen as flow direction shifts, some sweep their screens when flow gets too slow to provide ample throughput—barnacles are the exemplars here (Trager, Hwang, and Strickler 1990). Some use special collecting pipes to direct water through their screens—such as the ascidians and caddisfly larvae of figure 7.3. Some small creatures go colonial to get access to water that passes at decent speed, either sticking up on communal stalks (many hydroids and erect bryozoans) or making colony-wide currents by pooling their pumping (encrusting bryozoans and colonial ascidians). Some (sponges and bivalve mollusks, most notably) work at pumping, passing remarkably high volumes of water through themselves in channels of equally remarkably low resistance—recall the previous two chapters. Some take advantage of self-generated flow patterns, feeding from vortices upstream or downstream or above themselves—no sharp distinction can be made between feeding on oncoming suspended particulates and resuspended detritus, especially when change in flow triggers some creatures (Taghon, Nowell, and Jumars 1980) to switch between one and the other. And catching particles may depend on more than patience and trust in statistics. Where the catchers aren't orders of magnitude larger than the caught, watchful detection followed by grabbing and grappling come into the picture—copepods play this game with remarkable sophistication (Koehl and Strickler 1981; Strickler 1985), and in a simpler version, it's what's done by protozoans such as the *Vorticella* of figure 6.9 (Sleigh and Barlow 1976).

To add to the complexity, the physical mechanisms for separating food from flow turn out to be a diverse lot, underlying, again, our terminological shift from a term, filtration, to one, suspension feeding, with only minimal mechanistic implications. Here, though, the possibilities have been clearly delineated in two key papers—the first, by Rubenstein and Koehl (1977), who adopted and adapted the physical approach of aerosol engineering, and the second, by Shimeta and Jumars (1991), who made an important distinction between particle encounter rate and particle retention efficiency. Simplifying a little, they recognized no fewer than six different ways in which organisms might capture particles, the first five of which are shown in figure 11.6:

1. Simple sieving. A particle gets caught by a mesh or rake simply because it's too large to fit through the holes or between the elements. This, of course, is how we usually think filters work. Bigger particles will be captured, smaller ones will not, with a sharp size threshold.

2. Direct interception. A sufficiently large particle passes close enough to a collecting surface so it touches without deviating from its streamline. A very small particle is unlikely to touch because of the no-slip condition, so encounter rate will go up more or less steadily with particle size.

3. Inertial impaction. Flow may go around a collecting surface—it can't go through—but the particle, denser than the fluid, turns less sharply and

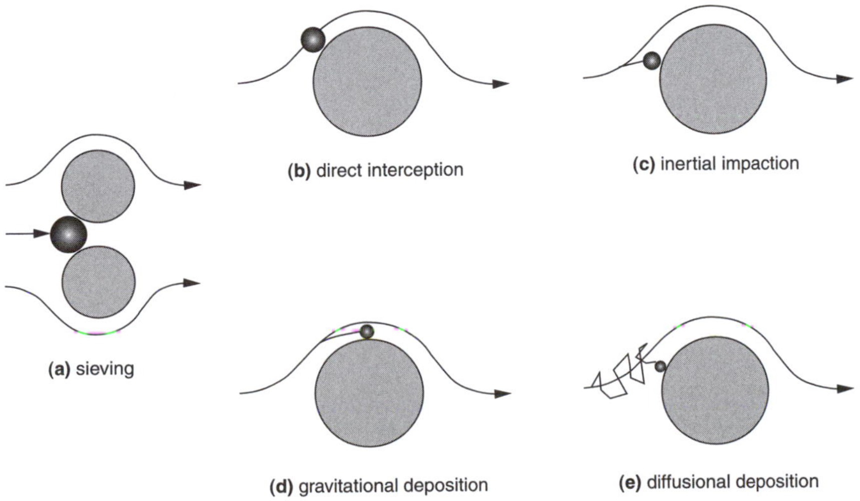

FIGURE 11.6 Five ways to suspension feed, shown diagrammatically.

may hit the collector. It has thus shifted from its streamline. Again, bigger is better for capture, with more mass and less surface area.

4. Gravitational deposition. Particles denser than the fluid will sink, so they continuously shift downward from the streamlines. Doing so may lead them to touch a collecting surface. Flow speed doesn't matter as much as it does for the preceding mechanisms. Once again, bigger particles are better, with more weight and less surface area.

5. Diffusional deposition. Very small particles get bumped about, setting them into Brownian motion. That can cause them to bump into a collecting surface. Here, obviously, the smaller the particle, the more likely it is to be captured. To Brownian movement one might add a macroscopic version based on random swimming activity.

6. Electrostatic attraction. Both particles and collectors can carry charge, unless short-circuited by seawater's conductivity. And opposite charges attract. The mechanism is unlikely to work in seawater and, it's only a bit more plausible for freshwater. But it may be of great consequence for pollination in air (Bowker 2002).

Paddles Versus Rakes

The appendages of a great many organisms—mostly but not solely arthropods—bear arrays of bristles. The arrays look relatively similar, usually sticking out in two dimensions, like combs, rather than in three, like bottle brushes. Figure 11.7 shows a few cases. What's fluid-mechanically peculiar is that these overtly similar appendages serve two radically different functions. Some get put to use for creating currents or for locomotion at low Reynolds numbers—they're paddles. As we've seen, drag depends little on orientation, although a recovery stroke near the body's surface offsets that awkward symmetry somewhat. Bristle-paddles can play another game, one not considered earlier. Drag still depends on exposed area—a big paddle has more drag than a small one. And a bristle-paddle can be made to fold back into a bundle with minimal exposed surface area for a recovery stroke. However they act, they depend on the main consequence of operation at low Reynolds number, reduction in the steepness of velocity gradients. A plane of bristles can behave as if it were a paddle with a continuous surface, and many small creatures swim or even fly by moving paddles that lack any continuous membrane over their bristly struts.

Some of these bristle-bearing appendages, by contrast, capture food particles with various rakelike actions. They may be held rigid while an animal swims through water or is exposed to a current, they may make repeated sweeps, or they may sweep when something edible and within reach is detected. However they act, functional efficiency depends on being leaky—on allowing water to pass between individual setae. For them, the viscous forces of low Reynolds numbers pose problems. As we saw back in chapter 6 when considering flow through silkworm moth antennae, making a fine mesh filter can be a fine way to persuade fluid to go around rather than through a structure. Viscosity becomes a bad thing if you're trying to filter out molecules or edibles.

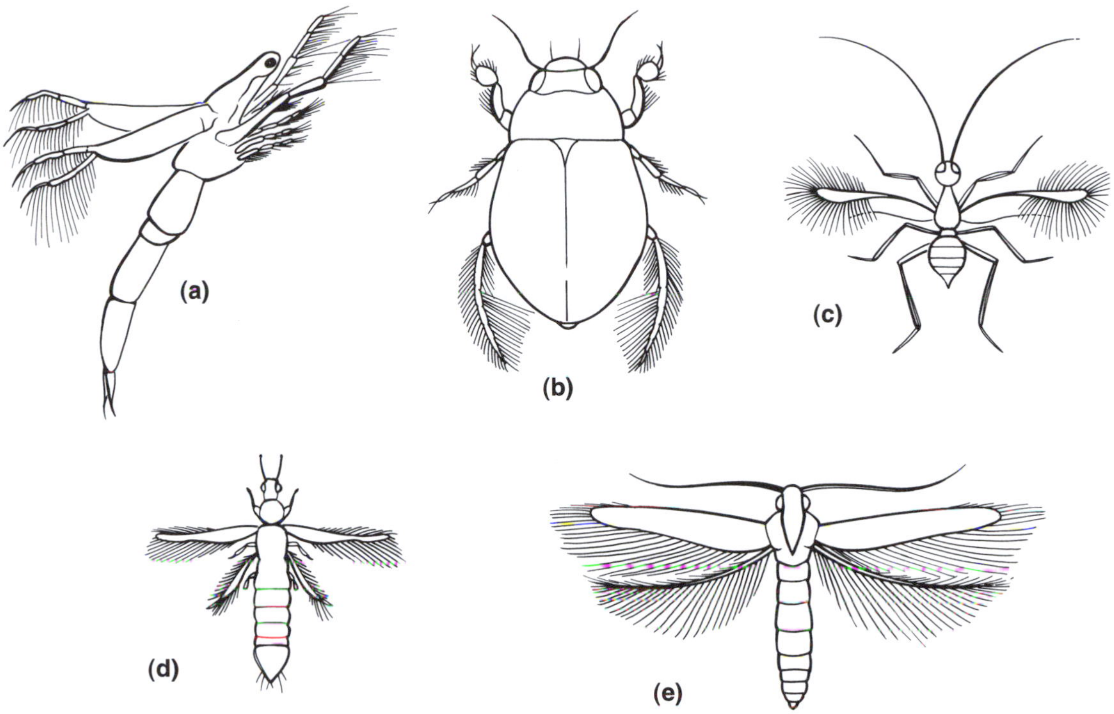

FIGURE 11.7 Bristle-bearing appendages used for locomotion: (a) a cladoceran crustacean, (b) a diving beetle, (c) a mymarid wasp, (d) a thrips, and (e) a moth. All but the beetle are quite small—the mymarid is about as small as flying insects get. The first two swim, the latter three fly.

One can't tell a paddle from a rake by appearance alone, as pointed out in a classic paper by Cheer and Koehl (1987). Even the ratio of bristle diameter to interbristle gap makes no unequivocal difference. What matters most isn't morphology but Reynolds number. These structures operate from Reynolds numbers as low as 10^{-6} (some protozoan pseudopodia) to a little above 1.0 (the cephalic fans of black-fly larvae), a range that exceeds a millionfold. Here we can't just lump everything into a generic bin labeled "low Reynolds number flow"—within that domain, low values permit paddling while higher values facilitate raking. "Leakiness" provides the simplest measure of the difference; Cheer and Koehl (1987) define it as I earlier did "transmissivity" for the antennae—the volume flow rate of fluid through a structure divided by the volume per time that they sweep, in effect the fraction of the fluid directly upstream that ends up going through the structure.

In practice, transitions can be more sudden than one might expect—stretching "sudden" to a change in Reynolds number of a single order of magnitude. Spacing between bristles does matter somewhat, as does bristle diameter. Model experiments and calculations give a good idea of the possibilities and of variations in the sensitivity of different arrangements to changes in their operating conditions. But reality remains complex, dependent on the specifics of morphology and, in particular, on how appendages move and on their proximity to adjacent surfaces—matters explored by, among others,

Hansen and Tiselius (1992), Loudon, Best, and Koehl (1994), and Loudon and Koehl (2000).

Flow in the Interstices

Chapter 9 talked mostly about internal, laminar flows; the present chapter looks at external, laminar flows. One subject of great biological relevance could be put in either—the subject of a chapter nine-sharp or eleven-flat. What about flow through porous media? Such flows are much more common than one might imagine. Where a stream flows across a porous bed—meaning anything except pure rock—water goes in and out of the bed depending on variations in the local hydrostatic and dynamic pressures, and water can flow for considerable distances within beds of sand or pebbles. One can get a good sense of the phenomenon by exploring around in a stream with a jar of dye and a syringe, perhaps with a length of catheter tubing on the needle. (Sodium fluorescein, or uranine, is a good choice; it's both nontoxic and visible at the most homeopathic of dilutions.)

Engineers, especially chemical engineers, have long been interested in the matter. Columns of particles do great service in separating the components of mixtures and solutions at both laboratory and industrial scales. The oil-extraction industry cares about percolation of viscous hydrocarbons, and hydrologists care about groundwater flow. Where I live, one's land has to pass a test before approval for a septic tank—the soil must "perk" adequately (as in a coffee percolator). The basic rule, Darcy's law (discovered by Henri Darcy, a French hydraulic engineer, in 1856), goes back almost as far as the rule of Hagen and Poiseuille. It makes the same assertion, that volume flow varies directly with pressure drop per unit length and inversely with viscosity. It needs only an additional factor or two to describe the porosity of a bed and the tortuosity of the channels through it (see, for a start Denny [1993] and Vogel [1994a]). In a macroscopic and physical sense, flows through porous media can be treated as internal flows.

Of especially bioportentous consequence are the flows within sandy beaches. A breaking wave causes swash, which runs up a beach as it converts some of its kinetic energy into gravitational potential. But not all its water runs down the beach again. Some, as you may have seen but not noticed, sinks into the sand—after all, the sand perks and the water is above sea level. Eventually the water returns to the ocean, but it may travel long distances in the interstices between sand grains, driven as a pulsatile, unidirectional flow by this gravitationally powered intertidal pump. We're talking about large-scale water movement—Riedl (1971) measured a typical speed of about half a millimeter per second and calculated that the resulting flow into the ocean exceeded that of all the earth's river systems combined.

For small organisms, flows through porous media represent external flows. Organisms live on and between the grains of substrata through which water flows—lots of organisms, all small, but dazzlingly diverse. Bacteria and fungi have long been recognized as important parts of this fauna; what may come as a bit of a surprise is just how much else lives down there. Riedl and his coworkers and successors (see Riedl and Machan [1972] and Higgins and Thiel [1988]) have documented the presence of oddly aberrant members of phylum after phylum of invertebrates—figure 11.8 shows a few members of

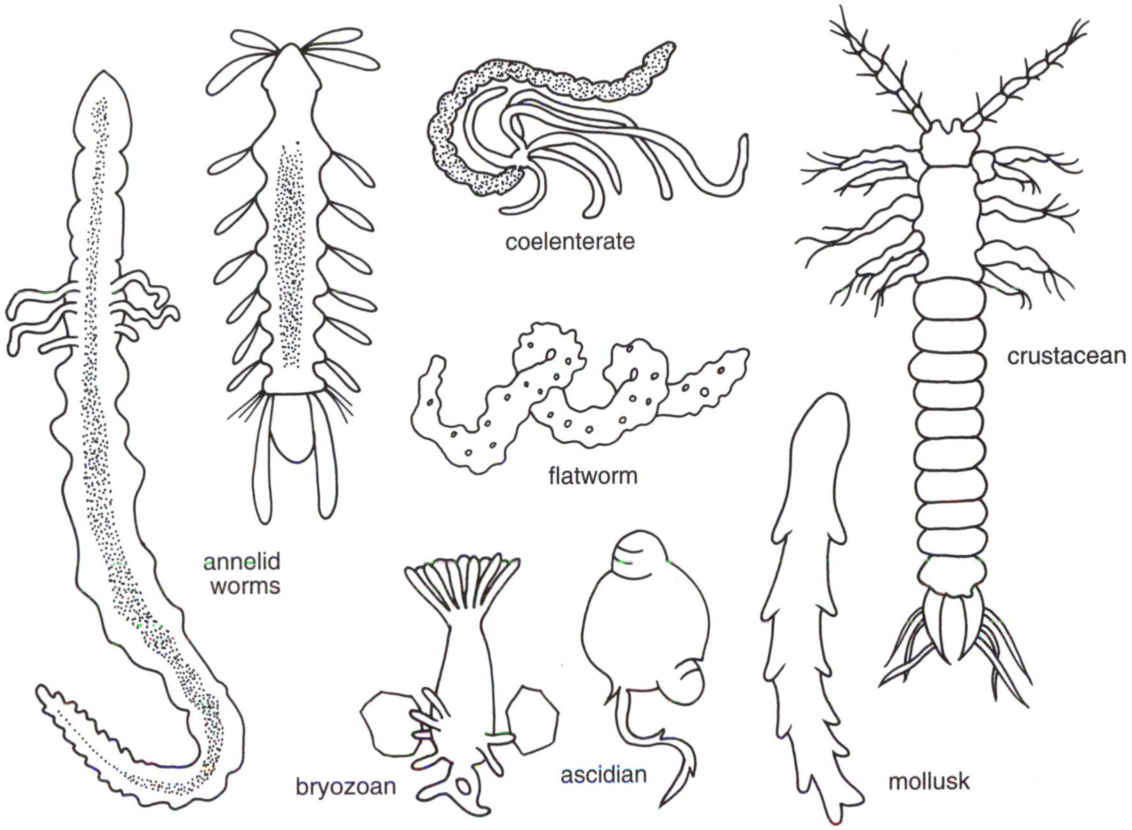

FIGURE 11.8 A variety of interstitial invertebrates, some quite divergent from the more familiar members of their respective taxa. Redrawn, with permission of Cambridge University Press, from Swedmark (1964).

this so-called meiofauna. Many decades have passed since their discovery (see, for instance, Swedmark [1964]), but we still have all too little information about the immediate physical environment of its members, the situations of creatures between the grains. With substantial pressures and flows, one can speak of the "stormy interstices," but that must come with the realization that these are completely laminar storms of a kind with which we have no direct experience. In the only hydrodynamic study I know of, Crenshaw (1980) showed that where the organisms are of the same size range as the grains and passages, their own presence profoundly alters the local flow. That gives them the option of a kind of hydrodynamic tuning.

CHAPTER

About Lift

The bounded waters
Should lift their bosoms
higher than the shores
And make a sop of all
this solid globe.

WILLIAM SHAKESPEARE, expressing his preference for fluid over solid mechanics

DRAG WAS A FINE MESS OF COMPLICATIONS, EVEN IN OUR LIMITED AND simplified treatment. Lift may be less complex, but it's even stranger; again we're onto something whose existence has been long known but whose physical explanation was only worked out in the early part of this century. The first thing to settle is a definitional matter—we'll use "lift" for any force at right angles to the direction of flow of the overall fluid. For a craft in level flight, fluid approaches horizontally, and it produces upwardly directed lift. The peculiarity of the definition emerges when flow isn't horizontal. Consider the local wind striking a glider that's slowly losing altitude—to the extent that the glider descends, the wind approaches along an upward slope. The lift force, then, no longer points exactly upward but now points slightly forward of the vertical. As we'll see shortly, that slight forward tilt of the lift vector is what keeps the craft from slowing down and eventually tumbling earthward. More extreme yet is the lift produced by a sail, not upward but horizontal.

More nonsense and misapprehension surrounds the mechanism underlying aerodynamic lift than any other topic we'll meet. In part, the problem stems from the peculiarity and counterintuitive character of the subject. If anything, the origin of lift exceeds even drag and streamlining in its distance from the reality of people who live in slow air and think mainly about solids. In part, the problem comes from a polite fiction told to the youth, one so widespread we no longer can even guess who ought to drink the hemlock. You may have heard that lift comes about because when a bit of fluid splits at the front of a wing, its upper half must take a longer path to get to the rear edge than its lower half. So for the halves to rejoin, the upper one must go faster, which, in turn means (invoking the sacred name of Dan Bernoulli) it has a lower pressure. Lower pressure above than below—voilà, lift. But why should the halves recognize each other and rejoin? In fact, they don't. The fluid on top ordinarily gets to the rear *before* that moving along the bottom. Were they to arrive at the same time, Bernoulli's principle would predict less lift than we measure and depend upon to keep our aircraft aloft. Put the other way around, if that mechanism described the origin of lift, to get its lift a wing would need a bigger hump than a camel's. Doesn't Bernoulli's principle explain the origin of lift? It explains lift, all right, but not by that particular scenario. Lower pressure above comes from higher flow speed; what's at issue is the origin of that higher flow speed above a wing.

One sometimes hears that we have to choose between an explanation that traces to Newton's laws and one that invokes Bernoulli's principle. Both matter—they just provide different levels of explanation. Newton must be satisfied. To get an upward force (assuming, for simplicity, level flight), a downward momentum flux must be created of exactly its magnitude. Recall that *mv/t* has the same dimensions as force, and recall how drag was introduced back in chapter 7 as the reduction in momentum flux of a fluid crossing an object. In that Newtonian sense, lift equals the rate of creation of momentum at a right angle to the overall direction of flow—whatever the mechanism.

Lift and Circulation

More important than definition and dimensional legality is its physical origin. Around the start of the twentieth century, Frederick Lanchester, in Britain, and Ludwig Prandtl, in Germany, proposed what we now recognize as the basic story. And an odd tale it turns out to be. Consider (ignoring the real chronology) a truly bizarre kind of sailboat, the invention, only a few years later, of a German naval engineer, Anton Flettner. This ship (figure 12.1) has one or more cylindrical, vertical funnels ("smokestacks") on the deck—they may be as much as 10 meters in diameter. The funnels are mounted on bearings and connected to motors below that cause them to rotate around their vertical axes. Because of the no-slip condition, the air near each funnel moves around with the funnel, with the air nearest the funnels going fastest. (Technically, the bits of air translate in a circle or "circulate" around the funnels rather than truly rotating—recall our talk about irrotational vortices in chapter 6.) If the ship encounters a wind from one side, and the wind moves across the deck in the same direction as the front of each funnel is turning, then the ship moves forward—"reaching," a sailor would say.

What's happening? We have superimposed a circulating motion of fluid (caused by the rotation of the funnels) and a unidirectional motion (the wind from the side). The combination produces a force at right angles to the wind, by our definition, lift; for the ship its forward direction means that the lift provides thrust. The phenomenon goes by the name "Magnus effect." If the wind came from the other side, the ship could continue to move forward if it reversed the rotation of its funnels. It can go backwards as well as forwards; the only thing it can't do is sail *directly* into the wind—it's still a sailboat. (Only the *Flying Dutchman* could do the latter and only while the wind whistled Wagner.) We're looking at the same thing that makes spinning spheres deviate from normal ballistic trajectories, deliberately, as when a baseball

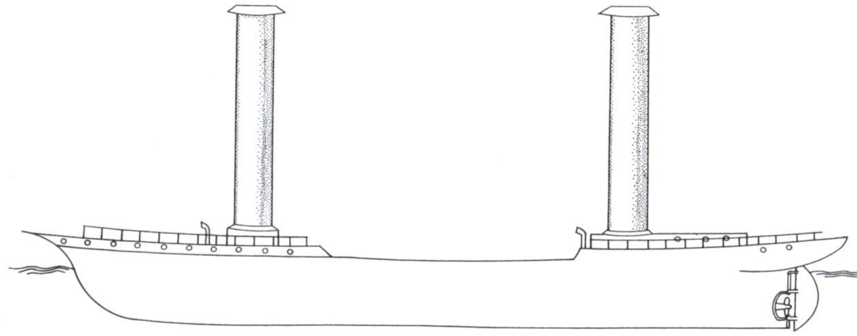

FIGURE 12.1 The 600-ton brig *Buckau* equipped by Flettner in the early 1920s with a pair of rotating funnels, each 16 meters (52 feet) high and 3 meters (10 feet) in diameter.

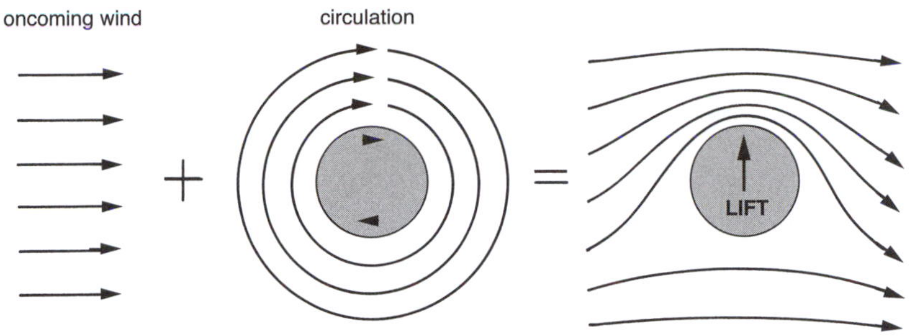

FIGURE 12.2 If a rotating body (here a sphere or cylinder) encounters a wind, the resulting asymmetry of flow around the body gives rise to an asymmetrical force—lift.

pitcher throws a curve (Adair 1990) or a tennis player manages a bit of top spin, or embarrassingly, as when a golf ball is sliced (Brancazio 1984). If the extra force is upward (lift in both senses) and comes with little or no extra drag, then a projectile may go just a little farther. The range of a tiny jumping mite gains a few percent because the animal rotates in flight, the lower part going into the oncoming air and the upper part going with the flow (Wauthy et al. 1998).

Anton Flettner's craft may have been a technical success, but it was a commercial failure, a sailboat that needed a motor. It needed no great bunkers of coal, hard to obtain in remote parts of the world, nor the large crew normally needed to manage sails, but its prospects sunk beneath a surfeit of ships, commercial stagnation, and the advent of cheap petroleum in the postwar 1920s. Boats with Flettner rotors surface from time to time, impelled by hope and journalistic hyperbole (wind, of a sort). In the same way, lengthwise-rotating cylinders can replace the wings of a functional airplane—the cylinders rotate so their tops move rearward and their bottoms move forward. Despite their complications and relatively poor performance, small models fly well enough to startle the conventionally minded among us.

But what's really happening? Where the wind and the flow induced by rotation of the cylinder go in the same direction (in front of the funnels), they add, so the overall fluid motion is more rapid than either alone. Where they're in the opposite direction, they oppose, and the net motion of the fluid is slower than either alone (fig. 12.2). By Bernoulli's principle, rapid flow correlates with lower pressure. So the ship with wind from the side will feel a pressure difference fore and aft, impelling it forward; the rotating pseudowings of the airplane will feel a similar one above and below, holding it up—in both cases, *lift*, as we've defined it.

That lift turns out to be proportional to the product of the surface area of the cylinders, to the density of the fluid medium (here, air), to the rate of rotation, and to the magnitude of wind (for a ship) or forward speed (for an airplane). All these are straightforward direct proportionalities—no exponents other than unity.

While as technology these Flettner rotors may be unattractive, they've evolved in nature on several occasions. You're probably familiar with the spiral descent of maple seeds (properly samaras in particular and fruits in general), in which the action of a lift-generating airfoil reduces sinking speed, as in figure 12.3a. That's also what happens in an autogyro or unpowered helicopter. The downward spiral of tulip poplar or ash samaras looks overtly similar to the descent of maple seeds. But a close look reveals a curious sym-

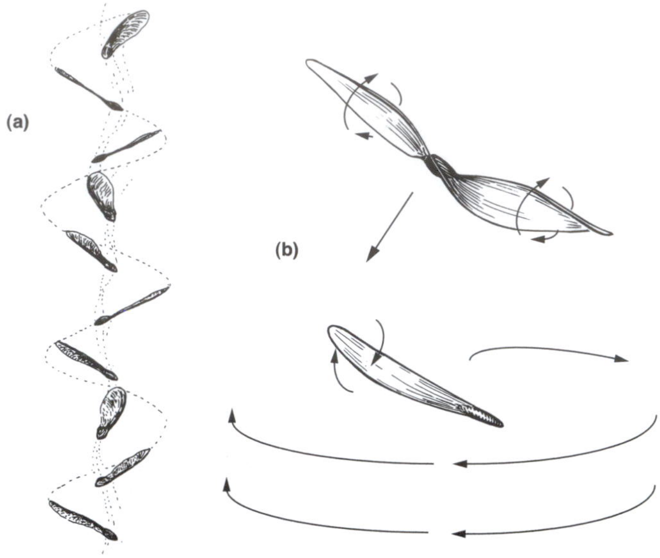

FIGURE 12.3 (a) The autogyrating descent of the samara (fruit) of a maple (*Acer*), essentially a glider taking a helical glide path. (b) The Flettner-autorotating samaras of the tree of heaven (*Ailanthus*) and of ash (*Fraxinus*).

metry about the long axis of these latter—unlike maple seeds, they don't have distinct upstream and downstream edges. And that, as McCutchen (1977) pointed out, goes along with the lack of any functional distinction between the edges—such a samara rotates around its long axis while more obviously spinning about a vertical axis of descent, as in figure 12.3b. One doesn't notice the autorotation because the spinning descent obscures it. But tulip poplar, ash, and some other samaras (such as those of the tree of heaven, *Ailanthus*) are indeed Flettner rotors. Azuma (1992) termed the motion "wheeling." See, for further examples, Augspurger (1988) and Greene and Johnson (1990).

If you're not convinced, flip a strip of index card (about 1 inch by 5 inches) and watch it descend. If it rotates, it will not descend vertically but will instead glide off with some horizontal component. Its top rotates opposite its direction of gliding, that is, with the wind it encounters, while its bottom rotates against its wind—the reverse of what something rolling on a solid surface would do. If you use a tapered instead of a rectangular strip, it will descend in a helix while rotating lengthwise just as do these particular samaras.

How Airfoils Work

Most lift-producers, however, use that familiar shape we call an airfoil, whether manifested as the cross section of an airplane wing, of a propeller blade in air or water, or of a billowing mainsail (fig. 12.4a). None of these has the habit of rotating rapidly around its long axis. So now what's happening? Such airfoils, convex above and concave or flat below, experience the same kind of circulatory flow pattern as did the cylinders and autorotating samaras, *but without themselves rotating*. Either because of their own asymmetry or by facing the flow at a slight inclination (or both), they cause the flow to be more rapid on one side (on the top of a well-behaved aircraft wing) than on the other.

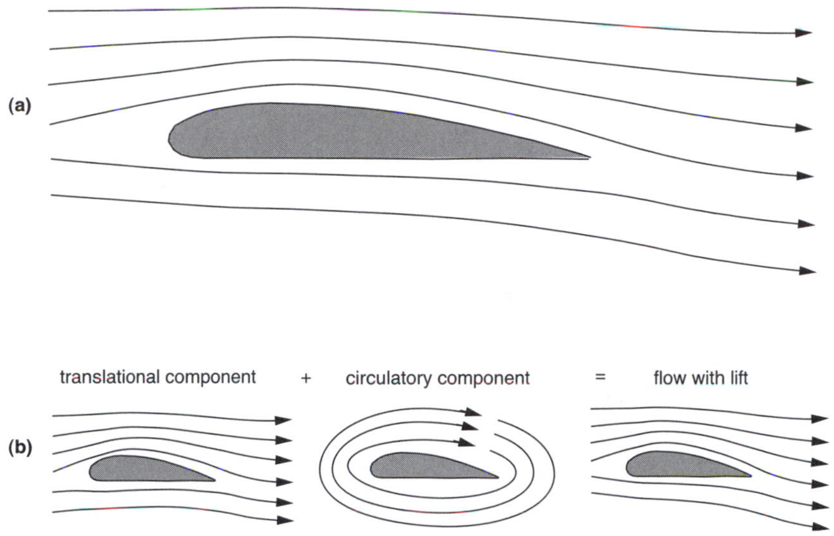

FIGURE 12.4 (a) The streamlines around the cross section of a lift-producing airfoil. (b) How those streamlines reflect a combination of translational and circulatory components of flow.

With airfoils, though, there's a telling quantitative difference. The lift again varies directly with the density of the fluid and the area of the wing (top view or "plan form" area is the usual area of reference). But it's now proportional to the *square* of the speed of the flow encountered by the airfoil rather than being directly proportional. Apparently, an airfoil makes the air around it circulate, as shown in figure 12.4b, just like Flettner's rotating cylinders. But here the translational wind interacting with the peculiar shape of the airfoil *causes* the circulation—instead of being proportional to the product of separate translational and circulating flows, lift now varies with the square of translational flow, or flying speed. In obedience to Bernoulli's principle, pressure drops below ambient above the wing, where flow goes faster, and it rises above ambient below the wing, where flow goes slower. (And, again, a bit of air going over the top typically arrives at the rear *before* the old mate with which it parted in the front.)

This notion of lift production as a consequence of circulation about an airfoil doesn't imply that air (or water, for a lift-producing hydrofoil) actually travels upstream beneath the wing. It doesn't deny that, were the vectors of translational and circulatory flows added at every point, all the resultant vectors would point quite happily downstream. But their combination wouldn't point exactly downstream—it would have a slight downward inclination representing the airfoil's lift. The circulation doesn't appear as a physical vortex quite like the ones we met in chapter 6—just a net vortex analogous to the vorticity in a velocity gradient above a flat surface. Nonetheless, that net vortex (called a bound vortex) amounts to more than a mathematical abstraction, and its presence helps to explain quite a few otherwise mysterious phenomena, including some biological ones. (The usual polite fiction about lift can't deal with these!)

Vortices don't form ends; recall how vorticity in a flow near a surface could tangle around a tree or other protrusion, forming what we know as a horseshoe vortex. A vortex really prefers to be toroidal, so it has no ends and manages to make ends meet. Of course the horseshoe vortex can't get its ends

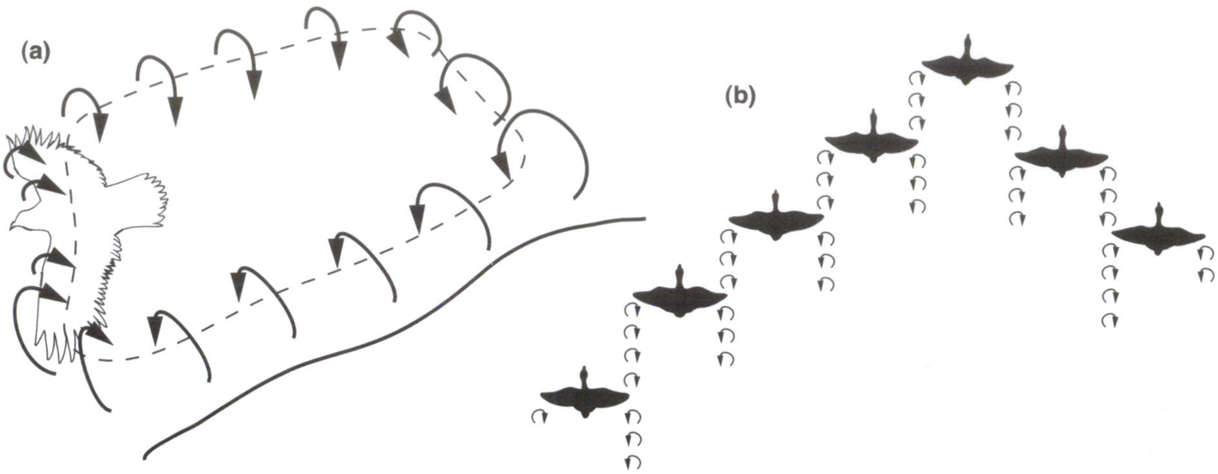

FIGURE 12.5 (a) The toroidal vortex system associated with a lift-producing wing, here a gliding bird—bound vortex, tip vortices, and starting vortex. (b) the V-formation of flying birds, showing how each sets its position (and the phasing of its wings) to get maximum benefit of the tip vortices of the one ahead.

together—viscosity eats its trailing ends as they go downstream. Consider that bound vortex on a wing. It extends beyond the ends of the wings, turning downstream like the sides of the horseshoe vortex, as in figure 12.5a. Sometimes condensation of moisture marks these tip vortices, making them visible from an airplane window or the ground. For an airplane, something called a starting vortex completes the misshapen torus. It remains behind at the airport, where it can make mischief for a small plane taking off behind a big one. Making more lift takes a stronger bound vortex, and the starting vortex has the same strength of circulation (albeit with the opposite spin) as the bound vortex—the heavier the airplane, the stronger the vortex it leaves on the runway.

The vortex pattern of a flapping wing gets still more complex, but a bird, for instance, still leaves a pair of tip vortices when flying rapidly forward. They just vary in strength over the course of the wingbeat, with the portions formed during a downstroke circulating more strongly. Large birds that fly steadily take advantage of those tip vortices, whether they're migrating geese seen as spots against the moon or a group of pelicans flying into the sunset. If the bound vortex on a wing goes rearward across its top, and net airflow behind the wing goes downward (momentum flux still counts!), the tip vortices must go downward on the sides facing the bird's body and upward on the sides away from the body. Upward flow shouldn't be wasted—it can offset gravity. And that's what the V-formation, shown in figure 12.5b, does, with each succeeding bird getting a bit of a boost from a tip vortex left by the one ahead. While the help may not be enormous, flying isn't cheap, and even the little assistance they get pays off in the long haul (Hainsworth 1989). Proper phasing of the wingbeats of each bird maximizes the payoff—look carefully at the next V-formation you see. The leader, as you'd expect, gets no benefit, but at least it pays no extra price; for long flights, birds often exchange the leadership role. Of course a symmetrical V isn't necessary, and we commonly see unbalanced ones and even diagonal Is.

Perhaps the most direct use of the tip vortex of an airfoil, certainly one that splendidly points up both its reality and the way a lift-producing airfoil makes it, occurs in a wading bird, the spoonbill (*Platalea* spp.). Spoonbills have elongate bills (fig. 12.6), wider near the tip than farther back, and rounded above and flat below. Airfoils? The birds would have either to fly sideways or fly with their heads turned 90 degrees. Weihs and Katzir (1994) showed that the bills were, in fact, hydrofoils. Spoonbills have the peculiar habit of sticking their bills into shallow water and sweeping them rapidly from side to side. A bill generates lift, of little direct use, but in doing so, it sheds a tip vortex for each sweep. The vortex interacts with the bottom of pond or stream, creating shear forces that stir up prey, which the bird then snaps up.

Airfoils—A More Quantitative View

For a craft that's neither accelerating upward or downward, lift must just equal weight. The proportionality of lift with the square of speed implies big trouble getting that critical amount of lift at low speeds. Two changes commonly compensate. First, the area of the wing can be changed. Birds certainly do this, using the broadest possible expanse when taking off, landing, or flying slowly; and large jet airplanes disgorge a remarkable amount of extra surface for takeoffs and landings. The other adjustment is in the angle between wing and oncoming air—the so-called angle of attack. Up to a point, increasing that angle gives a proportionate increase in lift. Still, raising the angle of attack raises two problems along with the angle. First, that up-to-a-point. At some not immense angle (typically about 20 degrees), flow separates from the top of the wing (as with the sphere mentioned much earlier), and the lift drops off catastrophically. So operating near the point of maximum lift, the "stall point," can be hazardous. Second, increasing the angle of attack increases the drag as well, so operating near the stall point isn't usually economical either. (Remember that power is drag times velocity.) The long-term best operating point, that giving the maximum lift-to-drag ratio, typically occurs at a much lower angle of attack.

The easiest way to view the behavior of airfoils consists of a so-called polar plot, a device introduced by that towering figure, Gustav Eiffel. To offset the

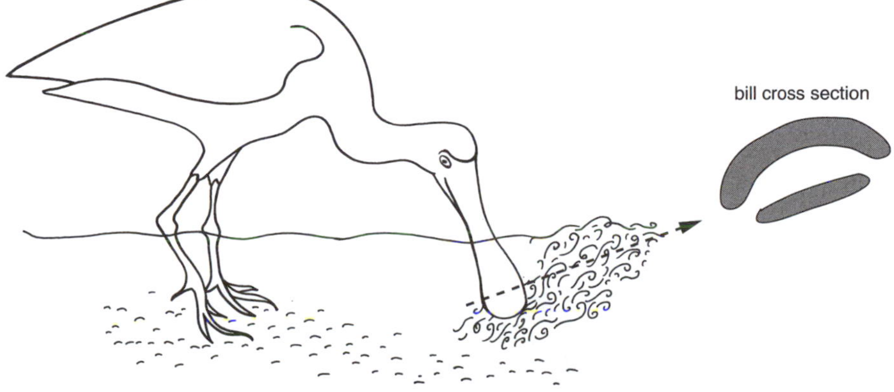

FIGURE 12.6 The way a spoonbill sheds vortices from its bill as it moves it from side to side, taking advantage of its profile as an asymmetrical hydrofoil.

effects of size one first defines a lift coefficient, C_l, comparable to the drag coefficient, C_d (equation (4.10) or equation (7.13)); the area used is the plan form or top view of the airfoil:

$$L = \frac{1}{2} C_l \rho S v^2. \tag{12.1}$$

One then plots, for a given airfoil, a graph of lift coefficient versus drag coefficient (based on the same reference area), noting the angles of attack on the curve (fig. 12.7). A straight line from the intersection of the axes is a line of constant lift-to-drag ratio. The steepest such line that touches the curve gives the maximum ratio for that airfoil configuration. The point at which that line touches marks the optimal angle of attack for economical steady flight. The highest horizontal line that touches the curve gives the maximum lift coefficient and the angle at which that coefficient occurs; that may be the best place to operate when taking off or trying to fly as slowly as possible.

Dramatic effects of size are evident on the polar plot, which, don't forget, has dimensionless axes. First, for the small fliers, that is, at low Reynolds numbers, the curve shifts far to the right—drag increases relative to lift. For streamlined airfoils, skin friction causes most of the drag, and skin friction gets worse as the Reynolds number drops, as we saw in the last chapter. Thus, the best lift-to-drag ratios drop. For an airplane wing the best ratio might be 25, for a locust wing it's about 8, for a fruit-fly wing it's less than 2 (Vogel 1994a). A model at $Re = 10$ couldn't do better than a lift-to-drag ratio of about 0.4 and at $Re = 1.0$ a lift-to-drag-ratio of about 0.2 (Thom and Swart 1940).

Second, at low Reynolds numbers, drag depends less on orientation, so inclining an airfoil to extract lift incurs a relatively lower additional drag. That has an odd consequence. A small flier not only has a lower best lift-to-

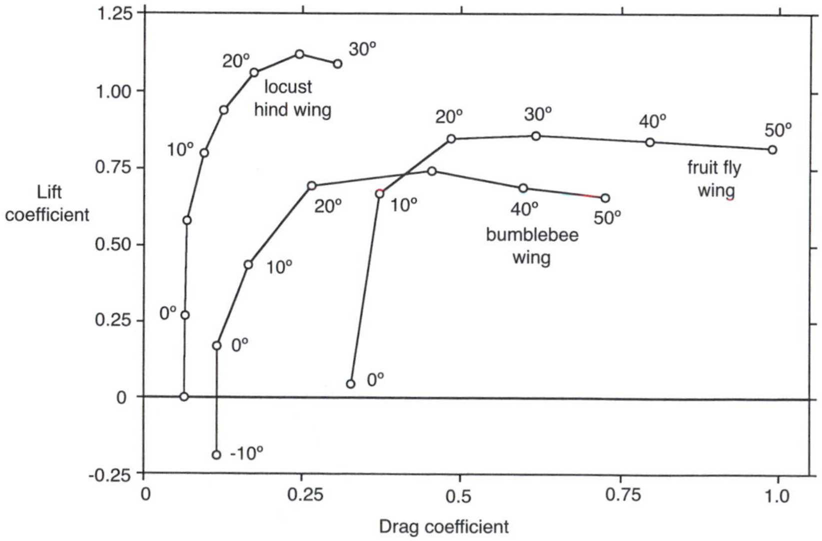

FIGURE 12.7 Polar plots for several insect wings: a locust hind wing (Jensen 1956), the linked fore and hind wings of a bumblebee (Dudley and Ellington 1990), and a fruit-fly wing (Vogel 1967).

drag ratio, but gets it at a very high angle—at the angle that just precedes separation or stall. For a fruit-fly wing operating at about $Re = 100$, that's around 20 degrees; for the models at *Re*s of 10 and 1.0, maximum lift-to-drag ratios occurred at about 45 degrees. And the fact that no microscopic creatures fly using the bird-bat-insect-airplane scheme falls into place—the great viscous forces of low–Reynolds number flows give inescapably high drag and thus impractically low lift-to-drag ratios.

Notice that size shifts things right or left on the plot more than it shifts them up or down. Maximum lift coefficients don't vary with size as much as do drag coefficients. Values of maximum lift coefficient for the wings used by either nature or human technology run around unity—up to low single digits, down a bit below unity. We're talking, of course, of the coefficients, not lift itself.

(Variation in size and form—diversity, in a word—is the hallmark of life, and a diversity of recent books tackle the diversity of nature's fliers. Tennekes [1996] gives a splendid introduction to these and other matters of scaling flying machines, human and natural; Norberg [1990] does a thorough job on vertebrate fliers and Dudley [2000] does on insects; Alexander [2002] gives the most general and accessible look at how and why plants and animals glide or fly; and Azuma's [1992] account is more technical and specialized.)

Both lift coefficient and lift-to-drag ratio depend on the shape of the airfoil as well as on the angle of attack, and a century of extensive work (with, it must be admitted, mainly military motives) has explored those relationships. Less obviously, perhaps, both variables depend as well on the relative length and width of a wing—on the ratio of these variables, which is called the "aspect ratio," familiar to many of us in the context of video displays. (Where wings taper, the ratio of tip-to-tip wing span squared over wing area serves as aspect ratio.) For a given area, longer and skinnier is better, which is why sport gliders have such extremely long and skinny wings. Structural considerations set the practical limit—one doesn't want wings either too heavy or too prone to deform.

We can look at aspect ratio from either of two viewpoints. One begins by observing that the longer, narrower wing intercepts more air and, consequently, has to give it less of a downward deflection to get the same downward momentum flux. More air, less speed (here downward)—we're talking in effect about Froude efficiency, and more m/t and less Δv improved matters (chapter 7). The ideal, of course, would be an infinitely long wing, which would give an infinitesimal deflection to an infinite mass per time. It would produce lift with no expenditure of power—$m\Delta v^2/2t$. Free lift may sound too good to be true, but recall that force (here lift) isn't work, and the chain holds up the chandelier with zero power consumption.

The other viewpoint looks at those tip vortices mentioned earlier. The greater the aspect ratio of the wing, the less the lift is taxed by its tip vortex—more wing, less tip. Incidentally, the infinite wing isn't quite as far-fetched as it sounds. In a wind tunnel, one simply runs a test airfoil from wall to wall so it has no tip that might shed a tip vortex. Only the spoonbill's bill, it seems, ought to prefer a low aspect ratio. It wants a strong tip vortex, never mind the lift, and the strength of that vortex varies inversely with the submerged length of the bill—its equivalent of wing span. (The game changes at speeds none of nature's fliers reach, so we're ignoring all those delta-wing and radically swept configurations.)

Lift's Multifarious Roles

The use of lift-producing airfoils and hydrofoils to produce thrust will get specific scrutiny in the next chapter. What matters here is illustration of the various other roles of these wonderful devices—spoonbill bills only begin the bill of particulars. Sand dollars, unimposing even among the unprepossessing echinoderms, provide several more cases. Such a disk-shaped creature normally lies flat (or partially buried) on the sand, feeding by drawing small particles into a mouth in the middle of its lower ("oral" of course) surface. Their upper ("aboral") surfaces are typically humped up a bit, so they're flat on one side and convex on the other—like a wing, but round. Alexander and Ghiold (1980) looked at one particular sand dollar, appropriately named *Mellita quinquiesperforata* because of five radial slots that go from top to bottom surfaces (fig. 12.8a). They found that the combination of the reduced pressure on the top, due to the convexity of the sand dollars, and the slots drew water up through the sand. That water contained edible particulates—recall the mention of interstitial fauna in the last chapter. Postfacto examination of animals whose slots had been plugged revealed empty stomachs.

The slots seem to do something else as well. Telford (1983) found that they reduced lift on the sand dollar. That sounds bad until we realize that sand dollars have no interest in flying but a great interest in not flying. Slot-lowered lift meant that a 20 percent greater flow was needed to dislodge individuals. For flatfish such as plaice and flounder, lift is a similar embarrassment. Lying on the bottom, such fish present convex upper surfaces and thus develop lift if the water moves across them. At flow speeds inadequate to push it downstream, a plaice begins to lift off and has to dig into the substratum in order to remain in place and not flounder (Arnold and Weihs 1978).

O'Neill (1978) described an even less obvious case, explaining yet another long-known but puzzling bit of behavior. Members of another species of sand dollar, *Dendraster excentricus*, a subtidal inhabitant of beaches on the Pacific

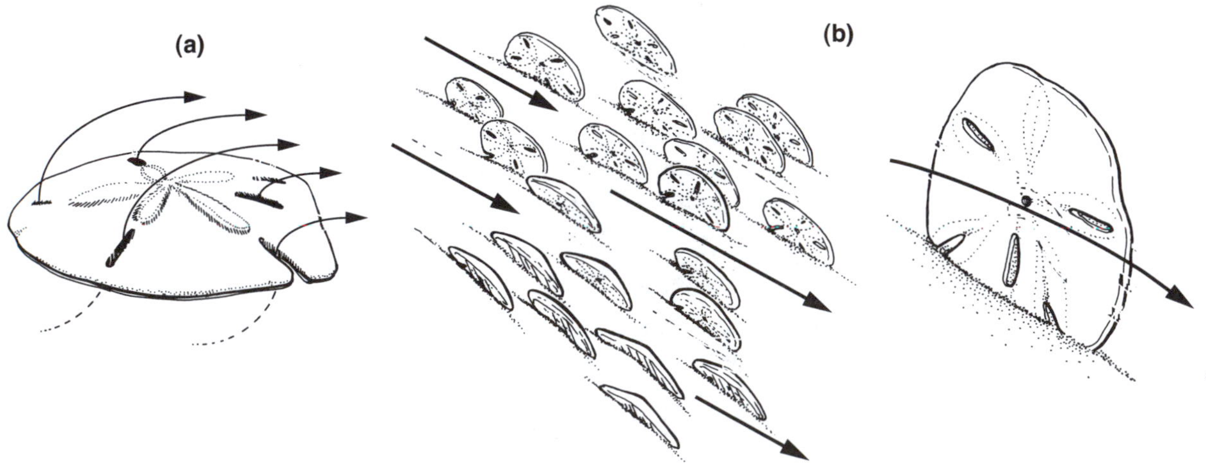

FIGURE 12.8 (a) Flow up through the five slots of *Mellita quinquiesperforata*, useful for reducing lift and drawing up food particles. (b) The way another sand dollar, *Dendraster excentricus*, stands on edge and uses communal lift to direct current closer to its mouth.

Coast of North America, have the peculiar habit of standing on their edges in patches containing hundreds per square meter, as in figure 12.8b. More peculiar yet, all individuals orient with their oral surfaces pointed the same way. She showed that neither drag nor lift has much importance. But circulation matters—it brings passing flood particles closer to the oral surfaces of the sand dollars. Aggregation enhances the effect, with the individuals interacting much like the wings of biplanes and triplanes. The system is remarkably refined. The sand dollars adjust the gaps between individuals depending on flow speed, and populations from more sheltered locations consist of slightly humpier (more cambered) individuals with greater lift coefficients.

How to Glide

According to Newton's first law of motion, a body remains at rest (or in steady motion) in the absence of any external force. For an aircraft flying steadily, the law implies that all forces on it must balance out to zero, taking into account both their magnitudes and their directions. Two forces act on the craft—the downward pull of gravity, experienced as its weight, and the rearward force of the wind it meets, felt as its drag. Somehow, the aircraft must do something to offset these forces, in other words, to produce a force equal to their resultant. In an airplane, the propeller or jet provides a horizontal force opposing drag and thus propelling the craft through the air. The forward motion then inveigles a pair of fixed wings into generating lift. In most flying animals, the functions of propeller and wings are combined, much in the manner of a helicopter. Most commonly, wings don't beat quite up and down in the manner of propeller blades that reciprocate rather than rotate. Instead, they beat in a plane inclined a little off the vertical, so they take air from above and in front and accelerate it downward and rearward.

But we'll defer to the next chapter further explanation of how an airfoil that experiences drag and produces lift can be persuaded to generate thrust. Only four lineages of animals have ever figured out how to do the trick; one has gone extinct (the pterosaurs) and three (insects, birds, and bats) remain with us. Staying aloft by making lift alone turns out to be far more widespread among organisms, both plants and animals. These others glide, doing so in remarkably diverse and sophisticated ways. In a formal sense, a gliding descent constitutes a source of work—a downward force, weight, together with motion in the direction of the force. So that work might be made to do something useful, such as slowing the descent itself.

It's perhaps easiest to approach the matter with a look at what's often called "simple gliding." That's defined as a situation in which a craft descends just fast enough to maintain a steady speed and a straight glide path. In simple gliding (fig. 12.9), an aerodynamic resultant force must be exactly equal in magnitude and opposite in direction to the weight of the flyer. The resultant has to be the combination of lift and drag with nothing left unbalanced to cause acceleration—with lift defined (as before) as the force at a right angle to the glide path and drag as the force along the glide path.

A curious and crucial bit of geometry jumps out of figure 12.9. Both the angle between the resultant and the horizon and that between the lift and the glide path are right angles sharing a vertex. Turning the resultant-to-horizon angle counterclockwise gives the lift-to-path angle. So the angle between the

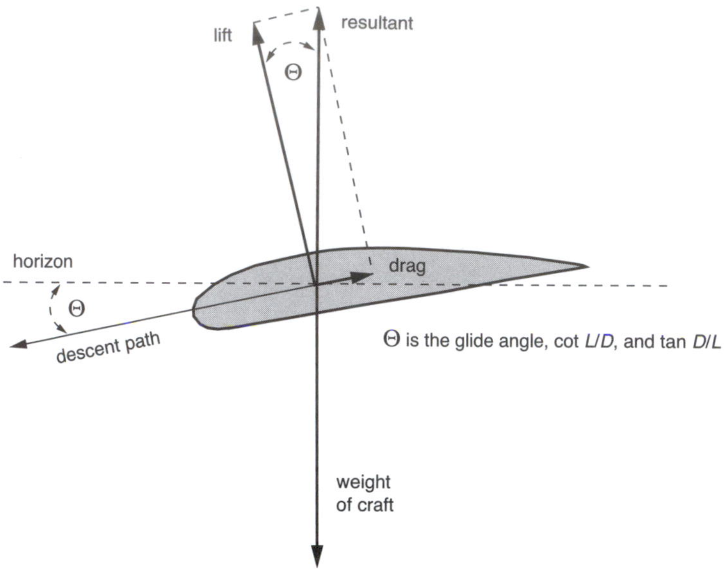

FIGURE 12.9 The forces and angles in simple gliding. As a result of its descending path, the craft (here represented by an airfoil section) meets an ascending wind. To move steadily, the craft has to descend just fast enough so forces balance—the resultant of lift and drag must equal weight.

resultant and the lift must be the same as the angle between the horizon and the glide path. The resultant-to-lift angle is the cotangent of the lift-to-drag ratio (or the tangent of the drag-to-lift ratio), while the horizon-to-glide path angle is the so-called glide angle. Thus,

$$\cot\theta = \frac{1}{\tan\theta} = \frac{L}{D} = \frac{C_l}{C_d}, \tag{12.2}$$

where θ is the angle between the horizon and the glide path. In short, the lift-to-drag ratio sets the glide angle—a ratio of, say, 40:1, means the craft will descend 1 meter for each 40 it goes horizontally. So for a good low angle or a nearly horizontal glide the craft needs a high lift-to-drag ratio, which, incidentally, 40:1 is. We see again why good gliders have long, skinny wings—tip vortices lower the lift coefficient and lift-to-drag ratio and inevitably steepen the best steady glide, so they have to be minimized.

This relationship places serious limitations on gliding among very small creatures—as we saw earlier, small size is inevitably concomitant with a much reduced best lift-to-drag ratio. If a fruit fly were to glide (it apparently does not) its best lift-to-drag ratio of 1.8 would require that it descend at an angle of almost 30 degrees to the horizon. Very small biological items carried on the wind—wind-dispersed seeds, baby spiders, and so forth—don't glide but rather go for drag maximization with fluff, silk, and such devices to slow their descent while they let wind currents move them horizontally.

The reader of especially instant insightfulness might notice an odd implication of equation (12.2). Both lift and drag coefficients include in their definitions a correction for speed, so they ought to vary only minimally with how fast a craft glides. Therefore, glide angle should vary only a little with speed. What then determines how fast a given gliding craft moves steadily downward? Nothing matters anywhere near as much as its weight. Why? Weight has to equal the necessary upward resultant force, the aerodynamic lift. Lift, in turn, varies very nearly in proportion to the square of speed. Result—wing characteristics may set glide angle, but weight sets glide speed. Corollary one—the heavier glider descends along the same path as the lighter one, merely moving faster along that path. Corollary two—for simple gliding from a fixed height, the glider carrying bricks can go just as far as the one carrying feathers. The military (at least during World War II) put great loads in gliders that carried troops into battle, accepting increased danger in landing to achieve greater payload and less time aloft and exposed.

Of course lift and drag coefficients do depend somewhat on speed. A given airfoil achieves its best ratio at a particular speed, dropping (all too literally) off at least a bit if moving either faster or slower. The easiest way to visualize the situation consists of a particular graph called a "glide polar." (Fortunately for most of us, neither polar diagram nor glide polar uses polar coordinates!) In it, as in figure 12.10, sinking speed—loss of altitude—is plotted against horizontal speed. Since one is talking about descent, the plot puts the zero line of the x-axis at the top. If the lift-to-drag ratio of an airfoil didn't vary with speed, its performance would be represented by a descending straight line from the origin, a line whose slope gave that lift-to-drag ratio. In practice, a particular airfoil gives a curved line with an optimum—the point to which a line from the origin is tangent. At that best lift-to-drag ratio or C_l/C_d, the craft can go farthest from a given release height in otherwise still air. At the correct weight (to set speed), the craft ought to descend at that optimum angle. Downward speed, of course can be read as the point at which a horizontal line from that tangent point reaches the y-axis.

Yet another optimum can be derived from the glide polar of an airfoil or a gliding aircraft. Minimum descent rate must occur at a speed slightly below the speed that gives the minimum glide angle. Minimum descent speed, on the

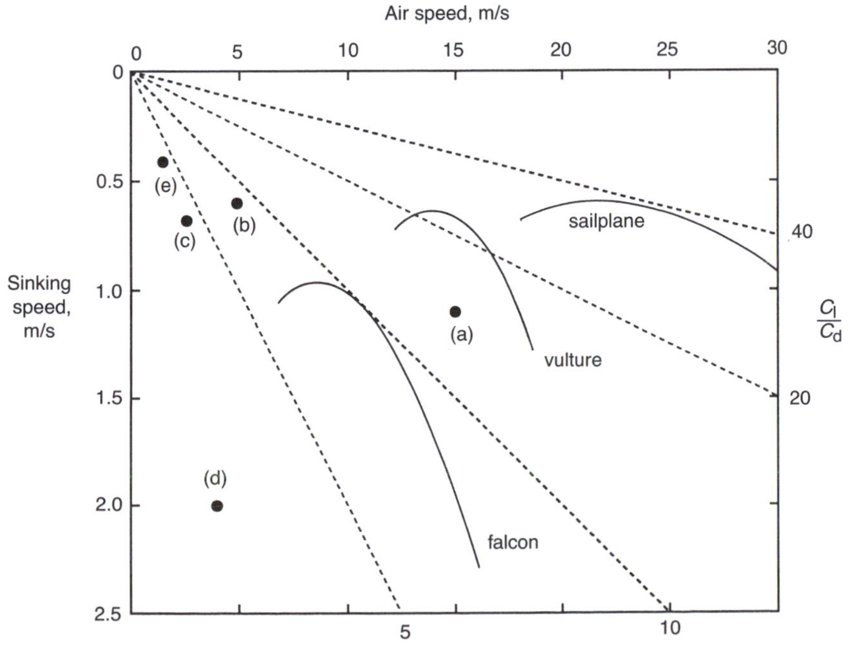

FIGURE 12.10 Glide polars for several craft—sailplane, vulture, and falcon (Tucker and Parrott 1970). Also shown are single data points for some other gliders: (a) Andean condor, (b) desert locust, (c) monarch butterfly, (d) gliding phalanger (or possum), and (e) Javanese cucumber seed-leaf.

plot, is the point to which a horizontal line is tangent. If a craft wants to stay aloft as long as possible, ordinarily to take advantage of atmospheric structure and motion—"soaring" as opposed to "gliding" in the conventional lingo—that's the optimum that's optimal and for which it should opt.

The distinction between those two optima has important biological consequences. Recall the polar plot of figure 12.7. Airfoils that worked at lower Reynolds numbers reached lower lift-to-drag ratios than those operating at higher Reynolds numbers. The problem becomes especially acute down in the size range of small organisms. It takes only a little work with models and a modest faith in biological design to deny that the problem lies with the poor design of the wings of small creatures. So (and as we see in the glide polars of fig. 12.10), organisms are worse than human-built gliders, and the smaller the creature the worse it does—in terms of lift-to-drag ratio. But organisms don't look so bad if the criterion for quality shifts to the slowness of descent. If the organism uses atmospheric structure and motion, again, descent speed matters more than glide angle. And that's as often as not what matters, whether a bird is trying to stay in an ascending thermal bubble of air or a samara is being carried laterally by a wind—as emphasized for the latter by Nathan and Muller-Landau (2000) and Nathan et al. (2002).

Scaling Descents: How, Who, and Why

That note about speed and glide angle has once again brought up that recurrent issue of scaling, which bears relevance to flight, both powered and unpowered, in other ways as well. First, the physical side, in particular for unpowered fliers—gliders and parachuters. Truly remarkable are both the number and diversity of airborne organisms that can't or sometimes don't use self-powered flight. Equally remarkable are number of times such passive aerial life has evolved. But life puts a high value on acquiring food, dispersing, and encountering mates, all activities that can be aided by aerial activities. Who are we talking about? Figure 12.11 gives some of the players. In all this diversity, though, we can recognize two general schemes for keeping aloft without active propulsion.

The ability of a wing to produce lift scales with its area, as in equation (12.1). The lift that's necessary to keep aloft scales with weight and thus, usually, volume. So, other things (i.e., shape) being at least roughly equal, the smaller creature has less difficulty staying up. The usual variable cited is something called "wing loading"—the ratio of weight, mg, to wing area, S:

$$W_l = \frac{mg}{S} = \frac{\rho V g}{S}. \tag{12.3}$$

Crucially, the variable isn't dimensionless but (ignoring density and gravity as size-independent) has a residual dimension of length. The smaller flier should have the lower wing loading. In short, it's lighter for its wing area and thus can (indeed, must) fly or sink more slowly. Its nicely low wing loading offsets its unfortunately low lift-to-drag ratio.

What we're looking at comes down to nothing less basic than the scaling of a surface-to-volume ratio, first introduced back in chapter 3—next to this, even Reynolds number pales. Big things fly faster, glide faster, drop faster,

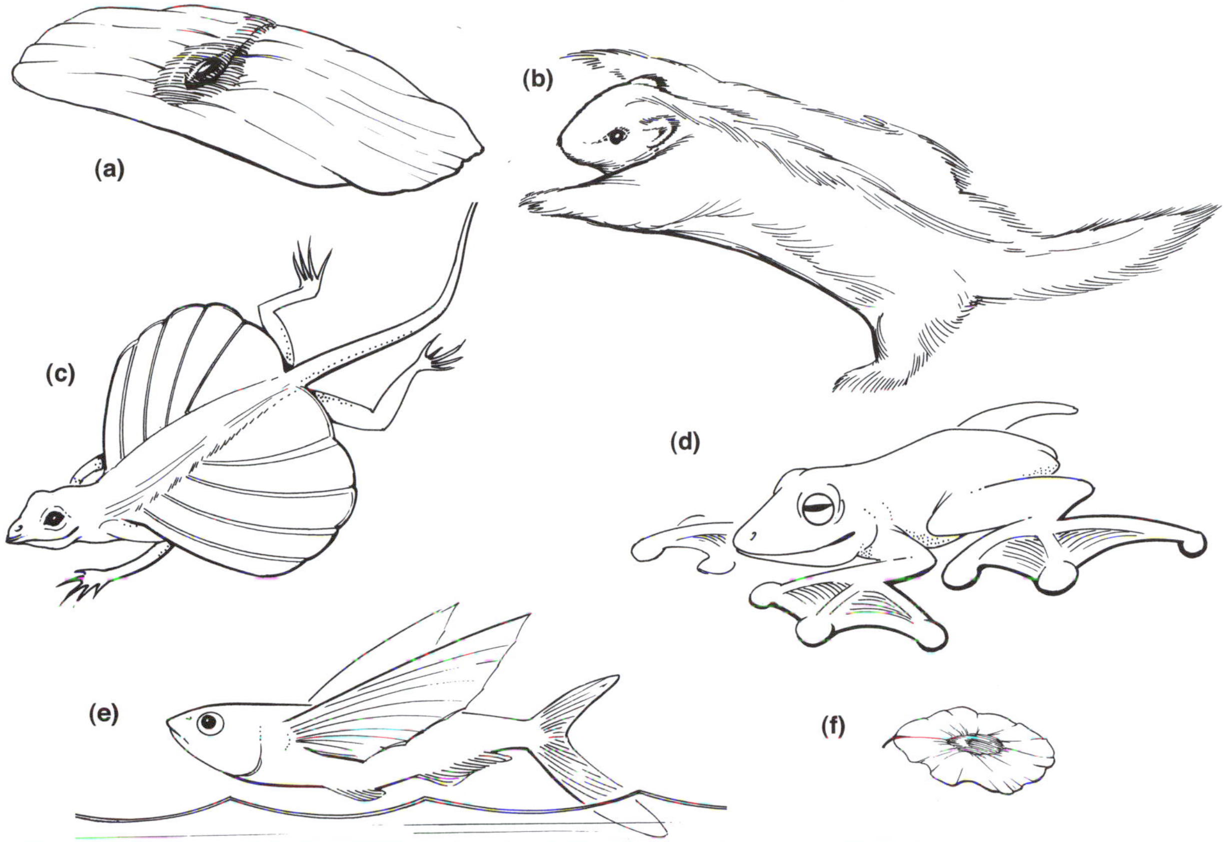

FIGURE 12.11 Various gliders, not drawn to the same scale: (a) the seed-leaf of the Javanese cucumber, *Alsomitra*; (b) a gliding phalanger (or possum), *Petaurus breviceps*; (c) a lizard, *Draco*; (d) a big-footed frog, *Rhacophorus*; (e) a fish, one of the Exocoetidae; and (f) an elm (*Ulmus*) fruit.

whether spores, pollen, seeds, flying animals, model airplanes, or human-carrying aircraft. So familiar is the generalization that we forget to notice it. But think for a moment—most model airplanes are slower than most full-sized ones and, even so, use proportionately larger engines (McMahon and Bonner 1983). Very slow aircraft that carry people, such as helicopters and hovering fighters, consume great quantities of fuel or, as human-powered craft, must be extraordinarily weight-efficient—Tennekes (1996) put this business with especial clarity.

Both nature and human technology use two fundamentally different schemes to slow down gravity-impelled descents—lift production, as described in the preceding pages, and drag production, as brought up in the last chapter. Here Reynolds number does come into the picture. Lift production, as already explained, gets problematic at low Reynolds numbers—that crucial variable, airfoil shape, gets to be a bit of a joke with all the semistagnant fluid obscuring it. Worse, drag gets relatively worse and, with the predominance of skin friction, it gets harder to circumvent. Conversely, the necessarily high drag at low Reynolds numbers can be made a virtue. And the Reynolds number contains the product of size and speed. So what we expect—and find in nature if not in our own devices—is a reliance on maximization of drag for small biological items and on maximization of the lift-to-drag ratio (implying, incidentally, drag minimization) for larger ones. Never mind ancestry—size sets sinking style.

Reynolds number enters in yet another way. How will speed of descent vary with the size of the descender, assuming as a base line, still air and isometric fallers of invariant density? For something that descends (recall the last chapter), the upward force must equal weight at terminal velocity. And that's true whether the upward force comes from the resultant of lift and drag or from drag itself. (Drag was defined, remember, as a force parallel to the local flow direction, and flow will be upward when you fall downward.) The scaling argument can be most easily framed in terms of drag.

Down in the size range to which Stokes' law applies, drag varies with the first power of both speed and length. Weight, with our assumptions, will vary with length cubed. Setting the two proportional, we get

$$lv \propto l^3$$

and

$$v \propto l^2.$$

To put things in the usual frame of reference (and since body mass is proportional to length cubed), speed must therefore scale with the two-thirds power of mass or weight, so

$$v \propto M_b^{2/3}. \tag{12.4}$$

Other things equal, bigger is faster, and fairly drastically so. This is the aerial world of spores and pollen, all those seeds with fibers sticking out to increase their drag, spiderlings with their silken threads, and so forth. It's also the aquatic world of sinking diatoms and a lot of other plankton. That severe dependence of speed on size may underlie an apparent pattern among drag-based descenders, at least the aerial ones. The smaller among them, such as pollen and spores, rarely deviate much from simple, unadorned spheres, while the larger ones—seeds, especially—add drag-increasing fibrous envelopes. Put in terms introduced in the last chapter, with increasing size comes an increasing divergence between the core radius and the nominal or Stokes radius. Or so it appears—I don't know of specific data.

At higher Reynolds numbers, drag varies with the second powers of both speed and length. Again setting drag equal to weight and putting the matter as a proportionality,

$$l^2v^2 \propto l^3$$

and

$$v \propto l^{1/2}$$

so

$$v \propto M_b^{1/6}. \tag{12.5}$$

Bigger is still faster, and smaller remains slower—although no longer quite so overwhelmingly. The case, in fact, doesn't rest on slowing descent by increasing drag. Lift varies with speed in nearly the same way as does drag, so the argument should apply to descent slowing by maximizing the lift-to-drag ratio, and thus to gliders as well as parachuters. Nor does it matter whether the descender glides in a helix, like a maple seed, or goes cross country, like a bird. For that matter, it doesn't matter whether it uses a conventional airfoil or a Flettner rotor.

The dividing line between drag maximization and lift-to-drag maximization turns out to be relatively sharp—at a mass of about 50 milligrams for seeds (Matlack 1987). These lift-to-drag maximizers then run up to about 10 kilograms (the largest birds). That 200,000-fold range runs from the smallest craft that might develop a reasonable lift-to-drag ratio to the largest craft that demands neither impractically large wing areas nor equally impractically high flying speeds. That range, over five orders of magnitude, should (by equation (12.5)) give a speed range of about sevenfold—which seems at least roughly to fit with what we see.

At the low end we meet mainly seeds—or more properly, flying fruits such as samaras that contain seeds. Most of them glide downward along a helical path, working more like autogyros than what we think of as gliders—maple seeds are representative of a scheme occurring in a great many families of both softwoods and hardwoods (gymnosperms and angiosperms). Some of the tropical autogyrating gliders (of the family Juglandaceae especially) get quite large and heavy—ounces, not grams. And one can build large models that work well—they descend only a little faster than small ones (equation (12.5)). Already mentioned are those, such as ash and tulip poplar, that use Flettner autorotating airfoils rather than conventional ones—but Flettner devices occur less widely than ordinary autogyrating gliders.

Much rarer are gliding pieces of plant that do not autogyrate, ones that travel in a straight line. The best-studied is the seed leaf of the Javanese cucumber, *Alsomitra macrocarpa* (Azuma and Okuno 1987). Gliding in a straight line may be unusually useful for *Alsomitra*, which descends in the relatively still air beneath the high canopy of southeast Asian forests. It achieves a lift-to-drag ratio of 4.6 at a Reynolds number of 4,000. That corresponds to a descent angle of 12 degrees. By contrast, a smaller autogyrating maple samara (of *Acer diabolicum*) does 3.3 at $Re = 1{,}400$ (Azuma and Yasuda 1989). Neither number is startlingly high, and most likely performance in terms of lift-to-drag ratio has been sacrificed for stability in a craft completely lacking active control. So asserting that lift-to-drag ratios have been maximized to slow descent overstates things just a bit. For that matter, a lot of botanical gliders turn out to be remarkably poor in aerodynamic terms. The most likely rationalization, far from original here, invokes that need for unusual stability in craft that totally lack active control.

Insects, at least the larger ones, sometimes do a bit of gliding, even if none goes long distances without active wing beating as well. But even big ones, such as locusts and butterflies, remain near the lower size limit for practical gliding. The trouble, of course, comes from the low lift-to-drag ratios at low Reynolds numbers. Not only would a fruit fly descend at a steep 30 degrees, but it would drop faster than a fluffy, drag maximizer. The large dragonfly-like Palaeodictyoptera of the Paleozoic, with wings much longer than any living insects, ought to have been much better gliders than contemporary forms, whatever their facility for flapping.

Birds and bats are by no means the only gliding vertebrates. Members of several genera of squirrels glide, ranging from the giant flying squirrels (*Petaurista*) that get up to 2.5 kilograms in mass (Scholey 1986) to small ones with adults of under 100 grams (Thorington and Heaney 1981). And flying-squirrel-like phalangers (properly, the possum genus *Petaurus*) occur among the marsupials of New Guinea. All use a patagium, a extension of skin connecting forelegs and hindlegs. Flying fish may be the best of the nonflying gliders. Two groups fly, one using only enlarged pectoral fins, the other both pectoral and pelvic fins in a staggered biplane configuration (see, yes, Fish [1990]). Some emerge from the water in two steps, taxiing along with fins in the air and the lower lobe of the tail beating furiously in the water. One (*Cypselurus*) can go as fast as 20 meters per second (45 miles per hour) with single flights of 50 meters and repetitive flights extending up to 400 meters. Less familiar are flying frogs (*Rhacophorus*), with great webbed feet (Emerson and Koehl 1990); a lizard (*Draco*), with rib-stiffened membranes extending to either side of its trunk; and flying snakes (*Chrysopelea*), with flattened bodies most of which they orient crosswise to their flight paths (Socha 2002).

None of these sets records for aerodynamic quality, and none gets lift from structures of close homology with the wings of birds or bats. Curiously, the world center for terrestrial nonflying gliders is that same tropical forest that supports the Javanese cucumber. Flying frogs, lizards, and snakes occur only there, and it has more forms of flying squirrels than anywhere else on earth, perhaps more than everywhere else together. Dudley and DeVries (1990) suggested that what's special is that that dipterocarp-dominated forest, perhaps a result of its volcanic rather than lateritic soil, is higher than other tropical forests. They pointed out that the greater vertical dimension beneath a dense canopy should increase the utility of gliding.

Zoologists have traditionally divided animals that descend through the air into parachuters and gliders based on whether they go downward at angles greater or less than 45 degrees off vertical (Oliver 1951). That division has little to recommend it. For one thing, the dividing line is completely arbitrary; it certainly doesn't mark a clear shift from simple drag maximization to lift production with drag minimization. For another, few of the relevant animals glide in anything close to a steady state. Most make a rapid initial descent at a fairly steep angle followed by a flare-out, sometimes adding a slight upsweep prior to landing. And that pattern makes fine functional sense. Making lift takes speed, and gravity can be counted on to augment any push off. All of these animals have relatively high wing loadings, so decent gliding demands speed. At the same time, speed isn't the handiest thing when landing—why not reinvest that speed in a bit of level or ascending flight prior to touching down or grabbing another tree? All it takes is increasing the angle of attack of the airfoil (from maximum L/D toward maximum L) just before landing, what every bird does every time it lands on a branch without a head wind.

Finally, going for distance may not be the only thing (or even the main thing) that matters for these arboreal creatures. If an animal jumps, it sets its trajectory at takeoff. Little adjustment can be made by postural shifts and so forth later on. Gliding, even at a poor glide angle, puts aerodynamic control into the picture, and that can be effective further along—perhaps to pick a

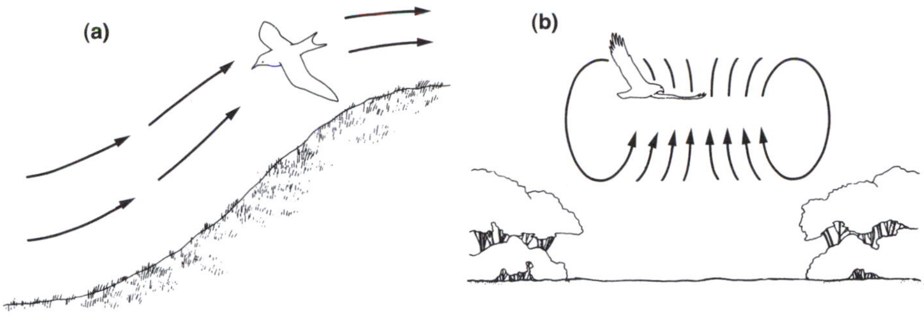

FIGURE 12.12 Types of static soaring: (a) slope soaring, and (b) thermal soaring.

good landing spot or to avoid some predator coming from behind or another with jaws agape at a landing site.

Soaring

One might say that soaring is just gliding with an altitude. Instead of depending solely on gravity and inevitably descending, it enlists atmospheric structure, from which it extracts the power necessary to stay airborne. In theory, almost any such structure (except perhaps an unmitigated downdraft) can be tapped—any temporal or spatial non-homogeneity, which is to say anything other than a steady and uniform horizontal motion. Active flight costs energy; soaring can be free; the advantage is obvious. But how to do it? Never mind theory—a look at what animals do shows at least four possible schemes. These can be most easily viewed in a simple hierarchy; fortunately we have a reasonably well-established terminology for them. Thus, in words and in figures 12.12 and 12.13:

- *Static soaring* comes in at least three varieties:

  1. *Declivity or slope soaring* takes advantage of ascending air on the upwind side of a hill. The trick can't be simpler—an animal glides in the

FIGURE 12.13 Dynamic soaring in a wind whose horizontal speed increases with altitude. The bird alternately ascends and descends, extracting energy from the gradient. It reverses its heading at the marked points.

ordinary fashion, but the upward motion of the air compensates for the downward component of the bird's motion, while the horizontal part of the oncoming wind compensates for its forward motion. So it stays in place, in a fixed position in front of, and often a bit above, the hill. I've seen sea gulls and hawks do it; many other birds engage in slope soaring as well. Hang gliders do just this when they can; under favorable circumstances they can stay aloft at their pleasure.

2. *Thermal soaring* is a little fancier. As a day wears on, the sun heats the earth a little unevenly; the earth responds by disgorging into the atmosphere bubbles of warmed air. Each rises, and the interactions between the sides of the bubble and the air it passes forms the bubble into a torus, which continues to ascend. In such a torus, the air on the outside of the ring moves downward and that on the inside of the ring moves upward with respect to the torus as a whole. If a bird can circle within the inner portion of the ring, it can descend with respect to the air around it but (a) ascend with respect to the earth and (b) remain within the torus, riding upward with it. Where I live, turkey vultures do most of the thermal soaring; during the summer one can often see several at once, each above a plowed field that's surrounded by forest. I wonder if highways running through forests or well-vegetated areas encourage thermal soaring—by providing a dependable series of thermals, a line of elevators, if you will, and, via roadkill, a source of food as well.

Gliding works better with long, thin wings, but birds that engage in slope and thermal soaring commonly have wings much shorter and wider than those that soar over the ocean (as below). Instead, they have separated primary (= tip) feathers, splayed out much like fingers. One explanation, and I see no problem with it, is that their design represents a compromise that improves the quick control needed for soaring in the less steady air over land. In particular, thermal soaring requires turning in tight circles to stay in the rising tori.

3. Less widely practiced or appreciated is *sea anchor soaring*. Withers (1979) called attention to the possibility and to its use by petrels. Here a bird floating on water and facing a dependable wind spreads its wings, which lift its body above the water. It leaves its feet in the water, arranged so they incur lots of drag, which keeps the body from speeding up. That keeps them from losing their wind—like a kite whose string has broken. Petrels get their name from the device. To mariners on ships they appeared to be walking on the water, like Peter, in the New Testament. The mariners couldn't easily see what the feet were doing—even very small waves would have obscured the way they worked as drogues or anchors, things familiar to sailors. Of course the birds move slowly backward as the feet move through the water, since nothing can have drag without moving relative to the surrounding fluid. But that may be an advantage, increasing their chances of encountering fish. Prions, another kind of seabird, seem to do the same thing (Klages and Cooper 1992), although they keep their heads in the water as they feed on swimming copepods.

- *Dynamic soaring* takes advantage of the velocity gradient in the lower portion of the atmosphere rather than any updraft or difference between media. At this point we have only one well-understood example, what's often called *gradient soaring*, illustrated in figure 12.13. To take advantage of the velocity gradient, a bird must go alternately up and down, which it does in great vertical loops that carry it, overall, in a downwind direction. How it works was first suggested, I think, by Cone (1962); decent observational data was provided later by Pennycuick (1982). But the basic motion has been nowhere better described than in a remarkable book about an even more remarkable journey. Escaping from their crushed ship in the Antarctic ice, in the Antarctic summer of 1915, Ernest Shackleton and a few others made an 800-mile trip through the wild southern ocean to South Georgia in what was little more than a sail-equipped life boat. The captain of the crushed ship, Frank Worsley, wrote an account of that journey; despite the cold, wetness, crowding, and buffeting, he noted and described the motions of the albatross in terms that no scientific paper could improve (Worsley 1977):

He sweeps before the gale with a mighty rush, then, turning sharply, lets the wind strike his underside, soars almost perpendicularly, and again turns with the wind, coming down in a long, symmetrical swoop, carelessly lifting a few inches to clear the top of a breaking sea, while conforming his flight to the surface of the ridges and hollows.

CHAPTER 13

Thrust for Flying and Swimming

> Birds fly, men drink.
>
> **motto of the Man Will Never Fly Society (www.manwillneverfly.com)**

WHILE GENERATING THRUST MAY NOT BE THE OLDEST APPLICATION OF lift, it can't be anything but the most important. We see the same underlying scheme whether we consider the wing of a hummingbird, the fluke of a whale, the forelegs of a sea turtle, the tail fin of a tuna, or the wing of a swimming penguin. It's not the only way to get sufficient thrust to make one's way through air or water, but, except at Reynolds numbers below about 10, thrust-from-lift has been the gold standard of both nature and recent human technology. Paddling and jetting, unless some secondary consideration supervenes, can't match its effectiveness and efficiency—recall the discussion and data on Froude efficiency in chapter 7.

While nowhere near as obscure as the origin of lift, getting thrust remains a less-than-obvious trick. We start with a device, an airfoil or hydrofoil, that when exposed to a flow experiences a force in a direction that's not quite the same as that of the flow. Although we divide the force into drag and lift, components parallel and normal to the flow, the device itself sees only the unitary resultant. How can that resultant of drag and lift make thrust—negative drag, a force opposing the flow?

The Source of Thrust

During the nineteenth century, some of the people trying to create aircraft recognized that greater efficiencies could be reached if the functions of staying up and progressing forward were divided, with fixed wings responsible for the first and a propeller directly responsible for the second and only indirectly for the first. The arrangement is rare in nature, although the forewings of beetles (the elytra) in some cases function as fixed wings. But it provides a convenience for explanation, and it has been quietly used to divide chapters 12 and 13. So we'll talk initially about propellers, only later adding the twin complications of biological airfoils, making both lift and thrust, and reciprocating rather than rotating.

Getting thrust, in essence, amounts to an extreme form of gliding. If the descent angle of the airfoil is even greater than needed to get a vertical resultant (as in steady gliding), then the resultant force of lift and drag will be directed forward of a vertical line through the craft. In short, it will have a component of thrust, as in figure 13.1. But how to get the fluid to strike the airfoil in a sufficiently nonhorizontal direction? Downward, upward, or

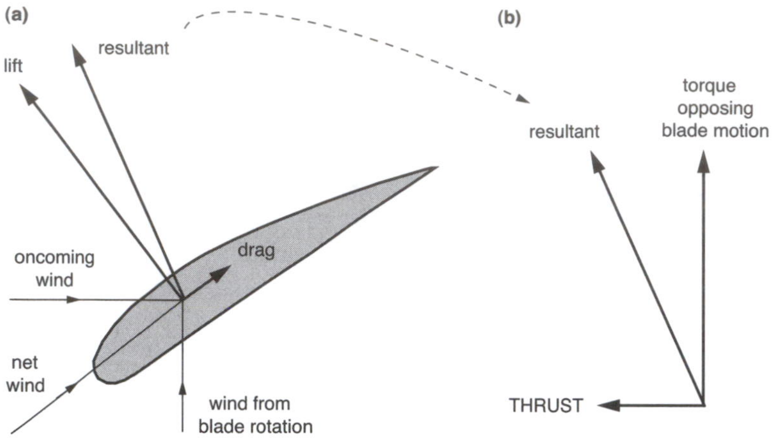

FIGURE 13.1 The origin of thrust. (a) If the oncoming wind strikes a blade element from a sufficiently nonhorizontal angle, then the aerodynamic resultant may be tipped forward (upstream) of the vertical. (b) If so, it has a component opposite the free-stream flow—negative drag, or thrust. The thrust component appears explicitly when the resultant is reanalyzed, as at the right.

sideways will do equally well—we just can't allow it to come from directly ahead or all we'll get is lift and drag. The trick consists of nothing more nor less than persuading the airfoil to move with respect to the craft—in short, to use some form of a motorized propeller. The flow striking this rotating (or heaving or flapping) airfoil will thus be something other than a direct head wind.

Most of the same considerations described for lift still apply—for instance, the quality of the airfoil, even if now a flapping wing or propeller blade, is still measured by its best lift-to-drag ratio. And, as in gliding, the scheme doesn't work at all well at very low Reynolds numbers, so the swimming of all microorganisms must be based on different mechanisms entirely.

Look again at figure 13.1a. A cross section of an airfoil moves downward, encountering an upward flow, $v_{\text{rotation of wing}}$, as a result of that downward movement. It also encounters a flow opposite that of the craft's forward movement, v_{forward}. The resultant relative wind, $v_{\text{wind, net}}$, of course, is what it really sees, and it meets this at an angle, the "pitch angle," which corresponds to its angle of attack as defined in the last chapter. As a consequence of its shape, pitch, and that relative wind, it feels lift and drag in the form of an off-flow resultant force. So we turn our attention (fig. 13.1b) to that resultant, dividing it (for our analytical convenience) into an upward force and a forward force. This last, directed opposite the motion of the aircraft as a whole, acts on the aircraft as thrust. Again, the propeller blade knows only lift and drag; the craft feels thrust, coaxed into existence by spinning the propeller with an engine.

What about that other force, the upward one? Upward sounds good, but it's offset, unfortunately, by a downward mate due to the motion of the opposite propeller blade. (Nor, obviously, can one evade the matter by using still more blades; and, if flapping, the upstroke that must follow every downstroke provides the same dismal compensation.) We recognize that other force as "torque per unit radius." It's what the engine sees and works against. We'll assume (pending chapter 22) that the engine, whether metabolic and muscular or thermal combustion can satisfy its demand by producing sufficient power to keep things turning.

Consider what happens as an aircraft increases its speed, assuming for simplicity that the propeller turns at a constant rate but adjusts its pitch to keep a good lift-to-drag ratio—as in figure 13.2. Initially (fig. 13.2a), the resultant points well forward since the propeller's turning provides all the net wind. So the craft gets lots of thrust relative to the demand on the engine. As the craft picks up speed (fig. 13.2b), the resultant begins to tip farther aft, which ought to reduce the resulting thrust. But if thrust drops it all, it doesn't drop much since, early in the process, the additional oncoming wind increases the resultant force enough to provide additional thrust—if at a cost in power, since torque per radius increases by a somewhat larger factor. As the craft gains more speed (fig. 13.2c), the resultant tips still further rearward, which now more severely decreases thrust. Of course the craft can't keep speeding up until thrust equals zero. It suffers drag, from the various sources of the latter, and it can't exceed the speed at which thrust drops and drag increases to the point at which they become equal.

Advance Ratios and Top Speeds

The point at which maximum thrust equals drag sets maximum flying speed. Drag can be reduced by streamlining the body of the aircraft and by using longer (and thinner, since area is what matters most for lift) wings. Thrust can most readily be increased by using a longer propeller (faster tip speed) and spinning it more rapidly. Putting the latter quantitatively, the relevant expression is something called "advance ratio," *J*—in essence how far the aircraft moves forward relative to how far the tip of the propeller moves in its circle:

$$\left(J = \frac{v_f}{nd}\right), \tag{13.1}$$

where v_f is forward speed, n is the rate of revolution of the propeller, and d is the diameter of disk swept by the propeller blades. For ordinary airplanes, advance ratios run below about 4.0. The importance and components of *J* ought to make sense after revisiting figure 13.2.

Here we need (perhaps parenthetically at this point) an additional factor for wings or fins or flukes that reciprocate rather than rotate. It's a factor,

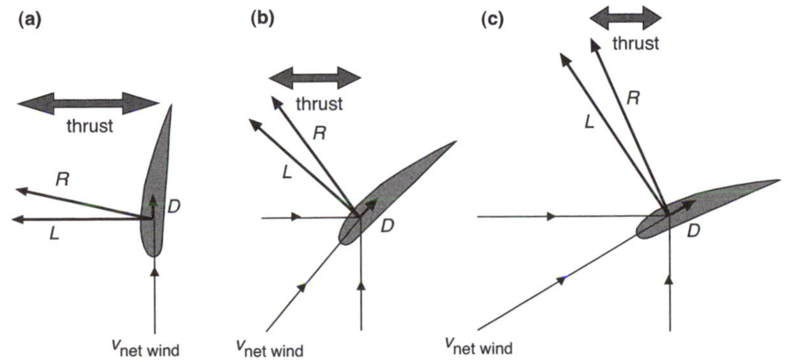

FIGURE 13.2 How thrust drops with increasing speed. (a) The propeller's turning provides all the relative wind striking it. (b) Addition of an oncoming wind increases the relative wind but shifts its direction, tilting the resultant. Depending on the circumstances, this may slightly decrease its forward component, thrust. (c) More oncoming wind shifts the relative wind still closer to horizontal, tilting the resultant closer to vertical, and further reducing thrust.

stroke amplitude, with no direct airplane analog. It's also, a matter of some importance, one that an animal can alter, one might say, at will. So a flying animal might turn left by increasing the amplitude of its right wing and reducing that of its left. An airplane with a single propeller can't do that. Conversely, an aircraft with two propellers can increase the revolution rate of one and decrease that of the other, something that no known flying animal does—left and right wings always beat at the same frequency. What's used (Ellington 1984) is an alternative, slightly fancier *J*, which we might call J_b, for the biological advance ratio.

$$J_b = \frac{v_f}{2\phi nR}, \tag{13.2}$$

where ϕ is the amplitude of the wingbeat (called, also, the "stroke angle"), in dimensionless radians, the arc the wing sweeps through in its stroke; *R* is wing length (the analog of the length of propeller blade); and *n* is wingbeat frequency. To get figures comparable to advance ratios cited for airplane propellers, one has to multiply J_b by π. (Sometimes equation (13.1) is given with π in the denominator, as $2\pi nR$, and is thus equivalent to equation (13.2).)

All too few good data are available for top *J*s, mainly because we have few reliable data for maximum flight speeds of freely flying animals. For insects, smaller size means a lower best advance ratio. Operating at lower Reynolds numbers, small insects have two things working against them. They have lower best lift-to-drag ratios and thus have to beat their wings more often relative to their flight speeds to get an adequate forward component on the aerodynamic resultant force. And they have higher body drag coefficients (recall chapter 7) for that thrust to overcome.

Advance ratio, in the form of equation (13.2), provides a handy tool for the skeptical biologist. One encounters wildly high and quite as wildly variable figures for the top speeds of animals in flight. For insects, at least, one can do better by assuming a reasonable *J*, a moderate amplitude—perhaps 2 (115 degrees), wing length as measured, and wingbeat frequency from Greenewalt (1962) or other source. The result may remain an estimate, but it's better than grazing in the literature or on the Web. Denny (1976) used it to good advantage to estimate the speeds with which insects could fly into spider webs.

What, though, is a "reasonable" advance ratio? I gathered a fair amount of data (much but not all from Azuma [1992]) for insects from fruit flies to dragonflies and applied some value judgments about biases and general reliability. From them I suggest the following formula:

$$J_b = 0.144Re^{0.235}. \tag{13.3}$$

(Butterflies and moths—Lepidoptera—with their very broad wings, achieve about twice the J_bs given by the formula. The heavy-bodied, relatively slow beetles—Coleoptera—get only up to about half of these J_bs.) Anyone who tries the formula will notice a slight problem—Reynolds number (based here on body length) depends on flying speed, just what one is trying to estimate. But it doesn't depend so strongly on flying speed as to preclude using a little trial-and-error iteration. For some idea of the range with which we're dealing, a large (2-milligram) fruit fly (*Drosophila virilis*) can fly at about 1.2 meters per second; a large (500-milligram) dragonfly (*Aeschna mixtra*) can do about 7 meters per second. Claims for much higher speeds for insects should be viewed with considerable skepticism.

With birds, the picture looks less tidy. We do have data for something close to the insect J_b—Pennycuick (1990) did extensive theodolite tracking of birds (mostly large ones, for obvious reasons) in steady, level, free flight. His data assume a stroke amplitude of one radian—birds don't sweep their wings through arcs as wide as is typical of insects, and that seems to be typical. Advance ratios run from 1.49 to 4.27, impressively high. If there's any pattern, it is that the biggest—pelican, eagle, and vulture—have the lowest J_bs, but further down in size, no clear correlation emerges. Any suggestion of an inverse relationship with size can't be considered seriously paradoxical, since Reynolds number effects ought to be minimal in this range, mostly above 100,000. The low-J_b fliers may just be relatively heavy-bodied creatures analogous to beetles.

Several other items while we're talking about advance ratios. Among flapping animals, skates and rays have especially extreme values. The underwater flappers among them (some use a different locomotory mode) seem to go farther forward than at all reasonable for each stroke. What they're doing is using their whole bodies as thrust-generating hydrofoils, with the wingtips beating to do more than develop thrust themselves—they beat as much to force the middle, in opposite phase, to heave up and down while pitching enough in the process to generate lift and thrust. A cownose ray easily exceeds an advance ratio of 2 (Heine 1992); by comparison a penguin, another flapping swimmer, does only about 0.5 (Hui 1988a).

Advance ratio applies as well to descending, autogyrating seed such as came up (or went down) in the last chapter. Only here the situation is reversed. The better seed is the one with the lower, not higher, advance ratio—that is, a lower speed of descent and a faster rate of rotation. That sounds as if operation at lower Reynolds numbers buys some advantage. Not so, though. To get a lot of rotation out of a slight descent takes a force vector directed well away for the descent direction, and that's of course helped by a higher, not lower Reynolds number. In other words, not one but two things switch—low instead of high J as advantage, and power extraction from the passing air rather than insertion into it. The Js, incidentally, run around 0.2 (Azuma 1992; Vogel 1994a), where amplitude (ϕ) in equation (13.2) is taken as 2π. These are satisfyingly lower than those for flying insects in roughly the same range of Reynolds numbers—one wants to maximize, the other to minimize, and both face the same basic design constraints. The Flettner-rotators, ash and tulip poplar, though, don't do quite as well, giving advance ratios about 75 percent higher.

Combining Wings and Propellers

So far we've talked only about propellers, whether rotating or reciprocating. While it takes a bit of mental agility, one is best off thinking about wings as wings (in the familiar technological sense) when the matter at hand is gliding but as propellers when what's needed is thrust. A flying insect or hummingbird, or even a more ordinary bird, resembles a helicopter more than it does an airplane. The wings push a single airstream; it's directed fully downward during hovering and at least slightly downward even at the fastest forward speed. Except during brief periods of acceleration, the rate of creation of downward momentum must, to repeat what was said earlier, equal body weight.

How do nature's fliers so effectively combine functions we still separate when we want to fly with decent speed or economy? Scaling, once again, provides the key. A variable introduced in connection with sinking earthward in the last chapter makes the difference. Wing loading is the ratio of weight, *mg*, to wing area, *S*:

$$W_l = \frac{mg}{S} = \frac{\rho V g}{S}. \tag{12.3}$$

Again, the variable (ignoring density and gravity as size-independent) depends on volume over area, which gives it a residual dimension of length—wing loading will be proportional to length or to the cube root of body mass. So the smaller flier (assuming isometry) has a lower wing loading, quite aside from aerodynamic variables such as Reynolds number. In other words, it's lighter relative to its wing area, and thus it needs less lift relative to that area. And lift depends on speed, in particular on the square of speed, as recognized in the formula for the lift coefficient (equation (12.1)). Result—the smaller flier doesn't have to fly as fast.

In effect, the argument repeats the one made in the last chapter for falling. Lift required varies with mass; lift available varies with area times the square of speed. So speed should vary with the square root of length or the sixth root of body mass. A 100-fold increase in body mass ought to mandate a 2.15-fold increase in flying speed—not overwhelming, but noticeable. If one looks at flying speeds over the full range of craft, from tiny insects to jumbo jets, that scaling rule holds fairly well. A 1-milligram insect flies at about 1 meter per second; a Boeing 747 of 350,000 kilograms cruises at 250 meters per second. The sixth-root rule predicts an 84-fold speed difference, not the observed 250-fold one, but that's pretty close where the weight range exceeds 10^{11}. Perhaps the smallest insects ought to go faster, but that's easily attributed to the deterioration of the lift-to-drag ratio at low Reynolds numbers. (In addition, the density of the 747 comes nowhere near that of animals—after all, it's hollow.) (Tennekes [1996] presents the story in especially lucid form.)

Now think back, yet again, to the argument behind the idea of Froude propulsion efficiency. It should cost nothing to produce the force to stay up—just give an infinite volume of air per unit time an infinitesimal downward push. That takes nothing more than a helicopter with blades infinitely long. But that's just the ridiculous extreme of the logic that the bigger the wing, the greater the efficiency of producing lift, the reason the most severely power-limited craft, human-pedaled airplanes, have huge wings and go slowly. Less lift per area, of course, but . . . No, wait—that's the implication of the lower wing loading that comes with small size! The smaller craft, with its lower wing loading, makes the better helicopter. It pushes air downward more slowly—closer to that zero-speed ideal. A typical insect pushes air downward at about 0.5 meters per second; a typical bird pushes air downward at no less than 8 meters per second (Azuma 1992). So the helicopter trick, which we do

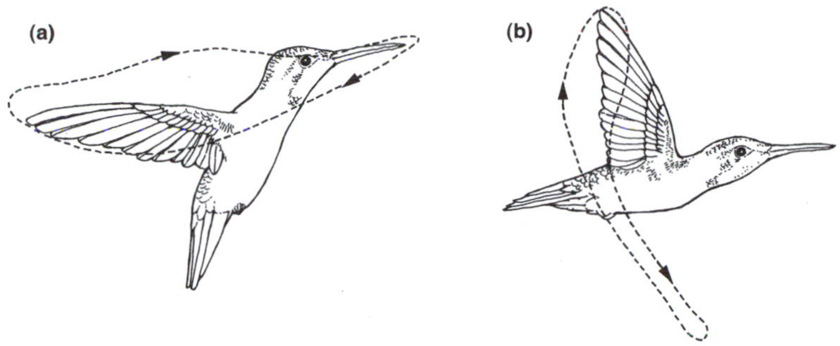

FIGURE 13.3 The stroke plane and body attitude of a hummingbird (a) when hovering in still air or (b) when flying rapidly forward (or hovering in a strong head wind).

only at the expense of economy (i.e., with low efficiency) gets more efficient as the craft gets smaller.

The bottom line, so to speak, is that the smaller the craft, the less will be gained by using propeller and fixed wings instead of combining the functions, helicopter-fashion. So the scheme used by flying animals makes sense on their scale, if not on ours. In fact, the larger birds hedge a bit, bending the wings in midspan as they beat them, so the inner part of each wing works mainly as a fixed wing and the outer part as a propeller. And the smaller the flying animal, the better it gets at zero-speed flight—hovering. Big birds typically need a running takeoff or a cliff or headwind; somewhat smaller birds can do a zero-speed takeoff; still smaller ones can hover very briefly; yet smaller ones can do sustained hovering.

Hovering, in most instances, happens like helicopter or "osprey" aircraft—the animal tilts the plane of beating rearward until it's about horizontal (fig. 13.3). Thus, the downstroke becomes a fully forward stroke and the upstroke rearward. One notices this most often in hummingbirds, occasionally when a small bird approaches a perch from below, once in a while when a hover fly is seen in a bright light against some dark background. Some small insects can hover with a vertical plane of beating, but they use more power in the process—the payoff comes when they suddenly buzz forward to lunge at prey (Ennos 1987).

Speed, Power, and Economy in Flight

There's no profit in spending more energy getting one's lunch than the lunch returns. Active flight isn't cheap, but an animal might at least choose the most economical speed. We've now seen that bigger means faster; now we focus on the individual flier, capable of making headway at a range of speeds, and inquire about what determines speed optima.

What matters in flight is drag. Drag increases with the square of speed, more or less, at least above the Stokes' law range, and power is drag times speed, increasing therefore with the cube of speed. So for best economy per unit distance, go slowly. Zero speed sounds best, or at least some speed that gets you to the lunch before you've spent too much on maintenance metabolism—keeping body and soul together.

That simple reasoning, though, is simply wrong. The trouble comes from the fact that, not only does the body of the animal suffer drag, but so do the wings, as a price for making lift. One component, "profile drag" behaves as we might guess—wings can't avoid skin friction (chapter 7), so none of the curves on the polar diagram of figure 12.7 touched the *y*-axis or zero-drag line. Profile drag increases with speed and is relatively worse at low Reynolds numbers. The other is the odd one. Wings suffer "induced drag" because they're not infinitely long, and they thus have to give passing air some finite deflection. What's odd is that induced drag gets worse and worse as flying speed drops—it's maximal during hovering, another view of the problem described a few paragraphs back. The hovering animal (recall fig. 13.2a) has no oncoming wind, so the wind speed over the wing is lower, so it has to deflect the wind further by operating at a higher angle of attack and a worse lift-to-drag ratio. Hovering, then, is hard work—a (briefly) hovering pigeon makes a lot of noise banging its wings together as it maximizes its stroke angle, and even hummingbirds hit maximal stroke amplitude when hovering. Paradoxically, perhaps, wings suffer more induced drag at low speed.

At the same time, higher speed does increase the drag on the animal's body. What's worse, in practice, is that it gradually reduces the animal's advance ratio and limits the power available from beating wings (or spinning propeller, for that matter), as we saw in figure 13.2c. Muscular power available may be some fairly fixed multiple of basal metabolic rate, but aerodynamic power available most certainly is not. It's maximal at some flying speed intermediate between zero and top. How might we put all of this together in handy form? Figure 13.4 plots drag against flying speed in a form that's at least qualitatively applicable to both flying animals and airplanes. Total drag combines the induced drag, the result of producing lift that predominates at low speeds, with the profile drag and the so-called parasite drag, or drag of all non-lift-producing structures, both of these latter predominant at high speeds. It's easiest to lump these last two, as pointed out by Alexander (2002).

Drag, though, isn't the best currency for further transactions. Multiplying drag by flying speed (not the speed of local flows over the wings) gives power, what the muscles must generate and what has then to be transmitted outward across the hinge points of the wings. So we can plot power required against flying speed, as done in figure 13.5. On this same graph, then, it's possible to plot power available. The animal can sustain flight wherever power available

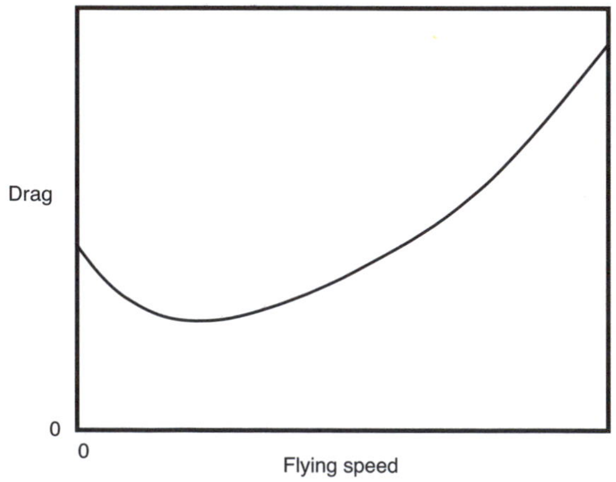

FIGURE 13.4 The overall drag of a flying animal versus the speed at which it flies. (The wings make some drag-causing wind even at zero speed.) This amounts to a measure of the force a propulsive system has to supply.

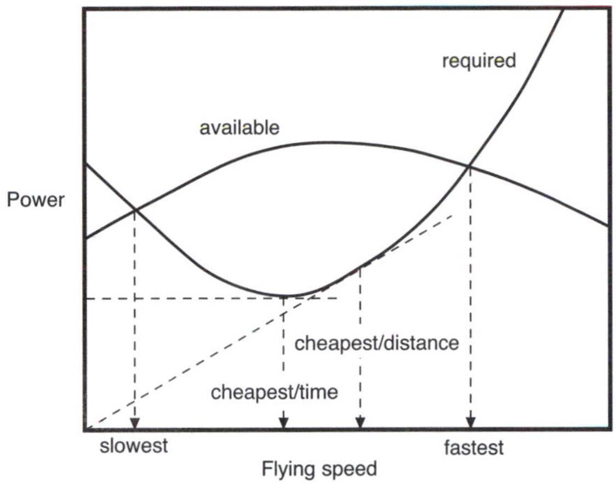

FIGURE 13.5 Power plotted against flying speed, and some consequences of the overall curve. Total power needed is the sum of parasite power, profile power, and induced power (from the three sources of drag); the shape of the curve for total power available reflects the changing effectiveness of thrust generation with speed (recall fig. 13.2). A flier can operate wherever availability exceeds demand—this one can't quite sustain hovering. In addition, the demand curve gives two further things—the speed for flight at minimum power and the speed that maximizes distance flown relative to fuel consumption.

exceeds power required. For instance, if (unlike the situation in the figure) power available exceeds power required at zero speed, the animal can hover.

We began by worrying about economy, and permissible speeds aren't necessarily economical speeds. The same graph, with only a slight increase in complexity, can deal with that as well. Here, as with the polar diagram for a wing (fig. 12.7) and the glide polar (fig. 12.10), two distinctly different optima can be recognized. First, a horizontal line tangent to the minimum point on the curve for net power required gives the minimum power for sustaining flight; a vertical line from that tangent point to the *x*-axis one gives the speed for cheapest flight. A sloping line from the origin that's tangent to the curve for net power required gives a different intersection point. A vertical line to the origin from this more rightward junction gives the cheapest speed per unit distance traveled—no mystery about why this occurs at a somewhat higher speed—and the corresponding horizontal line gives the power required for this minimally costly travel. For instance, for a starling weighing 70 grams the minimum power speed is 7.8 meters per second (18 miles per hour) while the maximum range speed is 13.3 meters per second (30 miles per hour) (Houston 1986). Rayner (1999) gives a more sophisticated view of the whole business. The fundamentals may be the same for animals and aircraft, but getting animals to reveal their curves taxes tempers and ingenuity.

Beating the Problems of Beating

Consider, again, nature's need to use reciprocating rather than rotating propulsors. She gains a few minor advantages—a propeller sends air helically backwards, giving a rolling as well as pushing force to an aircraft, and it's tricky to make one propeller on a fixed shaft blow preferentially on one side or the other. A pair of wings can be balanced side to side or run deliberately unbalanced. Dragonflies, in particular (D. E. Alexander 1986) play all kinds of games with their four wings, separately controlling almost everything about their separate strokes except their underlying frequency. As befits aerial predators, they thereby gain the maneuverability that a relatively large flier usually lacks in encounters with smaller ones. But for the most part, beating wings up and down or back and forth just makes trouble.

To start with, a wing has to be accelerated and decelerated twice during each stroke, a problem we encountered in connection with walking and running at the start of the first chapter. Elastic energy storage, as in running, can help quite a lot, and it's widely used. Insects make particularly great use of it with the help of a bit of nature's most resilient elastic, resilin (chapter 15), as a pad associated with the wing hinge or (in dragonflies) in series with the flight muscles. Insects, in a sense, meet the problem at its worst, on account of their high wingbeat frequencies. Even a large dragonfly typically does over thirty complete strokes per second; flies and bees exceed a hundred; mosquitoes do around four hundred, and some small midges push a thousand. That's *per second*, not per minute. Remember, wingtip speeds must exceed flight speeds by enough to achieve decent advance ratios, and small insects have both low maximum advance ratios and short wings. Bear in mind, too, the premium on speed imposed by a ubiquitously breezy world.

As a general rule, flying animals operate at nearly fixed wingbeat frequencies, using other variables for control. For the smaller insects, that clearly reflects their use of mechanical tuning to achieve elastic energy storage (more in chapter 23); for larger fliers something analogous is likely. Offsetting the lower frequencies of larger fliers are their longer wings, so wingtip speed can still rise a little with size, keeping the advance ratio appropriate for their somewhat higher flight speeds.

Next problem. A configuration appropriate for a wing in downstroke won't work at all well during upstroke. The wing has to be abruptly rotated lengthwise to shift the angle of attack between half-strokes and (except in the smaller kinds) the front-to-back curvature (camber) of the wing has to be reversed at the same times. In insects, the changes, often done hundreds of times per second, used to be attributed to small muscles in the thorax that attach to the bases of the wings. But Ennos (1989) showed that most of the shifts turn (one is once again tempted to say "cleverly") on their wings' structure and mechanical properties. Just the right tuning of flexibility enables them to take advantage of aerodynamic forces to do the job. That circumvents matters both of power and precise phase control, with the small muscles just exerting biasing forces over longer (if still very short) periods.

Next problem. If one accelerates a wing from rest, it doesn't begin to exert much lift until it has traveled forward several times its own width. The bound vortex—the circulation—responsible for lift just doesn't begin instantaneously. The phenomenon has been known since the 1920s; it's called the "Wagner effect." Does this mean that a wing can make no lift (or thrust, of course) during the first third or half of its cycle? Calculations done with that assumption tend to the conclusion that insects should remain earthbound. For that matter, calculations that ignore the Wagner effect don't always (depending on simplifying assumptions and on which insects provide models) get them off the ground either. The first really thorough experimental study, done in the 1950s, on the performance of intact desert locusts and on their isolated wings in a wind tunnel, implied that flight was okay, so the problem receded from view for a while. But that study didn't go beyond locusts, with relatively low wingbeat frequencies; and it couldn't rule out an internally canceling set of uncertainties. In short, all or a lot of insects, at least, could still be taking advantage of special, unsteady phenomena.

The first of these to come to light (Weis-Fogh 1973) depended on the way paired wings parted at the start of the downstroke. If they peeled apart starting with their leading edges, then the air moving into the space between them would initiate a pair of bound vortices, one then moving downward around each airfoil, foiling Wagner (fig. 13.6). In addition, it had long been recognized that if the angle of attack of an airfoil is rapidly raised, it can be briefly taken beyond the stall point without stalling. Recent work has shown that the rapid lengthwise rotation of insect wings during their stroke takes advantage of the effect. Quite a few other unsteady effects have been proposed and demonstrated in flapping models or through computational simulations—periodic flow outward along the wing, temporary vortices just above and behind the leading edge of airfoils at what would ordinarily be unusably high angles of attack, interaction of these leading edge vortices with the normal tip vortices, and others. See, for instance, Dickinson, Lehmann, and Sane (1999), Dickinson et al. (2000), and Birch and Dickinson (2001) along with the references they give.

Bird flight, in a sense, comes with more complications than does insect flight. Wing area and profile, even wing length, change more during each stroke; and flexible, jointed, feathered wings can't be dealt with as easily as wings of little more than inert chitin. Wind tunnel data on isolated wings tends to be incongruent with predictions from tracking records in the field. Birds fly faster and tolerate abuse less readily. They're easier, though, to persuade to fly unrestrained in the working section of a wind tunnel, something that has become almost routine in the decades since Tucker (1968) showed that it could be done. And they will fly through clouds of particles, which makes them fair game for recently developed techniques of digital particle image velocimetry.

What has been clear since Spedding's work in the 1980s (1987; and other papers in the *Journal of Experimental Biology*) is that birds, like insects, have much invested in unsteady flows. In particular, they leave peculiar vortices in their wakes, with different patterns at different relative speeds of flight—as if changing gait in some sense. When hovering (as in fig. 13.7a) a bird leaves a stack of toroidal vortices beneath itself, one for each half-stroke. As it gains speed (fig. 13.7b), that shifts to a zigzag pattern of obliquely descending and

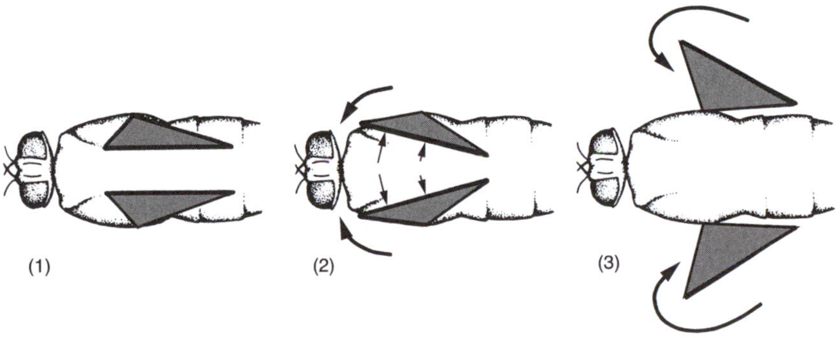

FIGURE 13.6 Weis-Fogh's "clap-and-fling" mechanism for the initiation of circulation through interaction of a pair of wings. At the start of the downstroke the two wings part, first in the front, and air passes around the leading edges to fill the gap. That air goes from what will be the undersides to what will be the tops of the wings, so the direction is appropriate for lift-producing circulation.

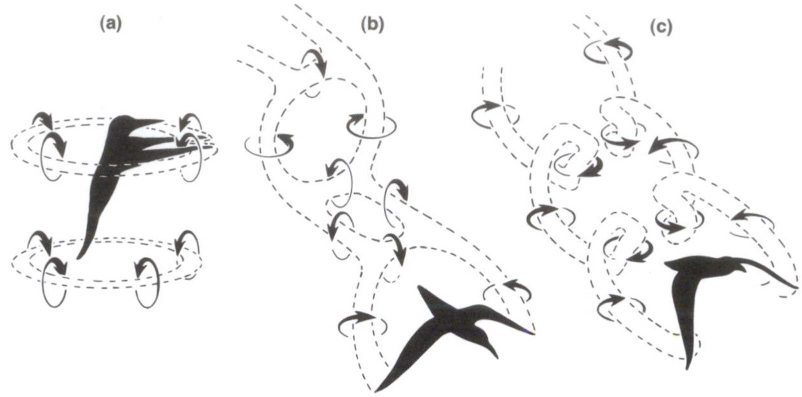

FIGURE 13.7 The vortices behind flying birds. (a) A hovering bird leaves a vertical stack of separate vortex rings beneath itself. (b) A bird flying slowly forward leaves a zigzag ladder of vortices, with tip vortices reversing direction between half-strokes and with adjacent rungs of the ladder having opposite directions of circulation. The wider vortex rings correspond to the lift-producing downstroke. (c) A bird flying rapidly forward leaves a pair of tip vortices that wobble inward when shed during the upstroke and outward when shed during the downstroke—since the downstroke produces more lift.

interconnected vortices—a bit like a downward staircase. Each downstroke leaves a wide vortex ring; each upstroke a narrower one. At yet higher speeds (fig. 13.7c), the bird leaves only the paired trails of tip vortices long familiar from airplanes, although even this pattern has some odd features—it zigzags both in and out and up and down in tune with the wingbeat.

The Diverse Ways to Swim

When considering the two ways to get about in continuous media, one notes at the outset that flight is hard and swimming is easy—the near-neutral buoyancy of biological material makes a world of difference. After all, whatever the biological risks of swimming poorly or stopping, it causes no major relocation or other physical hazard. That makes analysis sound easier. Not so. The very ease of swimming has permitted it to evolve on far more occasions than the mere four for powered flight. And whereas all flying animals use variants of the same physical scheme (butterflies may be a near-exception), we know of quite a number of almost completely different ways to propel oneself through water. Nor, as far as I know, do we have anything like the nice entry-level books of Tennekes (1996) and Alexander (2002) on flight. The best general introduction, if now a bit old, is Webb (1975).

Perhaps we ought to view swimming techniques in a hierarchical scheme. In the most general (= phylum) terms, we can recognize three common ways to do it, as follows.

Jetting

Squirt water backwards and you'll go forwards. The game is one of momentum conservation in its simplest form, and quite a lot has been said about it earlier in contexts such as propulsion efficiency (chapter 7) and pumps (chapter 9). Nature's jets are all pulse-jets rather than steady squirters, which raises the possibility of doing productive tricks with unsteady flow, something brought up by Weihs (1977). It may make a poor showing from an energetic point of view (except, as noted earlier, perhaps among some jellyfish), but it demands the least modification of conventional animal structure—what's more ordinary than chambers or tubes with muscular walls?

Drag-based propulsion

Arrange two strokes in opposite directions so one moves with greater drag than the other. The result will be a net thrust. That's, of course, what we saw in various forms of propulsion at low Reynolds numbers—cilia extending for a power stroke and bent down parallel to flow during their recovery, and flagella taking advantage of the same drag difference by moving sinusoidally or helically bent appendages so different portions met flow at different inclinations. In addition, appendages with planar arrays of bristles can splay them out during a stroke in one direction and collapse them together or back as they move in the opposite direction. As we'll see, drag-based propulsion finds use at higher Reynolds numbers as well. Perhaps we might view it as the default mode in the sense that just about any jointed or flexible appendage can be pressed into service to achieve some minimally effective swimming stroke. One thinks of the dog-paddle typical of normally terrestrial mammals and most of our own crude modes of swimming.

Lift-based propulsion

Already noted was the flightlike swimming of skates and penguins, sometimes called "aquaflying." The general scheme consists, as before, of generating thrust by using cunningly shaped airfoils (or hydrofoils) to make lift through circulation, and moving those foils so the lift (or a good portion of it) appears as thrust when the animal's body is the frame of reference. This familiar flapping of paired appendages located in midfuselage only touches the possibilities for lift-based propulsion—most of the schemes use devices less like conventional wings, and the devices vary far more than those used for flight. For efficiency, as the figures in table 7.1 attest, it can't be bested—at least if the Reynolds numbers aren't too low. Perhaps the absence in nature of any other mode of active flight provides the best evidence of its efficiency.

While fish provide the exemplars of lift-based swimming, one should bear in mind that not all fish use the mode—a few squirt their respiratory streams out as jets and a few use pectoral fins as oars that move fore and aft. And not all lift-based swimmers do so under all circumstances. Moreover, quite a few other vertebrate groups have come up with the scheme—some amphibians; several kinds of reptiles including, in particular, the now-extinct ichthyosaurs as well as marine turtles; several orders of mammals such as the cetaceans, the pinnipeds (seals and such), and the sirenians (manatees and dugongs); and a variety of birds.

Lift-based swimming turns up in unexpected cases as well; figure 13.8 shows a few. Pteropods (sea butterflies) are shell-less gastropod mollusks that flap their paired anterior appendages up and down (Satterlie, LaBarbera, and Spencer 1985). Chaetognaths (arrowworms) can do a form of lift-based undulatory

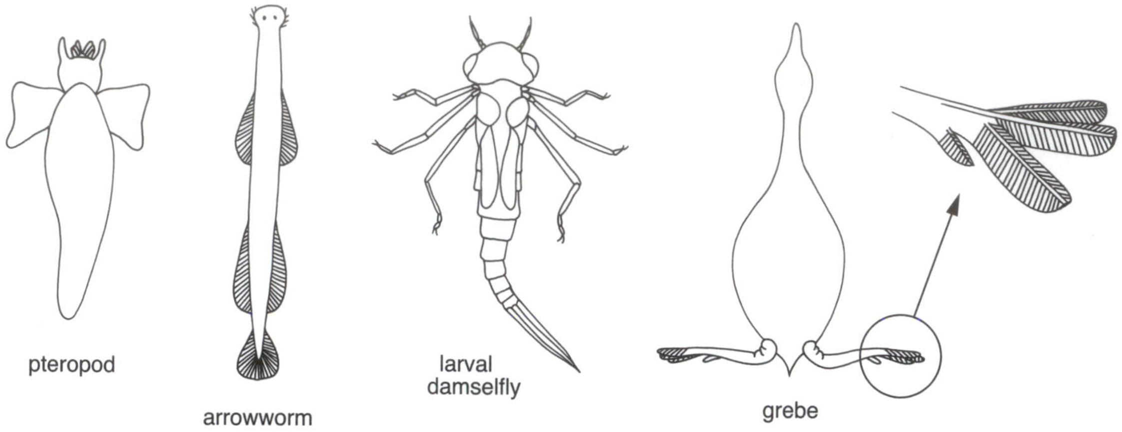

FIGURE 13.8 Less familiar lift-based swimmers: pteropod, arrowworm (chaetognath), damselfly larva, and grebe, with its airfoil toes. Grebe and toes redrawn from Johansson and Lindhe Norberg (2001) with permission of The Company of Biologists, Ltd.

locomotion (Jordan 1992). Larval damselflies swing what amount to caudal fins in quite a fishlike fashion (Brackenbury 2002). Diving grebes do it with their feet, using toes shaped and deployed as slotted airfoils like the splayed tip feathers of hawks and vultures (Johansson and Lindhe Norberg 2001).

As has proven useful for many decades, we can divide the continuum of lift-based swimmers into three (some sources split them further) slightly arbitrary general categories. Slowest and least efficient, for the most part, are the most undulatory swimmers, termed "anguilliform," after the eel, *Anguilla* (fig. 13.9a). Definitionally, these pass waves of bending down their bodies, with bodies flexible enough to encompass at least one full wavelength. They may or may not have fins, but none have especially protrusive ones. Generally faster and more efficient are what we regard as ordinary fish, who do what's called "carangiform" swimming, after *Caranx*, the jack (fig. 13.9b). A wave of bending moves rearward, but the stiffer fish is less than a full wavelength long. Most have substantial pectoral fins (paired forefins), and all have a fairly large caudal (tail) fin, although one that doesn't splay out too much further up and down than the fish's own height. The fastest and most efficient swim-

FIGURE 13.9 Modes of swimming in fishes: (a) the anguilliform motion of an eel, (b) the carangiform motion of a trout, and (c) the thunniform motion of a tuna.

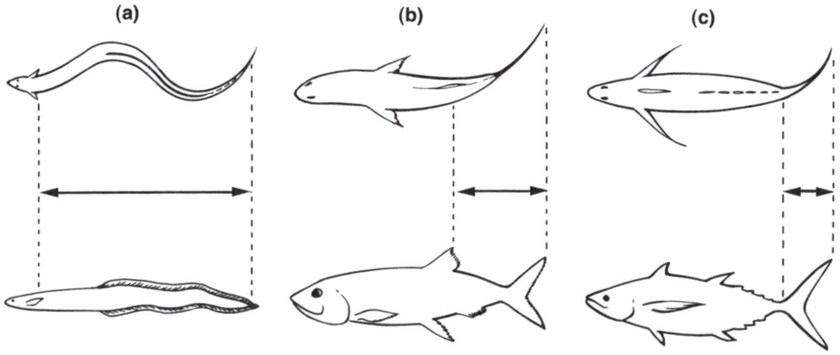

mers are those with stiff bodies and large, well-splayed tail fins, the "thunniform" fishes, after *Thunnus*, a tuna (fig. 13.9c). Their trunk musculature does little to bend their bodies, mainly swinging their caudal fins via long tendons going back through their caudal peduncles.

While the categorization smells of its ichthylogical antecedents, it accepts many other lift-based swimmers without profound problem or prejudice. Lots of worms and all sea snakes are anguilliform; whales are and the ichthyosaurs were thunniform. (That the cetaceans swing their tails up and down rather than side to side makes no difference, of course.) What it doesn't easily accept are the animals that beat anterior appendages in birdlike fashion—the pteropods, the sea turtles, the sea lions and walruses, the penguins, auks, and murres (Lovvorn 2001), and even a few fish such as ratfish (Combs and Daniel 2001). Propulsion with pectoral fins may not be typical of fish, but it's effective enough—Webb (1974) obtained a propulsion efficiency of 0.84. And these other animals include many excellent performers.

How Swimming Is Done

Thunniform swimming needs little explication beyond what was provided for flapping flight. It has some of the same problems, such as having to reverse the camber of the foil twice during each stroke, it uses half-stroke to half-stroke elastic energy storage to keep accelerative costs under control, and so forth. That they're mainly oceanic may reflect some reduced maneuverability or at least the awkwardness of a splayed tail in any confined place or where a fish might take advantage of fine-scale local flow variation.

Trickier are carangiform and anguilliform swimming. What seems to happen is that the waves of bending that pass, with increasing amplitude, from front to rear generate a train of alternating vortices in the passing water, vortices that are thrown backward in the wake. As noted back in chapter 6, such vortices rotate in directions opposite those of the Von Kármán trail behind cylinders, consistent with the injection rather than the extraction of momentum into the wake. That swimming fish leave a trail of vortices behind has been known for nearly half a century; in three dimensions, the trails (as in fig. 13.10) look similar to those left by birds in forward flight. Only recently, with the advent of laser sheet lighting to illuminate particles and of automated digitizing techniques to convert successive points into velocity vectors, have these trails of interconnected vortices begun to tell us just what a swimming fish does to the water passing it. A look at Lauder (2000) will give a good idea of how the approach can be used.

Vortex shedding, even if circulation directions are reversed, shows the same regularity as what happens behind cylinders—the value of the Strouhal

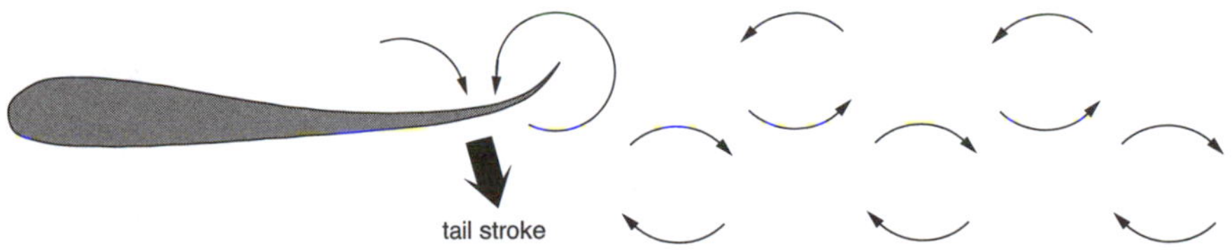

FIGURE 13.10 The vortex trail behind a swimming fish.

number (equation (6.7)) remains nearly constant (Triantafyllou, Triantafyllou, and Yue 2000). The only additional change consists of a slight redefinition of l, previously the cross-flow diameter, as A, the excursion of the tip of the tail, effectively the amplitude (linear, not angular) of the tail beat:

$$St = \frac{nA}{v}, \tag{13.4}$$

(where n remains the frequency, here of tail beat and vortex shedding; and v is the swimming speed). Over the immense length range of fish, the Strouhal number varies only between about 0.25 and 0.35. Given the differences in geometry, that's satisfyingly close to the 0.20 for vortex shedding behind cylinders. For that matter, the Strouhal number remains around 0.2, based on diameter, for both swimmers and fliers from flagellate protozoans and nematodes on up; this, according to Bschorr (1988), gives optimal energy transmission to the fluid medium.

It's not entirely clear how these vortices interact and move backward in the wake to achieve the remarkable efficiencies of, in particular, carangiform swimming. What they don't do is travel rapidly backward with respect to the water (as would happen with lower propulsion efficiency), but rather remain nearly in place. Nor do they waste much energy in moving sideways. Also not entirely clear (and this may sound odd), is just what drag a swimming fish experiences. With insects and birds, we could get at least a reasonable estimate of parasite drag by putting wingless bodies in wind tunnels. With most fish, we have no handy knife to separate thrust-producing and drag-imposing structures. Towing a dead or frozen fish gives numbers for drag, as do calculations from decelerations of gliding fish. However precise such numbers, they have uncertain relevance to actively thrusting fish. A variety of indirect approaches imply that fish must produce sufficient thrust to overcome drag in amounts between two and four times that from towing tests and decelerative calculations.

What must represent a kind of bottom line is that swimming, as done by fish, cetaceans, and such, gets mass from place to place with remarkable economy. Water may be dense, but streamlining is terrific, even allowing for the loss (as just above) from body oscillation. Watch aquatic mammals in an aquarium—they do a lot more moving around for no apparent reason than do terrestrial mammals. Quite a few fish, both bony fish and sharks, must swim continuously to drive water across their gills, a fool's errand were swimming more costly. A 1-kilogram fish pays 0.5 joules to move its weight 1 meter; a 1-kilogram bird pays about 2 joules; a 1-kilogram running mammal pays about 7 joules—each shift drives the cost up roughly fourfold (Tucker 1975)—look ahead, if you wish, to figure 24.5. (Naturally the bird goes a lot faster and must therefore maintain a much higher metabolic rate than mammal or fish, but that's something else again.)

You don't want to judge swimming by human standards. We're terrible swimmers. For a company interested in producing swimming-assisting gear, I once estimated that with good streamlining and good coupling of our locomotory muscles to a proper propulsive unit we could reach ten times our unassisted speed. I think I noticed gasps from several swimming coaches, folks who'd give an arm or a leg for a speed gain of a few percent. Even wearing the crudest of swim fins boosts one's speed impressively.

Drag Versus Lift As Ways To Get Thrust

About the middle of the nineteenth century, steamboats began a switch from paddle wheels to propellers, which meant a switch from a drag-based to a lift-based mode of propulsion. For almost all boats, we now rely on lift. From the very smallest, propeller-driven craft hold the speed records. (Take a look at the next paddle-wheel pleasure ship—you may notice that the paddle wheel is purely cosmetic, driven by the water rather than the other way around, with a propeller presumably submerged and out of view.) Nature doesn't make quite as clear a choice as does our technology. Semiaquatic turtles paddle in a drag-based mode; sea turtles, which swim long distances, flap using a lift-based mode. Ducks paddle, as do muskrats, large water beetles, a variety of worms, and various crustaceans.

What besides accidents of ancestry might be at issue? What doesn't much matter is the tax paid as the drag of the recovery stroke; again streamlining works superbly, so the drag of a paddle during the recovery stroke can be only a few percent of that of the power stroke. The time wasted by a recovery stroke ought to matter more, although it, too, isn't critical. More important, it seems, is something that goes back to the relationship between thrust and an animal's speed, discussed, with figure 13.2, early in the chapter. Thrust-from-lift increases slightly with speed and then decreases gradually as the pernicious effect of a high advance ratio comes into play. Thrust-from-drag, by contrast, starts high and then drops off rapidly. So, as speed picks up and demands more thrust to offset body drag, thrust drops. That militates against high speed, at least with respect to the speed of movement of an animal's propulsive appendages. What the paddlers have in common are their relatively low top speeds. The comparison has been explored by several people, most recently by Fish (1996) and Walker and Westneat (2000).

Why should any creature paddle? Lift-based swimming gets you there both faster and more economically—all measures of cost of transport or propulsion efficiency favor lift-based propulsion. Two factors may keep paddling, or its analogs, on the table. First, it's the gait that gets you out of the starting gate quicker—it wins in initial acceleration, in force generated at zero speed. An accelerating predator or ambushed prey might well find it preferable. To me, the pectoral and pelvic bones of some long-necked, small-headed, giant aquatic reptiles, plesiosaurs (fig. 13.11), seem more suitable as attachments for muscles that pull legs fore and aft than up and down—in short, for paddling rather than flapping. Perhaps that tells us that they were lunging predators, shooting the neck forward as the body went forward, and thus putting the bite on some fish that swam into the vicinity—rather than chasing anything down. (For a more serious look at swimming in Mesozoic reptiles, see Massare [1988].) Crustaceans such as crayfish give a great single, drag-based lunge by flexing their abdomens, speeding up from a standstill at five times gravitational acceleration (Webb 1979).

The second advantage of paddling emerges from its use of paddles rather than foils. Lift-producing airfoils or hydrofoils give decent lift-to-drag ratios (the critical thing, again, to get thrust) only if given quite specific and peculiar

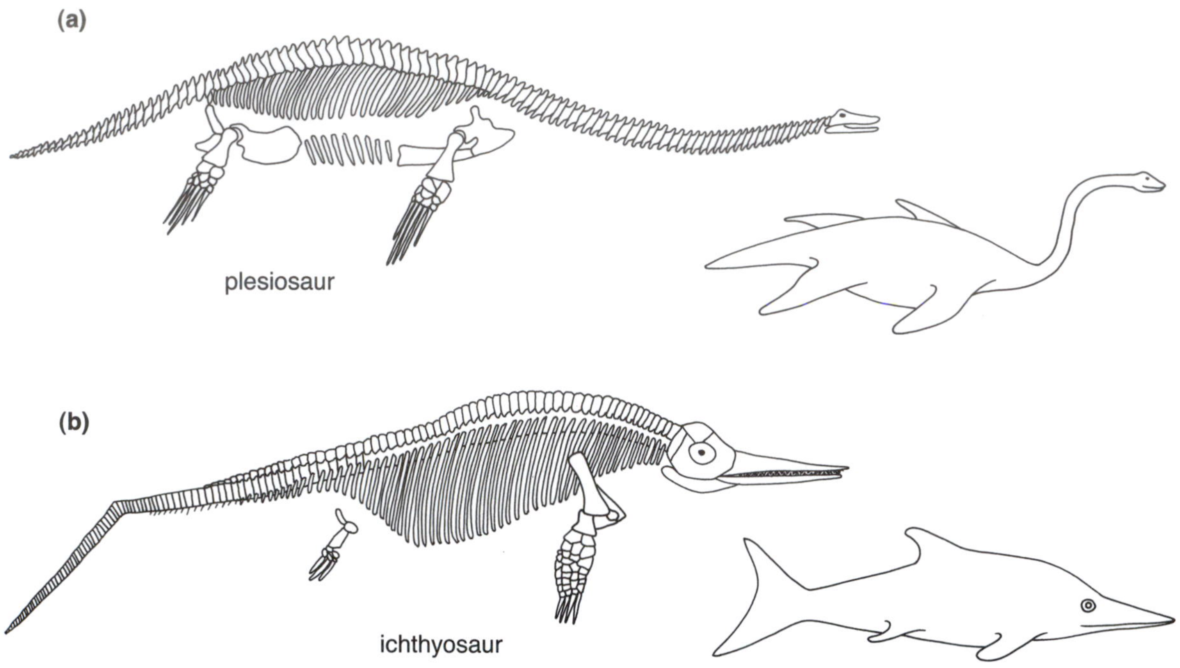

FIGURE 13.11 (a) The skeleton of a plesiosaur, looking a little like a snake threaded through a turtle. Note the lack of a downwardly protruding keel analogous to the breastbone of a bird (a turkey or chicken, for instance) to which muscles that pulled a limbs downward might attach. (b) Another Mesozoic aquatic reptile, an ichthyosaur, for comparison. It has an obviously propulsive tail.

profiles. They don't take kindly to multimedia use. Watch a seal or sea turtle—so graceful and swift in the water—stagger about on land. Semiaquatic turtles, frogs, ducks, muskrats—all these amphibious creatures paddle.

Still, it's possible to get the best of both worlds. Ordinary fish accelerate quite well, as anyone trying to net one in a large tank knows well enough. What they do is called a "C-start," bending into a single arc and then straightening out with a sudden motion of a body whose posterior is almost broadside to flow. That initial movement uses the drag of that broadside body to push directly rearward on the water, and it works almost as well as pushing off against a solid wall.

Swimming's bottom lines. For best efficiency when operating at moderate and high Reynolds numbers—use lift-based thrust generation. For propulsion with structures that must do other things as well, for operation at Reynolds numbers below 10 or so, for good initial acceleration—use drag-based thrust generation. For structural and evolutionary simplicity, for good initial accleration, for operation at moderate and high Reynolds numbers—use jet propulsion for thrust.

CHAPTER 14

Moving at the Air-Water Interface

They didn't think much of the Ocean:
The waves, they were fiddlin' and small,
There was no wrecks and nobody drownded,
Fact, nothing to laugh at at all.

Albert and the Lion,
origin obscure

THE PLANE OF CONTACT BETWEEN LIQUID WATER AND GASEOUS AIR LAST made an appearance in chapter 5, where the concerns (except for a few words about walking on water) were static phenomena. We've now been talking about locomotion, about flying through air and swimming through water. Far fewer creatures swim and fly near that interface than do so well removed from it. Still, some do, and both they and the phenomena affecting their motion are diverse. Moreover, yet other organisms, mainly sessile ones, live in a fully three-phase mechanical world, hanging on as they're intermittently struck by breaking waves.

Surface Waves

The surface of a liquid is like the brow of a thinker—a perturbation wrinkles it into waves. Surface waves on a liquid, though, persist only as long as they move, so one can't separate the subjects of waves and wave motion. We'll touch only superficially on the most relevant phenomena here, ignoring really large waves, the special problems of waves in shallow (relative to the distance between wave crests) water, the rules for the motion of groups of waves, and so forth. The interested reader might look at Bascom (1980) or Trefil (1984) for simple and engaging accounts or at Denny (1988) for both greater detail and a more biological slant.

The first and perhaps the most surprising observation about waves is that, while waves travel steadily, the water of which they're made travels only back and forth and up and down—a wave exists as an independent entity, divorced from some specific mass of water in motion. You've undoubtedly seen how waves pass under a floating object, which may oscillate quite impressively (especially if you've sitting on it!) both laterally and vertically, but whose overall spatial progress remains slight. What's happening is that individual bits of water beneath the waves move in nearly circular orbits. Beneath the crest of a wave, the bits of water move in the same direction as the wave as a whole. Beneath the trough, the bits of water move opposite the direction of wave movement. As a crest approaches, they move upward; as a trough approaches, they move downward. One can look at what's happening in at least two ways; a good sense of the phenomenon requires some degree of mental synthesis.

A diagram of representative orbits (fig. 14.1a) may look like a set of vortices, but it isn't. A set of instantaneous velocity vectors (fig. 14.1b) loses the

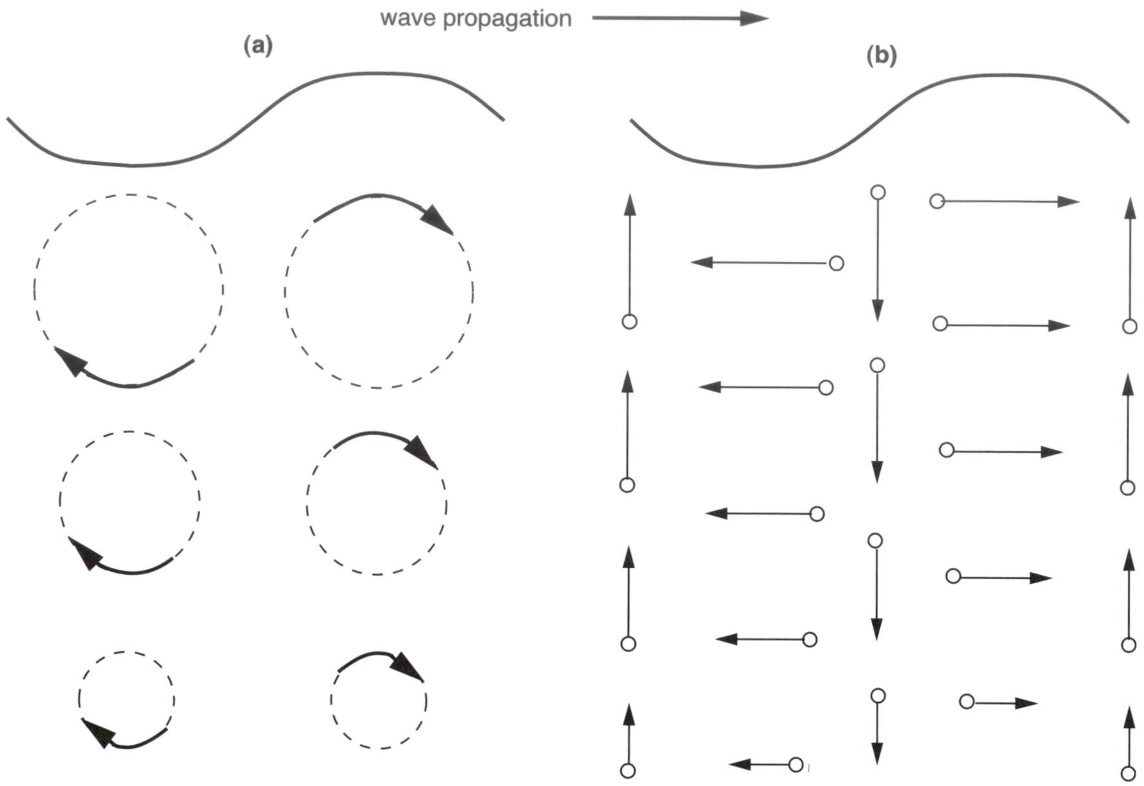

FIGURE 14.1 (a) The orbits of bits of water beneath deepwater waves over time—imagine you are looking while traveling with the waves. (b) The actual velocities (speed and direction) of bits of water beneath waves at an instant in time.

sense of orbiting bits, but it gives a better notion of the oscillatory, circularly translational motion of columns of water beneath the moving waves. "Columns," though, needs some qualification—the further beneath the surface, the lower the amplitude (both laterally and vertically) of the orbits. You may have experienced this depth-dependent decrease in wave motion when swimming underwater in the ocean. What happens in water shallow enough so a bottom interferes with orbital motion? That's the case (roughly, but by definition) where the depth is less than half the crest-to-crest wavelength. There the orbits flatten with depth, becoming ever more elliptical and at the bottom (ignoring the boundary layer) becoming purely lateral.

A special variable for wave speed avoids any confusion with water speed. The variable is called the celerity and usually is designated c; it put in a cameo appearance in chapter 10 in connection with pressure waves in blood vessels. Beyond that, waves are characterized by their heights, to which we won't pay much attention here, by their wavelengths, the distance between adjacent crests or troughs and symbolized by λ, the Greek lowercase lambda, and by their periods, t, the interwave interval at a fixed location. By definition, then, celerity is wavelength divided by period:

$$c = \frac{\lambda}{t}. \tag{14.1}$$

A surface wave survives only if moving under the impetus of its inertia. The air-water interface really prefers to be flat and horizontal—flat because of the surface tension of water and horizontal because of gravity. Since flat and horizontal amount to the same thing, a wave feels these two physically distinct destructive forces acting in concert. In practice, these forces due to surface tension and gravitational acceleration determine the relationship between the wavelength and the celerity of waves. Just which one dominates depends on the scale of operations, so how fast a wave moves varies rather peculiarly. Surface tension (causing "capillary waves") of course predominates at small scales, and gravitation (generating "gravity waves") predominates at scales more germane to human activities. The combination is clearer in the equation below and in a graph (fig. 14.2) than in words:

$$c = \sqrt{\frac{g\lambda}{2\pi} + \frac{2\pi\gamma}{\lambda\rho}}. \tag{14.2}$$

Note the presence of g (gravitational acceleration) in the first term within the radical and γ (surface tension) in the second term, both in numerators. Those are the wave flatteners of gravity waves and capillary waves—real waves combine the two. The equation does idealize a bit, presuming a mathematically tidy wave shape, water much deeper than the wavelength, and so forth. It was established by Lord Kelvin when, a century ago, Britannia ruled the waves.

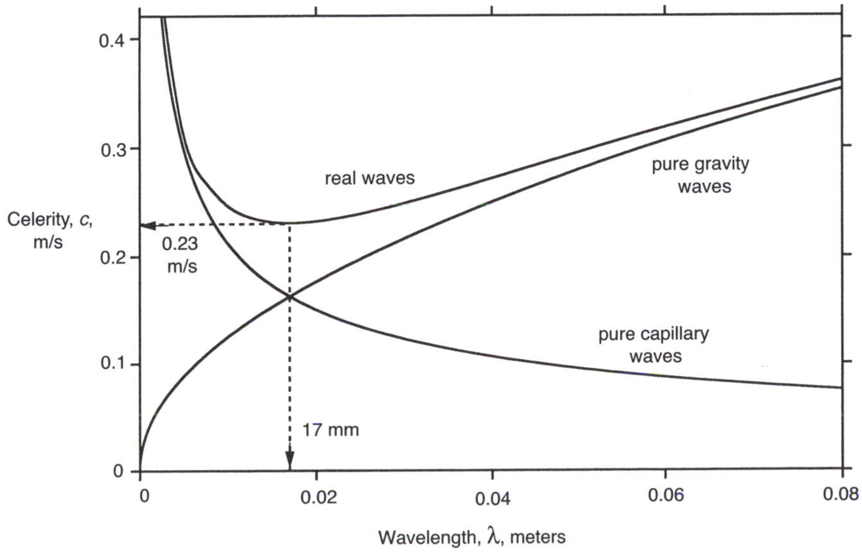

FIGURE 14.2 Wave speed, or celerity, as a function of wavelength—for capillary waves, gravity waves, and real (combined) waves.

(By the way, the ratio of the second to the first term within the radical sign, omitting constants, is the Jesus number, from equation (4.11).)

Notice also the odd way wavelength appears in the first numerator but the second denominator. For gravity waves, celerity goes up with wavelength; for capillary waves, celerity goes down with wavelength. That sets a minimum speed below which real waves cannot travel; it's about 0.23 meters per second (0.5 miles per hour) and occurs at a wavelength of about 17 mm (0.7 inch). In no earthly way (setting *g*!), can you persuade a wave of water beneath an atmosphere of air to move more slowly.

Of the biology that rides on surface waves, most cases involve large gravity waves and the consequences of their destructive dissipation on shores. A few studies, though, have explored the world of small waves and how organisms capitalize on them. One is an investigation of wave-making by the whirligig beetles (*Dineutus*) of ponds and slow streams (Tucker 1969). *Dineutus* swims on the surface, supported about half by its buoyancy (by gravity) and half by surface tension. Swimming slowly, it makes no waves—that minimum celerity is no abstraction in its world. Swimming faster, waves propagate outward from it, and Tucker calculated swimming speeds from photographs showing the propagating waves and the angle made by the arms of their "V" (Fig. 14.3). Incidentally, an analogous V is formed by shock waves in supersonic flight; the maximum design speed of a supersonic aircraft can be calculated from the sweepback angle of the wings. Which is why the military often censors photos showing top or bottom views of new designs.

A world of unfamiliar phenomena may be available to the fauna of liquid surfaces. For instance, whirligig beetles seem to be able to echolocate through reflection of their ripples—it must be a queer sort of sense that requires body motion to send the signal. And Wilcox (1979) has shown that males of at least one species of water strider can tell one sex from the other by sensing surface waves—males add on a high-frequency flourish, a kind of tremolo, that conveys the message to any male receiver that they're not suitable mates for each other.

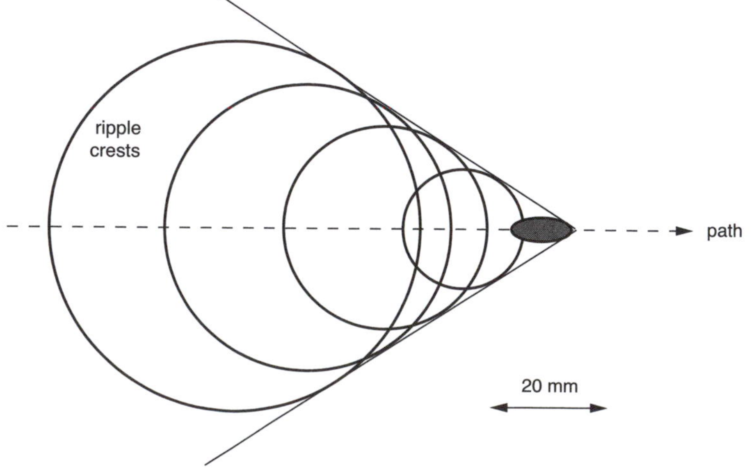

FIGURE 14.3 A whirligig beetle and the pattern of waves it makes when moving (for it) rapidly.

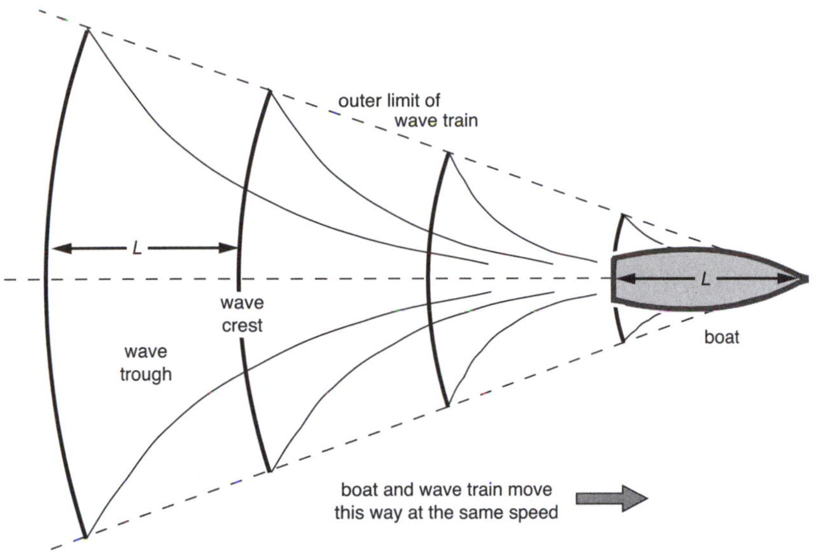

FIGURE 14.4 The pattern of waves produced by a surface ship or swimmer when going at its top practical (hull) speed.

The Speed Limit for Surface Ships

A conventional ship moving across a body of water leaves a trail of waves—although "leaves" can mislead. The pattern (as in fig. 14.4) consists of waves in that diagonal V noted for the whirligigs and waves running directly across the ship's path. From the stern of the ship, the waves appear fixed in place because they are fixed—with respect to the ship. We now know how fast waves will move, as a function of their crest-to-crest length (equation (14.2)), so we can tell how fast a ship is moving from the wavelength of those transverse waves. That information may have some occasional use; more important is the way those waves impose a practical speed limit on surface ships. The rule imposed on ships has wider biological ramifications than anyone guessed a few decades ago, so it deserves a little attention.

For ships longer than about 5 centimeters, only gravity waves make much difference, so we'll ignore capillary waves henceforth. Recall—its inertia keeps a wave going, gravity wants to extinguish it. What if we look, as has helped us in earlier instances, at these agencies in the form of the simplest possible ratio of two forces? That ratio (the Froude number, *Fr*) should set the relationship between the speed of a wave and its crest-to-crest or trough-to-trough wavelength. Inertial force can be put most simply as *ma*; gravitational force is just *mg*:

$$Fr = \frac{ma}{mg}.$$

For steady motion at least, acceleration can be taken as velocity per time or velocity squared per length, canceling *m*s, we get

$$Fr = \frac{v^2}{gl}. \tag{14.3}$$

Some specific value of the Froude number ought to characterize the movement of waves. In fact, equation (14.2) gives us the specific value, which turns out to be 0.16. (Occasionally, the square root of the expression above is used as the Froude number. If you're dimensionless, so are your roots. If so, that critical value of the Froude number becomes 0.4.) But there's more to the Froude number; next to the Reynolds number, the engineer has given the biomechanic no more useful a dimensionless number.

The Froude (pronounced as in "crude," not "crowd") number bears the name of William Froude (1810–79), the British engineer and naval architect who took the first productive steps toward a rational science of hull design. Nineteenth century ship designers faced two questions of types that remain familiar. First, to test the performance of hulls, it's easiest to make small models and tow them in a tank of still water. At what speed should a given model be towed to approximate the real thing best—that is, how can one ensure equivalence between model and reality? Froude recognized that in an equivalent model the square of the speed of testing ought to be proportional to the waterline length of the model. Thus, if a model is four times smaller than reality the square of its speed should be four times lower, or the speed itself should be halved. That, of course, is just what would be required to keep that ratio, v^2/gl, constant. Ship hull testing still cares about consistency of Froude number. It does have a problem beyond those of submarine and airplane testing—Reynolds number doesn't stay entirely submerged, and no convenient technique permits changing size and speed while leaving both Reynolds and Froude numbers unaltered. Froude, though, matters more, so the usual approach focuses on it while keeping size and speed close enough to reality so Reynolds isn't seriously offended.

We're not looking at something uniquely relevant to surface swimming. Equation (1.2), describing the maximum speed of a pendulum swinging through a given arc, matches equation (14.3) in all but name. So, not to be coy about it, the normal maximum walking speed, the walk-trot or walk-hop transition point, of legged animals happens at a particular Froude number (about 0.5). The reason is that the same ratio of forces, inertial to gravitational, matter. That will be important in chapter 24, which takes terrestrial gaits beyond their mere mention in the first chapter. Think back, as well, to the beetle that maintained a gas bubble under rapidly moving water, mentioned in chapters 5 and 7 (Stride 1955). Pressure reduction within the bubble follows Bernoulli's principle; it can be described by the equation we used to define the pressure coefficient (equation (7.12)). At the same time, the hydrostatic pressure forcing gas back into solution can be described by the equation for manometry or pressure unit conversion, $\Delta p = \rho g h$ (chapter 2). Putting one over the other, we get a ratio that okays the scheme if it's greater than 1:

$$\frac{\rho v^2 C_p}{2\rho g h} > 1.0, \tag{14.4}$$

where h is depth and C_p is the empirically determined pressure coefficient. Cancel densities and ignore constants and one sees, again, the Froude number. For small bubbles, C_p will be about 0.2 and the critical depth will occur at a Froude number (as in equation (14.3)) of about 10 (Vogel 1998b).

Back from towing tests to the speeds of real ships. A ship makes a train of waves that travel with it; its speed thus sets the wavelength of those waves. In the process, a ship makes a bow wave at its bow and one or more waves at intervals along the hull. The faster it goes, the more widely spaced are those waves. Eventually, it has only a single one, often called the stern wave. At that point, the ship has reached its fastest practical speed. Its problem comes down to a peculiar imbalance. Above that top speed (called the "hull speed"), the world starts to look as if it's uphill in all directions. To go faster, the ship has to climb up its bow wave with no help from any other waves, which takes a disproportionate output of power. Or it has to cut through, no cheap course either. For surface ships near maximum speed, wave-making costs more than offsetting ordinary fluid-mechanical drag. Which says, in effect, that the Froude number matters more than the Reynolds number.

As a consequence, the practical speed limit for a surface ship with a hull supported by buoyancy rather than planing depends on the Froude number, which provides an index to that point for ships of divers sizes. The practical maximum speed corresponds roughly to that value of $Fr = 0.16$ noted earlier, and the reason ought now to be obvious. In fact, few ships exceed $Fr = 0.12$—minimization of the cost of transport relative to distance happens at something lower than maximum speed. But, as you can see from equation (14.3), at any given Froude number the larger ship (with longer hull length, l) goes faster. Other things being equal (and there are several complicating factors), a ship four times longer ought to be able to go twice as fast. Figure 14.5 puts the matter graphically.

That rule has biological implications. Very few animals swim on the surface of the world's oceans, rivers, and lakes—a few waterfowl, a few semi-aquatic mammals, an occasional large whale. Curiously, even air-breathers such as seals and dolphins avoid the surface, staying submerged except when taking an occasional breath. Surface swimming, quite useful for our relatively large craft, turns out for animals to be either too slow or too expensive—except for the whales, animals are just too small. A ship 100 meters long can

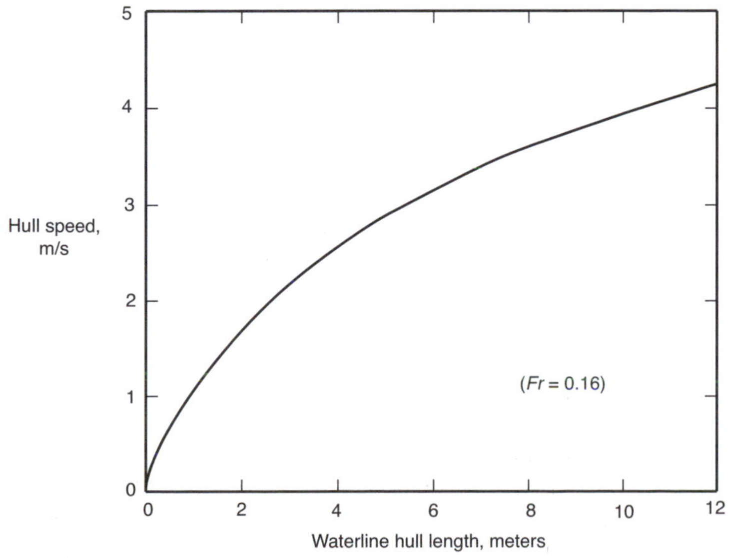

FIGURE 14.5 The maximum practical speeds ($Fr = 0.16$) for ships with conventional floating hulls.

exceed 12 meters per second (28 miles per hour or 43 kilometers per hour) before reaching the equivalent limit. By contrast, a duck, with a hull length of 0.33 meters, hits its limit at about 0.7 meters per second (1.6 miles per hour), which corresponds to a Froude number of just about the maximal 0.16 (Prange and Schmidt-Nielsen 1970). Not that a duck gets breathless—its respiratory system can handle the far greater demands of flight. What seems to limit speed is simple quantity of leg muscle; presumably natural selection long ago recognized that investing more in surface swimming machinery would be a fool's errand. (Ducklings, shorter than mama, can swim in the slower water of her wake, taking advantage of their shallow draft and short legs to stay above the wake of her paddling feet, according to Fish [1994].)

For a muskrat, with a hull length of 0.25 meters, that same Froude number gives a speed of 0.63 meters per second; Fish (1982) found that few muskrats on a pond exceeded that speed, and that the typical swimming speed was just a little lower—0.58 meters per second, or $Fr = 0.14$. Similarly, calculated hull speed and maximum swimming speed coincided (at the same speed as the muskrat's) for Australian water rats; like humans they can run about twice as fast as they can swim (Fish and Baudinette 1999). Humboldt penguins can swim with impressively speed, reaching 4.5 meters per second, but only when fully submerged. Above about 0.5 meters per second, surface swimming costs more, and as swimming speed increases, a penguin does an increasing fraction of its swimming fully submerged (Hui 1988b).

Small and particularly powerful boats have the option of a gait transition analogous to a switch from a walk to a run. They can rise up on the water's surface and hydroplane—in effect sliding along, taking advantage of the water's resistance to being accelerated downward. Hydroplaning boats make no such length-dependent wave train and thus evade the problem altogether. Small size helps mightily. For boats of similar shape and density, the area of water surface beneath their hulls will vary with the square of their lengths. How much weight they have to raise to hydroplane, weight otherwise supported by its displacement of water, will vary with the cube of their lengths. Double the length of such a boat and you double the weight relative to the length—to hydroplane, the boat must propel itself faster.

Aside from brief hydroplaning of large waterfowl during landings, the birds that do hydroplane are, not unexpectedly, of modest size. Grebes, for instance, whose slotted, hydrofoil toes were mentioned in the last chapter, appear to hydroplane. Grebes swim with especially effective, lift-based propulsion from legs extending behind them; their wings are relatively short, and some can't even walk—they're not sometime surface paddlers but highly specialized avian miniseals. At the Lincoln Park Zoo in Chicago, I once watched some razorbills plane, which they did with striking speed and agility. Nine-day-old mallard ducklings can hydroplane quite effectively, reaching 1.7 meters per second, or about four times their hull speeds and well over twice the hull speed (about 0.7 meters per second, again) of adults. Their small size must help, along with a broad and relatively flat lower surface (Aigeldinger and Fish 1995). Planing might matter a lot to them, because at that age they're not yet capable of flight and still dive poorly.

The rule changes, though, for much smaller craft. Consider again the whirligig beetles described earlier in the chapter. For a hull length of 10.6 millimeters, equation (14.3) predicts a top speed of only 0.13 meters per second, only a third of their relatively impressive observed maximum speed of about 0.4 meters per second. As it happens, the hull speed limitation set by a maximal Froude number doesn't apply, since the beetles are short enough to be in the range where wavelength is mainly determined by surface tension rather than gravitational forces. So one has to invoke the second term of equation (14.2) or look up the top speed on the graph of figure 14.2. Indeed, Tucker (1969) noted that the lengths of no common whirligig beetles exceed 17 millimeters, the length above which gravity waves begin to predominate.

Again, size imposes its constraint in determining the practicality of a way of living. But in one size-wise respect, the whirligigs live in a funny world. If you're a ship in the capillary-wave region, then smaller is faster, not slower—at least with respect to wave-limited top speed. Does this have practical ramifications? We don't know. Quite possibly very small and very fast ships of the whirligig's ilk would run into a less favorable aspect of surface tension. They might be immobilized, since surface tension depends on the length of the air-water-body line, and in small things that gets disproportionate relative to either area (proportional to muscle cross section and thus force) or volume (proportional to muscle volume and thus, perhaps, power). Tiny insects do appear to get trapped on the surfaces of ponds and puddles—apparently they can't muster sufficient force to free their wings.

One suspects they run into the problem of lacking sufficient inertial force to offset the force of surface tension. Another two forces, another dimensionless number. So the problem might be posed in terms of what's called the "Weber number," *We*, the ratio of inertial force to surface tension force:

$$We = \frac{v^2 \rho l}{\gamma}. \tag{14.5}$$

(The density here is that of the object, not the fluid.) Put this way, they're of insufficient size and can achieve insufficient speed to have much hope against the surface tension tar baby. That may be the problem, as well, of microscopic, planktonic diatoms, which, as noted in chapter 11, arrange to sink (albeit slowly) even though photosynthetic. The water's surface may present hazards to creatures other than storm-afflicted mariners.

Drag in Shallow Swimming

What happens if a normal swimmer, one of nature's submarines, moves along just beneath the water's surface? It ought to make a surface wave, at least a small one. And that implies working against gravity, quite beyond the problem of excessive buoyancy faced by, say, a duck with air in its feathers (chapter 5). Some years ago, Hertel (1963) reported that a streamlined body towed just beneath the surface could suffer up to five times the drag it felt when submerged at three body diameters or more. The peak occurred at minimum submergence—an average depth equal to half the body diameter—but the drag was still elevated threefold at a depth of a full body diameter. No details of the experimental conditions were given.

Those data have been cited repeatedly, in particular as an explanation for what's usually called porpoising behavior—sequential parabolic leaps out of the water while swimming at high speed (fig. 14.6). Restricting time spent just

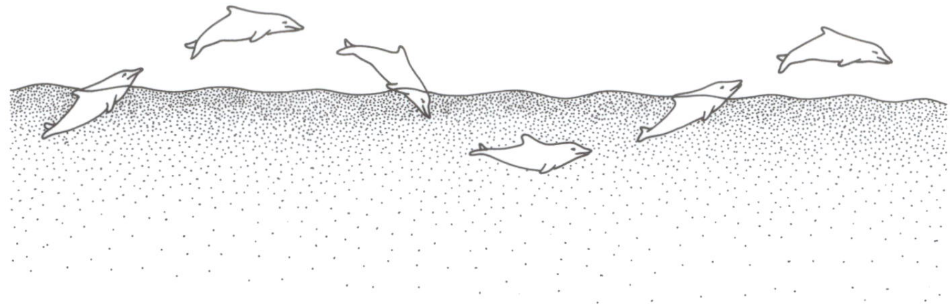

FIGURE 14.6 Successive positions of a dolphin that is porpoising.

beneath the surface could save energy, for instance, for porpoises and dolphins (Au and Weihs 1980). I think skepticism is in order. Hertel's book has so many loose and undocumented statements, miscitations, and such—it takes creativity far too far. Hoerner (1965; see also Azuma [1992]) reported far less drastic drag increases, and these mostly happen with partly immersed bodies at about the Froude numbers expected for induced surface waves, as we talked about earlier. More specifically, quite recently Goldman (2001) measured the drag of bodies moving just beneath the surface at Froude numbers relevant to swimming animals, and found no such increase in drag. Moreover, porpoising can be explained in other ways. Hui (1987b) found evidence that the behavior can minimize energy spent breathing in penguins; Fish and Hui (1991) make the same case for dolphins. They remind us that only air-breathers porpoise. It may also be the case that, while the smaller air-breathers when traveling rapidly want to stay close to the surface, they also prefer to avoid the peculiar water motion of surface waves. So no certain explanation of the porpoise of porpoising has yet emerged—not that there's any emergency.

Using the Energy of Surface Waves

Dolphins commonly travel considerable distances with what must be negligible effort by riding the bow waves of ships and the stern waves of small boats. They're not body surfing quite the way a human might, doing the trick while fully submerged with only the minimum surfacing needed for blowhole breathing. But they're taking analogous advantage of an interplay between their weight and the ascending water of the wave. The phenomenon has occasioned quite a disputatious literature. Sparing the history and details, they seem, as expected, to save energy with the trick (Williams et al. 1992). One might ask whether we're viewing some behavior that arose to take advantage of a newly available free ride, dolphins being nothing if not a bright and behaviorally versatile bunch. It appears more likely that they've long done an analogous trick using surface waves and even the wakes of much larger cetaceans.

Waves do represent energy rattling around that animals might do more than avoid by, say, porpoising. We know little about the operational details of any specific case, but, in addition to the evidence (largely indirect) that dolphins can take advantage of wave energy, a good case has been made that large whales can use their flukes as both passive and active wave-energy extractors (Bose and Lien 1990).

Flying birds may also capitalize on surface waves as an energy source. Pennycuick (2002) has described a mode of soaring ("gust soaring") by which petrels and albatrosses take advantage of winds deflected upward from waves and of the aerial vortices associated with them. Albatrosses can reach 20 meters and petrels about half that, both far higher than either could get from simple height-wise velocity gradients. He speculates that the peculiar tube-nostrils of these birds might be involved in the pressure detection needed to play the game.

Withstanding the Energy of Surface Waves

Consider the mechanical situation of some unprepossessing snail or limpet on a wave-swept intertidal rock. When emersed, it gets bashed by every wave that breaks against the shore. That obvious peril turns out to be the least of its worries—water rarely flies fast enough for any major hammer effect, and even a little air entrained in the water greatly reduces the impact by compressing momentarily. Being bashed by some solid object in the water represents a far worse, if much less common, hazard.

The creature also feels drag of the conventional sort. A storm surge velocity of 15 meters per second produces about the same drag as would a wind some twenty-nine times greater (the multiplier being the square root of the ratio of the densities of the two media). That's over 400 meters per second or nearly 1,000 miles an hour (ignoring the complications of supersonic air flow). And it feels lift; as mentioned in chapters 8 and 12, lift may present worse difficulties for staying attached than does drag—it's an upward force peeling a creature loose rather than a sideways one sliding it along.

In addition, an organism experiences yet another force related to the mass of water it displaces, a force related to the water's acceleration rather than to its velocity per se. Nothing intuitively obvious, this last force arises because the presence of the organism forces a volume of water to decelerate—or, which is the same thing, to accelerate in a direction opposite that of the wave surge. The phenomenon is called the "acceleration reaction," the extra mass is the "added mass," and the ratio of the added volume to the volume of the organism is the "added mass coefficient." The existence of this third force may say something about the scaling of at least the relatively rigid organisms that live on intertidal rocks. Drag and lift scale with area, as does the area of attachment of limpet, barnacle, or snail. So (except for protruding further out into the mainstream) they shouldn't determine any particular relationship between degree of exposure and size. But the acceleration reaction depends on volume, making the bigger the worser, relative to attachment area. That may explain (at least in part) why wave-swept rocks harbor mainly small creatures. The best general references on this distinctly peculiar business are Daniel (1984), Denny (1987, 1988), and Denny, Daniel, and Koehl (1985). Gaylord, Hale, and Denny (2001) provides an entry point to more recent work.

Slapping, and Other Interfacial Tricks

A few organisms have figured out another way to use that eight hundred-fold change in density at the air-water interface, at least intermittently. A moving object entering water from air suddenly encounters a great increase in drag; it may have been moving at a steady, terminal velocity, but now it decelerates drastically. That puts a lot of force on it ($F = ma$, once again), as anyone who has done a belly flop instead of a graceful dive needn't be reminded. It's a collision, and momentum must be conserved. So the decrease in momentum of the body entering the water has to be picked up as an increase in momentum of the water. The body decelerates, water must accelerate. Having mass, water resists acceleration forcefully since fluids as well as solids must follow Newton's second law. The severity of that acceleration depends on the suddenness with which the body enters the water—in large measure on its shape, as both belly flop and its cure attest. In effect, resisting acceleration means resisting entry of the body.

Several kinds of animals, most notably the basilisk lizards of Central America, make good use of that resistance of bodies to sudden, broadside entry into water. Basilisks can run across small bodies of water without sinking much beneath the surface, as in figure 14.7; they do so with their big feet, doing what Glasheen and McMahon (1996a, 1996b) called "slapping" locomotion. The feet do penetrate the undisturbed surface, and their descent pushes water aside—the acceleration of water just mentioned. That leaves an air-filled cavity as the feet go down. The cavity, though, doesn't persist, and they have to get each foot up again before inward water movement (gravity, again) makes the cavity collapse. The size range over which the trick can be done extends from 1 gram to 10 kilograms, a remarkable ten thousand-fold range. Small lizards do better than big ones, both by measurements of forces on model feet and judging by how far in (relative to body size) they penetrate. Inertia (the water's) fights weight (the lizards); not too surprisingly, the depth a leg or model penetrates (for its size) depends on the Froude number, which we've now met in yet another context.

And here ends the second section of the book. The reader—student or instructor or other—must be all too aware of how much ground has been covered. The writer prepares to move on, with that slight sense of relief (gratification remains in abeyance) attending completion of any lengthy task. He remains all too aware of what has not been said, since every paragraph has represented a decision, a guess, really, about what should be included and what quietly

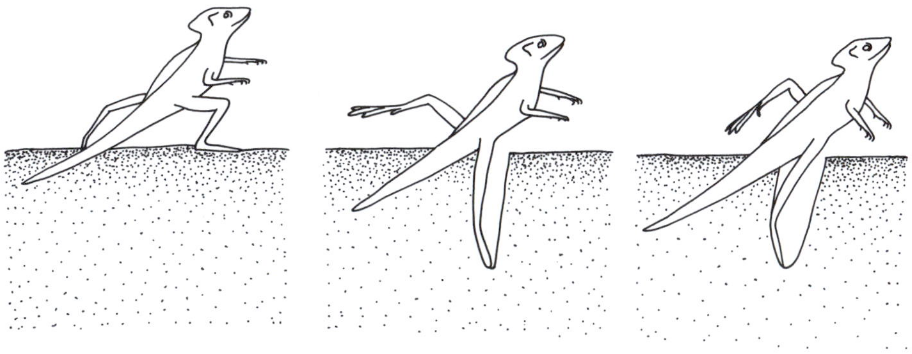

FIGURE 14.7 A basilisk, about as big as can manage the trick, running on water. Redrawn from Glasheen and McMahon (1996a) with permission of The Company of Biologists, Ltd.

omitted. And so you've heard only the most general outline of what we know of swimming and flying and of suspension feeding; even less, relatively speaking, of the low–Reynolds number world of cells and microorganisms. Moreover, you've heard nothing at all about flow of incompletely Newtonian fluids—snots, slurries, synovial fluids, and such. Beyond that, nothing has been said about thermal phenomena—convective flows driven either by temperature difference or external agency. I mention these matters of unarguable biological content lest one think we have come to the edge of the known and the known-to-be-relevant and omitted only matters whose importance remains conjectural.

PART THREE

Solids and Structures

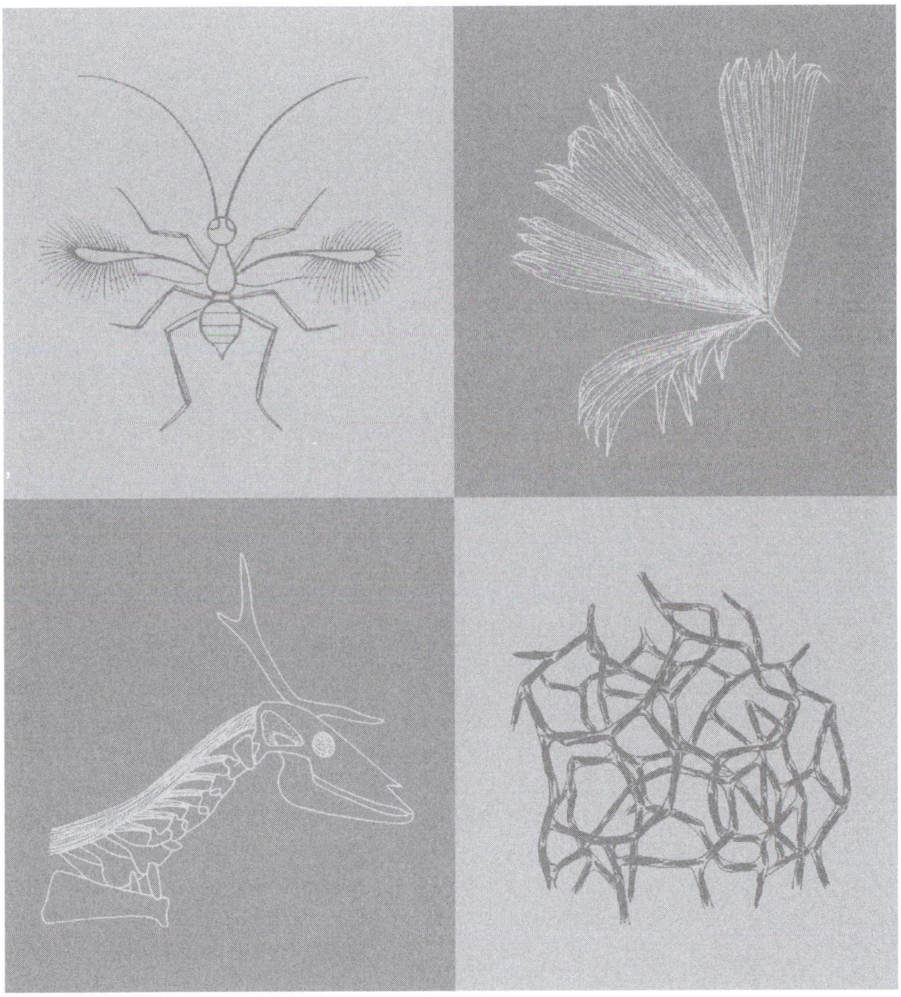

For the most part, organisms cope with the fluids they encounter; by contrast, they make solids and structures for their particular purposes. Only two fluids, air and water, affect them; the relevant solids are legion. Fluids have only a few bioportentous properties—density, viscosity, and surface tension; nature's solids make up a far more multidimensional lot. Still, our greater intuitive familiarity with solids more than offsets their trickier analytic problems. In short, we're comfortably closer to home.

CHAPTER 15

A Matter of Materials

> The first requisite of a good citizen of this Republic of ours is that he shall be able and willing to pull his own weight.
>
> **THEODORE ROOSEVELT, proposing tensometric testing of immigrants**

TO REMIND OURSELVES, GASES RESIST ONLY COMPRESSION, LIQUIDS RESIST tension as well as compression, but solids resist three independent kinds of stress—compression, tension, and shear. Again, we recall that gases and liquids, collectively fluids, resisted *rate of* shear, while solids resisted shear stress per se. Solids have shape, fluids don't. Jelly, yes; syrup, no.

In the next few chapters, attention will turn to solids. Within solids, in practice, we'll include materials ranging from the very soft (but still elastic) mucuses to the very hard (but still deformable) dental enamels and corals. Oddly enough, the degree of "solidity" in the vernacular sense is definitionally irrelevant—an ordinary word has again been pressed into service for an unfamiliar task.

Following the practice of engineering, the general subject of solid biomechanics can be divided hierarchically into materials, structures, and structural systems. Steel is a material, a beam is a structure, and a bridge is a system. Imperfect, perhaps, but useful. In nature, matters get messier since no functional benefit rewards an organism for paying attention to our tidy grouping. Bone is made of several materials in an orderly array; but is bone therefore a structure or merely a composite material? Still, bear in mind that the question is terminological, not real, and should thus deprive no one of any sleep. Just keep an eye out for the presence (or absence) of articles—a material, "bone," when written with no article, a structure when "a bone" or "the bone." And recognize the awkwardness and arbitrariness as part of a scheme that corresponds only imperfectly with reality. We'll begin here with materials; using the mechanical properties that describe how the materials behave, we'll build upward toward structures and systems in subsequent chapters.

The Sorts of Solids

Because organisms use a wide range of solid materials, we have to face the great bugbear associated with any kind of diversity, that of classification. The most straightforward classification is a hierarchical one based on chemical composition, and it makes a lot of sense if one wants to know how materials come to have their mechanical properties. Thus, we can recognize some general categories such as polysaccharides and proteins as well as subsidiary and somewhat less heterogeneous categories such as the generic proteins, silk and collagen. And then silks from silkworm moths differ chemically from that of spiders. Beyond that, their compositions differ among the silks of different silkworm moths and even among the silks used in different parts of the orb web of a single spider. So the hierarchy has quite a few levels and gets relatively complicated.

For us, the main problem isn't so much the complication as the fact that a classification of materials based on chemical composition doesn't correspond all that well with their mechanical properties. As an alternative, we might categorize solids by their mechanical behavior and concomitant biological roles. The commonest version of such a scheme, and the one we'll use here, considers three categories—*tensile*, *pliant*, and *rigid* materials; I think the arrangement was first introduced in a most useful book by Wainwright et al. (1976). A few other ways of dividing materials might be borne in mind as well. We can distinguish between *simple materials*, homogeneous accumulations of only one chemical species (with perhaps some water of hydration), and *composite materials*, nonhomogeneous combinations in which the mechanical properties reflect the interactions of two or more simple materials. And we can distinguish between *isotropic* and *anisotropic* materials: the values for the various mechanical properties of isotropic (not "isotopic!") ones don't depend on the direction in which the material is loaded, while for anisotropic ones direction makes a distinct difference.

On such matters I cannot go further without announcing my enthusiasm for two books by the late James Gordon—*The New Science of Strong Materials* (1976) and *Structures* (1978). Not only are they germane to the next few chapters, but I know of no better writing for nonspecialists about any area of contemporary science and technology. His *The Science of Structures and Materials* (1988), is prettier but not quite so stellar.

Tensile materials

These function by resisting being pulled upon—that is, by resisting *tensile* stress—basically they're biological ropes. But bear in mind that they don't always get used as ropes, often serving as components of complex composite materials subjected to other stresses. Organisms commonly use at least four biochemically distinct kinds.

1. The protein silk is produced in profusion by a variety of arthropods, of which spiders and silkworm moths make the best known silks. It serves a host of functions. In fairly pure form it resists tension in the webs of spiders; combined with a glue into a composite material it makes a tough cocoon for moths. It's used in a fine meshwork in the catch-nets of caddisfly larvae, mentioned earlier, as the world's smallest silk stockings. Tiny spiderlings and gypsy moth caterpillars take advantage of the drag of its threads. Some spiders line burrows with it to stabilize loose sand. We use the silkworm moth version by unwinding the thread of the cocoons of specially bred glue-deficient strains of *Bombyx mori* or by chopping the cocoons and dissolving the glue of some species of *Antheraea* ("wild" or "raw" silk).

2. Another protein, collagen, occurs in fairly pure form in our tendons, the ropes connecting muscles to bones. In combination with other materials, it forms a major component of composite materials with a wide range of properties and functions—skin, bone, muscle, and arterial wall are just a few.

It's present in almost all phyla of animals, but it's apparently absent in plants. In general, the price of edible muscle, red meat in particular, varies inversely with its collagen content. Collagen is tough stuff in both technical and everyday senses of the word "tough." But, as any competent cook knows, long treatment with heat or mild acid has a decently destructive action on it—different cultures choose different acids. The ballistae of the ancient Greeks and Romans (about which more later) used collagen in the form of cow tendons to store energy, but its excellent mechanical properties were offset by its cost and susceptibility to rot and decay in a warm climate. Of course the military mind has never been noted for economic fastidiousness.

3. A polymeric sugar (or polysaccharide), cellulose, is the basic fibrous, tensile material of plants. Plants don't seem to use it in any unadulterated state to withstand tensile forces—again it performs its mechanical function mainly as part (often the main part) of composites, usually as the thickened walls of individual cells. But cellulose does occur in fairly pure form in the external, drag-increasing fibers of wind-dispersed seeds such as those of cotton. We make extensive use of its ability to withstand tension, collecting it from various sources, spinning it into ropes, and often weaving the latter into fabrics—cotton, linen, and sisal, among others, are just cellulose fibers differing slightly in microstructure. We even extract it from wood, dissolve it, and reconstitute fibers by forcing the material through tiny holes to make the semisynthetic, rayon. Cellulose occurs only rarely in animals despite the ubiquity of the underlying monomer, glucose; I think only one group, the ascidians or tunicates (sea squirts), uses it as a structural material.

All organisms use proteins as chemical catalysts—enzymes—and use polysaccharides for energy storage (starch and glycogen are the commonest examples of the latter). But plants and animals appear to choose different default structural materials—plants have a predilection for polysaccharide while animals prefer protein.

4. A modified polymeric sugar, *chitin*, provides the nearest thing to an exception to that generalization. After collagen, it's the most widely used tensile stuff in animals; as a pure tension-resisting material it connects muscles to exoskeletons in insects, crustaceans, spiders, and such in the form of apodemes, their equivalent of tendons. (The claw of the Maine lobster, *Homarus*, has a piece of decent size, if you want an excuse to look at one. It's thermally stable, so the animal can be killed by immersion for a few minutes in boiling water.) Chitin occurs here and there in the animal kingdom and occasionally among plants. It mainly serves as part of the composite exoskeletons of insects and other arthropods. Our human technology hasn't put this fine material to use in any mechanical device.

Pliant materials

All materials deform under stress to some extent—solid never means perfectly rigid. For materials specifically designated as pliant, though, normal function depends on deformation. Where they differ among themselves is in the degree of their squeeziness and the extent to which and speed at which they recover their original forms after removal of a stress.

1. So-called natural rubber really isn't—the native substance of rubber-yielding plants is a liquid sap. A number of functional, mechanical rubbers do exist in nature; all are animal proteins. By a rubber we mean a material that can be rather drastically distorted in one or more directions yet remain capable of rapidly and forcefully returning to its undeformed shape. Perhaps the best of these *protein rubbers*, the one that returns the largest fraction of the imposed work following removal of the stress, is *resilin*. It's located in little pads or tendons next to the wing hinges of insects, where it has the important function of decelerating a wing at the end of, say, the downstroke and then using the work thus done on it to accelerate the wing at the start of the succeeding upstroke. Fleas use resilin to accumulate energy from muscles for quick release in their jumps. Another protein rubber is *abductin*, found in the shell-opening ligaments of bivalve mollusks. One or two adductor muscles hold the two halfshells or valves of a bivalve closed (the edible part of a scallop is one of these muscles). Closing compresses the ligament, so its elastic resiliency can reopen the shell if the muscles relax. Interestingly, scallops, which swim by repeatedly clapping their valves together, recover a greater fraction of the work done on their abductin than do clams and other more sedentary forms. Our own rubber, *elastin*, occurs mainly as a component of two composites, skin and arterial wall. The nearest thing to pure elastin is the nuchal ligament of large grazing mammals. It runs from a ridge on the rear of the skull back along the top of the neck to the thoracic vertebrae; it seems to act as a shock absorber as well as a support for the head. Dissected from a lamb neck (the local butcher can get you one of these), it's distinctly rubbery and responds quite differently to stretching than the collagenous tendons from, say, a lamb shank. Other pliant proteins have received less attention than these three.

One might expect some similarity in composition among these three proteins, but in amino acid content they're about as different from each other as proteins can be. At a higher level of organization, though, strong similarities appear—pulling on them amounts to pulling on molecular entanglements rather than on chemical bonds, which is why they're stretchier than the tensile materials.

2. The other pliant materials are pliant only as complete composites rather than as single chemical compounds; these *pliant composites* are a variable bunch both chemically and physically. At one extreme they grade off into true fluids, retaining only a little resistance to shear—the various mucuses that line pipes, coat surfaces, and even provide traction for snails and slugs, as well as odd materials such as the synovial fluid that lubricates our joints and the Wharton's jelly of the umbilical cord. Then there's the jelly of jellyfish and sea anemones (mesoglea), made up of sparse collagen fibers in a complex gel matrix. The cartilage of our ears, noses, and intervertebral disks is similar but stiffer—more collagen fibers and less gel. Skin and arterial wall are more complex still—pliancy is crucial, and its decrease with age helps neither appearance nor function.

Rigid materials

These resist stresses without allowing much deformation at all. None is a simple chemical compound—all are composite materials. Besides their obvious difference in stiffness, pliant and rigid materials differ in another way. Most of the pliant materials are to a substantial extent viscoelastic, meaning that their behavior is somewhat time-dependent in the manner of a fluid. A prolonged load or a load beyond some transition value results in something akin to flow—a permanent or semipermanent deformation. The rigid materials rarely show much trace of viscosity—whatever their other complexities, they are lawfully solid in a more important way than mere hardness. But they differ from most of the hard solids of human technology in that they're severely anisotropic, meaning (to repeat) that their mechanical response depends on the direction in which they're loaded.

What we're calling rigid materials includes a lot of familiar biological items. There's *arthropod cuticle*, a composite of chitin fibers in some proteinaceous material, with the addition of calcium carbonate salt in the larger crustaceans. In many instances, the fibers are arranged in sheets, each with a specific orientation, rather like plywood. There's *bone*, a composite of collagen fibers (about 50 percent of a bone's volume), some other proteinaceous material, and calcium phosphate salt. In most mammals, the collagen is in neat layers (lamellae)—again, the material is anisotropic, with mechanical response varying with the orientation of a stress. There's *keratin*, a composite of microfibrils in a matrix, both proteins; it's what hair, horn, and bird feathers are made of. We make use of it as tensile material in fabrics such as wool and mohair, but it's closer to a rigid material where nature uses it for mechanical purposes—horns, fingernails, hooves, feathers, and baleen plates. Horn resists compression well and stores a lot of the energy of compression relative to its weight. The old Asiatic bowyers knew that—they used it for the side of the bow that faced the archer, so drawing the bow compressed the horn on one face while stretching the wood on the other.

Which brings up *wood*, the cell walls of plants—cellulose microfibrils specifically oriented in an unstructured matrix mainly of a substance called lignin. Human technology derives no more important structural material from living systems; but, one hastens to add, the properties of fresh, tree-supporting wood differ substantially from those of the dried and sliced products of sawmills. Consider—when you want to check whether a tree or branch is living or dead, you do a mechanical test on a twig. If it snaps rather than bending, it's dead. Every woodcarver, woodcutter, and woodsplitter has intimate familiarity with the anisotropy of the material as well as with the variation of its mechanical behavior from species to species. For commercial, prepared wood, at least, Forest Products Laboratory (1999) offers a rich compendium of data.

There are yet other rigid materials, what Wainwright et al. (1976) refer to as "stony materials" and Vincent (1990) calls "biological ceramics." These are distinguished by being very heavily mineralized, with more mineral (some inorganic salt) than organic matter. Some are tiny, such as the spicules of even very large sponges—made mainly of either calcium carbonate or silica (roughly, glass). Others occur in massive pieces, such as the skeletons of the stony corals and the shells of most mollusks, both made of calcium carbonate with a small amount of organic matter. Stony materials occur here and there in the vertebrates—for instance, as the eggshells of birds (mostly calcium carbonate) and the dentine and enamel (mainly calcium phosphate) of our teeth. On the general matter of what occurs where and how it's put together, Brown (1975) remains a useful source.

Stresses and Strains

Not only do solids resist stresses such as torsion and shear to which fluids are indifferent, but they resist even a single stress in much more complex ways. Furthermore, they fail in a variety of ways—they may crack, crush, buckle, or collapse otherwise. So we need a complex of quantities to describe the behavior of even a simple solid—several elastic moduli, one for each kind of loading, as well as strength, extensibility, toughness, resilience, and so forth.

Perhaps the place to start is with the most straightforward way a material can be loaded—pulling on a sample at opposite ends to apply a tensile force. As we'll see, tensile testing holds relevance for loads other than tensile ones as well. The Procrustean testing apparatus is called a tensometer and may be either homemade or commercial—materials testing is standard practice among mechanical engineers. Tensile testing can be done in either of two ways. One can apply a graded series of forces to a material and simultaneously measure the extent to which it stretches, as in figure 15.1a. Or one can gradually stretch a material while measuring the force with which it resists extension, as in figure 15.1b. The first, manipulating force, may involve nothing fancier than hanging weights on something while measuring changes in length. The second may simply use a steady motor to stretch something that has a force transducer attached to its other end.

The procedure sounds just as routine as weighing out a pound of coffee, but it's all too commonly one of the most frustrating jobs in the masochistic world of experimental science. To test a metal, plastic, or ceramic, the engineer fabricates a sample of a deliberately convenient shape—large ends for attachment, gentle taper to the thin part that will take the most stress, and so forth. By contrast, the biomechanic most often insists on pulling on some anisotropic material of a shape determined by its own devious ends, one that may be wet and slimy, awkwardly small, rapidly deteriorating, and hard to separate from the rest of the organism. An extensive and profane oral tradition concerns just the design and use of grips for biomaterials. Finally, the biological data is only roughly reproducible—results vary more than we'd like between samples and usually depend more than we'd like on the details of the testing procedures. Statistics can further confuse things if used by rote, lumping variation due to the imperfections of the apparatus with underlying and significant biological variation.

(I see that I've used the word "biomaterials." Before going further, I must point out that the term has two quite different meanings, meanings that in practice one has to distinguish through context. I'll use "biomaterials" in the more literal sense of materials made and used by organisms—the stuff we've

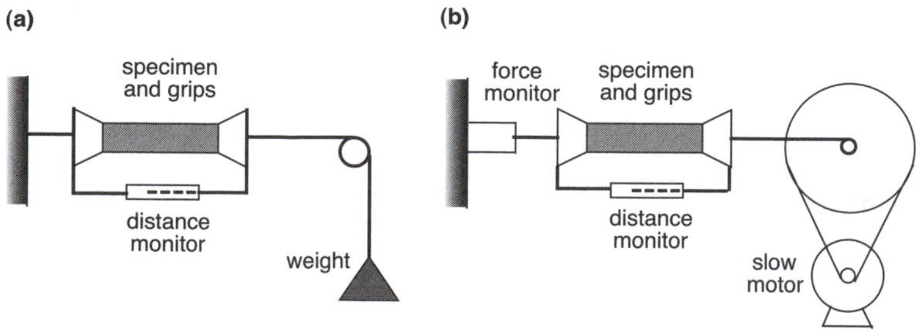

FIGURE 15.1 (a) Tensile testing by applying a steady force while measuring stretch. (b) Tensile testing by steady stretching while measuring force. Courtesy of the Center for the Deforming Arts.

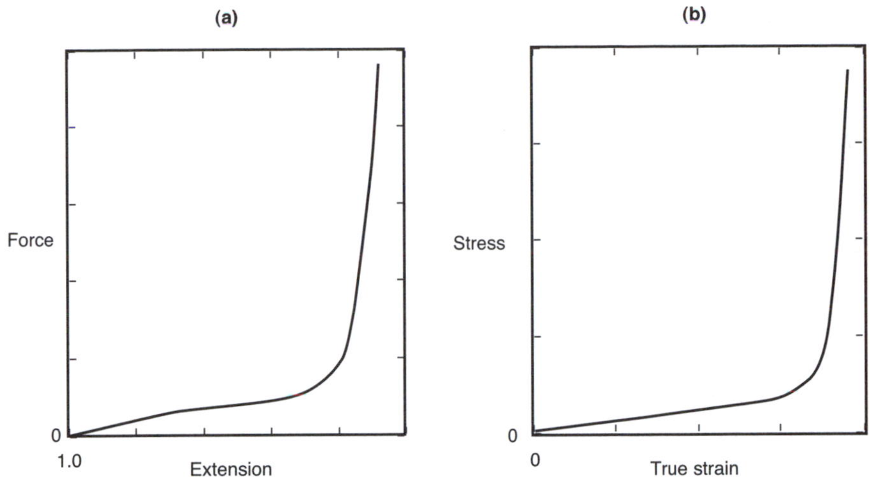

FIGURE 15.2 (a) A plot of force versus length. (b) Stress against strain, plotted for same data. Sometimes the two don't look all that different.

been talking about in this chapter, following the practice of Vincent [1990] and other biologically rooted people. The biomedical community and others doing human biomechanics most often consider as a biomaterial anything sufficiently compatible with human tissue to be suitable for prosthetic use.)

A successful tensile test produces a graph of force against length, as in figure 15.2a. Such a force-extension curve, though, describes the results of testing a particular specimen. Much more useful would be a curve that applies to the particular kind of material—"bone" versus "a bone" again. To get there, we manipulate the data on both axes to get the stress-strain curve of figure 15.2b. Every English-speaking neophyte biomechanic will tend to confuse those ill-chosen terms, "stress" and "strain."

To get stress (with dimensions of force over area and designated as lowercase sigma, σ), one divides each value of the force applied lengthwise by the cross-sectional area of the sample, as mentioned in chapter 2:

$$\sigma = \frac{F}{S}. \tag{15.1}$$

To get strain or fractional extension (dimensionless, as noted back in chapter 4, and given as lowercase epsilon, ε), one divides extension, Δl, by original length, l_o:

$$\varepsilon = \frac{\Delta l}{l_o}. \tag{15.2}$$

However the test has been run—whether the resulting strain has been monitored as stress has been manipulated, or whether stress is measured as strain is changed—the conventional presentation puts stress on the y-axis and strain on the x-axis.

Two minor matters afflict biomechanical stress-strain curves more than they do the ones with which engineers most often deal. Both are consequences of the greater extensibility (stretchiness) typical of biological materials. First, the value of the stress for a given force depends on the particular cross section chosen for a divisor. Trouble starts when the specimen has elongated by around 10 percent. If it retains anything close to constant volume—as is usual—then stretching one way causes significant shrinkage the other way. Just how much shrinkage occurs varies from material to material, so no universal correction can be made. In practice one needs to distinguish between two stresses, "engineering stress," which simply divides the force by the original, unstressed area, and "true stress," which divides the force by the particular cross section that corresponds to that particular force. True stress will exceed engineering stress for everything except the baseline value for a given test. Unfortunately, published data doesn't always state explicitly which one has been used, and where major ambiguity lurks, one may need a look at the methods section of the original paper.

Similarly, the value of the strain can take a simple or a more complex form. Engineering strain just uses the fractional extension of equation (15.2); and, again, it's reliable up to strains of about 10 percent. Above that, it diverges from true strain, which looks at extension as an incremental process. Here, at least, the two interrelate unambiguously, even if, once again, which is which may be ambiguous in published data. True strain, given in the box in equation (15.3), represents the result of integrating each incremental bit of strain as one goes from length l_o to length l, with a natural logarithm (a logarithm to the base e) in the result:

$$\int_{l_o}^{l} \frac{dl}{l} = (ln\, l - ln\, l_o) = \boxed{\varepsilon = ln \frac{l}{l_o}}. \tag{15.3}$$

Punching a few buttons on a calculator shows that at an extension of 10 percent, true strain is a mere 5 percent below engineering strain. But things then diverge more severely. At an engineering strain of 0.2, true strain is 10 percent lower; at 0.5 it's nearly 20 percent lower; at 1.0 (a doubling of original length), true strain is over 30 percent lower. Again, one must keep alert when looking at published data.

The Properties of Solids

Neither stress nor strain constitutes a property per se—one is what we do to a material, and the other is how it responds to that particular assault. But a stress-strain plot can yield at least six material properties. Figure 15.3 shows how various properties emerge from one.

1. The relation of stress to strain is the first of the material properties to concern us. It is, simply, stress divided by strain, or how much the material stretches for a given load, or the slope of the graph of σ against ε:

$$E = \frac{\sigma}{\varepsilon}, \tag{15.4}$$

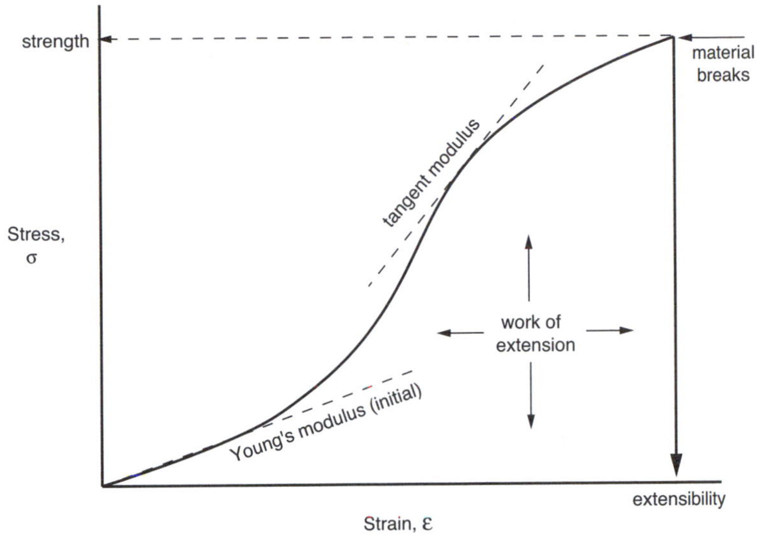

FIGURE 15.3 A graph of stress versus strain showing five of the six material properties that such a graph can display. (Compressive strength, appearing in the third quadrant, isn't shown.)

at $\varepsilon \cong 0$. It's called, variously, the "stiffness," the "elastic modulus," or, best, *Young's modulus of elasticity*, and values appear in table 15.1. Note carefully that the steeper the slope, the stiffer the material. So perhaps one should avoid using the term "elasticity"—in everyday use it implies precisely the opposite of what's meant here, which takes redefinition a bit far. Note also that the Young's modulus depends on how much the material has been stretched, except for materials for which the graph forms a straight line. Those materials are termed "Hookean" (after Robert Hooke, remembered also for aggravating Isaac Newton). Since metals, including metal springs, and cylinders of gas squeezed by pistons are substantially Hookean, a large literature in engineering considers that kind of behavior. But they're about all. Ordinarily biological materials should be presumed non-Hookean—thus cited values presume some "ordinary" degree of strain, all too often left unspecified. Taken strictly, Young's modulus implies the initial slope of the stress-strain curve, immediately beyond resting length; and, again strictly, any other value should be called the *tangent Young's modulus*, since it's the local tangent to the curve. Since strain is dimensionless, Young's modulus must have the same dimensions as stress—force divided by area.

Several points deserve comment. First, note that extraordinary range—no less than twenty millionfold in all, six millionfold for biomaterials. Notice also that bone, shell, chitin, and wood stand up to tension very well—their values are only about an order of magnitude below that of steel. And it's worth mentioning that the value for elastin is typical of rubbers (resilin, abductin, and butyl rubber are only two to four times stiffer); notice the two thousandfold difference separating it from collagen. That's the quantitative view of the comparison between a lamb's nuchal ligament atop the neck and the tendon from the leg, mentioned earlier.

Recognize, though, that stiffer needn't mean better. In an especially immediate biological context, excessive stiffness of one's aorta and other large ves-

TABLE 15.1 Examples of (Tangent) Young's Moduli of Elasticity, in megapascals, or meganewtons per square meter

Material	Young's Modulus of Elasticity (MPa)
Sea anemone mesoglea	0.01
Pregnant locust intersegmental membrane	0.2
Aorta, pig	0.5
Nuchal ligament (mainly elastin)	1.0
Dragonfly tendon (mainly resilin)	1.8
Abductin (bivalve hinge ligament)	4.
Cartilage	20.
Algal stipes and blades (variable)	20.
Mussel byssus thread (variable)	100.
Tendon (mainly collagen)	1,500.
Horn, bovid (mainly keratin)	4,000.
Tree trunks	6,400.
Silk (silkworm moth)	10,000.
Silk (spider orb frame)	10,000.
Wood, dry, with grain (typical)	10,000.
Locust apodeme (mainly chitin)	12,000.
Bamboo culm (stalk)	15,000
Teeth (dentine)	15,000.
Bone (large mammal)	18,000.
Mollusk shell (nacre)	50,000.
Coral skeleton	60,000
Teeth (enamel)	60,000.
Kevlar (synthetic fiber)	130,000
Steel (mild)	200,000.

Sources: Data from Wainwright et al. 1976; Waters 1980; Alexander 1983; Currey 1984; Dimery, Alexander, and Ker 1985; Gosline, Demont, and Denny 1986; Denny et al. 1989; Vincent 1990; Niklas 1992; Vogel 1995b; Bell and Gosline 1996; Gosline et al. 1999; Kitchener 2000.
Note: These values should be regarded as only rough approximations, with wide variations that depend on the rate of stretching, on where on the curve one takes the slope, and on no small amount of biological diversity for each item.

sels goes by the name atherosclerosis, and that's not what you want to have. To some extent it comes with age, and one gets a glimpse of its progress in the increasing difference between systolic and diastolic blood pressures—the *lack* of stiffness of one's arteries provides the main agency that smoothes out the pressures generated by the pulsatile performance of the cardiac pump.

In a different context, consider what can happen if you moor your yacht with a long, steel chain. If the boat drifts away from the dock, the chain will stop it—as a mooring cable must do. But the chain will hit full extension suddenly, which will cause a severe deceleration of the boat, which, in turn, will put a huge force on boat and dock ($F = ma$, yet again), damaging one or the other or else breaking the chain and letting the boat drift. By contrast, a stretchy mooring means lower deceleration and lower force—the hazard will be less, and a weaker mooring line will suffice. Closer to home, an inextensible automobile seat belt might be almost as bad as none, since its value depends on slowing you down as gently as possible (which isn't all that gently, of course) rather than stopping you abruptly.

2. Stretching the specimen in the tensometer to the point of breakage yields two further numbers. One is how much stress a material can take before it fractures, the maximum height of the graph; it's called the *tensile strength* of the material.* Like stress and Young's modulus, it has dimensions of force over area. We now have three items with the same dimensions, each measuring something distinctly different from the others. For each, the SI unit is the pascal or the newton per square meter, a very small unit—roughly, says a colleague, the stress of Newton's apple atop a garbage can lid. So one often encounters megapascals (MPa) or meganewtons per square meter or even gigapascals (GPa), a thousand times larger still. Data for the tensile strengths of a few materials are given in table 15.2. In looking at it, you must try to maintain a decent mental distance between stiffness and strength—some small bit of ribaldry can be contrived if need be.

The low value for arterial wall is interesting—presumably the occurrence of ruptured aneurysms has not been a major selective agent in mammalian evolution! Anything at or above 100 megapascals has to be considered a good tensile material—wood with the grain and collagen have about this value. Nylon is outstanding; you may have noticed that compared to other cordage, nylon ropes seem to have very generous load ratings relative to their diameters. And spider silk is superb—one can only wonder why, if one kind of creature can make a protein this good, the others, with the same synthetic machinery, don't do as well. About spider silk's special strength, though, more later.

Steel's high value may be a little deceptive, since strength has been expressed relative to cross section rather than relative to mass per unit length. Since steel is several times denser than anything else in the table, it doesn't stack up so well on that alternative basis. We persist in using steel tensile elements, though, for our suspension and cable-stayed bridges, in part because of just that lack of stretchiness upon which aspersions were cast earlier. Substituting Kevlar or nylon would generate an intolerable level of flexibility. As we've seen, nature typically takes a less disdainful attitude toward flexibility than we do.

3. A slight permutation of a tensometer can be used to measure *compressive strength* and stiffness, at least of fairly rigid materials. The test doesn't

* "Modulus of rupture" or its acronym, MOR, sometimes serves as a surrogate for strength. It's the maximum surface stress in a bent beam at the failure point. (Bending will await chapter 18.) Its values differ only a little from strength, differences reflecting the relative ease of crack initiation (chapter 16) in pure tension and bending; see, for instance, Ashby (1992) or Niklas (1992).

TABLE 15.2 Examples of the Tensile Strength of Materials

Material	Tensile Strength (MPa)
Barnacle cement	0.3
Aorta (arterial wall)	2.0
Algal stipes and blades (widely variable)	3.0
Wood (dry pine timber, across grain)	4.0
Brick, concrete	5.0
Abductin	10.
Teeth (enamel)	35.
Coral skeleton	40.
Teeth (dentine)	50.
Mussel byssus thread	50.
Wood (dry pine timber, with grain)	100.
Tendon (mainly collagen)	100.
Insect cuticle (mainly chitin)	100.
Wood (typical, fresh, with grain)	200.
Bone (large mammal)	200.
Hair, human (mainly keratin)	200.
Mollusk shell (mussel)	500.
Silk (silkworm moth)	500.
Cellulose fiber	750.
Nylon (synthetic fiber)	1,000.
Steel (typical)	1,000.
Silk (spider)	2,000.
Steel (high tensile)	3,000.
Kevlar (synthetic fiber)	3,600.

Sources: Data from Wainwright et al. 1976; Gordon 1978; Currey 1984; Gosline, Dermont, and Denny 1986; Craig 1987; Denny 1987, 1988; Bell and Gosline 1996; Gosline et al. 1999.
Note: Regard comparisons skeptically—what constitutes failure depends on judgment and convention.

have quite as wide applicability as a tensile test—for reasons that will be apparent when we take up structures such as beams and columns in chapter 18. Still, with a short enough specimen, one can measure how materials resist crushing. (Longer ones bend and buckle, making their effective loading regimes more complex, especially near the failure point). On the stress-strain graph, we're just looking at what happens in the third (lower left) quadrant rather than the first (upper right). Put another way, the line of the classic plot just gets extended (with data) downward and leftward. Stiffness rarely shifts much from the values obtained in a tensile test, but strength can be a lot different.

Notice how much better concrete does when resisting compressive rather than tensile loads—36 as against 5 megapascals. Other ceramic building ma-

TABLE 15.3 Examples of the Compressive Strength of Materials

Material	Compressive Strength (MPa)
Cartilage (human costal)	1.
Concrete	36.
Wood (dry ash timber)	45.
Bone, compact (lengthwise)	170.
Teeth (enamel)	200.
Teeth (dentine)	300.
Mollusk shell (nacre)	380.
Steel (typical)	1,000.

Source: Data from Wainwright et al. 1976; Gordon 1978; Oberg, Jones, and Horton 1984; Vincent 1990.

terials are similar, and that difference determines how we use them. It's why no one bothers to lay bricks with a decent glue—you're risking trouble if your structure subjects its bricks to much tension. It's why concrete structures depend critically on the amount and placement of steel reinforcing rods lying invisible (and skimped on by unscrupulous builders) inside. And it's why Gothic cathedrals, lacking metal reinforcement in roofs and buttresses, could not permit any element to experience tensile loading. That low tensile strength (plus a thick callus) makes it possible to break a brick with a chop of the hand—you just have to arrange a little unsupported area under the center of the brick so the bottom is put in tension. I once did this in a large hydraulic press; I screwed up my courage, shielded my face, and felt quite silly as brick after brick broke with only a tiny click and a very low load.

By contrast, wood is a little stronger—but only a little—in tension than compression, a point to which we'll return. Ferrous metals, except for very specialized steels, have about the same strength one way or the other. That, together with their similar Young's modulus in tension and compression, explains why I-beams are symmetrical, top to bottom, and can be installed upside down with impunity. Bone behaves like ferrous metals in this way. But bone is so anisotropic and bones as entities are so nonuniform in composition that similar tensile and compressive strengths more likely reflect functional demands than inevitable consequences of composition. Teeth look wrong, with the outside enamel weaker than inside dentine in either tension or compression. That may reflect the price we pay for the functionally critical stiffness, the high Young's modulus, of tooth enamel. Again, stiffness is not strength.

4. The other number that emerges directly from stretching to breakage is breaking strain or *extensibility*; it's the maximum horizontal extent of the plotted line. Be reminded that extensibility is not elasticity. As maximum strain, extensibility retains strain's dimensionlessness—distance extended divided by original length. And, like strain, all but low values depend on

TABLE 15.4 Examples of the Extensibility of Materials

Material	Extensibility
Coral skeleton	0.0003
Mollusk shell (nacre)	0.006
Bone (yield or giving strain)	0.006
Steel (high tensile)	0.008
Wood (typical)	0.01
Insect cuticle	0.01–0.02
Bone (ultimate, breaking strain)	0.02
Kevlar (synthetic fiber)	0.027
Horn (mostly keratin)	0.04
Cartilage	0.08
Tendon, rat tail (collagen)	0.1
Steels, ordinary	0.1–0.35
Silk (silkworm moth)	0.2
Algal stipes and blades	0.2–0.35
Keratin (wool)	0.5
Mussel byssus thread	1.0
Abductin (mollusk hinge ligament)	1.0
Elastin (nuchal ligament)	1.2
Silk (spider)	0.3–16.
Pregnant locust intersegmental membrane	15.

Sources: Data from Eshbach 1952; Wainwright et al. 1976; Currey 1984; Denny et al. 1989; Vincent 1990; Bell and Gosline 1996; Gosline et al. 1999.
Note: True values are given where above 0.1.

whether we calculate the "engineering" version or the "true" version. This quantity (table 15.4 gives some values) is less often quoted than either Young's modulus or strength for reasons related, I'd guess, to our peculiar technological predilection for nonstretchy stuff.

In practice, extensibility varies more or less inversely with the stiffness of a material so it at least follows the commonplace notion of elasticity even if not defined in that way. Thus, nuchal ligament, as one might expect, stretches further than tendon. While a stretchy nuchal ligament can act as a shock absorber, a stretchy tendon would undo the shortening of the muscle that it attaches to a bone. The current record holder, the intersegmental membrane of the pregnant female locust, achieved its renown in an investigation by Vincent (1975); it's made of about 12 percent protein and 12 percent chitin, with the rest water, and it undergoes something called "stress-softening." Mother locust, a creature of dry habitats, stretches these membranes between her abdominal segments to get her eggs about 8 centimeters underground—deep enough so the desiccated eggs have a reliable source of water. The eggs are kept fairly dry before being expelled, presumably so the locust can hold a large number and still fly.

5. Stretching something takes a certain force or stress. It also takes a certain work. But force is not work, and materials that take the most stress (the strongest) may not take the most work. On that difference depends a peculiar kind of fragility that afflicts a lot of strong materials, whose practical strength seems far lower than what tensometer data suggest. As it happens, the work needed to extend a material corresponds more closely to our everyday notion of strength than the property of strength itself.

Work, though, doesn't provide an adequately general measure since it depends on the amount of material being tested. What we'd really prefer is a material property, which we can get simply by dividing the work of extension by the volume of the sample of material. So just as we shifted from force to stress, we shift from work to work per unit volume. In this nice, orderly world, work per volume (joules per cubic meter in SI units) comes out dimensionally equivalent to force per area (yet again!). And we get force per area (or work per volume) by multiplying together the dimensions of the axes of the plot of stress versus strain. To what aspect of the stress-strain plot, though, does the quantity correspond? That multiplication gives the answer—the area under the curve gives the work per volume absorbed as the material extends.

Figure 15.4 gives the stress-strain plots of two materials of the same strength. The one with the higher curve absorbs more work; most often it is said to have the greater *work of extension* or *strain energy storage*. Table 15.5 gives the work of extension (or in some accounts "toughness" or "resilience") for a variety of materials under, unfortunately, nonequivalent test conditions. The main uncertainty, of course, is just how far to extend a material in a test—to the limit of reversible stretch (the maximum distance from which it will snap back), to some safe working limit before fracture, or to the breaking point itself.

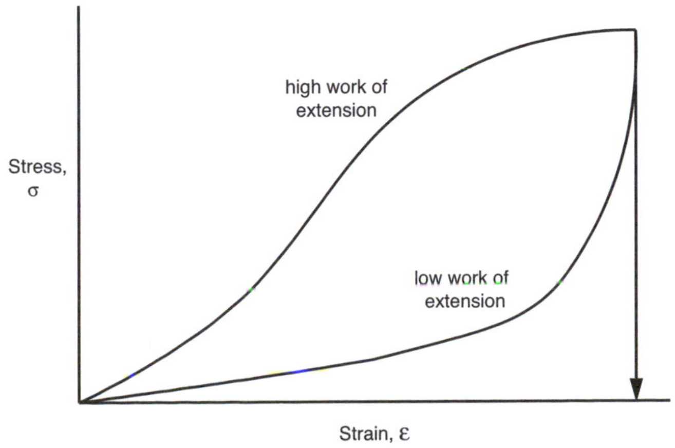

FIGURE 15.4 Stress-strain plot for a material with a high work of extension and one with a low work of extension.

TABLE 15.5 Examples of Work of Extension or Strain Energy Storage, Sometimes Called Toughness—Plus the Same, Expressed Relative to Mass Instead of the Usual Volume

Material	Work/Volume (MJ/m^3)	Work/Mass (J/kg)	At Strain	At Stress (MPa)
Coral skeleton	0.06	30	0.0003	400
Shell (mussel)	0.5	180	0.002	560
Wood (dry yew)	0.5	900	0.009	120
Aorta (arterial wall)	0.5	500	1.4	2
Steel (spring)	1.0	150	0.003	700
Horn (keratin)	1.8	1,500	0.04	90
Kelp (algal) stipe	2.6	2,000	0.4	9
Tendon (collagen)	2.8	2,500	0.08	70
Bone	3.0	1,500	0.02	150
Rubber (natural)	10.	8,000	3.0	7
Mussel byssus thread	13.	8,000	0.36	20
Abductin	13.	10,000	1.0	10
Kevlar (synthetic material)	40.	30,000		
Silk (spider frame)	200.	160,000	0.30	1,400

Sources: Data from Wainwright et al. 1976; Gordon 1978; Currey 1984; Gosline, Demont, and Denny 1986; Denny 1987; Abé, Hayashi, and Sato 1996.

Work per unit mass—even if not the standard quantity—may be a more reasonable basis for comparison between natural and technological materials than work divided by volume. Thus, collagen is only a little better than steel on a volumetric basis, but it far surpasses steel relative to the mass of a sample. Natural rubber is a peculiar case—it's weak (low strength) but stretchy (high extensibility), so while its stress-strain curve may be displaced from most of our materials, it has lots of area underneath. That's, of course, why it's useful for powering model airplanes! Spider silk is out of sight, about which more later.

6. A final property from these stress-strain graphs. Say you stretch a material and then allow it to shorten elastically back to its original length. During the test, you measure its stress. A real material will not recover along quite the same stress-strain curve that described the stretch; instead it will trace a downward curve somewhat below that along which it ascended (fig. 15.5). Therefore, the area under the curve representing the stretch will be greater than that representing recovery. What this means is that not all the work put into stretch comes out again in a mechanical form.

We call the fraction of the work that comes out relative to that that goes in the *resilience*; it's the ratio of the area underneath the recovery curve to the area underneath the loading curve, given as a percentage that must be below one hundred. The exception, of course, is active muscle, which does work

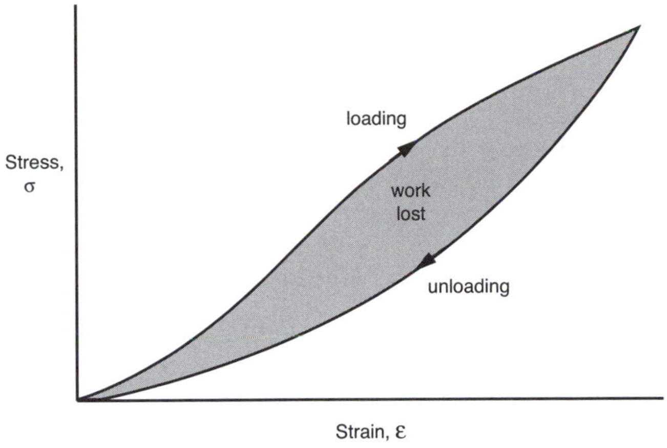

FIGURE 15.5 The stress-strain curves for an imperfectly elastic material that's first stretched and then allowed to shorten again.

rather than absorbing it, and which therefore goes counterclockwise instead of clockwise around its equivalent lengthening-shortening loop.

("Resilience" also has been used simply for the maximum possible strain energy storage before breaking, equivalent to work to fracture—Gordon [1978] used this alternative definition; to complicate things further, the phenomenon to which we are applying the present measure also is described by variously defined variables called, variously, elastic efficiency, hysteresis, energy loss per cycle, rebound resilience, loss coefficient, specific damping capacity, and coefficient of restitution.)

A few values are given in table 15.6; they're just for rough comparisons since all vary with state of hydration, temperature, and so forth.

The datum for collagen refers to operation at much lower strains than those for the other materials; it may be a much stiffer material, but under appropriate conditions it's very resilient. As we saw, spider silk absorbs a lot of energy; but the energy doesn't come out again in the form of mechanical work, and instead appears as heat. Resilin makes as good a rubber by this

TABLE 15.6 The Resilience of Some Biological Materials

Material	Resilience (%)
Silk (spider)	35
Elastin (nuchal ligament)	76
Abductin (sessile bivalves)	80
Abductin (scallops)	91
Collagen (tendon)	93
Resilin (insect wing hinge)	97*

Sources: Data from Wainwright et al. 1976; Denny 1980; Vincent 1990.
*Probably an overestimate.

criterion as the best of our synthetic stuff. Must it be that good? In terms of energy loss, the collagen with which a kangaroo jumps isn't that much worse than the resilin involved in wing beating—neither 7 percent nor 3 percent represent a high tax on performance. But the high resilience of resilin may be important in another way. As Vincent (1990) pointed out, all loss comes out as heat, and the high rate at which energy passes out through an insect's wing hinges may make local heating an especial problem. Imperfect resilience does make heat—feel the tires of a car after rapid driving even in cold weather and the heat from noncoincidence of loading and unloading curves will be immediately apparent.

7. One important property doesn't come directly off a stress-strain curve. Breaking something takes a certain stress, what we called its strength, a force per unit cross-sectional area. At the same time, it takes a certain amount of work per unit cross-sectional area to break the item—that's often called the *work of fracture* or, sometimes, toughness. (Again usage varies; toughness often is equated with strain energy storage or work *to* fracture. In practice, the two meanings can be distinguished by their dimensions—work per volume for work of extension to the fracture point; work per area for work of actual fracture.) The idea that it takes work to made new surface isn't unique to solids. Surface tension where immiscible fluids contact has dimensions of force over length, equivalent to work over area. It takes work to make new surface in a liquid just as in a solid—in any substance whose bits prefer a cozy communality.

Table 15.7 gives data for work of fracture. You may discern some similarities in the relative rankings of materials for strain energy storage (table 15.5)—the reason a single word, toughness, describes either property. The differences between strength and work of fracture deserve attention. In a material with a high work of fracture, cracks do not easily propagate, so where the two properties diverge, it's work of fracture to which we ought to pay most attention. As a general rule, the more jagged the break, the more work it took to make the break, and the higher the work of fracture of the material. A brick not only breaks at a low tensile strength but requires little work to make the break when it occurs. Glass has a high tensile strength but a crack, once started, similarly propagates with very little work. Mild structural steel has far less tensile strength than so-called high-tension steel, but the former may be safer inasmuch as it has a high work of fracture and thus is less apt to fail catastrophically. Work of fracture is the inverse of our ordinary notion of brittleness—cloth takes a lot of work to rip; glass breaks all too easily.

Lots of interesting items jump out of these data. Nicely stiff tooth enamel again shows its apparently concomitant brittleness. Wood's quite fabulous anisotropy appears nowhere more strongly—you can split a log with wedge and sledge lengthwise but not crosswise. Don't try the latter unless you're prepared to deal with a sharp, heavy, upwardly mobile wedge—I speak from the school of hard knocks. Skin's nicely high value means that a cut will not become a tear. We see why mild steel and wood are much more forgiving construction materials than concrete, brick, or stone—construction codes in earthquake-prone places make much of the matter. Biological ceramics such as teeth and shell do far better than natural ones such as stone. All the biologi-

TABLE 15.7 Work of Fracture, Also Sometimes Called Toughness

Material	Work of Fracture (J/m^2)
Concrete, brick, stone	3–40
Algal stipes and blades	80–800
Wood (with grain)	150
Mollusk shell (parallel to surface)	150
Teeth (enamel)	200
Teeth (dentine)	550
Mollusk shell (nacre)	1,000
Aorta, pig	1,000
Sea anemone mesoglea	1,200
Insect cuticle	1,500
Mollusk shell (crosswise)	1,650
Bone (cow leg)	1,700
Steel (high tensile)	10,000
Wood (across grain)	12,000
Skin, rabbit	20,000
Steel (mild)	100,000–1,000,000
Fiberglass	1,000,000

Sources: Data from Gordon 1978; Alexander 1983; Jackson, Vincent, and Turner 1988; Denny et al. 1989; Vincent 1990; Abé, Hayashi, and Sato 1996.

cal ones are composite materials that incorporate secondary, crack-stopping components that increase the work of fracture.

So we apply a stress and are rewarded with a strain and eventually a fracture, and we thereby encounter this complex spectrum of properties—(1) Young's modulus of elasticity, or stiffness; (2) tensile strength, or breaking stress; (3) compressive strength, another breaking stress; (4) extensibility, or breaking strain; (5) work of extension, or strain energy storage; (6) resilience; and (7) work of fracture. Note that numbers 1, 5, and 6 refer to how a material responds to stresses of any level, while 2, 3, 4, and 7 refer specifically to actual failure.

Several More Properties

All these measures have come from materials stressed lengthwise—we've loaded specimens in tension and compression. But recall from the description of solid materials in chapter 6 that part of what it means to be a solid consists of resistance to shear deformations. Not unexpectedly, quantifying the way solids resist shear provides us with another variable. A little more surprisingly, as we'll see in chapter 18, that measure turns out to have considerable importance for structures that resist twisting. But, then, we'll also find that Young's modulus of tensile stiffness matters when something bends. Never mind any suspense—ultimately we're going to use these material properties to talk about structures.

Shear modulus

Consider the block of material in figure 15.6a, a block with plates glued onto its top and bottom so we can subject it to a shearing force. As in figure 15.6b, the block has an area, *S*, in the plane of the shearing force, *F*; as a result of the force, it distorts by an angle γ (lowercase gamma). We call that force-per-area, τ (lowercase tau), the shear stress. Just as *E*, the Young's modulus, represented tensile stress divided by tensile strain, an analogous variable, *G*, the shear modulus or shear stiffness, represents the shear stress, τ, divided by shear strain, γ:

$$G = \frac{\tau}{\gamma}. \tag{15.5}$$

Dimensions give little trouble. Shear stress remains a stress, a force per unit area, although now using area, *S*, of the test block parallel rather than crosswise to the force. Shear strain is an angle, automatically (in radians) dimensionless so we needn't divide by any original length—or, put another way, the unsheared block has a shearing angle of zero. So the shear modulus has the same dimensions, force per area, as the Young's modulus or of shear stress itself.

Data for shear modulus don't quite slip off every page, but some do exist, both for technological and natural materials. I've put what I've found in table 15.8; notice that the values don't differ all that much, relatively, from those cited earlier for Young's modulus in table 15.1—*G*s run from 40 percent of the *E*s on downward to a bit under 10 percent. That relationship—*E* to *G*—will take on significance further along.

Poisson's ratio

If you make something get longer by pulling on it lengthwise, most often it gets back at you, not only by resisting your efforts, but by getting skinnier as

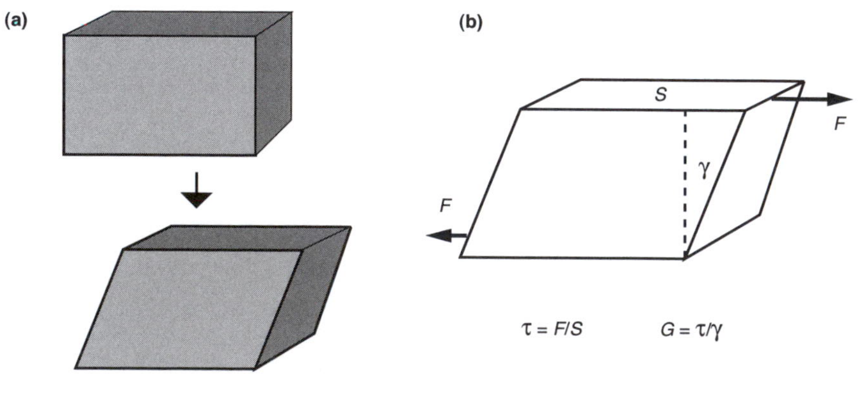

FIGURE 15.6 (a) A block of material loaded in shear. (b) The conventions for defining the shear modulus.

TABLE 15.8 Examples of Shear Modulus, Or Stiffness in Shear

Material	Shear Modulus (MPa)
Elastin (nuchal ligament)	0.060
Resilin (dragonfly "tendon")	0.064
Aorta, rat (arterial wall)	0.15
Cartilage, rabbit (articular)	0.35
Wood (fresh tree trunks)	450.
Bamboo (fresh culm, or stalk)	900.
Bone (compact mammalian)	5,000.
Teeth (enamel)	65,000.
Steel (typical)	77,000.

Sources: Data from Eshbach 1952; Wainwright et al. 1976; Currey 1984; Vogel 1995b; Abé, Hayashi, and Sato 1996.

it lengthens. How much skinnier ought a material get? Assuming a system of constant volume, we do a simple calculation. Cross section must decrease in the exact same proportion as length increases. Thus, the compressive true strain, or fractional decrease, in either of two crosswise directions (ε_y and ε_z) must equal half the tensile true strain, or fractional increase in the lengthwise direction (ε_x). Simple, but not often coincident with reality. So we need a new variable, a factor of lateral contraction, ν (lowercase nu), defined as:

$$\nu = \frac{-\varepsilon_y}{\varepsilon_x} = \frac{-\varepsilon_z}{\varepsilon_x}. \tag{15.6}$$

(The minus signs take account of the fact that a lengthwise length increase causes a crosswise length decrease.) That factor of lateral contraction has long gone by the name of Poisson's ratio, with no ichthyological implication—Siméon-Denis Poisson introduced it.

For the simple isovolumetric case (ignoring a complication or two), the ratio must be 0.5, as implied above. But, again, if that were the way the world worked, we'd not have bothered with the matter. Values rarely run above 0.5, at least for three-dimensional systems, although some odd-ball foams with higher values have been made—they thin down disproportionately when stretched. Mostly values run below 0.5. And that holds even for biological materials, where one might have expected the isovolumetric assumption to save the day. For our systems, though, overall volume may not change but fluid can move around relative to fibers and compartments, making even worse complications, about which Vincent (1990) and Niklas (1992) remonstrated.

Poisson's ratio has more immediate relevance than you might have guessed—so immediate that you already have a feeling for it. Consider the different behavior of tapered rubber and cork stoppers for bottles. A rubber stopper, with its high ratio, spreads sideways as you push it axially. (We're

TABLE 15.9 Examples of Poisson's Ratio, or Factor of Lateral Contraction

Material	Poisson's Ratio
Cork	0.00
Concrete	0.10
Cornstalk	0.23
Aorta (at 100 mm Hg)	0.24
Apple flesh	0.28
Cartilage, mammalian (articular)	0.30
Teeth (dentine and enamel)	0.30
Ferrous metals	0.30
Aluminum	0.33
Bone (compact mammalian)	0.40
Rubber, at low strain	0.50
Cross-weave cloth pulled at 45 degrees	1.00

Sources: Data from Den Hartog 1949; Eshbach 1952; Gordon 1978; Currey 1984; Niklas 1992; Abé, Hayashi, and Sato 1996.

loading in compression rather than tension, but that makes no great difference.) So as soon as you stop pushing, it will get skinnier and thus looser. You have to give it a little twist as you push, which puts the outside in tension and squeezes the inside enough to preshrink it. A cork stopper you can simply push to the desired tightness. Have you ever wondered why corks of cork are preferred for wine, champagne, and such? Cylindrical cork stoppers can be press-fit without fear of shrinkage following removal of the (com)pressing force.

Shear modulus versus elastic modulus

In almost every book on materials, one encounters a lovely formula for converting one elastic modulus into the other. All you need besides the first (whichever) is Poisson's ratio:

$$G = \frac{E}{2(1+\nu)}. \qquad (15.7)$$

It does seem reasonable that response to shear and tension ought to show some similarity—after all, what is shear from the point of view of a bit of material but a strangely oblique combination of tension and compression? The rule works fine for metals. And metals, we shouldn't forget, are the basic materials of materials engineers, with all else appearing almost parenthetically in the classic textbooks. A look back at the data in tables 15.1, 15.8, and 15.9, though, ought to provide a cautionary note. Steel comes out okay, but the G and E for bone imply a Poisson's ratio of 0.8, about twice the measured value. Worse still, the values for the tree trunks and bamboo stalks require

Poisson's ratios of 6 or 7, which beggar belief—as do standard data for dry timber, even with their higher *G*s and *E*s. And those for elastin and resilin imply ratios at least as high as well.

Moral of the story. For many purposes we can use the formulas and approaches bequeathed us by metal-minded material mongers, and we'd be silly to start from scratch. Sometimes, however, we're led far astray. Thus *E* can be more easily measured than *G*—should one measure the first, guess that wood is isovolumetric and calculate *G*? In one investigation, I would have drawn quite different conclusions if I had done so (Vogel 1995b)! I hasten to add that the same problems with classical and simplified material science afflicts the current generation of engineers themselves, as they make more use of plastics, composites, and an array of less rigid materials—Ashby (1992) gives a good sense of the problems.

Perhaps that's the right note on which to end a chapter—a note of caution. Or several notes. I find a text on materials for engineers talking about something called creep of materials, with a time scale of 10,000 hours. I turn to work on intervertebral disks and learn that I'm on the order of a centimeter shorter at night than in the morning. I look at work on sea anemones and learn that creep matters on the scale of a single wave, seconds, and on the scale of the tides, hours. I encounter a host of formulas, some of which will appear in pages to come, that presume Hookean materials. They're too attractive to be rejected and, mirabile dictu, sometimes give useful service for our particular problems. Still, some don't always, and others always don't. Biological materials seem complex along every imaginable axis—the wonder is that we can make any sense out of the subject at all.

CHAPTER

Biological Materials: Tuning Properties Properly

Our time has seen the
synthesis of Polyisoprene
And many cross-linked
Helices unknown
To Robert Hooke; but
each primordial Bean
Knew Cellulose by heart:
Nature alone of Collagen
And Apatite
compounded Bone.

JOHN UPDIKE

WE HAVEN'T QUITE FINISHED WITH MATERIALS. IN PARTICULAR, SEVERAL matters deserve attention, ones that make further links between all those properties and the biological role of specific materials. In talking about strain energy storage and resiliency, we alluded to the importance of the shapes of stress-strain curves; with a further look at those shapes we'll see a lot more functional significance. With an examination of composite materials, we can see how nature makes stuff that takes compressive, bending, and twisting loads, not just simple tension. And we need to look at the whole business from the opposite end—looking at the properties and roles of specific materials rather than at specific properties for a diversity of materials.

All the emphasis on tensile tests in both the previous and present chapters may strike the reader as justified only by the simplicity of tensile loading—loads in which only cross-sectional area, and not cross-sectional shape, mattered. True enough, properties in tension can be more easily defined. But tensile tests prove more useful than you might at expect. Often failure in loading modes other then tension ends up being attributable to failure in tension. Thus, a compressed cylinder typically bends before breaking; one side of the bend is in tension even though the ends of the cylinder as a whole are being pushed towards each other. And failure in tension, as we'll see, courts catastrophic consequences; a compressively loaded crack can commonly be ignored. Furthermore, local tensile loading is an inevitable component of overall loads described as shearing and torsional. But we're getting ahead of ourselves.

The J-Shaped Stress-Strain Curve

Metals are mostly a Hookean crowd, and human technology is hooked on them. As we'll see in this and subsequent chapters, they hold a host of very real advantages for us. That their linear stress-strain plots simplify our analyses simply comes with the territory—their Hookean behavior has little to do with their excellence, and such behavior most emphatically can't be taken as any kind of gold standard for material quality.

Common among nonmetals, both of organisms and organic chemists, is what has come to be called a "J-shaped" stress-strain curve, one that ascends slowly at first and then gets increasingly steep. Young's modulus, the stiffness, thus increases with extension, as in figure 16.1, for the nuchal ligament of a deer (Dimery et al. 1985). The J-shaped curve shouldn't be dismissed as a

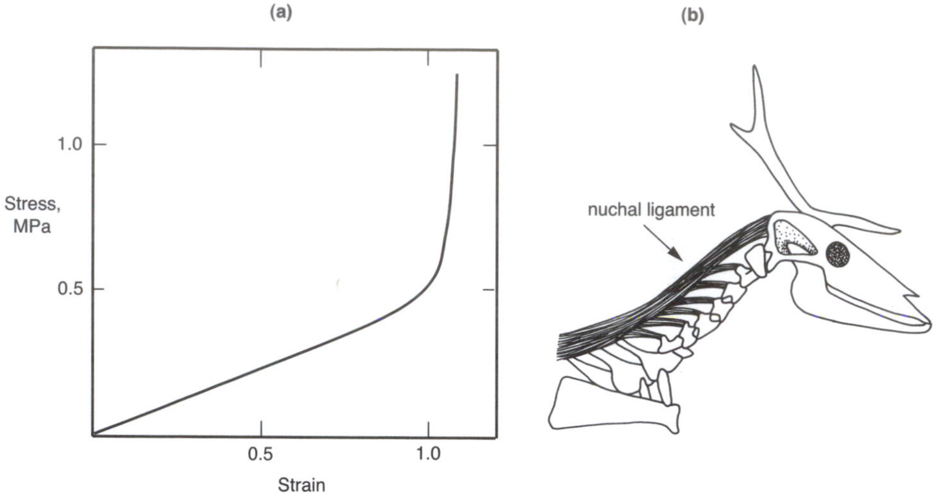

FIGURE 16.1 (a) A typical J-shaped stress-strain curve for a biological material, this one of the nuchal ligament of a deer (Dimery, Alexander, and Deyst 1985). (b) The location of the nuchal ligament.

mere accident of chemistry—even if sometimes it may be—but as something of functional significance. We can recognize three complementary advantages conferred by materials with such J-shaped curves on structures made from them.

First, getting stiffer as the failure point approaches makes a structure safer. Stretchiness may give useful buffering from bumps and bruises, the slings and arrows of outrageous fortune, but bumps and bruises tend to pale beside dismemberment when it comes to fitness. So it's nice when functionally risky extension requires disproportionate force. You may want your boat secured by a stretchy hawser, but you'd really prefer that at some point the boat comes to the end of its tether.

Second, the J-shaped curve provides safety in yet another way. Quite simply, the upward concavity reduces the area under the curve relative to that of a Hookean material. For a given combination of stress and strain, the concave curve indicates a lower work of extension, which translates into less energy that might be released upon failure. Energy release drives crack propagation, the latter almost never a mark of successful design. Skin stretches quite readily, as anyone who has skinned a chicken carcass can attest. But puncturing a stretched piece of chicken skin gives nothing like the dramatic result of puncturing a stretched piece of rubber. A bite or other injury may penetrate our skin, but even (as across a flexed elbow) where the skin is stretched, nothing particularly dramatic ensues.

To appreciate the third advantage, perhaps the most important, one has to think back to the discussion of Laplace's law in chapter 3. (If you've already read chapter 10, you'll notice that we're returning to a case it made; the issue also formed one of the teasers at the start of chapter 1.) In a curved vessel, internal pressure generates tension in the wall that can expand and ultimately even split the vessel. The more pressure, the more tension, obviously. But less obviously, the greater the radius, the more tension will be generated by a given pressure; equation (3.5), for a cylinder, repeated below, shows the direct proportionality. All of which comes straight from geometry and dimensional reasoning:

$$\Delta p = \frac{T}{r}. \tag{3.5}$$

Now put materials into the picture, and imagine a real cylinder with a thin, elastic wall. What happens if the pressure inside rises—where will the cylinder expand? If the material is Hookean, a well-expanded portion has just as much tendency to expand further as one that has expanded only minimally. The extra tension in the wider wall will just balance the increased ability of the internal pressure to produce further tension. Result—a cylinder with a stretchy wall can't be built of a material with a stress-strain curve that's convex above, and it can be made only with considerable risk and instability of a material that's Hookean. If rubber were Hookean out to the point of failure, rubber balloons could not be cylindrical. In fact, the stress-strain curve of rubber has an upwardly convex part at its start (fig. 16.2), and the inflation of cylindrical balloons inevitably begins with an aneurysm, one part expanding almost to the point of rupture before the rest follows suit. But the aneurysm doesn't rupture, since the curve ends with a upward Hooke before the breaking point—for high strains, it's concave upward.

Fortunately, our arteries do not have walls with linear stress-strain curves, nor ones even as initially Hookean as rubber. They have that nice concave-upward J-shaped curve. As do a lot of other pliant biological materials. Not that the curve is universal, as table 16.1 shows. About the only universal rule is that no common biological material shows as linear a stress-strain curve as do metals.

How does the J-shape come about? We haven't paid much attention to the molecular bases of mechanical properties, nor will we. The J-shape can, in fact, originate at almost any level of structural organization. It requires no more than some form of progressive recruitment of strain-resisting components. Where a pure material gives such a curve, one has to recognize that pulling on springlike interatomic bonds doesn't represent what's happening. Instead, in stretching something, one is forcing order on randomly writhing

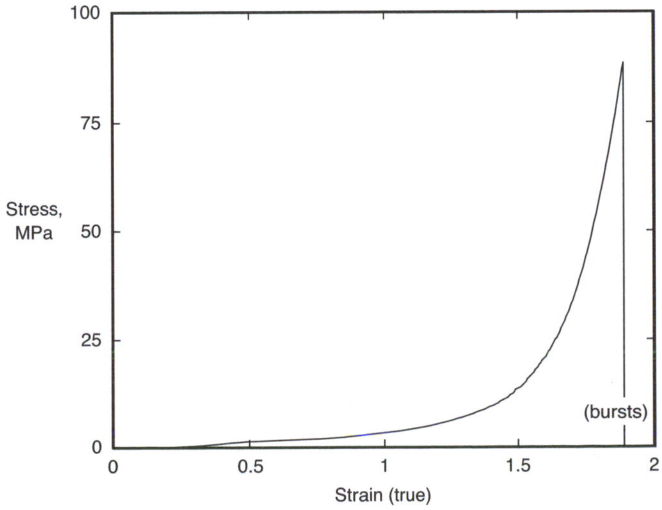

FIGURE 16.2 A stress-strain curve for natural rubber. This one comes from the same data used to generate figure 3.2b. The only additional data cranked in are the mass of the balloon (less the neck), 0.45 grams, the density of rubber, 1100 kilograms per meter cubed, and the fact that rubber is isovolumetric, with a Poisson's ratio of 0.5. Try the conversion—the formulas needed are in the text. Hanging increasing weights from a long, cylindrical balloon gives a similar curve.

TABLE 16.1 A Crude Division of the Shapes of Nonmetallic Stress-Strain Curves

Curve J-shaped	Curve not J-shaped
Rubber (near breaking)	Rubber (near start)
Collagen, tendon (near start)	Collagen, tendon (near breaking)
Nuchal ligament	Keratin, as wool, horn, and so forth
Locust intersegmental membrane	Mussel byssus thread
Spider silk, viscid	Spider silk, orb frame
Algal blades	Bone
Mammalian skin	Cartilage
Arterial wall	Glass

Sources: Data from Wainwright et al. 1976; Gosline, Demont, and Denny 1986; Alexander 1988; Denny et al. 1989; Vincent 1990; Bell and Gosline 1996; Vollrath and Knight 2001.

macromolecules and thus decreasing the system's entropy rather than increasing its internal energy. But these are abstruse thermodynamic concepts, and the deceptively ordinary terms mislead as much as they inform. While we'll touch on them again in the next chapter, for further information one should consult books such as those by Treloar (1975) and Wainwright et al. (1976).

For multicomponent materials, incorporation of fibers in a matrix gives another route to this nonlinearity. Stretching causes fibers that were initially random in orientation to take on, little by little, the direction of the pull—and thus to bear an ever-increasing fraction of the load. So the initial pull works mainly against a soft matrix, while further pulling works more and more against a stiffer fibrous component. To avoid aneurysms, our arterial walls use another version of the trick. Initially kinky fibers gradually pull out to full extension and take on an increasing fraction of the load, as was shown in figure 1.3.

At still higher levels of organization, some simple geometry can be recruited. Consider a weight suspended on a string between two supports, as in figure 16.3a; for simplicity, we'll assume a weightless, inextensible string.

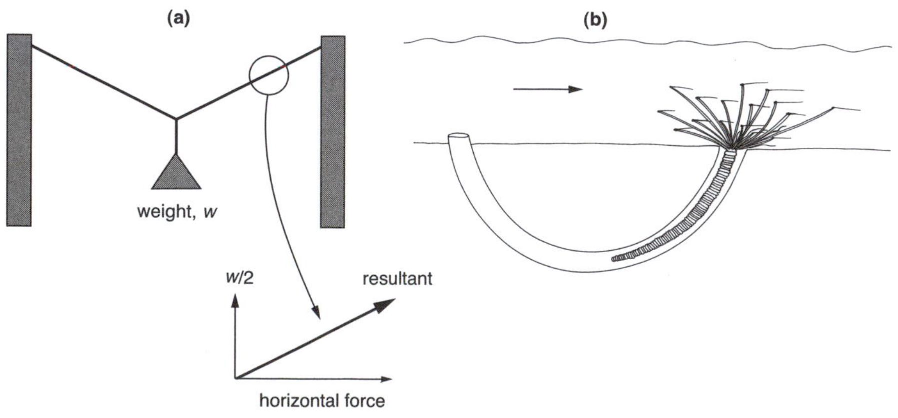

FIGURE 16.3 (a) A weight on a string between two supports and the way the force along the string depends on the dangle-angle of the string. (b) The tentacle set up of *Eupolymnia* (Johnson 1993).

Moving the supports apart raises the weight, but the weight exerts anything but a steady pull on the supports, as you can see from the diagram below. If the weight hangs vertically, it exerts no lateral pull on the supports. If the strings are pulled out horizontally, the weight exerts an infinite load on the supports! In fact, it can't quite be done if strings and weight have any weight at all. To quote a nineteenth-century mathematician, William Whewell (words once used as an example of inadvertent poetry), "And so no force, however great, can stretch a cord, however fine, into a horizontal line that shall be absolutely straight." So extension hits an asymptote, generating a J-shaped curve. A heavy, loose chain will give roughly the same result, so it can be used for mooring or anchoring a boat—if it's heavy enough or a weight hangs from it.

One little organism, at least, plays this game (or its converse) in an especially straightforward fashion. A marine polychaete worm, *Eupolymnia heterobranchia*, lives in a tube in a mud flat with its front end sticking out. It stretches out its long, flexible, sticky, feeding tentacles crosswise to flow, as in figure 16.3b—like the main cables of a suspension bridge but loaded by drag instead of gravity. According to Johnson (1993), extended straight from animal to anchor point, the tentacles could not withstand much drag. But by a combination of passive sagging and muscular activity, they extend ever farther as flow increases. Drag may increase with flow, but the tentacle-stretching tensile consequence of drag does not—tension in the tentacles remains almost constant over at least a sevenfold speed range.

Trading One Property for Another

In some hard-to-pin-down way, mechanical properties covary. For a given material, a high value of one property all too often goes along with a low value of another. Neither nature nor human materials scientists seem to have unlimited wiggle room within some hyperspace whose axes are the properties described in the last chapter. But both technologies have a host of tricks for making the best of what seem to be inevitable (if not quite rigid) trade-offs. Very stiff materials tend to fracture fairly easily, very strong materials rarely extend very far, the strongest materials aren't also the stiffest, and so forth. Ashby (1992) provides a host of trade-off graphs for materials of interest to engineers; not surprisingly, only wood as lumber is of biological origin. It's instructive to do the same kinds of comparisons with biological materials, and it gets even more interesting if we put some engineering materials into the mix. So:

1. *Young's modulus of elasticity, or tensile stiffness, against work of fracture* (tables 15.1 and 15.7). The biological materials fall reasonably close to a descending line on a log-log graph as in figure 16.4, suggesting that, indeed, nature has to trade off a high value of one property to get a high value of the other—stiff materials tend to be brittle. Comparison of two similar materials, the dentine and enamel of teeth, gives the same picture. Interestingly, the product of the two properties varies remarkably little. Its fourfold variation could be (but probably is not) accounted for by the uncertainties in measurements and calculations for biological stuff. We're talking here about wood,

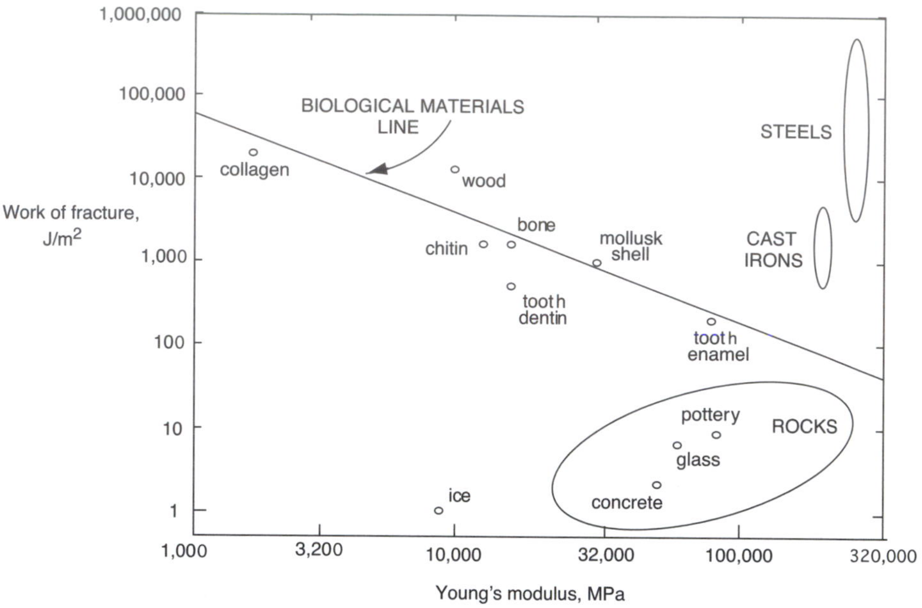

FIGURE 16.4 The empirical relationship between Young's modulus of elasticity and work of fracture.

chitin, the tooth materials, mammalian bone, and nacre of shells—diverse in composition, function, and phylogenetic position.

Is this product of stiffness and work of fracture (about 10×10^{12} square newtons per cubic meter, incidentally) inviolate for rigid solids? Were this the case, rigid nonbiological materials would give the same value. But steel, ordinary mild structural steel, has a product around 5,000 times higher. That nontrivial difference goes far to explain why we humans have gotten so wrapped up in a metallic (especially a ferrous metallic) technology. Lest you feel sorry for nature, though, consider some nonbiological nonmetals. Stone, brick, cement, and pottery are far worse—stiff but fragile, further rationalizing the predilection for metals of human designers. Solid water—ice—does still worse by another large factor, perhaps why we use cracked ice so casually and certainly why ice has limited appeal as a structural material even where melting and sublimation give no trouble. So we're left with questions both of why the biological rigid materials aren't better and why the rigid ones turn out to be such a homogeneous bunch.

2. *Young's modulus of elasticity against tensile strength* (tables 15.1 and 15.2). We find less regularity than in the previous comparison—and a few suggestive comparisons. A diverse lot of rigid materials have strengths in the range of 100–500 megapascals, with no obvious correlation (direct or inverse) with Young's modulus. But both tooth enamel and coral skeletal material push the modulus up around fourfold; both seem to pay with a twofold or threefold decrease in strength. Is there a limit beyond which nature has to pay in the currency of strength to increase stiffness? If so, it might explain why bone rarely runs above about 20,000 megapascals. Another curiosity is the great strength of the frame silk of orb-weaving spiders, four times that of the other biological materials, especially since it comes with a thoroughly decent

value of stiffness. In this sense, it's very special stuff, much better than the biochemically similar silks of silkworm moths.

But how far have organisms, with all their diversity and lengthy evolution, pushed the envelope? The great combination of strength and stiffness of spider frame silk pales beside that of Kevlar, another organic polymer. It's twice as strong as silk; at the same time it's no less than ten times as stiff. Steel does as well—while only half as strong as silk, it's fully twenty times as stiff. The comparison of stiffness and strength carries the same implication as that of stiffness and work of fracture. We get the hint of constraints on nature's materials, constraints somehow not incumbent on nonbiological materials. The greater fundamental diversity of nonbiological materials might—repeat, might—have some relevance here.

3. *Strength against extensibility* (tables 15.2, 15.3, and 15.4—for cartilage and mollusk shell nacre, compressive strength provides an adequate surrogate for tensile strength). Here, if anywhere, one ought to expect a high value of one property to come at the expense of a low value of the other. After all, a look at any stress-strain graph reminds us that the product of these properties ought to come close to maximum strain energy storage. For a Hookean material, the area under the curve is exactly half that product. A more complicated picture, though, emerges from the data. A look at the biological materials as a single set reveals little pattern. Some of the least extensible materials, such as coral skeleton, aren't especially strong; the most extensible material, a particular spider silk, happens to be the strongest by far.

Nonetheless, we can discern the kind of homogeneity we saw when comparing stiffness and work of fracture if we focus on materials commonly loaded in compression and look at the product of strength and extensibility. All the high values drop out! Wood, bone, insect cuticle, cartilage, and mollusk shell form a group, none far from 1 megapascal, and only the very inextensible coral skeleton drops well below that value. Furthermore, ignoring coral can be justified on biological grounds—while they often get smashed up in storms, breakage may be nonfatal, even functionally reproductive and disseminative. The high products of strength and extensibility turn out to belong to materials that don't get pushed around, that is, normally feel only tension—tendon, byssus thread, hair, and the silks. Here, spider silk truly shines, coming out as stronger than anything (except Kevlar) and at the same time especially extensible.

Which Properties Are Most Important When . . .

The specific synthesis of solid materials by organisms has a result more consequential than mere proliferation of kinds of materials. It's a matter we might simply call "tuning"—the alignment by natural selection of the properties of materials and their applications. At one level the subject is almost trivial—the availability of steel cable permits a different kind of bridge from anything that can be made of bricks; rubber can-openers aren't where you want to invest your money, and (for fans of modern sculpture) a soft, sway-backed engine block is strictly an Oldenburg, not an Oldsmobile. For a given task, some materials are more appropriate than others. But beneath that bald statement lurk substantial biological issues. For one thing, the materials are of interest in themselves—for instance, just what are the limits to the mechanical versatility of protein? Or what are the practical trade-offs among the properties, that is, what deficiencies must be suffered in one property in order to achieve a high value of some other?

Even without answers to such questions, a look at tuning can use the properties of materials to give us insight into the tasks to which they're put. As suggested toward the end of chapter 1, we assume good "design" and use the assumption as a tool to ask what really matters in an adaptive sense in the operation of an organism. Here, comparisons are exceedingly helpful—we can usefully ask how material properties change from species to species, from habitat to habitat, from place to place within a single organism. We can look for similarities of mechanical adjustments among very distantly related organisms subjected to similar environmental changes; we can investigate differences in properties of materials that interact harmoniously in some common function.

Consider our teeth (Waters [1980] has a good discussion of their properties), crucial to feeding ourselves, subject to stresses and abrasions over long periods, and (as we know too well from the disproportionate professional attention they require) none too reliable. Enamel, the outer layer on teeth (fig. 16.5a), is very stiff and resistant to compression—handy for biting and chewing–but it takes tension poorly and is exceedingly brittle—stiffness and toughness seem in practice as antithetical here as in the various commercial steels. Dentine, the next layer in (and the bulk of the hard stuff in teeth), is less stiff than enamel but not quite so brittle. An off-axis load might safely develop some substantial tension in the dentine—harder dentine would most likely lead to increased functional fragility. Perhaps there's a general principle involved—limit the hardest material to small pieces at the sites of wear, as we do in making carbide-tipped drills and saws. Teeth do crack without enor-

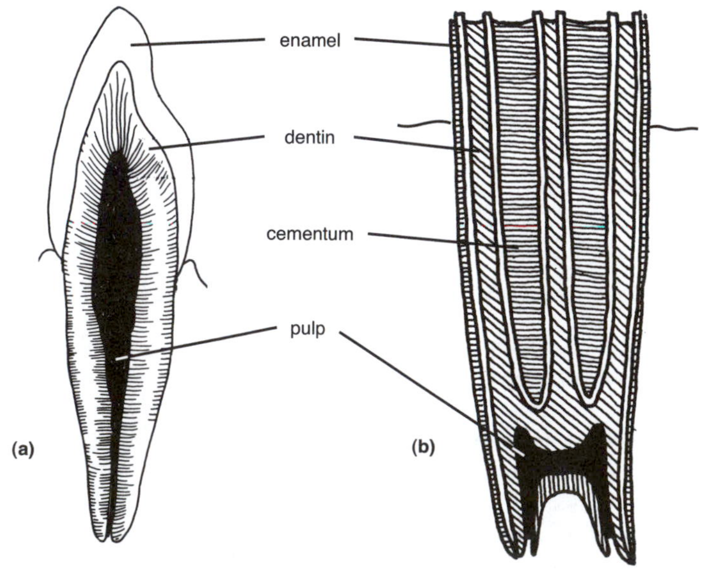

FIGURE 16.5 (a) The structure of a human tooth, a fairly typical mammalian example; and (b) the contrasting structure of a bovid tooth. Both are, of course, longitudinal sections.

mous provocation—their poor behavior in actual fracture is an unpleasant converse of the hardness needed for trituration. Trade-offs seem unavoidable, although it's no easy thing at this stage of our understanding to give them proper quantitative expression.

While the spectrum of natural materials lies a bit toward the less stiff end of the scale than the materials we humans have found most useful, nature (as the old evolutionists stressed) is red in tooth and claw, both items that need to be stiff as well as strong. Our teeth only nip the edge of that world. Still, mammals, even individual mammals, go in for more dental diversity than any other vertebrates—even in the face of our relatively stereotyped jaw mechanics. We humans make pointed teeth (canines) that penetrate tough skin and collagenous flesh, and we put them well outboard of the jaw's hinge and jaw-closing muscles. We make shearing teeth (incisors) that work like scissors or other shearing cutters. And we make crushing and grinding teeth (molars), which we put, for greater force, back near hinges and muscles. The diversity fits nicely with the omnivory of hominid primates. Other mammals extend their jaws forward, with canines well out in front for effective snapping. Others put shearing teeth as far out front for grazing and browsing, with a substantial gap between them and the grinding molars farther back.

Grazing has perhaps elicited the most dramatic dental specializations in mammals. About twenty million years ago, grasses and grasslands appeared on earth. Grass (and, incidentally, wood) provides poor fodder. It yields little energy relative to its mass, so a grazer has to process huge volumes. Much of that energy comes as chemically inert cellulose, which mammals hydrolyze only by enlisting symbiotic microorganisms in rumen or intestine. It's full of abrasive stuff like silicon dioxide and has lengthwise fibers that demand crosswise chewing rather than rapid tearing. Long-lived grazers, concomitantly, have especially special teeth, with their components typically layered side by side, as in figure 16.5b. This odd-looking arrangement ensures that, while teeth may wear down (hence judging a horse's age by looking at its teeth), they won't wear smooth. The harder material (enamel, most particularly) will continue to protrude as the softer materials (cementum and dentine) wear down between them. Elephants take the process a step farther, renewing their teeth throughout their lives by making them in the rear of their jaws and reabsorbing worn-down ones in the front. That does, of course, preclude tooth-type diversity, but elephants use trunks instead of front shearing teeth to detach vegetation from substratum.

Many invertebrates use crystals of metal salts to harden their cutting, rasping, and grinding equipment. (Metal salts aren't, of course, metallic, in the mechanical or metallurgical sense, so they're not examples of metallic materials in nature.) For instance, the radular ribbons with which chitons (a group of mollusks) scrape algae from rocks contain layers of iron, as magnetite (Fe_3O_4), within a chitin (not chiton) matrix (van der Wal et al. 1989). Another mollusk, the limpet *Patella*, backs the magnetite cutting tips of its radula with silicon dioxide (Runham et al. 1969). Incidentally, mollusks renew their radulas from the back as they wear near the front much the way elephants renew their teeth. The mandibles of herbivorous insects contain zinc or manganese salts (Vincent 1990). Glycerid polychaete worms use copper salts (Gibbs and Bryan 1980) in their jaws, concentrating the copper compounds near the tip; and nereid polychaetes do the same with zinc (Bryan and Gibbs 1980)—either kind can give a person a most unwormlike bite.

Varying the composition of a material from one place to another, as do those polychaetes, constitutes yet another dimension of tuning. Nature, synthesizing big things in tiny cellular factories, uses compositional gradients far more extensively than do we humans, who typically construct relatively small things with big equipment.

Local concentration, almost inevitably invariant in our materials, gets put to good use in such places as those metal-salt-hardened jaws of glycerid polychaete worms. Not only do the copper salts occur in higher concentration near the sharp ends of the fanglike jaws, but, according to Lichtenegger et al. (2002), copper-containing fibers are more concentrated in the middle of the jaws than nearer the surface. Thus, stiffness increases from outside inward, and the (relative) softness of the surface reduces the chance of crack initiation. As these investigators pointed out, a worm's jaws must often close on a bit of hard gravel rather than softer edible material, so reducing the chance of contact damage might matter a lot. Inadvertently bite on a pebble instead of a bean, and you're quite likely to break a tooth.

Cracks about Cracks

In the last chapter, the fact was casually put on record that most pliant materials and all rigid materials in organisms turn out to be composites of several components. And the work of fracture of a material proved to be related to the surface created as it broke. We're now in a position to make something of these items. First, a direct hands-on experience (fig. 16.6). Take a sheet of aluminum foil, fold it once, and make a small slit at right angles through the fold. Then pull lengthwise on the fold—with very little applied force the foil tears, starting at the slit. Now make a round hole at the far end of a similar slit with a paper punch and pull again. This removal of further foil should foil the further facility of the foil to fail.

What has happened gets us back to the difference between force and stress—the latter, force per unit area, is what matters most in the real world. The sharp end of the slit (or almost any crack) acts as a force concentrator and thus as a local stress increaser (fig. 16.7a). As you keep pulling, of course, the crack keeps advancing, so there's always a sharp edge. And sharp edges are bad news; smooth-edged voids, by contrast, may stop cracks (fig. 16.7b). Make the portholes and hatchways on your ships and the window and door cut-outs on your airplanes round-cornered to avoid locally high stresses.

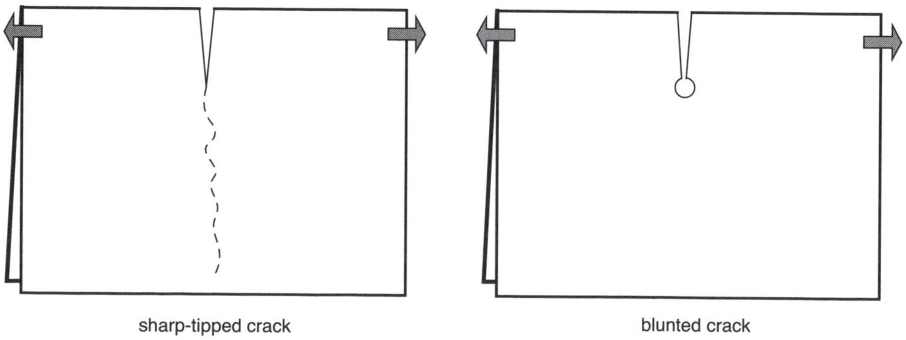

FIGURE 16.6 What happens when you pull on pieces of something like aluminum foil with and without a hole at the end of a transverse slit.

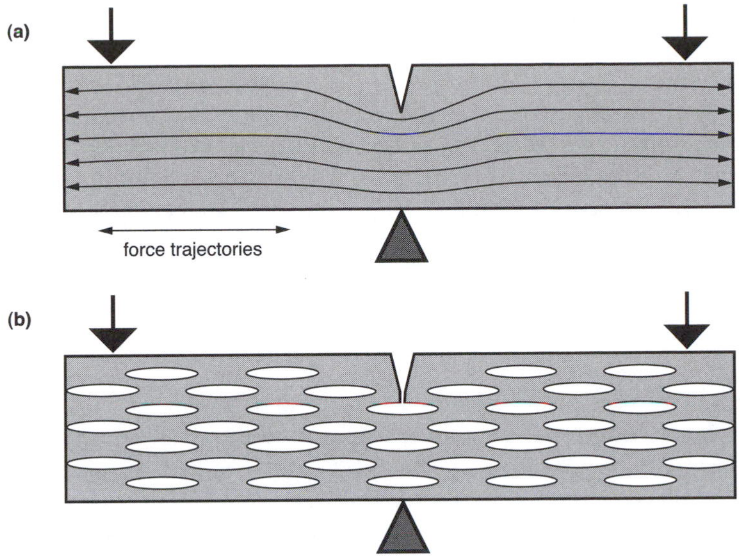

FIGURE 16.7 (a) Force concentration at advancing crack. The lines, force trajectories, behave rather like streamlines—bunching now indicates local stress increase instead of speed increase. (b) How rounded voids blunt the end of any crack; the material may be weaker by virtue of lower cross section, but stress hasn't increased much.

Round the edges of the grooves between the teeth on a gear. Put lots of little spaces (voids) in bone, in the tiny hard ossicles of a starfish, or in coral, but never let them have sharp edges. Fracture depends not just on the material but on the disposition of its components. The notion isn't new—I'm told that in 1545 Roger Ascham, tutor to Queen Elizabeth I, advised bowyers to prick small holes at the ends of fractures in longbows. And in 1964, Robert Chapman, machinist, advised the author to drill round holes at the proximal ends of the radial slits in a metal disk that would be made into a fan. Gordon (1976, 1978) gives an especially good account of the business of cracking, to the understanding of which he himself made important contributions.

One ought to be able to calculate the strength of a material from knowledge of its chemistry—of bond strength for crystalline materials and of intermolecular attraction for noncrystalline items. It rarely works, though. Consider glass. On the basis of intermolecular attraction it ought to have a strength of about 14,000 megapascals, but measurements on glass in bulk give 180 megapascals, nearly two orders of magnitude lower. What's behind this huge discrepancy? Nothing more nor less than crack propagation. A hint about the mechanism comes from measurements on very fine glass filaments, which are much stronger than bulk glass from the same melt. Moreover, the finer is the stronger, down to around a micrometer, where they come close to the chemist's calculation. Another hint comes from data we've already seen. Work of fracture measured what it took to make new surface. Rigid materials with high works of fracture, like bone and wood across the grain, break with jagged edges; materials with low works of fracture, like ceramics, break cleanly across. It costs energy to make surface; the more surface that has to be made, the more energy a crack demands.

And that's the first part of the key—whether a crack will propagate depends on the balance of two factors, on whether their algebraic sum comes out positive or negative. On the one hand, extending a crack takes energy in an amount determined by the area and nature of the surface that's made. That energy comes from the stress on the material and the extension at the crack, in dimensional terms force per area times distance, or work (or energy) per area, as in work of fracture. On the other hand, cracking relieves the force concentration as it proceeds. If relief of that force concentration (called, a bit misleadingly, "stress concentration") releases more energy than required for further cracking, then watch out—the crack can grow with terrible speed. A graph (fig. 16.8) tells the story. At a certain length (or depth, if you prefer), the net energy involved in making the crack propagate begins to rise, meaning that the process has become energy-yielding rather than energy-requiring. At that point, called the "critical crack length," the crack will keep advancing with no additional input of energy. The external force required has dropped to zero—at least approximately, since one can, if with effort and wastage, crack a piece of glass part way across.

The second part of the key comes down to a peculiar bit of scaling. The energy needed to move a crack along varies with the work of fracture, *energy per unit surface area created*. The energy available is that of stress and strain, force per area or *energy per unit volume affected*. Surface being created relieves force concentration throughout a volume. Energy per area over energy per volume has a dimension of length, and that's proportional to the critical crack length. Specifically, where W_f is the work of fracture (per area), and W_s is the strain energy stored (per volume) at the fracturing extension, the critical crack length, L_c, is

$$L_c = \frac{W_f}{\pi W_s}. \qquad (16.1)$$

Thin glass filaments get their great strength from their minute diameters, approaching the critical crack length for glass. For instance, for a work of fracture of 10 joules per meter squared and a strain energy storage of 1 megajoule per meter cubed (reasonable values for glass), the critical crack length is about 3 micrometers.

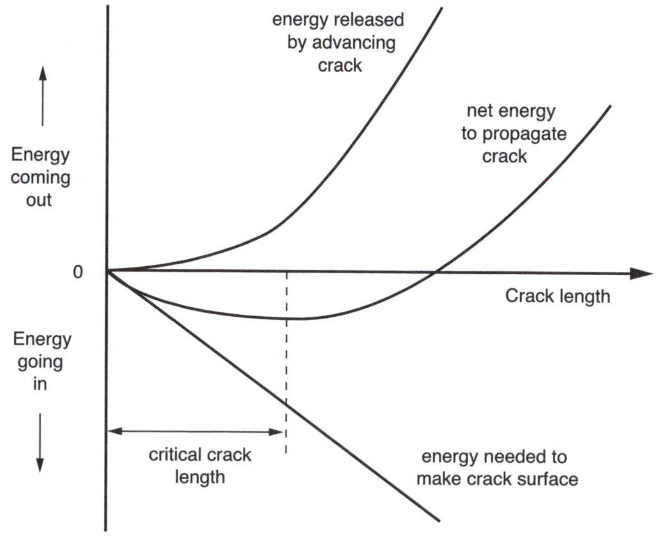

FIGURE 16.8 The energy input and output relationships as a crack propagates.

In the simplest case, the Hookean material usually mentioned in engineering texts, strain energy stored is half the stress times the strain ($\sigma\varepsilon/2$), so

$$L_c = \frac{2W_f}{\pi\sigma\varepsilon}. \tag{16.2}$$

How, then, can you avoid propagating cracks? The formulas suggest several possible approaches.

■ Use materials of high work of fracture—metals rather than ceramics if you're human, in any case something well down the sequence in table 15.7. Skin is good stuff—a puncture won't propagate even if you stretch that part. Fresh wood doesn't fracture across the grain for the same reason. Bones and teeth don't do as well.

■ Use materials with a low strain energy at fracture extension. Spider silk might present a problem were it not for the skinniness of its strands and its habit of immediately converting strain energy to heat. Rubber does give problems, as when one sticks a pin in a balloon, on account of its high strain energy and low work of fracture. Here, of course, is where the J-shaped curve comes into its own—the area under its concave curve has less area than that under a straight line of the same extreme values.

■ Use materials at low values of stress and strain, the reason I included equation (16.2). The fracture point in real materials and structures often depends on imperfection in manufacture or accumulated defects from wear or accidents. No surprise—the lower the operating levels of both stress and strain, the more intrusive an imperfection or defect must be to cause failure. Nature tends to go in for high strains (recall table 15.4), but she compensates with J-shaped rather than linear stress-strain curves and by going up to the highest strains in thin filaments loaded in tension. A nuchal ligament may be massive, but the elastin fibers within it have diameters of only about 7 micrometers.

That comment about thin elastin fibers brings up still another way not to crack, one implied in the words about thin glass filaments.

■ Use materials that are divided transversely to the direction of the load into thin fibers or filaments or, for bending loads (to come up again in chapter 18), into layered sheets. Most sponges, for instance, incorporate tiny elongate spicules of calcium salts or glass, which provide stiffness with good material economy and little risk of fracture. In using this device, organisms must profit by a lucky accident. For the values of the work of fracture and strain energy storage of the materials available to nature, fibers in the range of 1–10 micrometers in diameter correspond to the critical crack length range and so are especially appropriate. That's a size especially suitable for the synthetic machinery of cell-sized factories.

Recall the little exercise tearing aluminum foil. Aside from using a metal, you did none of the above but instead used still another crack-foiling scheme.

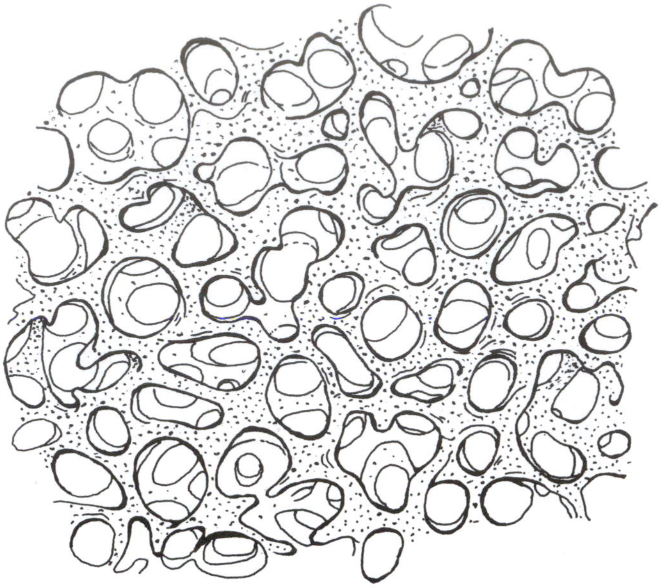

FIGURE 16.9 The holey ossicles of an echinoderm, here a starfish, a so-called cellular solid.

Ordinary cracks in stiff materials have very sharp ends. That's bad, since the sharper the crack's tip relative to the crack's length, the more concentrated the force at the tip. The relevant formula for the concentration factor is

$$C_{\text{stress}} = 2\sqrt{\frac{l}{r}}, \tag{16.3}$$

where l is the crack length and r is the radius of curvature of the tip. So we have another couple of crack-avoiding schemes.

■ Use materials that don't develop sharp tips on any cracks they might suffer. Some highly extensible materials use this device to minimize their fracture risk. Macroalgal blades are broad enough for tearing to be a real risk, and they have a low work of fracture. But their extensibility is high, and they use that extensibility to engage in spontaneous crack-tip-rounding (Denny et al. 1989)—as does Saran wrap in the kitchen, making it far harder to tear off the roller than aluminum foil or waxed paper.

■ Use "foamy" materials in which any threatening crack will in short order run into a hole. Not only does this reduce the chance of cracking, but it saves material—less can be more. Styrofoam works this way, erring perhaps on the side of excessive holeyness if cracking matters most, but gaining greatly in material economy and lowered thermal conductivity. The little hard bits of echinoderms, the ossicles, develop as single crystals, but they avoid the excessive brittleness typical of crystals by being especially holey, as in figure 16.9. Wood gains some material benefit from similar voids. Such materials come under the heading of "cellular solids," the term having no connection with "cellular" in the strictly biological sense—but in the sense that Hooke (him again) originally used the word for the microscopic holes in cork.

For a crack to start from some imposed stress or to be in danger of propagating after, say, a surface suffers a bite, the edge of the surface must be loaded in tension. Loose skin extends without fighting back. So we have still another way not to crack.

- Proliferate edge (if flat) or surface (if massive), deviating fractally upward from two- and three-dimensional form. The ruffled edges of some macroalgal blades may work this way; the edges of the leaves of many higher plants may do likewise—the two-dimensionally pressed leaves of herbarial specimens commonly fold but almost never split at the edges. Deliberate looseness of skin in animals is more problematical at this point. (Of course, we're talking about a structural rather than a material trick, a reminder of how arbitrary is our distinction.)

Incidentally, we now understand the correlation between maximum strain energy storage, or work to the fracture point, and work of fracture. To accumulate a lot of strain energy, a material must be resistant to fracture, which means that it must demand a lot of energy for cracks to propagate.

Composite Materials

But there's yet another and perhaps even better way to keep cracks from propagating disastrously. It consists of making a material of at least two components, one stiffer than the other. If a crack runs through a little fiber of stiff material and then reaches an unstiff (compliant) component, the latter will give a little, accommodate the crack, and reduce the force concentration at the tip of the crack (fig. 16.10). Result—the crack stops. It amounts to having self-rounding tips on cracks. While the material as a whole, the so-called composite, may be less strong than its stiff material, it now demands far more work to get it to the breaking point and still more work to fracture it—both the work to the fracture point (strain energy storage) and the work of fracture have been increased. Several tricks help here. First, the stiff stuff should be divided into small pieces—keeping the stiff material smaller than its critical crack length makes best use of it. Second, a goodly *difference* in properties between components makes it hardest for a crack to make the transition from one to another. Third, making a lot of internal surface at which components contact maximizes the chance of a crack hitting such an interface. Finally,

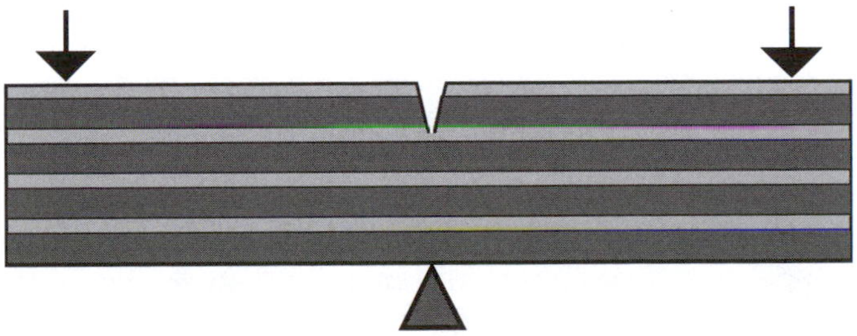

FIGURE 16.10 How composites minimize crack propagation. A crack in stiff, brittle material hits soft material and gets its tip dulled. (Think of this as the opposite of what happens when a knife penetrates—it gets dulled when it hits something hard.)

giving the components a specific, nonrandom orientation can increase the number of interfaces a crack caused by the anticipated load will hit and thus makes the crack run the most jagged (and work-demanding) possible course.

(We ought to note that composites further mock our distinction between materials and structures. While they have structure at a level well above the molecular, for convenience and convention we continue to treat them as materials.)

Except when we build things of wood, our technology makes only limited use of composites (and when we talk of them, they're always referred to as "advanced" composites!). For one thing, they're troublesome to fabricate; for another, we make great use of metals, which give excellent performance as pure materials, at least pure in the present sense, even if alloys. But glass rod and fiberglass rod differ enormously—far more than one would expect from a small fraction of relatively soft epoxy glue. Organisms don't use pure metals, and they use composites for all their rigid and most of their pliant materials—one is tempted to connect the clauses of the sentence with a "therefore." They inevitably divide their stiff stuff into small pieces that form components of composites. Even coral, mollusk shell, and teeth are composites. The sulfurous smell when the teeth are drilled comes from burning protein—enamel, awkwardly brittle, would be even more so were it not for its 2 percent of organic material; dentine is fully 29 percent organic (Waters 1980). You can drill a hole in a scallop shell with little worry that it might shatter (if you want to make earrings or other ornaments, for instance)—it's not as brittle as you might think.

The eggshells of birds are mechanically impressive devices, surprisingly resistant to external loading; Vincent (1990), though, complained about how little we understand them, muttering at "half-boiled notions" in the literature. They're mostly mineral but have a critical 2–4 percent of organic matter, making them into composites. Still, cracks can propagate, of which fact the chick takes advantage to get out—before pushing, it pecks around a circle so it can then break the egg along the dotted line.

Equally impressive is the composite character and consequent resistance to crack propagation of the leaves of grasses, also investigated by Vincent (1982). If a grass leaf is slit or notched it does tear more easily, but only (and fairly precisely) in proportion to its reduced cross section—there's just no sign of any significant stress concentration. Do your worst to a grass leaf—it just doesn't go along with attempts to tear it crosswise. The phenomenon has major implications for the grazing techniques of grass-eaters: grass must either be sliced almost completely across or torn up from the ground. A cow typically wraps its tongue around some grass and pulls; it doesn't take the time and trouble to chew the truly tough stuff. A lot of ecology and evolution must depend on such mechanical events.

Nature makes her composites from a diversity of materials. Some of the variation must reflect phylogenetic habit. Thus, higher plants go in for polysaccharides such as cellulose, while animals make more use of proteins, ones such as collagen and keratin. Some variation clearly reflects different functional demands—stony coral would make poor cartilage. Common to all, though, is the basic scheme of combining pieces of materials of diverse properties on a microscopic but not molecular level. Table 16.2 lists a baker's dozen of them, distinguishing (where possible amid the complexity) crystalline or fibrous components from softer, amorphous matrix.

TABLE 16.2 Some Natural Composites and Their Components

Material	Hard Component	Matrix
Wood	cellulose (polysaccharide)	lignin, hemicelluloses
Sponge body wall	calcareous, siliceous spicules, collagen	miscellaneous organic
Stony corals	aragonite ($CaCO_3$) crystals	chitin fibril network
Sea anemone mesoglea	collagen fibrils	protein/polysaccharide gel
Mollusk shell	calcite ($CaCO_3$), aragonite	protein, sometimes some chitin
Chiton radular teeth	magnetite (Fe_3O_4) crystals	organic, chitin plus protein
Echinoid teeth	dolomite ($CaMg(CO_3)_2$)	organic
Bird eggshell	calcite crystals	protein, some polysaccharide
Cartilage	collagen fibrils	mucopolysaccharide
Bovid horn	keratin fibers	wet, amorphous keratin
Bone	hydroxyapatite ($Ca_5(PO_4)_3(OH)$)	collagen, other organic
Tooth enamel	hydroxyapatite	organic
Arthropod cuticle	chitin fibrils + cross-linked protein + (sometimes) calcite	

Sources: Data from Brown 1975; Gosline 1971; Wainwright et al. 1976; Vincent 1990.

Nature uses those diverse materials in diverse arrangements—variations in microstructure underlie a world of anisotropic composite materials. We humans most often produce anisotropy of construction components at a macroscopic, structural level. Thus I-beams bend more easily side to side than up and down. But nature tinkers at what we're regarding here as a material level as well, orienting fibers, layering sheets, and so forth. Such variation in local concentrations and orientations of the components of composites determine, as much as any other factor, the range of values of every property mentioned so far for every natural composite.

It's worth noting again that the critical crack lengths of the 1- to 10-micrometer range for typical biological hard stuff set the size scale for the components of composites. That the range corresponds so well to the size scale of living cells must be coincidence—one can make better arguments for the size range of cells, as, for instance, the one in chapter 10. Coincidental as well must be its coincidence with the useful viewing scale for light microscopes. Alloyed metals, whatever their other attraction—and we use most of our metals as alloys—don't make use of the composites' trick. Cellular lengths are not molecular lengths, separated by as many orders of magnitude in lengths as separate cells from us.

Some Particular Materials

A full picture of how organisms use materials needs a more focused look at at least a few of the major things of which we're made. So we'll now do just that, picking four for scrutiny—wood, spider silk, collagenous materials, and mammalian bone. Each of these, of course, comes in a range of kinds whose properties vary considerably. The diversity of even the most homogeneous (probably spider silk) must be comparable to that of all the ferrous metals of human manufacture—even if the full spectrum of materials available to us exceeds what nature can employ.

Cellulose, in the form of wood

While both plants and animals do their syntheses within cells, they differ about as drastically as two cellular systems could. Not only are the cells of higher plants much larger than those of animals—around tenfold in length and thus about a thousandfold in volume, but they play a much more direct mechanical role. They don't just produce structurally important materials; they themselves constitute structurally important components. Some cells remain alive when working as structural elements; others die and leave only their hollow shells as parts of complex foams and fibrous composites. And this cellular system produces a notably hierarchical kind of material, wood, whose functional properties reflect components operating on different scales. Figure 16.11 gives a crude view of wood's organization at the (plant) cellular level.

At the base of the hierarchy are microfibrils of cellulose wrapped helically as layers of what are called secondary cell walls—"secondary" in a developmental but certainly not in a mechanical sense. The pitch of the helices may change if a cell elongates, as many do, and that pitch has an effect on Young's modulus. Wood, technically secondary xylem, represents only one of a host of types of mechanically significant tissue—again "secondary" carries no implication of mechanical afterthought. Trees (except palms and cycads) grow in girth through divisions of a layer of cells just beneath the bark. That cambial layer then produces on its outside various forms of phloem cells, which (among other things) transport photosynthetic products, and which contribute to the bark. The cambial layer also produces, on its inside, various forms of xylem cells, the water-transporting pipes discussed in chapter 5. Only the

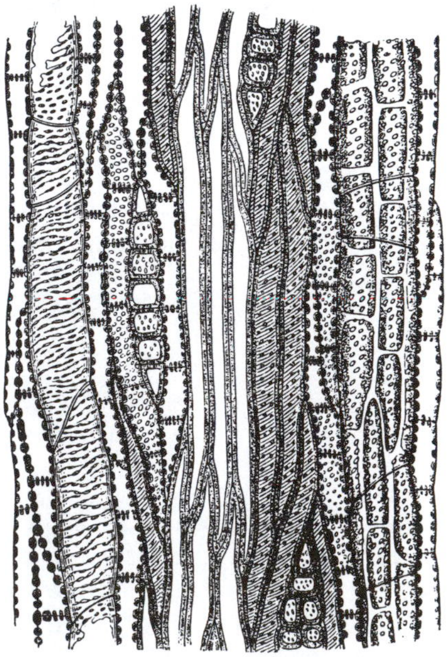

FIGURE 16.11 The complex organization of a typical wood.

outermost of these xylem elements do that job (and are dead by the time they're functional); the successive inner layers are, one is tempted to say, just so much dead wood, the secondary xylem that we saw up into timber.

That construction of lengthwise tubes with relatively modest cross-connections gives wood its spectacular anisotropy. Recall the figures from table 15.2—the tensile strength of pine timber runs 25 times higher with the grain than across it. Every lumber miller, furniture maker, and shipwright pays careful attention to the grain of every piece, knowing how poorly it bears loads across its grain. Even three-layered plywood retains serious anisotropy, bending much more easily across the grain of the outer layers—pay attention when you use unsupported pieces of it and don't go (as the old saw goes) against the grain. At the same time, even some notably hard woods such as that of the oaks split lengthwise with only the mildest of provocation. Crosswise, though, most woods resist fracture well, with the highest work of fracture of any rigid biological material; the orientational difference can be as much as a hundredfold (table 15.7). Not only can we use all kinds of intrusive fasteners such as nails and screws without initiating fracture, but a tree can be injured by a crosswise ax stroke and yet not crack in the next storm. A sawyer must cut almost all the way across the trunk before a healthy tree topples.

Looking at a grove of trees with a mechanical eye, one might guess that wood excels at withstanding compressive loads, so obvious is the static gravitational loading. Looking at wood at the level of hand lens, penknife, or modest microscope, one might guess the opposite from the conspicuous pull-resistant lengthwise fibers. Test data confirm the latter view. As tables 15.2 and 15.3 show, tensile strength exceeds compressive strength about twofold. That may confer functional advantage. Local failure of a foam in compression ought to produce little more than local squashing. Local failure in tension reduces the area left for support and thus increases the stress of the load; furthermore, it may produce a dangerously self-propagating crack. Still, that may be too superficial a view. Trees commonly prestress their wood in tension (should one say "intentionally"?), so initially a compressive load just relieves preexisting tension. A wind bending a tree will increase tension on one side and compression on the other. That, though, may not result in any actual compressive stress, but merely a reduction in tension. Incidentally, the prestressing may be responsible for a lot of irregular shrinkage, bending, and twisting as pieces of wood are cut from tree trunks. On the credit side, the prestressing opens the kerf produced by a saw rather than making it squeeze out against the saw blade.

I've heard of one peculiar situation in which the different consequences of compressive and tensile failure of wood can make big trouble. You find that the top surface of your wooden diving board looks worn and ugly and decide to turn the board over. It has suffered undetected local compressive failure on what was the lower surface but is now the upper. Those failed points now get loaded in tension, a dive drives a crack across, and. . . .

While there's a wealth of information on commercial timber (see, for instance, Bodig and Jayne [1982] and Forest Products Laboratory [1999]), few analyses have focused on functional, fresh, intact pieces of tree, or even freshly cut pieces of wood. The two states differ so markedly that (as mentioned earlier) we most often do a mechanical test to see if a twig or branch remains alive—if it snaps, it's dead; if it bends, it's alive. Most of the difference reflects no more than water content. Less water makes wood stiffer but often makes it stronger. Wainwright et al. (1976), Jeronimidis (1980), and Niklas (1992) provide a lot more information.

Silk, as made by spiders

Certainly the most extraordinary material among those tabulated here is spider silk (that of silkworm moths is substantially less extreme)—it has the greatest tensile strength, astonishing extensibility, and by far the greatest strain energy storage. It should be admitted that the figures are samples of diverse data—silks vary considerably in their properties, quite clearly tuned by natural selection to their particular tasks. We mainly hear about the silk of orb webs, but even that comes in no fewer than seven different kinds, each produced by separate glands and extruded through multiple spinnerets—Vollrath and Knight (2001) give an accessible account. A single aranaeid spider makes frame silk for the main members of its orb, viscid silk for the spiral threads that catch prey, cocoon silk, prey-wrapping silk, and so forth. Other kinds of spiders make other kinds of silks for other tasks.

The task of the silks of an orb web is itself extraordinary (see Denny [1980]). The fibers must be thin to minimize visibility—they're around a micrometer in diameter, far thinner than one of our hairs. They must use a minimum of material—protein is valuable stuff, and a web takes a lot of thread. Gosline, DeMont, and Denny (1986) noted that a 75-milligram spider uses only 180 micrograms of protein to make a web. Even so, it can make only a few webs before running out of material unless it feeds or eats its old webs. If a large bit of prey enters, it should be caught or at least should not destroy the web—extensibility can multiply the effect of strength by what literary folk might call the "tar baby effect"—if you can't stretch something *far* enough to stress it much, what does your own strength matter? And high extensibility may permit a spider to lower itself on a line only barely strong enough to support the animal's weight—impact loading can be largely ignored if the structure just gives upon impact. The poor showing in one property, silk's lack of resilience (fig. 16.12b), in fact may be a virtue. The work of prey against web has to be dissipated somehow—the device is, after all, a catchnet and not a trampoline!

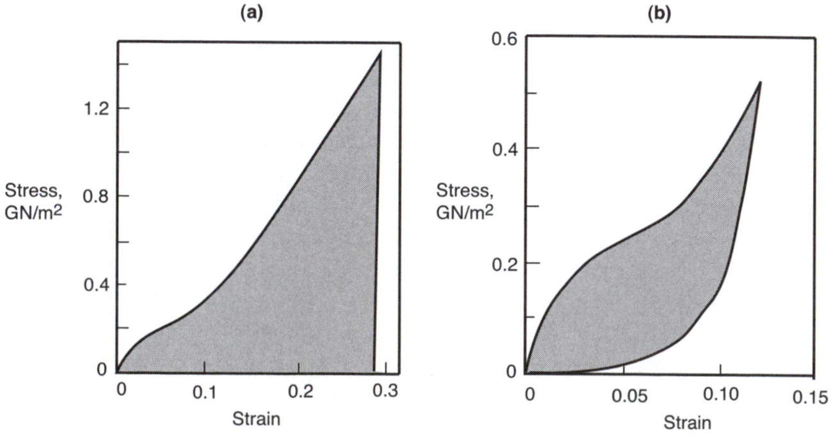

FIGURE 16.12 (a) Strain energy storage (the area under the curve) and (b) loading and unloading curves (whose difference points up the low resilience) of spider silk (Gosline, DeMont, and Denny 1986). Note the different scales on the two graphs.

An extensive literature, both popular and technical, contends that spider silk manages to outperform any materials of human manufacture. That's mainly nonsense. In neither Young's modulus nor tensile strength does it match high-tensile steel or at least one of our polymers, Kevlar. Its extensibility exceeds these, but we don't use these materials on account of their extensibility, and we can make materials of high extensibility (rubber, for instance) when it matters to us. In fact, in only one of the properties we've been considering do spider silks look truly remarkable—strain energy storage, as in figure 16.12a. That matters to a system that's trying to decelerate an impacting mass, and silk does it well—five times better than Kevlar, twenty times better than rubber. Spider silks do have an unusual combination of properties. But I know of no evidence that these can be achieved (if one wants them) only by a sequence-specific heteropolymer of amino acids, something unlikely to lend itself to cheap manufacture.

Collagen and things made of it

Except, perhaps, for the muscle proteins myosin and actin, animals contain no more abundant protein than collagen. By one measurement, collagen makes up no less than 20 percent of the entire body protein of a mouse. Like the silks, it's not a single molecular species but remains a recognizable generic. And like the silks, it serves a variety of functions—an even wider variety by most criteria. Most sponges, the macroscopic animals most distant from us, use it (renamed "spongin") to hook up their calcareous or siliceous spicules into large-scale three-dimensional arrays, structures that will reappear in chapter 18. It plays a crucial role in giving sea anemone mesoglea its viscoelastic character (chapter 17). We use it as a hard element in cartilage and as a matrix component of bone, as noted earlier, in table 16.2. As also mentioned earlier, it forms a critical component of the walls of our blood vessels, unkinking with their expansion to give the J-shaped stress strain curve of figure 1.3. The name, "collagen," refers to its role as a source ("-gen") of glue ("kolla-"); boiling the skin of large mammals, which contains lots of it, gives so-called hide glue.

Only in our tendons, of which it makes up 70–80 percent (of dry weight), do we use anything close to pure collagen. So most of what we know about collagen comes from work on tendon, and the properties of collagen and tendon differ little enough so we can use either as surrogate for the other. In stiffness (Young's modulus) and tensile strength it's undistinguished. In extensibility, it's remarkably low (10 percent) for a material normally used to take tensile loads. In strain energy storage, only spider silk does notably better; and, relative to mass, collagen beats steel. In resilience, only resilin beats its (typical) value of 93 percent. In short, it doesn't stretch far, but it stores a lot of energy in doing so, and it gives back almost all that energy when released.

What about the biological role of collagenous tendons? They usually run between the ends of a muscle and its attachments to bones, so the force (and shortening) of the muscle is transmitted to the bones. Work is stored—in a running person or hopping kangaroo about 40–50 percent of the work done on a leg in landing is recovered as it pushes off again (Alexander 1983). The leg tendons do most of that storage despite their low mass relative to bones, muscles, or the animal as a whole.

Elastic energy storage matters a lot for animals, a point brought up back at the start of the first chapter of the book. Our reciprocating appendages have to be decelerated and then reaccelerated, and the process shouldn't be done wastefully. Furthermore, we can't convert motion into chemical or electrical potential, so we're pretty much stuck with gravitation and elasticity as modes of storage. And gravity is a weak reed, next to useless in aquatic systems and dependent on the earth's unalterable gravitational value.

Using an elastic element of low extensibility may seem odd, but as its primary role, tendon must transmit to the skeleton the distance change as well as the force of shortening. So a conventional rubbery material, as used in insect wing hinges and bivalve shell hinges (or natural rubber), wouldn't do for tendon. These may store energy efficiently, but they do so at impractically great extensions. Fleas use resilin to jump while kangaroos use collagen, concomitantly both the anatomical arrangements and the time course of the process could hardly differ more. And the low resilience of spider silk, whatever silk's other virtues, rules it out as a muscle-to-bone connector or for any kind of battery for recoverable energy storage.

The use of collagen for storage and release of work in the ballistae of the ancients has already been brought up; Gordon (1978) has a good discussion of the matter. Ballistae, one of which is shown in figure 16.13, were probably the ultimate destructive machines of classical civilizations, products refined over centuries by the military research establishments of the Mediterranean powers—their history and associated lore have been collected by Marsden (1969, 1971) along with translations of the ancient artillery manuals that survive. As we've seen, collagen is excellent stuff for the purpose, high both in strain energy storage (especially relative to mass) and resilience, surpassed in the first only by rubber, of which the ancients knew nothing. The Romans tried bronze springs; after all, collagen is perishable and expensive. But the available metals just didn't give anything like the same performance on a weight-for-weight basis—nor do any of the fancier alloys we now make.

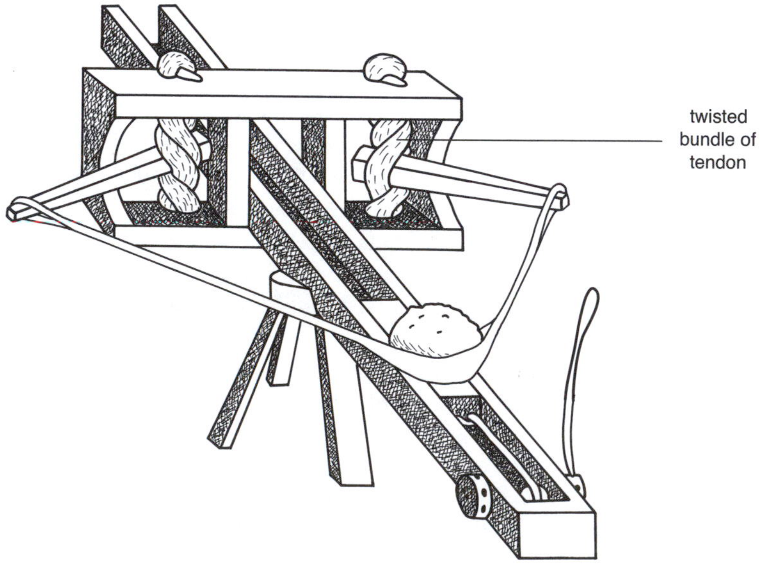

FIGURE 16.13 A ballista. Note how little mass besides that of the projectile has to be be accelerated.

Throwing a 40-kilogram (90-pound) projectile 100 meters, what a big ballista could do, is no small achievement. But a lesser mass would be much less effective against stone or masonry fortifications, and a shorter distance would put the artillery within the range (about 80 meters) of enemy archers. These ballistae were comparable in power with the early cannon, a millennium later. And they were more efficient by far than the larger medieval counterweight trebuchets. I'd guess that they were ultimately limited by their prime movers. Collagen only stores work; unlike gunpowder, it may be reusable, but the artillerymen had to crank in all the work expended in tossing each shot. A ballista couldn't have accommodated more than two artillerymen (although more might have worked in relays). According to my crude calculations (Vogel 2001), based on the power output of human athletes, this Herculean task must have taken some time and restricted the shooting rate to once every 5 or so minutes. (Since ballistae didn't use raw, slippery tendons, we're not too sure of the relevant elastic properties; still, they probably weren't better than those reported in table 15.5.)

One particular use of a collagenous rope gives a sense of the range of performance nature can get from this material—and of some of the peculiarities of applying data from standard analyses. As familiar to anyone who has poked around rocky, wave-swept shores, mussels don't dislodge easily. Each is attached to rocks by twenty to sixty stringy byssus threads of sufficient tenacity to resist extreme hydrodynamic forces. (Denny [1988] provides an especially good view of the origin of these forces.) At first glance, a collagenous material looks inappropriate for the mission. After all, low extensibility means that unless particularly well matched and faced with forces of invariant strength and direction, only some subset of threads will bear the load. Imagine hanging from a group of inextensible ropes each of slightly different length—having more than one will gain you nothing, since they'll break one by one instead of sharing the load.

A byssus thread consists of two mechanically distinct regions, called, for their distance from the shell, proximal and distal. The material of both regions proves to be unusually extensible for collagens, which sounds right and proper. But then, according to Bell and Gosline (1996) things get more complex. The proximal region can be strained to a greater fraction of unloaded length, but it never achieves the breaking strength of the distal region, as you can see from figure 16.14a. So it looks as if their proximal regions take care of distributing the load among the threads. Not so. The distal region of threads happens to be two to four times longer and only half as wide. So a given force will stretch it quite far. Replotting the data as force against extension for a whole thread as a structure, as in figure 16.14b, shows a nice match. Even better, it shows how the distal thread yields (the horizontal portion of its curve) just short of the breaking force—an extension that will permit threads to reorient closer to the direction of the applied force and to share increasing loads among an increasing number of threads.

Collagen makes meat tough, and culinary error can compound the problem. As noted long ago by the great physiologist and raconteur Albert Szent-Györgyi (and alluded to in the last chapter), steak house menus give a table of relative collagen content of different muscles—or, more precisely, its reciprocal, in the form of relative price. In short, muscle with the least collagen costs the most. For tough meat, the solution, in an unusually literal sense, consists of lengthy mild acid hydrolysis, as one might do when making glue

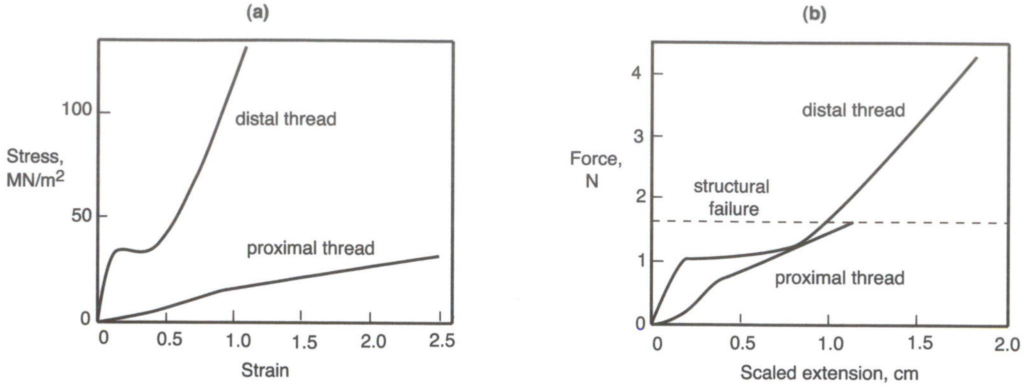

FIGURE 16.14 (a) Stress-strain curves for proximal and distal byssus thread, and (b) a force-extension curve for the structure, both from Bell and Gosline (1996). The shape of the curve for distal byssus may look bizarre to the engineer, but it's similar to those of keratins—what you'd get if you tested one of your hairs. Redrawn from Bell and Gosline (1996) with permission of The Company of Biologists, Ltd.

from skin. Choice of agent provides the acid test of a culture, with contenders including tomato, lemon, and vinegar. In some special cases, very brief treatment suffices to denature the muscle protein without much cross-linking of the collagen—squid mantle, according to one cookbook, should be merely "threatened with heat." I recommend, for rings of mantle, a dip of about 40 seconds into boiling water, followed by cooling and a dip into cocktail sauce.

Bone, mainly mammalian

Bone, although totally different in composition, isn't all that far from wood in mechanical behavior. It's similarly stiff but less anisotropic, which is to say that the direction of a load makes less difference. Also, its stiffness doesn't depends as much on whether it's loaded in tension or compression. The stiffness appears at first glance to have a similar significance—holding a massive body off the substratum. In fact, impact loading matters more than steady gravitational loading. Furthermore, muscles load bones, using the latter as levers, at least as much as does body weight. Whatever the origin of the load, an overly flexible femur would be as unhandy as a flexible tree trunk, rubber ax handle, or soft pry bar, whatever its strength or toughness.

Of all the materials we've considered, bone must have the most complex structure. One can speculate that compression-resisting materials will naturally be more complex than ones that resist tension, something noted in the last chapter when pointing out that nature used simple, noncomposite materials only to resist tension. Of the major compression-resisting materials, bone almost alone belongs to the cellular world of perpetual metabolic activity, continuous resynthesis, and load-dependent remodeling. By contrast, the wood of trees may be wet, but it's dead at functional maturity; arthropods shed their cuticles from time to time; mollusk shell gets augmented only by surface addition.

Bone also comes in several forms, even within a single mammal, indeed, even within a single anatomical bone. At a fairly macroscopic level, it divides into compact and cancellous bone. No surprise—compact bone has within it no spaces other than those occupied by capillaries, bone cells (osteocytes), and a few other items. Cancellous bone consists of struts and plates of roughly millimeter size in what's essentially a marrow-filled foam (fig. 16.15a). (Air fills most of the cancellous bones of birds.) At a finer level, about a hundred times finer than the millimeter level of structure distinguishing cancellous and compact bone, we get to the underlying composite itself, as in figure 16.15b. For the most part, it consists of lamellae about 4 micrometers thick with lengthwise parallel collagen fibrils and mineral crystals. The orientation of the fibrils varies between adjacent lamellae, and the lamellae may be organized into the complex Haversian systems described in histology textbooks. Or the fibrils may be randomly arrayed in what's called "woven bone." Each kind of bone, of course, has different mechanical properties. All this comes from the excellent books by Currey (1984, 2002).

Not only do the forms of bone have diverse properties, but the mechanical diversity of anatomical bones, even within a relatively homogeneous group such as large mammals, needs more than casual mention. Data cited in the last chapter should be considered the grossest of averages. Currey (1999) provides an especially nice set of cases in point, comparing the properties of the antlers of red deer, the femurs of cows, and the tympanic bullae of the ears of fin whales. (The latter are bones forming part of the base of the skull.) Table 16.3 gives his data.

Size and ancestry may differ only slightly, but these bones serve the most varied of roles:

- Antlers, elaborated annually, advertise age, size, and sex; as Currey points out, structures of cardboard would do the trick. They do get used, and sometimes broken, when males bang heads together. Good strength and work of fracture helps prevent such breakage. Low stiffness ought to be a virtue as well, minimizing transmission of impacts to the skull. High extensibility can be viewed either as part of the same coin or as the inevitable concomitant of high strength and low stiffness. Low density (from low mineral content) may give a slight advantage in carrying these huge sex symbols around, reducing both mass and turning moments.

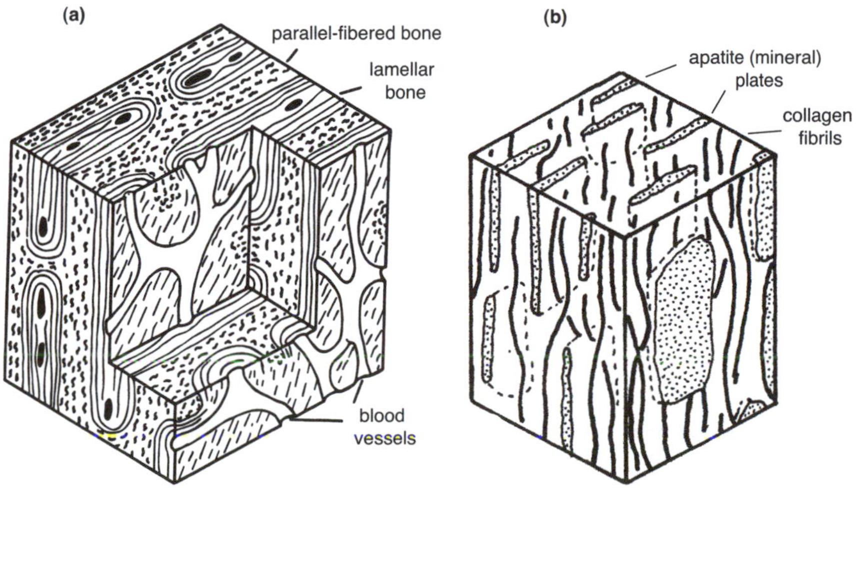

FIGURE 16.15 The complex organization of bone viewed at two levels—(a) a semimicroscopic view, and (b) one level of its underlying composite material.

TABLE 16.3 Five Properties of Three Bones

Property	Deer Antler	Cow Femur	Whale Bulla
Work of fracture (J/m^{-2})	6,190	2,800	20
Tensile strength (MPa)	179	148	33
Young's modulus (GPa)	7.4	26.1	31.3
Extensibility (%)	11.4	0.4	0.2
Density (kg/m^{-3})	1,860	2,060	2,470

Source: Curry 1999.

■ A femur (thigh bone), of course, get stressed when an animal walks, with peak stress about twice body weight (per cross-sectional area), and, when an animal runs, with peak stress about seven times weight. Here it provides an example of the ordinary.

■ The tympanic bulla of a whale has no mechanical function in the ordinary sense, but serves as a passive component of the auditory system. Its function requires that it be stiff; that stiffness comes from a high mineral content and brings with it low extensibility and the great brittleness that's reflected in the minute work of fracture. But what bully goes around fracturing whale bullae?

Bone's stress-strain graph, as in figure 16.16a, looks peculiar. At low stresses and strains, it behaves like a reasonably Hookean elastic, while at high strains, stress maxes out and becomes fairly constant up to the point of fracture itself. So the graph has an inflection, usually called the yield point. Below the yield point, a bone returns to its normal dimensions following removal of a load, and normal loads don't stress bone beyond that point. Beyond the yield point, bone may not fracture, but it deforms plastically, and

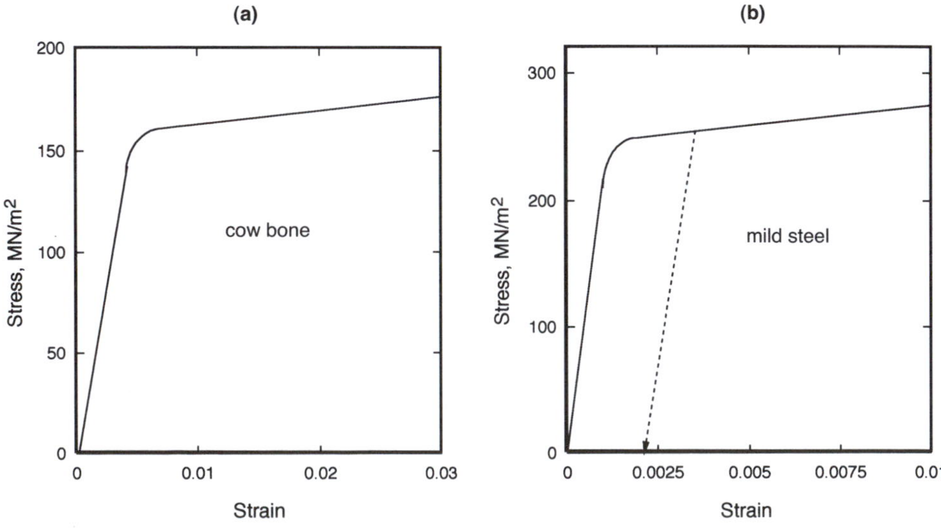

FIGURE 16.16 Stress-strain curves for bone (a) and steel (b), a bit idealized. Bear in mind both the safety feature of the plateau and the completely different bases of that plateau.

loading does damage. Deformation, though, may reduce the load by changing a bone's shape, or it may use up enough time so a transient load (a sudden impact) never takes it to the fracture point. In any case, it greatly increases the area under the stress-strain graph, which (recall the last chapter) increases the energy needed to reach fracture.

Whatever the advantage of plastic yield, the damage must be repaired for bone to reacquire both full stiffness and its normal unloaded shape—but repair comes naturally in such a living tissue. Plastic damage takes the form of microcracks—tiny dislocations between lamellae and at other such places. So bone that has been pushed into the plastic region becomes both misshapen and less stiff. Using it again without a period of rest (days, typically) can cause trouble at loads that would normally be tolerable.

The stress-strain graph for mild steel loaded in tension (fig. 16.16b) looks remarkably similar to that of bone—an elastic region, a yield point, and a plastic region. It's one of the reasons that mild steel proves as safe as we find it for structural purposes—it may stay awkwardly bent without actually breaking. What happens to deform steel plastically, though, differs. Dislocations travel through the material, but things seize up again in the new form. So full stiffness is restored, with only shape having changed. Following removal of the load, a steel specimen drops back down its stress-strain curve on a line parallel (same elastic modulus) to the original. When pressing pieces of steel into new shapes (stamping out curved panels, for instance) we take advantage of the shift of shape without change of properties—just push beyond the yield point to reshape a piece for good.

For ancient technology, at least, biological materials were crucial. Stone was about the only nonbiological thing in common use; and horn, tendon, the various woods and fibers were certainly more important than minor bits of metal and ceramic. But the *biological* role of biomaterials received very little attention even before the great metallic revolution of the nineteenth century and the plastic revolution of the twentieth century. Only within the past few decades has more than an occasional offbeat biologist done more than speculate about the subject. If this renaissance of activity tells us anything, it is that for almost any complex material property, there's probably some species making some special capital out of it. So we get insight into the relationship between material properties and effective applications. What we must not assume is that we're seeing everything possible, the best possible materials used in the most efficient ways. That's unnatural.

CHAPTER

More Complex Materials: Viscoelasticity

> If it is love that makes the world go round, then it is surely mucus and slime which facilitate its translational motion.
>
> **ROGER H. PAIN**

THE BASIC DISTINCTION BETWEEN SOLIDS AND FLUIDS NEEDS REITERATION before we move to materials that make a mockery of it. You may recall from chapter 2 that solids resisted shearing loads, while fluids didn't. And you may recall if you've read chapter 6 that fluids, instead, resisted shearing motion—not shear but rate of shear. As put back then, for solids, at least (we can now say) ordinary Hookean ones,

$$\text{shear stress} = \text{shear modulus} \times \text{shear strain.} \tag{6.1}$$

For fluids, at least (as we did mention then) ordinary Newtonian ones,

$$\text{shear stress} = \text{viscosity} \times \text{shear strain } \textit{rate}. \tag{6.2}$$

We did admit the existence of some funny fluids (at least funny liquids), noting (sanguinely) a bit of non-Newtonian behavior on the part of blood.

A pair of demonstrations might, to use a mechanical metaphor, shatter illusions. First, suspend some cornstarch in water, making a 100 or so milliliters (perhaps half a cup) of loose slurry. Gently stir the suspension, noticing how you get the feeling of a layer of more solid stuff on the bottom even though pouring it into another container reveals no such layer. Then stir much more vigorously. In a viscous liquid you'd have to work harder, and the liquid might move outward from the center. Here something more radical happens. The stuff solidifies, and cracks, attributes solely of solids, appear. Stop stirring and the suspension resumes its liquid character. What you've made is called a "shear thickening" substance, one to which the equations above are laughably inadequate.

Then pour into a container a suspension of some flexible fibrous material or long-chain polymer—the lengths of the suspended elements matters much less than their aspect ratios. Roofing compound (not roofing cement) with fibers will do, as will a concentrated suspension of DNA. Try to stir the suspension with a rotating vertical cylinder such as a dowel turning in the chuck of a handheld drill. Instead of moving radially, as would some ordinary liquid, the suspension creeps up the stirrer. You may even notice that you're "unstirring" rather than stirring, concentrating a higher viscosity fraction near the center. You're seeing what's called "strangulated flow" or the Weissenberg effect (Lodge 1960). Yes, you can stir up roofing compound, but you have to do something else such as paddling it back and forth.

These demonstrations (not models, but the actual thing, of course) may be a bit far from biological reality, as well as being closer to fluid than solid domains. So try something close to home, if a bit less dramatic. Carefully measure your height when you first arise in the morning and then repeat the measurement after an active day. You're shorter at night! The culprits, mainly, are your intervertebral disks of cartilage. Under constant compressive load they "creep," in this case compressing. Since the compression is noticeable relative to your overall height, it's really substantial relative to the thickness of the discs. There's a long-term creep as well, described by that closet biomechanic, T. S. Eliot: "I grow old . . . I grow old . . . I shall wear the bottoms of my trousers rolled."

Creep can be a subtly variable business. When looking at the resistance of tree trunks to bending and twisting (chapter 18; Vogel 1995b) I found that a given load gave a given bend, but a given load gave a slightly ambiguous amount of twist—trunks were creepy in torsion (so I had to standardize test durations) but not in bending. One woody vine, wisteria, though, crept in both twisting and bending. I suspect it elongates under steady or repeated loading, thereby reducing its possible load when a tree sways in the wind.

So we have to consider a world of time-dependent material properties, a world in which the values of the mechanical properties depend on the time course over which loads are applied. Put another way, many real materials that we consider solids have some of that characteristic fluid property, viscosity, and so care how *fast* as well as how *far* they're stressed. Many care, as well, about their own previous histories of loading and responses. It's the world of "viscoelastic" materials, ones that resist both shear and rate of shear, and that do so in astonishingly diverse ways, the world of things between a nice viscous liquid like glycerin or corn syrup and a nice soft solid such as a Jello salad or salmon mousse. The parent subject, at least for the more liquid-like cases, goes by the name "rheology" or, sometimes, "biorheology."

As we'll see, among living things the lower the Young's modulus the more viscoelastic the material is likely to be, with the jellies and mucuses showing the most complexly time-dependent behavior. Since nature makes a lot of structures out of relatively soft materials, viscoelasticity takes on considerable importance. This single chapter, which does little more than introduce the players, has two main intentions, in a sense considering constraints and opportunities. It provides a note of caution lest one get caught up in some sticky stuff (such as taffy, a fine example of a viscoelastic material). And it gives some sense of where in organisms viscoelasticity might have a positive role.

Models

Several now classical models ought to be useful at this point. (In this whole discussion, I'm largely following Wainwright et al. [1976] and Vincent [1990].) In each, a fixed load ("stress") is suddenly applied, maintained for a fixed time, and then removed. Meanwhile, length ("strain") is monitored. In addition, a fixed stretch ("strain") is suddenly applied, maintained for a fixed time, and then removed. At the same time, the responding stress is monitored. Bear in mind the severe idealizations of the models, especially their assumptions of perfectly Newtonian fluid elements and massless Hookean solid elements.

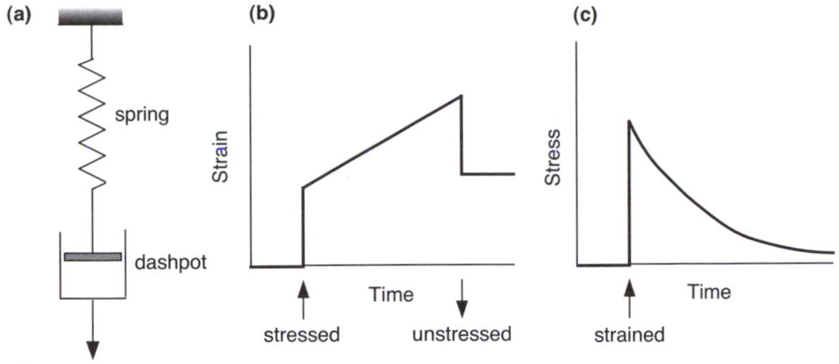

FIGURE 17.1 (a) A Maxwell model, labeling spring and dashpot icons. (b) Strain changes when a fixed load (stress) is first imposed and then removed. (c) Stress changes when a fixed strain (length increase) is imposed.

First, a so-called Maxwell model (fig. 17.1a), consisting of a spring (elastic element) and dashpot (viscous element) serially arranged. Consider what happens when a force such as a hanging weight is suddenly applied, maintained for a while, and then suddenly removed. The load initially stretches the spring in standard Hookean fashion, and the spring maintains its new length since the load remains constant. The same load hits the dashpot at the same time, causing it to lengthen steadily in standard Newtonian fashion. The graph of figure 17.1b shows how the load moves, merely calling the motion "strain" and making downward into upward. Removal of the load lets the spring return to its original length; the dashpot, though, remains elongated, so the system has a new rest length. It sounds at least reminiscent of how your intervertebral disks behave during the day. If, conversely, the model is suddenly strained by a fixed amount, the spring does its best to resist, initially applying lots of force, since the dashpot hasn't as yet lengthened. As the dashpot lengthens, the spring shortens until it exerts no force at all, the force decaying asymptotically as it approaches its unloaded length. The graph of figure 17.1c shows that rise and fall in force ("stress"). Bread dough behaves at least roughly like the Maxwell model. So does, I've heard, a 10 percent solution of hog gastric mucin in water; but that's not exactly a household item. Compression tests on asphalt give results close to the Maxwell model (Biggs 1965).

Second, a so-called Voigt or Kelvin model (fig. 17.2), consisting of a spring and dashpot in parallel. Now a fixed stress initially tries and fails to stretch the system, since the dashpot doesn't do anything in a dash. Gradually, though, the dashpot lengthens, its motion getting ever slower as the simultane-

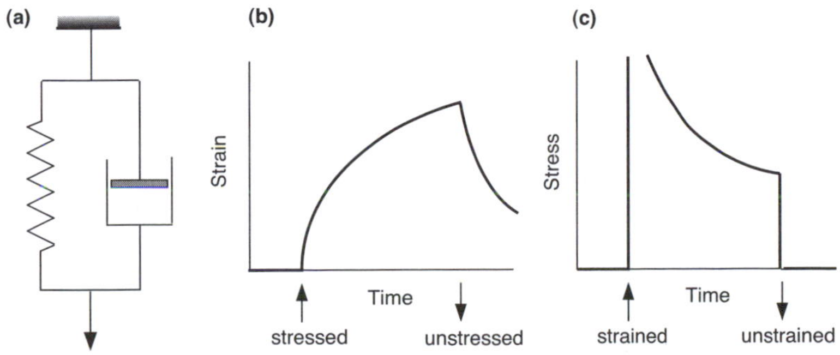

FIGURE 17.2 (a) A Voigt or Kelvin model. (b) Strain changes when a fixed load (stress) is first imposed and then removed. (c) Stress changes when a fixed strain (length increase) is first imposed and then removed.

ously lengthening spring springs into action. Eventually (if the load is maintained), the system will achieve a new, longer, equilibrium length determined by the stiffness of the spring and unaffected by the dashpot. Removing the load makes the system shorten, at first rapidly on account of the stretched spring, and then ever more slowly as the spring exerts less and less restoring force. And now a fixed strain can't be applied instantaneously—the dashpot won't permit it. Or, at least, the stress momentarily hits the ceiling and drops back to a value determined by the spring alone. We underpin our automobiles with springs and dashpots (shock absorbers) in parallel, essentially Voigt/Kelvin devices. So cars appear soft-sprung and compliant for slowly applied loads (curves, changes in contents) and hard-sprung and resistant for sudden loads (potholes). Egg white (albumen) apparently fits the model fairly well, as do the results from tension tests on rayon (Biggs 1965).

One can look at more complex models; indeed one must if one wants to find a fit with more than the occasional biological viscoelastic. A "standard linear solid," for instance, consists of a second spring in parallel with a Maxwell model. These idealized physical models lend themselves to simple mathematical formulations. But ultimately they just abstract some possible behaviors, inviting (but not providing) molecular explanations and inviting (perhaps deceptively) identification with the behavior of specific materials. As a step toward reality, we turn to the behavior of a generalized polymeric material.

The "Master Curve"—Relax with a Creep

How to test a material? We might perform practical versions of the two thought-tests already mentioned. One consists of applying a constant tensile stress and watching how the material elongates, or strains, as a function of time. That's usually called a "creep test." The other consists of stretching a sample by a fixed amount relatively quickly, and watching how the stress exerted by the material on its mounting changes over time. That's usually called a "stress-relaxation test." Either (or the two together) can tell a lot about both how the material behaves and how it can be used by organisms.

Creep testing for any material that extends more than a small fraction of its rest length—which means all soft biological materials—presents a slight problem. Stress isn't force, and true stress for a sample of anything decently extensible will change as a given force stretches it. That much should have been made clear in chapter 15. How, then, can we apply a constant stress? Hang a weight from a specimen and the stress will gradually increase as it strains; in practice that tends to give little action followed by sudden extension to the breaking point. Figure 17.3a shows about the simplest possible creep tester. It resulted from a discussion I once had with an undergraduate student, who described (I had never seen one) an exercise machine that made you lift a weight suspended from a spiral pulley (Vogel and Papanicolaou 1982).

Say one hooks a piece of polymeric material to the device. It gradually elongates, so strain increases with time. Since stress is constant, moduli can be calculated (putting stress over strain) at different times; the whole thing can be plotted as elastic modulus against time. (Sometimes the reciprocal of the modulus, *creep compliance*, is used as the dependent variable instead.) For most such materials, the initial modulus is high; it then drops and levels, as in figure 17.3b. One can, if convenient, speak of an unrelaxed modulus at the

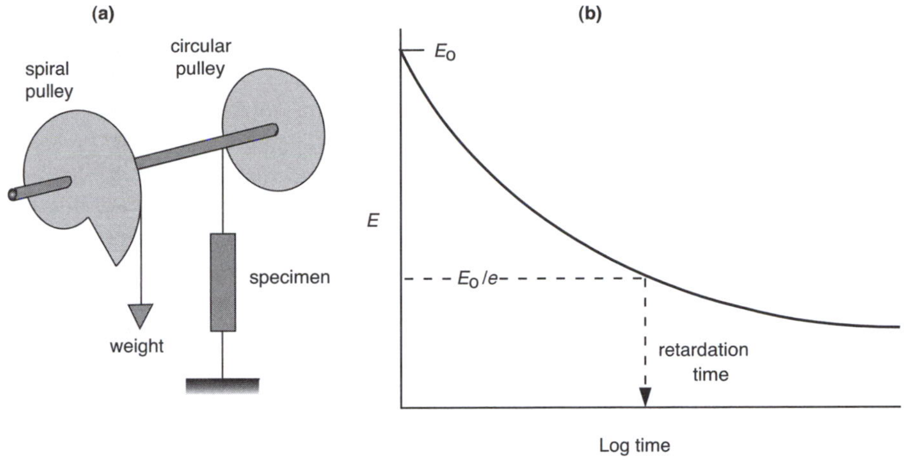

FIGURE 17.3 (a) A creep tester, using a spiral pulley so the weight exerts an ever-decreasing force as the specimen lengthens, thus keeping imposed stress constant. (b) Young's modulus of elasticity versus time, the result (after adjusting for specimen dimensions) of a creep test.

start, and a relaxed modulus further along. An index to how fast things change is the *retardation time*, the time necessary for the modulus to drop from its initial value to $1/e$ (or 0.368) times its initial value. Of course the use of e implies that the process can be treated as an ordinary exponential decay, to which it's often a pretty rough fit. Still, using any other single number to describe a decay process would be as bad.

A stress-relaxation test has its own practical problems for biological materials. In particular, gripping becomes critical, since any slippage over time corrupts the results in a way that's harder to deal with than slippage in a creep test (Shadwick [1992] gives a good sense of the realities here). Its outcome can be treated similarly, generating a modulus (usually called the *relaxation modulus*) that decreases with time much like the one determined by a creep test. And the curve of its decrease can be described (again ignoring a lot of reality) by an analogous variable, the *relaxation time*, τ, the time needed for the stress to fall to $1/e$ times its initial value. It turns out that the relaxation time represents the ratio of viscosity, μ, to Young's modulus of elasticity, E:

$$\tau = \frac{\mu}{E}. \tag{17.1}$$

And yes, it does indeed have dimensions of time. Note, parenthetically, that a viscous liquid, with its lack of resistance to stretch ($E = 0$), can't be subjected to a proper stress-relaxation test. A simple elastic solid will have a relaxation time of infinity, or a zero rate of relaxation, since it doesn't flow and $\mu = \infty$.

Figure 17.3b needs more than a glance. For one thing, it effectively plots stiffness against time, not simply length or strain or even stress. Stress, of course, is constant, and as strain increases, stiffness decreases—the modulus is, after all, stress over strain. More peculiar is the abscissa, not time but the logarithm of time. So one gets a view of how the material behaves where times run from perhaps a hundredth of a second ($\log t = -2$) to perhaps a day (around $\log t = 5$). The funny time scale presents some challenges to the experimentalist, and often data from several separate runs have to be combined. Polymer chemists have a trick for compressing the practical time scale, one, though that can't be casually adopted for work on biological polymers. Working at higher temperatures has the same effect as working at longer times, with a standard rule playing a role analogous to the way fluid mechanists can shift scales for model work by maintaining the same Reynolds number (chapter 7).

So the curve can be lengthened, sometimes out to log $t = 8$, or 3 years, as in the "master curve" of figure 17.4. The usual practice matches the logarithmic x-axis with a comparable y-axis to accommodate the now wider modulus range. The initial plateau commonly is spoken of as the "glassy" state; the material then makes a "glass transition" to a "leathery" state in the second to hour or day range; it then goes into a "rubbery" state. Following that it may either level out at that rubbery stiffness or continue to soften, becoming ever more liquid.

Dynamic Testing

One can view another important aspect of the complex world of viscoelastic materials by looking at what they do under cyclic loading. A sinusoidally varying strain is the easiest cycle to impose—one just has to connect a long arm (a "pitman arm") between the specimen and a revolving wheel that cranks it. A force-measuring device either in the arm or on the opposite attachment then tracks stress, as in figure 17.5a. For a material that's perfectly Hookean (fig. 17.5b), stress will track strain, giving a sinusoidally varying response for both, with no phase shift. Plotted as a stress-strain diagram to eliminate time the data appear as a straight line—data points march repeatedly up and down the line. Or, if the material is non-Hookean, but still perfectly elastic, the line will curve. Either way, staying on the same line means that the system gives back elastically all the energy that has been put in. For a viscoelastic material (fig. 17.5c), responsive stress will precede imposed strain—stretch something and it fights back hard at the start and then less hard as the strain persists. So the stress-strain plot develops a loop, with data points marching clockwise around. Energy has been lost, and the energy per unit volume lost per cycle can be read from the area within the loop.

That loop ought to be familiar on at least two accounts. First, think back to Maxwell and Voigt/Kelvin models—in both, stress change in effect antici-

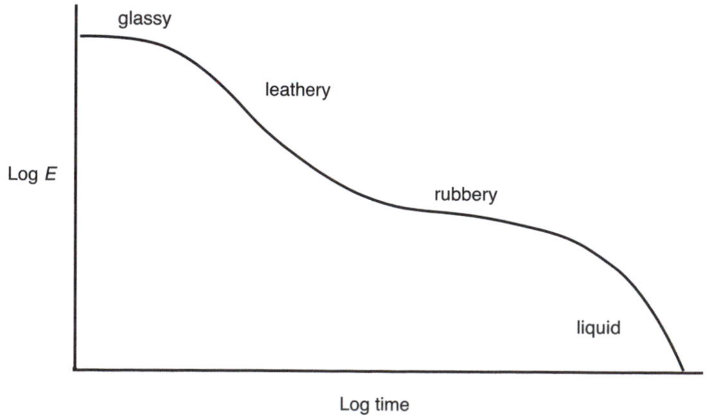

FIGURE 17.4 Master curve, a plot of log modulus versus log time, with the different states through which the material passes.

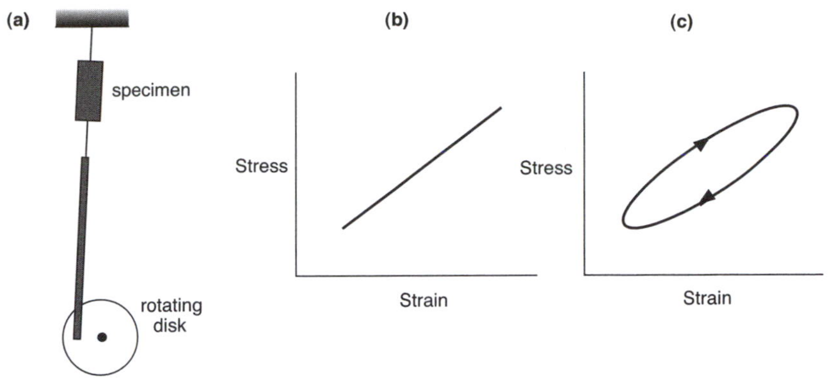

FIGURE 17.5 (a) Push-pull dynamic testing apparatus. (b) Stress-strain "loop" for a Hookean material. (c) Stress-strain loop for a viscoelastic material.

pates strain change. Then think back further, to chapter 15. We got the values of resilience from just such loops. Stress change has to precede strain change if the system is to absorb and then dissipate energy. Looking forward, instead, we'll encounter counterclockwise looping, so-called work loops, when we look at muscle in chapter 22.

Putting what the loop says more formally, we can define the strain we apply as

$$\varepsilon = \varepsilon_o \sin \omega t, \tag{17.2}$$

where ε_o is the maximum strain in each cycle, ω is the angular velocity, and t is the specific point in time. The stress that responds can then be described as

$$\sigma = \sigma_o \sin (\omega t + \delta), \tag{17.3}$$

where σ_o is the maximum stress in each cycle and δ is the phase angle between stress and strain.

A bit of manipulation produces a pair of moduli:

$$E' = E_o \cos \delta, \tag{17.4}$$

and

$$E'' = E_o \sin \delta, \tag{17.5}$$

where E' is the *storage modulus* and provides a measure of the energy stored elastically in each cycle, and E'' is the *loss modulus* and gives the energy dissipated in each cycle.

One more variable. The tangent of the phase angle is commonly used as a measure of the lossiness of the system, the viscosity relative to the elasticity of the system:

$$\tan \delta = E''/E'. \tag{17.6}$$

Sometimes it's called the *damping*. (By the way, the parent infinitive form is "to damp" and not "to dampen.") Bear in mind that all these properties—storage modulus, loss modulus, and damping—depend on the frequency (and to some extent amplitude) of the cyclic load. That means one ought not draw conclusions about function from tests done under the wrong conditions.

Real Material—The Uses of Viscoelasticity

All real materials are to some extent viscoelastic. But for some systems, that viscoelasticity is no more than a curiosity or trivial nuisance while for others it holds major functional significance. Biological systems in particular take advantage of it, and their diverse applications capitalize on the multidimensional character that so complicates the present account. Given the extent of both material and biological diversity, perhaps the best tactic here consists of describing a series of illustrative cases.

Beating on bone

We've already noted that bone, stressed, reaches a yield point, straining further with little or no increase in resistance to stretch. Unloaded out in that plastic region, it initially returns to dimensions somewhat closer to its loaded dimensions than was its original state. Depending on the severity of the load, it may or may not ever regain its original dimensions—some residual strain may persist. But that's only one aspect of the time-dependent behavior of a material that looks on the face of things like an ordinary stiff material. We animals load our bones both steadily and impulsively. Something stiff that's loaded impulsively may at least briefly develop very high levels of stress—in a sense, that's the price of being stiff. So a bit of time-dependent behavior can help matters. The tensile strength of bone, in fact, increases with strain rate, that is, with increasing impulsiveness of the loading regime. And it does so well within the rates that animals normally encounter (Currey 2002).

Still, even if one can say that even bone is viscoelastic, one has to recognize that it isn't severely viscoelastic; Vincent (1990) cited figures for tan δ for its osteons and lamellae no higher than 0.01, indicating a storage modulus about a hundred times higher than the loss modulus—a relatively undamped system.

Sea anemones and their mesoglea

Consider a solid material with properties and role about as distant from bone as a supportive, compression-resisting material can be. The body wall of a sea anemone—which can be quite substantial in size—consists of inner and outer surface layers separated by the thick mesoglea. One doesn't go far wrong viewing the system as a tall can of seawater whose walls are mostly made of jelly. That jellylike stuff, mesoglea, was the least stiff stuff in table 15.1. Fortunately for the biomechanical trade, it's far less brittle than jelly, so the old definition of an impossible mechanical task, nailing jelly to a tree, doesn't apply. (Recall how easily jelly fractures—it really is brittle.) Mesoglea has about the toughness of aorta or insect cuticle (table 15.7), hard as it may be to imagine something (or somebody, for that matter) so tough that's so soft.

The toughness comes from collagen fibrils embedded in the soft matrix—it's a proper composite (Gosline 1971).

A typical anemone has a rare facility for changing shape, ranging from a low barrel to a tall cylinder with a few flourishes in between, over times ranging from seconds to hours, as in figure 17.6. Obviously, its mesogleal stuffing must participate in the process. Muscle drives some of the shape changes, in particular the sudden expulsion of water in the central cavity from its single apical opening. But tracts of cilia drive other changes, such as reinflation by pumping water back in. You may recall, if you've read chapter 9, that ciliary pumps produce exceedingly low pressures, and here we're asking that they pump up creatures that may reach half a meter in height and live in moving water.

Alexander (1962) showed the crucial role of mesogleal viscoelasticity for anemones. In creep tests on samples, strain increased from an initial value of about 0.2 to a final level ten times that, achieved after around 10 hours. That means the mesoglea has a lot of viscosity relative to its elasticity—it's hard to make it do anything fast but fairly easy to make it change shape slowly. It has a retardation time (recalculated by Biggs; see Vincent [1990]) of a little under an hour. How nice! The pulsating or reversing flows of waves passing above won't sweep it about very much, but after it has hunkered down, the low-pressure ciliary pump will be adequate to pump it back up again, albeit slowly. It can stand up to a single wave but deflect in a tidal current that imposes the same drag. Furthermore, the anemone's body wall can resist the stresses of its own short-term muscle contractions, so it can bend or straighten without getting an aneurysm wherever its muscles aren't active (Koehl 1977a).

So mesoglea is a lot more viscoelastic than bone, and the viscoelasticity isn't just an accident of soft-tissue construction but plays a large role in its life. To put a number on it, taking the most commonly investigated species, *Metridium senile*, tan δ runs between about 0.3 and 0.4 (Koehl 1977a)—a lot more viscous dissipation, a lot less storage of any imposed load than in bone.

FIGURE 17.6 The shapes that a sea anemone, *Metridium*, can assume. Don't try this at home.

Why slugs are slimy

A terrestrial slug doesn't so much crawl as slide across its world on its belly (technically its foot), leaving behind a trail of slime—its pedal mucus, consisting of water with some polysaccharide and protein. It's neither a cheap nor a rapid way to go, a slug being the very paradigm of sluggishness. It is (or, strictly, was) a mysterious business. Slugs use neither cilia (as push the mucus lining our respiratory passages) nor peristaltic motion (as do, for instance, earthworms). How they propel themselves turns out to depend on that mucus, in particular on viscoelastic behavior more complex than anything we've seen so far.

The key can be seen if, with just the right lighting, one watches from beneath while a slug walks across a glass surface. Some kind of waves pass lengthwise; they're about a millimeter or so in wavelength and travel at a couple of millimeters per second (fig. 17.7). It certainly looks as if the slug gets around by pushing back on its mucus, and that's indeed what it does. But consider—that will work for one "step" only. Any operating system that pushes needs to recover if it's to push again. Imagine that the resistance of the mucus varied in proportion to the speed of pushing. No recovery would be possible since, with recovery stroke necessarily the same length as power stroke, the product of force and time, the impulse (equivalent to momentum, mv, as in chapter 2), of the two strokes will be equal and opposite. That's the same problem encountered by reciprocating locomotion at low Reynolds numbers (chapter 11). So—'snot possible—the slug must be doing something more subtle as it shears its mucus.

For small shear deformations, the mucus may be viscoelastic, but it's not radically so. A typical tan δ is around 0.07, well below that of mesoglea, indicating better elastic storage and lower viscous loss. It gives a linear and reversible stress-strain line at strains around 0.1 but a more lossy (plastic) recovery at strains around 4. The shear modulus of a big banana slug (*Ariolimax columbianus*, which looks like a yellow and black overripe banana) runs between about 100 and 300 pascals. That may be over three orders of magnitude lower than anything in table 15.8 but still resistant enough to shear to be pushed against, at least gently.

A harder push, a strain sufficient to raise the stress to 400 or so pascals, provides a transforming experience; figure 17.8 shows what happens when mucus is strained in increments. At strains above 5 or 6 (higher values at higher strain rates), the mucus abruptly yields, transforming in a flash into a gooey liquid with a viscosity of 3–5 pascal-seconds—three thousand to five thousand times the viscosity of pure water. Stress falls abruptly, dropping by half, and the stress-strain curve goes flat and horizontal—stress now depends only on strain rate, as befits a fluid. Thus, the mucus can be pushed backward

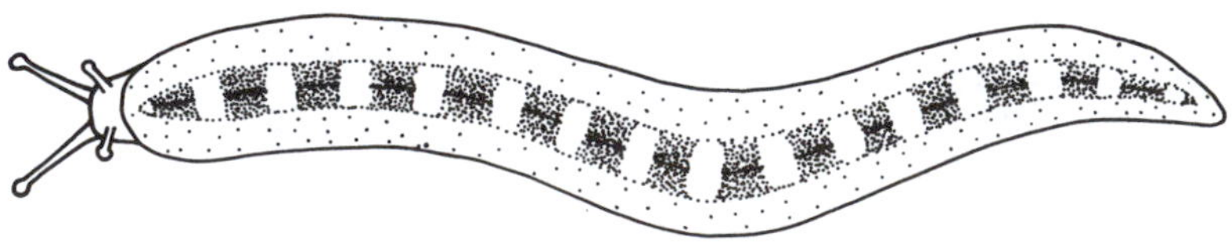

FIGURE 17.7 The waves beneath a slithering slug.

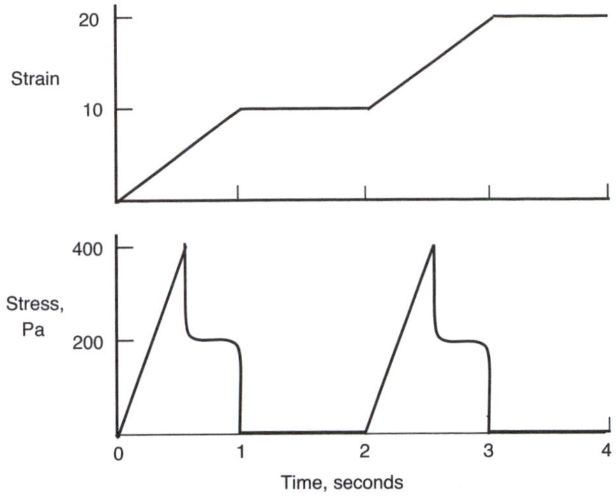

FIGURE 17.8 Experimentally imposed strain (above) and the responding stress (below) for the mucus of a slug.

upon, propelling the slug forward, but when nudged just beyond that yield stress it turns liquid. Then a bit of slug can slide forward across it and be ready to push backward again. Meanwhile, and critically, the mucus has to recover, to "heal" back into a viscoelastic solid, which it obligingly does forthwith. In life, a "step" is about a millimeter in length and the mucus layer is about 10 micrometers thick, so the strain gets up to around 100 during the recovery phase of each pedal wave. Beneath a given slug at a given time, all stages of the operation are taking place, with mucus in different states under different points along the animal. All this from Denny and Gosline (1980) and Denny (1984).

Et al.

Viscoelastic materials abound in nature, and one can go on with such stories, some in which we can make confident connection between complex material behavior and biological role, and some in which the connection, if any, remains unknown or uncertain. For instance, mesoglea stuffs jellyfish, as it does sea anemones (who share their phylum). *Jellyfish jelly* plays a somewhat different role, though, storing mechanical energy between pulsating contractions of the surrounding muscles. Concomitantly, it has decent resilience (58 percent) and a tan δ around 0.2 rather than the 0.3–0.4 for anemone mesoglea (DeMont and Gosline 1988a). *Synovial fluid*, the lubricant in our joints, behaves more like a fluid than a solid, with a tan δ well over 3. So it flows nicely—at least at low rates of shear strain. As the rate (or, in dynamic tests, frequency) increases, solidity increasingly exerts itself, and tan δ drops to about 0.1, indicating a predominantly elastic material (Gibbs et al. 1968). High shear strain, of course, is what happens when interbone compression thins the layer of fluid and motion of one bone relative to another bone shears it. So the stuff stays properly positioned between the articulating bony surfaces during activity, capitalizing on behavior almost exactly opposite that of slug slime.

In fact, the biological use of complex materials transcends the normal world of viscoelasticity and shear-dependent solid-fluid transitions. A totally different kind of yield underlies the operation of the pit of a larval ant lion. An unbound pile of solid particles will form a slope no steeper than some "angle of repose," a phenomenon of great interest to both highway engineers and designers of automatic parts for feeding and packaging equipment. Lucas (1982) showed that the ant lion carefully maintains the greatest possible slope. The addition of a wandering ant then precipitates a miniature avalanche, with a large and efficient set of jaws barely buried at the bottom of the pit.

CHAPTER

18

Simple Structures: Beams, Columns, Shells

Q. What's the difference between girder and joist?
A. One wrote *Faust*, the other *Ulysses*.

Heard on BBC program, *My Word*

IN EXPLORING THE GAMUT OF MATERIALS FROM THE ENAMEL OF TEETH to the jelly of jellyfish we found that the values of their properties "made sense" in terms of their uses by organisms. To get such nice coincidence between materials and applications, though, we limited our view to the simplest modes of loading—tension and, almost parenthetically, crushing compression. These loads share a unique characteristic—as stresses they elicit strains that, while depending on the material at hand, are almost wholly indifferent to how that material is arranged. Only the cross-sectional area of the material matters, and that gets taken care of by defining stress as force per unit area. Thus, we can specify, for instance, a value of the stiffness, E, that characterizes a particular material and that shifts only slightly (relative to the range of Es we meet!) with the degree of strain.

By contrast, responses to other modes of loading—compression beyond simple crushing, bending, shear, torsion—depend on something else in addition to the properties of the material. That other factor is the *arrangement* of the material, a topic touched previously only when we considered composites. The need for an additional factor is, I think, easiest to see if we consider what happens when we bend a reasonably stiff object. Bend two cylindrical rods of the same length and amount of material, one solid and the other in the form of a hollow (but not excessively thin-walled) tube. The solid one will bend more easily than the hollow one. Their tensile strengths and stiffnesses are the same—there's no percentage in making ropes hollow—but the tube better resists bending (as well as compression and torsion, for that matter). Obviously we need something beyond the material properties of the last two chapters to describe this effect of *shape*. So—yet more properties; these, though, properties of structures rather than merely of materials.

Flexural Stiffness

Resistance to bending, as we just did or imagined doing with the two rods, can be described in terms of something called "flexural stiffness," the analog for bending of Young's modulus of elasticity for tensile loads. The variable turns out to be the simple product of two factors, one reflecting the material present and the other its arrangement. The first is simply Young's modulus

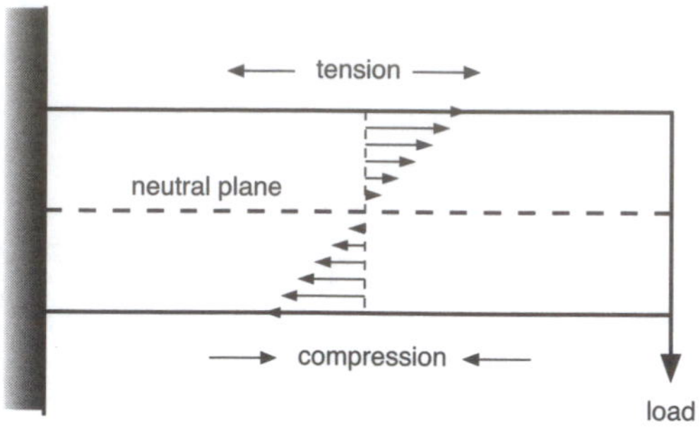

FIGURE 18.1 Bending something subjects one side to compressive stress and the other to tensile stress. Somewhere between the sides is a neutral plane that feels neither; stress increases linearly with distance to either side of that plane. If bending can be in any direction, we speak of a neutral axis instead of a plane.

itself, E. The second is given the name "second moment of area,"* which amounts to an explicit definition for those familiar with such matters; it's designated I. So flexural stiffness (a.k.a. "bending modulus" or "flexural rigidity") is just EI; it has no common symbol of its own, and only the I needs further explanation.

Second moment of area and neutral plane

If you were to bend an infinitely thin rod or sheet nothing would be either compressed or extended—the process would consist of unadulterated bending. But as soon as an object has some finite thickness, then bending does other things. It stretches the surface on the outside of the curve and compresses the surface on the inside. Still, if one side is stretched and the other squeezed, some plane running down the middle must be completely unstressed in either tension or compression. It's called the "neutral plane," and, as ought to be obvious from Figure 18.1, the stress on an object's material increases with distance to either side of that plane. Not that the neutral plane is truly unstressed—it's quite clearly sheared by the opposing stresses to either side. Thus, two apposed strips of wood bend less readily if glued together than if unattached—the glue joint prevents shearing at the line of contact. Still, material near the neutral plane can contribute little to the object's antipathy toward bends. In short, the further a bit of material lies from the neutral plane the more it helps resist bending—moving material away from the neutral plane makes an object stiffer. That's why a tube is stiffer than a rod where both have the same cross-sectional area of material, and that's why the joist under a floor is made higher than it is wide rather than square in cross section.

In fact, material contributes to the I of flexural stiffness in more than direct proportion to its distance from the neutral plane, which is why we talk about second moment rather than first. The tube's stiffness exceeds that of the rod disproportionately, because two factors combine in its favor. The matter is perhaps best explained using a rather contrived model. Consider a structure, a kind of beam, that protrudes outward from the wall to which it is attached—two easily bent (but nonsagging) struts, each of which can change length by

* The second moment of area is sometimes called the "moment of inertia," a name also used for a variable that includes mass. We don't need that potential for confusion.

stretching or compressing a spring (fig. 18.2). The springs are linearly elastic—they obey Hooke's law in either tension or compression—and each strut is the same distance, d, from the neutral plane. The free end of the beam is depressed a distance y by a weight F_w that simultaneously generates a force F_s stretching the top spring a distance Δl. We need only worry only about the top spring since the bottom one just mirrors in compression its stretch.

Now the stretch of the top spring will be proportional to the product of how far it is from the neutral plane and the deflection of the whole beam:

$$\Delta l \propto dy.$$

By Hooke's law, the stretch of the spring is proportional to the force exerted on it, so

$$F_s \propto \Delta l.$$

Combining these, we see that the deflection of the whole beam will be proportional to the force on the spring divided by the distance from the neutral plane:

$$y \propto F_s/d.$$

At the same time, by the notion of moments of forces, as introduced in chapter 2, the force on the spring will be proportional to the weight on the end of the beam divided by that same distance from the neutral plane—think of the neutral plane as the pivot and the vertical end-piece of the beam as the board of a seesaw:

$$F_s \propto F_w/d.$$

What we're looking for is something proportional to how much force it takes to get a given deflection of the beam—but that's now easily obtained by few substitutions from these relationships:

$$F_w/y \propto d^2.$$

In that exponent of 2 lies the disproportionate stiffness of the thicker beam. Put in verbal form, the material of an object being bent resists stress in

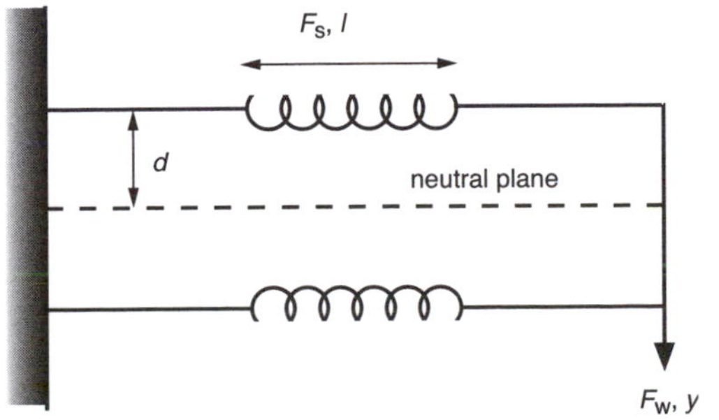

FIGURE 18.2 A beam composed of two inextensible but bendable struts, each with an inserted spring. The right-angle attachments of either end of the struts are assumed fixed. Thus, a weight at the end bends the beam by stretching the top spring and compressing the bottom one.

proportion to its distance from the neutral plane. At the same time, the applied force causes an effective force on the material in inverse proportion to that distance from the neutral plane—the more distant material both resists stress more effectively and feels less stress from a given weight.

But F_w/y refers to the force on a particular element of our particular example. To get the overall second moment of area, I, itself, we have to take each little element of area of a cross section of the object being bent and multiply its area by the *square* of the distance, d, from the neutral plane—we have to integrate outward from the neutral plane. But just where is the neutral plane? Imagine a cross section of the object transformed into a board of uniform thickness that preserves the shape of that cross section—the neutral plane is the hinge-line about which it balances. If bending can occur in any direction, than the neutral plane is replaced by a line running the length of that plane, the "neutral axis." For symmetrical forms such as cylindrical rods or tubes, the neutral axis is typically the center line. (At least that's the case if tensile and compressive elastic moduli, the Es, match, an additional condition beyond the assumption of Hookean behavior.) For an object of some less simple cross section this neutral axis corresponds to the "centroid" of the section. The latter is similar to the center of gravity (chapter 2)—the centroid is the center of gravity of a cutout of the section using a material of uniform thickness.

How does this second moment of area vary with the cross-sectional size and geometry of an object? Consider a set of cylinders, including both hollow and solid ones, of a given kind of material, in three simple situations. First, assume a given amount of material per unit length of cylinder, the situation with which we began. The second moment of area, I (and hence EI), will be proportional (very nearly) to the *square of the average radius*—with increasing radius the material works with disproportionately increasing effectiveness. Second, assume a tube with a given thickness of wall. Now the cross-sectional area of material will increase in proportion to the average radius, so EI will be proportional not to the square but to the *cube of the radius*. Third, assume isometry, where an unchanging shape requires that wall thickness increase in proportion to the radius. Wall area will now be proportional to the square of the average radius, and EI will be proportional to the *fourth power of the radius*. Flexural stiffness thus increases even more drastically than does weight! Unfortunately for big sorts such as ourselves, this does not make creatures relatively sturdier as they get bigger—bones get longer as well as fatter with increases in size, so the big beast is, as pointed out back in chapter 3, worse off than the small one. We'll get back to scaling later in the chapter.

The situation can be summarized by citing some useful formulas (given also in fig. 18.3) for the second moment of area. For a hollow tube of outer radius r_o and inner radius r_i or a solid tube for which the latter is zero,

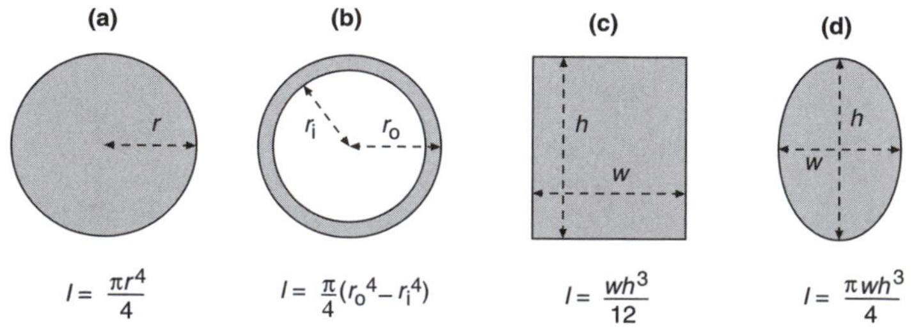

FIGURE 18.3 Cross sections and formulas for the second moment of area, I, for (a) a solid cylindrical rod, (b) a hollow cylindrical tube, (c) a beam of rectangular section, and (d) a beam of elliptical section.

$$I = \pi\frac{(r_o^4 - r_i^4)}{4}. \tag{18.1}$$

For a solid rectangular beam of height (sometimes called "depth") h and width w, loaded in the height plane,

$$I = \frac{wh^3}{12}. \tag{18.2}$$

For an elliptical rod, with similar height and width designations,

$$I = \frac{\pi wh^3}{64}, \tag{18.3}$$

about 60 percent of the moment of area of the rectangular beam. One pays a considerable price for the rounded ends and the resulting greater concentration of material nearer the neutral plane. In compensation, one might make the top and bottoms of the elliptical walls thicker than the sides (assuming upward or downward force, of course). Which organisms commonly do.

Onward to flexural stiffness

So much for I; E was defined three chapters back as the slope of a stress-strain plot. Quite obviously, I has dimensions of length to the fourth power; E, you'll recall, has dimensions of force per length squared. Thus, EI is force times the square of length or, in SI units, newton-square meters (not "per" anything).

The basis for the preference of both nature and human technology for hollow tubes over solid rods ought to be quite obvious, whether you prefer to think of bicycle frames or bamboo poles or the exoskeletons of insects or our own long bones. One wants to get that radius-to-the-fourth-power as high as practical—the greater its value, the more effectively one uses a given amount of material. We generate red blood cells in the spaces in our bones; birds fill them mainly with air (presumably in the interest of weight economy); insects put their muscles and everything else on the inside; aircraft may run electrical or control cables through them or use them for fuel storage—all secondary uses of structures made hollow for mechanical reasons.

We now have a quantity, EI, the flexural stiffness, that gives the resistance to bending of a *structure*. A not-so-parenthetical note of caution, though, must be sounded lest one take formulas for deflections based on EI as completely trustworthy. For the results of data for E and calculations of I to correspond to practical reality, quite a few conditions must be met. The material must be homogeneous and isotropic. It must be linearly elastic—Hookean—which means that its stress-strain curve must be uncurved. It must deform equally under tensile and compressive loads. It must not deform very far—typically less than 10 percent. And its shape must not change appreciably as it is loaded. Finally, a calculated I anticipates functional reality only for certain shapes—it sometimes misleads. Stiff metal structures, as we ordinarily use them, meet these conditions pretty well. For biological materials and

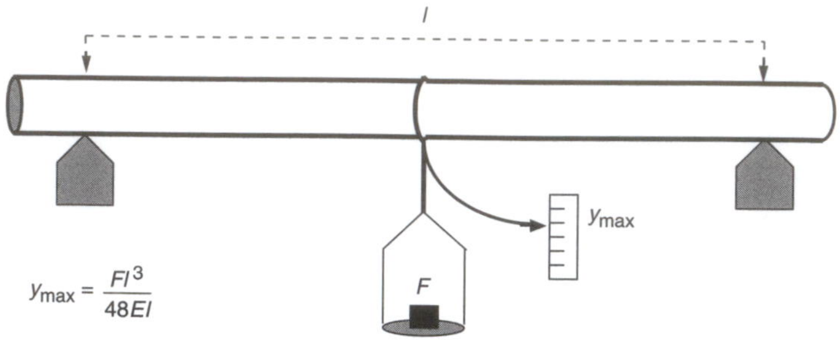

FIGURE 18.4 How to measure *EI* by using a three-point bending test.

shapes, the conditions are often much less reasonable and frequently quite disabling. So what do we do? Two things. We retain the quantity *EI* but we measure it as a single variable in some kind of bending apparatus rather than calculating it from its components. And we take even these results with a grain of salt when drawing conclusions from them.

In fact, *EI* turns out to be one of the easiest of all mechanical properties to measure—you can easily improvise a testing apparatus with a pair of supports (sawhorses will do), weights (a bucket to which water can be added), and a tape measure. The specimen, some elongated object, is laid across the supports and the bending stress is applied to the middle (perhaps with a rope), as in figure 18.4. For small deflections, the deformation downward in the middle, y_{max}, is given by Gordon (1978) (and standard textbooks of structural engineering) as

$$y_{max} = \frac{Fl^3}{48EI}, \tag{18.4}$$

where F is the weight (not the mass) applied and l is the distance between the supports. One minor problem with thin-walled or soft specimens comes from the sharp loading points; that can be minimized by loading along lines rather than at points, by hanging the beam from loops of thread or rope. Equation (18.4) deserves examination as well as use—it states that the deflection of such a beam under constant load is proportional to the cube of its length, making a long beam a very bad business. But it's inversely proportional to the flexural stiffness, the latter proportional to the square of some linear dimension. So thicker is much better.

Bending Beams

Equipped with a measure of stiffness in bending we can move on to somewhat more specific structures. First, then, beams. By a beam we usually mean some more-or-less horizontal girder subjected to up-and-down or side-to-side loads (downward and gravitational most often) that tend to bend it. The joists beneath a floor or flat roof are typical beams; so are the roadways of most bridges; sometimes, too, is the entire bridge. And so is the protruding neck of a bird or quadrupedal mammal, the petiole that holds a leaf away from a branch, a horizontal branch itself, the vertebral column between forelegs and hindlegs of a quadruped, the tubular abdomen of an insect, and so forth.

The two kinds of beams that will most concern us have in fact made their initial appearances. A "simply-supported beam" sits on supports at each end, as was the one with which we measured *EI*. Bending downward loads its top in compression and its bottom in tension. A "cantilever beam" protrudes outward from a single support, as was the one used still earlier to rationalize the behavior of *I*. Bending its free end downward loads its top in tension and its bottom in compression. A third kind, a "fixed beam," has its ends rigidly fixed to supports at each end—two supports, like a simply supported beam, but with fixed ends, like a cantilever. This last will be of little concern here. For any of these, all but the very top and bottom surfaces experience shearing loads. (As a matter of custom and verbal convenience, we're assuming gravitational loading.)

Simply supported beams

We use these more, I'd guess, than does nature; but they do occur as arched feet, backbones between pectoral and pelvic girdles, and an occasional part of a plant. Equation (18.4) applies, strictly, to a simply supported beam in what's known as three-point bending (fig. 18.5a) with the load in the middle. That's what an archer does when drawing an arrow and bending the bow—to pick an example in which the load isn't gravitational. Alternatively, the load can be off center, as in figure 18.5b.

Loads often distribute themselves along a beam instead of being specifically located on a weightless structure. So yet another formula comes into play. For a uniformly loaded (or self-loaded) beam (fig. 18.5c):

$$y_{\max} = \frac{5Fl^3}{384EI}. \qquad (18.5)$$

Since the variously caused deflections are nicely additive, at least if the sum remains small, one can, for instance, use equation (18.4) for an applied load together with equation (18.5) for self-loading.

Comparing equations (18.4) and (18.5) tells us that a given load will make such a beam sag in the middle 1.6 times as far if centered than if distributed uniformly along the beam. Equation (18.5) holds practical advice for us bookish academics—combined with equation (18.2) it tells us what happens when we put books on bookshelves supported (but not attached) at their ends:

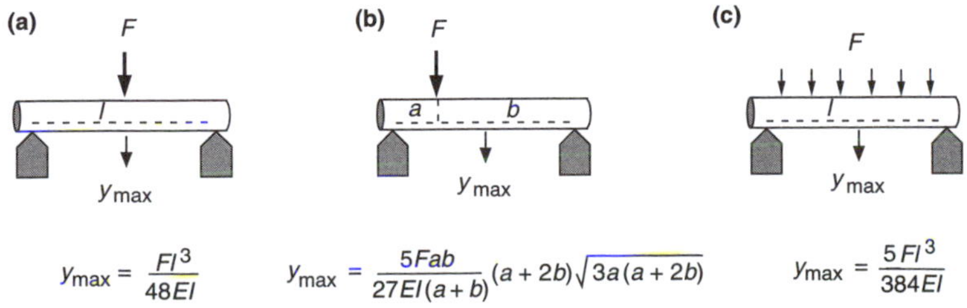

FIGURE 18.5 A fixed beam, with formulas for maximum deflection when (a) point-loaded in the middle, (b) point loaded anywhere along its length, and (c) uniformly loaded along its length.

$$y_{max} = \frac{5Fl^3}{32Ewh^3}. \qquad (18.6)$$

Stiffer is proportionately better, assuming sag is bad, although one rarely has much choice of material. (You can get a datum for E from table 15.1.) Wider (w) is proportionately better as well. Thicker (h) is disproportionately better—glue together two boards and you'll reduce the sag eightfold. (Without glue or screw, though, the boards can shear on the neutral plane, and you'll gain only the additive twofold in sag reduction.) Longer is disproportionately worse. Going from a 2-foot to 3-foot shelf length will increase sag not 1.5 times but nearly 3.5 times. In fact, that understates the real problem. Since the load of books will ordinarily increase in proportion to length, force will be proportional to length, and sag will scale with the fourth rather than the third power of length. Double the length and the sag increases fully sixteenfold. A shelf 36 inches long will sag over four times as far as one of similar construction that's only 30 inches long. Buy short shelves or support them at narrow intervals!

Cantilever beams

Back to cantillating about cantilevers. Consider a cantilever beam whose length is decently greater than its vertical thickness (figs. 18.1 and 18.2, again), with the load concentrated at the unsupported end. Formulas for the deflection of its outer end look much like those given for fixed beams–equations (18.4), (18.5), and (18.6). For a weightless, end-loaded beam of uniform cross-sectional shape and area,

$$y_{max} = \frac{Fl^3}{3EI}. \qquad (18.7)$$

For a force distributed uniformly along the length of the beam,

$$y_{max} = \frac{Fl^3}{8EI}. \qquad (18.8)$$

So an end load makes such a beam sag about 2.7 times farther than a uniformly distributed load.

As already mentioned, tensile forces will increase with distance up and compressive forces will increase with distance down from the neutral plane. In addition, the forces will steadily increase as one moves inward toward the support from the load. In effect, the loading force acts as a turning moment (clockwise in the figure) with the attachment point of the beam as the pivot—the beam can be viewed as one big lever or half of a seesaw. Closer to the pivot the distance of deflection will be less while the force will be more—the product of force and distance has to remain constant. Clearly, a beam of uniform cross section puts too much material out near its free end where forces are lower. Put another way, the stresses on the material near the free end will be wastefully low.

So what should be the shape of an efficient cantilever beam, that is, one of uniform strength, that's loaded at its outer end? If the deflecting moment decreases with distance from the support, then to make best use of material, the height ("depth" in the usual jargon) of the beam ought to decrease from its support to a point where the load is attached. For a beam of rectangular cross section and constant width that weighs much less than the load whose load is located at the outer end, the taper should be parabolic (fig. 18.6a). That's enough of a nuisance so we often approximate the parabola by a linear taper that doesn't quite come to a point. A crane used to lift and swing containers from the decks and holds of cargo ships may come close to the weightless, end-loaded condition. But it must be uncommon in biological systems—perhaps a few stems hold sufficiently heavy fruits horizontally.

If, by contrast, the load distributes uniformly along the beam (still assumed rectangular in section, of constant width, and weightless), then a linear taper for beam depth (fig. 18.6b) comes out best. That's also a bit unusual in living systems. Organisms most often use beams that are circular (or elliptical) in cross section, and for these the common engineering handbooks (such as Oberg, Jones, and Horton [1984]) don't give such direct solutions. Denny (1988) faced the matter in admirably direct fashion. Some degree of taper, though, is virtually universal, for the branches of trees, for long necks and upheld tails, for archy's long, thin cockroach antennae as well as for the cat's whiskers of mehitabel (Marquis 1927).

One can continue such analyses, with increasing sophistication and analytical complexity, for other loading regimes, looking at, for instance, the best taper for a beam that's entirely self-loaded. That might apply, in part, to a horizontal branch of a tree. But not only do biological beams have oddly shaped cross sections, they withstand peculiarly mixed loads. For instance, a branch in a wind will face, in addition to gravitational load, a sideways load. Worse, it will most likely face a torsional load as well, assuming that the leaves generate most of its drag and that the leaves don't combine to apply a perfectly torqueless sideways force. For such situations, the computer-aided optimization described by Mattheck (1998) may provide the best approach. It, though, involves its own assumptions, mainly an omission of the anisotropy and place-to-place variation of the properties of the underlying material.

In any case, silly figures result from treating, say, the branch of a tree as a simple, gravitationally loaded cantilever designed for material economy. Consider the permissible length of a horizontal branch. A reasonable figure for the Young's modulus of fresh wood might be 9000 meganewtons per square meter; one can assume a density slightly less than that of water, say 800 kilograms per cubic meter. Assuming a maximum deflection of a tenth the length of the cantilever, one can calculate that maximum length for a given cross

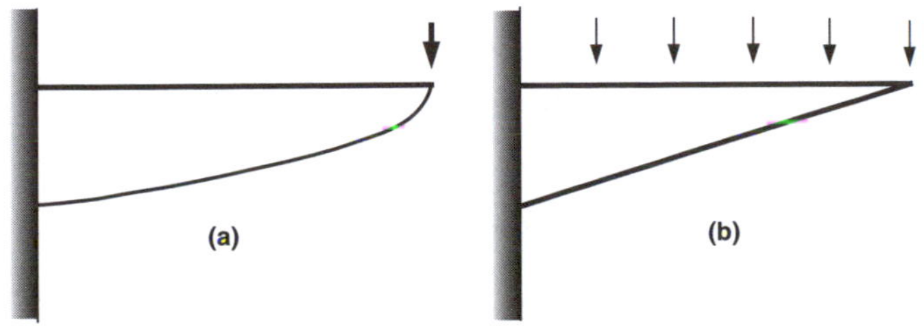

FIGURE 18.6 Cantilever beams of optimal shapes: (a) the parabolic taper of end-loading, and (b) a linear taper for a load uniformly distributed along its length.

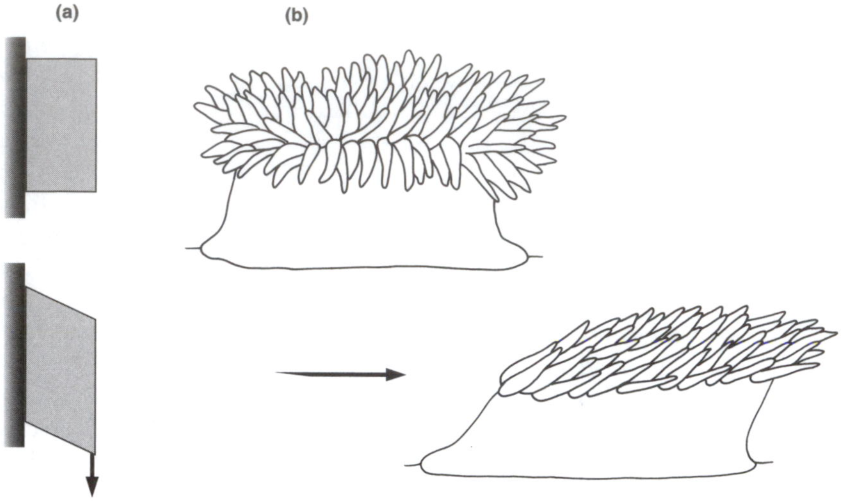

FIGURE 18.7 (a) A short thick cantilever that responds to a load by shearing rather than by bending. (b) *Anthopleura*, sheared by a water current.

section. Using the far-too-conservative equation (18.7) for an untapered cantilever gives a length of 6 meters (20 feet) for a branch 10 centimeters (4 inches) in diameter. Using a more realistic taper permits a branch that's even more extreme.

Furthermore, one ought to bear in mind a particular limitation of this focus on bending moments. Consider an extreme, a very short, thick cantilever (fig. 18.7a). It won't bend as do these beams, but instead it will shear downward, changing its side-view form from a rectangle to a parallelogram. Strictly, all cantilever beams want to shear downward as well as bend; we've just been focusing on situations where bending poses the predominant problem. Organisms know both—Koehl (1977b), for instance, found that a tall sea anemone, *Metridium senile*, bent in a current, while a short, squat anemone, *Anthopleura xanthogrammica* (fig. 18.7b) sheared.

Cross-sectional shape

For that matter, equation (18.3) tells us that branches shouldn't be circular in section if designed as economical cantilevers faced with gravitational self-loading. They should instead be severely elliptical, pushing ellipticity to the point where other factors limit the height-to-width ratio. Plenty of pieces of plants do have elliptical sections, and radially asymmetrical growth of wood represents a common response of trees to long-term unidirectional forces. So while we're not talking about anything morphogenetically radical, most horizontal branches are close to circular in section, as if to underline the point that gravitational self-loading can't be what matters most.

But the example has brought us back to cross-sectional shape, which does matter in many instances, both in nature and human technology. The simplest examples go back to what equation (18.2) implies—beams should be high but not wide. We make timber beams in just such a manner—the only common exceptions are diving boards and leaf springs, where substantial deformation is positively essential. Wooden floor joists have nominal sizes of 2 by 8, 2 by 10, even 2 by 12—the width of about 4.5 centimeters (1.75 inches) stays fixed, while the height of the beam increases. One can do even better by putting material mainly at top and bottom, since material on the neutral axis takes no tensile or compressive stress and only a little shear. That may be awkward if slicing beams from tree trunks, but we do it routinely with metals. If the material withstands tension and compression with equal stiffness (as do metals), the same amount ought to be used top and bottom. We've now invented that most basic structural member, the so-called I-beam, with the serifs on the printed "I" representing the heavy top and bottom flanges of the beam. Of course, some material is still needed between top and bottom flanges to prevent any shearing strain that might bring the flanges closer together and ruin everything.

An I-beam makes good use of metals, and it works well for a technology heavily concerned with gravitational loading. But biological materials most often differ in their responses to tension and compression. And this account has repeatedly pointed out that static gravitational loading much less commonly provides the main challenge even to terrestrial organisms than one (with the biases of exposure to human technology) might expect. So the beam may be present, but the I-form rarely appears. Where gravity does matter, most often we find tension-resisting stuff concentrated at the top (tendons and other fibrous materials) and stuff that resists compression (bone and so forth) at the bottom. Elsewhere, hollow tubes predominate.

Consider a broad leaf on a tree. The greatest forces on its petiole ("stem") and midrib probably occur as it's pulled by the drag of the blade in a wind storm, but these forces are tensile and thus easy to resist. Without wind, it's a beam faced with the task of keeping its blade in a position to intercept sunlight, which, on the average, comes from above. So its design, as in figure 18.8, ought to reflect gravitational loading. Which it does, but more by using internal material anisotropy than externally obvious cross-sectional specialization. It uses thick-walled, liquid-filled cells along its bottom, which resist compression well, and long cells with lengthwise fibers along the top, which act as ropy tension resistors. The petiole and midrib are as truly cantilevers as any protruding I-beam, but internal structure—anisotropy at various levels—matters at least as much as overall cross section in efficiently dealing with gravity. And the rest of the leaf blade, an extension of the cantilever, faces much the same mechanical situation. Veins protrude downward to get some

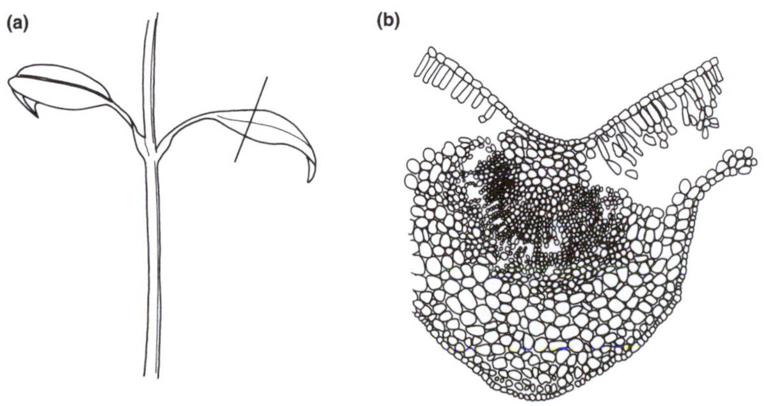

FIGURE 18.8 (a) A leaf and its petiole act as a cantilever beam, with the upper part in tension and the lower part in compression. (b) A cross section through the midrib of a leaf. The large cells near the bottom are nearly spherical and liquid-filled as appropriate for bearing compression, while the smaller ones further up are elongate and ropy.

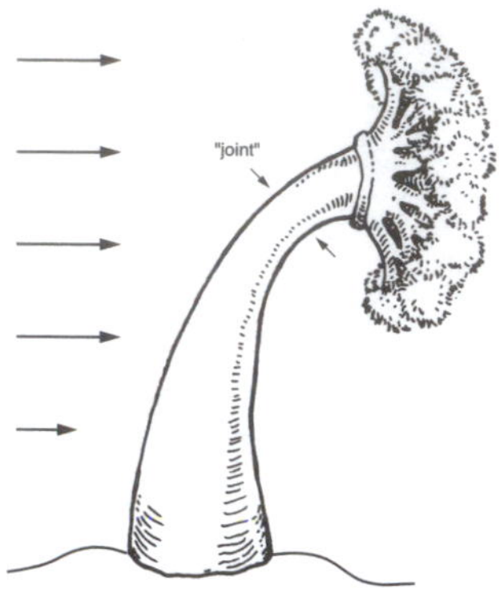

FIGURE 18.9 The sea anemone, *Metridium*, bent by a current of seawater from which it can then suspension-feed. Figure 17.6 gives the erect form.

height to the beam and to continue the compression-resisting material of petiole and midrib. The blade is always at the top—a flat sheet can take tension, but it's almost as bad in compression as a rope. An insect's wing is likewise made up of a membrane and veins, but it's stressed about equally on up- and downstrokes; concomitantly, and in contrast with leaves, the veins commonly protrude equally above and below the membrane.

What if the beam must withstand loads from all directions, not just downward? One obvious solution is to use a double-I girder, one with a cross section in the form of a Maltese cross. But that shape just begs to have its corners connected for a little more stability (especially in the face of torsional loads); with corners connected, the internal webs that connect opposite flanges become unnecessary, and we have reinvented something long known as a "box girder" or "box beam." Robert Stephenson used just such a beam in his Britannia Bridge of 1850—the trains ran through the center of a horizontal iron box (Billington 1983). One can do even better by making more corners. Stresses will be high at the corners of the box, which therefore need reinforcement and which have no specific function anyway, but the more corners, the more obtuse each can be. And the more obtuse the corner, the less the problem of excess stress. No surprise—the ultimate beam for withstanding forces from all directions turns out to be a hollow, cylindrical tube. As noted, these predominate among nature's beams—but, as we'll see, for other reasons as well as their excellence in dealing with omnidirectional bending forces.

In some cases, nature capitalizes on the way hollow tubes bend. A tall sea anemone, *Metridium*, has an area of its columnar body just beneath the crown of tentacles that is narrower than anywhere else, as in figure 18.9. The material properties of the stalk don't vary, but when a gentle water current is present, the stalk bends at this point rather than at the bottom, and the tentacles are exposed broadside to the flow in the best position for feeding on suspended matter (Koehl 1977b). It doesn't take much narrowing to concentrate the bending—I, as equation (18.1) shows, depends strongly on the radius.

Columns: Making Ends Not Meet

Complementary to beams are columns. In the usual engineering applications, beams run horizontally and withstand crosswise forces that would bend them, while columns run vertically and withstand lengthwise forces that would push their ends into closer proximity. In biological applications the same relative force directions apply, so the general distinction works as well—we just take the overall orientations, horizontal and vertical, with smile or smirk. A tree trunk in a wind or a sea anemone in a current represents as real a cantilever beam as any horizontal crane or jib. The upshot is that the hazards of column-type static loading occur less commonly in nature than in human-built structures. But nature does use columns, even if their loads are more often impulsive than static.

How columns respond to loads differs a little from what we saw for stressed beams. With beams we worried about bending, with shearing almost always a secondary consideration. With columns, we have to worry more about true failure and less about mere deformation. And we have to distinguish among three modes of failure—crushing, bowing out to one side, and crumpling—which gives the designer more to worry about.

The initial effect of a weight on a column is purely compressive, and the analysis is just that of a material rather than a structure. Thus, chapter 15 (especially table 15.3) provides enough to go on, at least for a start. Short, fat columns do indeed fail in pure compression, either crushing their material or shearing, with an oblique line of failure marking instances of the latter. (Sometimes one can produce an example of that shearing by squeezing a short piece of wood end-to-end, or with the grain, in a vice.) Crushing collapse mirrors tensile failure—only cross-sectional area, not shape, matters.

Lengthen the column, though, and it bends before it fails. We say that it buckles. Buckling can happen in two distinct ways; which one occurs depends more than anything else on the relative thickness of the walls of the column. Either form constitutes a form of degenerative collapse in that all is well up to a particular load, after which the more a column buckles, the more it will yet buckle under the same load. So what matters for either is the critical force that initiates the destruction.

The first form of buckling, "Euler" (pronounced "oiler") buckling, generates a smooth bend, with compressive forces concentrated on the concave side and tensile forces on the convex side. It's what happens when you push together the ends of a strand of uncooked spaghetti or a long, thin dowel. After all the talk about beams, no one should find it strange that an overall compressive load can generate tensile forces, and that the initial failure for many materials may be a tensile one. Making a column fail in tension by adding weight to the top does admittedly seem counterintuitive. Still, it further rationalizes the emphasis on tensile properties in the last three chapters.

For Euler buckling, we have a convenient formula for that critical force, designated F_E:

$$F_E = \frac{n\pi^2 EI}{l^2}. \tag{18.9}$$

Once again, the composite variable flexural stiffness, EI, plays a key role. In equation (18.9), l is just the length of the column; n, though represents the kind of structural consideration we run into in this business. Its value (dimensionless, at least) depends on how the ends of the column are restrained. With both ends merely pinned—that is, free to bend at the support, n is 1; if one end is firmly fixed (inserted into the support or firmly braced or rooted), n is 2; if both ends are firmly fixed, then n is 4. You can persuade yourself that the end conditions matter, and do so about as these values indicate, by pressing dowels end-to-end, with and without ends inserted into tight holes in blocks of wood, as in figure 18.10.

The condition of having one fixed end is of particular biological interest—it's the situation of long, slender plant stems such as those of dandelions, grass, bamboo, and others. Before making serious use of equation (18.9), though, one must recognize that it applies to isotropic materials, which plant stems are anything but. (We'll see just how drastic the effects of anisotropy can be when, shortly, we look at torsion.) As emphasized by Schulgasser and Witztum (1992), their anisotropy greatly increases the risk of buckling for plants that use thin-walled tubular construction. Mainly, the tubes, normally circular in cross section, go somewhat oval just prior to buckling, and that reduces

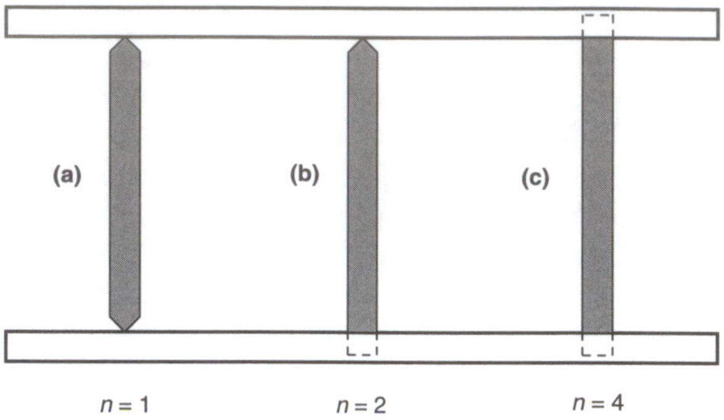

FIGURE 18.10 Three different ends for columns sensitive to Euler buckling and their *n*-values (equation (18.9)). (a) One with both ends pinned, that is, free to bend any way with respect to its top and bottom. (b) One with the upper end pinned and the lower one firmly rooted so it can't rotate. (c) One with both ends firmly attached to end pieces that can only move up and down—that can't themselves rotate.

the critical force. Preventing that ovalization may be one of the roles of the periodic transverse bulkheads so conspicuous in, for instance, bamboo.

Anyone who has squashed an empty metal can knows about the second form of buckling; it's called "local buckling" or "Brazier buckling." Unsurprisingly, it's mainly a problem for columns (and beams) with very thin walls—it's the main practical reason why the advantage of tubular over solid cylinders can't be pushed to extremely thin-walled, extremely wide tubes. The usual formula for the critical local buckling force, F_L, for a thin-walled cylinder in compression turns out to be a more rough-and-ready empirical affair than the one for Euler buckling:

$$F_L = K\pi(r_o - r_i)^2 E. \tag{18.10}$$

Here $(r_o - r_i)$ or outside minus inside radius is the cylinder's wall thickness; and K is a semiempirical constant between 0.5 and 0.8, with the specific value depending (inversely) on extent to which the cylinder has imperfections at which buckling might start (Wainwright et al. 1976). Notice that neither radius per se nor second moment of area make any difference to the initiation of local buckling.

Local buckling does occur in biological columns—it's certainly involved in the lodging of slender crop plants in wind storms, and it can be deliberately induced in any dandelion stem. A low-density foam core reduces susceptibility, and many plants (but not dandelions!) have such cores. But most biological cases of local buckling involve bending of such structures as cantilever beams. One of the initial reviewers of this book pointed out that a bent sea anemone, as in figure 18.9, buckles a little on the concave side when it bends in a current. So the same consideration of minimum acceptable wall thickness plays a role in bending as well. Of course local buckling happens even in solid structures—it's the initial event in the compressive failure of a wooden beam, which, to remind ourselves, is weaker in compression than in tension.

How might a cylindrical tube be designed to balance the tendencies to fail by Euler and local buckling? It's a simple matter to equate the two and notice that Young's modulus drops out. So the optimum doesn't depend much at all on the material of the tube but comes down to purely geometrical considerations. But the specific optimum will vary with both the end conditions (n) and the imperfections in the surface (K).

Twisting—Distortion by Torsion

Materials take tensile, compressive, and shearing loads. Simple structures such as beams and columns take, in addition, bending and twisting—called, respectively, flexion and torsion. So to complete the present story, we turn to torsion. Bending a cantilever beam, recall, put the top in tension, the bottom in compression, and the middle in shear. What about twisting a protruding shaft or column? As figure 18.11 shows, the outside feels shear and tension, while the inside feels compression. Twist a wet towel to get a feel for these stresses—the outside is clearly sheared and stretched, and compression of the inside squeezes out water, which is why one ordinarily twists a wet towel. Less happily, bones sometimes fracture under torsional loading, as when we attach a long board to a shoe (something called a ski) and twist; the internal compression makes the bone quite literally explode, with much messier results than ordinary bending fractures.

Where does torsion matter? For human technology, rotating shafts provide the most important place; and elaborate tables of values help in choosing shafts, capstans, and other devices for which the ability to transmit torque defines their worth. Transmitting the torque of motors to the downstream components of a machine demands adequate resistance to twisting distortion or actual torsional failure. Nature uses relatively few rotating shafts, but, as we'll see shortly, she seems to care about torsion (even before we took up skiing), if with a somewhat different approach to the phenomenon.

Torsional stiffness

For flexion, the key factor was *EI*, the flexural stiffness. For torsion, the analogous composite variable is *GJ*, the torsional stiffness. Here *G*, the shear modulus, or resistance to shearing, replaces *E*, Young's modulus of elasticity, or resistance to stretching. And *J*, the polar second moment of area outward from a neutral axis of twisting, replaces *I*, the second moment of area lateral to a neutral plane.

We met *G*, the shear modulus, back in chapter 15, where it was defined as the shear stress, τ, divided by shear strain, γ, nicely constant for a Hookean material:

$$G = \frac{\tau}{\gamma}. \tag{15.4}$$

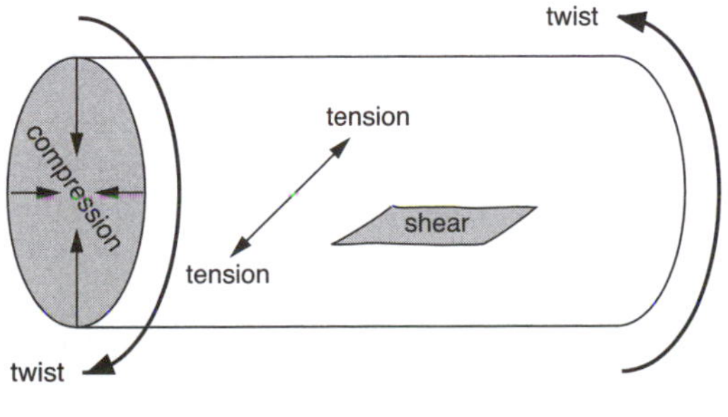

FIGURE 18.11 Shearing, tension, and compression in something subjected to torsional loading.

To reiterate, shear stress is a force per unit area, the latter the area being sheared and thus parallel rather than crosswise to the force. Shear strain is an angle, automatically (in radians) dimensionless and starting from zero rather than some original length. So the shear modulus has the same dimensions, force per area, as shear stress itself or as Young's modulus.

For the materials in ordinary engineering practice, shear modulus tracks Young's modulus with but a slight correction for Poisson's ratio, ν; a formula,

$$G = \frac{E}{2(1+\nu)}, \qquad (15.6)$$

was given earlier. (The internal compression caused by the external tension of twisting explains the involvement of Poisson's ratio.) The malediction about use of the formula ought to be repeated as well. It assumes a homogeneous, isotropic material such as a metal or an ordinary plastic. It should not be assumed applicable for any biological material even in the way we used Hookean behavior as a first approximation for beams and columns—separate measurements of *E* and *G* must be made (or on record) before buying in.

The polar second moment of area, *J*, differs only slightly from *I*, the ordinary second moment of area, both conceptually and practically. One multiplies each bit of cross-sectional area by the square of its distance from the neutral axis and sums the result—it's the most ordinary of integrations. For a hollow circular cylinder (or a solid one, setting $r_i = 0$), then,

$$J = \pi \frac{(r_o^4 - r_i^4)}{2}. \qquad (18.11)$$

The value of *J* turns out to be just double the value of *I* given by equation (18.1) because of nothing more obscure than the Pythagorean theorem. Just as thickening or deepening a beam gained bending stiffness disproportionately, so an increase in the size of a cylinder greatly increases its resistance to twisting—in the commonest application its ability to transmit torque. Double a solid shaft's diameter and it will handle sixteen times the torque.

For other shapes, things get trickier, and, in practice, formulas from analogous integrations must be regarded as maximal values—the rules for integration don't change, but resistance to twisting doesn't necessarily follow in proper obedience to *J*. A numerical example should drive home the point. Consider the angle to which a given torque twists two circular tubes. The tubes have wall thicknesses a tenth of their radii; they're identical except that the wall of one has been slit lengthwise for its full length. Using the formulas for torsional stiffness cited in Eshbach (1952) and elsewhere, the tube with a slit will twist about 260 times as far as the one without a slit! Nevertheless both, of course, have the same value of *J* according to equation (18.11). Thin the wall tenfold further, and the one with the slit twists almost thirty thousand times as far as the one without. Even the tolerant biologist finds such errors intolerable.

The bottom line is that both *G* and *J* have sneaky problems that can catch the biologist, who deals with irregular shapes and anisotropic materials. As with E and I, the practical answer lies in measuring the combination as the single, composite variable, GJ. Instead of, say, a three-point bending test (fig. 18.4) for EI, one tests in torsion. Torsion tests prove almost as easy as flexion tests, adding only two minor bothers. One needs to prevent residual bending from contaminating the results, and, unlike a three-point bending test, one can't avoid rigidly gripping at least one end of the specimen. Figure 18.12 shows two simple ways to set up a torsion test. One uses a specimen glued into some cylindrical capstan (a gelatin capsule or a plastic test tube, for instance), with off-axis pulls from opposite directions to avoid simultaneous bending. The other, suitable for larger objects, takes advantage of the torsionally rigid grip of the locked headstock of a metal lathe, with a bearing (a "live center") in the tailstock to prevent bending. The formula for dealing with the results turns out to be quite as simple as those (such as equations (18.4) and (18.7)) for testing flexural stiffness:

$$\theta = \frac{Frl}{GJ}, \tag{18.12}$$

where θ is the angle of twist (in radians) resulting from the application of a force, F, tangent to the structure of torsional stiffness GJ and length l at a radius r.

Where torsion matters

Again, while we borrow our formulas, models, and analytic techniques from the mechanical engineers, we have to keep our distance from their materials, practices, and objectives. Human technology has by long and pragmatic tradition focused on stiff structures. Nature builds more flexible things, resorting to stiff stuff only when some special situation, such as staying erect against gravity, justifies its usually greater cost in material.

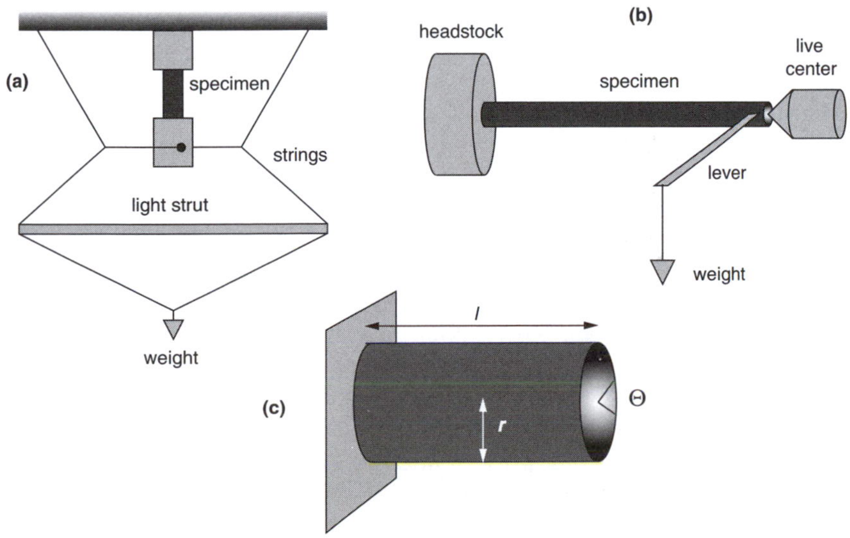

FIGURE 18.12 Two ways to perform a torsion test: (a) embedding the ends of a small object, fixing one end and twisting the other; and (b) using a metal lathe, lateral arm, and weights to twist larger objects. (c) The variables for use in equation (18.12).

That difference is especially evident in structures subject to torsional loads. Low resistance to twisting must be avoided in our artifacts. A wing that twists too much when aerodynamically stressed may part from the rest of the airplane—as did the wings of the Fokker D8 during World War I (Gordon 1978). The Tacoma Narrows suspension bridge dropped into Tacoma Bay in 1940 in part because of excessive torsional flexibility. More recently, the John Hancock Tower in Boston turned out to turn through an uncomfortably large angle in high winds and had to be retrofitted with a pair of 300-ton damping blocks, at the cost of an entire floor (Wiesner 1988). We make great use of I-beams, which have decidedly low values of GJ, but you'll rarely see anything built of a single one—we use them in groups where one offsets another's torsional flexibility.

Nature, by contrast, takes a less disdainful attitude toward torsion—in some applications adequate resistance matters, but in many others function depends on having sufficient torsional flexibility. A bird's wingtip feathers must twist in one direction during the upstroke of the wings and in the other direction during the downstroke to keep the local wind striking the wing at an appropriate angle to generate lift and thrust (chapter 13). The turning could be done at the base, with a completely inflexible feather; the aerodynamics are improved and material saved if the local flow forces twist the feather by just the right amount. Ennos (1988) showed that insect wings take even greater advantage of aerodynamic forces combined with limited torsional stiffness. Propellers have to be twisted lengthwise to make the local wind direction, varying along the propeller's length, strike the blade at a constant angle of attack. For an insect wing, the lengthwise twist that works for an upstroke will be exactly the opposite of what's needed for the downstroke. Small muscles at the bases of the wings had been thought to do the required phasic adjustments, no mean task with wingbeat frequencies in the hundreds per second. In fact, aerodynamic forces plus sufficient torsional flexibility minimize the need for any such active control.

Another use of torsional flexibility, perhaps less sophisticated, happens on a larger scale. Wind on a tree will twist it unless everything (including the wind) is perfectly symmetrical about the trunk. But twisting brings bits of tree closer to a downwind orientation and brings the bits into closer proximity to each other. Both should reduce the tendency of the tree to bend over. Clever—lowering torsional stiffness ought to reduce the requirement for flexural stiffness! While we don't have data for any intact tree, the effect has been shown for clusters of leaves (Vogel 1989), and casual observations in storms suggest that it works at larger scales. Tree-level use is consistent with the relatively low values of torsional stiffness of fresh samples of tree trunks and bamboo culms (Vogel 1995b). On a smaller scale, Ennos (1993a) found that sedges swing around in a wind rather than bending over, doing so with stems of remarkably low torsional stiffness. A banana leaf, pushed sideways, twists rather than bends, again using a structure, its petiole (or leaf stem) of very torsional stiffness (Ennos, Spatz, and Speck 2000). And daffodil flowers, borne off to one side of their stems, swing around similarly, reducing their drag by about 30 percent in the process (Etnier and Vogel 2000). Twisting in the wind isn't just a slogan left over from the Nixon presidency. Daffodils appear to "dance" in the wind, as noted by the poet, William Wordsworth, because down near ground level, winds are especially puffy.

A measure of relative flexibility in torsion

A dimensionless ratio facilitates comparisons of what we've been loosely calling relative torsional stiffness. Both *EI* and *GJ* have dimensions of force times distance squared, and they come by those dimensions in analogous ways. Putting *EI* over *GJ* gives a measure whose numerical values conveniently come out between about one and twenty or so. Length dimensions cancel out, so the ratio has no awkward size bias. An engineer concerned with maximizing stiffness in one or the other modes might call it the ratio of stiffness against bending to stiffness against twisting loads. The biologist, dealing (we believe) with controlled and deliberate flexibility, does better calling it the "twistiness-to-bendiness ratio." Converting from resistance to change to ease of change inverts the words without changing the formula:

$$\frac{\text{flexural stiffness}}{\text{torsional stiffness}} = \frac{\text{twistiness}}{\text{bendiness}} = \frac{EI}{GJ}. \tag{18.13}$$

How does this variable behave? Consider, first, structures of ordinary isotropic, homogeneous materials of forms for which the formulas for *I* and *J* can be trusted. The ratio *E*/*G* will depend on Poisson's ratio; for an isovolumetric material, $\nu = 0.5$, and *E*/*G* (equation (15.6)) will be 3.0. (For steel, $\nu = 0.3$, and *E*/*G* will be 2.6.) For solid or hollow cylinders, *I*/*J* is 0.5 (equations (18.1) and (18.11)). Thus, a cylinder of an isovolumetric material will have a twistiness-to-bendiness of 1.5. (A steel cylinder will give the slightly lower value of 1.3.) Deviation from a circular cross section will raise the value—twisting will get easier, relative to bending. Table 18.1 gives some values from standard engineering handbooks.

So if we're looking for places where nature contrives to raise the ratio, we might look for elongate structures whose cross section isn't circular. They certainly abound—figure 18.13 gives a few. A primary wing feather usually has a lengthwise groove along its central shaft (the rachis). Lots of stems and petioles are grooved, flattened, or of some other noncircular form. These structures ought to have twistiness-to-bendiness values over 2.0.

But what of reality? Most experimentally determined values come out far in excess of these, again emphasizing that we're using the language of conventional engineering to talk about a world that's anything but. Table 18.2 gives some examples.

TABLE 18.1 Twistiness-To-Bendiness Ratios of Isovolumetric Objects of Different Cross-Sectional Shapes by Conventional Calculations

Shape	I/J	EI/GJ
Circular section	0.5	1.50
Square section	0.59	1.77
Equilaterally triangular section	0.83	2.49

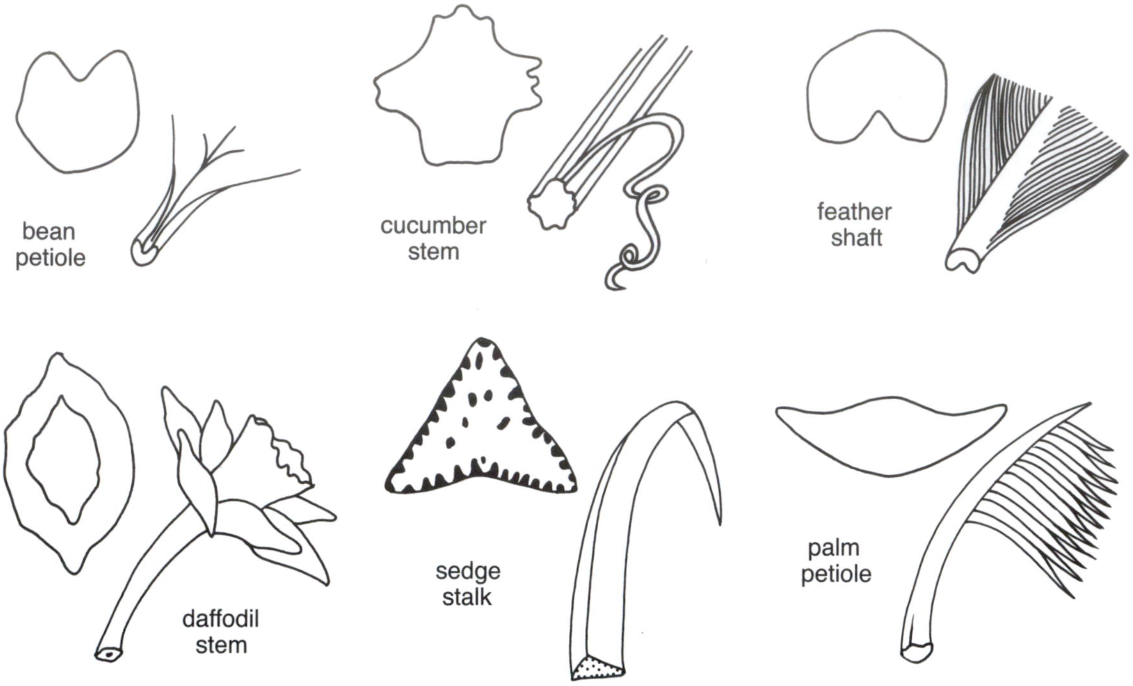

FIGURE 18.13 A variety of biological structures with noncircular cross sections that may be torsionally loaded.

Evidently both shape and material anisotropy matter to the *EI/GJ* ratio. And given the minor effects of shape alone (table 18.1), anisotropy matters most. All those highly oriented composites and that hierarchy of microstructure apparently contribute to these dramatically high values. Shape, observable casually, hints at where *EI/GJ* might be high—without making the main contribution to it. The contrast with typical engineering materials is sharp.

Most importantly, the values (excerpts from a larger set) "make sense" in functional terms, suggesting that this particular variable gives insight into biological design—even if one that merits little attention from people engaged with human technology.

■ An argument was made that twisting might be advantageous for tree trunks; the values for trunks of all three lineages are high. But those high values don't just characterize wood per se, since vines and roots yield lower ones. Sliced timber, interestingly, gives about the same ratio even though its separate values for *E* and *G* differ markedly from fresh wood. (Bodig and Jayne 1982). (Similarly, the ratio for banana petioles varies far less than either *E* or *G* alone, which drop over tenfold outward along the length of the petiole, according to Ennos, Spatz, and Speck [2000].)

■ Having petioles that twist easily permits leaves to cluster and reduce their drag in storms (recall chapter 7), but to work as proper cantilevers and hold their blades out, they mustn't bend too easily. So high values are reasonable. Ones with grooves along their tops have higher ratios than ones without them.

TABLE 18.2 Twistiness-to-Bendiness Ratios of Objects as Tested

Object	Cross Section	EI/GJ
Steel spring wire	circular	1.3
Pine root, fresh	circular	2.3
Grape vine, fresh	circular	2.7
Pine trunk, fresh	circular	6.1
Maple trunk, fresh	circular	8.3
Bamboo culm, fresh	circular	8.6
Maple petiole	circular	2.3
Sweet-gum petiole	top groove	4.5
Banana petiole	U-shaped	45–100
Tomato plant stem	circular	3.8
Cucumber plant stem	cruciform	5.8
Sparrow primary feather	bottom groove	4.8
Green bean petiole	top groove	4.9
Tulip flower stem	circular	8.3
Daffodil flower stem	elliptical	13.3
Sedge stem	triangular	25–100

Sources: Data from Vogel 1989, 1995b; Ennos 1993a; Ennos, Spatz, and Speck 2000; Etnier and Vogel 2000.

■ Stems of vines such as cucumber must reach upward to grab successive support points; twisting might permit leaves to be swung around into the lee of the stem and reduce the chance that an initially unsupported part might be bent over by gravity or wind. Of course once attached, neither torsional nor flexural stiffness will count for much.

■ In cross section, feathers look like grooved petioles upside down. Again, that makes functional sense. If an elongated structure must have a groove to raise *EI*/*GJ*, the groove should be on the side that's loaded in tension. That location won't increase the structure's tendency to buckle, since tensile loading is nearly shape-indifferent. A leaf blade bends its petiole downward; its aerodynamic loading bends a feather upward—leaf blades hang from the ends of their petioles; flying birds hang from bases of their wing feathers.

■ Daffodils have flowers that extend to one side of their stems; the stems twist in a wind, and the drag of the flowers drops as they swing around (like the sedges). Tulips, by contrast, have flowers that sit squarely atop their stems, so they should feel little if any torque from a wind; at the same time, the drag of their symmetrical flowers shouldn't be orientation-dependent.

But not all yet drops into place. By my measurements, tree trunks differ little in their twistiness-to-bendiness. Yet splitting wood for burning leaves me painfully (literally) aware of what looks like a relevant structural difference.

Some trees—where I live, sweet gum, sourwood, and sycamore—resist splitting with accursedly interwoven fibers; others such as the oaks, pines, and tulip poplar part obligingly with minimal provocation. As a general rule, the ones that don't like to split often grow with nonvertical trunks, so they should be subject to strong torsional stresses in winds. Those that split readily almost always grow quite upright, and torsional problems should be much less. Is something of mechanical interest going on? Do tests that impose only minimal deflections give the whole story or does extrapolation from them mislead us? In addition, we still have only the least information on how the material anisotropy of any of these biological items lowers G relative to E.

Shells and Domes

In addition to cylindrical beams and columns, nature uses lots of spherical—or at least multiply curved—structures. Some, such as many nuts and all avian eggshells take the form of complete spheres or ellipsoids. Others, such as our crania and turtle shells, come as partial spheres or ellipsoids. Still others, such as the various logarithmic spirals of mollusk, brachiopod, foraminiferan, and other shells, come in more complex forms.

Shells and domes face three kinds of loads—piercing, transmural pressure, and crushing. Piercing loads, such as caused by bird beak or squirrel tooth, don't much interact with their multiply curved shapes, so the usual responses apply—tough, thick, crack-resistant material and enough "give" to accommodate as much as possible of the piercer's travel. Some resist net internal pressures, if very few take net external pressure; we'll defer a look at curved structures sustaining transmural pressure differences to chapter 20. Many face crushing loads, either from being stepped on or having something dropped on or from some deliberate squeeze on opposite sides, as when a crab squeezes a shelled mollusk.

Applicable information in the literature on structural engineering comes mainly under the heading of spherical domes and mainly considers structures that make up less than complete spheres or ellipsoids. Partial spheres, though, do reasonably well for our purposes. After all, the stresses on a half shell sitting on a horizontal surface and pressed down from above match those of a full shell squeezed top to bottom. A shell or dome big enough to hold a large number of people sees significant gravitational self-loading, but that complication matters little for nature's structures.

How does a load on top stress a thin-walled shell or dome? If the wall is sufficiently thin and reasonably stiff, we can ignore beamlike bending, and the tension, compression, and shear that vary from outside to inside of the wall. What matters, then, are two remaining stresses. *Meridional stresses* run up and down the walls, following what would be lines of longitude (like the meridian of Greenwich) on a globe. *Parallel stresses* run circumferentially, following lines of latitude (like the forty-ninth parallel that separates Canada and the United States). The load on the top generates compressive meridional stresses and tensile parallel stresses, as shown in figure 18.14. That's intuitively reasonable—pushing down ought to produce compression, and preventing outward buckling of a circle of already curved meridians ought to cause tension.

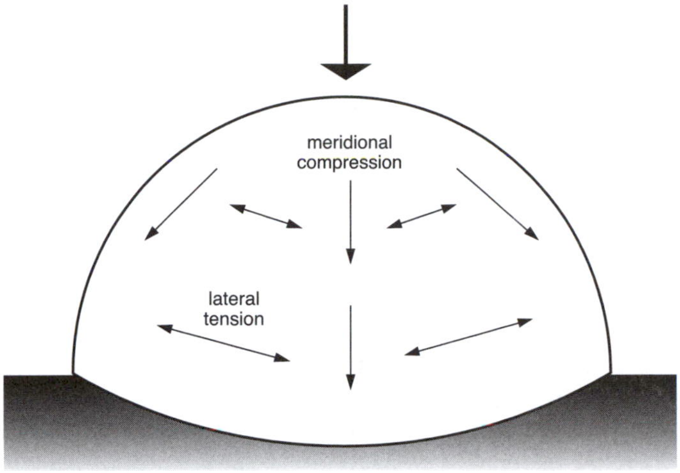

FIGURE 18.14 The stresses on a dome that bears a weight on its top.

(Brunelleschi, in his famous dome for Santa Maria del Fiore, in Florence, Italy, completed in 1436, took the then-radical step of using iron elements to aid in taking the circumferential tension. Tension clearly gave architects headaches before the advent of cheap steel in the nineteenth century.)

As should be clear by this point, compressive stresses pose greater problems for biological structures than do tensile stresses. So it's worth asking about the magnitude of the compressive meridional stresses on a dome; Salvadori (1971) gives the following formula:

$$\sigma_m = \frac{F}{2\pi r_o (r_o - r_o) \sin^2 \phi}, \tag{18.14}$$

where F is the weight pushing down on top, r_o and r_i are the dome's outside and inside radii respectively, and ϕ is the local angle between a meridian and the horizontal—zero degrees at the top (North Pole) and 90 degrees at the equator. (A more extensive analysis of domes and shells can be found in engineering sources such as Timoshenko and Woinowsky-Krieger [1959].)

At the equator the stress comes out to the loading force divided by the total cross section of wall, which makes sense, since nothing matters except the downward push of the load. At the very top, the formula implies an infinite stress, which seems less sensible. All it implies, though, is that a dome that takes a downward force cannot have an infinitely thin membrane, since the top has to act like a simply supported beam. Thus, the meridional, compressive stress drops from some high value at the top to just that of the weight at the equator, if the dome extends down to an equator; for any extension of the dome below the equator, the stress rises again. Since domes used in human architecture worry as much as anything about self-loading, the problem at the top isn't too worrisome—it's the point of minimum self-loading. And since they rarely form full spheres, the equivalent bottom problem only rarely matters. (We sometimes use large, full spheres to store compressed gases; for these, the main stress comes from pressure differences, not self-loading or crushing.)

Of concern here (as usual) is scaling. What happens to the stress in the material as size changes? For an isometric series of domes or shells, both radius and wall thickness will vary with length to the first power. If we assume a constant external force, F, stress thus varies inversely with length squared. That suggests that bigger is better. But the external force will more likely vary with volume or length cubed, that is, we scale both external and internal worlds. So the meridional stress on the material of dome or shell will vary directly with length—being bigger will be more stressful. Unless different materials are used, the bigger dome or shell will need a disproportionately thick wall. That may underlie what to me seems nature's preference for using domes and shells at relatively small scales, reserving big ones for cases such as marine mollusks where the material is cheap and thick, heavy walls don't compromise mobility.

Some of the few relatively large shells with thin walls are those of sea urchins and other echinoid echinoderms. They resemble pressure-supported structures (chapter 20), but they lack the requisite internal pressures (Ellers and Telford 1992), so they have to have proper shells, at least in the engineering sense. For the biologist, they have "tests" rather than "shells," and the latter distinction isn't just our usual terminological proliferation. Tests, unlike shells, are growing structures of articulated hard elements. For some, at least, collagen-swathed sutures permit significant local deformation, which should reduce impact loading and thus offset some of the hazards of a thin shell (Telford 1985). Nonetheless, they do smash easily—one can pick sea urchins off rocks, hit them against the same rocks, and, without benefit of tool, eat their gonads, in some species (especially *Strongylocentrotus droehbachiensis*) accounted a delicacy. Doing the same to mollusks is a lot harder. The best rationalization I can offer for why sea urchins tolerate such fragility is that the wave forces don't provide either piercing loads or a sudden, hammering impact.

CHAPTER 19

Less Simple Structural Matters

> The knee is a structure of such mechanical implausibility that it must be designed to do something very well. Unfortunately, I do not know what that something is.
>
> JOHN CURREY

CHAPTERS 15 THROUGH 17 CONSIDERED MATERIALS; BY ADDING SHAPE, chapter 18 brought us to structures. But it talked only about structures made of single materials. Here we add two additional elements—structures made of several materials and structures with joints or articulation points. With these additions, we find ourselves somewhere around the frontier between structures and structural systems. First, though, we need a look at how size matters for structures. After all, we have to deal with structural elements ranging from microtubules to the trunks of sequoia trees.

Scaling Simple Structures

Most often, bigger means weaker. But the operative scaling rules hold surprises, relationships that matter greatly to the engineer building a bigger bridge or the biologist worrying about the relative designs of organisms spanning a hundred-millionfold length range. Thus, we saw that fixed rectangular beams scaled oddly—sag varied with the first power of the width, the third power of the height, and the third power of the span. Beams of circular section behave similarly, with sag scaling with the fourth power of the radius or diameter, the second power of cross section, and the third power of the span.

What's most important here is that structures that normally experience purely tensile loads face scaling rules that differ from those feeling compressive or combined loads. Pull on a rope, and the force you must exert to break it depends entirely on its cross section and not at all on its length. The longer rope will stretch farther before breaking—why we normalized extension by dividing by original length—but it won't be weaker. D'Arcy Thompson (1942), in his famous chapter on scaling, comments that "trees should not bear heavy fruit on slender branches" but he says nothing about the lengths of stems. Stems, tendons, muscles live in this forgiving regime, and one gets the strong suspicion of its relevance in the long, slender arms of gibbons that brachiate, swinging from handhold to handhold.

Push on a long column, and it responds quite differently. Consider what happens if one enlarges a cylindrical column without changing its proportions—its diameter and wall thickness remain the same fraction of its length. Perhaps it faces failure at a critical Euler buckling force. If so, by equation (18.8), that critical force will be proportional to I/l^2; since I is proportional to the fourth power of length, the force will scale with the square of length. Perhaps it faces local buckling. If so, by equation (18.9), the critical force will be proportional to the square of wall thickness, again the square of a length.

Those squares sound auspicious, with critical force rising faster than length and at the same rate as cross-sectional area. Not so. The larger structure faces the larger load, both from self-loading and from whatever it encounters from environment or activities. Gravitational force scales with weight, thus volume, and thus length cubed. That more than eats up a critical force, in essence strength, that scales with length squared. And other forces scale in still worse ways than gravitational ones. Think of impact forces—the taller you are, the farther you fall, and the faster you hit. You hit faster whether from the accelerational increase in velocity with distance or from a higher terminal velocity from your lower surface-to-volume ratio. Think of drag—area may scale with length squared and drag with area, but most often bigger and thus higher means exposure to stronger winds and water currents. Faster flow means more drag, in proportion to speed if you're small and flows are slow; in proportion to the square of speed if you're big and flows are rapid.

Tensile loading, in fact, constitutes only a partial exception. Breaking force varies with cross section, thus the second power of a linear dimension. Isometric size change implies that the mass or weight of a load will vary with the third power of a linear dimension. So the larger remains the weaker—except, however, if the tensile element is elongated without change of load. In this special case, strength doesn't depend on length, or, put another way, it scales with length to the zero power. As Galileo (1637) pointed out, the leg bones of large animals must be disproportionately thick. We noted back in chapter 3 that skeletal mass increased faster than body mass—the scaling exponent for terrestrial mammals was about 1.08—above one, but not enough to compensate completely. By contrast, muscle takes only tension, and all of us mammals invest about the same 40 percent of our mass in muscle.

So the bigger is the weaker for anything except the load-wise direction in tensile loading. Small, terrestrial animals such as most birds have spindly legs; elephants, rhinos, dinosaurs, and titanotheres have massive ones. The difference strikes one more strongly when looking at, say, ants, which spend their lives running around on legs that make pins and needles look fat.

Nonetheless, the extreme slenderness of many relatively long arthropod legs remains striking. In group after group, the legs of some or all members appear truly disproportionate—among the arachnids, the harvestmen (often called "daddy longlegs" in the United States) especially; among the insects, the crane flies (daddy longlegs elsewhere); among the crustaceans, the spider crabs; among the myriapods, the scutigerid centipedes. Figure 19.1 shows some of these creatures. And one whole obscure marine group, the pycnogonids, have almost nothing but legs—with so little torso, their intestines run out into their skinny legs. Some of these walk, some fly; some are terrestrial, some are aquatic. As arthropods, they put their muscles inside those skinny tubes, which minimizes the leverage possible at joints and requires that the muscles shorten by the tiniest fraction of their lengths. The best suggestion that I've heard—from several people—is that these superlong legs make their owners less terrain-sensitive, reducing the effective bumpiness of the world across which they walk. A harvestman walks in leisurely fashion across a patch of English ivy in my yard. An ant would need great speed to cover the same overall distance in the same time.

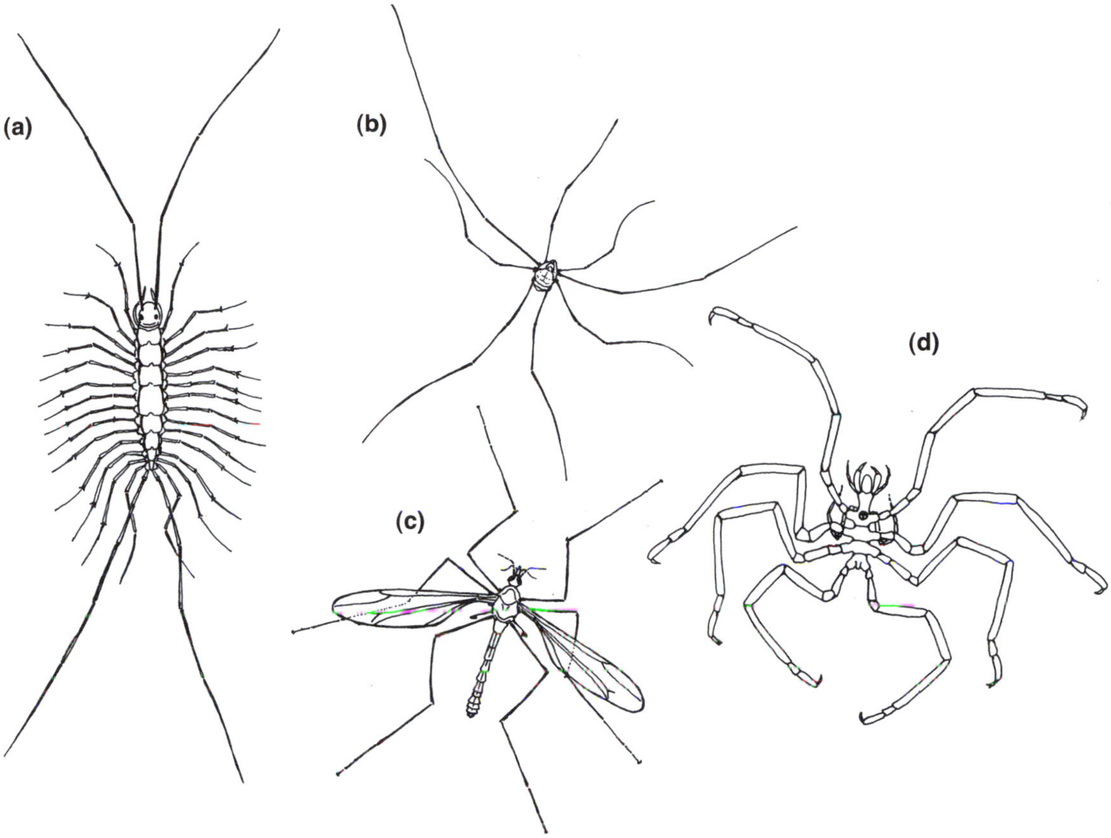

FIGURE 19.1 Arthropods with extremely spindly legs: (a) scutigerid centipede, (b) harvestman or daddy longlegs, (c) crane fly, and (d) sea spider or pycnogonid, considerably enlarged.

One more matter of scaling, one might say the extreme height of silliness. Consider the possibility that an untapered column might crush the material at its base as a result of simple self-loading. One just equates compressive strength (from the data of table 15.3) with weight (not mass) per unit cross-sectional area and solves for length:

$$l = \frac{\sigma_c}{\rho g}. \tag{19.1}$$

For something of wood (a hypothetical tree trunk, perhaps), with a strength of 50 megapascals and a density of 800 kilograms per cubic meter, the maximum height comes to over 6,000 meters! One could go a little higher with bone, a little lower with concrete. The bottom line is that crushing as a result of self-loading can be safely ignored as a design criterion for our kind of situations.

But having gone to this extreme, I can't resist mentioning a suggestion (Isaacs et al. 1966) whose practicality depends on the tensile analog of that calculation. We make earth satellites that stay above specific locations over the equator by putting them up at an altitude, about 35,786 kilometers or 19,323 miles, where the period of their revolution corresponds to that of the earth's rotation. That location, though, need be occupied only by the center of mass of the satellite, not by the whole thing. So we could make one that stuck out above and, more importantly, below, that geosynchronous altitude. If the satellite dangled down to the earth, one could get into orbit by climbing up it—the same energy would be required in the end, but the power needed could be reduced to any level by sufficiently slowing the ascent. The fly in the ointment—no material has anything close to the requisite tensile strength relative to its density. Try a calculation yourself, using equation (19.1) and tensile rather than compressive strength. While trees needn't worry about crushing the bottoms of their trunks, geosynchronous satellites cannot lower functional skyhooks.

Trusses

Consider, again, an I-beam working as a gravitationally loaded cantilever. The top flange takes tensile forces, the bottom flange takes compression. The middle keeps top and bottom properly distant; it takes mainly shear. Sometimes that middle part is referred to as the "webbing," which sounds like something other than a continuous surface. That's a good term in a mechanical sense—one can punch holes in the middle part of an I-beam with little loss of bending resistance and a useful reduction in weight—the side walls and wings of aircraft hide large numbers of such perforated beams. Round holes work well, since they resist initiation and propagation of cracks; round holes with their own circumferential flanging are even better.

The business can be pushed still further, as shown in figure 19.2. We can replace the top flange with a set of pieces joined end to end that take tension, replace the bottom flange with another set that take compression, and replace the webbing by a literal webbing of diagonal elements that take shear. We call the result a "truss," and it holds at least two advantages over an ordinary

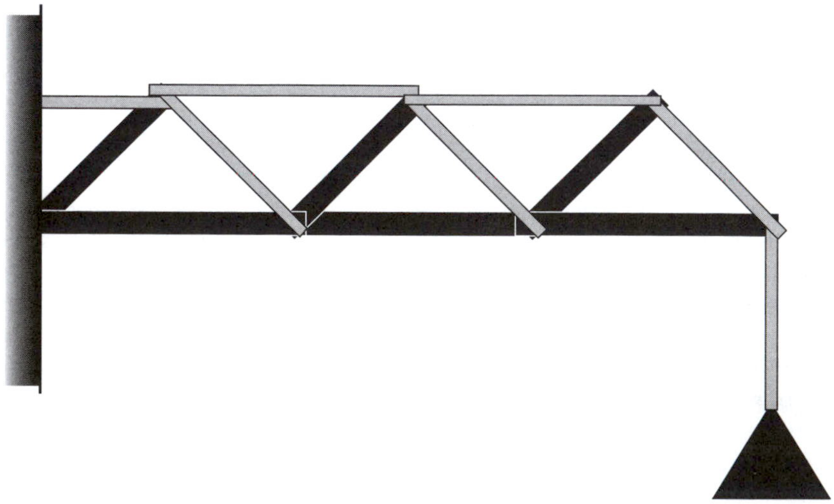

FIGURE 19.2 A cantilever truss. The compressively loaded elements are shown darker and wider.

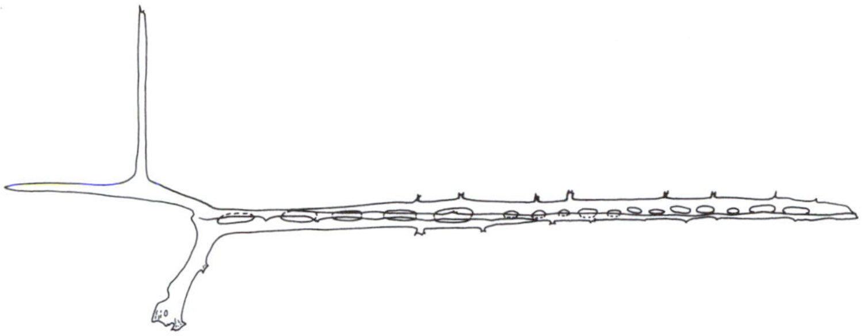

FIGURE 19.3 In some kinds of sand dollar larvae, the spicules that stiffen the arms are fenestrated, forming a truss instead of a simple beam. In this one, of *Dendraster*, the truss has three longitudinal members; the structure is about 0.3 millimeters long.

beam. The first, of course, is the simple weight economy of using diagonals instead of a continuous plate as webbing. The second, more subtle, is that the same resistance to bending can be obtained with this cleverly arranged set of short pieces. For our building projects, it reduces the weight of individual parts that have to be brought to the site. For animals (trusses being rare in plants), it adds articulation points that, properly equipped with muscles and bearing points, can confer mobility.

In short, a truss is just a structural framework that functions as a beam, a set of interconnected elements that do the task we previously assigned to a single piece. Again, it may be supported at several points or at opposite ends or it may extend as a cantilever outward from a support. Note that a bridge may be built as two cantilever trusses reaching toward each other so the result looks superficially continuous. Or a bridge may be designed to work as a pair of cantilevers during construction and as a fixed beam thereafter. A little playing (good exercise, too) with the formulas in the last chapter will show you that the final fixed beam can safely bear more load than the paired cantilevers—but, then, the bridge need only be self-supporting during construction.

The obvious feature of a truss is the arrangement of its elements as sets of triangles. In a well-designed truss, each element is called on to resist only compression or tension under the intended loads. For trusses of metal the two sorts of elements may look just the same, so we're not immediately aware of their opposite functions. Nineteenth-century engineers tried, it seems, every conceivable kind of truss, and the arrangements—Pratt, Warren, and others—still bear the names of these early explorers. Analogous trusses are present but relatively uncommon in living systems. D'Arcy Thompson (1942) pointed out that the bracing of the metacarpal bone of a vulture's wing takes the form of a Warren's truss; a similar arrangement is found in the arms of the free-swimming larvae of a sand dollar, *Dendraster* (Emlet 1982; fig. 19.3).

The rarity of natural trusses similar to those of engineers almost certainly traces in large measure to a basic difference between biological materials and metals, one we've repeatedly noted—biological materials aren't equally resistant to compression and tension. Nature uses an "ordinary" truss only when the direction of loading varies and its elements must withstand either compression or tension—as with a wing that must beat both up and down. With a consistent direction of loading, she goes a step further in specialization and employs different materials to withstand the different stresses. A mammalian backbone, as D'Arcy Thompson (1942) eloquently argued, is a truss working as a beam (fig. 19.4); the neck and thoracic regions are best analyzed as paired cantilevers extending each way from the forelegs. Tensile forces are concen-

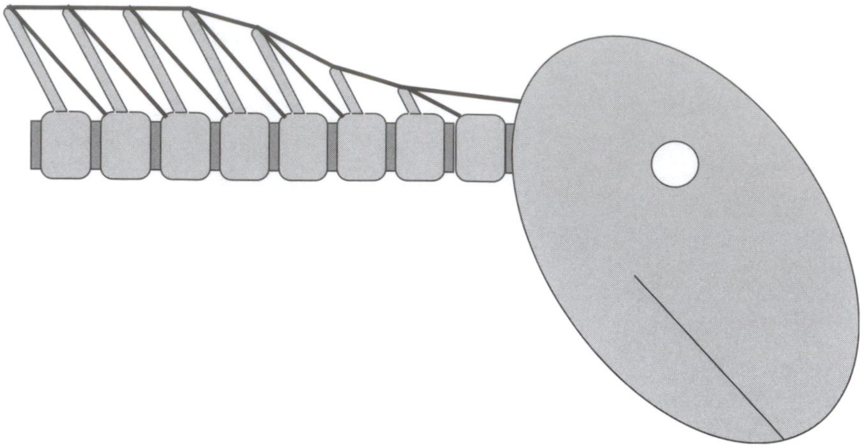

FIGURE 19.4 The truss formed by the cervical (neck) part of the vertebral column of a mammal. The combination of vertebrae, intervertebral disks, muscles, ligaments, and tendons support the head.

trated at the top, where they are resisted by muscles and tendons running between the vertical struts (neural spines) of the vertebrae; compressive forces are concentrated at the bottom, and there the vertebrae are appressed, with alternating bony centra and cartilaginous intervertebral discs. And the neural spines and diagonal tensile elements run crosswise to each other, the former taking compression and the latter tension. Again, using muscles as one form of tensile element gives options quite beyond any common technological analog—such a living truss permits moment-to-moment adjustments of both shape and stiffness. Using tensile elements of different elastic properties—such as making a nuchal ligament mainly of elastin and other connective tissue mainly of collagen—provides further versatility.

Braced Frameworks

Trusses are only one of a general class of structures sometimes referred to as "braced frameworks"—systems of struts. In such systems, by definition, loads are accommodated by changes in the lengths of individual members rather than by changes in their geometric interrelationships. A well-developed theoretical branch of engineering addresses the design of stress-resisting frameworks using minimal amounts of material; a good reference is Parkes (1965). About the only place where the results appear congruent with biological design is in the internal bracing (the so-called trabeculae) of bones. D'Arcy Thompson (1942) first noted the similarity between a cross section of the head of a femur and the head of a theoretical crane (derrick, not bird), and Currey (1984, 2002) gave the matter some attention. Figure 19.5 gives an example of the phenomenon. It might be mentioned as an item of biological hubris that if a bone is broken and reset in a different orientation or if the stresses on a bone change, the trabeculae reorganize to adjust their stiffening to the new situation. Since trabeculae fossilize, they might preserve information about how a bone of some extinct animal was put to use. The leg bones of saber-toothed tigers from the LaBrea tar pits in Los Angeles have especially splendid trabeculae.

Why, beyond a few obvious trusses and the trabeculae of bones, are braced frameworks in the classical engineering sense so notably rare in organisms?

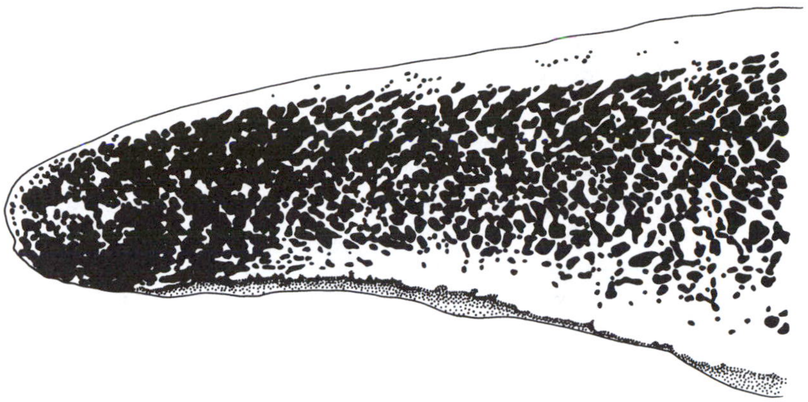

FIGURE 19.5 A cross section of part of the ischium of a horse. The free end of this rearmost bone of the hip gets pulled downward by leg muscles. The dark parts are voids in the solid; notice the generally diagonal arrangement of the trabeculae. You can make a similar preparation by grinding down lengthwise the upper portion of a well-dried lamb or pig femur.

Wainwright et al. (1976) suggested (and I quite agree) that such frameworks accomplish something organisms ordinarily avoid—making structures of great stiffness; lurking here is that general biological preference for strength over stiffness, for accepting rather than preventing deformation. For instance, geodesic domes are impressively efficient but stiff structures; in nature they seem to appear only as the shells of viruses (and there more likely as a matter of informational simplicity than for any mechanical advantage).

The subject of braced frameworks, though, still holds biological interest. It describes a useful distinction between nondistortable forms, or "statically determined structures," and distortable ones, officially known as "mechanisms" (a peculiar appropriation of a most ordinary word!). A simple test tells whether a structure is statically determined and thus a proper braced framework. Does knowing the length of each element give you an unambiguous picture of the shape of an array? Figure 19.6a quite obviously shows a mechanism—the struts can rotate with respect to each other at the pins. In figure 19.6b, by contrast, even though the joints may allow free rotation, the shape of the array remains fixed—it's statically determined. The distinction between these simple cases rests on the different consequences of three versus four or more joints in a loop—all of which amounts to a fancy way of saying that triangles are intrinsically rigid but quadrilaterals aren't.

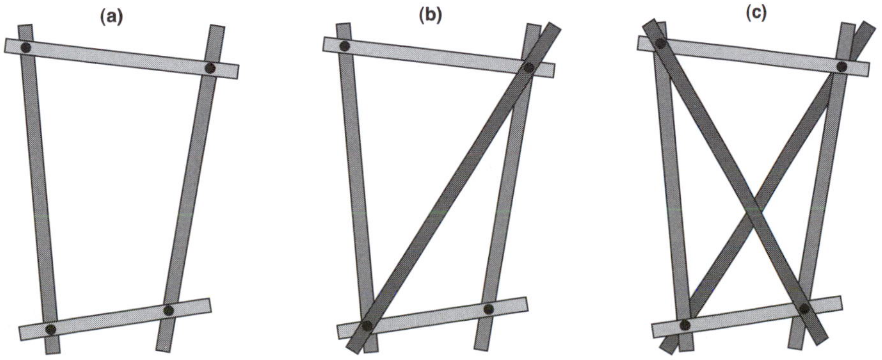

FIGURE 19.6 Three ways of pin-joining struts: (a) is a mechanism—its shape is indeterminate; (b) is a statically determined structure—it has a definite shape; and (c) is also statically determined but it's redundant—any single strut can be removed without changing its shape.

Three points need to be kept in mind. First, specific rules based on the number of joints and struts determine whether structures are statically determined or flexible, rules that work for structures considerably more elaborate than those illustrated here. (See Alexander [1983] for a quick look, or Hunt [1978] for the full works.) Second, neither arrangement should be thought of as superior to the other in any simple sense—the two just play different roles in mechanical contrivances. The skull and jaw of a fish or a snake, with (as we'll see) their complex mechanisms, serve one function; the trusses already mentioned do something quite different. Finally, any mechanism can be made statically determinate by sufficient reinforcing of the joints. The result is almost always a bad design, involving stress concentrations and excessive material—but "bad" remains relative, potentially outweighed by other considerations.

One of the peculiarities of statically determined structures is that a minimal structure usually works best, not only for simple material economy but also for both ease of construction and effective operation. We normally regard redundancy as a Good Thing—as I write I save my prose on two different disks and take comfort in the knowledge that I have access to a second word processor. But consider a nonredundant design to which a redundant element is added (fig. 19.6c). The nonredundant design of fig. 19.6b tolerates minor variations in the length of each strut—it can shift its geometry by rotation at the joints so the pieces will fit. One class I taught built a geodesic jungle gym of 120 struts joined by single bolts at the apices. We put it together without tightening the screws, and found the last piece apparently 5 centimeters short. We gave the structure a vigorous shake, and the final hole dropped onto the final screw as if, one might say, they had been made for each other. We hadn't made such precisely measured struts—the structure was just comfortably nonredundant. To fit, any *additional* (121st) strut would have needed real precision or an oversized hole in one end.

And the resulting redundant structure would most likely be weaker, not stronger. Look again at figure 19.6c—that the final strut strengthens the structure deludes the designer. Unless the struts are very carefully measured or very elastic, the last strut may carry all the load, it may carry none, or the load may be partitioned in some way that depends (if there's any stretch) in a nearly unpredictable way on its own magnitude. But (and perhaps organisms can teach us a lesson), only the stiff need be uptight about these problems of redundancy and load sharing. The automatic remodeling of the trabeculae of bones in the face of changing loads may be remarkable; as remarkable, for organisms, is its necessity only in a few unusually stiff structures. Flexibility, whether in tension, compression, shear, bending, or torsion, results in load-sharing—not necessarily optimal, but much better than nothing. Nature can build a big structure without worrying about such things as expansion joints.

Structures of Struts and Ties

Backbones put in an appearance earlier in connection with trusses. One peculiar feature of a backbone as a truss, noted at that point, is that some of its elements, rather than being made as flexurally stiff bones, are instead tension-resisting tendons and ligaments. We need to make this distinction explicit, and to do so let's call structures that resist all manner of stresses "struts" and those that can only be pulled on "ties." Quite obviously, a complex structure can always be built out of some combination of struts alone—structural struts can work as functional ties since they can take all possible stresses. By contrast, only under the special circumstances of solely tensile loading, can a structure be built only of ties. Spider webs exemplify this latter situation, and they nicely direct attention to the advantage of using ties. A given load can almost always be borne more cheaply (in terms of material) when it acts in tension rather than in compression. No great mystery about why—tensile elements need not worry about failing by either Euler or local buckling.

That freedom from buckling underlies the difference in the way struts and ties scale, a topic that has already come in for attention. Recall that while in general bigger is weaker, tensile elements evade the general rule as far as length goes. We rate string and rope by safe maximum tensile load; no allowance need be made for length, and 100-pound test is still 100-pound test no matter how long. The implication is that quite different guidelines apply to the best forms of compression- and tension-resisting elements—the first ought to be kept short, but for the second length matters little.

Should the astute designer, then, try to minimize the use of struts while maximizing the use of ties? On scaling grounds, certainly. But using ties has particular consequences.

- Ties, in one sense, may be safer than struts. For both Euler and local buckling, the key formulas equations (18.8) and (18.9)) gave a critical force; exceed that force even briefly and a column will collapse. Once started, buckling demands little to keep it going. Stretching an element takes an ever-increasing force if the force against length curve (and usually the stress against strain curve) slopes upward. The system gets unstable only when the piece begins to elongate so severely that the decreasing cross section makes a given force produce an increasing stress. In addition, any load-induced sag of a tensile element stretched between two points will reduce the force on the element by the geometric argument made in chapters 3 and 16 and illustrated in figure 16.3.

- But using ties places severe demands on joinery. Compressive elements can be stacked or fitted against one another, as with bricks or the centra of vertebrae. Until the invention of Portland cement, in 1824, mortars and cements had next to no tensile strength. Even now, the mortar between bricks mainly assures a good area of contact and so minimizes crack initiation. By contrast, traditional shipwrights long enjoyed greater pay and prestige than carpenters—one has to pay more attention to tension when making wooden sailing ships that will withstand waves. And we have special receptors (Golgi tendon organs) that, in part, inhibit muscles that might pull tendons off bones.

- Tensile elements, in part because they take only tension, tend to produce less rigid structures than do compression elements. Suspension bridges move a lot more from winds and loads than do arch bridges or cableless truss bridges. We stopped using suspension bridges for heavy railroad use back in the middle of the nineteenth century—they could carry the payloads with less self-loading, but their liveliness raised excessive risks of derailments.

Consider, though, how nature builds her structures. The last two consequences of using tensile elements may matter little to her. At least we have little evidence that tensile joinery presents a special challenge, whether hooking muscles via tendons to bones or the upper sides of branches to tree trunks—although tensile failure can be especially vexatious, which may be why we have those Golgi tendon organs. Nature glues barnacles, limpets, and macroalgae onto rocks, where they withstand the monumental tensile forces of breaking waves. And she makes few, if any, static structures that, like Gothic cathedrals, depend for their integrity on stable stacking.

Moreover—and the point keeps coming up in this account—she does a lot with flexible structures, more often building to a criterion of adequate strength than (as do we) of ample stiffness. The stiffness of particularly stiff biological structures, such as the trunks of hardwood trees, the bones of large vertebrates, mollusk shells and hard corals, and so forth, usually have some special rationalization. Moreover, muscle works only as a tensile element. Since one can't make much of a structure that will take only tension, any structure than incorporates muscle, meaning almost any macroscopic, motile animal, must have at least one compressive element as well.

Spicular Reinforcement—Grit in the Jello

Should a braced framework or a mechanism that acts in a supporting role be made of a few large struts or of a lot of little ones? The problem is the now familiar one of scaling, in particular of the scaling of the tensile and compressive elements of a framework.

R. Buckminster Fuller (1962) recognized this distinction as well as the economy of using tensile elements—as mentioned, where buckling can't happen, less material is needed to resist a given load. He coined the term "tensegrity" to describe structures in which the tension elements form a continuous lattice into which short, isolated compression elements fit—"islands of compression in a sea of tension" is his metaphorical phrase. While the resulting structures are esthetically satisfying (there's a neat mast by Kenneth Snelson outside the Hirshhorn Museum in Washington, D.C.), they suffer the disadvantage of not being particularly stiff unless built with huge tensile stresses on their cables. But if, as we've argued, nature isn't usually as interested in stiffness as are we, then such latticeworks of tiny struts should abound in nature. Which they don't, particularly. The main limitation of the scheme, I'd guess, has to do with motility, as noted just a few paragraphs ago. Concomitant with self-propulsion is usually a set of muscles pulling on solid elements. For an organism of a given size, the smaller the compression elements, the greater the number of muscles needed to work them; indeed starfish, with very short ossicles, have a fabulous number of muscles (Eylers 1976).

There is, though, a group of animals that use very short struts (called "spicules")—typically a few tenths of a millimeter in length—and yet achieve heights of up to 2 meters and withstand severe environmental forces. These are sponges, with variously shaped spicules of brittle material laced together by a form of collagen. Their spicular skeleton is (as are sponges generally) all too often dismissed as primitive, but for sessile animals nearly devoid of contractile tissue it's an eminently sensible arrangement. Indeed, even the largest sponges devote only a few percent of their wet weight to supportive material, both spicules and collagen together. (Bath sponges are aberrant sorts with no spicules and extra collagen instead.)

At one stage I had the idea that the spicules and collagen of sponges ought to be Fuller-on-earth, with geodesic lattices, tensegrities, octet trusses, and all of his imaginative and efficient structures. I looked at quite a number of sponge skeletons (and illustrations in old accounts) without finding much coincidence with Fuller's designs; I was about ready to admit that sponges were truly primitive in the ordinary sense. But then in Bermuda, on the edge of the ocean, I found the skeleton of a tiny sponge that (being my own) I could abuse (fig. 19.7). I compressed it nearly flat (between fist and desk) whereupon it promptly sprang back to its normal shape. Examination revealed a three-dimensional array of mechanisms without a single statically determined set of spicules. Sponges seem to be consummate specialists in accepting deformation—tiny, brittle spicules, too short and isolated to be at much risk of bending, laced together with as good a set of inextensible but bendable straps as nature has devised. One is tempted to call these mechanisms "unbraced frameworks."

If mechanisms rather than statically determined structures are acceptable, then yet another possibility arises. At a crude level of analysis, a statically determined structure is something specific, while mechanisms are everything else. Putting small pieces of brittle material into a pliant matrix gives a composite called a "filled polymer"—it amounts to a kind of random array of mechanisms. Koehl (1982) looked into the extent to which the connective tissues of animals that had embedded spicules behaved like proper filled polymers—embedded spicules are fairly widespread, not just in sponges, but in some coelenterates, echinoderms, mollusks (the chitons), arthropods (stalked barnacles), and ascidians. She took isolated animal spicules of various kinds and concentrations, embedded them in gelatin (raspberry flavored), and performed various mechanical manipulations on the products. Since the normal function of spicules is to stiffen tissue (although we're still considering relatively unstiff structures), she used that as a criterion of effectiveness. Even a relatively small proportion of spicules dramatically increases stiffness; more spicules or more elongate or irregularly shaped spicules give more stiffness, and small spicules are more effective than are large ones for a given added mass. One factor that matters a lot is the area of contact between spicules and matrix, not unlike

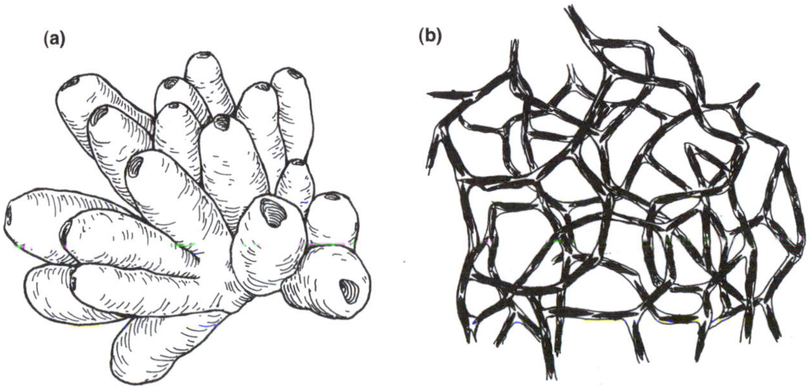

FIGURE 19.7 A sponge (a) and the skeleton of a sponge (b), showing how tiny, stiff spicules are embedded in an easily bent collagenous meshwork.

other composites. So whether the spicules are in a specific framework or in a random array, roughly the same rules seem to apply.

Mechanisms

Nature, of course, makes mechanisms of much greater sophistication than these unbraced frameworks and random arrays of compressive elements—our arrays of muscles and bones, for instance. The parent subject in engineering goes by the name "kinematics," and it's critical in the design of every mechanical appliance—a look at any textbook of kinematics (such as Gans [1991]) will give a sense of the subject's scope. The engineering rules may still hold, but nature's willingness to use elements of varying degrees of stiffness, to bend them repeatedly, and to incorporate multiple contractile engines as tensile elements yields mechanisms of quite different appearance and functional character. A look at some particularly extreme mechanisms ought to emphasize the point as well as give a look at how structural elements can be combined.

Multibar linkages

Consider how an element of a mechanism behaves when its ends are connected in what's called a multibar linkage, as in figure 19.8. The easiest way to envision what's possible is to assume one fixed bar and then examine how the others might move. The minimum number of bars has to be four, since three would constitute a statically determined structure rather than a mechanism. In such a four-bar linkage (fig. 19.8a), any part of each of the three movable bars can shift along a fixed line (or, really, arc)—we say it has one degree of freedom. A five-bar linkage (fig. 19.8b) constrains its members less. The ones attached to the fixed bar can only swing in a simple arc, but the others can both move in a plane and rotate. Addition of yet more bars in the loop will further reduce the constraints or—perhaps a better way to put it—confer on its members additional elements of permissible motion.

We mammals make no great use of multibar linkages, but a lot of other vertebrates depend on them. The most famous are snakes that can swallow items of prey whose diameters and cross sections exceed those of themselves. How pythons (genus *Python*) manage was carefully analyzed by Frazzetta (1966), who regarded their skull and jaws as linkages with no fewer than

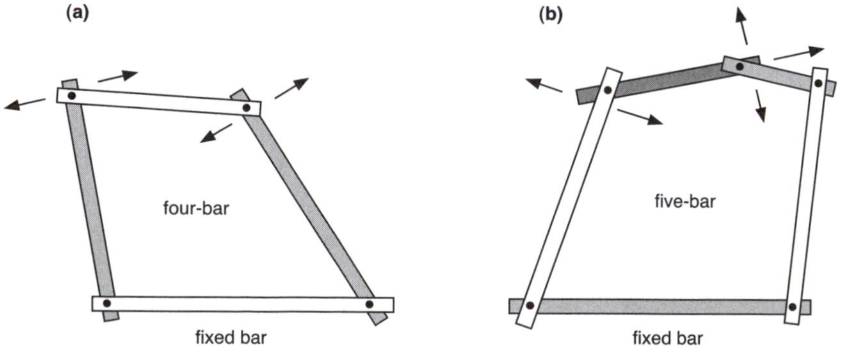

FIGURE 19.8 Mechanisms made up of multibar linkages: (a) four-bar, and (b) five-bar.

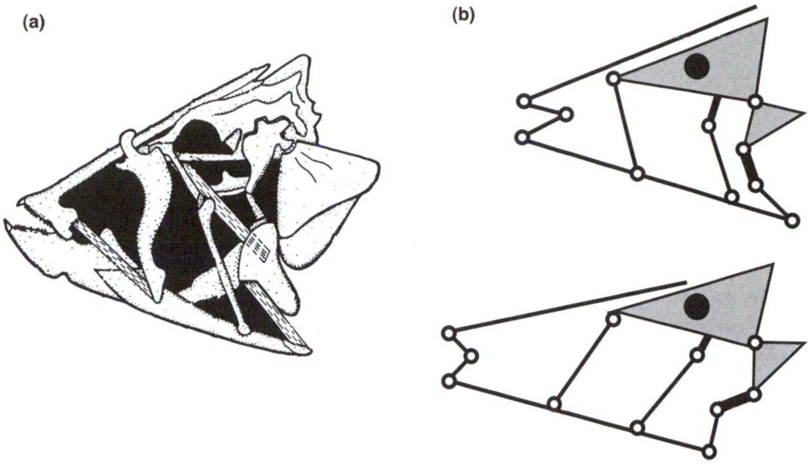

FIGURE 19.9 The head of a sling-jaw wrasse, normal and protruded—somewhat simplified here but still remarkably complex—seen as (a) anatomy and (b) mechanical diagram. Part (a) has been adapted from Westneat, M. W., 1991; *J. Exp. Biol.* 159: 169 with permission of The Company of Biologists, Ltd.

eight bars. Such snakes use two such linkages, one on each side of the head, with a lot of flexibility in between. The setup permits the mouth to gape sufficiently to accommodate huge prey, which then get digested at leisure. The members of another genus of snake, *Dasypeltis*, of quite a different lineage (colubrids rather than boids) eat only eggs, which they swallow whole and break in their esophagi. The eggs, like python prey, exceed the snakes' cross sections. Eggs don't fight back, and these oophagous snakes construct lighter skulls. But their bones are more firmly attached, and their impressive gape depends on a jaw with three hinge points along its length on each side (Gans 1974).

Nonetheless, the jaw mechanism of even a fancy snake looks simple next to what some fish do with multibar linkages in their heads (Westneat 1991). The most extreme must be the sling-jaw wrasse, *Epibulus insidiator*, which shoots out an otherwise unnoticeable snout to snag prey. According to Westneat and Wainwright (1989), who've analyzed the biomechanics of the system, this wrasse can protrude its jaw by a length equal to 65 percent of normal head length. Protrusion takes only about a thirtieth of a second; acceleration exceeds 100 meters per second squared, or 10 g; and snout speed hits 2.3 meters per second, or over 5 miles per hour. The components—bones, ligaments, and muscle—may be ordinary, but their arrangement is anything but. Figure 19.9 shows the system from the viewpoint of an anatomist (fig. 19.9a) and a kinematic analyst (fig. 19.9b).

Why might we mammals be so ordinary? Perhaps relatively larger brains limit the degree to which head bones can specialize; we certainly have highly fused cranial cases. Perhaps more importantly, we do more with specialization of tooth form, function, and position. Neither fishes nor snakes do much mastication, and the teeth of an individual vary little from front to back.

Joints and junctions

What can we say in general terms about more ordinary joints and linkages, the kind with which we're equipped? Conventional kinematic analysis recognizes six possible motions at joints, that is, how one member can move when the other is fixed. These consist of translational motion in three orthogonal directions (up-down, back-forth, left-right, one might say) and rotational motion about three axes, as in figure 19.10. Rotation in a vertically longitudinal plane, thus about a left-right axis, is *pitch*. Rotation in an horizontal plane, thus about a vertical axis, is *yaw*. Rotation in a vertically transverse plane, thus about a longitudinal axis, is *roll*. Joints and junctions can have between zero and six degrees of freedom, depending on how many of these motions they permit. (Alexander [1983] and Currey [1984, 2002] give good introductions to the subject; McGowan [1999] considers our own joints more specifically.)

For the joints of animals, that's unnecessarily elaborate. We don't go in for joints that make adjoining members do translational sliding (except within the joints themselves). For back and forth movement, protrusion and retraction, the way of a piston in the barrel of a hypodermic syringe, for instance, nature's devices (as will appear in the next chapter) manage without recourse to discrete joints—except, stretching the point slightly, the telescoping segments of some insect abdomens. And we make minimal use of joints that move slideways from side to side or up and down. So the main motion of all common joints consists of rotation about one or more axes—one or a combination of pitch, yaw, and roll. Pitch and yaw we normally call bending, and the distinction between them can be a bit arbitrary or nonuseful. Roll we normally call twisting—move your hand and forearm about your wrist and elbow until you have these well distinguished in your mind. That means one, two, or three degrees of freedom (fig. 19.10, again). We can put the possibilities hierarchically:

A. *One degree of freedom*

■ *Bending about one axis.* Our hinged elbows and knees work this way, at least superficially, as do the leg joints of arthropods, and the hinges of bivalve mollusks. And, of course, door hinges and innumerable items of human technology. Bending can be accomplished by making one element slide or roll about the end of the other, as do door hinges, knees, and such. Viewed close up, many more of nature's bending joints truly bend than do the ones humans make, although as we become more adept at using less stiff materials we're making more hinges whose material does indeed bend.

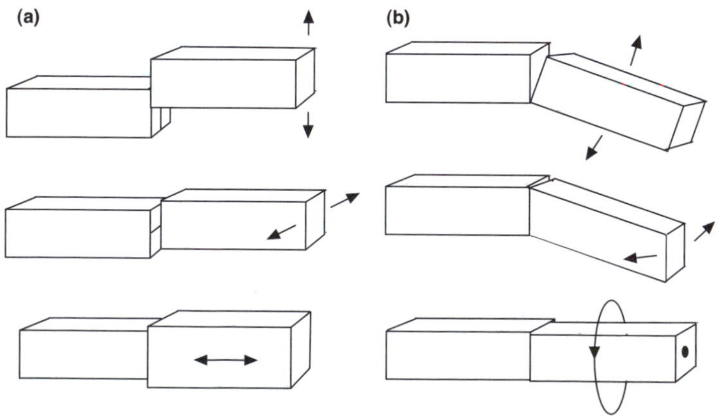

FIGURE 19.10 The six possible degrees of freedom at a joint: (a) translational movements; and (b) the more important rotational movements, from top to bottom pitch, yaw, and roll.

■ *Twisting about a lengthwise axis only.* This kind of joint, rare in nature, turns up in sea pens (*Ptilosarcus*), which stand unbent on bay bottoms and rotate so they face the current (Best 1988). The ascidian, *Styela*, mentioned in chapter 7 (and shown in fig. 7.3b) does much the same thing (Young and Braithwaite 1980)—and for the same reason, taking advantage of ambient currents for suspension feeding. Pushing up the twistiness-to-bendiness ratio of a cylinder, of course, restricts bending relative to twisting and thus encourages just this kind of specific torsional flexibility, but without an absolute prohibition on hinge action and without a discrete joint. Etnier (2001) considered twisting in systems of stacked rotatable elements, the stalks of crinoids (sea lilies, a once major group of echinoderms) and the antennae of decapod crustaceans. She found that while they twist, they also bend, with twistiness-to-bendiness ratios no higher than 6.6, not notably different from tree trunks, feather shafts, and other unjointed systems.

B. *Two degrees of freedom*

■ *Bending both up-down and left-right.* Our wrists and knuckles do both, although with greater up-down than left-right excursion. (Ankles, of course, can twist a bit as well; hands get most of their twisting from rotation of the forearm.) Arthropods gain two degrees of bending freedom by putting two 1-degree bending joints next to each other, each oriented at a right angle to the other. Examine the legs of a lobster or crab (boiled is fine) the next time you get a chance. The classic work on such cases was done by S. M. Manton, in the 1950s and 1960s; as put, with a long list of references, by Wainwright et al. (1976), "The accounts of her researches in this field constitute a monument in the study of mechanical design of the most mechanically diverse group of organisms that have ever lived." The technological analog of the pairing are the universal joints used (among other things) to link a car's engine, attached to the body, with its wheels, sprung beneath the body. Bending must be allowed, but freedom to twist wouldn't permit transmission of the engine's output. Shift levers (the old kind) and joy sticks work the same way.

■ *Bending either up-down or left-right plus twisting about a lengthwise axis.* If bending can happen in only one plane and twisting about one axis, a joint will have two degrees of freedom. Currey (2002) pointed out that a human knee, flexed at around 90 degrees, can twist as well as bend, even if the knee joint only flexes when fully extended. Try it—and note that when your leg is stretched out, you can twist only at hip and ankle. A technological case isn't obscure—the lever to the right of the steering wheel on many cars goes up and down to activate the windshield wipers and twists to set the speed of the intermittent setting.

C. *Three degrees of freedom*

■ *Bending both ways plus twisting about a lengthwise axis.* These abound in the skeletons of vertebrates such as we. Our hips and shoulders have this versatility; our ankles do to a slightly more limited extent. The joints between our cervical and lumbar vertebrae do so as well, so you can bend and twist both your neck and your torso. While our wrists don't twist much, relegating that to the forearm, that hands the hand no limitation. Twisting joints are oddly rare, at least relatively speaking, beyond the vertebrates. Arthropods don't go in for rotating ball joints that would permit twisting, something we take advantage of when disarticulating a crustacean leg with a joint-dislocating twist. Some degree of twist can be accommodated by serially multijointed systems, such as the antennae of decapod crustacea and the stalks of crinoids already mentioned and, of course, our backbones, but each joint allows only slight angular motion. And insect wing hinges can twist a little—but only a little, with most twisting taking place out on the wing's surface itself as a result of transarticular pulls on the various veins and of (as noted in the last chapter) aerodynamic forces.

We have, in this sense, only five main kinds of joints among organisms. That's far fewer than the possible permutations of six kinds of motion (Currey [1984] noted a dozen practical possibilities); indeed it's a surprisingly narrow range of possibilities. But it only implies simplicity for biological joints by a single criterion, albeit one that looms large in our own technology.

Once again, human technology turns on stiff materials. Joints almost always either move freely or don't move at all. An unintended degree of freedom may be tolerable, if minor, or it may represent undesirable wear, poor design, or actual danger. Stops often restrict the range of motion in one direction or around one axis or another. Nature's joints may constitute a narrower range of kinematic possibilities, but specifying their motion requires additional variables besides degrees of freedom. Where a linkage involves materials spanning a wide range stiffness, at least two other ones need attention.

One, of course, is the range of permissible motion. Except for the rotation of bacterial flagella, which are true rotary engines, all motions approach limits. And "approach" conveys the sense of what happens better than "hits." Limits vary with the force involved, with a joint's previous history, with whether some element moves actively or passively, even with the state of the joint with respect to some other avenue of mobility—as noted for knees. Very limited motion gets amplified by an additive series of joints, as with those between our vertebrae or between the segments of a long crustacean antenna. We have joints that we don't even think of as joints, cartilaginous and ligamentous connections that permit small but crucial movements.

Nor must joints resemble the spatially discrete structures of human technology. We noted earlier (chapter 18), when talking about flexural stiffness, how a slight reduction in cross-sectional area of the stalk of a sea anemone just beneath its tentacles produced what amounted to a semilocalized, passive joint—to which we can now assign two degrees of bending freedom. And to which we can now append yet another note. Animal hinges mainly bend, but some do a combination of bending and sliding—most of ours, for instance. Knees look like simple hinges. But they're not, and replacement knees present a much greater challenge to the designers of prostheses than do replacement hips, for all the overtly greater freedom of motion of the latter.

The other variable is the ease of motion, again highly variable, with every degree of both resistance and resiliency represented. Bending incurs more resistance than sliding, but, as in insect wing hinges and scallop shell hinges, highly resilient materials deform and provide both energy storage (chapter 15) and muscle antagonism (chapter 23). Sliding, as within our freely-movable synovial joints like hips and elbows, happens between well-lubricated surfaces—dissect the muscles off a fresh lamb leg and you'll find surfaces that slide over each other with remarkable ease. McGowan (1999) gives a good introduction to matters of friction and lubrication of joints.

So for animal joints, degrees of permissible motion tell only part of the functional story. The lovely rules and analyses of engineering kinematics turn out to be in at least one way too general and in others seriously inadequate for the joints animals make. To leave you (one might say) with something to chew on, consider your jaw—vastly simpler than those found among fishes and snakes. On a skeleton one sees what appears to be a single, discrete hinge, rotating about a crosswise horizontal axis. But that would only bite. You're quite capable of masticating food, in part by moving the lower jaw side to side, rotating about a vertical axis. And you can move it back and forth, translating in a horizontal plane. Nor are these all—by preferentially contracting one rather than both of the big jaw muscles you can chew on one side or the other, thus rotating about a lengthwise horizontal axis. Adequate analysis takes the best skills of the functional morphologist, familiar but not in thrall with the formal kinematics of ordinary mechanisms.

CHAPTER 20

Hydrostatic Structures

> If you can't demonstrate it with balloons, it's probably not important anyway.
>
> **STEPHEN A. WAINWRIGHT**

ORDINARY STRUCTURES, AS THOSE OF THE PAST TWO CHAPTERS, HAVE to resist both compression and tension—they need both struts and ties. We distinguished between elements, such as bones, that could do either and elements, such as tendons, that could serve only as ties. Flexibility in its various guises and the standard mechanical properties may have varied, but all structures consisted entirely of solid materials, as solids were defined back in chapter 2. But think back (or look back) to the practical distinctions among the states of matter. Fluids as well as solids resist imposed forces, at least selectively. They can't handle shear (as opposed to shear rate), and liquids, although not gases, will take tension—at least under special circumstances. Both solids and fluids, though, resist compression, the definitional task of a strut. Question—might gases and liquids be put to use as compression-resisting elements in practical structures? Could they serve as, or somehow in lieu of, struts?

We all know balloons. They give children their the first evidence that air is a real substance, something that can transform a limp piece of rubber into a large, well-shaped object—or the lack of which can do the opposite, as found by Piglet in *Winnie the Pooh*. We use inflated tires, inflated blimps, and sometimes even inflated buildings; all are structures best described as internally pressurized. All carry to an extreme the use of separate materials to withstand tension and compression. They resist tension with their outer membrane, which may, as in a tire, be quite a thick and complex piece of material. They resist compression with their internal fluid; in these cases the latter is either air or helium, both gases. The gases are lightweight, cheap if air and usefully buoyant if helium, and (a mixed blessing) somewhat compressible. Since balloons, tires, and blimps have shapes that withstand mechanical abuse, gases must obviously be materials that can be used to withstand compression in competent structures.

Many organisms carry bags of air within lungs, seedpods, and the like, but only rarely can the air be regarded as a structural material. The floats of some large algae and of the Portuguese man-of-war (a colonial coelenterate) are perhaps the best natural analogs of balloons and blimps. Life is wet, and water provides the common working fluid of nature's internally pressurized systems. It's functionally incompressible (which is to say that its Poisson's ratio is 0.5), so it takes compressive stresses far better than any gas, even relative to its density. While dense, it's no denser than the other materials of organisms. To the benefit of aquatic organisms, it's neutrally buoyant at any depth. It's also cheap—its acquisition entails none of the inevitable metabolic costs of making protein or polysaccharide.

So how might one make a structure based on an internally pressurized aqueous system, a submarinating blimp perhaps? The overall shape will most likely be a circular cylinder, a sphere, or an ellipsoid, for reasons that go back to Laplace's law (yet again); we'll look mainly at cylinders. Where pressure is higher inside a structure than outside, sharp corners tend to round themselves, and the tension on flat surfaces makes them bow outward. Put another way, radii of curvature tend toward constancy for a given structure if the external membrane merely takes tension.

Since these internally pressurized structures can be deformed either by environmental or internal activities, we ought to consider not just shape but the possibilities for change of shape. At the start, we have to distinguish among three somewhat idealized general arrangements. First, the structure—call it a "hydrostat"—might have a constant volume, so any increase in length is matched by a compensatory decrease in cross- sectional area. This most obvious constraint (essentially just an incompressible system) is central to the functioning of a peculiar group of structures collectively named by Kier and Smith (1985) "muscular hydrostats," to which we'll return later in the chapter. Here some of the situations will involve constancy of volume; others will not. Second, such a system might have constant surface area, an outer membrane that can be changed in shape but not size. As far as I know, no biological case of the latter has been described.

Hydrostats of Fixed Fiber Length

The third possibility is the least obvious. *A system might have some tension-resisting fibrous component in the membrane, one whose fibers can't change length.* (It may take you a moment's thought or some cross-hatching on paper to persuade yourself that the second and third possibilities are actually distinct.) Systems with inextensible fiber reinforcement in their walls come closest to what we most often find in organisms. Their behavior, not at all self-evident, will take a bit of explaining.

For a start, fibers might reasonably be arranged in a surface membrane in two fundamentally different ways. Either some might run lengthwise and some circumferentially, as in the sleeve of a shirt. Or they might run helically, with separate left- and right-hand helices extending down the length of a cylinder, as in a good-quality necktie or a dress cut on the bias. The two possibilities (shown in fig. 20.1) turn out to respond quite differently to the various stresses to which structures might be subjected. If you want to get a feeling in the most immediate sense for the difference, just inflate cylindrical balloons in tubes of

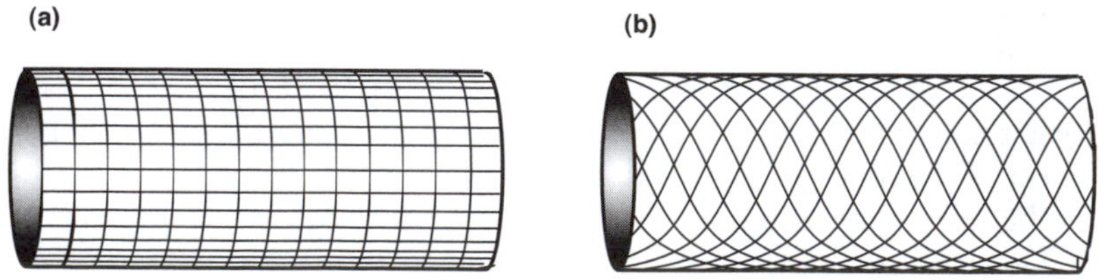

FIGURE 20.1 Two contrasting ways to arrange fibrous reinforcement in the membrane of an internally pressurized cylinder—fibers may run lengthwise and circumferentially (a), or in right- and left-hand helices (b).

loosely woven fabric, with one tube made with the threads running lengthwise and circumferentially and the other with the threads running helically.

The model with longitudinal and circumferential fiber reinforcement does not deform in response to pushes or pulls—compression or tension—on the whole cylinder. For a push the circumferential fibers prevent expansion in girth, and for a pull the longitudinal fibers prevent increase in length. So if inflated to any significant degree of pressurization, maintaining constant fiber length means maintaining constant shape as well. Nor does it like to bend—it has a relatively high flexural stiffness (Koehl, Quillin, and Pell 1995). When it does deform in response to a bending stress, it easily develops the kinks of local buckling. Only twisting—torsional stress—elicits a smooth deformation. And that doesn't alter the cylindrical form.

By contrast, the helically reinforced model lengthens and shortens smoothly as it is pulled and pushed. It bends more easily than the other, and it's less prone to local buckling. Only torsional stresses aren't accommodated with a noticeable deformation—each of the two sets of helical fibers resists twists in one or the other direction. In terms introduced earlier (chapter 18), it has a low twistiness-to-bendiness ratio, pushing the ratio in the opposite direction from that done by lengthwise grooves and other bits of noncircularity.

With few exceptions, nature uses the second arrangement of fibers for her internally pressurized, water-filled cylinders. These structures (often termed "hydrostatic skeletons" or "hydroskeletons" as well as "hydrostats") have helical reinforcing fibers. And this particular arrangement is no rare or once-evolved thing. It occurs in the stems of young herbaceous (nonwoody) plants such as sunflowers; it provides a wrapping for flatworms (platyhelminths and nemerteans), roundworms (nematodes), and segmented worms (annelids); it stiffens the body wall of sea anemones; it determines the response to muscle contraction of the outer mantle of squids; and it's a major functional component of shark skin. The material of the fibers varies widely, the functions of these hydroskeletons are even more diverse, but the wrapping is almost always helical.

Stresses and Scaling

The relationship between the structure and the behavior of these hydrostatic beams and columns is perhaps best developed with a hypothetical and idealized mathematical model expressed as a graph (fig. 20.2).* "Fiber angle" is the angle between the reinforcing fibers and lines running lengthwise along a cylinder. Stretching out a cylinder decreases the fiber angle—the fibers run more nearly lengthwise; compressing a cylinder end-to-end increases the fiber angle—fibers run more nearly circumferentially. The graph represents the results of taking a cylinder and either compressing it into a disk (moving left) or stretching it into a line (moving right). If the cylinder isn't fully inflated, its volume can be lower than the maximum given by the curve; since a cylinder can't be fuller than full, the area above the curve can't be occupied. At full extension or compression, of course, the cylinder has a volume of zero, so maximum volume must be reached somewhere in between. "In between"

*I've replotted the graph given elsewhere to put length rather than fiber angle on the x-axis; the result is a little like a mirror image of the usual version, which I have always found confusing.

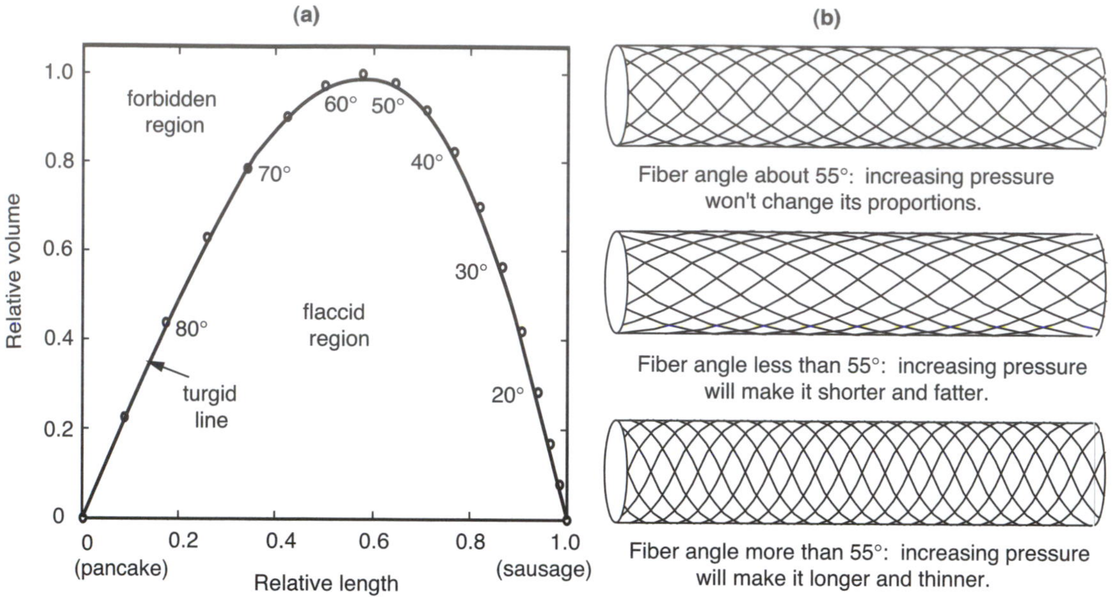

FIGURE 20.2 (a) The changes in volume and fiber angle as the length of a helically wound cylinder varies. The length of each fiber (and thus of the fully extended cylinder) has been arbitrarily set at 1.0. Beneath the curve the cylinder isn't fully inflated and so isn't circular in section; above the curve it would be impossibly overinflated. (b) The appearance of representative cylinders, and the practical consequences of their different fiber angles.

turns out not to be something ordinary, such as 45 degrees or 60 degrees, but an angle of 54.73 degrees, or, for practical purposes, 55 degrees. Bear in mind the idealized conditions assume, in particular, perfectly inextensible fibers, perfectly incompressible fluid inside, and rigid caps on the cylinder's ends.

The trigonometric derivation of this point of maximum value can be found in Wainwright et al. (1976). Its origin matters more here. Recall the truism that one can't push on a rope—tensile elements such as the present inextensible fibers withstand only tension. And recall from chapter 3 the different stresses in the different directions in a thin-walled cylinder. Or, better, we repeat the relevant (and easily derived) formulas that relate stress to pressure difference, Δp, radius of curvature, r, and wall thickness, Δr. For the hoop stress in a thin-walled cylinder:

$$\sigma_h = \frac{\Delta p r}{\Delta r}. \qquad (3.7)$$

For the longitudinal stress in such a cylinder, we get

$$\sigma_l = \frac{\Delta p r}{2\Delta r}. \qquad (3.8)$$

Notice that the hoop stress is exactly twice the longitudinal stress. That's why the reinforcing fibers must, if they're not asked to take anything but tension, run more nearly circumferentially than longitudinally—at that angle of 55 degrees. We're talking about geometry, not material science or structural engineering. Mathematical reality has priority; design must fall into line.

These equations have implications for how internally pressurized, hydrostatically supported structures ought to scale. Consider the hoop stresses in the wall of a transverse slice of a long cylinder, one whose material can tolerate no more than a certain stress. If, by equation (3.7), the slice has a wall of constant thickness—a standard membrane, for instance—then Δr as well as σ_h will be fixed, and tolerable pressure difference, Δp, will vary inversely with radius, r. Small tubes do withstand higher pressures than large ones, as we know from experience with the tires of racing bicycles and large trucks and from the pressures (both positive and negative) of many atmospheres inside the cells and vessels of plants. Small aerosol cans, at least shaving cream cans designed for travelers, appear to take advantage of that bit of scaling. A can of half the radius ought to take twice the pressure, and (assuming constant length) it should hold half as much rather than, as one would expect from its volume, only a quarter as much as a large one. I haven't measured total volume such cans yield, but I do have to push with care when using a small one lest I get out far more than I want, and I'm surprised at how long the small cans last.

If, alternatively, wall thickness stays in proportion to radius, the simple isometric case, then the tolerable pressure difference ought to be indifferent to the size of the system. That sounds reasonably auspicious for biological systems.

What of biological reality? From the work of Quillin (2000), we have some information on the forces that earthworms can exert against the walls of their burrows. The worms push hard, with radial forces running about seven times the anchoring forces involved in crawling in preexisting burrows or in resisting extraction by a robin. We noted that if membrane thickness remains constant, tolerable pressure (still assuming constant breaking stress) will vary inversely with radius. Pressure, of course, is force divided by area, and thus is proportional to force divided by radius and cylinder length. So outward force per unit body length should be independent of size or, put in the usual terms, it should scale with body mass to the zero power—$F \propto M_b^{0}$. If, by contrast, membrane thickness scales with radius, then tolerable pressure remains constant. That implies that outward force should vary directly with radius or with body mass to the one- third—$F \propto M_b^{0.33}$.

So what does happen? For earthworms ranging from 0.01 to 8 grams, Quillin found that $F \propto M_b^{0.4}$, reasonably close to the assumptions of a thickness proportional to radius and constant material strength. The bigger is the more forceful, but more closely proportional to diameter than to mass, so relative to mass, big worms are wimps. Earthworm hatchlings can push at a monumental 500 times their weight; large adults can push at (only) a still impressive ten times their weight.

More Reality

With hydroskeletons we can't turn (as we most often do) to the designs of engineers to show us the range of possibilities—the scheme is more nearly unique to living structures than almost anything we've previously considered. At this point we just have an abstract graph; still, implicit in the graph (fig.

20.2) is a surprising diversity of biological cases. A brief description of several follows.

■ *Herbaceous stems.* If you make a tiny lengthwise slit in the outer layer of a very young sunflower stem, the cut edges pull apart. You've relieved some of the tension in the outer layer, so it gets a little shorter. At the same time you've relieved some compression in the core; but, being mostly water, the core doesn't expand noticeably. The young sunflower amounts to a flagpole built of water—no other compression- resisting material is present in any great quantity until woody fibers begin to be laid down. That internal pressure contributes to support can be shown by an even simpler manipulation. Soak long pieces of carrot root in salty water and watch them go flabby as water is drawn out osmotically—the amount of water lost can (at least roughly) be checked by initial and post hoc weighings. Without muscles or motility these systems don't "do" anything as do the animal schemes we'll come to—they're constant-volume arrangements, at least in the short term. But if such helically wound structures grow by elongation, then fiber angles will change over somewhat longer periods.

The compressive core holds especial interest, at least to me. In a freshman botany course I dutifully memorized the "fact" that the inner, thin-walled, cells (parenchyma) of such a stem had a particular shape—a so-called ortho-tetrakaidecahedron, a fourteen-sided solid with eight six-edged faces and six four-edged faces (fig. 20.3). Only years later did I read D'Arcy Thompson (1942) and get an idea of the significance of that shape. If a set of distortable spheres of equal size are squeezed together so they completely fill some large volume, each will (ideally at least) take on this particular shape. It's the shape that permits each to expose the minimum surface area. So the peculiar shape of these cells supports the idea that they're being squeezed and thus that they function as a compression-resisting core. Only thin cell walls are needed—

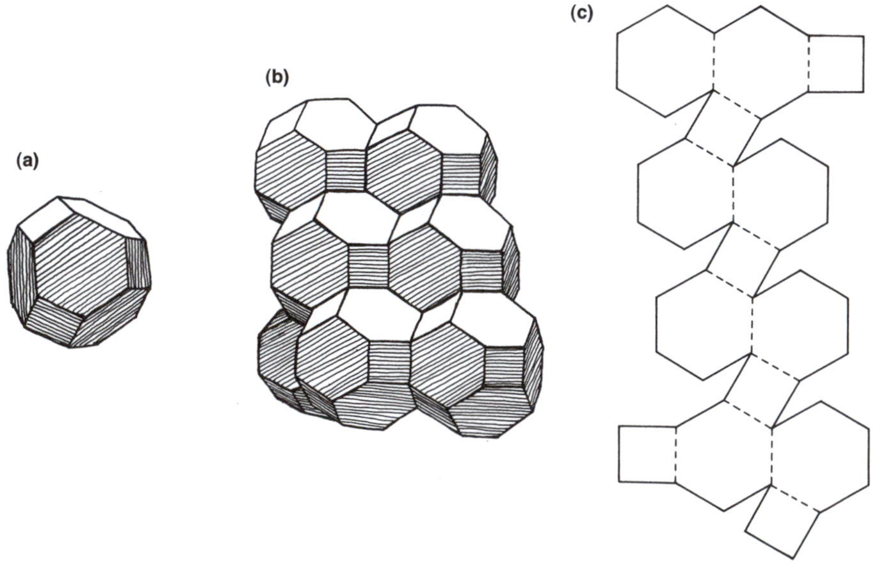

FIGURE 20.3 (a) An orthotetrakaidecahedron. (b) A group of these, packed together. (c) A cutout that can be folded on the dotted lines to produce the solid. If you make a few of these with paper and tape, you can demonstrate that they pack together without leaving voids between.

pressure shouldn't differ among the individual cells, and most of the motile animal systems lack any partitions at all. Still, these flimsy-looking cells have a crucial supportive function. The fourteen-sided shape, incidentally, idealizes somewhat—the real cells actually vary a bit (Hulbary 1944). Wainwright et al. (1976) mention this system, using the *Nasturtium* petiole (leaf stem) as an example.

That the peripheries of stems commonly experience predominantly tensile stresses and their cores mainly compressive stresses has been known since the nineteenth century. Plants, in fact, may include nature's most complex hydroskeletal arrangements, since individual cells typically have substantially elevated internal pressures, since the cellulose of the walls of individual cells forms crossed-fiber helical fibrillar networks, and since these cells often contribute to larger-scale hydrostats. (The internal pressure is usually referred to as "turgor" pressure inasmuch as it keeps the cells turgid.) In addition, growth not only involves changes in fiber angle, but fiber angle determines, in large measure, the allometry of growth—whether mainly in girth or length. Niklas (1992) gives a good general account. The basic paper I know of on individual cells is by Hettiaratchi and O'Callaghan (1978); its logic has been extended to whole plants, both extant and extinct, by Vincent and Jeronimidis (1991).

Nor are vascular plants the only ones to employ hydrostatic support. Fungal hyphae develop high internal pressures—as will get attention later, one fungus, *Pilobolus*, uses the pressure to launch its ballistic sporangia (spore clusters). And such pressures play crucial roles in both supporting and determining the growth forms of other fungi (Money 1997).

■ *Limp worms.* Cowey (1952) seems to have been the first to recognize the importance of fiber arrangement in hydroskeletons, looking at a nemertean worm—a long, flat, and unsegmented creature. (A more general treatment followed a few years later, as Clark and Cowey [1958].) Normally these worms form severely flattened cylinders; the fibers in their surface layers lie at angles to their long axes somewhere in the 40–70 degree range. Since they're not circular cylinders, they contain less volume than they might, and they live beneath the curve of figure 20.2, in the region labeled "flaccid." If such a worm contracts longitudinal muscles it will get shorter, moving to the left along a horizontal line. Sufficient contraction will bring it up against the curve, where it finds itself circular in section and more turgid to boot. If the worm contracts circumferential muscles it gets longer, moving to the right, but again it gets more nearly circular and more turgid. Not all of these worms can change shape enough to hit the limiting line, but they all use this curious scheme in which shape and stiffness are predictably interrelated. Internal pressures, though, are never very high—even when actively locomoting, these creatures are a limp lot.

■ *Some stiff worms.* Nematodes, or roundworms, contrast sharply with the various sorts of flatworms. They have a strong external cuticle, a normally round cross section, and longitudinal muscles only. A grotesquely large species, the intestinal parasite *Ascaris*, was studied by Harris and Crofton (1957). *Ascaris* has a normal fiber angle of about 75 degrees; being unflattened, it lives just on the curve of figure 20.2, part way down the left-hand slope.

Contraction of its muscles can only shorten it further. But shortening can only happen if it decreases in volume to move down the slope, which it can't, or if the fibers stretch, which they do very little. Mainly, muscle contraction makes it much stiffer, generating internal pressures up to 30 kilopascals—around a third of an atmosphere. The hypertensive worms do get shorter, but only by about 10 percent. Nematodes can bend by contracting muscles on one side only; and this, too, increases internal pressure. Circumferential muscles are quite superfluous—the resilience of the cuticle antagonizes the action of the longitudinals. Or, looked at another way, with a pressurized core as strut, muscle on one side can antagonize muscle on the other side just as do the biceps and triceps muscles of our upper arms. The scheme permits nematodes some unpleasant life-styles such as burrowing through our flesh.

■ *Starfish feet.* Something similar happens in echinoderm tube feet—small, soft, unjointed, and exceedingly numerous organs used for locomotion, handling food, and similar functions, noticeable when a starfish creeps up the glass wall of an aquarium. The walls of these organs have, like nematodes, longitudinal muscles and connective tissue fibers in crossed-helical arrays, as in figure 20.4. Their fiber angles are high—about 67 degrees for fully extended feet—so tube feet also lie on the left side of the curve of figure 20.2. Contraction of muscle tries to shorten a foot and increase that angle still further, pushing it downward on the curve. The high fiber angle minimizes fattening of the tube foot, so contraction should produce little actual shortening and considerable pressure rise. That is, unless the system can expel fluid.

But the system does expel fluid, so things get more complex than in nematodes. Above each tube foot is a bulbous chamber, the ampulla, equipped with circular muscles and reinforcing fibers at right angles to that muscle. So contraction of foot muscle forces fluid into the ampulla, extending its muscle. That couples the muscle of foot and ampulla in a hydraulically linked antagonism (McCurley and Kier 1995) much like muscles on opposite sides of a bending nematode; more about such systems later in the chapter. The whole thing hooks onto the water- vascular system of pipes, so its overall volume can vary. At the same time, a one-way flap valve prevents contraction of either foot or ampullary muscle from simply forcing water back into those pipes (Maerkel and Roeser 1992).

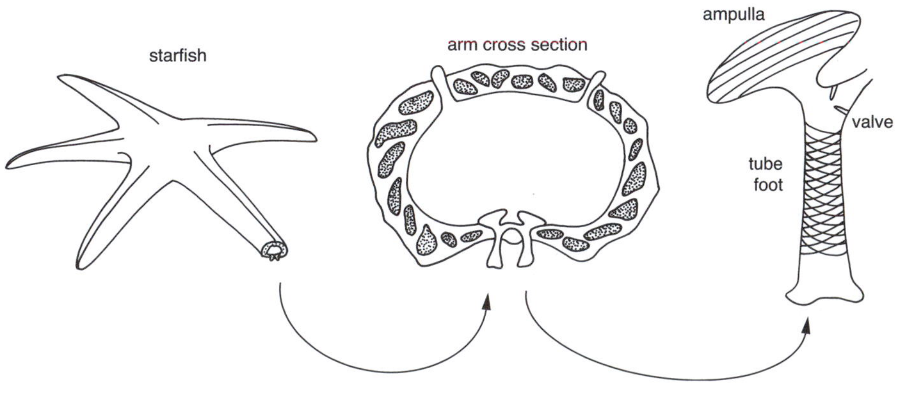

FIGURE 20.4 The layout of the tube foot and ancillary gear of a starfish or other echinoderm.

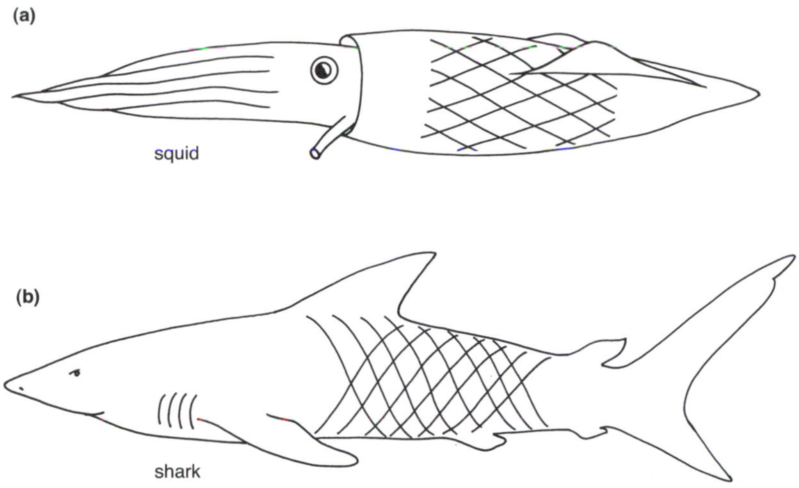

FIGURE 20.5 (a) A squid and its outer mantle, showing the fiber arrangement. (b) A shark, showing the fiber arrangement and its asymmetrical tail. Part (b) redrawn from Wainwright, Vosburgh, and Hebrank (1978) with permission of the copyright holder, American Association for the Advancement of Science.

■ *The squid's mantle.* How a squid refills its mantle cavity was mentioned in chapter 7; here we look at another piece of the story of jet propulsion in squid. A squid's mantle lacks longitudinal muscles but is well equipped with circumferential ones. Fibers of collagen in the mantle make an angle of about 25 degrees with the squid's long axis, as in figure 20.5a, so the system is on the right-hand slope of the curve of figure 20.2 (Ward and Wainwright 1972). Contraction of circumferential muscles tends to extend the mantle. But, with such a low fiber angle, extension implies a reduction in volume—contraction lowers the fiber angle and thus pushes the mantle down the right side of the curve. Here the reduction in volume does indeed occur, and very forcefully too. Water squirts out the siphon, and to conserve momentum the squid must accelerate in the opposite direction.

■ *Shark bodies.* It came as more of a surprise than it should have when Wainwright, Vosburgh, and Hebrank (1978) showed that sharks utilized a hydroskeleton. Sharks, after all, have conventional skeletons, albeit somewhat less calcified ones than that of most other vertebrates. Shark skin is sturdier stuff than fish skins generally (it has been made into leather on occasion), and it has the crossed helical fiber array typical of these hydrostatic arrangements, shown in figure 20.5b. The system, though, is more complex than those described previously—muscles attach directly to the skin, which thus acts both as external, pressure-resisting membrane and as an external, whole-body tendon. Fiber angles, not unreasonably, vary with location on the fish—the unstretchy skin must transmit the forces generated by the body musculature back toward the tail. During locomotion, the pressure inside the body of a shark rises to as much as 200 kilopascals—twice atmospheric and as high as that inside an automobile tire. So sharks are just shark-shaped balloons with teeth.

In sharks, the peculiar resistance of crossed helical fibers to torsion may have functional significance, at least of a negative kind. Sharks lack swimbladders and thus sink if inactive. Swimming must produce a little lift in addition to thrust, and a shark gets that lift, in part, by beating a tail fin that extends farther dorsally than ventrally, as you can see in the figure. That asymmetry might make a shark uselessly twist lengthwise, reversing twist twice during each full tailbeat—were it not for a torso that, while flexible in bending, resists twisting.

■ *Mammalian penises*. The intromittent organ of male humans, as well as those of a lot of other male mammals, depends on hydrostatics for its functionally critical stiffening. (Still other mammals, from raccoons to whales, use a bone, the os penis, either instead of or in addition to hydrostatics.) It came a surprise when Kelly (1997) announced that the fibers involved in hydrostatic mammalian penises broke the more- or-less general rule that nature always reinforces her hydrostatic cylinders with helical fibers. By contrast with the other systems, the fibers in penises go around circumferentially and extend longitudinally. Perhaps we should have guessed. Recall how cylinders with hoopwise and lengthwise fibers behave. Their flexural stiffnesses run higher than those with helically wound fibers, which sounds appropriate for an intromittent organ. And they have constant shape as well as volume. In particular, their hoopwise fibers resist lengthwise compression, which must be critical for such an organ. Kelly (2002) gives further details on the mechanics.

One ordinarily assumes that the stiffness of hydrostatic penises reflects the pressure of the blood that inflates them. Yes, they're inflated by blood pressure, via restriction of venous outflow, to a constant volume, shape, and size, set (in a way similar to what happens in arterial walls) by unkinking of collagen fibers. But muscle action near their bases applies a further squeeze, which raises the pressure well beyond systolic blood pressure. Pressures up to around 1,000 kilopascals (10 atmospheres) have been measured in the corpus cavernosum of a dog penis during intromission. That further squeeze, of course, causes only a minimal minimization of overall volume in this virtually incompressible system. Not that the dog tickles the edge of trouble—rupture pressure is well over an order of magnitude still higher. Indeed, that high implicit safety factor of about 11 (we'll talk a lot about safety factor in chapter 25) makes one wonder whether the instrumented dogs used by Purohit and Beckett (1976) were properly single-minded about their task.

■ *Coelenterates*. We've been assuming a thin layer of continuous and inextensible fibers, typically of collagen in animals and cellulose in plants. No law says that organisms must stick with these conditions, nor do they. Sea anemones such as *Metridium* embed discontinuous collagen fibrils in a matrix in the mesogloea of their body walls, and the fibrils can slide past one another to the extent (and at a rate) permitted by their attachments to the molecules of the matrix (Gosline 1971)—we considered the role of viscoelasticity when they pump water into their gastrovascular cavities to change shape in chapter 17. Anemones generate only modest pressures, usually under 100 kilopascals (a thousandth of an atmosphere) as they pump with their cilia. But with fibrils running more nearly circumferentially than longitudinally, *Metridium* extends lengthwise several fold as it inflates (fig. 17.6). Simpler polyps such as *Hydra* play analogous games as they change body dimensions and extend tentacles, again using a combination of variable body volume, discontinuous fibers in a matrix, and muscle and ciliary action (Wainwright et al. 1976).

The present account emphasizes the phylogenetic and functional diversity of hydrostatic skeletons. But it gives short shrift to the actual mechanics of their operation. For one thing, bending gets mentioned only as something that happens in nematodes, echinoderm tube feet, sharks, and such. That subject has been given proper attention by Alexander (1987, 1988). Also, we've glossed over partially hydrostatic systems. For instance, when you lift something heavy, you may hold your breath and tighten your abdominal muscles, increasing the pressure within your torso and thus taking some of the compressive stress off the backbone. (That's not an especially good practice, since the increased pressure reduces venous blood return to the heart. Don't, repeat, don't, ever do it when breathing heavily from activity—I once damaged a ventricle doing this so-called Valsalva maneuver.) Human heel pads and other mammalian foot pads make use of hydroskeletal support; our pads, which provide impact damping, some energy storage, and protection for bones, work as fluid-filled cushions—see, for instance, Ker (1999). They're complexly viscoelastic—if you want a stable reading of your height, you should stand for almost two minutes to allow your pads to creep into stability (Foreman and Linge 1989). Our intervertebral disks also use hydroskeletal support, with a tension-resisting periphery and a compression-resisting core (Hukins and Meakin 2000).

If helically reinforced hydrostats occur so widely, why don't we see them in our immediate surroundings? Any answer has to be taken as nearly pure speculation—we have trouble enough deciding what determines why something does occur, never mind making a case for why something doesn't. Still, one can suggest possibilities. We're large and both inertially and gravitationally challenged, which puts a premium on flexural stiffness. Too much reliance on tension yields structures of relatively low stiffness; and that's compounded when the compression component is stiff only in a volumetric sense. So even lengthwise-hoopwise reinforcement using heroic pressures doesn't produce the stiffness of a bone. And the more versatile helical reinforcement has that low twistiness-to-bendiness ratio alluded to a few pages back. Standing upright while challenged by horizontal forces goes best if a structure (think again of a tree) resists bending while accommodating twisting; these hydrostats do the opposite, resisting twisting while accommodating bending.

Muscular Hydrostats as Antagonists and Levers

The biceps and triceps muscles of your upper arm antagonize each other—contraction of one results in reextension of the other. The mantle of a squid has short, radial muscle fibers that thin the mantle and thereby reextend the circumferential muscles. The high internal hydrostatic pressure and slightly stretched cuticle reextend the longitudinal muscles of a nematode as a whole and permit contraction on one side to effect reextension on the other. Remember that, while muscle fibers actively shorten, they don't lengthen except as pulled upon by some external force. The piston of an internal combustion engine provides a close analogy—burning fuel pushes on the piston; something else, whether another piston or a flywheel, must alternately push it in the opposite direction. In organisms, though, who antagonizes whom may be less than obvious.

Another problem. Where they use discrete hinges and solid beams, muscular mechanisms can be analyzed in terms of comfortably familiar levers. But muscles aren't always used in arrangements so close to our technological world—

we've already mentioned nematodes, sea anemones, squids, and other nearly or entirely boneless beasts. Nonetheless, we could treat these systems as combinations of muscles and fairly simple passive components. What about a system almost entirely composed of muscle, with no body of fluid taking compression and with only a little passive tension-resisting stuff dispersed here and there? What supplies the crucial antagonism, and what might confer enough leverage to get long movements from that short-stroke engine, muscle? Stick out your tongue—nothing hypothetical about such systems.

The human tongue (and lots of other tongues) uses one of these "mostly muscle" systems; others are the highly extensible tentacles and less extensible arms of squid and the trunks of elephants. Yet another is the prehensile snout of the manatee, a large vegetarian aquatic mammal (Marshall, Clark, and Reep 1998). As mentioned earlier in this chapter, Kier and Smith (1985) have coined the term "muscular hydrostats" for these arrangements. They (the hydrostats, not necessarily Kier and Smith) can perform an impressive range of motions—they can extend, contract, bend, and twist; and they can radically alter their stiffnesses with little if any change in shape. Moreover, the motions needn't be slow—the tentacle of a squid extends by 70 percent in a thirtieth to a sixtieth of a second (Kier 1982).

What these systems can't do is change their volume: like the water of which it's mainly composed, muscle remains for all practical purposes incompressible. And that's the crucial feature that permits muscles to act as both tensile and compressive components of a solid beam or column, that permits muscles to antagonize each other without either discrete struts or an internal bag of water. Imagine a solid muscular cylinder in which muscle fibers run in various directions—contraction of different sets of fibers can cause the cylinder to change length and radius, one (of course) at the expense of the other. Calculating the relationship between radius and length for a fixed volume gives the result plotted in figure 20.6a. Implicit in the curve lies a remarkably versatile scheme to get variable leverage as well as antagonism—figure 20.6b gives a pair of cases.

Consider, first, a short fat cylinder of the proportions of a tuna fish can, with a radius of 1 unit and a length (or height) of 1 unit. The circumference is then twice pi. An outer circumferential muscle contracts 16 percent, reducing its length by what we'll called a unit. What will happen to the length of the cylinder? It will increase not by a unit but by only 0.4 units. Since distance increases by a factor of 0.4, muscular force will increase by 1/0.4, or 2.5 times. (The matter of leverage doesn't care that shortening has converted to lengthening.) Now consider a long skinny cylinder of the same volume as before, one with a radius a twentieth of its length. Again the circumferential muscles shorten by 16 percent, but that's now only a third of a unit. What will happen to the length? It will again increase by 40 percent, but the latter is now nearly a 3-unit extension—lots of distance but little force. We can calculate a 0.13-fold force increase (really a decrease) and a 7.9-fold distance increase. This second situation differs markedly from the first—instead of a short but forceful push we now have a long, easy, and probably rapid push. It gets real distance out of what little shortening a muscle can do. A squid tentacle, for instance, need decrease its diameter only 23 percent to get its 70 percent extension. And its long axial thrust comes without benefit of a single bone or other hard element for any muscle to pull against.

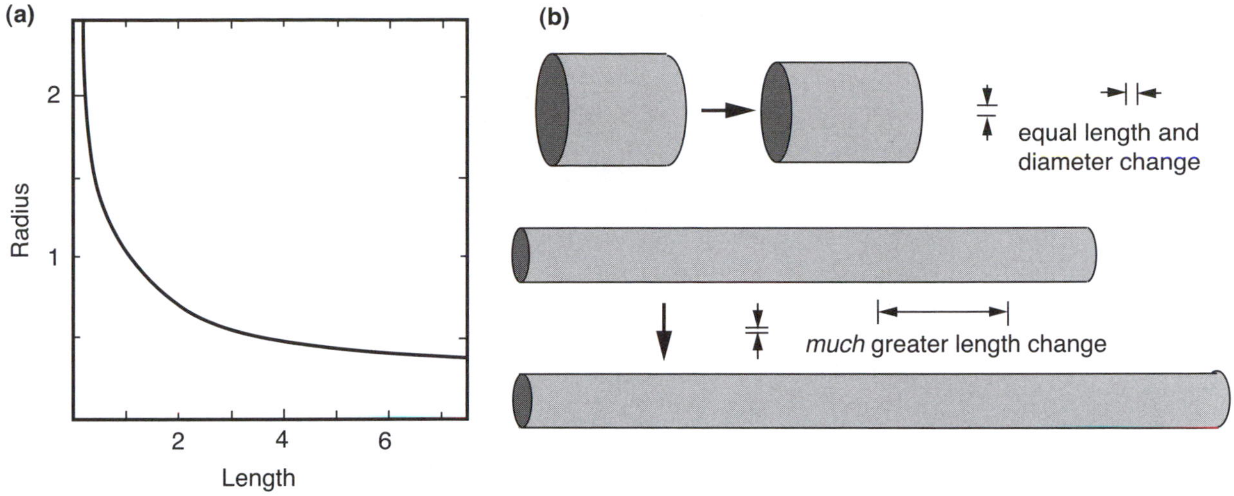

FIGURE 20.6 (a) Radius versus length for cylinders of fixed volume, here set at p units. Thus, the radius in this example equals one over the square root of the length. (b) The practical consequences of the relationship. In a short, fat one, a reduction in diameter gives a comparable length increase. By contrast, in the long, thin one, a given decrease in diameter produces a great increase in length.

Squid tentacles and lizard tongues—the latter are more impressive mechanically than the mammalian organ—operate with great amplification of speed and distance. Squid arms and elephant trunks are slower and extend less far, but they act more forcefully. Despite the fact that tongues, tentacles, and trunks (the alliteration, for once, is not mine) must represent separate evolutionary origins of muscular hydrostats, both their mechanical capabilities and anatomical arrangements are strikingly similar; figure 20.7 shows the three drawn to the same scale. All have muscles oriented (1) perpendicular to the long axis of the appendage, that is, crosswise or circumferentially; (2) parallel to the long axis and thus longitudinally; and (3) obliquely or helically around the long axis. Contraction of perpendicular (radial or circumferential) muscles causes overall elongation; contraction of longitudinal muscles causes overall shortening; contraction of perpendicular muscles on one side causes bending one way; contraction of longitudinal muscles on one side causes bending the other way; contraction of one or the other sets of helical muscles

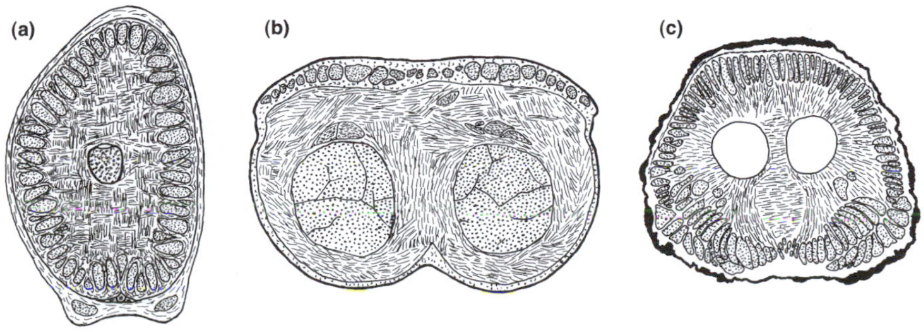

FIGURE 20.7 Cross sections of (a) squid tentacle, (b) lizard tongue, and (c) elephant trunk. In these tracings from micrographs, dots mark lengthwise muscle; dashes indicate crosswise, helical, and circular muscles.

causes twisting. In addition, contraction of both sets of helical muscles causes either lengthening or shortening depending on whether, given the shape of the cylinder at a particular moment, their fibers run at angles greater or less than 55 degrees to the long axis, as on the graph of figure 20.2. Kier (1992) gives a good general account.

There's a lot more biomechanical diversity and complexity in the ways animals use predominantly muscular structures. We noted earlier the way slugs passed waves down their feet as they walked—those feet are mainly made of very complexly arranged muscle. And that's characteristic of the feet of gastropods—snails, whelks, and such (see, for instance Voltzow [1990])—and of bivalve mollusks as well, some of which can burrow with remarkable rapidity. Vertebrate tongues as well turn out to be a mechanically diverse bunch of organs. Even among just the amphibians (where the diversity has received especial attention), tongue elongation mechanisms include both hydrostatic and ballistic (inertial) systems (Deban, O'Reilly, and Nishikawa 2001).

Hydraulic Linkages

Apply inward pressure on one wall of a fluid-filled container and you exert the same pressure outward on all other walls. If the fluid is a nicely incompressible liquid and if the container is stiff, you don't have to push inward through any appreciable distance to exert the outward pressure. Moreover, if the inward push makes fluid move outward, the volume that moves outward matches that of your inward push. If the movements aren't too rapid, viscous losses through shearing near walls will be minimal. With negligible work done either compressing or moving the fluid, pressure-volume work will be conserved. You do some work, the applied pressure difference times the volume moved, and you get out almost the same work. If the fluid moves outward over a greater area than it moves inward, then the outward movement must extend over a shorter distance but exert a greater force. (Force, remember, is pressure difference times area.) Figure 20.8 illustrates the principle, using a model made of two hypodermic syringes, some flexible tubing, and a soap solution. The commonest technological example is, of course, the braking system of an automobile, with a master cylinder replacing one syringe and the slave cylinders of two wheels the other. Notice that the idea of leverage applies here also—one just picks the relative cross-sectional areas of the cylinders to get the desired adjustment of force and distance, with speed going along with distance.

Hydraulic linkages occur in a wide variety of biological systems, many of which have already cropped up in these pages. The jetting of a squid, the operation of the tube feet of starfish, the crawling of worms, the erection of a penis have already been mentioned. In addition, one can point to the preda-

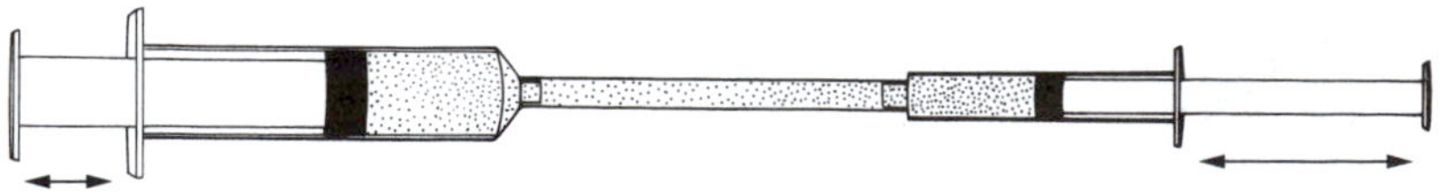

FIGURE 20.8 A hydraulic linkage made from two hypodermic syringes. The piston of the smaller syringe moves faster and farther than that of the larger one.

tory strike of dragonfly larvae, which lunge forward by squirting water out their anuses (Tanaka and Hisada 1980), and the expansion of the wing veins and thus the wing membranes of newly emerged adult insects (Glaser and Vincent 1979; Elliott 1981). We even use a hydraulic scheme to make contraction of our left ventricles filter blood through the glomeruli of our kidneys. In short, hydraulic action drives virtually all pressurized systems in which motion occurs.

One particular hydraulic device is worth a little more attention here, partly because its existence comes as yet another surprise and partly because it achieves antagonism for contractile muscle in an unusual way. The eight legs of a spider differ little from the six of an insect, but a curious special feature of spider legs has been known for almost a century. While properly equipped with flexor muscles (ones that decrease the angle between one segment and another), they lack the antagonistic extensor muscles (ones that increase that angle toward 180 degrees). Biologists casually assumed that elasticity of the interarticular membranes provided the antagonistic force, not on the face of it an unreasonable idea. But Ellis (1944) remembered that spiders die with legs severely flexed. If elasticity did the extension, they would more likely die with legs extended or at least not so flexed—as do insects. He found that cutting off the tip of a leg prevented reextension until the tip was resealed; and he found that mild exsanguination reduced a spider's ability to extend any of its legs. He suggested that extension in spider legs was hydraulic, not muscular or elastic. The idea was confirmed by Parry and Brown (1959), who measured resting pressures of 6.6 kilopascals and transient pressures of up to 60 kilopascals (over half an atmosphere) in spider legs. An isolated leg could lift more weight as the pressure inside it was increased, and the spiders turned out to have a special mechanism to seal off a joint that prevented fatal depressurization when a leg was lost.

The system works well enough, and its actions don't differ noticeably from those of insects. Some spiders (salticids) even jump well, some do dexterous tasks such as rotating a dead insect as they wrap it in silk, some walk rapidly and without misstep across delicate webs. Perhaps the hydraulic arrangement permits larger flexor muscles and the latter permit grappling with larger prey—spiders are preeminently predators on creatures of around their own size. But my adaptationist biases show through—present advantage needn't be decisive when looking at a problem of origins.

A tiny fungus, *Pilobolus*, uses a hydraulic system in quite a different way and for quite a different purpose. It's a tiny fungus that grows on the dung of grazing mammals (on "cow-pies") and bears a cluster of spores (a sporangium) on a cylindrical stalk or hypha (fig. 20.9a). The hypha grows toward a light source and develops substantial internal pressure by an osmotic mechanism. At maturity, the sporangium gets blown off the top of the hypha, and it goes on its way atop a cylindrical jet of cell sap. The initial ("muzzle") speed of the sporangium can be as high as 20 meters per second, or 45 miles an hour. But its high drag relative to its mass (being only about 0.3 millimeters in diameter) restricts its range, in practice to about 3 meters. To complete its life cycle, a spore must pass through the intestine of a grazer; since cows, at least, will not graze close to their own excrement (there's a "region of repulsion") the fungus has contrived a way to propel its spores away from the endangered feces. The hypha, incidentally, grows toward a light and thus aims the sporangium—if a culture is placed in a chamber illuminated only through

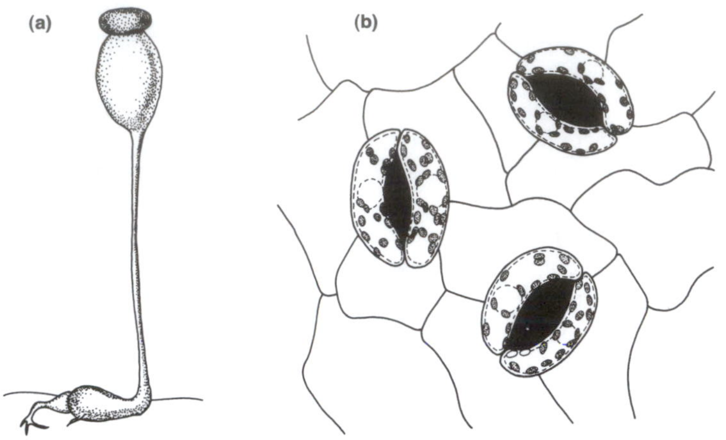

FIGURE 20.9 (a) *Pilobolus*, with its projectile, the sporangium, on top, with the subsporangial swelling the stalk below. (b) Stomata (shown dark) and their flanking guard cells on the lower surface of a leaf.

a hole, all the hyphae will shoot their sporangia through the hole. Mycology instructors often make *Pilobolus* do the trick, which goes back, I think, to Buller (1934).

The best-studied hydraulic engine may be the one that regulates the opening of the apertures (stomata) on the surfaces of leaves. Typical stomata are flanked by pairs of so-called guard cells, with shapes somewhere between kidneys and knackwursts (fig. 20.9b). A complex mélange of environmental and internal triggers and control systems combines to set the internal pressure (via the potassium concentration) of a guard cell. The width of the aperture is nearly proportional to the internal pressure, but the details of the coupling remain controversial, including the mechanical role of cells adjacent to the guard cells (Niklas 1992). If you inflate an annular balloon (as when an automobile inner tube is grossly overinflated), the increase in pressure reduces the area of the central opening. But increasing the pressure in guard cells increases the analogous area, so the tube isn't a good model. Either of two schemes seems reasonable. First, the inner walls of the guard cells might be less elastic (higher Young's modulus) than the outer walls, so inflation with water stretches the outer walls more and ends up bending the inner ones more sharply outward. Second, the lengths of the guard cells may be constrained, so increased inflation causes bending; since there's already a little outward bend, further bending of the Euler sort will increase the curvature. Either way, outward expansion pushes on other epidermal cells—the pressure in a guard cell of the leaf of a bean plant is about 3.6 megapascals (36 atmospheres!), which is well above the 0.6 megapascals of adjacent cells (Raschke 1979).

Other hydraulic engines in plants take part in the process by which the tendrils of climbing plants curl around a support and in the reorientation of leaves in those species that have different day and night postures (Satter 1979).

CHAPTER 21

Structural Systems

> An octopus once started
> a string quartet.
> But the act fell flat on its
> face, since it
> was left without a leg to
> stand on.
>
> **perpetrator anonymous**

THE GUIDING HAND OF EVOLUTION SHOULD BE MOST EVIDENT AT THE level of systems. Although the most complex level of mechanical design in living things, it's thus the level at which we can inquire most directly about the biological consequences of design. After all, success in propagating progeny depends more immediately on a good supportive system than it does on any particular structural component or material; indeed a structure or material has quality only in the context of the system that uses it. Evolution is sometimes referred to as a process of optimization; I agree with the general notion, but I'm not enthusiastic about using the word "optimization." Part of my hesitancy is semantic—something can't be *more* optimal just as it can't be *more* unique, so "to optimize" is synonymous with "to perfect." We mean, of course, "to improve" or "to make more nearly optimal", but a battle cry that needs a footnote is too easily misunderstood. Still, if natural selection doesn't push organisms in the direction of some sort of functional improvement, then modern biology rests on an awfully shaky foundation. Concomitantly, if one of our mechanical analyses focuses on factors other than those demanded by normal function, then the analysis probably barks up the wrong tree.

In the present chapter we'll look at some issues connected with structural systems—not that we haven't before—moving from a relatively specific problem, intermittent adhesion, then to a more general one, keeping a large plant upright, and finally to an attempt to see hierarchical order in supportive systems as a whole.

Staying Put—Adhesion

Consider how organisms can avoid being swept away by winds and water currents or dropped by gravity. We tend to care more about locomotion, perhaps as mobile creatures ourselves, and the biomechanical literature reflects that concern. But one might make a devil's advocate argument that staying put is more important in the overall scheme of nature; certainly in a forest of trees or in the wave-swept intertidal zone, fixation dominates fitness-relevant mechanics. And fixation involves a wide array of problems. How might the parts of an organism be interconnected so they don't unfasten themselves when pulled, pushed, twisted, or sheared? How might an organism attach itself to a substratum, whether rock, soil, beach, prey, or host? And so on.

Often gravity is either unavailable or inadequate. Or worse—for a tree frog on a pane of glass or a fly on a ceiling, gravity occupies the wrong side of the balance of forces needed to abide by Newton's first law. Still, organisms stay stuck, and they do so using a great diversity of schemes for attachment and adhesion.* Attachments may be permanent—structural materials are commonly nailed, glued, riveted, or welded; the bones of our skulls never part at their intermeshing sutures; a barnacle ordinarily stays glued to its rock. Or they may be releasable—plugs can be unplugged from sockets, catches and latches disengage, a starfish can walk by alternatively attaching and detaching its hundreds of tiny tube feet. Attachments may be substantially rigid, as with bolts and welds or the connection between a coral and its substratum, or they may be flexible, as with buttons, zippers, hinges, and adhesive labels, or with most of our joints and the connection between the copulatory organs of those species so engagingly equipped.

Wherever nature makes a structure, biologists busy themselves painstakingly describing it—but too often paying little attention to mechanical functions. Nachtigall (1974) has gone the next step, gathering a vast collection of structural schemes for attachment, classifying them by function, and comparing each with its technological analogs. Among interlocking joints he recognizes miters, rabbets, dovetails, and mortises and tenons; under releasable attachments he describes plugs and sockets, hooks and eyes, snaps, vises, forceps, anchors, suction cups, and others. The diversity simply defies summarization. Even Velcro has its biological analog—what Nachtigall calls "probabilistic" attachments, coatings of burrs invented repeatedly as a dispersal device for the seeds and fruits of plants. All that these latter require is gentle contact with a sufficiently irregular surface (fur perhaps originally but cloth works at least as well) and enough burrs attach to provide a surprisingly strong connection. Velcro, in fact, represents one of the rare cases of successful bioemulation; its inventor, Georges DeMestral, deliberately worked from the seed barbs of cocklebur or burdock (Vogel 1998a).

And biochemistry plays varied roles in attachment. Organisms manufacture good glues, most of them proteins; indeed until recently most glues used in our technology were mildly modified proteins obtained from blood, hooves (the old horse consigned to the legendary glue factory), bone, skin, fish remains, and so forth. As noted earlier, the similarity of the words "collage" and "collagen" is no mere coincidence—the protein was named for its use as a source of glue. Indeed, the highly hydrated proteinaceous mucuses and the stiffest of biologically produced proteinaceous glues are simply the extremes of a chemical and mechanical continuum. The chemical connections range in specificity and strength from direct covalent bonding through a variety of ionic and hydrogen bonds to the relatively general and weak "van der Waals force" between molecules.

Adhesion—loads and failure

A joint or an attachment that's subjected to a load may fail in any of four different ways—compression, shear, tension, and peel, listing them from least

*A distinction is made between "adhesion," in which two items are held together, and "cohesion," in which a single item (such as water) resists being pulled apart.

to most difficult to withstand. Pure *compression* between two decently mated surfaces requires, of course, no glue at all; one can argue about whether the actual joint actually fails. Mortarless walls have stood for thousands of years. Yet cement is barely adhesive or even cohesive, with a tensile strength an order of magnitude lower than its compressive strength. So it makes a very weak joint with even a rough brick. Cement does, though, distribute stresses and thus minimize the chance of initiating cracks through local tension. Resisting *shear* takes only a little irregularity in the mated surfaces—a few holes in bricks or blocks together with a layer of cement automatically provides sufficient mating. The attachments between our vertebrae and the intervertebral discs do quite well, at least in pure compression and shear.

Tension proves trickier, but the same chemomechanical schemes that make tensile elements themselves can be mobilized to make tensile attachments. In addition, tensile joints can be made into predominantly shearing joints by arranging for the mating surfaces to interdigitate. Wood workers make great use of the subterfuge, with dovetailed joints, inserted disk-shaped "biscuits," and so forth. Not only does tension mostly convert to shear, but the area of contact increases as well. Currey (1984) noted that the fibers at the crucial joints between tendons and bones interdigitate sufficiently so that one can't say exactly where one material stops and the other begins. He also reminds us that joining structures (such as tendon and bone) of quite different elastic moduli generates particular problems. Gordon (1978) gives a good view of how human technology makes use of such lapped joints and inserted joints, together with an explanation of the strange force distributions they generate.

Perhaps the worst problems of failure come from asymmetrical loads, loads that, through bending, cause high local stresses, what in simple systems is termed *peeling*. A piece of tape may adhere strongly to a surface, but peeling can remove it with only a small force. What makes such trouble is that old bugbear, force concentration and its consequent high local stress. In effect, the applied force acts over a very small area—the line of peeling times a short distance in the direction of peel—so a small force makes a large stress. And peeling may be initiated by loads that don't obviously run across the surface of contact. Consider a tensile load applied not quite perpendicular to a surface. That load will act almost entirely at one edge of the area of contact (fig. 21.1a), and failure will occur by peeling. And the stiffer the glue, the worse the situation—a stiffer glue reduces the effective area over which the force acts and thus raises the local stress.

Loads that might cause peeling occur widely in nature. Neither the holdfast of a large marine alga nor the byssal thread of an intertidal mussel nor the foot of the wave-challenged snail can be assured of tensile forces perpendicular to its attachment surface. All respond with some tapering disk of attachment that's increasingly flexible peripherally (fig. 21.1b)—the whole thing distorts a little to avoid stress concentrations at an edge. Thus, a good shape for a glued joint may resemble (and may also function as) a suction cup, even though the two mechanisms are physically distinct.

Adhesion—attachment mechanisms

More numerous than the modes of detachment are the physical devices for staying attached. Of especial interest are the diverse ways by which organisms

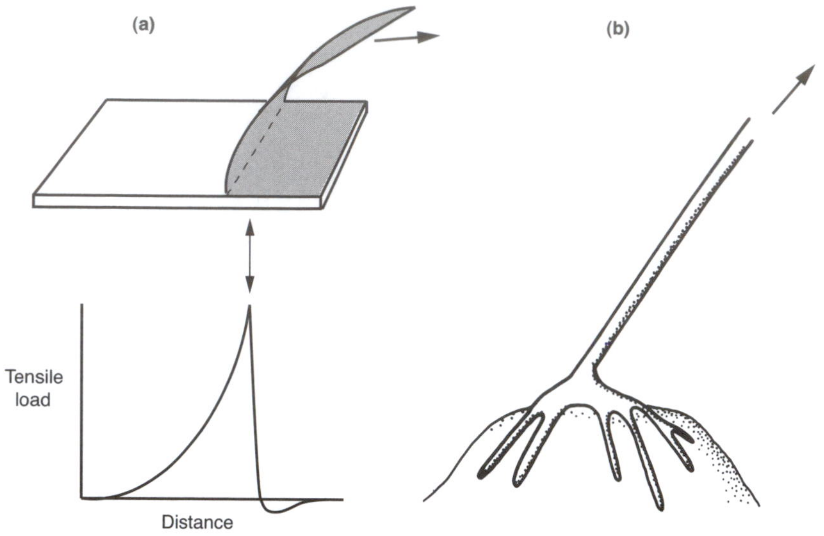

FIGURE 21.1 (a) The local increase in tensile stress at the separation line when one surface is peeled from another. (b) The stipe and holdfast of a large alga attached to a rock—the arrangement minimizes the chance of peel failure by making it unlikely that peeling will get started.

(or their parts) achieve temporary—that is, readily reversible—attachment. We're back to the frog on the glass, the fly walking across the ceiling.

■ *Hooks*. These abound, from ones revealed only with scanning electron microscopy to those of large tree-climbers such as sloths. Mechanically, they're curved, end-loaded cantilever beams. Interest in hooks goes back at least to Robert Hooke's (hah!) *Micrographia* of 1667, perhaps the first atlas of microscopy, a book much admired by the famous diarist and proto-yuppy, Pepys—who bought a microscope and found it difficult to use but a fine household decoration. Even the tiniest hooks don't work on glass, and cases that look like hooking sometimes prove otherwise, as Stork (1980) found in some small beetles.

■ *Suction*. While we usually refer to this scheme for resisting tension as suction, external pressure really does the heavy lifting. If a flexible surface presses against a rigid one, if some force then distorts the flexible surface so its center pulls away from the interface, and if a seal is maintained so fluid cannot move into the region of reduced pressure—then the external atmosphere will push the surfaces together. That's how a suction cup works. Unless one manages to put an internal liquid in tension, the maximum stress holding the surfaces together can't exceed the local atmospheric pressure. But tension does develop in at least one case—Smith (1991) showed that an octopus can, remarkably, suck harder than atmospheric pressure; that is, it can load the water beneath a sucker into tension (chapter 5). Maintaining the seal is crucial, which limits the use of suction-based devices on rough or porous surfaces. Organisms that use such suction commonly use a bit of mucus as well; the latter does duty both as sealant and as additional adhesive.

As one expects, the maximum force that suction adhesion can resist follows the area of contact within the seal. And the maximum force ought to increase with depth, since (ordinarily at least) ambient pressure—atmospheric plus hydrostatic—sets its limit. One study put limpets in a hyperbaric chamber to look at the possibility. As it happens, they don't take particular advantage of depth, with attachment strength increasing only slightly and never rising much above half an atmosphere (Smith, Kier, and Johnsen 1993).

Besides mollusks, echinoderm tube feet make use of suction adhesion, as do a wide variety of other aquatic systems—either as the only attachment mechanism or in combination with others. Among terrestrial systems one thinks first of wet ones—frogs, for instance. But the mechanism finds use even in arboreal mammals. Several of the smallest bats, for instance, use it to cling to smooth leaves, with disklike structures on wrists and ankles. In the 3.5-gram *Thyroptera tricolor* of Central America, suction provides the main mechanism; these bats' minimal reliance on other schemes such as the two kinds of wet adhesion that follow limits their ability to cling to anything but smooth surfaces (Riskin and Fenton 2001).

■ *Stefan adhesion.* This particular form of adhesion works best against shearing loads. Recall that the viscosity of a fluid indicates its resistance to rate of shear (chapter 6). If two surfaces are separated only by a thin layer of liquid, then even slow motion of one across the other will give a high shear rate. If, in addition, the liquid between is very viscous, then movement may take appreciable force applied for respectable lengths of time—the process is called "Stefan adhesion." It may be a temporary thing, but so are, say, the wave forces in rocky intertidal areas.

I know of no demonstrated case in which Stefan adhesion is the exclusive mode of attachment, but one suspects its involvement when, as Crisp (1960) showed, a steady force will slowly slide barnacles (some at least) across smooth surfaces. Wigglesworth's (1987) observations of flies clinging beneath glass sheets raises a similar suspicion. He noticed that the legs slowly slide centripetally across the glass, periodically detaching and swinging outward again. But shear of two surfaces isn't the only way to load the liquid in shear—a tensile stress between two objects with fluid between will force inward flow between the surfaces, and the no-slip conditions at the liquid-solid interfaces will ensure the occurrence of shear. Thus, Stefan adhesion can play a role in resisting the initial tensile separation of two smooth and well-mated surfaces. It's of course a player in the mucus locomotion of slugs, snails, and such, described in chapter 17.

■ *Capillary adhesion.* Another physical mechanism makes things adhere in the face of tensile loads in nonsubmerged situations. If a thin layer of water separates two smooth hydrophilic surfaces, then any attempt to pull the surfaces apart creates additional air-water boundary in the form of a curved interface (fig. 21.2). Surface tension opposes making surface and so resists the

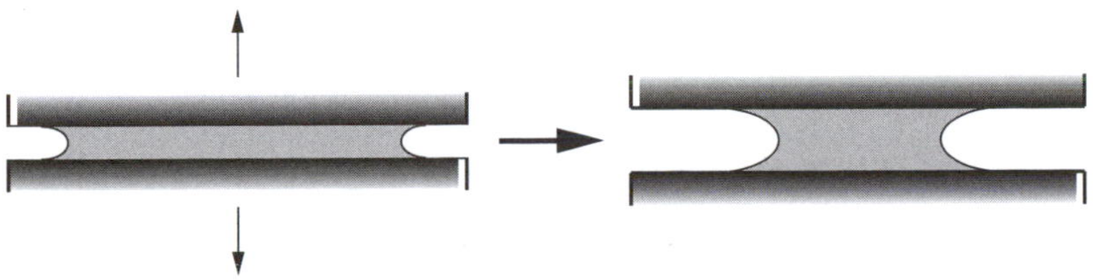

FIGURE 21.2 Adhesion based on surface tension. Pulling apart two hydrophilic surfaces separated by a water film (shaded) increases the area of the air-water interface, which takes work.

separation. Thus, in an aerial system, water can act as an adequate glue by a mechanism commonly referred to as "capillary adhesion." Aphids appear to use it to adhere to surfaces (Dixon, Croghan, and Gowing 1990). And it almost certainly helps a tree frog rest placidly on a vertical pane of glass, about which more shortly. Denny (1993) gives a good account of both capillary and Stefan adhesion, examining each for potential biological relevance.

Capillary adhesion ought to be proportional not to the area of contact but to the wetted perimeter, as with the other phenomena associated with surface tension. Since it involves force (or pressure), surface tension, and a curved surface, Laplace's law ought to apply. Here (by contrast with a sphere or ellipse) two radii work against each other. Consider, as in the figure, what happens when a small circle of liquid separates two smooth and parallel plates being drawn apart. The overall radius of the liquid helps push the plates apart—the liquid would like to minimize its surface. But, assuming full wetting, the radius of curvature of the periphery of the liquid (half the liquid's thickness) pulls the plates together—that peripheral surface would also like to minimize its area. Thus, for the pressure change within the liquid,

$$\Delta p = \gamma\left(\frac{1}{r_o} - \frac{1}{r_e}\right), \tag{21.1}$$

where γ is surface tension, r_o is the overall radius, and r_e is the radius of the curved edge of the liquid. Under most circumstances, r_o will be much larger than r_e, so the first term in the parentheses can be ignored.

■ *Molecular adhesion.* Besides glues, of course, the principal player here is a kind of weak, long-range (by molecular standards) attraction called the van der Waals force. This force may be responsible for a certain amount of macroscopic mischief. If two surfaces are brought sufficiently close together they most often show some tendency to stick together. One rarely notices—few surfaces can be brought into sufficiently intimate contact since their irregularities preclude adequate mating. A pair of scrupulously clean microscope slides behaves as if stuck together by a weak glue, but a few specks of dust from the air are sufficient to minimize the nuisance. Organisms, though, can arrange adequate contact with a rigid surface by using an extremely large number of very small and flexible contact points

That use of lots of contacts proves critical when a gecko uses its toe pads to climb a smooth wall. Each foot has nearly 500,000 very thin hairs, each roughly 0.1 millimeters long. Each of these has hundreds of smaller setae; thus, the four-footed gecko has of the order of a (U.S.) billion contacts, each 0.2–0.5 micrometers in diameter and adding up to 100 square millimeters of pad area per foot. Autumn et al. (2000), whose figures these are, measured the adhesive strength of individual foot hairs. While the average stress measured on whole feet was about 100,000 pascals (about an atmosphere), they found that individual setae do ten times better—not all setae operate at once. In a subsequent paper, Autumn et al. (2002) provided direct evidence for van der Waals force as the principal mechanism involved. It seems that adhesion strengthens merely from subdividing contact area into small elements, quite beyond the issue of fit of contacts to an irregular surface. A similar implication comes from Stork's (1980) work on small beetles (cited earlier), in which most other possible mechanisms were either ruled out or (for the two mechanisms of wet adhesion) showed to be minimally involved.

Interestingly, the area of the toe pads of a gecko increases with increasing size of animal more drastically than does not only body surface but even body volume—the bigger gecko has disproportionately large pads (Maderson 1964, 1970). But falling off a wall depends not just on mass but a turning moment about a pivot at the bottom of the attachment (consider a picture hung loosely on a wall). So the bigger animal will be not only heavier but will project further out from the wall. Thus scaling more drastically than with volume sounds entirely appropriate.

Adhesion—pulling apart some mixes

Naturally, nothing precludes an organism from using some combination of adhesive mechanisms; and determining relative contributions can be sticky. The sucker of an octopus works by suction almost exclusively. But limpets use suction mainly when they need mobility, relying on glue rather than continuous muscular effort for longer term attachment (Ellem, Furst, and Zimmerman 2002). They don't, though, use Stefan adhesion (Smith 1992).

The tube feet of starfish and other echinoderms are another messy case. Paine (1926) found that about 40 percent of the adhesion was done by glue rather than by suction. Thomas and Hermans (1985) claimed that at least one fairly ordinary starfish species attaches entirely by using ionic interactions and films of secreted protein—in short, with glue. But showing that interference with the protein prevents attachment may just reveal the protein's crucial role as a sealant. Moreover, starfish have the proper retractor muscles for generating a pressure difference—can these possibly be nonfunctional? Still, starfish can attach to wires too fine for their tube feet to act as suction cups, so they most likely use a mix of glue and suction, as Paine suggested long ago.

The toe pads of tree frogs provide another mixed and slightly controversial case. Emerson and Diehl (1980) argued that suction wasn't important, because reducing the external pressure had no effect on attachment. But, as Smith (1991) pointed out, the watery secretions that fill the space between feet and substratum may be able to sustain pressures below zero (recall chapter 5), reducing the strength of the argument. They also found that the frogs stuck about as well to leaves, glass, or Teflon. That indifference to surface texture and hydrophilicity argues against capillary adhesion, but that argument isn't strong since these frogs make chemically active mucus and thus don't depend entirely on water. Besides, surrounding a frog on a glass surface with water makes it fall off, which certainly points to capillary adhesion. But capillary adhesion may just hold the frog close enough to the surface to permit Stefan adhesion, so eliminating the first would inevitably disrupt the second as well. One might expect to resolve the matter with a look at scaling. Both Emerson and Diehl (1980) and Green (1981) determined that the strength of attachment varied directly with the area of toe pads—does this rule out capillarity, which should vary with perimeter? Not quite. If (as it appears) a toe pad consists of a varying number of flat contacting surfaces of constant size, then the length of edge should be proportional to overall area—just as the total edge length of standard floor tiles varies directly with the area of the floor.

Even less clear is the basis of "wax-running," by which some ants get around on the slippery epicuticular wax of plants. Federle, Rohrseitz, and Hölldobler (2000) measured attachment forces by making ants do their thing under varying amounts of extra gravity in a centrifuge. Scaling pointed to detachment force as dependent on contact area, but, paradoxically, ability to wax-walk turned out to vary inversely with attachment strength to a smooth acrylic (Plexiglas/Perspex) plastic. Whatever the mechanism, arboreal ants hang on spectacularly; attachment forces from 50 to 150 times their weights exceed those of all other insects yet investigated.

Two general notes about temporary attachment in animals. First, a creature must detach as well as attach; and for a small runner detachment can't dawdle. For suction adhesion, release of whatever muscular or hydraulic force generated the suction provides easy detachment. For the other mechanisms, the usual solution puts peel failure to positive use—a dimensional reduction from two to one means working along a line rather than over an area. (Note that a zipper depends on an analogous reduction, holding along a one-dimensional line while attaching or detaching at a zero-dimensional point; you do the same thing when lacing or unlacing boots.) Autumn et al. (2000) described how gecko foot hairs use peel failure for release.

Second, devices for intermittent adhesion in animals make extraordinary use of multiple contacts. The billion contacts of the gecko's feet may not be exceptional. Each of Stork's (1980) 5-microgram chrysomelid beetles had over ten thousand setae. William Kier found that even octopus suckers turn out to use tiny projections, pegs about 3 micrometers in diameter (Pennisi 2002). Using a lot of contacts must give some useful redundancy; more important, probably, are an improved fit to unpredictably irregular surfaces, easy ability to adjust adhesive strength, better resistance to shear forces, and (with interconnected space between them) useful pressure equalization. When attachment projections get down to the micrometer range, leakage of air between them (for suction, for instance) can't be too much of a worry—with a little moisture, either viscosity or surface tension ought to provide an effective seal.

Staying Put—Standing Stably on the Ground

In a world in which gravity is the only external force and attachment strength is negligible, the game has only one rule—to be stable the center of gravity, the point at which one's mass behaves as if concentrated (chapter 2), must be directly above some effective contact with the ground. Static stability requires at least three points of contact with the ground, with the center of gravity located above the triangle whose apices are those contact points. With two legs or one, the breadth of the foot or feet becomes crucial, with postural reflexes involving sensors and muscles that keep the system obedient to the rule. A wading bird can stand on a skinny leg because it ends in a wide foot; while you can walk on pointed stilts, you stand only with great difficulty and continuous adjustment.

We disobey at our peril, and it takes a substantial act of will (at least for an adult) deliberately to tip over. If some lateral force shifts one's center of gravity, the complex neuromuscular machinery that runs our postural reflexes directs the system to move the legs so the center of gravity is again directly above the ground contact. A persistent sideways force can be managed as well, so we can walk in a wind or haul a load.

What about an organism that can't deliberately move its contacts with the ground as its center of gravity wanders around? A tree is the extreme case; the schemes used by trees to avoid falling over ("windthrow" seems to be the bit of jargon) are varied and, as far as I can tell, surprisingly poorly known. Gravitational loading, as argued in chapter 7, doesn't present the worst challenge. The more serious problem for a tree comes from the drag induced by strong winds—recall chapter 1. Leaves incur most of that drag—leafless trees do blow over, but they're typically either rotten or deprived of the shelter of adjacent, leafed trees. Chapter 7 described how leaves reconfigure in winds into forms with low drag relative to surface area. But they can't avoid drag entirely, and photosynthesis depends on exposed area. The location of the leaves compounds the problem. Big trees in forests typically compete with their neighbors, with winners shadowing losers to get their place in the sun. So trees are stuck with not just a lot of drag but with a great lever arm for that force and thus a huge turning moment around their bases, as in figure 1.4. To avoid uprooting, they have to oppose this turning moment with one of equal magnitude and opposite direction.

Let's try now to put some pieces together, with, I should emphatically declare at the start, considerable elements of speculation. First, failure comes in two types—snapping and uprooting; which predominates varies greatly from species to species and forest to forest (Putz et al. 1983). Snapping represents failure of the tree's material itself, something that came up in chapters 16 and 18. Where I live, failure of an entire healthy tree seems most often to involve uprooting. Grabbing the ground is the weak link, which comes as no real surprise—selection for better-built trees cannot deal directly with the poor tensile strength of soil. Uprooting as failure implies adequately strong trunks, suggesting, therefore, that trunk strength isn't the critical variable. How, then, can trees avoid uprooting from the turning moment caused by the product of their heights and the drag of their leaves?

While no systematic study has yet been done, at least four distinct schemes seem to be used to keep roots and soil in decent contiguity. Combinations of more than a single scheme certainly occur, and a given tree may use different schemes or a varying mix of several as it grows from a sapling. (Mattheck [1991] considers some aspects of the tree's problem; Ennos and Fitter [1992] provide information on anchorage in small plants or very young trees; Ennos [2000] gives a good general view of the situation.)

■ Consider, first, what we might refer to as "compressive buttressing." Uprooting most often involves elevation of a large weight of roots and associated soil. Increasing the work necessary to achieve that elevation decreases the chance that a tree will blow over. That can be done by developing a stiff, wide base, and thus moving the pivot point or axis of turning well to one side of the center of the trunk and root mass (fig. 21.3a). The key components, then, are a stiff and massive trunk and a broad, upwardly tapered base that acts primarily as a set of buttresses on the downwind (compression loaded) side, pushing the pivot point laterally and thus increasing the work needed for turning. On the upwind side the broad base will contribute to the weight that the turning tree must lift.

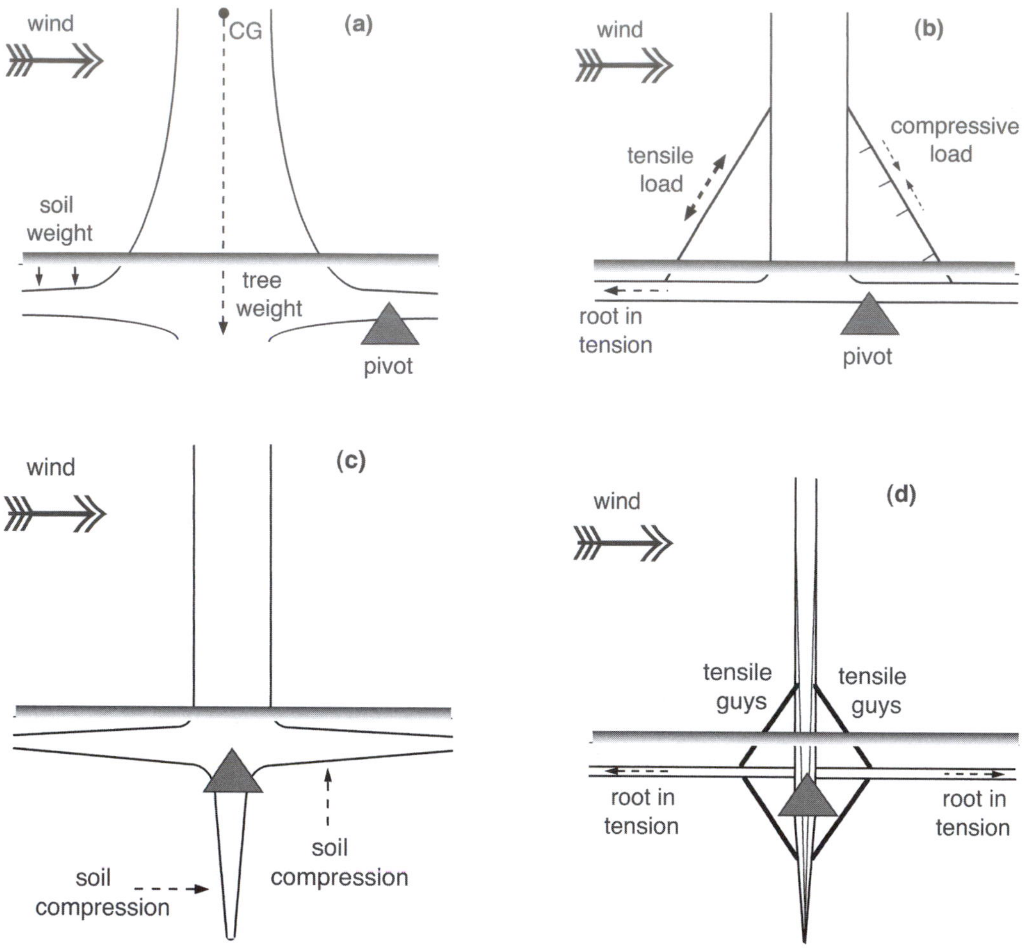

FIGURE 21.3 Four schemes to keep a tree from blowing over: (a) compressive buttressing, (b) tensile buttressing, (c) taprooting, and (d) diagonal guying. The center of gravity (CG) is indicated in (a).

Partly burying the broad base improves matters by using the substratum to increase the weight that must be lifted. Soil and stone are conveniently dense material, so for a root-soil plate even a small volume goes a long way. The substratum beneath the tree feels compression, which under most circumstances it will resist well. A stiff trunk will improve the effectiveness of the arrangement by minimizing downwind drift of the center of gravity in the wind, as will minimization of the weight of branches that can shift around. The paradigmatic example of such compressive buttressing might be a large specimen of an oak such as *Quercus alba* or *Q. robur*. This scheme may be the most important one for the large angiosperms of temperate North America, and it is certainly not uncommon among gymnosperms that lack vertical tap or striker roots, at least judging from work on Sitka spruce (Coutts 1983; Blackwell, Rennalls, and Coutts 1990).

■ The trees that we usually describe as buttressed, which occur mainly in tropical rain forests, seem to make little use of such compressive buttressing. Their conspicuous buttresses are simply too high and broad relative to their thickness for them to take major compressive loads without buckling. Instead, the buttresses work in a way just opposite the role played by the masonry buttresses of Gothic architecture with which they're visually analogous. They're tension-resisting structures, as shown in figure 21.3b (Mattheck and Bethge 1990; Ennos 1993b; Crook, Ennos, and Banks 1997)—the scheme, then, might be called "tensile buttressing." These tensile buttresses transmit and redirect the forces on the upwind side of a trunk to the roots. That loads the upwind roots in tension, almost certainly more strongly than those of trees using the first scheme. Soil, of course, has almost no tensile strength of its own. But in practice, a root should be able to pull hard against the general tangle of roots just beneath the soil surface in a rain forest. Stabilizing the center of gravity with a very stiff trunk should matter less than in compressive buttressing, a significant factor in permitting the tall, thin trunks of the trees that form the canopy of a rain forest. One might guess that trees mainly relying on tensile buttressing won't be suitable for planting in isolation.

■ An alternative scheme capitalizes on little more than the ability of soil to withstand compressive force. If the trunk is continued downward beneath the soil as a stiff taproot, and if ramifying lateral roots near the soil's surface fix the location of the tree, then pushing the trunk in one direction will push the taproot in the other (Edelin and Atger 1994). Soil, especially when beneath a layer of superficial roots, ought to resist this sideways push quite well; the scheme, which we might just call "taprooting" is shown in figure 21.3c. Taprooting depends on good resistance of the taproot to bending as a cantilever—a high level of flexural stiffness—as is sufficient broadside area to push against so as not to slip sideways through soil. (Additional substantial vertical "striker" roots, according to Perry [1982] and Crook and Ennos [1996], may supplement the mechanical role of taproots.)

A tree that uses the scheme without a healthy taproot is crippled. In over 25 years only one of over seventy loblolly pines (*Pinus taeda*) around my house has blown over with less than really severe provocation; that one had a rotted taproot. (Taproots normally break when a tree uproots, so they're easily overlooked.) My casual observations of several excavated pines suggest that taproots may develop noncylindrical cross sections in response to wind from a prevailing direction. But when poking around the bases of large uprooted Douglas firs, I was struck by the small size of the taproot breakage points. Of the schemes here, the relative importance of taprooting is the least certain; the best documentation of its role comes from work on larch, by Crook and Ennos (1996).

■ One further scheme, rare in true trees, can certainly stabilize woody structures of comparable height. The culms of bamboo use what amounts to a variation on tensile buttressing that we can call "diagonal guying," shown in figure 21.3d. Once again, tensile forces on the upwind side run downward and outward through tension-resisting structures to lateral roots. Once again lateral roots must withstand substantial tensile forces, taking advantage of a dense tangle of other roots in the superficial layer of soil. Those lateral roots, I assure the reader from hard personal experience, go far and, with their pelage of outgrowths, don't easily separate from the soil. The diagonal guying of bamboo looks relatively symmetrical above and below the lateral roots, with a taproot and a set of guying roots below as well as above. The scheme seems especially elegant in the way it uses ropes rather than solid buttresses for guying. In a tensile buttress, of course, the outermost region will carry almost the entire load, so the inner part is mainly wasted. But the otherwise admirable ability of dicotyledonous trees to grow in girth probably renders ropes impractical—guying ropes would need not only to thicken but would have to move ever further outward from the base. Bamboos, like palms, just don't grow girthwise.

A few final comments on large, stiff, hardwoods such as oaks. As we weekend woodcutters know well, the main part of the mass of an oak tree is concentrated in the trunk—the branches are conspicuous but yield much less fuel. And the base is wide, stiff, and shallow—uprooting leaves the lower (now horizontal) trunk well above the ground, and the hole created is a shallow one. Perhaps the stiffness of such tree trunks might more often be turned against them when we clear land. Around my home the first step in forest fighting consists of mowing down everything with chain saws. That's easy and satisfyingly destructive. Then a bulldozer deals with the stumps slowly and awkwardly. Why not attach a cable 10 or 20 meters up each tree and then to pull with a winch, the opposite end of which could be attached to the base of another tree? In most cases the tree should uproot with only a modest applied force—the trunk provides a long, stiff lever arm that we're perhaps too quick to amputate.

Trees, or at least treelike tall, woody, terrestrial plants, have evolved independently quite a few times (Niklas 1997). Wood supports—or supported—all. As Niklas pointed out, the obvious alternative, hydrostatic support, depends more critically on a reliable water supply and produces structures of greater density and thus greater supportive requirement. And we earlier noted the unfavorable twistiness-to-bendiness ratio of ordinary hydrostats for erect, gravity-defying structures. But at least one erect marine alga shows that treelike mechanical behavior isn't nature's only option.

Postelsia palmaeformis—the scientific name makes the same allusion as the common name, "sea palm"—lives in the lower intertidal zones of rocky, wave-swept shore on the west coast of North America. The plant, shown in figure 21.4, never reaches a meter in height, so it's not exactly a big tree. But, like trees, it has three parts, an attachment (holdfast), a column (stipe), and photosynthetic laminae (fronds). It stands against gravity, getting higher in dense stands, and it bends in response to lateral force, for it, waves. Unlike any tree, though, it responds to lateral force by bending until almost prostrate and then springing back upright. Concomitant with that untreelike behavior are a set of most untreelike material properties, set out in a lovely paper by Holbrook, Denny, and Koehl (1991). Table 21.1 lists these, along with typical values for wood.

Postelia's "trunk" certainly looks wimpy next to wood. It gives rather than standing tall in the face of fierce force, getting from a high second moment of area just enough flexural stiffness to stand erect at all. Only in work of extension does it play in the same league as wood, although it gets its high value by quite a different tactic—high stretchiness instead of high strength. Regaining its erect posture depends on reinvesting that work of extension, which requires a high resilience, which it has. But, as Holbrook, Denny, and Koehl

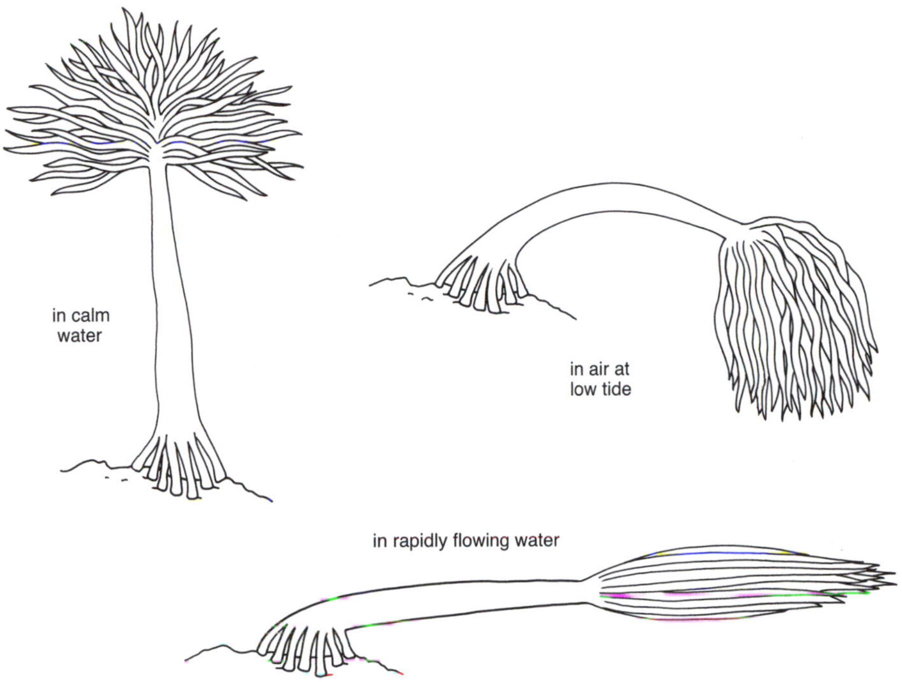

FIGURE 21.4 The postures of a sea palm, *Postelsia palmaeformis*, in still water, still air, and bent over by a wave.

emphasized, the energy storage underlying that resilience undermines its toughness. *Postelsia* is notably brittle and sensitive to scratches—its stipe may be soft, but cracks propagate all too readily. Before dismissing this alga as just a lowly tree wannabe, bear in mind that it has extraordinarily high photosynthetic productivity and that it invests far more, proportionately, in photosynthetic organs than any ordinary tree. Its fronds represent no less than 38 percent of its energy content or 35 percent of its weight (Lawrence and McClintock 1988)—compare this with the trivial weight and burning yield of the annual leaf-fall of a tree that yields a cord or so (perhaps 2 metric tons) of dry firewood.

The differences between *Postelsia* and all terrestrial treelike plants, extant and extinct, raise the paired questions of why erect marine plants aren't more treelike and why no large *Postelsia*-like plants occur in terrestrial habitats. My own guess amounts to the assertion that sea palms live on another planet.

TABLE 21.1 Sea Palm (*Postelia*) Versus Wood: Mechanical Properties

Property	Postelsia Stipe	Wood	Units
Young's modulus of elasticity	5–10	10,000	MPa
Tensile strength (breaking stress)	1.0	200	MPa
Extensibility (breaking strain)	0.2–0.25	0.01	—
Work of extension	100	500	kJ/m^3
Work of fracture (toughness)	400–900	12,000	J/m^2

Sources: Data from Holbrook, Denny, and Koehl 1991; tables in chapter 15.

The ratio of the lateral force, drag, to the axial force, gravity, is vastly higher in their world. Gravity remains gravity, at least when the tide is out, but water is eight hundred times more dense than air. Torrential water flows may be slower than storm winds, but not by a factor anywhere near the square root of 800, or almost 30, the factor that would produce the same drag. Four times slower flow might be more reasonable, which translates into almost an order of magnitude worse drag relative to area for *Postelsia*. And that ignores the extra accelerational forces associated with wave impact, discussed in chapter 14. Trees already invest far more, proportionately, in trunks, so an intertidal wooden tree would probably not even be able to hold up a trunk, much less deploy leaves. What about marine trees in quieter habitats? Places where flows remain gentle rarely offer substrata suitable for attachment. Conversely, the size scale on which trees operate and the temporal irregularities of the winds they face would probably make a terrestrial *Postelsia* impractical—although grasses come close to playing its game.

Categorizing Supportive Systems

Just as the difference between materials and structures became rather arbitrary when composites came up, so the difference between structures and systems lacks a decent dichotomizing distinction. In introducing materials, quite a few pages back, the point was made that a creature earned no selective advantage by adhering to some logical arrangement that we might find convenient. The point still applies. Some subtle advantages may underlie the distinction between the simple spicules that stiffen the arms of the larva of one species of sand dollar and the fenestrated spicules of a closely related species. Should nature care that a few perforations change a beam, which we consider a structure, into a truss, which we might prefer to regard as a supportive system?

Nonetheless we do need to reduce our cerebral jumble, even if that takes a degree of arbitrariness—the game of science, in the largest sense, consists not just of studying and describing nature, but of looking for order and patterns in her diverse entities and phenomena. And the way we've looked at structures leads to a piecemeal view of systems. So at this point a classification ought to prove useful. Still, the reader must recognize this particular hierarchy has only the author's authority, not nature's.

First, though, a few parenthetical notes. (1) Where loading generates only simple stresses (e.g., tension or crushing compression), any distinction between structure and material becomes minor—in effect we jump directly from material to supportive systems. It's only with bending, shearing, and torsional loads that the tripartite scheme of material, structure, and system proves properly applicable. (2) Several types of supportive systems may coexist in an organism either in different places or as systems and subsystems. Thus, a strutted system may have hydrostatic elements—recall that a person lifting a heavy weight tenses the walls of the entire torso and thereby assists the vertebral column with some internal pressure. Similarly, some supportive systems categorize less tidily and definitively than do others.

A. *Tensile systems* (fig. 21.5)

A modern technological example is a boat anchored in a current or a tethered balloon. Of course, humans have long made suspension bridges from vines—

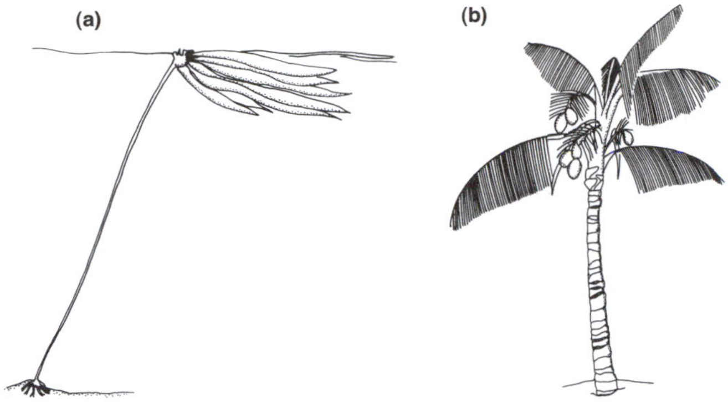

FIGURE 21.5 Tensile systems. (a) The bull kelp *Nereocystis* (a very large alga) is loaded in tension by almost any water movement. (b) The stems of the heavy fruits of coconut palms are loaded in tension by gravity.

vines, after all, are normally loaded in tension, so they ought to provide reasonable sources of material. In nature, tensile systems are reasonably common and certainly diverse in biological affinities. Among lower plants, the stipes of some large algae are notable. In the Pacific Northwest some algae whose holdfasts are attached to rocks and whose fronds are on the surface have stipes between them well over 100 feet long, stipes that are loaded in tension by waves and tidal currents. Among the higher terrestrial plants we see tensile systems in various epiphytes, tendrils, and dangling fruits. Mussels attach to rocks entirely through the use of tough byssus threads, mainly made of collagen. Even the setae by which the gecko hangs beneath a rock or ceiling are proper tensile systems.

Tensile systems ought to scale in an unusual way, something first pointed out by Galileo. Strength doesn't depend on length, so the thickness of your fishing line doesn't have to be adjusted for the depth at which you hang the bait. Cross-sectional area, though, ought to be proportional to load, so diameter ought to scale with the square root of load—load to the 0.5 power. Peterson et al. (1982) looked at several structures normally loaded in tension—fruit stems, algal stipes, and toe-pad setae. They found that in practice, diameter scales with load to the power 0.3 rather than 0.5; cross section thus goes up in proportion to load to the power 0.6, not the 1.0 that seems intuitively reasonable.

B. *Strutted systems, with or without external skins*

1. *Single or branched strut* (fig. 21.6). We use these in many forms—frame buildings (their collections of joists, studs, and rafters), bicycle frames, support towers for electrical transmission cable, and so forth. In nature we notice trees, branching corals, and other fairly stiff items. In these systems, all of the struts join in a common lattice, and no motion is permissible at joints—they're simple, statically determined systems. If our systems branch, they're usually equipped with lots of triangular elements, although some crude cases (frame houses) use an array of mechanisms braced against any possible deformations by a structural skin of plywood or something similar. In nature, more often than not, the branches of a system diverge without rejoining, although struts are sometimes joined into trusses—the arms of some sand dollar larvae and some bones in the wings of large birds have already been mentioned. Some cacti (such as the cholla), though, have supportive columns of branched

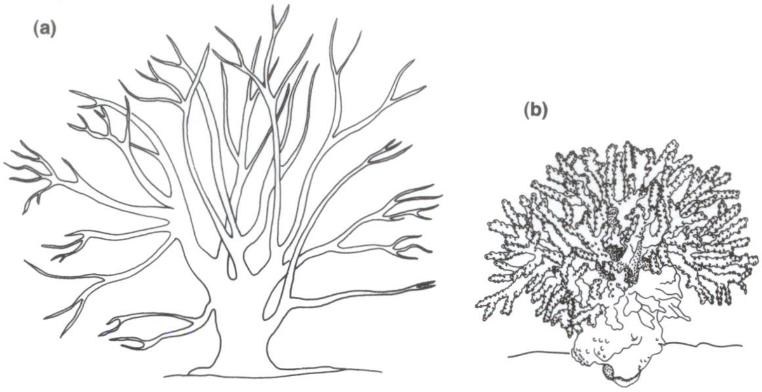

FIGURE 21.6 Two systems composed of branched struts. (a) The trunk and branches of a live oak, and (b) a hard coral from a tropical reef.

and rejoining struts—trusses in the form of hollow cylinders with perforated walls.

2. *Articulated strut* (fig. 21.7). These share the common lattice of compression-resisting elements, but their joints (articulations) permit motion. We use them infrequently, but we do deliberately build joints into many bridges, for example, so the resulting mechanisms can distort safely under changing wind loads, varied "live" or functional loads, or thermal size changes. Nature often uses the arrangement—major portions of vertebrate skeletons can be best viewed as mechanisms of articulated struts. The hard elements (ossicles) and their connections in echinoderms such as starfish provide another example.

Systems built around articulated struts combine nicely with muscles; sometimes, as in insect skeletons, the muscles are on the inside, but the principle is the same. Among the best features of these systems is their ability to alter shape or overall mechanical properties rapidly without having to change the properties of specific materials. You drastically increase your stiffness when changing from a supine to an upright posture by increasing the tension of a very large number of muscles—with a non-trivial increase in your metabolic rate. But even tensile tissues other than muscle may sometimes change properties fairly quickly in response to some chemical signal. These alterations have been studied most extensively in the so-called catch connective tissue of echinoderms (Motokawa 1984; Wilkie 2002). A starfish undergoes an impressive mechanical transformation as it shifts from being limp enough to crawl with its tube feet on an irregular substratum to being stiff enough so the same tube feet have adequate anchorage when pulling open the shell of a clam.

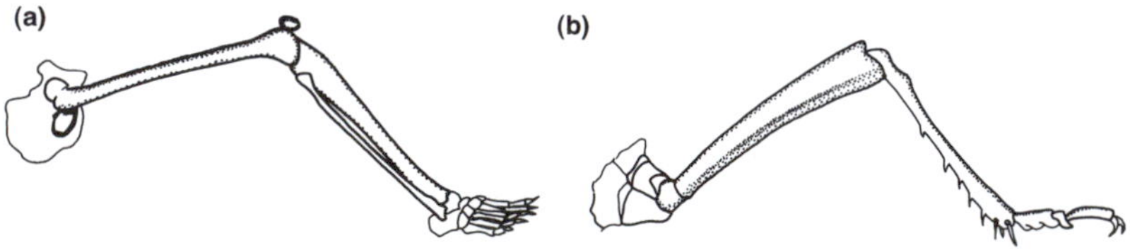

FIGURE 21.7 Two systems of articulated struts. (a) The endoskeleton of a vertebrate leg, and (b) the exoskeleton of an insect leg.

3. *Dispersed strut.* In our technology, these find use still less often; to most of us they're quite unfamiliar. We don't find it self-evident that decent support can be achieved in a system where struts don't directly join with each other but link only through tensile elements. So-called tensegrity domes and towers have been noted earlier as dispersed strut systems, but for us the scheme serves mainly as an art form. A few tents used discontinuous strutting, with the struts inserted into pockets of fabric just as are the battens of a sail. But I'm surprised so few of them use this style of construction, which ought to be very weight-efficient and avoid the troubles of making rigid joints between short pieces of pole.

In nature, the scheme is commoner but still far from widespread—the clearest example, noted in chapter 19, is the spicular skeleton of sponges, in which tiny rigid elements are laced together by collagen (fig. 19.7). And there are occasional forays in this direction among sea anemones (coelenterates) and sea cucumbers (echinoderms). It ought to be reemphasized that the arrangement is not intrinsically flawed in some way; the limitation is more likely to lie in problems of compatibility with attachment surfaces for muscles.

C. *Nearly rigid surfaces*

1. *Flat surfaces.* It's difficult to make a surface that will stay flat under appreciable stress, but we like flat floors in our houses or on bridges and are willing to invest a lot of material to achieve the result. Astonishingly, we even make flat, horizontal roofs—they not only require sturdy supports but have to be waterproof rather than working as rain guides as do shingles, slate, and thatch. To get a flat surface, castings must be thick or the flat plate well braced and trussed beneath—heavy girders run up to our luxuriously flat floors; and strong joists, trusses, or flanges run just beneath them.

The main use of flat surfaces in nature consists of photosynthetic structures such as leaves. These also are well braced beneath; most leaves seem to circumvent problems of loads perpendicular to their surfaces simply by flexing or reorienting in winds (fig. 1.5). Quite a few leaves of various lineages (mostly monocots) use a slight longitudinal V-fold to get adequate flexural stiffness, which must also give them nicely low torsional stiffness. Other leaves use another deviation from flatness, crosswise fan folding, discussed by Niklas (1992). Figure 21.8 shows a few such schemes.

Insect wings provide yet another example of braced, flat surfaces—cylindrical cantilever beams (veins) support a thin membrane. A pound of fruit-fly wings laid end to end would stretch about 500 miles, a *very* low mass per unit length—a steel wire to go so far would have about the same diameter as a red blood cell. Yet in each second of flight the tip of a wing moves several meters and reverses direction four hundred times. Other paddles and fins are fairly flat as well, as are some feathers, the book gills of horseshoe crabs, and a scattering of other stiff structures. In all these cases, though, flatness suits functions other than support. From a mechanical viewpoint the flatness of these systems, however impressive, is perhaps best regarded as a necessary evil—and their designs incorporate features that offset their intrinsically low flexural stiffness.

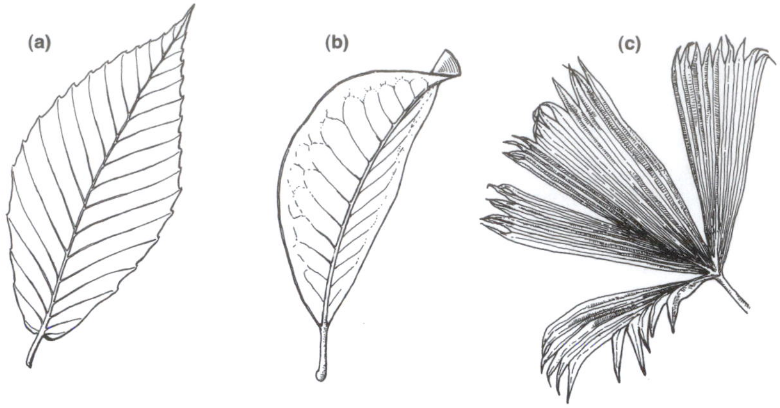

FIGURE 21.8 Thin leaf surfaces avoid bending in various ways. Veins may provide supporting trusses (a), the whole leaf may be cambered lengthwise (b), or pleats can make a ridge-and-valley self-trussing system (c).

2. *Hollow cylindrical tubes.* The way these give high flexural and torsional stiffness with minimal material hasn't been lost on either nature or engineers. We use them as subsystems when building bicycles and racing cars and as entire systems in so-called monocoque aircraft fuselages, cylindrical storage tanks, glass jars, and metal cans. Nature also uses them in diverse places—bamboo stems; vertebrate long bones; insect, spider, and crustacean appendages; the wing veins of insects; and the feather shafts of birds. Sometimes they contain the entire organism, as in lots of threadlike algae, although it's unclear how much of the stiffness of these last comes from fluid pressure rather than from their tubular solids—hydrostatic systems and hollow cylindrical beams aren't mutually exclusive. Microtubules (fig. 22.3), stiffening elements in cells, are also hollow cylindrical beams, although they may derive additional stiffness from ordered water molecules at their surfaces.

My colleague, Stephen Wainwright (1988), argues forcefully that the cylinder is the default shape in the design of organisms. Indeed, nature makes far more cylinders than those that make up the present category—solid struts and internally pressurized containers appear elsewhere in this classification.

3. *Spherical and ellipsoidal shells.* Cylinders curve in one direction, circumferentially—their lengthwise curvature is zero. Here we're talking about structures with their surfaces curved in two directions; and, as pointed out earlier (and implicit in Laplace's law), that double curvature gives greater strength and stiffness for a given investment of material. Maximization of internal volume for a given surface compounds the material economy—for that nothing beats a sphere, and a slight egg-shapedness (spheroidicity) doesn't make things much worse. We make domes, spherical storage tanks, skylight covers, bowls, and amphorae. Sometimes skin and structure are integral (cast domes); sometimes a thin skin covers a strutted structure (geodesic domes, for example).

Nature puts this shape to use in a large number of places. Our skulls are nearly spherical domes—and the light and thin bone needs only minimal internal bracing. Similarly, a turtle's shell is a light, strong dome, as are the shells of many bivalve and gastropod mollusks; the thoraces of many insects, spiders, and crustaceans; the eggs of birds; and nut shells. Smashing the inner wall of a coconut takes quite an effort, and the resulting pieces don't weigh a lot. Still, domes have several disabilities. Localized loads can be troublesome, and resistance to local penetration may demand enough material to offset most of their cheap resistance to uniform transmural pressure differences. Perhaps the main limitation of domes lies in their uncompromisingly regular geometry. An organism has a lot to do in life besides resisting the slings and arrows of outrageous fortune, and the turtle's shell looks like a liability when the animal takes arms against a sea of troubles.

D. *Internally pressurized systems*

1. *Hydrostats.* In a hydrostatic system, an internal pressure greater than that outside loads a skin in tension; fluid inside feels compression. As mentioned in chapter 20, we use this in nonrigid airships (blimps), pneumatic tires, and occasional inflatable buildings. In these cases, the internal material is air. Nature makes very extensive use of hydrostatic systems, but her preferred stuffing is water, sometimes together with living tissue. Water, of course, resists compression with less strain than any gas, so these systems can be quite sturdy. Sharks, worms, squid mantle, many plant cells, and herbaceous (nonwoody) stems are all examples that have been discussed earlier.

2. *Muscular hydrostats.* Here any distinction between systems accomplishing support and ones devoted to motility gets thoroughly blurred. These devices consist almost entirely of muscle and with the same machinery do both functions; they were recognized as a distinct and consistently constructed category of supportive system relatively recently, by Kier and Smith (1985). Naturally enough, we have no known nonliving or botanical examples. In effect, contraction of one group of muscles forces extension of another group, not due to some solid, bony linkage, but on account of the constancy of volume of muscle itself. Squid tentacles and arms, elephant trunks, and reptilian and some other tongues (including our own) work in this manner.

CHAPTER

Achieving Motility

> The function of muscle is to pull and not to push, except in the case of the genitals and the tongue.
>
> **LEONARDO DA VINCI**

ORGANISMS DO MORE THAN JUST ACCOMMODATE FORCES, ACCEPTING some amount of deflection or displacement. Most of them move, whether by doing work on their surroundings or by forcing one internal bit to change position relative to another. One can argue that the ability to make rapid motions was what elicited the evolution of the quick-acting sense organs and the elaborate information-processing equipment of animals—that nervous systems, memory, behavior, thought, even emotion arose because we could move. The last chapter worried about how creatures could stay put; this one and the next will focus on how they do just the opposite.

For convenience, we'll separate this brief discussion of motility into two parts, prime movers and linkages. It's the distinction between an engine and a transmission, and it comes with the same understanding that the character of the first constrains or even determines the design of the second. For the prime movers in particular, we need a disclaimer at the start—much is known about their molecular mechanisms, much of that bears on their operation, and the volume of current work on these systems at the molecular level probably exceeds that of all macroscopic biomechanics combined. Nevertheless, I assert the author's prerogative and declare the subject one largely beyond the scope of the present macroscopically oriented book.

Engines

The most conspicuous natural device that confers motility is, of course, muscle. But nature has other machines in her armamentarium—they're just not so fast, massive, and good to eat.

Hydration devices

If you take some dry legume seeds and immerse them in water they swell up. If they're put in a closed chamber into which water can pass but out of which they can't squeeze or ooze, then their swelling will push outward on the chamber with surprising force. Try it—there's an old device called a "potato ricer" that can be filled with split peas and its handles tied with lots of elastic bands; expansion of the elastic gives an indication of force (you might even calibrate the rubber with weights). Alternatively, a common type of garlic press, one with a pair of handles on one side, will work. The affinity of many polysac-

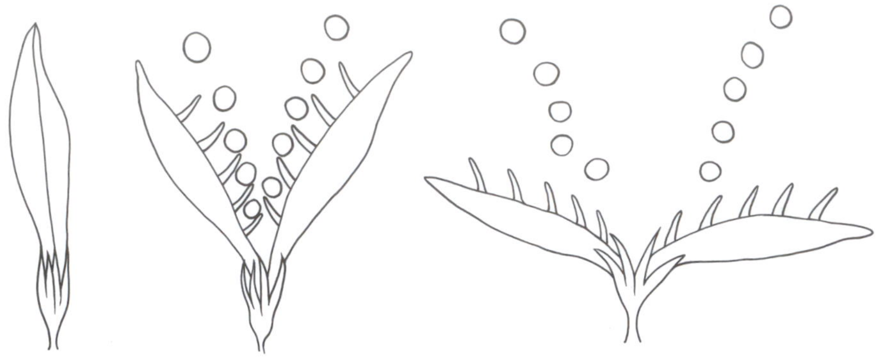

FIGURE 22.1 Seed expulsion as a dry seed capsule splits and its halves bend outward.

charides and proteins for water is extreme—the resulting pressures may reach thousands of atmospheres. (Salisbury and Ross [1969] give references to experiments and calculations.) Wetting of a small quantity of dry seeds accidentally left underneath can lead to the fracture of concrete pavement.* As far as I know, all extracellular mucuses and gels are secreted in concentrated form and then take up water—a conspicuous example is the jelly mass in which the eggs of a frog are suspended. When swollen, the mass is typically larger than the volume of the gravid female. Slime production by hagfish, according to John Gosline, provides an even more spectacular case.

The phenomenon, officially termed "imbibition," is sometimes used to move material. It's certainly involved in the growth of plants, where substantial amounts of material are lifted upward; the 40-centimeter upward growth I measured over the course of a single day on a bamboo plant in my yard must have resulted from hydration rather than from new organic synthesis. It has been put to further use in some plants—the relatively rapid (for a plant) folding of the leaves of the sensitive plant, *Mimosa*, uses the scheme, as does the closure of the trap of the Venus's flytrap and a few other leaf-folding arrangements among the grasses.

Some seed capsules have a particularly neat way of achieving motility by varying water content. Witztum and Schulgasser (1995) showed that the two halves of the capsule (as in fig. 22.1) store energy as they dry; at a certain dryness (or, in some, when the top of the capsule is wetted) the seam joining them gives way suddenly. The capsule halves spring outward, which not only releases the seeds within, but expels them explosively at speeds up to 12 meters per second (27 miles per hour); they travel up to about 3 meters. The springing mechanism, as they point out, matches that of the bimetallic strips we use for thermometers and thermostats—differential expansion of longitudinally joined layers produces the stress that springing relieves.

At least one case of motion driven by hydration occurs in animals—the mechanism with which the males of at least one genus of mosquitoes erect the hairs on their antennae (Nijhout and Sheffield 1979). Presumably, an antenna with recumbent hairs has less drag than one with erect hairs, but only with erect hairs can a male detect the hum of a female in flight. (Females,

*In one of C. S. Forester's novels, a leak in a ship's hold and consequent swelling of the cargo of rice splits the hull and sinks the ship.

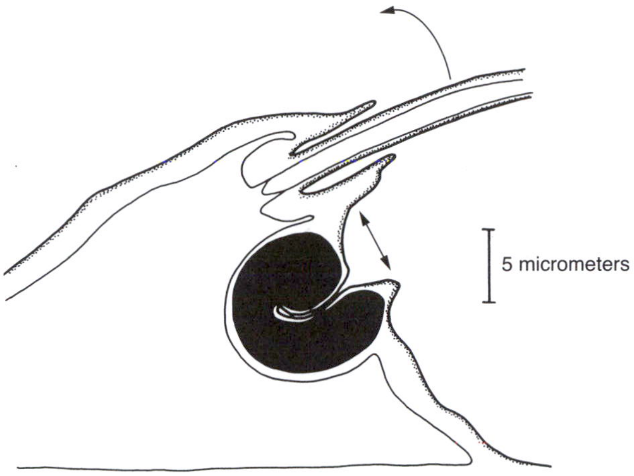

FIGURE 22.2 The annular pad (black) near the base of an antennal hair absorbs water, swells, and causes the hair to protrude outward from the axis of the antenna.

being larger, have a lower wingbeat frequency and thus buzz at a lower pitch; a tuning fork humming such a siren song will attract males.) Adjacent to the socket of a hair is an annular pad of homogeneous protein (fig. 22.2); the angle of erection of the hair tracks the angle of unfolding of the pad. During unfolding and erection the pad increases in volume by 25–30 percent while the cells just beneath it decrease in size.

Osmotic devices

Diffusion, as in chapter 8, moves (on average) material away from a location in which the material is more concentrated than elsewhere. Diffusion of water doesn't differ; it merely takes a slight mental readjustment to view a region as having a high solvent (water) concentration instead of having a low solute (sugar, for instance) concentration. Water, in short, moves toward regions in which it contains more solutes. For practical purposes, the water concentration decreases in proportion to the number of particles of the solute, whether these are molecules, as of sugar, or ions, as with the sodium and chloride ions that result from dissolving salt.

Consider the apparatus shown in figure 22.3. Water diluted by a solute

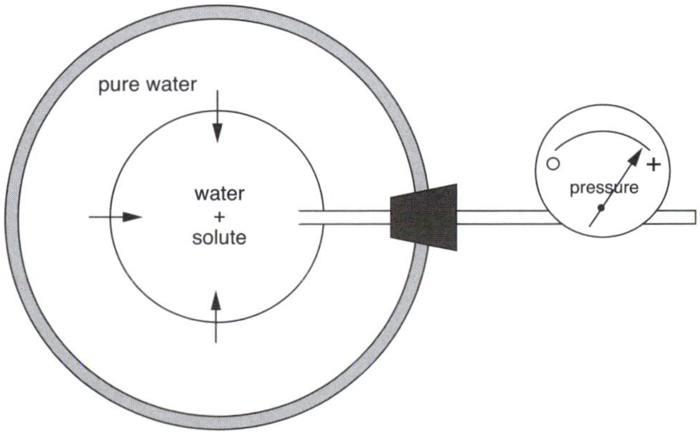

FIGURE 22.3 An osmotic engine: a net movement of water into a water-permeable bag occurs if the bag contains a solute as well as water. That movement can generate large pressure differences.

fills a bag with an inelastic wall, which is immersed in a large container of pure water. The wall of the bag has a peculiar property—water can diffuse through it but molecules or ions of the solute cannot (or do so much more slowly). That peculiarity results in a net inward movement of water that would, if the walls could stretch, increase the volume of the bag. Instead the pressure inside increases. Now, however your teacher or textbook of introductory biology may have confused matters, this pressure is a thoroughly ordinary hydrostatic pressure of the type that pervaded chapters 5 and 20. How much pressure is generated? In this somewhat idealized setup, a unit concentration of solute produces a pressure of 2.26 megapascals (over 20 atmospheres). The unit, though, is peculiar—it's a concentration equal to 6.0×10^{23} molecules (or ions) added to a kilogram of water, a so-called 1 molal solution.

Matters are made easier in practice if a bit more devious in concept by a scheme for counting out the molecules. You look up the so-called gram-molecular weight of the solute (for table sugar, sucrose, 343) and add that number of grams of the solute to the kilogram of water to get the molal solution—the fact that different molecules have different masses has then been dealt with. For fully ionizing substances, the pressure produced by the molal solution must be multiplied by the number of ions into which each molecule divides in solution—for table salt you multiply by two to account for one ion each of sodium and chloride. (The ions don't act quite independently, so a small but well-known correction must be applied as well.)

A solution capable of producing a hydrostatic pressure of 1 megapascal with pure water on the other side of a barrier is said to have an "osmotic pressure" of 1 megapascal. Osmotic pressure thus just designates a capability rather than a manometrically measurable pressure, the main reason why it confuses people. No kidding—these osmotic pressures correspond to (or can generate) huge hydrostatic pressures. About 30 grams of table salt in a kilogram of water (not all that concentrated) can generate 2 megapascals—so a closed bag must have a *very* strong wall if it's to avoid explosion. Or the bag might be tiny (like a cell), so the high pressures yield only modest tensions stretching its wall—equation (3.6) still applies.

To get work from pressure takes only a concomitant change in volume; that pressure-volume work forms the basis of operation of combustion engines whether the burning of fuel is external (steam engines) or internal (most liquid-fuel engines). Generating the pressure osmotically makes the arrangement no less useful as an engine, although, with diffusion playing a role, only a minuscule engine can run rapidly. Among organisms the scheme is confined largely to plant cells, where the relative status of water inside and outside determines the internal pressure (sometimes called "turgor") of individual cells. The interplay of structural constraints, internal pressure, and external forces then determines the shape of a cell at a particular time. The usual scheme for increasing internal pressure involves the release by a cell or the transport inward from specialized adjoining cells of ions, mainly those of potassium. (Plants are high in potassium, as people on low-sodium diets are informed.) The cell then passively (osmotically) absorbs water. Conversely, loss of water lowers internal pressure, with wilting the visible consequence. You can make vegetables wilt by putting them in a saline solution, since that results in a net water outflow. We've already met one impressive osmotic engine—recall the sporangium-shooting ballistic fungus, *Pilobolus*, in chapter 20.

Microtubules and microfilaments

Muscle is only one of a large group of mechanisms for achieving motility that depend directly on the properties and interactions of proteins. Within cells, microtubules (fig. 22.4) not only serve as supportive elements of the cytoskeleton but are strikingly common wherever cells undergo the chromosome movements associated with division or undergo the shape changes of development. Microtubules make up the most conspicuous components of cilia and flagella (except those of bacteria), organelles most of which do locomotory tasks as they send waves of bending outward from their attachment at the cell surface. Under some circumstances, microtubules alone can work as engines, pushing things such as chromosomes around. They exert their force and move things (hence doing work) by polymerizing, adding monomers of their dimeric protein, tubulin, to their ends. With no load, a microtubule can grow at 20 nanometers per second; with a load of 3–4 piconewtons, one can grow about 3.3 nanometers per second, according to Dogterom and Yurke (1997). (That's a power output of a bit over 1×10^{-20} watts.) They also prefer not to bend, appearing stiff and straight when photographed. As structures that serve by resisting bending, they're probably the world's smallest functioning beams. The articles in the book by Hyams and Lloyd (1994) provide a good introduction.

Still smaller organelles, microfilaments, seem to be associated with movement as well—especially that curious internal movement of the contents of plant cells called "cytoplasmic streaming" or "cyclosis" (see, for instance, Allen [1974] and Vallee [1998]). The speeds of cyclosis, about 5 micrometers per second (a few much faster cases are known) look far from trivial to a viewer with a microscope (which magnifies space but not time)—we're not talking about something evident only in time-lapse movies. Microfilaments are made of a protein called actin; actin and another protein, myosin, apparently occur in all living eucaryotic (nonbacterial, loosely) cells. As we'll see, these two proteins are the basic constituents of muscle. The details of the mechanisms by which microtubules and microfilaments generate motion aren't entirely clear, and not for lack of investigators or literature either.

Cilia and flagella

These organelles will be familiar to anyone who has taken a general biology course. They're not especially different—cilia typically occur in large numbers

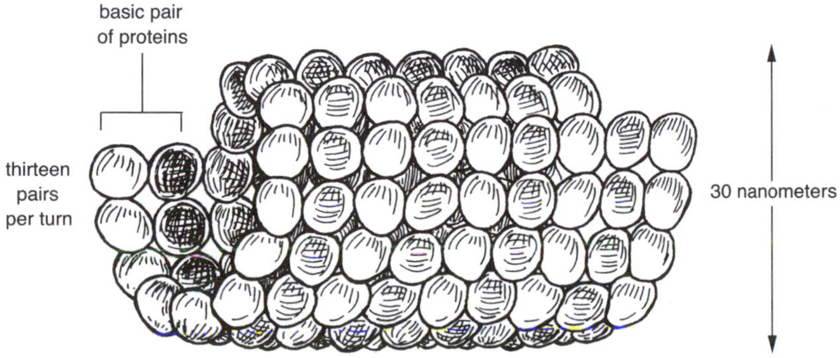

FIGURE 22.4 The arrangement of subunits in a microtubule. Two types of protein are wound in a long helix about 30 nanometers in diameter; each turn of the helix takes thirteen molecules of each of the proteins.

per cell and flagella as only one or a very few per cell. Otherwise, both make up an example (not the only one) of natural engines that work though the interaction of a pair of proteins, dynein and tubulin. They're the locomotory organs of a wide diversity of organisms—ciliated and flagellated protists, a host of different invertebrate larvae, and a wide range of spermatozoa. None of these is large. For fluid-mechanical reasons, large creatures can't be made to swim efficiently by ciliary action—one might imagine pumping water with them through a constricting pipe which forms a jet, but both the biology and physics of the arrangement are inauspicious. One group of invertebrates, the Ctenophora, or comb jellies, combines cilia into so-called comb plates and gets up to centimeters or inches in diameter—but ctenophores swim slowly relative to their sizes.

In one way or another, every group of animals except, oddly enough, the arthropods, uses motile cilia. In large creatures, they push things around internally—water over the gills of clams, eggs through fallopian tubes, water and particles across suspension-feeding equipment, mucus along the linings of various tubes. But they don't scale up as does muscle, remaining as linings in a two-dimensional world of surfaces along which fluid moves or across which material gets transported.

In typical cilia and flagella, a set of nine doublet microtubules with some specialized ends surrounds a pair of single microtubules, as in figure 22.5. Tubulin, as in other microtubules, makes up all of these. Each doublet bears a pair of arms (at least arms in cross section) made of dynein. The dynein of one doublet and tubulin of the adjacent one ratchet along each other; if this happens on one side of a cilium, the cilium bends. Depending on how bending propagates, cilia and flagella may sweep fluid or mucus sideways with an extended power and flexed recovery stroke, or they may propagate waves of planar or helical bending out from their bases. They may sound simple, but the processes of bending and propagation require a host of other proteins, an energy source, lengthwise transport of fuel and wastes, and so forth. An integrated picture of the process hasn't yet emerged, again despite great amounts of interest and effort.

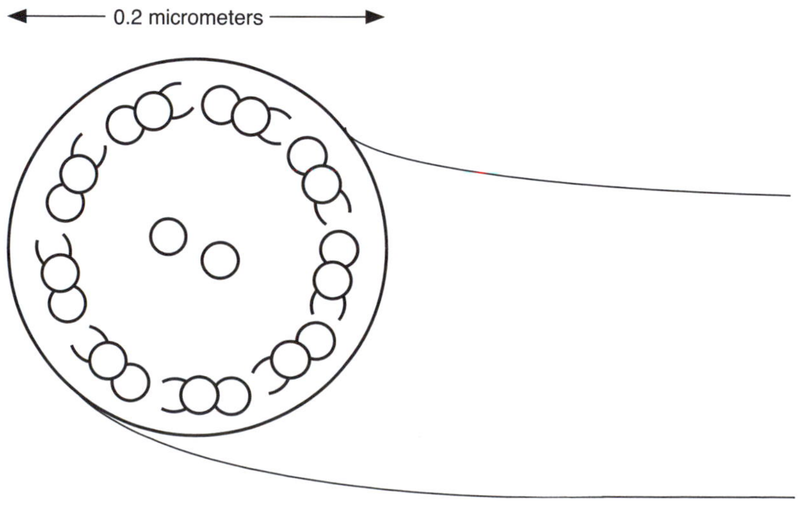

FIGURE 22.5 A cilium cut transversely (imagine looking at the end of a sliced strand of spaghetti), showing the typical set of nine doublet microtubules (with their side arms) surrounding a pair of singlet microtubules.

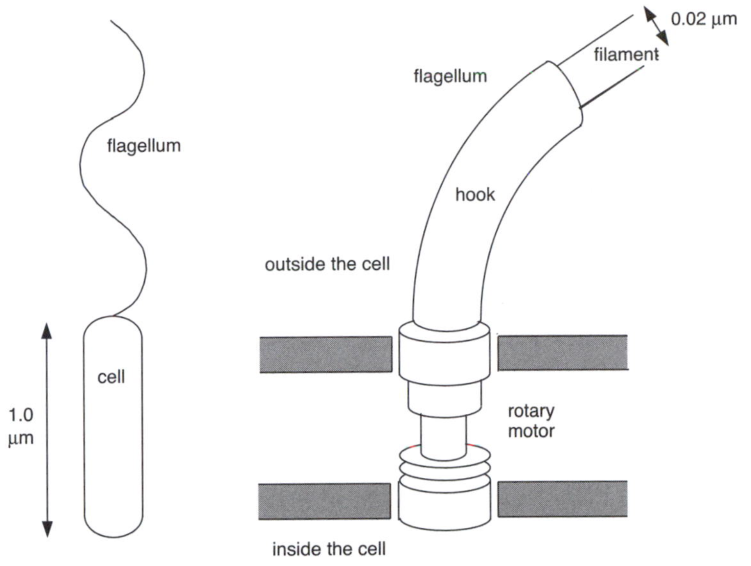

FIGURE 22.6 A bacterium with a flagellum, and the motor and base of the flagellum as presently understood.

Bacterial flagella

The cilia and flagella just described occur in eucaryotes—organisms whose cells have proper nuclei. An entirely different kind of propulsive unit, confusingly, goes by the same name, flagellum, in the bacterial procaryotes. Eucaryotic flagella run about 0.2 micrometers across and have that nine-plus-two pattern of internal tubules. Bacterial flagella are much smaller, only 0.02 micrometers in diameter, and lack that internal structure. In electron micrographs, bacterial flagella look suspiciously like rigid structures bent into fixed corkscrewlike waves. Which they turn out to be. As Berg and Anderson (1973; Berg 1974) discovered, they work as rigid, rotating propellers, driven by rotary engines in the bacterial surface beneath, as in figure 22.6. In a sense, they're not engines at all, leaving that to their basal motors. The combination consists of the only true rotary engine and propulsive unit known in the living world—a proper wheel and axle mechanism (Dusenbery 1996).

They're far more efficient, at least in terms of power relative to weight, than ordinary flagella or even muscle, but they don't seem to scale up, and nature hasn't used them elsewhere. Or at least she hasn't manufactured them elsewhere, since some higher organisms symbiotically appropriate bacteria for use as locomotory organelles.

Other biological motors

Up her sleeves, nature has a few more ways of achieving motility.

A mechanism based on the surface tension of water is used by a fern that forcibly discharges its spores from the sporangium—evaporation decreases the volume of water and increases the surface curvature in a series of cuplike dead cells until the sustainable cohesion is exceeded; water then vaporizes and tension is relieved by a movement analogous to that involved in throwing a spear (Steward 1968). Surface tension gets put to use by a nematode, which leaps from one soil particle (through air) to another. It holds itself in a U-bent spring with a drop of water; failure of the droplet to stay together straightens it suddenly enough to catapult the nematode upward and laterally (Dusenbery 1996). (And the phalarope of fig. 5.8 uses surface tension to make a droplet of water ascend the length of its bill.)

Amoeboid (pseudopodial) motion occurs in a wide variety of cells, from amoebae and slime molds to our own white blood cells. It utilizes, in some way, the same paired proteins, actin and myosin, as do our muscles.

A variety of microscopic algae (diatoms, desmids, and others) engage in gliding motion across surfaces; the motion is usually associated with the secretion of slime, but the source of the shearing force between organism and substratum is unclear; it's certainly different from any of the mechanisms we've considered so far (Halfen 1979; Melkonian 1992). Leaving a slime trail may push up the cost of transport, as the work on slug locomotion indicates.

One kind of marine slime mold, *Labyrinthula*, restricts its slime to communal membrane-bound "slime ways," through which its autonomous cells glide (Dusenbery 1996). The engine involves interactions between the same two proteins, actin and myosin, as does that of muscle (Dietz and Schnetter 1999).

Gliding motion across surfaces, usually with slime—whether or not by the same scheme—occurs in procaryotic organisms (bacteria and their kin) as well. It's based on either of two mechanisms. Bacteria are often covered with tiny hairs, pili; retraction of one type (designated IV) through their outer membranes can move them around. Alternatively, they can secrete carbohydrate slime rearward to get a push (Kaiser 2000; Merz and Forest 2002). One cyanobacterium, *Synechococcus*, can swim as well, doing so at up to 25 micrometers per second without benefit of visible locomotory organelles (Waterbury et al. 1985; Stone and Samuel 1996).

At least one odd actinopod, heliozoan protozoan, *Sticholonche*, rows along with oars made of bundles of microtubules that emerge from the nuclear membrane through microfibrillar oarlocks. These form the central axes of external cytoplasmic protrusions—oars of a sort (Cachon and Cachon 1978; Margulis and Schwartz 1998).

In a very real sense, the process of growth itself constitutes an engine. That may sound like some kind of academic sophistry, but work-requiring events are well known in the early development of animals. More impressive, perhaps, is something I've heard that a tree can be persuaded to do. One bends a low branch down to the ground; one then embeds the end of the branch (pruned as necessary) in a bucket of cement or ties to it a substantial rock. Over time (on a botanical rather than a zoological, but not a geological scale) the branch will lift the cement or rock.

Muscle in Particular

We know much more about the mechanism of muscle contraction than about how nature's other proteinaceous engines work. We know, and one really has to start with the point, that muscles don't contract! "Contraction" misleads in three separate senses. First, on a molecular level, nothing literally shortens—no molecule exerts significant force by going from long (and perhaps thin) to short (and concomitantly fat). We'll get back to the matter shortly.

Second, on a macroscopic, whole-tissue level, muscle may get shorter, but it doesn't get smaller; it's an isovolumetric material whose increases in cross section exactly track its decrease in length. Contraction in one direction entails expansion in another as it must in a constant volume system. Try taking enough air in your lungs so you just barely float in a pool; then tense as many of your muscles as you can. Result—nothing; you don't sink as a result of any increase in density. Conclusion—muscle doesn't really contract. What it does is to pull what's attached to one end toward what's attached to the other, while swelling in the middle.

Finally, we often use muscle under circumstances in which it develops tension under loading regimes that prevent it from shortening or even extend it while it pulls. We speak of "isometric contraction," literally contraction without length change and about as oxymoronic as a term can get. Not only can muscle operate without shortening, it develops greater force while held at constant length and still greater force while being stretched. In this last circumstance, our vernacular language diverges from strict physical definition. If a muscle develops tension (and consumes energy) while being stretched rather than while shortening, it "works" in the ordinary sense; in strictest physical sense it does *negative work*. Work, after all, equals force times the distance the force moves something—in the direction of the force.

The contractile mechanism

We can do no more than summarize the aspects of the basic molecular mechanism that bear relevance to muscle's performance at higher levels of organization. For a more elaborate story, one should consult textbooks of cell biology, state-of-the-art summarizations such as those of Holmes and Greeves (2000), A. F. Huxley (2000), and H. E. Huxley (2000), or (for a *much* more superficial treatment) Vogel (2001). The contractile unit, the "sarcomere," shown in figure 22.7, consists of a set of thinner filaments extending outward from a crosswise attachment disk (the "Z-disc") and a set of thicker filaments midway between adjacent disks. In vertebrates, at least, sarcomeres vary little in length from a standard of about 2.5 micrometers. (Burkholder and Lieber [2001] give a wealth of data.) Contraction consists of the interdigitation of the two sets of filaments, of the proteins actin (thin) and myosin (thick). As they slip in between each other, then, the ends of the thick filaments approach the Z-discs and the ends of the thin filaments approach each other. Ordinarily the thin filaments go no farther than touching each other, but in some cases they may slide across each other to give further shortening.

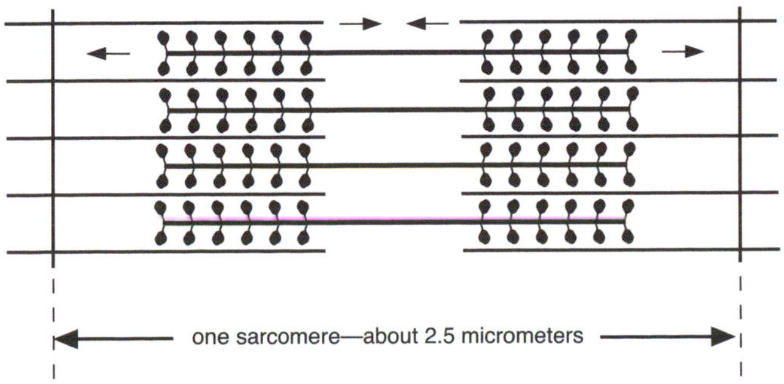

FIGURE 22.7 The basic contractile unit of striated, skeletal muscle. Cross bridges on the thick myosin filaments alternately attach and detach from sequential sites on the thin actin filaments. That makes the two sets of filaments interdigitate farther, shortening the whole apparatus.

While "sliding filaments" describes what happens at one level, the filaments don't so much slide as ratchet along one another. They don't, it seems, doing anything radically different from the interaction of dynein and tubulin in cilia. Cross bridges stick out from the thick myosin filaments, and these repeatedly attach to the thin actin filaments, swing, detach, and swing back, as in figure 22.8. They move in the manner of enzymes, forming enzyme-substrate complexes, changing configuration, and then detaching; indeed, myosin acts as an enzyme in every normal sense. It is, specifically, an adenosine triphosphatase, and uses the energy of hydrolysis of adenosine triphosphate (ATP) to adenosine diphosphate to push on the other filament. That ATP represents the ultimate energy input for muscle contraction. The unusual features consist of the arrangement and ultimate function of these enzymatic sites. Putting a lot of sites in series amplifies the tiny molecular motions typical of individual enzyme molecules to a scale appropriate for bending a whale's tail. Putting a lot in parallel amplifies the tiny molecular forces typical of individual enzyme molecules enough to move water with that tail. Just as arrays of cells make up large organisms, arrays of enzymes make up muscles. Function differs, though—most enzymes use motion to do chemistry; these, one might say, use chemistry to get motion.

These long chains of sarcomeres (myofibrils) are bundled into fibers, and a group of fibers operated by a single nerve cell (neuron) constitutes what's called a "motor unit"—with "group" in a functional as opposed to any anatomical sense. The name is apt; a single nerve impulse triggers a standard twitch of all the fibers in such a unit. To get more frequent contractions, the nervous systems sends more frequent impulses. At sufficiently high frequencies (or with less antagonism from other muscles and such), the units don't reextend fully between twitches and thus they shorten still farther. The large muscle of the rear of one's calf, the gastrocnemius, contains about a million muscle fibers; at the same time, it consists of about five hundred motor units. Thus, the nerve supplying the gastrocnemius has about five hundred axons running down its length, and, on average, two thousand fibers make up each motor unit. Sending unsynchronized impulses to the different motor units reduces the jerkiness of muscle contraction, although at near-maximal loads, some synchronization commonly appears, something you may notice as a slight twitchiness when trying to exert maximum force. Final point—muscles normally called on to do precise work have more motor units relative to their sizes. (A good source on such matters is McComas [1996].)

That's the vertebrate scheme; it's not the only way to run the machine. In particular, arthropods use a different triggering scheme, with fewer but functionally more versatile nerve cells.

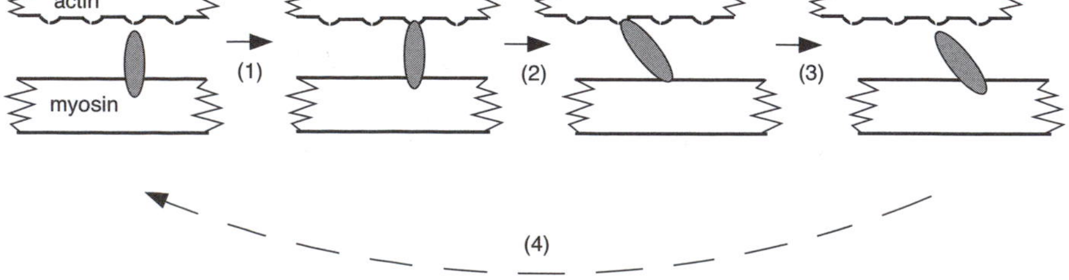

FIGURE 22.8 The action of muscle's cross bridges.

The variables of performance

What can we get out of this sliding filament mechanism? Three variables matter most—speed, force, and power; the last, of course, is the product of the first two. The various versions of the main kind of muscle of present concern, so-called striated, or skeletal, muscle, diverge only in a few quantitative particulars from the description just given. But the performance of striated muscle varies a lot more—nature has been able to tweak this basic device quite remarkably. Muscles themselves vary enormously in size, from ones far below a millimeter in length that work microscopic, but still multicellular, organisms such as microcrustacea to ones well over a meter in length in whales. That's a length range about a hundred thousandfold, and a volume range of the order of 10^{15}.

Speed, per se, does an awkward job of characterizing such a range. If the filaments of each unit (the relevant unit is a half-sarcomere) creep over each other at the same rate, whatever the length of the whole muscle, then the overall shortening will be proportional to the number of units in series, that is, to the length of the muscle. So speed has to be corrected for length (resting length, specifically) in much the same way strain involved correction for original length on a stress-strain graph. The only difference is that dividing speed by length leaves a residual dimension of reciprocal time. So our length-corrected speed measure, rather than being dimensionless, will have units of seconds-to-the-minus-one. In practice, we define an "intrinsic speed" as the speed of contraction (distance per time or meters per second) divided by the length of the muscle, and we referred to intrinsic speed in "lengths per second." Thus, if filaments creep across one another at 5 micrometers per second and a half-sarcomere is 1.25 micrometers long, the intrinsic speed of either the half-sarcomere or the whole muscle comes to 4 lengths per second (4 seconds^{-1}). Put another way, the muscle would contract by its own length in one fourth of a second. Of course, a muscle actually contracts by a shorter fraction of its length—40 percent is an extreme degree of contraction and 20 percent is a more normal operating contraction. With a light load, the latter would take about a twentieth of a second (0.25 times 0.20).

Note, incidentally, that the speed of filament creeping (5 micrometers per second) is roughly the same as that for cytoplasmic streaming, cited earlier—the similarity in speeds ties in agreeably with the notion of a similar underlying macromolecular machine.

Besides a fairly wide variation among the muscles of an individual animal, intrinsic speed varies systematically with differences in the size of at least terrestrial mammals, as predicted by Hill (1950). The logic is straightforward. Large animals move faster than small ones, although not proportionately so—speed goes up roughly in proportion to the square root of body length (as one might predict from the argument at the start of chapter 1). Assuming constant amplitude, speed varies with the product of leg length and frequency. So, assuming that body length tracks leg length, the frequency of leg swinging or footfalls varies inversely with the square root of body length (Pennycuick 1975). If body proportions are anywhere near similarity, then the muscles of small animals must contract a little more often (inverse square root) than those of the larger ones. But the muscles of the small animal are a lot shorter (proportional to body length, not its square root) than those of large animals, so muscles ought to contract at a speed proportional to square root of body length. The smaller ones, though, have proportionately fewer half-sarcomeres in series. So intrinsic speed, *IS*, should vary with the square root of body length divided by the first power of body length, or inversely with the square root of body length, or inversely with the sixth root of body mass:

$$IS \propto M_b^{-0.167}. \tag{22.1}$$

And that's about what happens—although both the particular exponent and the constant of proportionality vary a bit from muscle to muscle (see Close [1972] and McMahon [1984]; the latter is a particularly insightful book).

Of course any muscle can contract at a range of speeds (including, of course, zero and negative speeds). That matters little to the exponent in the proportionality above, since the speed at which a muscle puts out its greatest power remains in the vicinity of a third of its maximum, unloaded speed, whatever the size of the animal. And locomoting animals use their muscles at speeds close to what gives maximum power—as we do by changing gears on our bicycles.

What about actual intrinsic speeds? About the fastest on record is that of the extensor digitorum longus (finger extender) of a mouse, working at 22 lengths per second. By the equation above, an analogous speed demon of a muscle in a 5,000-kilogram elephant would have an intrinsic speed of about 2.8 lengths per second. More figures will emerge further along. This scaling argument assumes constant sarcomere length, which is okay for vertebrates but emphatically not a general condition.

Force

For speed comparisons we normalized to intrinsic speed. To compare the force production of muscles of different sizes we find it useful to do a similar normalization by using stress rather than force, just as we did for materials. Data for so-called isometric stress—maximum stress against an immovable load—have been accumulated from a wide variety of animals. The results don't vary too widely—striated muscle can pull at about 200 kilopascals, whether it comes from worms, mollusks, insects, or vertebrates (McMahon 1984). The figure, incidentally, is about 500 times lower than the tensile strength of collagen, explaining why fat muscles can be hooked to bones via skinny tendons.

Not surprisingly, the force a muscle can produce depends on the degree of overlap between thick and thin filaments. Peak force production doesn't happen exactly at rest length, but rather when a muscle shortening has proceeded to the point of maximum overlap between thick and thin filaments, as shown in figure 22.9. It hits zero when the filaments are stretched just to the point at which they stop overlapping and again when the everything is maximally interdigitated. The curve of force relative to filament interdigitation (Gordon, Huxley, and Julian 1966), in fact, provided an important piece of evidence confirming the sliding filament theory. An animal that wanted to get a bit more force from a muscle might work it at lengths shorter than resting, taking advantage of the extra overlap between filaments—the trick would be useful as long as the muscle was shortening by a fairly small fraction of resting length. That's what happens in the flight muscles of at least some birds, where length during activity runs from about 2.2 to 1.8 micrometers, well below the normal sarcomere length of 2.5 micrometers (Cutts 1986).

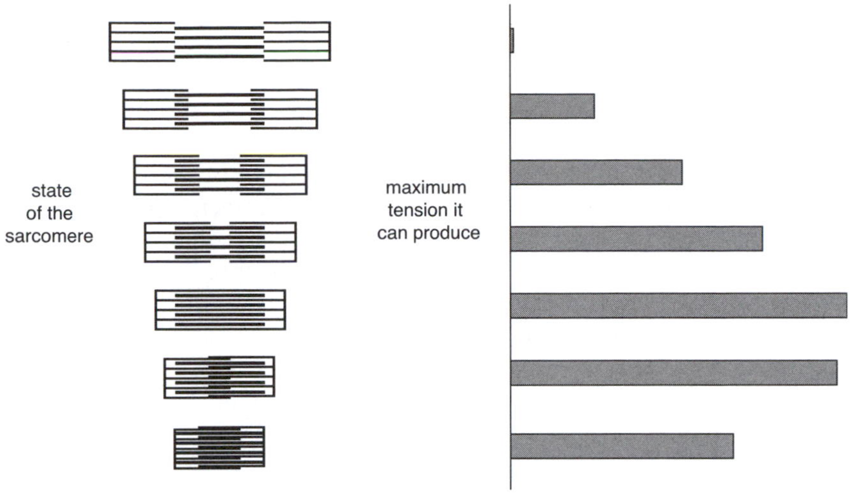

FIGURE 22.9 The force a muscle can exert relative to the degree of interdigitation of the filaments of its sarcomeres.

Maximum force varies with several other items. Muscle with which we can do continuous work, so-called red muscle, pulls a bit less hard than muscle used only in brief episodes, our white muscle—to consider the extremes of what's really more like a spectrum. Red muscle contains more mitochondria and other machinery for energy processing (the color reflects its greater content of myoglobin, an oxygen storing and transport-facilitating molecule). That leaves less space for sarcomeres. Insect flight muscle suffers an especially bad version of the same problem, with a large space allocation for mitochondria and, in many, for the triggering equipment. It minimizes the resulting tax on force by increasing the relative overlap between thick and thin filaments. That limits contraction distance, and flight muscle shortens by only a few percent. But high operating frequencies permit only limited shortening anyway since pushing intrinsic speed up ultimately compromises power as well as force.

Vertebrates may keep sarcomere length constant, tuning their muscles by adjusting other variables, mainly biochemical. Other animals recognize no such constraint. Where sarcomere length does vary, force covaries. Longer sarcomeres trade lower speed for greater force. How that happens takes a few words, a diagram (fig. 22.10), and a little thought. Imagine a muscle with one

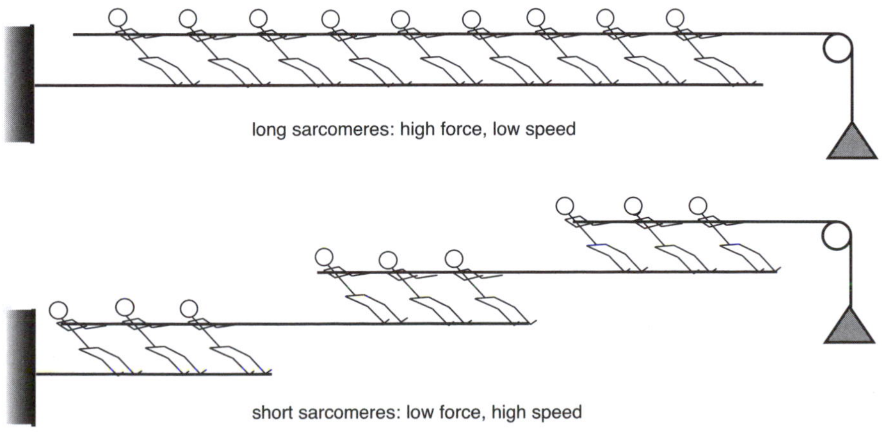

FIGURE 22.10 Pulling with long versus short sarcomeres.

half-sarcomere running its entire length—thick filaments attached to some fixed point and thin ones to some movable load. Contraction speed would be the rate at which filaments interdigitated. Now imagine the same muscle with a full sarcomere. The basal interdigitation will now pull, not directly on the load, but on the interdigitation of the other half-sarcomere. So the latter will be moving basally as it adds its contribution to the movement of the load, and the speed of its load should double. Put two sarcomeres in series and the second is moved basally by the first; speed will again double. That sounds as if shorter sarcomeres will always be better. But no, speed costs force. With one long half-sarcomere, every cross bridge pulls directly on the load. That's a lot of cross bridges pulling in parallel. Every increase in the number sarcomeres in series reduces the number of cross bridges pulling on the load, since an ever-increasing fraction of them just pull serially on one another. So more and shorter sarcomeres mean faster speed; faster speed means less force. Power output changes little, remaining proportional to overall volume.

A squid shooting out its tentacles (recall the muscular hydrostats of chapter 20) goes for speed and uses the shortest of sarcomeres in the muscles doing the task—about 0.9 micrometers (Kier 1985). By contrast, a crab attempting to crush a mollusk shell goes for force, caring little about either distance or time, and uses sarcomeres of around 20 micrometers (Taylor 2000). Bear in mind that final output force depends on larger-scale anatomical factors as well. Output force depends on the relative apportionment of muscle volume to force-producing fibers and components such as sarcoplasmic reticulum and mitochondria—recall the distinction between red and white muscle. It also depends on whether a muscle has longitudinal or pennate fibers, on its leverage, and so forth, subjects about which more in the next chapter.

Power

An isometric contraction produces zero power output—all force, no speed. A muscle contracting at top speed also produces zero power output—all speed, no force. Peak power happens in between, as shown on the graph of figure 22.11, at 15–40 percent of the maximum speed of shortening—muscles vary (Rome 1998). That graph turns out to say quite a bit about how whole muscle-powered systems operate. Indeed, one gets the same curve if an appendage (say an arm) of a person provides the data by working as hard as possible against a range of loads, as first done by Wilkie, in 1950. You do best if working muscles at neither excessively high nor excessively low speeds. Again, that's what underlies gear changing in bicycles—although the actual curve for that activity isn't peaky enough to justify, in and of itself, the great numbers of gear ratios we now supply. Incidentally, the casual cyclist does best at about one full turn of the pedals per second; for racing, rates about 50 percent greater work somewhat better.

What the graph doesn't say may be as important. Zero shortening speed gives no power output, but it still costs metabolic energy. For some reason we can't lock up our muscles and make them work like tendons. Isometric contraction may cost less than work-producing shortening, but it still costs. Among other consequences, that implies a drop in power efficiency—output over input—as contraction gets slower, hitting zero efficiency at zero shortening speed, and with a peak much like that for power itself. We'll come back to efficiency shortly.

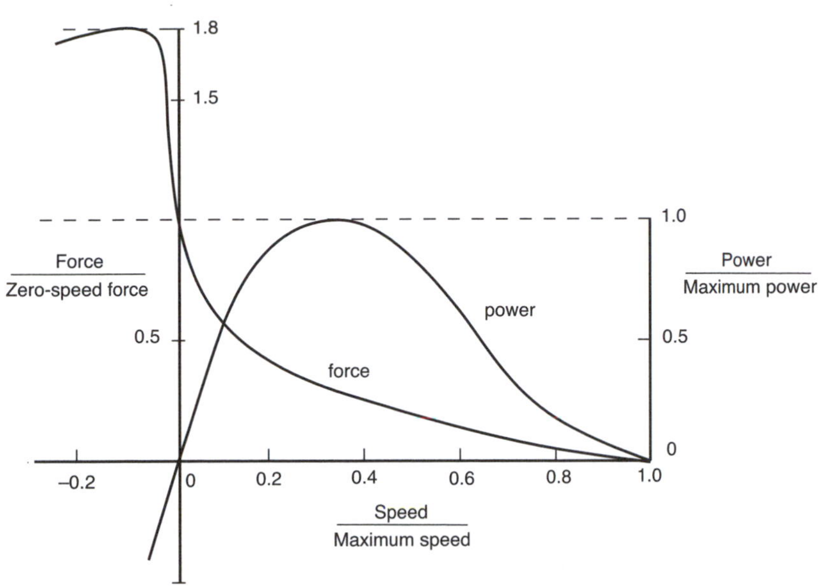

FIGURE 22.11 The force and power a muscle can produce as a function of the speed at which it shortens—or lengthens—under load. Power output, of course, goes negative if the muscle lengthens, since the system is then acting as a brake and absorbing work.

How much power can the muscular engine put out? As with force, muscles vary from one to another; specific muscle morphology entails a trade-off between force and power maximization. A good power-producing muscle generates a maximum output relative to mass of 200 watts per kilogram (0.6 horsepower per pound). But few muscles can sustain that level of output, and a whole human certainly cannot. The supply of fuel and oxygen and elimination of heat and wastes ultimately limit output rather than the capabilities of the actin and myosin filaments, as a quick reductio-ad-absurdum calculation demonstrates. The datum of 200 watts per kilogram implies that a 70-kilogram human devoting 40 percent of that mass as muscle should be able to put out 5,600 watts, or 7.5 horsepower. The practical maximum for an adult male laborer, working all day, has been recognized as about 100 watts since the eighteenth century; champion athletes can sustain 300 or so watts for a few hours—over an order of magnitude less than 5,600 watts. For power output, muscle seems to be the biological champion (at least above the bacteria); the runners-up, eucaryotic cilia, can do only 30 watts per kilogram. Still, muscles by no means match what we can get out of the engines we build, as table 22.1 shows—and our engines can sustain their high outputs.

Our ability—or inability—to sustain power-producing activities turns out to be a most peculiar business. In mechanical terms, a human (and lots of other animals) works like a huge spinning flywheel driven by a tiny motor. The limitations, again, are metabolic much more than muscular, but that makes them no less significant biomechanically. The flywheel, of course, represents our anaerobic capacity. It's equivalent to the energy we'd produce if consuming about 18 liters of oxygen; a liter of oxygen generates 20,000 joules. The resulting 350,000 joules amounts, since we're talking in terms of power, to as many watt-seconds of energy. Or a hundred watt-hours or (as we purchase electricity) a tenth of a kilowatt-hour.

The tiny motor means that the longer we're asked to sustain a task, the lower the power level at which we have to operate. According to Wilkie (1960),

TABLE 22.1 The Power Output Relative to Mass of a Variety of Engines

	Output	
Engine	**Watts Per Kilogram**	**Horsepower Per Pound**
Eighteenth-century steam pump	10	0.005
Cilia	30	0.02
Striated muscle	200	0.12
Electric motor	200	0.12
Automobile engine	400	0.24
Aircraft engine, turbine	6,000	3.6

Sources: Data from Nicklas 1984; French 1994; Vogel 1998a.

a trained athlete can sustain outputs of up to 4500 watts for up to a second, of 1,500 to 370 watts for about five seconds to 5 minutes, of 370 to 300 watts for up to a few hours, and of perhaps 150 watts all day—about 2 watts per kilogram. He generated the graph of figure 22.12 to help set the requirements for the first Kremer competition for human-powered aircraft. Incidentally, so limited are we by metabolic rather than muscular factors that the graph applies about equally well as an outer envelope of the possibilities for just about any efficient use of a human as a steady power source. It also predicts rather well the relative winning speeds for runs of different lengths.

Put another way, in a 100-meter dash, a runner relies entirely on anaerobic reserve; a 200-meter run is 82 percent anaerobic, a 1,500-m is 50 percent, a 10,000-m only 10 percent (Wilt 1968).

We humans do well as sustained workers by mammalian standards—we have good hearts and lungs. But we're not even in the running when compared to flying animals on the basis of power relative to body weight. (That may,

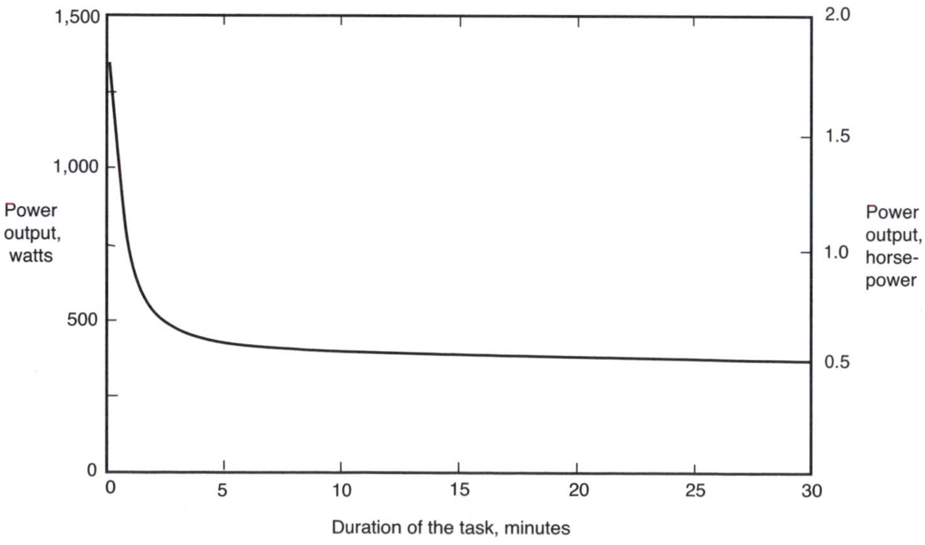

FIGURE 22.12 Power output as a function of duration for maximum output of especially fit human males. Note that duration refers to how long the person can work steadily at tasks that impose different demands—not how output diminishes with time as a person does a single task.

of course, be a peculiarly biased comparison since fliers must be far more fanatical weight-watchers.) A 0.1-kilogram bird uses oxygen in flight at a rate equivalent to 95 watts per kilogram (from birds flying in wind tunnels as summarized by Butler [1991]); assuming an overall system efficiency of 25 percent (more on this shortly), that's 24 watts per kilogram output. Bats do about as well as birds—so our inferiority isn't peculiarly mammalian. Insects do still better, with bumblebees consuming oxygen at a rate equivalent to up to 400 watts per kilogram or (again assuming 25 percent efficiency) 100 watts per kilogram (Ellington, Machin, and Casey 1990). For that matter, all the fliers fit about the same scaling relationship—smaller is simply better, with metabolic power (input), P_{in}, rising more slowly than body mass:

$$P_{in} \propto M_b^{0.72}. \tag{22.2}$$

Swimming costs a lot less than flying, either relative to distance covered or time. A salmon swimming at the speed that minimizes cost of transport, 0.55 meters per second (1.25 miles per hour) consumes energy at a mere 0.66 watts per kilogram (Brett and Glass 1973). Again making the wild guess of 25 percent system efficiency, that's an output of 0.165 watts per kilogram, over a hundred times less than what the bird put out (and thirty times less than a top-notch human does). Of course the salmon goes about thirty times more slowly than the bird, so the overall cost of transport isn't quite so much lower. Put another way, in flying, the speed achieved compensates in large measure for the intensity of the work. The comparison is given some relevance by the fact that both salmon and bird can sustain long migratory journeys.

Work loops

Think back (or look back) to a variable, resilience, introduced in chapter 15. It reflected—inversely—an area on a stress-strain graph, that between the loading and unloading curves. Stretching makes something ascend one curve, and releasing makes it descend along a curve a bit lower, with the area between representing the energy loss per unit volume each time you do so. Repeated cycles retrace the clockwise loop. In another context in chapter 17, we saw how that area within a stress-strain loop gave the loss per cycle of a viscoelastic material. Cyclic engine operation, whether the reciprocating pistons of a steam engine or the repetitive contractions of a muscle can be viewed in much the same way. Axes can be designated force and distance or pressure and volume change instead of stress and strain without changing the general principle. An engine, though, runs counterclockwise instead of clockwise around its loop—it puts out mechanical energy rather than dissipating it. Heat becomes work rather than the other way, to use an apt word, around.

Only recently has the approach become at all common as a way to look at the performance of muscles, thanks mainly to the work of Josephson (1985; Josephson, Malamud, and Stokes 2000). In practice, one attaches a freed end of a muscle to an ergometer that measures both force and position as a function of time. The muscle is then stimulated at a frequency matching that of its normal operation. Replotting to eliminate time (treating time parametrically), gives a so-called work loop, as in figure 22.13. The area inside the loop represents work (or, with stress and strain as axes, work per unit volume) per cycle. Multiplying by frequency and density then gives power output per unit

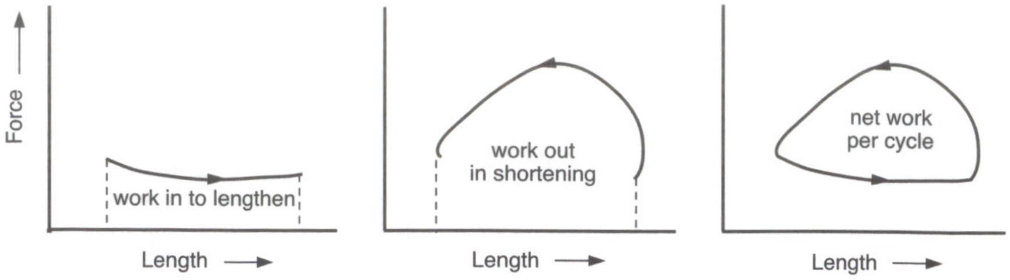

FIGURE 22.13 A so-called work loop for a muscle going through a cycle of contraction and reextension. Note that this power-producing system progresses counterclockwise around the loop, by contrast with the resilience loop for a passive, power absorbing material as in figure 15.5 or the stress-strain loop for a viscoelastic material in figure 17.5c.

mass. Adjusting the loading regime produces different loops. Thus, one can look at how power output is affected by the host of variables that together form the load.

Power efficiency

When you walk on level ground, you do no net work beyond the losses of your locomotory equipment (accelerating and decelerating limbs, for instance) and a trivial amount of air resistance. Almost nothing has worked against an external resistance. So, by definition, your efficiency approaches zero. But we do engage in activities that produce external work on many occasions—we tow the barge (against drag) and lift the bale (against gravity). If we know the drag of the body of a flying animal, its flying speed, and its oxygen consumption (the latter a proxy for power input), then we can calculate an efficiency. Similarly, if one of us climbs steadily on mountain or stepping machine, calculating an efficiency presents no great problem. While the results don't lack for awkward assumptions and uncertainties, the peak efficiencies of muscle-powered systems doing sustained work show a fair consistency. The benchmark figure is that used earlier, 25 percent—external work done per unit time over metabolic rate in excess of basal rate. Enough data has accumulated to suggest strong skepticism for any figure above this. But, and the point needs emphasis, most systems most of the time don't do that well. Our dismal efficiency when walking on the level shouldn't be forgotten.

As a quick example, consider what happens when I walk on an inclined treadmill. At a 10 percent slope, I sustain 1.75 meters per second (read as 4 miles per hour); the treadmill's calibration suggests that my 61 kilograms (135 pounds) requires an energy input of 698 watts. The speed and slope translate into an output of 108 watts, so I'm working at an efficiency of 15.5 percent. If I increase the slope to 15 percent and decrease the speed to 1.33 meters per second, which gives the same input, my efficiency rises to 17.5 percent. Both data are suboptimal (of course efficiency isn't the point of treadmill use!) but no surprise. Minetti (1995) found that humans climb best (most height relative to cost) at a slope of 25 percent, corresponding to the slopes at which mountain trails begin to use switchbacks to keep from getting steeper. So even with the steeper slope, I'm going too fast and not steeply enough.

Negative work

Notice that the lines on the graph of figure 22.11 extend to the left of the origin. Those extensions correspond to force production while lengthening and thus to negative work, both alluded to earlier. Force production hits a maximum, not under the isometric conditions usually used in the laboratory, but during lengthening. While that may startle at first, it does fit one's experience—lowering something at a constant speed seems easier than holding something at a given height even though the tasks ask for the same force. Negative work may seem a strange notion, but it's no trivial matter in everyday life. You accelerate and decelerate your limbs; deceleration requires negative work. When walking down a flight of stairs or descending on a fairly steep path, you do almost entirely negative work. When walking on the level, you do both negative and positive work. Only when ascending on a slope of above about 15 degrees does all your work appear on the positive side of the ledger (McMahon 1984).

(And the greater force produced by muscle doing negative work—called "eccentric contraction" as opposed to shortening, "concentric contraction"—can make trouble. You're much more likely to tear muscle or tendon when landing on a foot than when taking off. And "delayed onset muscle soreness," DOMS, comes more often after doing something such as descending a steep slope that requires sustained and severe negative work. Negative work, though, plays a peculiarly positive role if you want to build up muscle, according to Lindstedt, LaStayo, and Reich [2001]. Not only can you lower a heavier weight than you can lift, but the action stimulates net muscle synthesis more effectively. Weight lowering does more good than weight lifting, at least if one arranges matters to take advantage of one's ability to lower more weight than one can lift. Especially effective in increasing muscle strength and size is a peculiar bicycle ergometer, one with a motor that turns the pedals backwards. The trainee then tries, and persistently fails, to make them turn in the right direction.)

The power efficiencies of positive work run below about 25 percent—what of those for negative work? They're negative, of course, severely so, but here the lower the better! One figure (Margaria 1938) gives –118 percent. Now an efficiency of –100 percent implies that it costs (metabolically) just as much energy as is being absorbed (+100 percent means that as much energy is put out as expended). Thus, –118 percent says that 18 percent more energy is absorbed than the energy paid to absorb it. That means negative work can be done with much greater negative efficiency than positive work can be done with positive efficiency. One must emphasize, of course, that this isn't something for nothing. Indeed, one can ask why any energy should be needed to absorb energy—why isn't the efficiency minus infinity? After all, a completely passive device could lower something at a constant speed. At this point we can simply assert that no rule requires nature to do things in the most efficient possible way. One can ask, more pointedly, about where that energy goes. That depends on how the muscle is used. When walking downhill it gets transformed into heat, warming the body significantly. Under other circumstances such as running and hopping, some gets stored as recoverable elastic energy (Lindstedt, LaStayo, and Reich 2001).

Other kinds of muscle

A few variants on the basic (or textbook) muscle ought to be mentioned, if only to alert the reader to gaps in the present account.

Some insects—most notably the Diptera (flies and mosquitoes), Hymenoptera (bees, ants, and wasps), and Coleoptera (beetles) fly with muscle that oscillates at frequencies set by the mechanical properties of the flight apparatus rather than that set by incoming nerve impulses. Their so-called *asynchronous muscles* usually have been explained as requisite for achieving the high wingbeat frequencies of small insects. These can be in the hundreds per second; the record, for a midge with truncated wings, is 2,218 per second. A more recent argument (the two aren't fully alternative) suggests that self-triggering avoids the high fraction of muscle volume otherwise devoted to sarcoplasmic reticulum and transverse tubules—to the triggering machinery that normally occupies an ever-increasing fraction as frequency of muscle use goes up. Josephson, Malamud, and Stokes (2000) provide a good review.

Some mollusks manage to get around one of the peculiar disabilities of muscle, the fact that it must consume energy to exert a force even when nothing moves and no work is done. We can't pull something and then (as when raising a flag or working a winch) lock it in position. But a clam can clam up, pulling its half shells together, and then lock its adductor muscles, by use of so-called *catch muscle*. Mussels use the stuff to pull on the byssus threads with which they're attached to rocks. Catch muscle wins no prizes for speed or efficiency when shortening, but by virtue of a specialized kind of myosin and some other tricks, it can lock (Watabe and Hartshorne 1990).

Some other mollusks use *obliquely striated muscle* in addition to the usual transversely striated kind. Such muscles are especially slow ones, useful for sustained force generation, and they're used at lower intrinsic speeds than the cross-striated muscles of comparable animals (Millman 1967). In cephalopods (squid, cuttlefish, octopus, and nautilus), obliquely striated muscle is the rule while cross-striated muscle is exceptional. Squid run cross-striated muscle across their pair of tentacles, using it to reduce tentacle diameter and get speedy elongation (recall chapter 20). The same squid run obliquely striated muscle across their eight arms, using it to permit bending of the arms by resisting the lengthwise compression that would otherwise be caused by contraction of longitudinal muscles (Kier 1985).

We take the slow-and-steady business a step further with the *smooth muscle* of our digestive systems, bladders, bronchi, uteri, and blood vessels. It has an even less-regular alignment of actin and myosin filaments than molluscan obliquely striated muscle. It's an especially slow kind of muscle, typically capable of large length changes and prolonged exertion of force.

Somewhat similar, it seems, to smooth muscle are the various cell-level "unconventional" myosin motors, which occur in such disparate places as the microvillar brush border of the cells of small intestines, the cells of vertebrate brains, and in yeast.

And our hearts (but not all hearts) use muscle that self-triggers, but with pacemaking cells and a conduction system rather than the mechanical triggering of insect asynchronous muscle. That's why a heart can be transplanted without requiring neural reconnection or a pacemaker. *Cardiac muscle* has a number of other peculiarities as well.

CHAPTER 23

Using Muscle: Tuning and Transmissions

A horse divided against itself cannot stand.

ABRAHAM LINCOLN, biomechanist (without the usual typographical error)

RED VERSUS WHITE MUSCLE, ASYNCHRONOUS AND SYNCHRONOUS MUScle, cross-striated versus obliquely striated versus smooth muscle, long versus short sarcomeres—muscles don't lack for diversity. But that diversity pales into insignificance next to the diversity of mechanical tasks that muscles power. So we need to examine the concomitant diversity of schemes by which organisms make their motors actually *do* the mundane and profane tasks of daily life, the cleverlittledetails that provide the crucial grist to grind exceeding small in all those doctoral dissertations. In a sense, the limited range of engine options—muscle being muscle—demands a wide range of coupling devices. By analogy with the devices of our technology, we might lump these devices under the general heading of "transmissions."

Not that transmissions represent something new in the present account. We've already looked at hydraulic couplings, muscular hydrostats, jaw protrusion mechanisms, and others. But, for one thing, yet more deserve attention; for another, the subject itself deserves a more general view rather than the scattershot attention we've given it so far.

At this point, a little additional physical background is necessary, together with a few items of terminology.

Levers

These made a brief appearance in the second chapter but have kept a low profile since. First, let's remind ourselves that no conservation law applies directly to force. Nature conserves work (finally, energy), not force. A large applied force can, without evident inefficiency, press on something with much diminished force; and a small force can be persuaded, without fear of perpetual motion or divine retribution, to exert a greater force against some load. That persuasion justifies the existence of the various arrangements of pulleys, gear trains, hydraulic couplings, and, simplest of all, levers.

Recall the situation in which a force has a line of action lying to one side of an axis of rotation (fig. 2.5). We call the shortest, or perpendicular, distance between the force's line of action and the axis the "moment arm" or "lever arm" of the force; we call the product of force and moment arm the "moment" of the force. (The dimensions may be force times distance, but moments do not equate with work—in the latter force and distance run along the same line.) For all lawful cases of rest or steady motion, moments must balance out to zero, where one direction of rotational motion (clockwise, say) is taken as positive and the other as negative.

Any simple lever consists of a rigid beam, along which three elements are identifiable—(1) a fulcrum, pivot, or hinge point; (2) a point at which a force, the "effort," is applied; and (3) a point at which the force moves something, the "load." We evaluate the effectiveness of a lever by calculating its "mechanical advantage"; the latter is simply the ratio of the force exerted against the load to the force of the effort—if by pushing with a force of 10 newtons you indirectly apply 100 newtons then the mechanical advantage is 10. The mechanical advantage turns out to be equal to the ratio of the moment arm of the effort (the distance between effort and fulcrum) to the moment arm of the load (the distance between load and fulcrum)—the ratio is a consequence of the rule that moments must add up to zero.

For our purposes, the term "mechanical advantage" can be misleading, since one's inclination is to take "advantage" literally. That inadvertently casts muscular machines in a bad light. We'll replace it with the more literally meaningful and less disparaging "force advantage," the factor by which the applied force is multiplied by the lever or transmission. We can contrast the latter with its reciprocal, "distance advantage," the ratio of the distance the load moves to that moved by the effort. (Distance advantage corresponds to "speed advantage" as well—if an action takes a given time, then going further means going faster.) Muscles produce a lot of force but shorten by only a fraction of their lengths, so putting them to work almost always requires distance amplification.

Levers are commonly divided into three classes, unimaginatively called "first class," "second class," and "third class." The labels have little functional meaning, and, gallingly, the three don't form a single tripartite division. We're better served by the division shown in figure 23.1 between force-increasing and distance-increasing levers. The first have force advantages above one; the second have distance advantages above one.

Force-increasing levers abound in our everyday lives—the nail-pulling claws of hammers, pliers, toggle switches, drink can piercers, and so forth. In the traditional taxonomy, all second- and some first-class levers increase force at the expense of distance. All have effort arms longer than their load arms, whether the two arms are on the opposite sides of the fulcrum (first class) or on the same side (second class). Distance-increasing levers are less common—

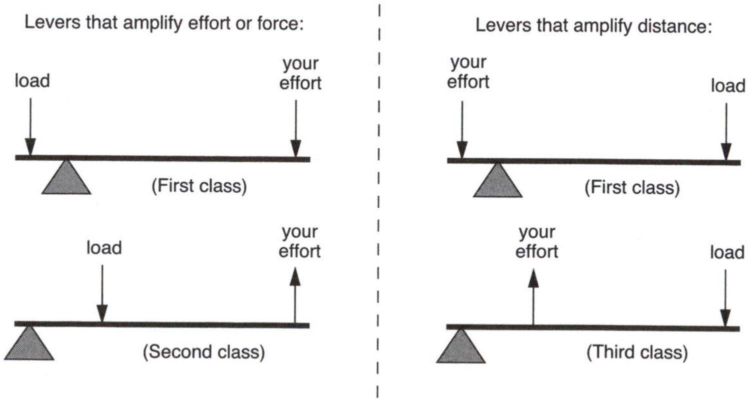

FIGURE 23.1 Force-increasing and distance-increasing levers.

one example is a gate latch where you lift the latch between the hinge and the catchment; tweezers and salad tongs are others. These have effort arms that are shorter than their load arms; again the two arms can be on opposite sides of the fulcrum (first class) or on the same side (third class).

Muscles and Skeletal Elements as Levers

A muscle, again, is relatively good at producing force and relatively bad at getting shorter. At least a device that gets only 20 percent shorter will have to operate with a substantial distance advantage to move a limb through an angle that may approach 180 degrees. For the good distance advantage it naturally will suffer a poor force advantage—the product of the two must be unity. Agility and strength are functionally antithetical in the design of a skeletomuscular system, whatever the changes we effect through training.

Quite commonly a first- and a third-class lever (both with distance advantages above one) are used together, with one muscle "antagonizing" the other. Thus, each muscle acts as the lengthener of the other. Muscle, be reminded, is a pulling engine; if it succeeds in shortening, some other device must intervene to stretch it back out again. Thus, the biceps muscle of the upper arm, which we use to raise the forearm, operates between its bones as part of a third-class lever (fig. 2.7)—at one end is the pivot; a little way out on the forearm is the attachment of the muscle, defining the effort arm, and at the far end of the forearm is the hand, the site of the load. On the opposite side of the upper arm is the triceps muscle, which attaches to the opposite side of the pivot, just above the elbow. The triceps, then, operating as part of a first-class lever, antagonizes the biceps.

What sorts of advantages actually result from such arrangements? For the biceps of the human arm, Currey (1984) gives an effort arm of 55 millimeters and a load arm of 300 millimeters; the force advantage is thus 0.18 and the distance advantage is 5.45. He comments on the contrast between a human and a mole—the latter has huge flanges on its forearms to get the attachment of the biceps further away from the pivot and thus to increase the force advantage in a manner appropriate to the mole's fossorial habit. Our triceps, by the way, has a much lower force advantage of about 0.05 and a higher distance advantage of 22—you're not too good at lifting forearm weights while upside down! More realistically, you will find it difficult to support your weight from your hands if you grab supports that are more than a meter apart. Bear in mind that the advantages vary from moment to moment as motion proceeds. Muscles don't ordinarily wrap around pulleys or capstans—more often they pull on protrusions. And when their tendons do follow a curved surface, that surface rarely has a constant radius. Alexander (1983) and McGowan (1999) give good discussions and a lot of additional information on such systems.

A few more figures for advantages, these from Tricker and Tricker (1967). For the quadriceps femoris, the large muscle of the front of the thigh, whose tendon runs across the kneecap and which extends the lower leg, the force advantage is 0.06 and the distance advantage is 17.5. This is, of course, the muscle that mutters nasty remarks when you try to hold a partial knee-bend. Further down the leg is the gastrocnemius on the back of the calf; it tapers down to the Achilles tendon, which runs across the heel. To stand on tiptoe, you contract this muscle. In the process, half your weight loads each leg, with the muscles operating as parts of first-class levers. But concomitant, perhaps, with a limited range of motion (a few inches of height gained) and high load, we give them a relatively great force advantage of 0.43 and a low distance advantage of 2.3. The main muscle for raising the lower jaw is the temporalis, located just in front of the hinge point of the jaw. It's part of a third-class lever with the effort arm fixed and the load arm determined by which teeth are in action. Thus, you bite with incisors at a distance advantage of 3.5 but chew with back molars at a distance advantage of only 1.7. The latter is a particularly low distance advantage, so you can exert forces hazardous to life, limb, and dentures.

The opposite extremes characterize the mechanical arrangements of arthropods. Building exoskeletal bodies, perforce, puts muscles on the inside, with little distance per force—thus distance advantages must be huge. An almost imperceptible shortening of a muscle must, as you can see from figure 23.2, swing an appendage through a wide arc. Inasmuch as a considerable number of arthropod lineages have come up with extremely skinny-legged creatures (recall figure 19.1), they've obviously figured out how to make muscles shorten by even lower fractions of rest length than they normally do and, at the same time, produce even greater forces. We'll get back to the scheme, one that we ourselves use, if not in such extreme form.

These relatively straightforward examples only scratch the surface of the arrangements and applications of living levers. Consider how a small insect might be arranged to beat its wings. The more rapid the beat, the smaller can be the wings, which means less to synthesize, less to injure, and less to attract predators. The intrinsic speeds of muscles are nothing extraordinary, and going for high intrinsic speed compromises the force they can produce—recall what was said in the last chapter. One way to increase the rapidity of motion is to limit contraction to a small fraction of muscle length and then to use linkages with heroic distance advantages. Another problem—flight demands very hard work; and, since the maximum power per unit mass of muscle seems to be unalterable, it inevitably requires a lot of muscle. The limited contraction fraction helps here, permitting more overlap between thick and thin filaments.

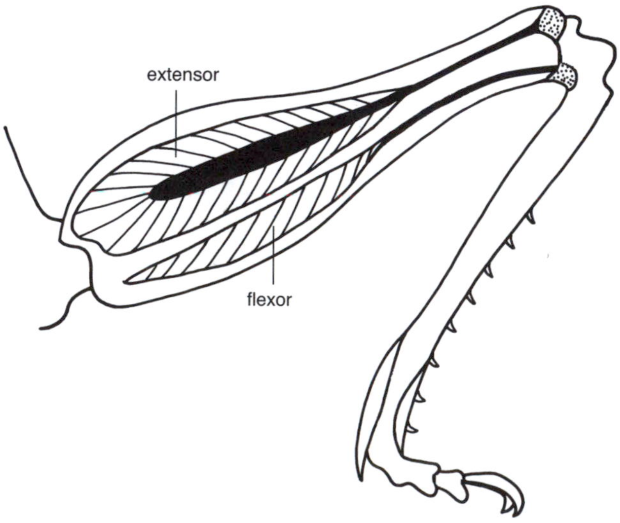

FIGURE 23.2 How the main jumping muscles in the upper hindleg of a grasshopper connect with its outer segments. Confinement within a tubular exoskeleton keeps the lever arms extremely short. Notice that the muscle that extends the leg is a lot bigger than the one that flexes it, as expected from the demands of jumping.

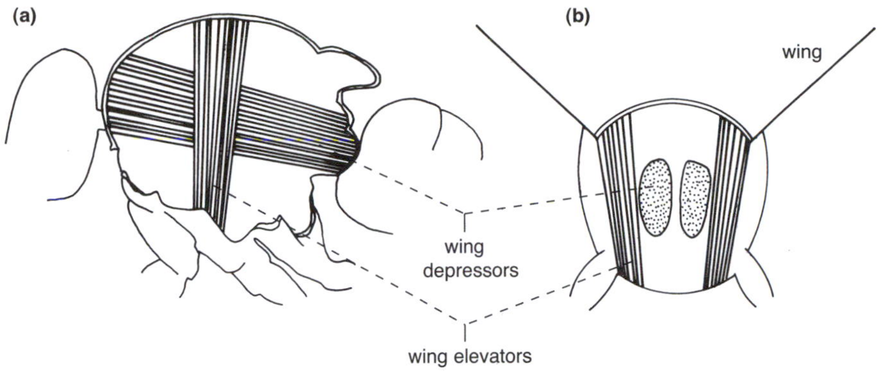

FIGURE 23.3 The main flight muscles in the thorax of a fly, in cutaway side view (with head to the left) (a) and in cross section (b).

Still, about 25 percent of the weight of a fly, for instance, is flight muscle, and the breast muscles of birds are impressively large. To add yet another complication, with its muscles on the inside, an insect has few ways in which a large mass of muscle can be hooked with appropriate leverage to a moveable structure. (And birds, with muscles outside, can't afford much compromise of their low-drag profiles.)

Most insects use an arrangement in which two pairs of muscles run across the inside of the thoracic box (fig. 23.3). Dorsoventral (vertical) muscles drive the upstroke and longitudinal muscles the downstroke, but neither pair actually attaches to the wing base, and both pairs run from wall to wall without tapering to tendons (or, being arthropods, chitinous apodemes). Instead they distort their thoracic boxes just a bit, and that distortion is linked through a complex hinging arrangement to the wings. Using the exoskeleton itself as linkage permits a lot of muscle to be fitted into a small container; the particular arrangement of the hinge means that no massive beam has to extend a wing far into the thorax to receive the muscles. And, while the muscles contract only a very short distance, the wing can swing through a broad arc.

For a generic fly (which has asynchronous flight muscle), the dorsoventral muscles might be about 2.5 millimeters long and contract 4 percent of that length while the center of the wing is about 5 millimeters from the hinge and swings through a third of a circle—about 120 degrees. As you can easily calculate, that comes out to a distance advantage of 100 or a force advantage of 0.01, quite extreme values! Even so, intrinsic shortening speeds are relatively high for these "indirect" flight muscles. If that fly beats its wings up and down 120 times each second, then the center of the blade moves 3.2 meters per second; the muscle must be shortening 32 millimeters per second, which is an intrinsic speed of 10 lengths per second. Weis-Fogh and Alexander (1977) give an intrinsic speed of 9 lengths per second for locusts, which fly with conventional muscle; calculations from data in Dudley (2000) indicate that such numbers are typical.

Incidentally, you may recall from the last chapter that intrinsic speeds can get much higher—one mouse muscle ran at 22 lengths per second. Flight seems to be just too power-demanding to push intrinsic speeds up that far, whatever the benefits of high wingbeat frequencies. Still, these figures for insects exceed what birds do—a pigeon, for instance, operates its pectoral muscles at 4.4 beats per second in level flight and at 6.7 beats per second during takeoff (Dial and Biewener 1993).

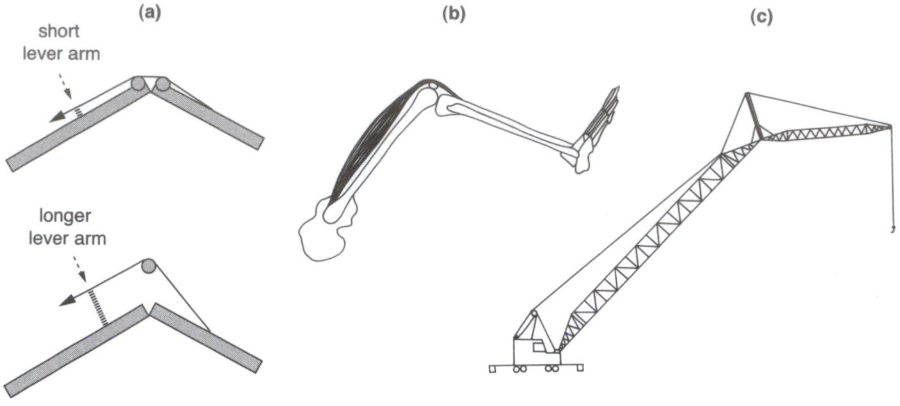

FIGURE 23.4 The uses of a jib. (a) The lever arm of a force is increased by running a cable over a more remote pulley instead of pulling close to the hinged members. (b) In a human leg, the tendon of the muscle that extends (straightens) the lower leg runs over an extra bone, the kneecap, which provides a lever-arm-increasing jib. (c) An articulated crane runs the support cables for its outer, hinged member over a jib to get the same advantage.

The generic insect scheme has some limitations—none of the (up to) four wings can beat at different rates or very far out of phase with any other. Dragonflies and damselflies (order Odonata) use a different setup from other insects, powering the wings with "direct" muscles that attach to the wings just inside the hinges; the mechanism permits greater flexibility in the phase and amplitude of movement of each wing. As a consequence, they seem to gain unusual maneuverability. It's no trivial matter to be large but still adequately maneuverable to catch small, flying prey. A dragonfly, for instance, can make an unbanked sudden turn when hovering by using its wings in the manner of a rower who turns a boat by pushing on one oar while pulling on the other (D. E. Alexander 1986). Odonata, though, have long wings and relatively low wingbeat frequencies (about 20 beats per second).

One shouldn't forget that the lengths of effort and load arms must be measured at right angles to the direction of the forces. A muscle or tendon may attach to a bone or piece of cuticle quite far from a joint, but if the attachment angle is something other than 90 degrees, then the actual effort arm is less than the distance between attachment and joint, as shown in figure 23.4a. For the examples earlier, cases were picked where the effort arm (at some point in the action) was in fact the shortest distance between attachment and hinge—biceps, triceps, gastrocnemius, and temporalis are nice, tidy cases. But with most muscles in our appendages running along the bones, muscles will most often pull obliquely.

An articulated crane has the same problem of a short effort arm. How can a cable from the cab elevate the outer, hinged part of the crane? If the cable just runs over a pulley whose axis is the hinge then the line of action will be minute—just the radius of the pulley. Most cranes add an extra strut, a jib nearly perpendicular to the two beams of the crane that does nothing more than increase the length of the effort arm (fig. 23.4c). Currey (2002) pointed out that the extra bone at the knee joint, the kneecap or patella, is an analog of that jib (fig. 23.4b). Its role is to keep the tendon that runs downward from the front of the thigh away from the hinge line of the knee. We're not great kickers, but we'd be worse without that little bone. For the joint between upper arm and forearm we use an extension of the forearm, the elbow, to accomplish the analogous task for the equivalent extensor muscle.

Leverage doesn't require morphologically obvious levers. We encountered a case earlier, when considering muscular hydrostats (chapter 20). The tentacle of a squid operates at a high distance advantage—it extends far and fast when attacking prey. The trunk of an elephant operates at a lower distance advantage—it can't change length by much, but it can pull with remarkable force. The hydraulic linkages mentioned in the same chapter also involve trade-offs of force and distance (or speed), although specific numbers are hard to find.

Pennate Muscles

Back to the problem of very short effort arms in arthropod legs and elsewhere. Compensation might be achieved with an arrangement similar to that of the flight muscles, mentioned earlier—a short, wide muscle produces relatively more force but shortens less far than does a long, narrow one of the same volume and work output. But how to stuff a short, fat muscle into a long and skinny leg? The usual solution uses something called a "pennate" (for "featherlike," also "pinnate") muscle. In such a muscle, as in figure 23.5a, the fibers run obliquely inward from the exoskeleton or one tendonous sheath to another tendon or apodeme running out to the load. The apodemes of crab and lobster claws that you pick out before eating the muscle form parts of such arrangements. For animals with exoskeletons, the extra short but forceful contraction permitted by pennateness facilitates operation at high distance advantages.

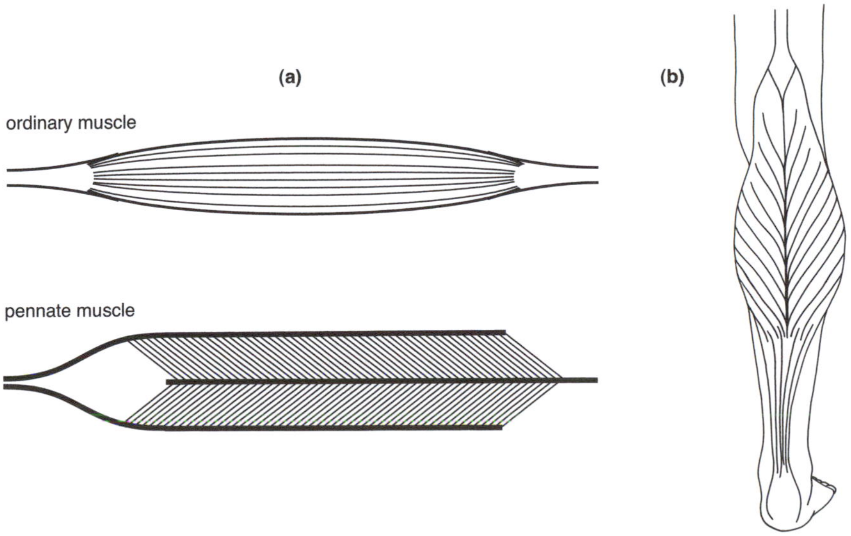

FIGURE 23.5 (a) The arrangement of fibers in ordinary and pennate muscles. (b) The human gastrocnemius, a pennate muscle, in situ.

Arthropods gain another advantage by using pennate muscles. As a constant-volume system, a normal muscle must become fatter when it shortens. That's not easy if muscle fills an exoskeletal tube. Not only don't pennate muscles change volume when they contract, but (perhaps counterintuitively), they don't change their cross sections either—they don't get fatter. When they change shape, as figure 23.6 shows, they take advantage of the rule that the area of a parallelogram equals the length of its base times its height. Sliding its top edge parallel to the bottom edge changes neither base length nor height, so it has no effect on area, or, in three dimensions, on volume. For an example, consider a ream of paper resting on a table. Pushing on one end converts the stack's cross section from a rectangle into a parallelogram. Same volume; same height.

While vertebrates, with endoskeletons, don't depend quite so much on pennate muscles, they (we!) use them in applications demanding especially high force per se. Gans (1974) describes how an amphisbaenid (a limbless, burrowing "worm-lizard") uses pennate muscles back in its trunk to aid the head in tunneling operations. We use them ourselves in such places as the gastrocnemius muscles of our calves (fig. 23.5b). These were mentioned earlier as having the particularly high force advantage appropriate for muscles that have to lift the weight of the body when we stand on tiptoe and must produce still more force when we run. The body muscles of a typical fish amount to a much more complicated case of the same underlying scheme—you've certainly noticed how they flake apart into short fibers after cooking hydrolyzes their septa. All these short-fibered trunk muscles lie close to the long axis of the body, and their shortening is conveyed to skin and skeleton by great sheets and (near the tail) bands of tensile tissue such as those septa—a fish integrates motor and transmission in a complex and as yet imperfectly understood way (Westneat et al. 1993).

Jumping frogs provide another example of the use of obliquely oriented muscle fibers—here a pair of muscles does the job, but they work just like a single pennate muscle of the kind shown in figures 23.5 and 23.6. These are the muscles that connect the peculiar bone, the urostyle, that extends a frog's vertebral column with the very long iliac bones of the frog's pelvic girdle (fig. 23.7a)—an arrangement seen in lab by almost every biology student but about whose functional oddity most laboratory manuals remain silent. The trouble with looking at a skeleton comes from the attachment of those iliac

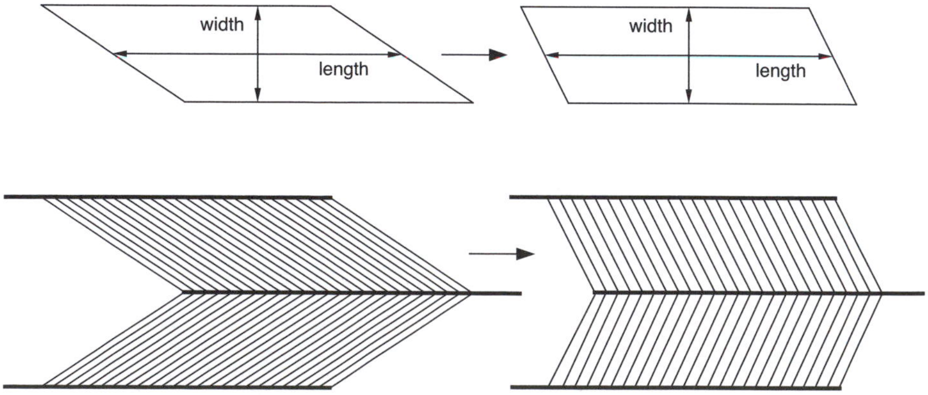

FIGURE 23.6 How pennate muscle works as a constant-volume system: contraction leaves width and length—and hence area and (in three dimensions) volume—unchanged.

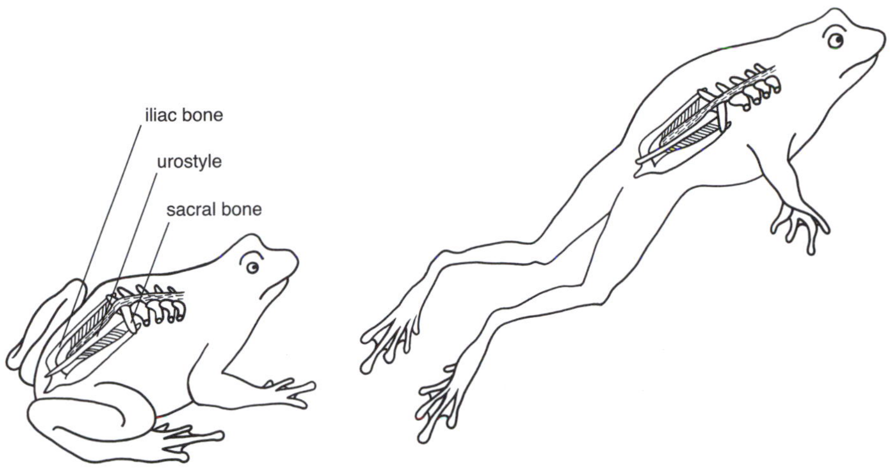

FIGURE 23.7 (a) The arrangement of iliac bones, sacral bones, and urostyle, in a frog. When it jumps, contraction of the diagonal muscles between urostyle and ilia straightens out the frog's back and thus adds additional extension to what the legs provide. Redrawn from Jenkins and Shubin (1998) with the permission of the Society of Vertebrate Paleontology.

bones to the lateral projections of the sacral (last) vertebra—the joint looks deceptively rigid. If rigid, those oblique muscles couldn't do much. In fact, these coccygeo-iliacus muscles (to name them properly) transmit the thrust of the legs from the ilia of the pelvic girdle to the urostyle of the vertebral column. Rigid joints might do the same, but they would force a frog to sit in a strange posture—its body wouldn't be able to fold just behind the sacral vertebra with its hindlegs underneath. As Jenkins and Shubin (1998) pointed out, the arrangement permits a frog both to get its legs beneath its abdomen to sit and to extend its legs behind it when it jumps—as shown in figure 23.7b. Moreover, the oblique direction of these muscles gives them typically pennate behavior—lots of force, little distance. That's consistent with limited range of motion of the joints between the ilia and the sacral vertebra.

Motors, Brakes, and Batteries

Anatomy books call muscles positioned to reduce the angle between two bones "flexors" and, conversely, muscles positioned to increase that angle "extensors." We're coming to realize just how misleading that terminology can be if taken as anything more than a key to location, that is, if treated as functionally descriptive. In a way, the neuromuscular physiologists knew that all along, but few of them cared about the details. After all, when walking, a normal person doesn't simply reextend a flexed foot until it hits the ground. Instead, it's decelerated just before footfall, so you tread softly rather than clomping around. That might be done with some passive damper or elastic—tendon, for instance. But you're better than that, getting just the right deceleration for different loads, slopes, surfaces, and footgear. What you're doing depends on using muscles as active, tunable brakes.

Treadmills, wind tunnels, image and motion analysis equipment, and electromyographic gear have become ever more widely available. These have allowed functional morphologists to give proper attention to quite a few specific cases, looking at how muscles really work at the various tasks we ask of them. Almost all propulsive systems used to move animals face a basic problem—however steady the motion of the animal as a whole, propulsive devices work unsteadily.

In a sense, that's a consequence of the general absence of wheel-and-axle engines. All propulsive devices—legs, fins, wings, even the chamber-squeezers of jets—have mass. So, once started, all prefer to keep going according to Newton's first law and have to be decelerated to a stop by some application of Newton's second law. Moreover, even for whole animals, steady motion is more the exception than the rule, as pointed out with splendid perspective by Dickinson et al. (2000).

So even if the central nervous system instructs an animal to move appendages, elaborate sensory feedback and auxiliary coordinating muscular machinery normally participate in determining what really happens. Sit on the edge of a table and swing a leg back and forth. You'll find that it much prefers to swing at a definite frequency, one determined by its pendulum characteristics. Doing what seems so smooth and easy takes considerable dynamic control—a lot of unobtrusive but effective machinery. Put on a heavy boot and your leg swings just as smoothly but at a lower frequency. Something must be making all those muscles readjust contraction rates and intensities!

Consider an example. Big cockroaches can easily be persuaded to run under decently controlled conditions with electrodes in various of their muscles. The same muscles can then be removed and work loops (recall the last chapter) determined under equivalent conditions. Full et al. (1998) and Ahn and Full (2002) looked in particular at two leg extensor muscles, shown in figure 23.8a. Only one produced power; the other operated as a brake, with its work loop cycling clockwise, like those of passive material, rather than counterclockwise, as would a proper engine (fig. 23.8b). In effect, only one was a proper extensor, the other worked as a brake on flexion. Nonetheless, the muscles had similar intrinsic properties—on the basis of microanatomy, location, and performance when isolated, one would have declared that they did the same thing and represented redundancy. In practice, the motor put out a respectable 28 watts per kilogram; in working (if the word isn't to be taken quite so literally), the brake absorbed 19 watts per kilogram.

Nor is this the only additional function we have to assign to muscle. The gastrocnemius muscle of a running turkey, according to Roberts et al. (1997), works two quite different ways. When going uphill, it does the ordinary muscular thing, putting out power, with an ordinary (if a bit strangely shaped) work loop. When the turkey runs on level ground, the gastrocnemius doesn't change length much at all. It produces just enough force to hold its own length in the face of a cycling load—an active element has been told to pretend that it's a passive element, a mere strut. Meanwhile, its tendons stretch and recoil, storing and returning energy. Beyond that, muscles can be persuaded to pretend that they're either supportive elements of variable stiffness or passive springs. We usually think of structures of elastin, collagen, resilin, and abductin as springs; we then recognize that wood can work as a spring; we may have heard that the wishbone (furcula) of a flying bird bends as a spring (Jenkins, Dial, and Goslow 1988). Muscle as virtual spring comes as more radical notion, but some wing-steering muscles in blowflies work that way (Tu and Dickinson 1996).

When muscle does negative work, absorbing energy, where does the energy go? It can't all be stored elastically for reuse—we saw in the last chapter than doing negative work still cost energy, although the energy absorbed was greater (in absolute terms, since the signs are opposite) than the energy spent

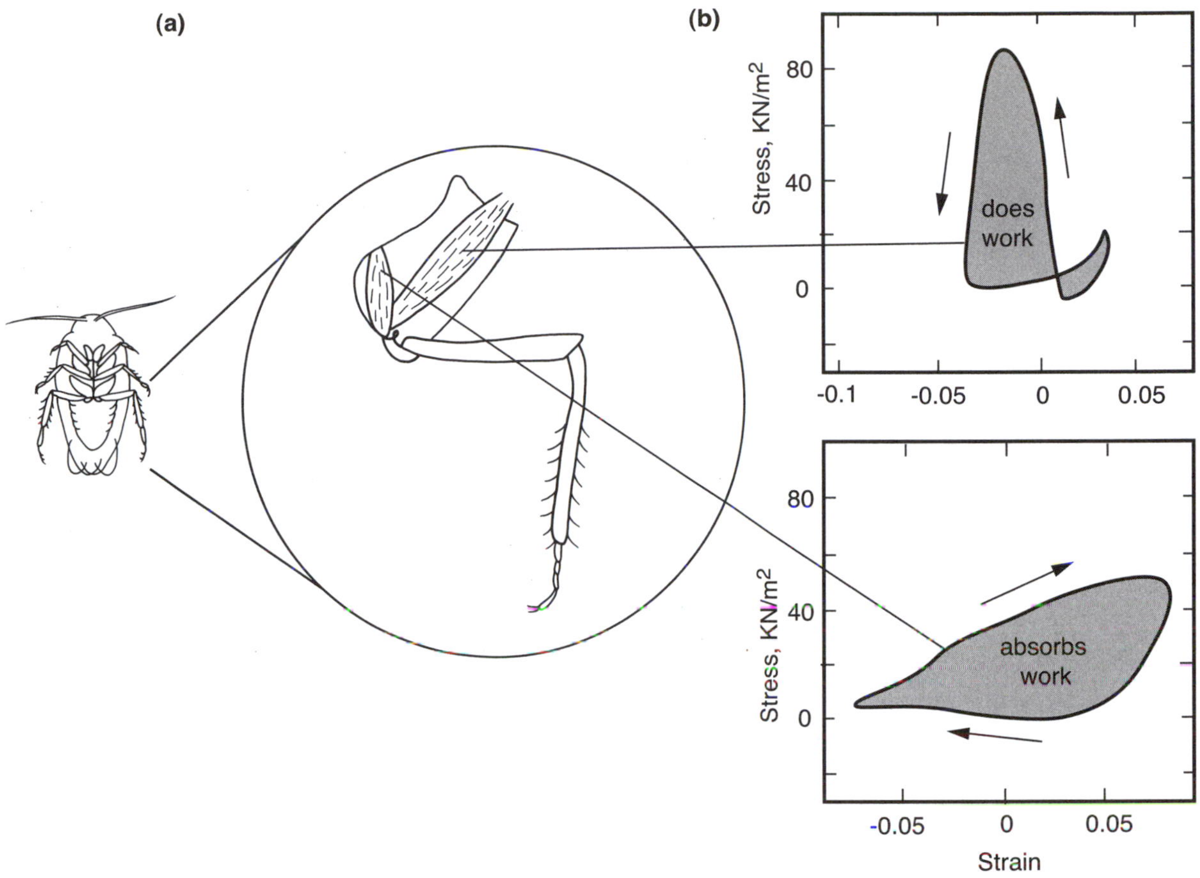

FIGURE 23.8 (a) Two extender muscles in the hindleg of a cockroach. (b) The work loops for those muscles. One turns out to be a flexion-resistor rather than a functional extender. Redrawn from Ahn and Full (2002) with permission of The Company of Biologists, Ltd.

to absorb it. Still, that result at least raises the possibility of some elastic storage rather than total thermal dissipation. That does indeed happen. An especially large protein, titin, whose molecules run the length of each half-sarcomere accounts for some, perhaps most, of the storage (Tskhovrebova et al. 1997). The collagen fibers of the surfaces of the muscle cells seem to provide additional storage (Tidball and Daniel 1986).

Still, tendon, with its splendid resilience and work of extension, appears to be the principle site of stride-to-stride energy storage, at least in vertebrates. Yet muscle can, on occasion, play a supporting role as a device to minimize slack or vibration in tendons. In the legs of horses, some tendons are fully a hundred times as long as their associated muscles (Dimery, Alexander, and Ker 1986). When the animals run, the tendons stretch but the muscles hardly change length at all. Long viewed as vestigial, the muscles seem to function as vibration dampers, working as fairly ordinary viscoelastic elements to protect tendons and bones from fatigue and impact failure (Wilson et al. 2001). Horses, of course, run on hard, hypertrophied toenails, not footpads.

Jumping When Small—Tiny Batteries

Energy storage plays an even more important role under circumstances other than in cyclic motion—in the acceleration of tiny creatures that lunge or jump, a situation mentioned briefly back in chapter 2. Speed of movement generally varies with the size of the organism. Thus, large fish swim faster than do small ones. In part that's because drag is roughly proportional to surface area while the amount of locomotory machinery increases with body volume. A large creature falls more rapidly than a small one of the same shape and density, again a matter of surface and volume. A large bird flies faster than a small one; the surface-volume relationship of drag and muscle permits it, while the surface-volume relationship of wing loading demands it.

What about acceleration? Consider the business of jumping—as Galileo first told us, in the absence of drag, jumpers of any size should achieve about the same maximum height. A rough-and-ready explanation goes as follows. The mass of an animal's muscle is nearly proportional to its body mass (chapter 3); the work its muscles can do (force times distance) in a contraction should be proportional to their mass, so the work a jumper can do ought to be proportional to its body mass. Put another way, its work per unit mass should be constant. Work equals force times distance, so it's equivalent to mass times acceleration times distance, and so work per unit mass should be proportional to the product of acceleration and distance. Assuming constant acceleration for any animal (not between animals), that product, though, is the square of velocity—the latter (and thus velocity itself) should therefore be constant:

$$\frac{W}{M_b} \propto ad \propto v^2 = \text{constant}. \tag{23.1}$$

In short, given muscle of some standard capability, all jumpers should achieve about the same takeoff velocity by an impulsive contraction of their muscles—the distance over which the acceleration takes place has dropped out.

That means that they should achieve the same maximum height. In the absence of drag, initial upward speed determines maximum height—after leaving the ground the organism or projectile just decelerates under the influence of gravity or, put the other way, it accelerates steadily downward even as it travels upward. Maximum height, h_{max}, is simply given as the square of the upward speed at the end of acceleration (termed, militaristically, the "muzzle velocity"), V_{muz}^2, divided by twice the acceleration of gravity:

$$h_{max} = \frac{v_{muz}^2}{2g}. \tag{23.2}$$

It must be admitted that even in the absence of appreciable drag, organisms may not quite follow our logic—jumping distance increases somewhat with size at least in frogs and toads, according to Emerson (1978) and a Useful Fact undoubtedly familiar to the residents of Calaveras County.

So what about acceleration? Bear in mind that once a nonflying creature leaves the ground it cannot apply force to the ground to cause acceleration of its mass. If muzzle velocity is constant, then acceleration (still assumed constant during a contraction) must be *greater* for the smaller creature—the speed has to be achieved over a shorter period of time because of the effectively shorter muzzle length, l_{muz}:

$$a = \frac{v}{t} = \frac{v^2}{2l_{muz}} \tag{23.3}$$

(Note that equations (23.2) and (23.3) express the same basic relationship.) A shorter leg cannot push a body to as great a height; indeed acceleration ought to be *inversely* proportional to body length.

What of reality—does it follow this logic? Table 23.1 gathers a few figures for acceleration for systems spanning a wide range of sizes and a diversity of propulsive arrangements. It shows that the values of acceleration in small-scale systems reach remarkably high values, even when we leave the limited world of aerial jumpers. A human can withstand accelerations of only about 100 meters per second squared (ten times gravity) or so; apparently these small systems are doing something rather special. But the look deceives us, a consequence of looking at a quantity once removed from stress, all that the material of the system really experiences. It's much the same sort of scaling illusion as we saw with respect to the high pressures that generated only ordinary stresses in the cell walls of plants (chapter 3).

TABLE 23.1 Maximum Acceleration for a Variety of Systems

System	Maximum Acceleration (m/s^2)
Car, 0 to 20 m/s (45 mph) in 8 s	2.5
Man, near record standing jump	15
Antelope, jump	16
Lizard, dash	30
Trout, sudden start	40
Spider (salticid), jump	51
Lesser galago, jump	140
Locust (first stage), jump	240
Squid tentacle, extension	250
Click beetle, jump	400
Rat flea, jump	2,000
Pilobolus, sporangium discharge	100,000
Hydra, nematocyst discharge	400,000
Rifle bullet, 3,000 m/s muzzle speed	500,000

Sources: Data from Parry and Brown 1959; Page 1964; Evans 1973; Bennet-Clark 1975, 1977; Webb 1976; Huey and Hertz 1984; Schmidt-Nielsen 1984; Kier, personal communication.

Note: To convert to multiples of gravitational acceleration on earth, divide by 9.8.

But something odd appears if we put the present discussion in the context of what was said in the last chapter about muscle contraction speed. Intrinsic speed, contraction velocity divided by muscle length, scaled roughly with the inverse square root of length:

$$IS \propto M_b^{-0.167} \propto l^{-0.5}. \tag{22.1}$$

As noted just above, though, to get the same final velocity as length decreases, acceleration would have to be inversely proportional to the *first* power of length, not its square root. So the increase in intrinsic speed as animals get smaller, while impressive, remains inadequate, the exponent about half of what's necessary.

And that's where energy storage comes in. Down to the size of a trout or a squid tentacle, unaided muscle can do a decent job with nothing more than ordinary leverage. Below that, muscle needs help; in practice, energy is put in slowly and stored elastically. Some kind of trigger then releases it at a higher rate. Work and energy may be conserved, but power gets amplified. Consider what a flea is up against. If a good muscle can put out about 200 watts per kilogram and if a flea is 20 percent jumping muscle, it has available only about 40 watts per kilogram of body mass. According to Bennet—Clark (1977), the observed performance of a flea requires 2,750 watts per kilogram—to jump, a flea simply must have a power amplifier. What it does is contract the muscles that push the hindlegs downward for about 50 milliseconds before it jumps; their contraction deforms small pads of that excellent protein rubber, resilin. The resilin then powers the actual jump during the 0.7-millisecond period of acceleration. The flea doesn't get something for nothing—in the process a little work leaks out; but work done slowly and thus at a lower power level is released rapidly and thus at higher power. A locust or grasshopper jumping with its hindlegs stores up work in chitinous apodemes and gets a tenfold power amplification (Bennet-Clark 1975). A click beetle stores up work by deforming the external cuticle; its power amplification is fully a thousandfold (Evans 1973). Each of these creatures has some kind of a mechanical catch to prevent premature extension while the work is being put in; the specific arrangements, though, are different for each case.

Larger animals use energy storage when they run around, as was mentioned at the start of chapter 1 and to which we'll return in the next chapter. But they're using it to bank decelerative work for subsequent acceleration where appendages must be cyclically stopped and started, gaining economy rather than speed or throw-distance. Energy storage prior to jumping in the interest of power amplification doesn't amount to much—at least compared to the insects just described.

A final note, prompted by inclusion of the nonmuscular systems of *Pilobolus* and *Hydra* in table 23.1. For one-shot systems too small for real-time muscular propulsion, power output (as in table 22.1) shouldn't matter much. With long-term energy storage no more costly than short-term storage, and with almost unlimited possibilities for power amplification, osmotic, water-absorptive, and other slow systems can be splendidly effective. Seed expulsion by desiccation came up in the last chapter; many more examples of explosive seed, spore, and sporangial discharge occur in terrestrial fungi and plants.

CHAPTER 24

Getting Around on Land

The centipede was happy quite,
until a toad in fun
Said, "Pray, which leg goes after which?"
That worked her mind to such a pitch,
She lay distracted in a ditch, considering how to run.

MRS. EDWARD CRASTER (1871)

GETTING ABOUT ON LAND, THUS AT THE SURFACE OF A SOLID MEDIUM, presents radically different problems than does swimming or flying through a continuous medium. So different are the biomechanical challenges that we take up one when talking about fluid flow and the other when considering structures and mechanisms of motility. For swimming and flying, the game consisted of passing rearward momentum to the surrounding medium while counting on the equal and opposite momentum for forward motion against one's drag. For terrestrial locomotion, pushing rearward still does the trick, but viewing it in terms of momentum transfer doesn't help a lot—that you slow the earth's rotation when walking eastward has little practical or analytical utility. So we find of little use the notion of Froude propulsion efficiency.

And precious little of your effort goes into offsetting your drag. So we find it simplest to assume that the external work done when moving steady on a level surface is zero. Which isn't as crazy as it sounds. After all, an enormously long train can be moved along a track by a modest engine, taking advantage of the low friction of steel rolling on steel—if the track is level and no one is in a hurry to get up to speed.

But the complications, in the final analysis, are no worse than those bedeviling swimming and flying, and we can recognize generalizations of at least equal potency and immediacy. Which was why chapter 1 began with a few pages on walking and running. We return to those gaits now, equipped with a lot more information about their locomotory tasks and the equipment they have available to drive them.

Gaits—The Possibilities

Humans use two principal gaits, walking and running. We may speak of ambling, shuffling, striding, jogging, trotting, and so forth; but these are flourishes on two themes. So distinct are they that you're never in doubt whether at any moment you're walking or running. In walking (fig. 24.1a), each leg contacts the ground for more than half the total stride time—we say that the duty factor exceeds 50 percent. And you nearly vault over a straight leg as each leg passes beneath your torso, so you're highest in midstride. When running (fig. 24.1b), your legs have duty factors below 50 percent, which is to say that you're completely airborne twice during each full stride (or pace). Since the average downward component of force must equal your weight, you

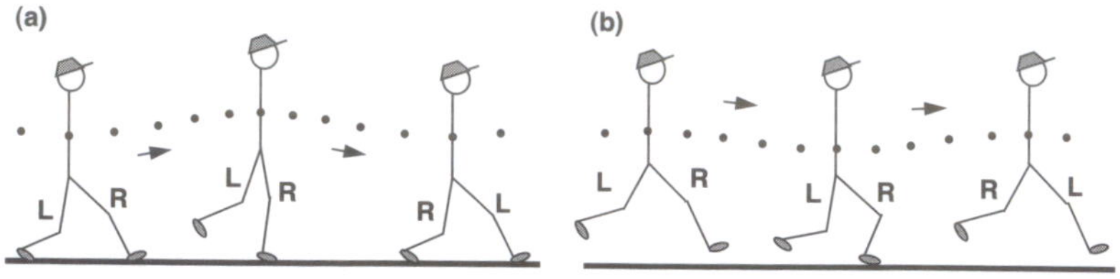

FIGURE 24.1 (a) The sequence of footfalls in a walking half-stride, and (b) the sequence of footfalls in a running half-stride.

must press harder on the ground when contacting it when you run. (And impact adds to this, so ground contact force hits about twice the peak level in running as it does in walking.) Neither leg is fully extended downward at any time, so you're, on average, closer to the ground. Furthermore, you're lowest rather than highest at midstride. Within each gait, you for the most part adjust speed by changing stride length and only minimally by changing stride frequency. While we're able to hop, skip, and jump, they don't take us far.

Things get a little more complicated when a mammal goes around on all fours. Walking has a remarkably standard leg-fall sequence, animal to animal. The order goes right front, then left hind, then left front, then right hind. Watch a dog or cat or horse. Or try it yourself, moving briskly on hands and knees. I watched my second grandson, just after he mastered crawling—he took up the proper mammalian pattern without prompting. (You can adopt other patterns deliberately without evident instability or other problem, but you ordinarily don't.) Duty factors vary with the size of the animal, as does the degree to which legs are bent during the process. Small animals may have lower duty factors for each leg, and they bend their legs more, but neither aspect changes dramatically. Trotting by quadrupeds (fig. 24.2a) uses a footfall pattern similar to that of walking, but it synchronizes the diagonals. Thus, left front and right hind move together, alternating phase with right front and left hind. It's spoken of as a "symmetrical" gait. For most animals, trotting includes a pair of airborne periods in each full stride. But as animals get

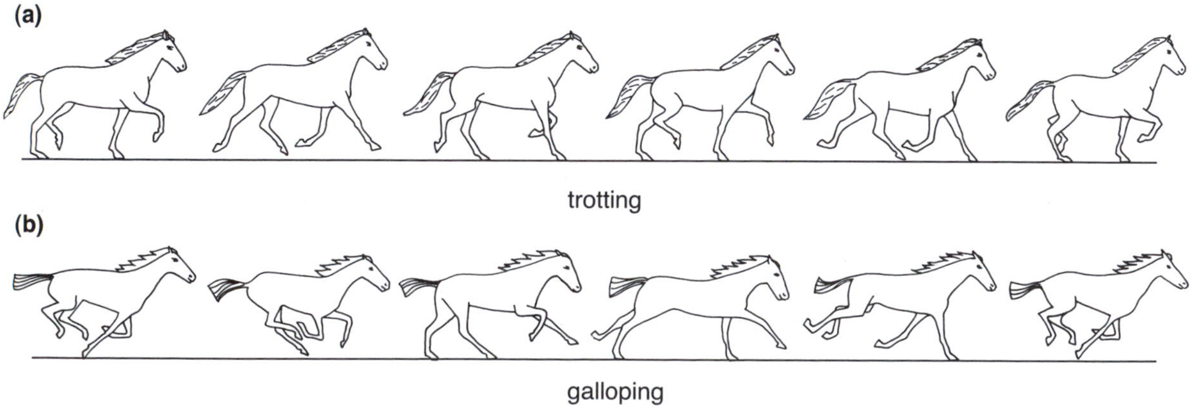

FIGURE 24.2 (a) The footfall sequence of a trotting horse, and (b) the sequence of footfalls in galloping. The pictures are redrawn from the pioneering high-speed photographs of Eadweard Muybridge, done around 1877—the first conclusive evidence that a trotting horse sometimes had no leg on the ground.

larger, the relative length of the airborne periods decreases; trotting elephants don't ever get all feet off the ground at once.

Quadrupeds add to these two a third principal gait, galloping (fig. 24.2b). In galloping, the two front legs work simultaneously, as do the two back legs, rather than diagonal pairs working in synchrony. As a result, it has a single airborne period rather than a pair of them. The practical consequence is that the back can bend and straighten during the stride, lengthening it and (as we'll see) storing elastic energy from limb deceleration for subsequent acceleration. Paired legs don't work quite together, with one always striking just before the other, so galloping is an "asymmetrical" gait. Thus, one animal might lead with the left forefoot and another with the right forefoot. For that matter, the word, "galloping" (or its musical equivalent, Rossini's overture to the opera *William Tell*) has an onomatopoeic suggestion of that asymmetry.

Gaits—Walking's Speed Limit

Some questions posed at the start of chapter 1 could stand reiteration. Why do you make this abrupt shift between walking and running? At what speed do you make the shift? Do other legged creatures shift gaits at the same speed or do they at least shift at a speed determined by the same rule we feel compelled to obey?

We saw back then that the mode of stride-to-stride energy storage shifted in the walk-run gait transition. And we asserted that this change from gravitational storage to elastic storage was the crucial element in that transition. Gravitational storage meant some form of pendulum—in fact an odd inverted one was invoked—and pendulums run into practical limits on their amplitudes. So the formula for the period of a pendulum on earth could give us the upper practical speed for walking; with speed, v_{max}, in meters per second and hip-to-ground leg length, l, in meters,

$$v_{max} = 2.2\sqrt{l}. \qquad (1.3)$$

The constant of proportionality assumes that animals walk in a particular dynamically similar fashion, with particular weight distributions along the lengths of their legs, and with the same particular upper limit on the amplitude through which a leg can be swung. What we're looking at, again, is the practical upper speed limit for gravitational energy storage, stride to stride.

The same formula can be reached by several other routes, using reasoning that's now a lot more familiar than it was back in chapter 1. Consider, for one, the forces on an appendage swinging back and forth as a pendulum or an alternately rising and falling body, again acting as a pendulum. What keeps it going is what's often called its inertial force—given as mass times acceleration. What makes it stop as it swings upward is gravitational force—mass times the earth's gravitational acceleration. We can write the two as a ratio, cancel masses, and reduce acceleration to speed and length (tacitly assuming steady motion). Sticking to loose, dimensional terms,

$$\frac{ma}{mg} = \frac{v^2}{gl} = Fr. \qquad (14.3)$$

What we get is the dimensionless Froude number, perhaps familiar from chapter 14, where it determined the maximum practical speed for a boat making gravity waves on the surface of a body of water. (Sometimes the square root of v^2/gl is called the Froude number—always check.) All that differs is the definition of length—there, the hull length of the boat, here, the hip-to-ground distance of a leg.

The same reasoning that suggested that equation (1.3) should have a particular numerical constant, together with the same data set that gives that constant, yield a value for the Froude number of about 0.5. (The first paper, Alexander [1976], gave 0.6 based on more limited data.)

Alternatively, think of the way a pendulum system interconverts kinetic energy (energy by virtue of motion), $mv^2/2$, and gravitational potential energy (energy by virtue of distance from the center of the earth), mgh. We can look at the two as another ratio, this one (absorbing the constant of ½ into the Froude number):

$$\frac{mv^2}{mgh} = \frac{v^2}{gl} = Fr. \tag{24.1}$$

In informal terms, it says that g and l limit the amount of kinetic energy that can be put away in gravitational storage from stride to stride. Longer legs or a more massive (or dense) planet ought to permit faster walking. In practice, differences in gravity may not matter much, but experimentally altering g tests the robustness of the rule. Kram and his collaborators (Kram, Domingo, and Ferris 1997; Griffin, Tolani, and Kram 1999, Chang et al. 2000b) have done just that, with such maneuvers as simulating reduced gravity for subjects walking on a treadmill by attaching to them strings that went up and over pulleys and led to long springs. Their results satisfyingly confirmed the transition value of the Froude number.

In practical terms, the cost in power of walking hits the cost of power of running at just about that transition point. Going slower, walking is cheaper; going faster, running is cheaper, as in figure 24.3, from Alexander (1996), but as data going back to Margaria (1938). The actual transition point, the particular Froude number, comes from looking at what animals really do, so animals appear to pick their most economical speed for making the transition. You can do a quick verification on a track with nothing more than a stopwatch—for timing your walking and running and for measuring your pulse. Figure 24.3 holds two other lessons. First, an ascending straight line that extrapolates back to the origin indicates a constant cost of transport in terms of energy expended relative to distance. When running, we humans gain distance, in this sense, with a speed-independent cost; by contrast, the curved line reflects an increasing price relative to distance for walking—slower is cheaper. Second, walking is better for slow transport, and running is better for faster transport. If you need to move, on the average, at the transition speed, the best tactic consists of shifting back and forth in both gait and speed, alternately walking at a speed just below normal transition and then running at a speed just above it.

As an inverted pendulum, you rise and fall twice during each full stride. As in figure 24.1a, you're highest in midstride, when you move forward less rapidly, and lowest at the extremes of the stride, when you're going fastest.

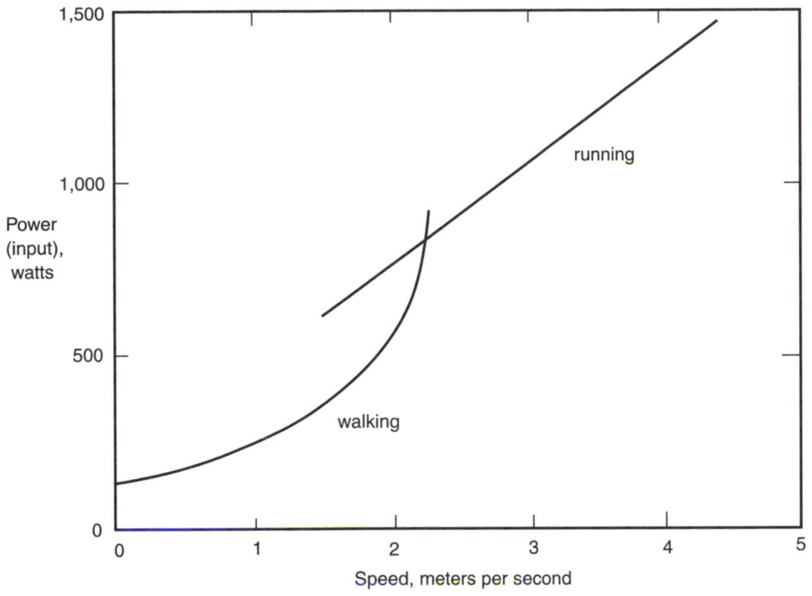

FIGURE 24.3 Energy expenditure per unit time—power—versus speed, for human walking and running.

Thus, you trade kinetic and potential energy. How good is gravitational energy storage, anyway? Not especially, it turns out. The changes in the body's center of gravity, both up and down and fore and aft, can be estimated fairly well. A 56-kilogram (123-pound) person rises and falls about 31 millimeters (1.2 inches) when walking. Full vaulting over a stiffly extended leg would require six times as much up and down movement (Lee and Farley 1998). And we're unusually straight-legged walkers.

Gaits—Galloping Slowly

What happens above the switch point? Instead of storing energy as gravitational potential, an animal stores it elastically—by deforming part of its body. Three places suggest themselves—bone, muscle, and tendon—and all do participate. But the major player turns out to be the least massive, tendon (Alexander 1988). When we run, when a horse trots or gallops, when a crow or kangaroo hops, all repeatedly stretch the tendons and other elastic tissue of feet and legs. That demands a different mode of coordinating the movement of the parts of body and appendages, so a run or hop looks unmistakably different from a walk. It takes advantage of the excellent energy-storage ability of our main elastic material, collagen, noted in chapters 15 and 16.

While we make only a single shift, that from walking to running (or jogging or trotting, much the same thing), most quadrupedal mammals can then shift to a third gait, going from a trot to a gallop. Again, the transition comes suddenly and unambiguously. But what's happening in present terms turns out to be a bit odd. Transition happens within a narrow range of Froude numbers around 2.5—not as narrow as the shift from walking to running, but still narrow relative to the size range of the animals. Why should Froude number work at all? Both trotting and galloping use elastic rather than gravitational energy storage, differing mainly in when and where tendon gets stretched.

Most likely, the Froude number's utility in defining the transition point between trotting and galloping comes from a different role of gravity (Vogel 1998b). We often encounter cases where a given dimensionless ratio, with its variables revamped, works for a different physical situation—demonstrating nothing less than the ubiquity of a few kinds of forces in setting the limits of biological design. What sets the transition point may be not so much how fast an animal can trot but how slowly it can gallop. A galloping animal spends a fair fraction of each stride off the ground. And that, in the absence of anything to give it aerodynamic lift, means that it's in free fall. So gravity can reasonably reenter the picture.

An animal in free fall for a time within each stride ought to tolerate a fall of a fixed fraction of leg length, l. Using d for fall distance, and the usual rule for distance in free fall,

$$d \propto l \propto gt^2. \tag{24.2}$$

According to Heglund and Taylor (1988), running speed at transition (v_{trans}) is proportional to leg length times stride frequency or leg length divided by the time for a single stride—something intuitively reasonable:

$$v_{\mathrm{trans}} \propto l/t. \tag{24.3}$$

Putting these together by substituting l/v_{trans} for each t in equation (24.2), we find that gl/v^2 comes out constant. So the Froude number rises again—if you're a dimensionless constant, so is your reciprocal. And its involvement suggests that we're on the right track when we assume that what sets the lower speed limit of galloping is the fixed fraction of leg length that an animal can permit itself to fall before again getting a pair of feet back on the ground.

The Energetics of Walking and Running

We're a little unusual in having that linear relationship between speed and cost for running shown in Figure 24.3. Figure 24.4a gives the equivalent relationship for horses, from data of Hoyt and Taylor (1981), with the same

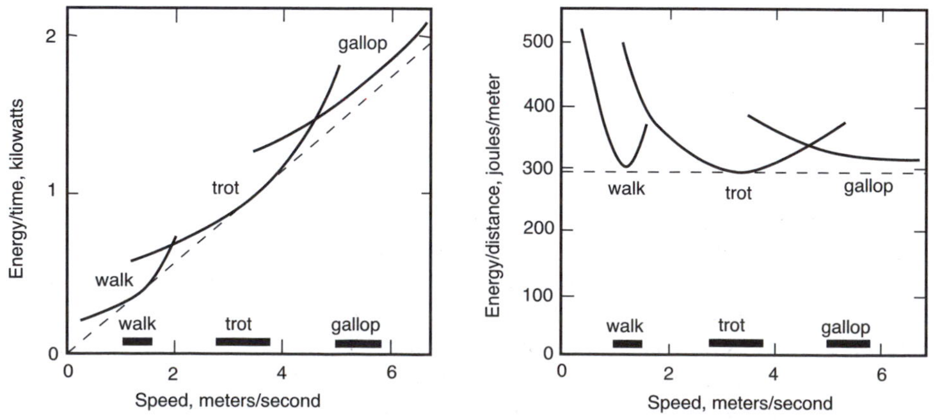

FIGURE 24.4 (a) Energy expenditure per unit time versus speed for the three principal gaits of horses. (b) The same data, now plotted as energy expenditure relative to distance. Bars at the bottom give the preferred speed for each gait.

information replotted as energy per unit distance rather than per unit time in figure 24.4b. Since horses have three basic gaits rather than two, the graph has three lines. None of them, though, has the linear, zero-extrapolated, long line of human running. Carrier (1984) has suggested that our ability to run with the same effectiveness over a wide range of speeds allowed primitive humans to get a leg up on prey by chasing them at speeds that were, for the prey, less than optimally efficient. Tested on a treadmill, Hoyt and Taylor's 100-kilogram ponies had three nearly identical minima, about 300 joules per meter, for the cost of moving themselves about. When free to choose their speeds rather than being constrained on a treadmill, the ponies picked one or another of those minima.

Which sounds as if the ponies could somehow monitor their cost of transport, adjusting speed to stay at a minimum. As it happens, they probably don't do just that but rather use something else as a proxy for cost. That something else appears to be force—as stress in bones, muscles, and tendons. Horses with added weights shift gears from walk to trot to gallop at lower speeds, deviating from the speeds that correspond to cost minima (Farley and Taylor 1991). Biewener (1990) saw a general rule—all animals hit about the same peak compressive stress in their leg bones. The compressive strength of bone is about 200 megapascals; peak loads during locomotion are about 50–100 megapascals. (That gives a safety factor, about which more in the next chapter, of two to four). Animals over about 300 kilograms (700 pounds) accept relatively lower locomotory performance and, in addition, build disproportionately thick bones. (That view agrees satisfyingly with the suggestion that maximum tolerable stride-to-stride fall distance sets the minimum speed for galloping.)

Legged locomotion, even with stride-to-stride energy storage, wins no prizes for efficiency. You can reduce the cost of going from one place to another by attaching 12 or 15 kilograms of bicycle to yourself, using your muscles in a biologically unusual (if carefully chosen) manner. That cost reduction isn't trivial—at 2 meters per second (4.5 miles per hour or 7.2 kilometers per hour), going a given distance by cycling takes 2.2 times less energy than walking; at 4 meters per second (9 miles per hour or 14.5 kilometers per hour), cycling takes 3.7 times less energy than running. Put another way, investing 15 watts of metabolic input per kilogram of body weight can move you at 8.5 meters per second (19 miles per hour or 30.6 kilometers per hour) if cycling but only 4 meters per second (9 miles per hour or 14.4 kilometers per hour) if running (Whitt and Wilson 1982). Or one can view the matter in a more general and biological context, looking at the energy needed to get a given mass a given distance—as in figure 24.5, from Tucker (1975). The data behind these lines (they're not theoretical) says that walking and running cost several times as much as flying, which costs several times as much as swimming.

As in so many places, the energetics of terrestrial locomotion depend on size. Peculiar to terrestrial locomotion (at least at constant speed on the level) is the dominance of internal factors. Drag, the bugbear of swimming and a serious impediment to flying, doesn't amount to much. Consider a human, running at 5 meters per second (11 miles per hour) and using 1,400 watts of metabolic power. Assuming the human can be approximated by a broadside cylinder of 0.5 square meters frontal area ($C_d = 0.47$; see chapter 7) gives a power expenditure of 18 watts. That's barely 1.3 percent of the input or a roughly 5.2 percent equivalent cost assuming an input-to-output efficiency of

0.25. (Pugh [1971] gives 7.5 percent for a speed of 6 meters per second, exactly the same as that 5.2 percent if drag varies with the square of speed.) Yes, it's easier to run with the wind than against it, but for drag at winds of the speeds induced by running itself, we're not talking about a major component of the cost of running.

Bigger is better, as evident from the line for running in figure 24.5. Over the huge size range between ants and horses, the minimum metabolic cost of transport—what an animal pays relative to mass and distance at its cheapest speed—scales with body mass to the power −0.32:

$$C_{\text{trans/mass}\cdot\text{dist}} \propto M_b^{-0.32}. \tag{24.4}$$

The size difference traces, in part, to a divergence between the metabolic cost of transport and its component, the mechanical cost of transport. An ant pays a metabolic cost of about 1,000 joules per kilogram-meter; a human pays about 4 joules per kilogram-meter (Alexander 1999). Alexander estimated the mechanical cost of transport, from raising and lowering and speeding and slowing its mass, as about 1 joule per kilogram for a mouse (1 per 1,000) and 2 joules per kilogram for a person (2 per 4). So stride-to-stride energy storage can't do as much good for the mouse as for a larger animal. Whether bounding kangaroos or running humans, energy storage can save about half of that mechanical cost for big creatures.

Bigger is better for some other (not entirely independent) reasons as well. Minimum metabolic rate—basal or maintenance metabolism—scales with body weight to the power −0.25, while cost of transport (relative to mass and distance) scales with the power −0.32. So the larger animal pays less, relative to the cost of getting somewhere, for keeping body and soul together. The effectiveness of muscle in converting chemical to mechanical energy is another factor. Recall, from chapter 22, that if animals are proportioned similarly and

FIGURE 24.5 How the cost of transport (energy to move a given mass a given distance at the most economical speed) for three forms of locomotion varies with body mass—and a few other vehicles for comparison.

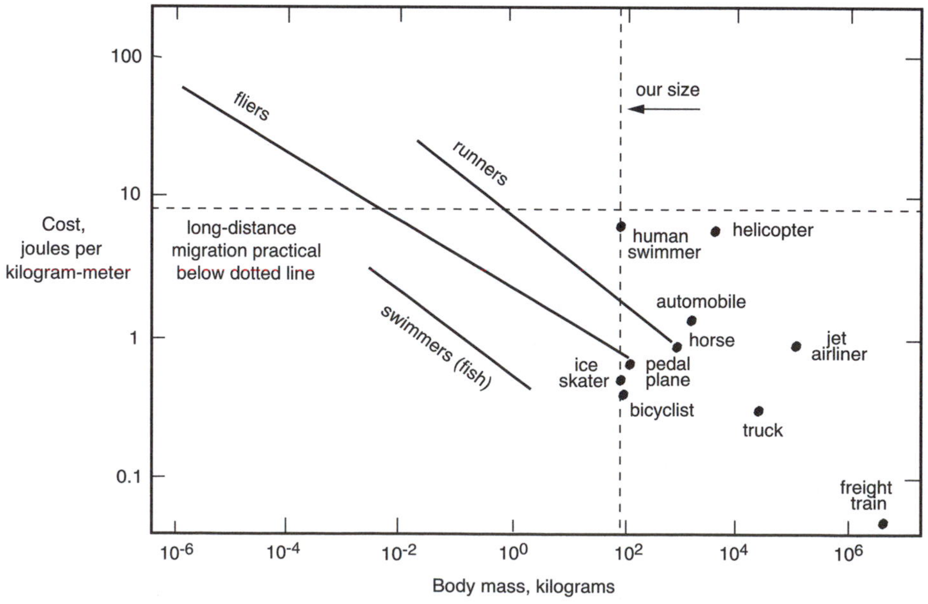

follow the pendulum rule for leg-swinging frequency, the smaller ones have to run their muscles at higher intrinsic speeds—*IS* scales with body mass to the power –0.167. And muscles work less efficiently at higher intrinsic speeds.

Yet another factor is the need for muscle to use energy to produce force, even when it does no work, when a rope (or tendon) would do as well. That empirical scaling exponent of –0.32 can be viewed another way. It's equivalent to saying that cost of transport relative to distance but not mass scales in proportion to body mass to the power 0.68:

$$C_{\text{trans/dist}} \propto M_{\text{b}}^{+0.68}. \qquad (24.5)$$

Now with muscle making up a fairly constant fraction of body mass, overall muscle cross section will scale with body mass to the power 0.67—indistinguishable from 0.68. Why might muscle cross section matter? Alexander (1999), again, has suggested that total energy cost for running is, roughly speaking, the sum of the cost of doing work and the cost of generating force, and that, for muscle, this second is surprisingly high. So in looking at how cost scales, we're looking mainly at the cost of force production. Odd, perhaps, but not unreasonable considering the remarkable inefficiency of running as a way to get from place to place.

Realizing that producing force imposes the greater part of the cost of terrestrial locomotion makes sense of some other things. An animal has to get purchase with the ground to push forward—it's hard to make decent progress on wet ice; and a dog slips awkwardly when trying to accelerate forward, stop suddenly, or corner sharply on a polished floor. As an animal goes faster, it requires a greater downward force. But the time-averaged static force it can exert can't exceed its weight. What to do? Adjust the duty factor, decreasing the fraction of the time any leg makes contact with the ground. That's exactly what happens—when walking, your legs have duty factors over 50 percent, so you're never without a leg on the ground. In all versions of running, factors run under 50 percent, so you're entirely airborne at times. The same thing happens to quadrupeds, if a bit more complicatedly. (Sir James Gray's classic books of 1953 and 1968 still stand as lucid explanations of what happens.) Paradoxically—at first glance—the faster an animal goes, the lower the fraction of the time any leg can make contact with the ground.

Another complicating factor is the need for an animal to exert force to stay upright. That would be a force proportional to its mass if nothing else changed. With both basal metabolic rate and cost of transport decreasing with size, that sustaining force would be an ever-increasing tax on performance. But large animals adopt different resting postures from small ones, standing straight-legged and thus using passive bone more and active muscle less.

Beyond that, we're looking at a general problem encountered by animals attempting terrestrial locomotion. A swimmer or flyer can push directly rearward against the medium—an effective line of action of the net force goes through or nearly through its center of mass. A walker or runner has to push rearward against a substratum while holding its center of mass above it, so either the push is wastefully oblique or the animal must deal with an awkward turning moment. You lean forward when pulling something, a posture that only emphasizes your deviation from verticality when just walking or running. Ground may be pushed against with negligible slip—unlike air or water. But it's in the wrong place.

Other Gaits

We humans can get around using more than our ordinary two gaits. Unless very old, very young, or very corpulent, you're probably capable of moving arm-over-arm, that is, brachiating like a gibbon or some other primates. Again, most of the cost comes from generating force, and arm muscles pay a lot more in power than do leg muscles (Glasheen and McMahon 1995). Brachiating can use either a continuous contact gait or one with a purely aerial phase—the shift to such richochetal vaulting happens at a Froude number between 0.3 and 0.6, so it's oddly analogous to the walk-trot transition (Chang, Bertram, and Lee 2000a). But both forms use gravitational energy storage, pendulum fashion, rather than elastic storage (Bertram et al. 1999). For even the most accomplished brachiators, gibbons, its cost exceeds that of quadrupedal walking, both literally and figuratively a more pedestrian gait (Parsons and Taylor 1977).

We can also skip (and hop), although only children between around 4 and 5 years old skip spontaneously. Minetti (1998) looked at our skipping gait and found it an expensive way to go. It uses the elastic energy storage and low duty factor of running at what are normally walking speeds. Skipping, though, has a peculiar use. Consider the problem of walking on a planet with a lot less gravity than the earth's. The Froude number remains a reliable predictor of transition speeds, which are therefore slower (Cavagna, Willems, and Heglund 2000, Minetti 2001). Drop gravitational acceleration by a factor of four and the walk-run transition will decrease twofold. At the same time, running works less well—how much one can push forward depends on how much one can push downward. Astronauts on the moon, with a sixth the gravity of the earth, found that skipping worked fairly well. With its low duty factor for its speed, it gave adequate downward force. Where gravity is above that of the earth, walking gets more useful than here.

Quadrupeds have lots more alternatives. Horses can (or can be taught to) walk by moving first the two right legs and then the two left legs—the gait goes by various names, including pacing and racking. That results in severe lateral unbalance, since the horses don't bend their feet inward sufficiently to put their footfalls under the paths of their centers of gravity. As a result, a pacing horse sways from side to side, to the discomfiture of a rider. Pacing, though, is no mere curiosity. Camels and giraffes normally pace—camels are called "ships of the desert," supposedly because of that swaying. The usual explanation for their odd gait is that they have especially long legs that aren't especially far apart front to back. Thus, the normal footfall sequence of walking would restrict the angle through which a leg could be swung without bumping into the one ahead or behind.

Kangaroos, of course, hop. In a sense they have gaits that can be described as bipedal, tripedal and pentapedal, depending on whether the gait entails just hindlegs as when hopping at speed, hindlegs and tail, or forelimbs as well as when moving at slow speeds. They store energy elastically when hopping in much the same manner as other mammals do when galloping (Alexander and Vernon 1975), with about the same consequent saving of about half the cost of locomotion. What's a little odd is that their oxygen consumption decreases a little as speed increases (Dawson and Taylor 1973)—the explanation, according to Alexander and Vernon (1975) is probably an increase in effectiveness of elastic energy storage.

Insects typically have six legs, which permits locomotion by lifting alternate sets of three and thus never losing static stability—at no point would the center of gravity be anywhere but within the triangle formed by the three points of contact between legs and ground. In fact, they don't seem to make use of that stability, even if it might have been of relevance to some early, massive, clumsy forebear.

Terrestrial crabs typically get around using a sideways octopedal gait. An especially fast one, the ghost crab *Ocypode*, has three gaits that use energy-conserving mechanisms remarkably similar to those of mammals. It can walk, using the same gravitational pendulum action. And it has two running gaits, loosely analogous to trotting and galloping—a slower one with no fully aerial phase and a fast one in which all eight legs can be off the ground at once (Blickhan and Full 1987). So putting the muscles inside, upping the number of legs, and doing without collagenous tendons ends up making less difference than one might expect.

What about walking under water? That amounts to an extreme reduction in gravity and an extreme increase in drag—about the worst combination of circumstances for a pedestrian gait. Martinez, Full, and Koehl (1998) compared what a rock crab, *Grapsus*, did in air and in water. In air it found an ordinary octopedal walking gait practical. In water, with its weight (effectively, gravity) reduced tenfold, that wouldn't have been effective and wasn't used. In going under water, neither stride frequency or stride length changed; what did change were the duty factor and the number of legs making contact with the substratum. Rock crabs used a novel gait, "punting," in alternately pushing against the substratum and gliding through the water. They maintained about the same unimpressive speed in air and water, about 0.1 meters per second. For comparison, a ghost crab, about the same size, can go at least ten times faster.

Penguin go long distances with a peculiar waddling gait, and they suffer a cost of transport about twice as high as mammals of similar size. Is waddling bad? Griffin and Kram (2000) found, interestingly, that waddling makes the best of a bad deal—it permits use of the same kind of gravitational energy storage as the more familiar fore-and-aft inverted pendulum. The penguin's problem comes down to its short legs, requiring short, quick strides and higher intrinsic speeds of operation of its leg muscles. We cite allometric equations for cost versus body mass that work because leg length varies systematically with body mass—leg length represents the more relevant variable and body mass the proxy. Peculiar proportions pose the penguin's problem—it has the leg length of a much smaller animal, one that normally pays more for transport.

Still other gaits can be found here and there. Springbucks (or springboks) (*Antidorcas*) get their common name from their habit of "pronking" or "stotting," putting all four legs down at once and then bouncing high—up to 3.5 meters (10 feet) in the air. It seems to play a role in looking around for predators and warning nearby conspecifics of lurking danger as well as suggesting to the predator that, now having been detected, it might spare itself the trouble of futile pursuit (Caro 1986). Some mammals "bound," using paired forelegs and hindlegs alternately. We had a cat that would do it when going through snow whose depth was roughly the same as the cat's height. Figure 24.6 summarizes the footfall patterns of a whole array of gaits for quadrupedal locomotion. Finally, "slapping" locomotion of basilisk lizards on the surface of streams was mentioned in chapter 14—it amounts to bipedal trotting.

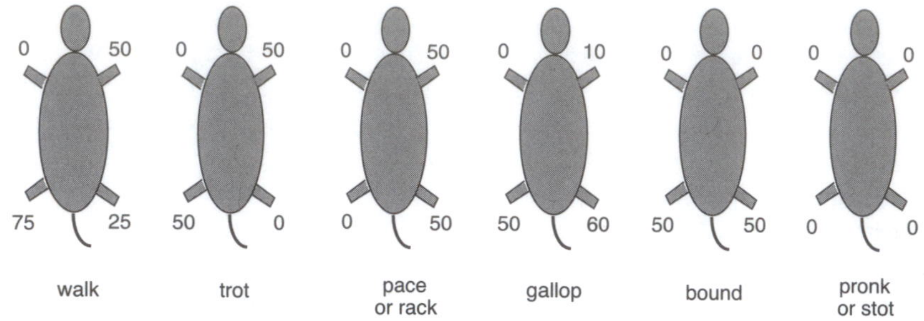

FIGURE 24.6 Footfall patterns for quadrupedal gaits. The numbers on each cartoon animal refer to the point during a complete cycle that the leg leaves the ground; the cycle starts at time = 0 and ends at time = 100.

Doing It without Legs

Snakes, some lizards, some amphibians, and various worms get around on the ground without any legs at all. They do so by using several different schemes, described in Alexander (1992) and lots of other sources.

The basic worm trick consists of stretching and squeezing alternate parts of a long, cylindrical trunk, moving each region of stretch or squeeze rearward, as in figure 24.7a. Despite the rearward progression, the scheme can't do much without one more component. Thus, a worm on a smooth and lubricated surface makes negligible progress—we're not looking at an analog of anguilliform (eel-like) swimming. The trunk needs some device so it slides more easily forward than rearward. For earthworms, setae, short rearward-pointing bristles, provide that crucial asymmetry.

A few snakes, such as boas and pythons, can move forward when extended lengthwise in a wormlike fashion rather than with their bodies bent into curves. They're not fast, and the behavior seems to be used for a stealthy

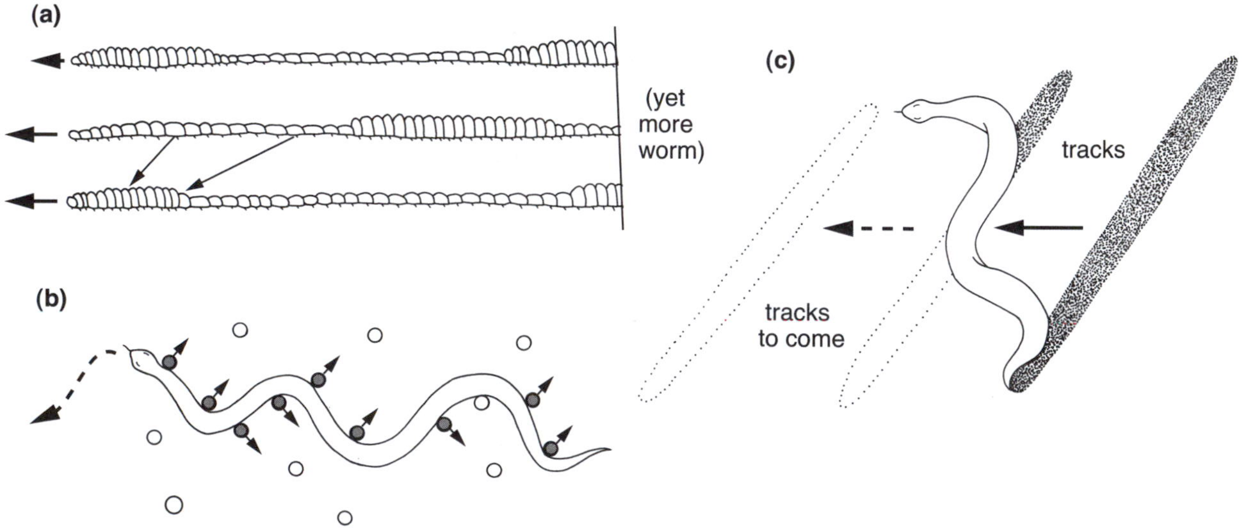

FIGURE 24.7 (a) Rectilinear locomotion by alternately sliding and grabbing the ground—here by an earthworm. (b) Locomotion by lateral undulation (serpentine motion), done where the substratum has sufficient irregularities to push against. (c) Sidewinding on sand, as done by a few snakes.

approach to prey. With backbones fixing their lengths, they can't work exactly like earthworms—they instead use waves of shortening that pass rearward along a powerful lengthwise band of ventral muscle. So the top of the snake moves steadily. Meanwhile, its bottom has a stop and start motion as successive segments are fixed to the ground while the ones behind the fixed ones are shortened and those in front of the fixed ones are lengthened. Moving segments are raised slightly above the fixed ones—it looks as if the snake is walking on its ribs, although it is not. Gray (1968) and Gans (1974) give good descriptions.

The more common scheme in snakes uses one or another form of serpentine (the name of obvious origin) body curvature—sometimes called "lateral undulation" to distinguish it from the "rectilinear" motion just described. As shown in figure 24.7b, the entire body moves all the time; what are fixed to the substratum are its curves. So each point along the snake's body alternately bends one way and then the other. The directional motion originates in the way each curve of the snake pushes against irregularities on the substratum and slides forward because those irregularities resist sideways forces and thus lateral motion much more effectively than they resist shearing forces and thus forward motion. The process doesn't differ greatly from what a flagellum does (chapter 11)—it depends on a similarly greater resistance to lateral motion than to axial motion. In one sense, the snake has it better than the flagellum, since points on the substratum may remain completely fixed rather than perpetually slipping. While not anywhere near as fast as it looks to a startled observer, this characteristic kind of serpentine locomotion does manage reasonable efficiency compared to moving on legs, mainly because no part of the body need to exert force to support itself against gravity.

Snakes can also engage in what's called "concertina" movement, in which one or two curves pass down the length of an otherwise straight animal. This works well for an animal confined within a channel just a bit larger than itself (such as a rodent's burrow). At least one snake, *Bitis caudalis*, the South African desert viper, can jump, getting entirely airborne and moving a distance almost equal to its length (Gans 1974).

Some desert vipers, such as a rattlesnake, *Crotalus cerastes*, of the southwestern United States, move across sand by what's called "sidewinding." The name refers to perhaps the oddest aspect of the motion, shown in figure 24.7c, the way the overall course of motion takes the snake at nearly a right angle to the heading of its head. As in normal serpentine movement, the snake propagates waves of bending rearward. But it doesn't slide along a serpentine path that traces the line of contact of those curves with the ground. Instead, it alternately fixes part of the body to the ground, pushing sideways against the sand, and lifts the adjacent part. So a given location on the snake never slides but repeatedly lifts and sets down. The process leaves a set of indentations in the sand at right angles to the snake's progress. In this way the snake never has to push against anything but a broad expanse of substratum—a good tactic since a sandy substratum doesn't resist being pushed upon at all well.

Another way to deal with sand brings us to nature's nearest approach to a macroscopic wheel—but not, to be up front about the matter, a wheel-and-axle combination like a bacterial flagellum or wagon wheel. A stomatopod crustacean (*Nannosquilla decemspinosa*), essentially a stretched-out shrimp, of the Pacific coast of Panama moves across sand when out of water by doing backwards somersaults. For about 40 percent of each turn it rolls like a wheel.

A bout of rolling lasts about half a minute; during that time it makes between twenty and forty revolutions and goes about 2 meters, doing 1.5 body lengths per second. The advantage appears to be the same as the sidewinder's—distributing one's locomotory force over a large area of sand—rather than any wheel-like economy. In fact it does about as much accelerating and decelerating as any legged creature. All this from Full et al. (1993).

Going Uphill and Downhill

So far we've been on the level. And so far bigger has been better, for several reasons discussed earlier. But what happens when an animal ascends a slope? Cost of transport relative to distance scales with body mass to the power 0.68 (equation 24.5). Body mass itself, of course, scales with body mass to the power 1.00. And in ascending, the cost of raising body mass against gravity must be added to the other costs. So bigger can no longer be better but must be distinctly worse. And that's just what we see. A horse is seriously slowed by even a slight slope, a dog is less sensitive to slope, a squirrel runs up a tree with ease and swiftness, an ant seems totally indifferent to whether it's going vertically uphill or downhill.

Experimental data nicely confirm theory and observation. Full and Tullis (1990) found that cockroaches did pay for going uphill. They consumed energy at twice their level best when going up at a 45-degree slope and three times as much when the treadmill was fully vertical. But those are the most modest of multipliers. A chimpanzee going up at a mere 15 degrees raises its energy consumption by 50 percent (Taylor, Caldwell, and Rowntree 1972). Another way to look at the picture, again from Taylor, Caldwell, and Rowntree (1972), comes from a comparison of the cost of gaining height at 2 kilometers per hour (0.56 meters per second or 1.25 miles per hour) at an animal's best speed and slope, relative to its energy consumption rate when moving on the level:

Mouse	1.23 times
Chimp	2.89 times
Horse	7.30 times

Of course, we don't quite minimize our cost of transport when walking a level course, as you can see from figure 24.8. The most economical gradient is a somewhat downhill one, about minus 5.8 degrees, or 10.2 percent. On or near the level, we do a mix of positive and negative work; above 8.5 degrees we do no more negative work—gravity takes care of any necessary braking action. Below minus 8.5 degrees we do no more positive work—gravity takes care of any necessary acceleration, and we need worry only about slowing our body parts (Minetti, Ardigo, and Saibene 1993). Positive and negative work come close to balance in level walking, which is as things should be since on average no load moves up or down. The difference between balance at level and the greater economy when going downhill at minus 5.8 degrees traces back to the different energetic cost of getting positive and negative work out of muscle, something we noted two chapters back. Thus, cheapest is a mix in which negative work predominates. Running differs from walking only inasmuch as negative work persists at inclines up to 17 degrees and

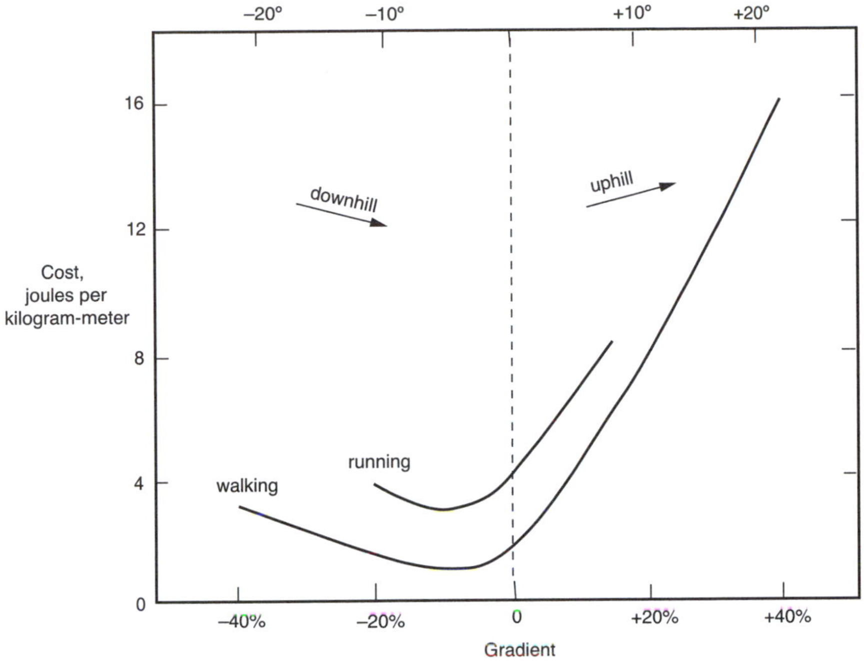

FIGURE 24.8 How our cost of transport varies with grade. The data refer to the energy relative to mass and distance needed to move along the grades at the optimal (cheapest) speed for each.

positive work at inclines down to minus 17 degrees (Minetti, Ardigo, and Saibene 1994.)

From such data, Minetti (1995) predicted the optimal gradient of mountain paths for humans interested in gaining altitude. Steeper gets you up more directly, but you go more slowly and use more energy. The best gradient turns out to be about 14 degrees (25 percent). That corresponds reasonably well to the gradients of real paths in the Italian Alps—where a hill was less steep than 14 degrees, paths took the most direct route; where a hill's slope exceeded that incline, paths deviated with switchbacks and such. In theory at least, one could tell the size of some kind of animal from the maximum slopes of its habitually used trails, although in practice the field data might be too untidy for confident analysis.

Carrying Loads

Most animals most of the time transport only themselves. A few flyers carry prey, a few terrestrial creatures carry young. We've persuaded a small number of domesticated animals to carry or pull, things none would do in the wild—nature knows precious few beasts of burden. Among mammals, we may be the most habitual pack animals of all. Few, if any, primitive cultures don't carry substantial loads for substantial distances, and our present affluent youngsters voluntarily carry worrisome loads. Indeed, the cost of carriage may be the oldest serious biomechanical concern. Nothing focuses the attention of governments more than military imperatives, and an interrelationship between load per soldier, speed of movement, and condition upon arrival has been recognized as long as armies have existed (Lothian 1921–22).

An extensive literature, mainly with military objectives, tells us that the cost of carrying a load by a human male differs in no significant way from the cost of self-transport. Shoulder a load equal to 20 percent of your weight, and you'll consume about 20 percent more oxygen in transport—there's no free ride. Intuitively reasonable, yes, but not the whole story, perhaps not even the important story. Taylor and his colleagues, set up to work large animals on a treadmill in Nairobi, Kenya, persuaded some local women to be tested with and without loads (Maloiy et al. 1986; Taylor 1986). These women could carry loads of up to 20 percent of their body weights with no increase at all in oxygen consumption. Carrying heavier loads, they did consume more, but only then in proportion to their weights. So they maintained that initial advantage, that 20 percent free riding load. A particular special posture could not be invoked as explanation, since some of the women carried the loads on their heads and others carried them, with the aid of a headstrap, on their backs.

We have no certain explanation of what these Kenyan women were doing that soldiers don't do. Quite likely early experience counts for something, with genetics providing the wiggle room. Thus, horses and dogs, carriers only at our instigation, give the same results as do human soldiers. One wonders if the present fashion among the young for carrying the paraphernalia of affluence in backpacks might translate into more efficiently pack-carrying adults. One needn't regard the trick as verging on magic. Bear in mind the costliness of terrestrial transport by legged animals compared to almost any other way of getting from place to place, and don't forget that we're paying in power to sustain force, which no basic mechanical rule mandates. So adjustment, neural or muscular, from early experience and sustained practice, seems reasonable enough.

Furthermore, horses, dogs, and inexperienced humans don't exhaust the relevant zoology. Kram (1996) tested rhinoceros beetles on a treadmill with weights of up to thirty times their own; he found that they expended only a fifth as much power moving added weights as they did moving their own weights. (Of course, as small creatures, self-transport comes at a fairly high relative price.) Interestingly, rhinoceros beetles do not ordinarily lug things around at all. By contrast, leaf-cutting ants, which carry loads that often far exceed body weight, pay the full price—carrying body and load costs the same (Lighton, Bartholomew, and Feener 1987).

A hermit crab does lug something around in air—the mollusk shell in which it lives. According to Herreid and Full (1986), a crab pays a lot for carrying a shell of a mass similar to its own—at 14 millimeters per second, the cost of transport doubles. But the cost of carrying the shell remains a constant increment on an increasing cost of self-transport, so at 50 millimeters per second (these are not fast creatures), that doubling of mass raises the overall cost by only about 50 percent. Lifting the shell seems to extract its price (a stationary crab doesn't stand erect, apparently), one that moving doesn't increase. The best tactic for such an animal, just what you may see crabs doing at the shore, consists of either sitting or running as fast as possible.

So we're in good company as far as the cost of load-carrying goes, but some animals do better for reasons still not entirely clear. One note ought to be added about the location of the load. Since the price doesn't vary with where we carry it, nothing except comfort matters. But there's one exception to that—carrying anything on your feet extracts a price about five times greater than carrying anything elsewhere. Lightweight footgear pays substantial dividends.

A general problem for locomotion of animals appears in starkest form when we consider getting around on land. An appendage can be flexed and extended, a trunk can be bent and unbent. Neither will give net transport in and of itself. That depends on arranging some kind of cycle in which motion of a body part in one direction propels the animal while the recovery motion doesn't do the reverse. Without proper rotary machinery, contriving propulsive schemes without some nonuseful body motion presents a major problem. Something can translate in a circle, as does a helical, eucaryotic flagellum, but that's not put to use (and one has a hard time envisioning how it might work) for terrestrial locomotion—hoop snakes remain mythological. So, to the problem of starting and stopping appendages, that of a recovery stroke must be added. It's normally done by lifting an appendage off the substratum—as in walking, running, even side-winding. But starting and stopping happen even without such disconnection. Thus, a snake doing normal serpentine locomotion can't push against the substratum with its entire body at every instant. Perhaps we ought not be surprised that the costs of all forms of terrestrial locomotion exceed those of swimming and flying. What seems so easy and obvious—pushing rearward against an immovable substratum—turns out to be awkward and inefficient.

PART FOUR

The Contexts of Biomechanics

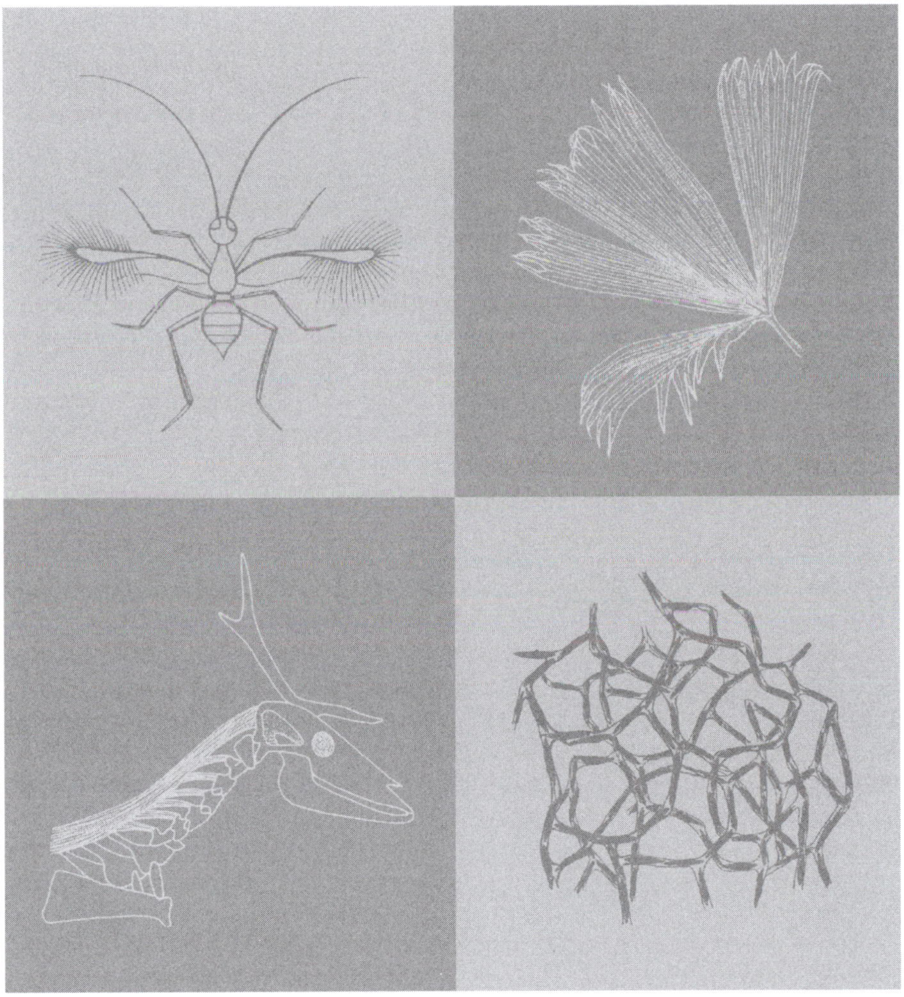

Where in the general scheme of biology does biomechanics lie? What does it draw on, and to what does it contribute? In a still wider context, what can work in biomechanics—and its way of thinking—reveal about worlds beyond academic biology? What can it tell us about how humans have done things? And what can it suggest in the way of biomimetic technology? These are big-ticket items, easy to lose sight of in all the preceding detail. They're also questions that don't lend themselves to answers of even the limited precision of everything we've so far talked about. Furthermore, they're distinctly more

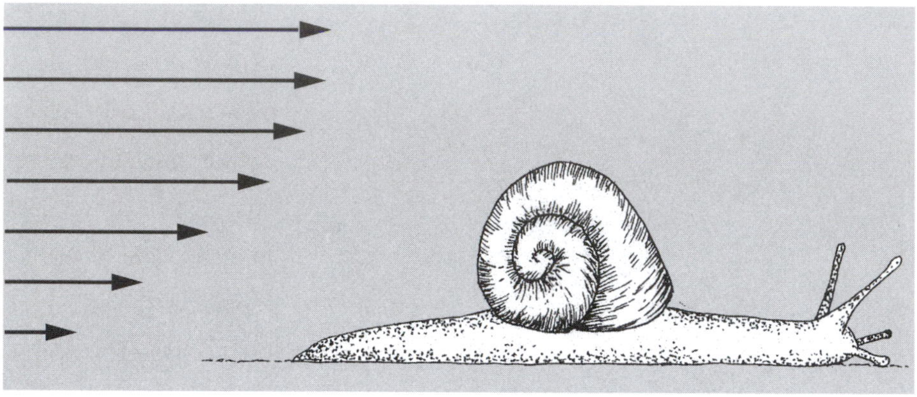

contentious, and I write of them with even less sense of personal authority. What I do believe is that they're issues we ought at least to keep in the back of our minds.

So I end with a lengthy essay that raises more questions than it answers, a look at several of these larger issues with some suggested generalizations. But don't forget that, except in rare and wonderful instances (one thinks of Newton and Darwin), the more general the statement, the further it departs from data and the larger looms its component of personal opinion.

CHAPTER 25

Loose Ends and Perspectives

From here on down, it's uphill all the way.

WALT KELLY (*Pogo*)

EVOLUTION IS A DESIGN PROCESS THAT, AS IN THE FIRST CHAPTER OF THE Book of Genesis, makes something and then looks at it to see if it's any good. We noted back in our own first chapter that "design" had linguistic problems, that it's more tolerable as a noun than as an active-voice verb, that we tacitly assert at one level the paradoxical notion that we have designs that were not designed. I bring up the point here because it will haunt most of what follows.

How Good is Good Enough?

Safety factors in supportive systems

One of the variables in any process of practical (by which I exclude strictly esthetic) design is what we call "safety factor." By this term we usually mean the magnitude of the load that just causes failure of a device divided by the maximum load the device is expected to experience. For instance, if a swing is normally used by no child of above 50 kilograms but accepts 500 kilograms before failure, then the swing has a safety factor of 10. And real factors do run that high—you'll be comforted by Diamond's (1998) figures of 7.6–11.9 for elevator cables. Of course where the extra material itself constitutes a significant load, lower factors are accepted—as low as 2 for a steel bridge or building. The factor must be evaluated for each structural component that bears a load, but the supportive system as a whole shoulders the final cost of failure. Gordon (1978) prefaces a sobering chapter on accidents with the comment that factors of safety are often no more than factors of ignorance and that engineering design amounts to applied theology. More generally (and equally readably), Petroski (1985) has devoted an entire book (and large parts of several others) to the failures of mechanical designs.

And organisms do suffer mechanical failures in their supportive systems. Alexander (1981) and Currey (1984) got the biomechanical ball rolling as they took initial stabs at the matter of safety factors, which turn out to be both more important and more complicated than one might guess. Not only do modern, nonaged humans break bones and pull tendons loose in various athletic activities, but skeletal remains from members of earlier societies show that a large fraction of the individuals suffered and recovered from fractures. The skeletons of wild-caught vultures, baboons, gibbons, and macaques often have healed fractures (Currey 2002). And the data are likely to understate the incidence of breaks if any significant fraction of the breaks had fatal consequences. Zebras and antelopes may break legs when racing ahead of predators. Hard corals and trees suffer breakage from their drag during storms. Even the crushing strength of different foraminiferans—tiny, shelled protozoa—has been shown to correlate with the coarseness of the sediment in which they live (Wetmore 1987), so mechanical failure may have been a selective factor.

How strong should a system or a component be? Whether we consider the technology of humans or of nature, two classes of considerations bear on the question. First, uncertainties—uncertainty as to what the maximal load might be during the lifetime of the system, uncertainty arising from inconsistent quality of the materials, and uncertainty as to the correctness and completeness of the mechanical analysis. Second, costs—the incremental (marginal, if you've taken an economics course) cost of making the system a bit stronger, incremental maintenance costs associated with more material (but perhaps a negative increment from a more lenient maintenance schedule), and the cost associated with failure.

Uncertainties. An internally pressurized system, whether a tank of gas or the cell of a stem, feels a fairly constant load. A bone in the hindleg of a kangaroo or a strut on a short vehicular bridge is subjected to much more variable loads. The forearm of a brachiating (locomoting by successive handholds) gibbon probably experiences still more variable loads. Lowell (1985) interpreted his data on the prying force a crab or bird had to exert to dislodge an intertidal limpet as showing that safety factors increased with increases in environmental unpredictability. Clearly, maximal loads in nature raise complex statistical issues of the kind dealt with by Denny and Gaines (2000). And materials are variable—for critical applications, samples of each batch are tested before use of the rest of each batch. I work in a building to which extra reinforcement was added when it was discovered that samples of concrete from batches already poured were below specifications. The Brooklyn Bridge has miles of subpar cables within its main catenaries—a supplier faked some tensile tests, but the fraud was uncovered in time to make adjustments. But even proper material has an acceptable range of variation. Living materials must also be variable, although, as Currey (1984) pointed out, the evident variability in existing data is more likely a result of difficulties in testing than of real variability in the material. Uncertainty in the analysis, though, applies only to the engineer—nature neither analyzes nor tests but goes right to production and field use!

Costs. Foremost, of course, is the cost of failure. Is it the economic cost of downtime and replacement of a transmission tower after an unusually large storm or do human life and such economic imponderables have to be included? We accept cars that pose perils to existence quite unacceptable in airplanes in which large numbers of people travel at once. What increase in cost of material or fabrication results from a given reduction in the likelihood of failure? An answer involves analysis of materials, design, the relationship of the magnitude of the load to the probability of failure, and the statistics of the uncertainty of the loading regime.

For organisms we at least have an applicable currency in which to measure costs and benefits—it's called "fitness" and it's a measure of an individual's contribution to future generations through the production of reproductively competent progeny. Neither life, liberty, nor the good of spouse or species have relevance except as they contribute to fitness. The trouble is that the currency isn't a convertible one—I think no one has yet linked a nonpathological change of a specific amount in the strength of, say, the Achilles tendon with a change in reproductive success. But fitness does at least provide a very good qualitative guide—we're not really surprised to find (as did Currey) that a short-lived, rapidly reproducing barnacle exhibits shoddier workmanship than a species with the opposite kind of life history. For that matter, we do the same on occasion, economizing in anticipation of a short working life.

And breakage is certainly common where the cost is clearly low. Some corals break readily in storms, but they grow rapidly, and breakage is secondarily an asexual reproductive device. Trees often lose branches in storm, but that tithe of Boreas is only occasionally serious. Also, by reducing drag, loss of a branch must reduce the stress on the trunk and base and thereby lessen the change of a really catastrophic failure. Thus, *local* failure may be one mark of fit design. It's altogether too easy to stay within our own perspective as organisms seriously discommoded by the loss of more than a digit or two. Autotomy, the self-amputation of an injured appendage, is common among creatures from spiders (legs) to lizards (the tail). As an arachnid, a harvestman has eight legs, but one needn't examine many to find a specimen with fewer, still making its way as a splendid all-terrain vehicle. For that matter, the tails of some lizards may function mainly as items with which to propitiate predators.

We do have some estimates of safety factors for some specific structures; table 25.1 gives a few values to serve as the basis for discussion.

In chapter 5, mention was made of the incompressible gas-filled buoyancy chambers of cephalopod mollusks such as cuttlefish and nautilus. Both the constancy and low values of their safety factors for resistance to implosion are remarkable—pressure at a given depth must represent an unusually invariable load.

The stems of garlics and other annuals grow to about half of that height at which they'd buckle and trees to a quarter. Niklas (1992) noted biological rationality in the concomitant difference between the safety factors of the stems of annual plants and of tree trunks—a longer life span demands a higher factor to achieve the same probability of avoiding failure. Not only does the cumulative chance of extreme events rise with time, but, as explained by Denny and Gaines (2000), the longer the time, the more extreme will be the environmental extremes that should be expected. And environment matters—as we'll say more about shortly, chronic mechanical disturbance produces a sturdier organism.

For mammalian bones, Alexander (1981) gives values ranging from about 2 to 6, with slightly lower values for loading in tension than in compression where the same bone normally experiences both. (Human weight lifters push their safety factors down to unity, but their fanaticism must be maladaptive.) Currey (1984) noted that the cost incurred as a result of an increase in skeletal mass will depend on how an animal spends its time. The cost must be especially severe for animals in which the value of the acceleration of the body as a whole is critical to getting food or being eaten. By contrast, for a marine mammal, buoyancy from air in the lungs needs to be offset, so heavier bones may even be a net benefit, not a net cost. For a slow runner, extra mass will be bad but no major factor on level surfaces. For fast runners, though, the work needed to accelerate and decelerate limbs becomes appreciable, and this factor makes extra mass a more serious impediment. But the cost of mass depends on the location of the mass. Alexander (1998) pointed out that frac-

TABLE 25.1 Some Safety Factors for Biological Structures

Structure	Load/Failure Mode	Factor
Cuttlefish, buoyancy chambers	pressure	1.3–1.4
Squid, shell (pen)	bending	1.3–1.4
Spider, dragline	tension	1.5
Reptilian hindlimbs	bending	5.5–10.8
Reptilian hindlimbs	torsion	3.9–5.4
Mammalian bones (general)	bending	2–6
Horse, leg bones in galloping	bending′	4.8
Ostrich, leg bones in running	bending	2.5
Bird wing bones	bending	2.2
Bat (microchiropterans) wing bones	bending	1.4
Bat (megachiropterans) wing bones	bending	3.9
Pigeon (wing) humerus	bending	3.5
Pigeon humerus	torsion	1.9
Pigeon wing feather shaft	bending	6–12
Tendons	tension	2–8
Teeth, human	(tensile) strain	15
Tree trunks	critical buckling	about 4
Stems, annual plants	critical buckling	about 2
Garlic, grown in windy field	critical buckling	1.78
Garlic, grown in greenhouse	critical buckling	1.03

Sources: Data from Alexander 1981; Currey 1984; Swartz 1991; Niklas 1992; Swartz, Bennett, and Carrier 1992; Kirkpatrick 1994; Biewener and Dial 1995; Biewener 1998; Corning and Biewener 1998; Diamond 1998; Blob and Biewener 1999; Niklas, Varne, and Berglund 1999.

tures occur more often in the longer, more distal leg bones than in the more medial ones, and he suggested that lower safety factors characterize bones that must be more severely accelerated. Fliers, of course, ought to be the most mass-sensitive creatures and should accept the lowest safety factors—they have to stay aloft as well as reciprocate their appendages. One smells such matters in the data of table 25.1, but one wouldn't want to bet the farm on currently published numbers.

A variety of estimates can be cited for the safety factor of tendons in vertebrates and of at least one insect apodeme. Some are lower than for bones—under 2.0 even. Low values might be explained in at least three ways. First, these structures are loaded in tension; tensile loads are simpler to withstand than any other loads, and they are less likely to be unpredictably increased by changes in the time course or direction of loading forces. Second, tensile loads are either applied by the animals themselves through muscles of definite strength or they are transmitted through muscles that by their own ability to stretch provide protection against overloads. The third suggestion, another of Alexander's, is the most interesting. Tendon and apodeme take little enough material—tension, after all, is the easy load, so making a bit extra won't cost much at all. But these structures are used to store strain energy, so what matters is the area under the stress-strain curve. Operating such a structure very conservatively amounts to working low on the curve, where area underneath is lower for each unit of extension. Strain energy storage improves if the material is stretched out to where the area underneath is greater for each increment of extension; that means working the material less conservatively and living with a lower safety factor.

But what, then, should we make of tendons with high safety factors? Biewener (1998) suggested that many tendons can't be permitted to stretch much if they're to serve their role of transmitting the length changes of muscle to skeletal elements. Double the cross section of a tendon and you much reduce its stretch for a given load—that you've increased its safety factor becomes an irrelevant epiphenomenon.

All of which views safety factor with the eyes of an engineer or a traditional physiologist. Bear in mind that that view may have little relevance—or be misleading—for systems produced by the evolutionary process. In a more immediate sense than anything that has come up in previous chapters, safety factor implies an anticipatory designer rather than a post hoc adjuster. How can natural selection generate designs (intentionally avoiding "design" as a verb) that cost more than they need to because they're built better than they need be? Put perhaps more sharply, how much reproductive potential should be sacrificed for contingencies that don't occur? Put with even more focus, if safety factor normally accounts for errors in calculation and design data as well as those in guesses about eventual use, how can the notion have meaning for a system that depends solely on past experience for direction?

The issue goes beyond small stuff such as what allowance to make for the differences in functional adequacy—as between, say, a young tree deferring seed production until assured of a place in the sun, and a postreproductive human? It gets down to the very meaning of "safety factor" in organisms. Adapting the engineer's concept gives numbers that "make sense" in an adaptive context; but conceptual justification remains elusive.

Such larger questions aside, data for safety factors may hold special value in biomechanics. Several chapters back, we raised the question of which mechanical properties nature cares about in different situations—we're always looking at design post hoc, if in a different sense than does natural selection. But safety factor may hint at what has mattered in particular cases of evolutionary adaptation (or optimization, if you prefer, which I don't). Thus, Blob and Biewener (1999) argued (from good, in vivo measurements) that the very high safety factors for the hindlimbs of alligators and iguanas in bending are incidental to their more functionally significant torsional loading, and Biewener and Dial (1995) made the same point from data on pigeon wings. Similarly, Corning and Biewener (1998) attribute the extreme values for the strength of feather shafts to a function that demands stiffness (recall our talk about *EI/GJ* in chapter 18). High strength just comes with the territory, something to which J. E. Gordon alerted us in another context decades ago.

Tolerances

A related matter should give less trouble but, unfortunately, remains largely unexplored—although this business of tolerance can be discerned as a background presence in Weibel et al. (1998). In designing things to be produced by mass-production techniques, an important element (as nicely described by Boorstin [1965]) consists of figuring out just how badly the components can be made and then making them no better than they need be. In this sense, the best overall design is the one that can be put together from components that need the least precision in order to go together and work together. The criterion can only rarely be discerned in the final products, in part because we produce many components to unnecessarily close tolerances more or less incidentally. A rigid mold produces standard castings even with respect to aspects needing no such standardization.

What about natural designs—how similar are two peas in a pod? In certain respects they must be more similar than (in the naturally selective sense) they need be. Protein molecules often tolerate substantial amino acid substitutions with no noticeable alteration of function, but the synthetic machinery of an organism's cells turns them out without such permissible sloppiness. That's part of the rationalization for the commonness of selectively neutral mutations. At a level of more immediate concern here, the matter might be put in terms of the cost of deviance—how much compromise of functional adequacy relative to a given amount of variation in structure.

Organisms of a given species living in a given habitat certainly vary among themselves, but they vary for a number of reasons. Some variability may be functionally irrelevant sloppiness, whether genetically or developmentally based—natural selection cares nothing for functionally irrelevant features. Some variability depends on personal experience in a particular habitat. Genetically identical trees won't look the same when grown in even slightly different places. The extreme must be oysters, with each one contingent on local rock shape. (I once heard of an attempt to build an oyster-shucking machine that was frustrated by shear morphological diversity.) But some variability, the interesting part, must depend on that cost of deviance.

My own experience suggests great variation in the cost of deviance. When looking at how much air passes through the antennae of silkworm moths, I measured light passage as an indicator of nonviscous transmissivity (Vogel 1983). Every individual of several species passed close to 44 percent of the incident light. A particularly sharp optimum seemed to obtain for an antenna with (presumably) the same kind of sensilla and overall structure that would maximize air transmissivity at similar wind speeds. Too open to wind and a passing scent molecule will be less likely to be captured; too tight-meshed and a scent molecule will be less likely to pass through in the first place. In our look at twisting and bending flower stems (Etnier and Vogel 2000), we found not just a difference in the EI/GJ ratio between daffodils and tulips but a difference in the standard deviations of the ratios. The daffodils, where we guessed that the variable matters, had a closer tolerance (13.27 ± 0.96) than the tulips (8.33 ± 3.20), where the ratio ought to be of less consequence. Similarly—and in a completely different system—Currey, Jayes, and Alexander (Currey 1984) looked at average mismatches between equivalent bones of the left and right limbs of birds. Mechanical behavior of bones of gulls living in the wild (on a garbage dump) matched much more closely than that of bones of commercial egg-laying hens, where natural selection for bone symmetry could not have been such an immediate matter.

Like safety factor, tolerance means something different in organisms than in our components and products, although here the difference is more subtle. For the most part (but not always or entirely!) our products work similarly because they're made of similar parts, and the more similar the parts, product after product, the more similar the assembled result. Nature doesn't work quite that way. As one looks more closely, the differences often become greater. We don't name or describe the routing of small blood vessels—they differ among individuals, even genetically identical ones. Major portions of early embryos can in many cases be killed without obvious effects on the resulting organism. Development works within a tangle of feedback loops, looking at outcomes and adjusting processes. The old embryologists coined a name for the phenomenon, calling it "equifinality" (Nijhout 1990).

Tuning Structure and Mechanical Properties to the Habitat

The subjects of safety factor and tolerance represent only two of the places where our mechanical analyses interact with such areas as morphology, ecology, and evolution. To take another example, consider the tuning of design to the mechanical details of particular environments. One can view this tuning through interspecific comparisons, through comparisons of individuals of a single species living in different habitats, or by subjecting organisms to varied mechanical regimes during their development. Each of these approaches has been taken, and each conveys the same general message as well as analogous specifics.

Koehl's (1977b) study of a pair of species of sea anemones of different structure and different habitats is now classic; she found (among other things, some already noted) that structural changes compensated for habitat changes with respect to some mechanical properties. As a result, the material of a tall anemone inhabiting sheltered bays (*Metridium*; figs. 17.6 and 18.9) experienced about the same stress as that on a low, wide anemone in the torrential flow of tidal surge channels (*Anthopleura*; fig. 18.7). More striking, in a sense, than the many cases now known of habitat-sensible interspecific variation (after all, a main reason to speciate) are intraspecific variations. Sponge morphology depends to a notorious degree on local habitat, especially on local hydrodynamic conditions, something decried in every work on sponge identification. Nor are sponges unusual. For instance, Armstrong (1987) showed in a species of large cabbagelike algae that not only could plants from a low-current habitat not tolerate transplantation to an exposed locality, but that algae from the latter place found a sheltered habitat inhospitable as well—they shed silt poorly and fared badly in competition with the local conspecifics.

Thigmomorphogenesis

Still more striking are appropriate readjustments of individuals faced with changed mechanical circumstances. And the actual mechanics of adjustment to different habitats still holds surprises. Telewski (Telewski and Jaffe 1986; Telewski 1995) looked at pines from windy and sheltered places; he found that he could produce the same structural differences within a group of genetically homogeneous plants by subjecting some to periodic mechanical perturbation as seedlings. These stressed trees gave higher flexural stiffness values than unstressed trees—one guesses that where winds are consistently present, a photosynthetic machine must maintain a good, sky-exposed orientation up to higher velocities. More interesting yet was his dissection of the changes in flexural stiffness, *EI*, into variation in *E*, Young's modulus, and *I*, the second moment of area—the material factor and the geometric factor. The wind-stressed or perturbed trees proved to have a lower *E* but more than made up for it with a higher *I*. Does this overtly paradoxical result reflect some inescapable correlation of increased material stiffness with brittleness and functional fragility? Put more simply, are inhabitants of a more stressful environments less tolerant of stiff materials? This widespread kind of tuning in plants (recall the data for garlic stems in table 25.1) was given the name "thigmomorphogenesis" by Jaffe (1973), as mentioned in chapter 7. The title of his paper defines it succinctly: "the response of plant growth and development to mechanical stress."

Interestingly, Barbara Best ran into the same phenomenon when she compared sea pens (erect, sessile, suspension-feeding coelenterates mentioned in chapter 19) from bay bottoms having different speeds of tidal flow—higher flow was associated with higher flexural stiffness but lower material stiffness or Young's modulus. But Charters, Neushul, and Barilotti (1969) found increases in both *E* and *I* in an erect marine alga in a more stressful environment, so we can't assert a consistent rule. Moreover, differences can be either genetic, epigenetic, or some combination of the two.

Adaptive remodeling

Adaptive remodeling during the life of an individual goes beyond altering growth trajectories; we might regard it as more remarkable were it not so familiar. We all know that functionally useful changes continue beyond our initial development and growth. Muscles obviously change in positive response to patterns of use. We exercise both to increase our mass of muscle and to adjust its mix of fiber types. Do weight-lifting and similar force-dependent activities and you increase your so-called white fibers, anaerobic force-exerting specialists. Do power-dependent activities such as distance running and you increase your so-called red fibers, aerobic power-producing specialists.

If we run, our bones mineralize. If we persist in using the same arm for a stressful task, its bones enlarge and our bilateral asymmetry increases. Loss of bone mass presents a serious problem for people who spend long periods in spacecraft. (And as I write this, I read in the paper that exercise works far better than calcium supplementation and other medication in preventing osteoporosis—I'm surprised anyone is surprised.) Consider what must be second nature to people who look at human fossils and associated prehistoric artifacts. How might they tell that a bone was well stressed by what an individual did to make a living? The bone will be larger, relative either to ourselves or, perhaps, to its opposite member. How might they tell how a tool was used? The tool will diminish through wear in some revealing pattern. Tools wear down; bones wear, one might say, up.

Adaptive remodeling: trees and the constant-stress model

Remodeling remains remarkable, and it remains a bit mysterious as well. Vertebrate bone, a living, dynamic, cellularized tissue, may do it well; but it doesn't represent a particularly extreme or uniquely sophisticated case. Load a tree and it responds thigmomorphogenetically. But it can do much more than a general alteration of proportions and stiffness. A tree can add wood in an asymmetrical pattern to offset a nonaxial load on a trunk or branch. The phenomenon has a curious phylogenetic correlate. The softwoods (gymnosperms) reinforce with "reaction wood" especially good at resisting compression, adding it on the side subjected to increased compression. The hardwoods (dicotyledonous angiosperms), by contrast, add tensile reaction wood on the side where shortening will restore a tree's original posture, as well as producing compression wood (Niklas 1992). Prevent normal loading, and a tree becomes fragile, a well-known consequence of guying trees—prevention of windthrow with guy ropes renders the tree more susceptible to windthrow thereafter (see, for instance, Jacobs [1954] or Osler, West, and Downes [1996]).

Both the trunk and the branches of most trees, for those of you who escaped a classical course in botany, grow continuously in girth. A layer of dividing cells, the cambium, lies beneath the bark (phloem) and wood (xylem), and contributes progeny to both. (Palm trees, monocotyledonous angiosperms, lack this cambial layer and grow only in height.) Variation in the size or character of these new cells over the course of a growing season produces the familiar concentric rings of mature wood. In order to make reaction wood, that cambial layer must somehow detect a need, that is, it has to include a sensory system. The requirement may be certain; how the system works remains much less so. In addition, it has to know what's normal—it requires a set point that, when restored and detected, will trigger cessation of reactive activity. (We're talking, of course, about a conventional negative feedback servosystem.)

Controversy surrounds the nature of that set point. Back in 1893, Metzger suggested that trees reconstruct themselves so that all points on the periphery of trunk and branches (in other words, the cambial cells) see the same stress. Neat, tidy, and morphogenetically irresistible. Assuming stress can be sensed, the cells just have to respond to abnormally high stress by producing more wood. Whether they respond to extra compression or extra tension matters little to the restorative character of the system. Nor does a cambial cell need to know anything about its location. If the entire tree is built of wood with very nearly the same mechanical properties, then the tolerable stress ought to be about the same anywhere. Besides, bending has long been known to stimulate cambial activity.

What stresses matter? Metzger argued that trees responded more to the stresses of wind-induced drag than to those of weight, that they scaled in a way appropriate for cantilever beams rather than axially crushed columns. One might take issue with the rationale, using what's been said in chapter 18 about the dangers of axial loads on columns. But the basic point that worse trouble comes from wind than from gravity still looks good—except for middle-sized pines during severe icing, I've never seen healthy trees fail as axially loaded columns.

What about restoration of constant stress as the goal of remodeling? At least superficially, it seems to hold (Niklas 1992). When examined closely, though, at least two kinds of problems emerge. First, to what wind does the set point correspond? One can argue that setting it just below what would cause damage won't work, since trees won't often enough experience sufficiently high winds to trigger growth and therefore won't be set for a real blow. Remodeling to substantially lower winds might give inappropriate results. And what about seasonal effects, where loading varies depending on whether the tree—and those around it—have leaves? And what about interactions of unsteady winds and elasticity?

The constant-stress hypothesis should be testable, which raises the second kind of problem. From extensive and elaborate calculations, Morgan and Cannell (1994) concluded that trees go ("design themselves?") for constant stress under average conditions—that is, modest winds. By contrast, under extreme conditions, stress varies from place to place, implying a degree of maladaptiveness that one only reluctantly attributes to long-lived and widely distributed organisms. Still, I wonder about the basis of their conclusion, since their calculations assumed that drag (mostly leaves) varied with the square of wind speed; my own results (Vogel 1984, 1989) suggest that a first-power dependence is more likely. Niklas and Spatz (2000) found against the defendant, obtaining from their calculations order-of-magnitude variation in stress and safety factor along the lengths of the trunks of cherry trees. Nonetheless, as they pointed out, ontogenetic shape change does certainly reduce wind-induced stresses. Their calculations, though, apply to leafless trees. That may facilitate estimating how drag changes with speed but might limit the applicability of the results. They make the attractive suggestion that in severe winds, trees by "design" fail in adaptively useful ways, sacrificing small branches by breaking near junction points and by such self-pruning reduce drag, bending moments, and stresses on larger branches and trunk. And self-pruning shouldn't be limited to the genus *Prunus*, the source of their data.

I'll go out on a limb and predict that in some form, constant stress will survive. But it may be the suboptimal moderate-wind version, with self-pruning rather than total collapse in big storms—as the late Tom McMahon once pointed out, an organism doesn't learn from failure (except, one should add, indirectly by natural selection). I just don't think that organisms are built like the deacon's perfect shay (in the poem by Oliver Wendell Holmes) that lasted exactly a century and a day and thereupon failed in every way at once. One can glue strain gauges onto trees (Laura Miller, personal communication), and these will yield stress data if one knows the Young's modulus of the wood. But that doesn't declare easy any properly definitive experimental study—one needs to follow the strains at a variety of locations on a variety of trees over a long enough period to include some severe weather. I can imagine a paper beginning, "It was a dark and stormy night."

Adaptive remodeling: rebuilding bones

Back to mammalian bone, which behaves in a similar manner, at least as far as the end results. Increase the load on a bone and it lays down more material. Bone, as a living tissue (which wood, once formed, isn't), not only increases in size, strength, and so forth, but it decreases in all the same ways if not routinely loaded—as in the weightlessness of space flight or lengthy bed rest. The general shape of a bone may be determined genetically, but the final character of the bone depends a lot on its mechanical situation. We're not looking at some minor subtleties here. In one experiment (Goodship, Lanyon, and McFie 1979), the ulna, one of the paired bones of the lower segment of the forelimb, was removed from a pig. That about doubled the strain on the remaining bone, the radius. After a few months, the radius altered enough so its strain returned almost to normal. A variety of other experiments (cited by Currey [1984]) have shown that neither the Young's modulus nor the tensile strength of the material of such bones changed in these manipulation. In other words, bone remains bone, with little change in mechanical properties—remarkable, considering the bone-to-bone range of these properties that we noted in chapter 16.

Such adaptive remodeling needn't involve large strains. In normal activities, adult bones strain about 0.002–0.003 (Rubin and Lanyon 1982), and strains only a little higher stimulate positive remodeling. Which they ought to, since strains under 0.005 can produce detectable damage. Routine strains that don't exceed 0.001, by contrast, lead to loss of bone (Lanyon 1990). Analogous to the constant stress hypothesis in trees is what's been called the theory of dynamic strain similarity, originally proposed by Rubin and Lanyon (1982) but having its origins (again like the business about trees) back in the nineteenth century as "Wolff's law of bone transformation." In a roughly Hookean world, of course, strain correlates with stress, so we're mainly looking at the irrationally independent histories of botany and zoology. Lieberman and Crompton (1998) give a good introduction to the basic process and issues on the bone side of the fence.

The sensory basis for remodeling remains murky, most likely because it's complicated. Straining bones produces electrical signals, which seem to be involved. And there's a complex interplay between genetic factors, hormonal status, and loading, as well as the temporal pattern of loading. Loading regimes don't quite rationalize normal bone shape, suggesting, perhaps, the kind of secondary considerations that might underlie self-pruning failure in trees. For some bony structures, such as skulls, minimizing strain may not be the relevant criterion of good function. "Failure" lumps disabilities that span too wide a range of degree and tolerability. Furthermore, bones don't just get fatter. We've already noted the reorganization of the trabeculae of long bones (see fig. 19.5) following an alteration in the direction of loading.

The wood of trees, the sarcomeres of muscle, the complex composite that makes up bone don't exhaust the cases of mechanically adaptive growth and remodeling. Both cartilage and tendon show mechanically adaptive responses to deformation of their cells. Our blood vessels do likewise—excessive shear forces on the endothelial cells from abnormally fast flow stimulate endothelial division followed by expansion of tissues further outward from the long axes of vessels. (The particular significance of this last phenomenon came up in chapter 10.) Perhaps the default assumption at this point ought to be that any cellular structure serving a mechanical role incorporates provision for adjusting its organization in response to changes in stress and strain.

The engineering analog

Consider the business of adjusting structure in response to stress change with a goal of achieving a constant and predetermined maximum functional stress. That sums up what all these biological systems are doing. It's also something that can be done very nicely with finite element modeling on computers. Claus Mattheck (1991, 1998), a German engineer, has been doing just that, using trees as his initial and paradigmatic material. Using a constant-stress model and a variety of other design parameters, his system generates nicely proportioned and lifelike trees and their components. Bone simulations do well enough (if not giving quite as tidy results as trees) to give the approach real promise for analyzing prostheses.

But the lovely designs that emerge need at least one cautionary note. Human design ordinarily assumes material of uniform mechanical properties, material, therefore, for which constant stress means least material. Concomitantly, the modeling scheme tacitly assumes that structural behavior can be altered only by changing the amount and location of material. But organisms are made of heterogeneous material, which they alter as much as they do geometry. One large alga, the giant kelp *Nereocystis*, adjusts both its shape (and thus drag) and its material properties to keep constant the ratio of breaking stress to normal habitat stress, whatever the habitat or stage of growth (Johnson and Koehl 1994).

The main difficulty comes not from overall alteration in material properties but from place-to-place differences—spatial anisotropy. Mattheck's (1998) tiger claw, based on an assumption of constant stress, turns out to be unrealistically blunt. Stresses on a real claw thus increase from base to tip, and, given their particular shapes, the same most likely holds for lots of other claws, fangs, and even (in invertebrates) jaws. While the constant stress model may generate very biological-looking structures, it may also mislead us. Consider the metal content of the jaws of nereid and glycerid polychaete jaws, mentioned in chapter 16. Increasing the proportion of metallic compounds out toward their tips almost certainly generates increasing stiffness (and perhaps strength as well). So a material-minimizing design won't be one that assumes constant stress—stress must be made (or at least allowed) to increase tipward. And that permits a more realistically sharp jaw. Nature plays with her elastic moduli, E and G, within individual components; she's not limited to adjusting the distribution of material, thus to playing with the geometric factors I and J.

Adaptation and Convergence

As my colleague Steve Wainwright once put it, in nature, the good designs eat the bad designs. The culling edge of natural selection is functional superiority, whatever other factors might complicate the picture. But equivocation, even embarrassment, commonly follows any statement that structures exist to serve some function. On the one hand that declaration reverberates with the heresy of purposiveness, with the "argument from design" long used (and again quite recently) by people arguing against evolution by natural selection. On the other hand, it embodies the central notion of natural selection itself—that functional advantage (measured in the currency of fitness) drives evolutionary change. The point was made in chapter 1; we're now in a position to get some more mileage out of it.

This notion of functional advantage can provide an analytic tool for studies of function, a tool that's unique to biology and especially potent for the kind of organism-level questions that come up in biomechanics. The basic argument goes as follows.

Reproductive fitness is an organism-level variable, so natural selection acts more directly on organisms than on their microscopic and molecular constituents. Thus, the assumption of efficient design, whatever its dangers, is more useful in studies of function close to the organismic than to any other level of organization. How might we make best use of that assumption without falling into the traps it sets for us? The worst trap is one's all-too-easy assumption that the function one cares about happens also to be nature's sole concern as well. Drag minimization may not matter for a limpet on a rock (Denny 1989; chapter 8), and a high safety factor in a tendon may have nothing to do with any chance of failure (Biewener 1998, as noted earlier in this chapter). But even the assumption of function itself may not be correct—you may be looking at some irrelevant accident. That the autorotation and consequent slowing of an elm tree's samara's descent has a fitness-related role seems reasonable. That the autorotation and consequent slowing of the descent of a willow oak's leaf in the fall matters to its fitness seems, at best, improbable.

How, then, can we separate what's significant from what's incidental? Here's where natural selection can come to (at least in part) the rescue. The key phenomenon is *convergence*—the appearance in different lineages of common structural arrangements where the nearest common ancestor to the lineages lacked that arrangement. (Evolutionary biologists commonly distinguish convergence from a less radical phenomenon that they call "parallel evolution"; they often lump the two under the heading "homoplasy.") Convergence, then, is about, not the shared, *derived* characters upon which we base our phylogenies, but shared *nonderived* characters. For phylogenetic analyses, the independent evolution of similar arrangements is a nuisance, a corruption to be circumvented. "What matters, converges" is a curse afflicting the phylogeneticist, who traditionally looks at multiple characters and minimally functional characters to minimize exposure to the problem.

Conversely, convergence should be accorded especial attention in investigations of function—traits that have converged have thereby shown themselves to be functionally significant. "What converges, matters" is evolution's gift to functional studies, a gift of greatest value for organismal-level work. Furthermore, the common aspects of habit and habitat of the organisms that have converged direct attention to the nature of that functional significance. Of course, the explicit use of convergence as an analytic tool implies acceptance of a strong form of adaptationism, at which accusations of the heresy of purposiveness are leveled. The best course, perhaps, is to accept the adaptationism and dismiss as irrelevant any implications of teleology, deliberate design, or purpose. The bottom line is that if adaptation indicates functionality (even if not necessarily vice versa), and convergence indicates adaptation, then convergence provides a finger that points to function. To investigate function is to ask the tactical "how" and not the strategic "why," so one is free to proceed, just observing some terminological amenities.

Why might convergence be most common at the organismic level? Mainly because that's the level closest to the operation of natural selection. It does occur at other levels of organization, as noted (with lots of examples) by Moore and Willmer (1997). While convergence (or something similar to it—the significance of similarities among members of different lineages comes in for some argument) occurs in biochemical pathways and other characteristics, the vast preponderance of common subcellular features of different organisms certainly represent common ancestry rather than convergence. Sequences of nucleotide bases, of course, don't converge at all; if nothing else, the degenerate nature of the genetic code makes the possibility overwhelmingly unlikely.

Nor is the relative commonness of convergence the only difference in the consequences of natural selection at different levels of organization. The composition of DNA changes relatively regularly—hence the use of "molecular clocks" in estimating the antiquity of lineage divergences. Thus, systematics ought to be done using characters most distant from the organismic level, as is now being done with noteworthy effectiveness using nucleotide sequences. Even amino acid sequences converge sometimes, due to that same degenerate character of the genetic code, and amino acid-based phylogenies can conflict with those derived from nucleotides (Simmons, Ochoterena, and Freudenstein 2002). Molecular techniques now make it unnecessary to seek out those near-oxymorons, functionless anatomical features.

By contrast, at the organismic level evolution is at least temporally irregular and often thoroughly episodic—between convergence and punctuational events, natural selection busily obscures its own history. Even the integrity of a species, it now appears (see, for instance, Lande [1980]) is often maintained by natural selection rather than gene flow. Thus, isolated populations may not diverge phenotypically if they face similar environments—despite continued genotypic divergence.

Convergence appears ever more ubiquitous. Image-forming eyes have evolved between forty and sixty times, the tracheal systems of arthropods have appeared four times, and massively complex instances of convergence characterize inhabitants of caves and commonly patterned mimics (Moore and Willmer 1997). We have more to work with than an older biologist might have anticipated. The phylogenies I learned and loved have turned out to be extensively corrupted by convergence. Such things as coeloms and segmentation with which the animal kingdom was divided turn out to be polyphyletically convergent. Grebes and loons, to me visually indistinguishable, prove only distantly related, but flamingoes, which look like neither, are close kin to the grebes (Van Tuinen et al. 2001). Upsetting? No. We now have license to look for functional hints in the common features of different coeloms, perhaps in connection with hydroskeletons, we're better positioned to ask about locomotor imperatives that might push segmentation, we can examine grebes and loons with greater confidence for modifications associated with planing.

In the present book, I've repeatedly invoked convergence to see underlying rules. The occurrence of regular patterns transcending lineage formed the basis for arguing that twistiness-to-bendiness ratios were functionally significant. We made even more of the way most hydrostatic systems used crossed-helical fiber winding. Murray's law for the diameter ratios as circulatory pipes branched wouldn't even have been mentioned did it not occur in widely disparate lineages. Even the frequent use of scaling rules tacitly built on convergence—whatever the lineage, flyers had a cost of transport on the flying line, swimmers on the swimming line. One smells functional significance rather than fidelity to lineage in such commonality.

What's critical for use of convergence as a tool for detecting adaptation and function is reliable phylogenetic information. Similarity of structure among organisms isn't sufficient; equally critical is the assurance that their most recent common ancestors lacked that similarity. Again, shared derived characters may form the basis of phylogenetic systematics, but shared nonderived, and thus convergent, characters are the basis for inferring function from structural similarity. Fortunately, the reliability of phylogenetic information has taken a dramatic upward leap in recent years.

One complicating matter shouldn't be passed over, though. "Functional significance" covers a lot of sins, and convergence can be driven by a number of different agencies. Perhaps it's handiest to divide biological driving factors from nonbiological ones. Among biological factors we find all those behavioral or sensory factors underlying mimicry, pollinator attraction, and so forth (not that physics doesn't play some role, of course). We also run into internal biological factors, genetic and developmental. Thus, the highest levels of sociality have evolved repeatedly in the wasps and ants, with their peculiar haplodiploid genetics, and only once in any other insect group (the termites). These biologically driven convergences (or parallels) are of less present interest than those driven by physical factors—all those physical and geometrical constraints on what organisms can get away with.

What Nature Doesn't

Biomechanics has an applied side, biomimetics. The two areas relate in a way analogous to English and journalism or biochemistry and chemotherapy. Not that it's the only academic field that contributes to biomimetics, but biomimetic yield has become an increasing impetus (and funding rationalization) for what we do. I have no basic problem with using the results of science here or elsewhere, although I do worry about the pernicious influence of the greater funding always available for applied as opposed to pure science. A more specific matter, though, needs to be ventilated (in several senses of the word, including that of crime novels).

Biomimetics has built, quite unnecessarily, in my opinion, on some counterfactual mythology (= bull). From Aristotle, 2,500 years ago, to the present, statements about the absolute superiority of nature's methods and devices abound—I have a decent-sized file of them. Nature, we're told again and again, defines the right way to go. She does the most with the least. Everything we come up with has already been invented by nature—meaning, in context, the evolutionary process. Artists and antitechnologists disdainfully disparage any claim of true novelty in the artifacts of human industry. But such knee-jerk naturalism simply doesn't withstand scrutiny. Indeed, biologists, who might just know something about nature's technological achievements, rarely proclaim her preeminence. Quite simply, human ingenuity has come up with many things that nature either never does or does only crudely and sporadically.

Elementary biology courses once asserted that nature had never invented the wheel. I've forgotten what moral followed that particular statement; in any case, it's a fine example of the trap set by inductive reasoning—the statement can't really be proved but only disproved. In fact, as we saw in chapter 22, it's false. Bacterial flagella (fig. 22.6) are not only smaller but do not wave and wiggle as do the cilia and flagella of other organisms. They truly rotate and are driven, as mentioned earlier, by a rotary engine just inside the cell wall.

The fact remains, though, that wheels remain uncommon in living systems; if we take a purist's approach and look only for arrangements in which a wheel and axle rotates with respect to the rest of an organism, the bacterial flagellum remains the only clear case. Plenty of structures *translate* in a circular path—most of a large group of tiny creatures, the rotifers or "wheel animalcules", come equipped with a pair of ciliary organs that wave around in circles; the filtering fans of some aquatic fly larvae (fig. 6.12a) are similar but larger. Neither, though, truly rotates—as far as we know, nature makes no macroscopic wheel.

Some years ago, Gould (1981) devoted one of his engaging essays to the subject of the scarcity of wheels, arguing the difficulty of evolving them de novo—moving through nonfunctional intermediates is not nature's style, and it's hard to see the beneficence of a partially evolved wheel. He also noted the difficulties of arranging nutrient supplies, axle seals, and so forth—matters of little concern on the scale of bacteria. He asserted that "wheels are not flawed as modes of transport; I'm sure that many animals would be far better with them," and that "animals are debarred from building them by structural constraints inherited as an evolutionary legacy." The goodness of wheels is obvious—modern technology without wheels is unimaginable.* Add a bicycle to a person and you reduce threefold the work needed to go a given distance—and that despite the extra dozen or more kilograms of machinery (chapter 24). A predator on a bicycle should have a formidable advantage!

Or is there something wrong with this most magnificent device? My fractious friend LaBarbera (1983) wrote an article as a rejoinder to Gould, arguing that "the concept of the general superiority of wheels as a mode of transport is false, and the limitation which constrains the evolution of such systems . . . lies in the limited utility of rotating systems in most natural environments." LaBarbera argued that while wheels might be wonderful on flat, hard terrain, they did badly elsewhere, and that appropriate surfaces for effective use of wheels are few and far between—the smaller the organism, the bumpier its world. Ascending a curb takes a lot of power, and a curb of height greater than the radius of a wheel presents an absolute impediment, whatever the power of the engine. The covered wagons of the North American prairies needed enormous wheels even for use in flat, dry country, and all wheeled cultures are road-building cultures. Wheeled vehicles can't turn as sharply as legged ones unless they're extremely complex—societies that didn't use wheels could get away with narrow, tortuous streets. And, given such streets, the introduction of wheeled vehicles would hold little appeal. Furthermore, on even slightly spongy terrain wheels lose any power advantage. Propelling wheelchairs on soft carpet takes great effort—to be traversed by the handicapped, carpets should have short, dense, unpadded pile.

Gould's point is not disproved—in a sense it's not subject to strict disproof. But his point is less self-evident than it has usually been regarded—the dominant constraint on the use of wheels in nature is as likely to be physical as it is to be due to some flaw in the adaptive ability of the evolutionary process. For that matter, Gould and LaBarbera argue only about why nature doesn't turn to wheels as final locomotory drivers. But wheels don't just roll our vehicles around. Consider the marvelously versatile appurtenances of our rotary technology—gears, chains, belts, pulleys, cams, and crankshafts, to mention just a few. Nature makes none of these, nor any roller bearings, waterwheels, windmills, or turbines either. In none of these cases do their arguments apply. Do these omissions trace to difficulties of synthesis, sustentation with nutrients, nervous control, or repair; or do we just glimpse evolution's difficulty with adaptive gaps? Or do we glimpse what some paleontologists call "the privilege of incumbency" and economic historians (less pretentiously) call "lock in"—where an inferior design preempts the functional niche and blocks proliferation of a superior, or potentially superior, one?

---

*How well the Amerindians did without them depends on one's point of view. They apparently knew about wheels and had them on toys (Ekholm 1946) but never used them in transport.

Wheels, though, form only the tip of an iceberg of large proportions. What else should we put on a list of devices almost or entirely unknown in nature? And which of such other omissions can we attribute to physical maladaptiveness as opposed to evolutionary oversight?

Other instances. We considered metallic materials in chapters 15 and 16. Each of us, and perhaps every living creature, can synthesize compounds containing metals—the iron-containing hemoglobin of our blood, for instance. Yet, ironically, no organism is known to use a piece of metal for any mechanical purpose. As noted earlier, some worms and mollusks harden the cutting edges of their teeth with metal salts, in much the same way that we use carbide-tipped saw blades, but these aren't true metallic materials. Nature even detects magnetic fields with metal oxides, not metals—here and there from bacteria to bird brains, we find chunks of magnetite, Fe_3O_4.

In place of metals, nature builds all her stiff structures out of composite materials. But the complex composites that serve organisms lack two features that make metals especially handy. If a metal is lightly loaded, it deforms elastically, springing back into shape when the load is removed. If it's loaded heavily, it begins to deform plastically; that is, it retains a new shape when unloaded again. So we can extrude or press standard metal stock into new and useful shapes. With composites, material and structure must most often be made together. Furthermore, metals don't decompose when heated above their melting points, so we can concoct alloys and cast beams, engine blocks, and even tin soldiers. Composites, whether bone, wood, or fiberglass, can't be pressed, stamped, forged, or cast. And recall the graph in figure 16.5 of tensile stiffness versus work of fracture—steels lie well above the line to which biological materials seem constrained.

Consider ropes. In both the Old World and the Americas, humans long ago figured out how to get long, flexible cordage of decent tensile strength out of relatively short fibers. It's a fine trick, twisting ("spinning") an array of fibers laid lengthwise but otherwise placed randomly against one another. The harder you pull on the twisted array, the more the fibers bunch together, and the more their resistance to sliding along each other keeps the rope from parting. We even make rope that doesn't spontaneously untwist itself—by spinning batches of fibers one way and then twisting several of the resulting strands in the opposite direction. Naturally occurring fibers make fine ropes. Until recently all our threads and ropes were made of minimally altered natural material—wool (the protein keratin) from a few animals and cellulose fiber from a wide variety of plants. But even though her vines, spider silks, and tendons bear loads in tension, nature never combines fibers into ropes in quite this way. The collagen of nature's ropiest material, tendon, does run helically—the protein is a triple helix—but its molecules are long, and the strength of tendon doesn't depend on the way strands resist sliding.

Weaving, likewise, probably emerged from multiple episodes of human creativity. Woven cloth for clothing may have been preceded by woven netting or basketry—the principle is the same. Threads of spun or stripped fiber run crosswise, maintaining the integrity of the fabric by weaving over and under each other. Threads alone do the job, with no need for glue at intersections or some binder to keep a feltwork of fiber together. As we've seen (chapter 20) crossed-fiber sheathings abound in nature, covering pressurized cylinders in everything from nonwoody plant stems to worm bodies, shark skin, and the human penis. Some aquatic insect larvae—caddisflies—make catch-nets of right-angled arrays of silk strands, and the orb webs of spiders have four-sided cells. But quite unknown is the in-and-out arrangements of crosswise threads with which we make flexible fabrics that retain their integrity without recourse to stickum. At their intersections, the silk threads of spider webs must stick to each other; and silk making up cocoons is bound into a composite with a proteinaceous glue. Strictly speaking, weaver birds don't weave, nor do spiders spin.

The same story holds for more complicated contrivances. Nature does make an occasional surface ship of the kind we've used since antiquity, a ship that floats by displacing a volume of water equal to its own weight. Ducks, muskrats, and beavers cruise on the water's surface as proper displacement hulls (chapter 14). But these swim intermittently, and none with the speed and grace of a fully submerged salmon, squid, porpoise, or penguin. Even humans, mediocre swimmers at best, go faster underwater. Odder still, even air breathers such as porpoises and penguins avoid the surface, either coming up only to breathe or porpoising up and down across the interface.

More fluid-mechanical motion. Parachutes—simple, lightweight things—let us descend unharmed from any height or let us stop a fast-moving vehicle without depending on friction with a surface. Nature gets nearest to our kind of device in a few odd creatures such as the flying frog with its large, webbed feet and the flying snake that's a bit flattened back-to-belly. And these do rather better than free fall as they descend in the forests of southeast Asia. Yet nature makes the stuff of parachutes—perfectly good strings of both protein (silk and collagen, again) and cellulose (all manner of plant fibers) as well as thin membranes.

Nor—while on lightweight fliers—does nature make lighter-than-air flying machines. Neither flexible blimp nor rigid airship has a natural analog. The omission is odder than one might think. Since craft that are lighter than air need no power to stay aloft, they would seem perfect for spreading wind-borne seeds and certainly better than the long silk threads used by spiderlings and some first-stage caterpillars. Helium may be unavailable, but hydrogen cannot present serious problems of either basic supply or generating mechanism. The process of photosynthesis, done by every green plant, begins by using solar energy to split water into oxygen that enters the atmosphere and hydrogen that will reduce carbon dioxide to carbohydrate. Nature even makes membranous bags, from internal lungs and swimbladders to the beautiful floats of the otherwise unbeautiful Portuguese man-of-war. Nowadays we use few blimps, but nature uses none.

For the past few centuries, we've made fire do much more than keep us warm and aid our chemistry and metallurgy. We make wide use of the way burning releases gases and high temperature makes them expand. Our lives would be vastly different without thermal expansion engines—whether external or internal combustion, piston or turbine. But nature makes no engines that extract energy from temperature differences, even without combustion. Not that nature lacks decent engines, as noted in chapter 22. The best of her engines, muscle, compare well in weight economy to electric motors and automobile engines. But none comes within shouting distance of the efficiency of the steam turbines with which we generate electricity or of the power-to-weight ratios of modern aircraft engines.

Flywheels, in which the persistent spinning of a heavy mass stores kinetic energy, may be the most ancient of short-term batteries. A good potters' wheel needs only an occasional kick to keep it going, and even a primitive one needn't be pushed steadily. Potters' wheels have been used for over 5,000 years—at least in the Old World; in the Americas pottery was made quite differently. We use flywheels to smooth out the action of our engines, whether piston engines or (in turntables and tape recorders) electric motors; and we use them in toys such as tops, yo-yos, and tiny model cars. Occasionally one hears suggestions that automobiles might use flywheels together with smaller internal combustion engines that run at more efficient high and steady speeds—the flywheels would intermittently supply energy for acceleration and hill climbing.

Once again, we look in vain for natural examples. Nature stores energy gravitationally—we act as pendulums when we walk, saving for the next step some of the energy we might otherwise lose when a leg decelerates. She also stores energy elastically—we stretch tendons when we run, using a different battery for the same short-term energy storage. Human technology uses both gravitational and elastic storage—counterweights and springs—but to these we add flywheels, unknown in nature.

So why doesn't nature use all these fine devices—and a lot of others not specifically mentioned here? No single, all-encompassing explanation proves persuasive.

Some of the gaps—or differences, to be fair—reflect size. What's sauce for a large goose may not be worth a gander for a small one. Surface swimming must be just such a case, as argued in chapter 14. A hull that displaces water makes waves that travel with it, and the faster the travel, the longer their wavelength. A boat shorter than the wavelength set by its speed has to work extra hard to climb up or cut through the wave at its bow. So for economical travel, smaller means slower. For a duck, the speed limit for efficient surface swimming is well less than a meter per second (2 miles per hour). Above that speed the pond slopes uphill in every direction. A 100-foot-long surface ship can do 7 meters per second (16 miles per hour) before it runs afoul of the same trouble. A microblimp may have too high a surface-to-volume ratio—an adequately light membrane to contain its hydrogen might need to be impossibly thin. Conversely, nature's small aircraft have little to gain by adding fixed wings to their propeller-like beating wings since the airflow behind beating wings is already quite slow.

Some gaps turn on methods of manufacture and on cultural needs. Perhaps ropemaking proved necessary for a technology that only had access to short fibers and that valued stored cordage of indefinite length. Spider and silkworm moth extrude continuous multistranded thread, so individual fibers are as long as necessary and the problem doesn't arise. We build things from standardized and stockpiled components—bricks, boards, screws, hinges, and so forth. Nature, by contrast, grows things for direct use from molecules. Metal sheets and beams, rolls of cordage, and bolts of fabric pour in and out of our factories; none have the same intrinsic utility for nature's synthetic methods.

Some gaps must be consequences of others. Electric motors depend on wires, and wires require materials of high conductivity. To get the same resistance per unit length, a pipe of a highly conductive salt solution—one that might be contained within organisms—would need almost ten million times the cross section of a copper wire. If materials can't tolerate great extremes of temperature, then heat engines will be impractical. The bottom-line Carnot efficiency of an engine with an input temperature of 1000 degrees (Celsius) and an output of 100 degrees is 71 percent. For one dropping from 40 to 0 degrees, about the greatest range an organism might manage, the comparable maximum comes to a mere 13 percent.* And flywheels without wheels are hard to imagine.

But many of the differences must emerge from the enormously conservative and constrained character of natural design—compared to which human engineers are unfettered free spirits. In nature's domain, tradition oppresses with a heavy hand. Anything but slight, step-by-step change is almost impossible. And the criterion of utility applies in the harshest fashion—profit (for nature, reproductive success) must be immediate rather than some promissory note to be redeemed by future generations. The absence of metals and the scarcity of wheels might just reflect such limited scope for innovation. Perhaps the lighter-than-air flier doesn't happen because only a well-developed version could reach the threshold of profitable buoyancy. Copying from another lineage? No copying, licensing, adapting—you can only modify what your ancestors bequeathed you. Making a parachute by combining a membranous leaf with threads of silk would require a mighty act of miscegenation.

The general point—nature neither holds nor should be expected to hold any natural superiority, and she provides neither comfort nor example for engineer-bashers. Our technology does all kinds of things with few or no natural analogs, even things we think might be quite useful to other organisms. Even ancient humans repeatedly invented things without benefit of natural models—metal implements, baked ceramics for pots and bricks, clothing of fabric, and so forth. We shouldn't view our suspension bridges as intrinsically less effective than nature's vines, our electric wires than her nerve axons, our turbines than her muscles, the propellers of our ships than the tails of her fish.

But neither should we ignore what organisms have evolved. We've much to learn simply because nature, for whatever her reasons, has done things differently. Even if her superiority is a myth, such a vastly different technology can only be a gold mine of stimulation, even inspiration. So what if her splendid achievements in multicomponent composite materials were mothered by her ignorance of metallurgy? Bone, tooth, wood, and the outer skeletons of insects still help us understand how mechanical behavior derives from chemical composition and supermolecular structure. That birds, bats, and insects could fly demonstrated as no theoretical analysis could have that active flight through the air could be done. Profit and knowledge might just emerge from asking how nature manages her mechanical affairs so well when she lacks so many of our favorite devices. Where she does what we do—tubular construction, lift-producing airfoils, and such—one suspects intrinsic technical superiority. Where she does something different—as in her uses of flexibility and her hydroskeletons—one gets the hint of multiple worlds, even ones as yet explored by neither technology.

In the final analysis, though—which means examining the basis of my own fascination with the subject—biomechanics explains aspects of one's immedi-

*The formula for Carnot efficiency is $(T_{source} - T_{sink})/T_{source}$.

ate world as satisfyingly as any area of contemporary science. Never mind whether it tells us how to build profitable widgets—it talks about a person's everyday views and experiences, telling about why you switch from a walk to a run at a certain speed, how your blood flows with least effort, how trees avoid blowing over. It *explains*—with the least possible recourse to some recondite and counterintuitive revelation.

LIST OF SYMBOLS

A	tailbeat linear excursion
a	acceleration
a	radius (cylinder, aperture, sphere)
a	scaling constant, from $y = ax^b$
BMI	body mass index
b	scaling exponent ("factor"), from $y = ax^b$
c	cost of transport
C_d	drag coefficient
C_{dx}	local drag coefficient
C_l	lift coefficient
C_o	orifice coefficient
C_p	pressure coefficient
C_{stress}	stress concentration factor
CF	condition factor (fish)
CI	corporal index
c	celerity (propagation speed) of a wave
c	concentration (mass per volume)
D	diffusion coefficient
D	drag
DA	distance advantage
d	diameter
d	distance
d	zero-plane displacement (in log boundary layer)
E	energy
E	Young's modulus of elasticity (tensile stiffness)
E'	storage modulus of viscoelastic materials
E''	loss modulus of viscoelastic materials
e	efficiency (%)
F	force
F_E	critical Euler buckling force
F_{in}	input force
F_L	critical local buckling force
F_{out}	output force
FA	force advantage
FI	flatness index (Vogel number, Vo)
Fr	Froude number (v^2/gl)
f	friction factor (pipe flow)
G	shear modulus
g	gravitational acceleration, 9.8 m/sec^2
H	total head (of pressure)
h	height, depth (vertical distance)
I	second moment of area
IS	intrinsic speed, muscle contraction
J	advance ratio
J	polar second moment of area
Je	Jesus number
K	local buckling initiation correction
k	miscellaneous constants
L	length, the dimension
L	lift
L_c	critical crack length
l	length
l'	entrance length (pipe)
l_o	original length
ln	natural logarithm (base e)
M	mass, the dimension
M	first moment of area
M_b	body mass
m	mass
mv	momentum
n	column end-condition adjustment factor
n	frequency
n	number of anything
P	power
P_{in}	power input
$Pé$	Péclet number (lv/D)

p	pressure
p_d	dynamic pressure ($\rho v^2/2$)
p_s	static pressure
Q	total flow (volume/time)
R	resistance (pressure/total flow)
R	universal gas constant, 8.314 J/mole-°Kelvin
R	wing length
Re	Reynolds number ($\rho lv/\mu$)
Re_x	local Reynolds number
r	radius, radial position
S	area, surface
Sh	Sherwood number (l^2/Dt)
St	Strouhal number (nl/v)
T	temperature
T	tension (force/distance)
T	thrust
T	time, the dimension
t	time, period
t	wall thickness
V	volume
Vo	Vogel number (flatness index, *FI*)
v	velocity, speed
v_*	shear velocity (in logarithmic boundary layer)
W	work
We	Weber number ($v^2\rho l/\gamma$)
W_f	work of fracture (work/area)
W_l	wing loading (weight/area)
Wo	Womersley number
W_s	strain energy stored (work/volume)
w	width
x	distance, flow direction
y	deflection of beam under load
z	distance, normal to surface or above ground
z_0	roughness length (in logarithmic boundary layer)
γ	shear strain
γ	surface tension
Δ	difference between two values
δ	boundary layer thickness
δ	phase angle between stress and strain
δ_u	unstirred layer thickness
ε	strain (fractional extension or compression)
η_f	Froude propulsion efficiency
θ	glide angle (off horizontal)
θ	miscellaneous angles
κ	constant (in logarithmic boundary layer)
λ	wavelength
μ	viscosity
ν	Poisson's ratio
π	pi, 3.1416
ρ	density (mass/volume)
σ	compressive or tensile stress or strength
σ_c	compressive strength
σ_h	hoop (circumferential) stress
σ_l	longitudinal stress
σ_m	meridional compressive stress (shell)
τ	relaxation time
τ	shear stress
ϕ	amplitude (of appendage movement)
ϕ	meridian-to-horizontal angle for dome
ω	angular velocity
$\propto$	is proportional to
$\tan \delta$	damping (of viscoelastic materials)

REFERENCES AND INDEX OF CITATIONS

Abdel-Malek, A. K., D. Mukherjee, and A. F. Roche. 1985. A method of constructing an index of obesity. *Hum. Biol.* 57: 415–30. (*70*)

Abé, H., K. Hayashi, and M. Sato, eds. 1996. *Data Book on Mechanical Properties of Living Cells, Tissues, and Organs*. Tokyo: Springer-Verlag. (*316, 319, 321, 322*)

Adair, R. K. 1990. *The Physics of Baseball*. New York: Harper & Row. (*247*)

Ahn, A. N., and R. J. Full. 2002. A motor and a brake: two leg extensor muscles acting at the same joint manage energy differently in a running insect. *J. Exp. Biol.* 205: 379–89. (*472, 473*)

Aigeldinger, T. L., and F. E. Fish. 1995. Hydroplaning by ducklings: overcoming limitations to swimming at the water surface. *J. Exp. Biol.* 198: 1567–74. (*292*)

Alexander, D. E. 1986. Wind tunnel studies of turns by flying dragonflies. *J. Exp. Biol.* 122: 81–98. (*275, 468*)

———. 2002. *Nature's Flyers: Birds, Insects, and the Biomechanics of Flight*. Baltimore, MD: Johns Hopkins University Press. (*253, 274, 278*)

Alexander, D. E., and J. Ghiold. 1980. The functional significance of the lunules in the sand dollar, *Mellita quinquiesperforata*. *Biol. Bull.* 159: 561–70. (*254*)

Alexander, R. McN. 1962. Visco-elastic properties of the mesogloea of the body wall of sea anemones. *J. Exp. Biol.* 39: 373–86. (*361*)

———. 1976. Estimates of speeds of dinosaurs. *Nature* 261: 129–30. (*480*)

———. 1981. Factors of safety in the structure of animals. *Sci. Prog. Oxf.* 67: 109–30. (*497, 499, 500*)

———. 1983. *Animal Mechanics*. Oxford, UK: Blackwell Scientific Publications. (*97, 310, 319, 345, 396, 402, 465*)

———. 1984. Stride length and speed for adults, children, and fossil hominids. *Am. J. Phys. Anthropol.* 63: 23–27. (*6*)

———. 1986. The ideal and the feasible: physical constraints on evolution. *J. Linn. Soc. Lond.* 26: 345–58. (*11*)

———. 1987. Bending of cylindrical animals with helical fibers in their skin or cuticle. *J. Theor. Biol.* 124: 97–110. (*417*)

———. 1988. *Elastic Mechanisms in Animal Movement*. Cambridge, UK: Cambridge University Press. (*5, 328, 417, 481*)

———. 1989. *Dynamics of Dinosaurs and Other Extinct Giants*. New York: Columbia University Press. (*55*)

———. 1992. *Exploring Biomechanics: Animals in Motion*. New York: Scientific American Library. (*488*)

———. 1996. *Optima for Animals*, rev. ed. Princeton, NJ: Princeton University Press. (*480*)

———. 1998. Symmorphosis and safety factors. Pp. 28–35 in E. R. Weibel, C. R.

Taylor, and L. Bolis, eds., *Principles of Animal Design*. Cambridge, UK: Cambridge University Press. (*499*)

———. 1999. *Energy for Animal Life*. Oxford, UK: Oxford University Press. (*484, 485*)

Alexander, R. McN., and A. S. Jayes. 1983. A dynamic similarity hypothesis for the gaits of quadrupedal mammals. *J. Zool. (Lond.)* 201: 135–52. (*5*)

Alexander, R. McN., and A. Vernon. 1975. The mechanics of hopping by kangaroos (Macropodidae). *J. Zool. (Lond.)* 177: 265–303. (*486*)

Allen, N. S. 1974. Endoplasmic filaments generate the motive force for rotational streaming in *Nitella*. *J. Cell Biol.* 63: 270–87. (*447*)

Anderson, E. S., and M. E. DeMont. 2000. The mechanics of locomotion in the squid *Loligo pealei*: locomotory function and unsteady hydrodynamics of the jet and intramantle pressure. *J. Exp. Biol.* 203: 2851–63. (*143*)

Andersen, N. M., and J. T. Polhemus. 1976. Water striders (Hemiptera: Gerridae, Veliidae, etc.). Pp. 187–224 in L. Cheng, ed., *Marine Insects*. New York: American Elsevier. (*108*)

Armstrong, J., W. Armstrong, and P. M. Beckett. 1992. *Phragmites australis*: venturi- and humidity-induced pressure flow enhance rhizome aeration and rhizosphere oxidation. *New Phytol.* 120: 197–207. (*149*)

Armstrong, S. L. 1987. Mechanical properties of the tissues of the brown alga *Hedophyllum sessile* (C. Ag.) Setchell: variability with habitat. *J. Exp. Mar. Biol. Ecol.* 114: 143–51. (*503*)

Arnold, G. P., and D. Weihs. 1978. The hydrodynamics of rheotaxis in the plaice (*Pleuronectes platessa* L.). *J. Exp. Biol.* 75: 147–69. (*254*)

Arya, S. P. 1988. *Introduction to Micrometeorology*. San Diego, CA: Academic Press. (*175*)

Ashby, M. F. 1992. *Materials Selection in Mechanical Design*. Oxford, UK: Butterworth-Heinemann. (*311, 323, 329*)

Au, D., and D. Weihs. 1980. At high speeds dolphins save energy by leaping. *Nature* 284: 548–50. (*294*)

Augspurger, C. K. 1988. Mass allocation, moisture content, and dispersal capacity of wind-dispersed tropical diaspores. *New Phytol. 108*: 357–68. (*248*)

Autumn, K., Y. A. Liang, S. T. Hsieh, W. Zesch, W. P. Chan, T. W. Kenny, R. Fearing, and R. J. Full. 2000. Adhesive force of a single gecko foot-hair. *Nature 405*: 681–85. (*428, 430*)

Autumn, K., M. Sitti, Y. A. Lang, A. M. Peattie, W. R. Hansen, S. Sponberg, T. W. Kenny, R. Fearing, J. N. Israelachvili, and R. J. Full. 2002. Evidence for van der Waals adhesion in gecko setae. *Proc. Natl. Acad. Sci. USA* 99: 12252–56. (*428*)

Azuma, A. 1992. *The Biokinetics of Flying and Swimming*. Tokyo: Springer-Verlag. (*248, 253, 270, 271, 272, 294*)

Azuma, A., and Y. Okuno. 1987. Flight of a samara, *Alsomitra macrocarpa*. *J. Theor. Biol.* 129: 263–74. (*261*)

Azuma, A., and K. Yasuda. 1989. Flight performance of rotary seeds. *J. Theor. Biol.* 138: 23–54. (*261*)

Bagenal, T. B. 1978. *Method for Assessment of Fish Production in Fresh Waters*. London: Blackwell Scientific Publications. (*71*)

Baker, C. J. 1997. Measurement of the natural frequencies of trees. *J. Exp. Bot.* 48: 1125–32. (*134*)

Barry, P. H., and J. M. Diamond. 1984. Effects of unstirred layers on membrane phenomena. *Physiol. Rev.* 64: 763–872. (*184*)

Bascom, W. 1980. *Waves and Beaches*, 2d ed. Garden City, NY: Anchor/Doubleday. (*285*)

Beers, M. H., and R. Berkow, eds. 1999. *The Merck Manual of Diagnosis and Therapy*, 17th ed. Whitehouse Station, NJ: Merck & Co. (*70*)

Bejan, A. 1993. *Heat Transfer*. New York: John Wiley. (*180*)

Bell, E. C., and J. M. Gosline. 1996. Mechanical design of mussel byssus: material yield enhances attachment strength. *J. Exp. Biol.* 199: 1005–17. (*310, 312, 314, 328, 347, 348*)

Bennet-Clark, H. C. 1963. Negative pressures produced in the pharyngeal pump of the blood-sucking bug, *Rhodnius prolixus*. *J. Exp. Biol.* 40: 223–29. (*103*)

———. 1975. The energetics of the jump of the locust *Schistocerca gregaria*. *J. Exp. Biol.* 63: 53–83. (*475, 476*)

———. 1977. Scale effects in jumping animals. Pp. 185–201 in T. J. Pedley, ed., *Scale Effects in Animal Locomotion*. London: Academic Press. (*39, 475, 476*)

Bennet-Clark, H. C., and E.C.A. Lucey. 1967. The jump of the flea: a study of the energetics and a model of the mechanism. *J. Exp. Biol.* 47: 59–76. (*39*)

Berg, H. C. 1974. Dynamic properties of bacterial flagellar motors. *Nature* 249: 77–79. (*449*)

———. 1993. *Random Walks in Biology*, 2d ed. Princeton, NJ: Princeton University Press. (*178, 179, 181, 207, 229, 237*)

Berg, H. C., and R. A. Anderson. 1973. Bacteria swim by rotating their flagellar filaments. *Nature* 245: 380–82. (*449*)

Bertram, J.E.A., A. Ruina, C. E. Cannon, Y.-H. Chang, and M. J. Coleman. 1999. A point-mass model of gibbon locomotion. *J. Exp. Biol.* 202: 2609–17. (*486*)

Best, B. A. 1988. Passive suspension feeding in a sea pen: effects of ambient flow on volume flow rate and filtering efficiency. *Biol. Bull.* 175: 332–42. (*403*)

Biewener, A. A. 1990. Biomechanics of mammalian terrestrial locomotion. *Science* 250: 1097–103. (*483*)

———. 1998. Optimization of musculoskeletal design—does symmorphosis apply? Pp. 70–77 in E. R. Weibel, C. R. Taylor, and L. Bolis, eds., *Principles of Animal Design*. Cambridge, UK: Cambridge University Press. (*500, 501, 509*)

Biewener, A. A., and K. P. Dial. 1995. *In vivo* strain in humerus of pigeons (*Columba livia*) during flight. *J. Morphol.* 225: 61–75. (*500, 501*)

Biggs, W. D. 1965. *The Mechanical Behaviour of Engineering Materials*. Oxford, UK: Pergamon Press. (*355, 356*)

Billington, D. P. 1983. *The Tower and the Bridge*. New York: Basic Books. (*376*)

Birch, J. M., and M. H. Dickinson. 2001. Spanwise flow and the attachment of the leading-edge vortex on insect wings. *Nature* 412: 729–33. (*277*)

Bishop, C. M. 1999. The maximum oxygen consumption and aerobic scope of birds and mammals: getting to the heart of the matter. *Proc. R. Soc. Lond. B* 266: 2275–81. (*62*)

Blackwell, P. G., K. Rennolls, and M. P. Coutts. 1990. A root anchorage model for shallowly rooted Sitka spruce. *Forestry* 63: 73–91. (*432*)

Blickhan, R., and R. J. Full. 1987. Locomotion energetics of the ghost crab. II. Mechanics of the centre of mass during walking and running. *J. Exp. Biol.* 130: 155–74. (*487*)

Blob, R. W., and A. A. Biewener. 1999. *In vivo* locomotor strain in the hindlimb bones of *Alligator mississippiensis* and *Iguana iguana*: implications for the evolution of limb bone safety factor and non-sprawling limb posture. *J. Exp. Biol.* 202: 1023–46. (*500, 501*)

Bodig, J., and B. A. Jayne. 1982. *Mechanics of Wood and Wood Composites*. New York: Van Nostrand Reinhold. (*343, 384*)

Boorstin, D. J. 1965. *The Americans: The National Experience*. New York: Random House. (*502*)

Bose, N., and J. Lien. 1990. Energy absorption from ocean waves: a free ride for cetaceans. *Proc. R. Soc. Lond. B* 240: 591–605. (*294*)

Bowker, G. 2002. *The Possible Role of Electrostatic Forces in Wind Pollination*. Ph.D. dissertation, Duke University, Durham, NC. (*240*)

Boys, C. V. 1902. *Soap Bubbles: Their Colours and the Forces Which Mold Them*. New York: E. and J. B. Young and Co. Reprint New York: Dover Publications, Inc. 1959. (*109*)

Brackenbury, J. 2001. Locomotion through use of the mouth brushes in the larva of *Culex pipiens* (Diptera: Culicidae). *Proc. R. Soc. Lond. B* 268: 101–6. (*143*)

———. 2002. Kinematics and hydrodynamics of an invertebrate undulatory swimmer: the damsel-fly larva. *J. Exp. Biol.* 205: 627–39. (*280*)

Brancazio, P. J. 1984. *Sport Science: Physical Laws and Optimal Performance*. New York: Simon and Schuster. (*247*)

Brennen, C., and H. Winet. 1977. Fluid mechanics of propulsion by cilia and flagella. *Annu. Rev. Fluid Mech.* 9: 339–98. (*237*)

Brett, J. R., and N. R. Glass. 1973. Metabolic rates and critical swimming speeds of sockeye salmon (*Oncorhynchus nerka*) in relation to size and temperature. *J. Fish. Res. Board. Can.* 30: 379–87. (*459*)

Bridgman, P. W. 1961. Dimensional analysis. *Encyclopædia Britannica*, 14th ed. 7: 387–387J. (*66*)

Briggs, L. J. 1950. Limiting negative pressure of water. *J. Appl. Phys.* 21: 721–22. (*101*)

Brown, C. H. 1975. *Structural Materials in Animals*. New York: John Wiley & Sons. (*305, 341*)

Brown, J. H., and G. B. West. 2000. *Scaling in Biology*. New York: Oxford University Press. (*59, 222*)

Bryan, G. W., and P. E. Gibbs. 1980. Metals in nereid polychaetes: the contribution of metals in the jaws to the total body burden. *J. Mar. Biol. Assoc. U.K.* 60: 641–54. (*333*)

Bschorr, O. 1988. Propulsive mechanisms in animal swimming and flying locomotion. *Aeronaut. J.* 92: 84–90. (*143, 282*)

Buller, A.H.R. 1934. *Researches on Fungi*, vol. 6. London: Longmans, Green, Ltd. (*422*)

Burkholder, T. J., and R. L. Lieber. 2001. Sarcomere length operating range of vertebrate muscles during movement. *J. Exp. Biol.* 204: 1529–36. (*451*)

Burton, A. C. 1954. The relation of structure to function of the tissues of the wall of blood vessels. *Physiol. Rev.* 34: 619–42. (*51*)

———. 1972. *Physiology and Biophysics of the Circulation*, 2d ed. Chicago, IL: Year Book Medical Publishers. (*194, 222*)

Bushnell, D. M., and K. J. Moore. 1991. Drag reduction in nature. *Annu. Rev. Fluid Mech.* 23: 65–79. (*163*)

Butler, P. J. 1991 Exercise in birds. *J. Exp. Biol.* 160: 233–62. (*459*)

Cachon, J., and M. Cachon. 1978. *Sticholonche zanclea: a reinterpretation of its phylogenetic position based upon new observations of its ultrastructure. Arch. Protistenkd* 120: 148–68. (*450*)

Calder, W. A. 1984. *Size, Function, and Life History*. Cambridge, MA: Harvard University Press. (*59, 87, 209, 218*)

Canny, M. J. 1993. The transpiration stream in the leaf apoplast: water and solutes. *Philos. Trans. R. Soc. Lond. B* 341: 87–100. (*213*)

Carey, D. A. 1983. Particle resuspension in the boundary layer induced by flow around polychaete tubes. *Can. J. Fish. Aquat. Sci.* 40(Suppl. 1): 301–8. (*135*)

Caro, C. G., T. J. Pedley, R. C. Schroter, and W. A. Seed. 1978. *The Mechanics of the Circulation*. Oxford, UK: Oxford University Press. (*216, 221, 223*)

Caro, T. M. 1986. The functions of stotting: a review of the hypotheses. *Anim. Behav.* 34: 649–62. (*487*)

Carpita, N. C. 1985. Tensile strength of cell walls of living cells. *Plant Physiol.* 79: 485–88. (*53*)

Carrier, D. R. 1984. The energetic paradox of human running and hominid evolution. *Curr. Anthropol.* 25: 483–95. (*483*)

Cavagna, G., P. A. Willems, and N. C. Heglund. 2000. The role of gravity in human walking: pendular energy exchange, external work and optimal speed. *J. Physiol.* 528: 657–68. (*486*)

Chance, M. M., and D. A. Craig. 1986. Hydrodynamics and behaviour of Simuliidae larvae (Diptera). *Can. J. Zool.* 64: 1295–1309. (*135*)

Chang, Y.-H., J.E.A. Bertram, and D. V. Lee. 2000a. External forces and torques generated by the brachiating white-handed gibbon (*Hylobates lar*). *Am. J. Phys. Anthropol.* 113: 201–16. (*486*)

Chang, Y.-H., H.-W.C. Huang, C. M. Hamerski, and R. Kram. 2000b. The independent effects of gravity and inertia on running mechanics. *J. Exp. Biol.* 203: 229–38. (*480*)

Chapman, R. F. 1982. *The Insects: Structure and Function*, 2d ed. Cambridge, MA: Harvard University Press. (*108*)

Charters, A. C., M. Neushul, and C. Barilotti. 1969. Functional morphology of *Eisenia arborea. Proc. Int. Seaweed Symp.* 6: 89–105. (*504*)

Cheer, A.Y.L., and M.A.R. Koehl. 1987. Paddles and rakes: fluid flow through bristled appendages of small organisms. *J. Theor. Biol.* 129: 17–39. (*241*)

Cheng, J.-Y., and M. E. Demont. 1996. Jet-propelled swimming in scallops: swimming mechanics and ontogenetic scaling. *Can. J. Zool.* 74: 1734–48. (*143*)

Christiansen, P. 1999. Scaling of mammalian long bones: small and large mammals compared. *J. Zool. (Lond.)* 247: 333–48. (*61*)

Clark, R. B., and J. B. Cowey. 1958. Factors controlling the change of shape of certain nemertean and turbellarian worms. *J. Exp. Biol.* 35: 731–48. (*413*)

Clements, J. A. 1962. Surface tension in the lungs. *Sci. Am.* 207(6): 120–30. (*53*)

Close, R. I. 1972. Dynamic properties of mammalian skeletal muscle. *Physiol. Rev.* 52: 129–97. (*454*)

Combs, S. A., and T. L. Daniel. 2001. Shape, flapping and flexion: wing design for forward flight. *J. Exp. Biol.* 203: 2073–85. (*281*)

Cone, C. D. 1962. The soaring flight of birds. *Sci. Am.* 206(4): 130–40. (*265*)

Corning, W. R., and A. A. Biewener. 1998. *In vivo* strains in pigeon feather shafts: implications for structural designs. *J. Exp. Biol.* 201: 3057–65. (*500, 501*)

Coutts, M. P. 1983. Root architecture and tree stability. *Plant Soil* 71: 171–88. (*432*)

Cowey, J. B. 1952. The structure and function of the basement membrane muscle system in *Amphiporus lactifloreus* (Nemertea). *Q. J. Microsc. Sci.* 93: 1–15. (*413*)

Craig, C. L. 1987. The significance of spider size to the diversification of spider-web architectures and spider reproductive modes. *Am. Nat.* 129: 47–68. (*312*)

Crenshaw, D. G. 1980. *How Interstitial Animals Respond to Viscous Flows*. Ph.D. dissertation, Duke University, Durham, NC. (*243*)

Crick, F. 1970. Diffusion in embryogenesis. *Nature* 255: 420–22. (*186*)

Crisp, D. J. 1955. The behaviour of barnacle ciprids in relation to water movement over a surface. *J. Exp. Biol.* 32: 569–90. (*171*)

———. 1960. Mobility of barnacles. *Nature* 188: 1208–9. (*427*)

Crook, M. J., and A. R. Ennos. 1996. The anchorage mechanics of deep-rooted larch, *Larix europea* × *L. japonica. J. Exp. Bot.* 47: 1509–17. (*433*)

Crook, M. J., A. R. Ennos, and J. R. Banks. 1997. The function of buttress roots: a comparative study of the anchorage systems of buttressed (*Aglaia* and *Nephelium ramboutan* species) and non-buttressed (*Mallotus wrayi*) tropical trees. *J. Exp. Bot.* 48: 1703–16. (*433*)

Culik, B., and R. P. Wilson. 1991. Energetics of underwater swimming in Adelie penguins *Pygoscelis adeliae. J. Comp. Physiol. B* 161: 285–92. (*143*)

Currey, J. D. 1984. *The Mechanical Adaptations of Bones*. Princeton, NJ: Princeton University Press. (*310, 312, 314, 316, 321, 322, 349, 394, 402, 404, 425, 465, 497, 498, 499, 500, 502, 506*)

———. 1999. The design of hard tissues for their mechanical functions. *J. Exp. Biol.* 202: 3285–94. (*349, 350*)

———. 2002. *Bones: Structure and Mechanics*. Princeton, NJ: Princeton University Press. (*349, 360, 394, 402, 403, 468, 497*)

Cutts, A. 1986. Sarcomere length changes in the wing muscles during the wing beat cycle of two bird species. *J. Zool. (Lond.)* 209: 183–85. (*454*)

Czirók, A., I. M. Jánosi, and J. O. Kessler. 2000. Bioconvective dynamics: dependence on organism behavior. *J. Exp. Biol.* 203: 3345–54. (*138*)

Daniel, T. L. 1984. Unsteady aspects of aquatic locomotion. *Am. Zool.* 24: 121–34. (*295*)

Darwin, C. 1845. *Journal of Researches by Charles Darwin into the Natural History and Geology of the Countries Visited During the Voyage of H.M.S. Beagle Under the Command of Capt. Fitz Roy, R.N.* London: J. Murray. Reprint, 2d ed., New York: The Heritage Press, 1957. (*137*)

Davenport, J., S. A. Munks, and P. J. Oxford. 1984. A comparison of the swimming of marine and freshwater turtles. *Proc. R. Soc. Lond. B* 220: 447–75. (*116*)

Dawson, T. H. 2001. Similitude in the cardiovascular system of mammals. *J. Exp. Biol.* 204: 395–407. (*62, 63*)

Dawson, T. J., and C. R. Taylor. 1973. Energy cost of locomotion in kangaroos. *Nature* 246: 313–14. (*486*)

Deban, S. M., J. C. O'Reilly, and K. C. Nishikawa. 2001. The evolution of the motor control of feeding in amphibians. *Am. Zool.* 41: 1280–98. (*420*)

DeMont, M. E., and J. M. Gosline. 1988a. Mechanics of jet propulsion in the hydromedusan jellyfish, *Polyorchis penicillatus*. I. Mechanical properties of the locomotor structure. *J. Exp. Biol.* 134: 313–32. (*363*)

———. 1988b. Mechanics of jet propulsion in the hydromedusan jellyfish, *Polyorchis penicillatus*. II. Energetics of the jet cycle. *J. Exp. Biol.* 134: 333–45. (*203*)

Den Hartog, J. P. 1949. *Strength of Materials*. New York: McGraw-Hill Book Co. Reprint, New York: Dover Publications, 1977. (*322*)

Denny, M. W. 1976. The physical properties of spider silk and their role in the design of orb-webs. *J. Exp. Biol.* 65: 485–506. (*270*)

———. 1980. Silks—their properties and functions. Pp. 247–72 in J.F.V. Vincent and J. D. Currey, eds., *The Mechanical Properties of Biological Materials: Symposium of the Society for Experimental Biology* 34. Cambridge, UK: Cambridge University Press. (*317, 344*)

———. 1984. Mechanical properties of pedal mucus and their consequences for gastropod structure and performance. *Am. Zool.* 24: 23–36. (*363*)

———. 1987. Life in the maelstrom: the biomechanics of wave-swept rocky shores. *Trends Ecol. Evol.* 2: 61–66. (*295, 312, 316*)

———. 1988. *Biology and the Mechanics of the Wave-Swept Environment*. Princeton, NJ: Princeton University Press. (*285, 295, 312, 347, 373*)

———. 1989. A limpet shell shape that reduces drag: laboratory demonstration of a hydrodynamic mechanism and an exploration of its effectiveness in nature. *Can. J. Zool.* 67: 2098–106. (*171, 509*)

———. 1993. *Air and Water: The Biology and Physics of Life's Media*. Princeton, NJ: Princeton University Press. (*93, 108, 184, 238, 242, 428*)

Denny, M. W., V. Brown, E. Carrington, G. Kraemer, and A. Miller. 1989. Fracture mechanics and the survival of wave-swept macroalgae. *J. Exp. Mar. Biol. Ecol.* 127: 211–28. (*310, 314, 319, 328, 338*)

Denny, M. W., T. L. Daniel, and M.A.R. Koehl. 1985. Mechanical limits to size in wave-swept organisms. *Ecol. Monogr.* 55: 69–102. (*295*)

Denny, M. W., and S. Gaines. 2000. *Chance in Biology: Using Probability to Explore Nature*. Princeton, NJ: Princeton University Press. (*498, 499*)

Denny, M. W., and J. M. Gosline. 1980. The physical properties of the pedal mucus of the terrestrial slug, *Ariolimax columbianus*. *J. Exp. Biol.* 88: 375–93. (*363*)

Dial, K. P., and A. A. Biewener. 1993. Pectoralis muscle force and power output during different modes of flight in pigeons (*Columba livia*). *J. Exp. Biol.* 176: 31–54. (*467*)

Diamond, J. M. 1998. Evolution of biological safety factors: a cost/benefit analysis. Pp. 21–27 in E. R. Weibel, C. R. Taylor, and L. Bolis, eds., *Principles of Animal Design*. Cambridge, UK: Cambridge University Press. (*497, 500*)

Dickinson, M. H., C. T. Farley, R. J. Full, M.A.R. Koehl, R. Kram, and S. Lehman. 2000. How animals move: an integrative view. *Science* 288: 100–106. (*277, 472*)

Dickinson, M. H., F.-O. Lehmann, and S. P. Sane. 1999. Wing rotation and the aerodynamic basis of insect flight. *Science* 284: 1954–60. (*277*)

Dietz, C., and R. Schnetter. 1999. Interaction of two myosins with microfibers causes locomotion in *Labyrinthula* sp. *Protoplasma* 206: 97–104. (*450*)

Dimery, N. J., R. McN. Alexander, and K. A. Deyst. 1985. Mechanics of the ligamentum nuchae of some artiodactyls. *J. Zool. (Lond.) A* 206: 341–51. (*325, 326*)

Dimery, N. J., R. McN. Alexander, and R. F. Ker. 1986. Elastic extensions of leg tendons in the locomotion of horses (*Equus caballus*). *J. Zool. (Lond.) A* 210: 415–25. (*310, 473*)

Dixon, A.F.G., P. C. Croghan, and R. P. Gowing. 1990. The mechanism by which aphids adhere to smooth surfaces. *J. Exp. Biol.* 152: 243–53. (*428*)

Dixon, H. H., and J. Joly. 1894. On the ascent of sap. *Ann. Bot.* 8: 468–70. (*102*)

Dodds, P., D. H. Rothman, and J. S. Weitz. 2001. Re-examination of the "3/4-law" of metabolism. *J. Theor. Biol.* 209: 9–27. (*62*)

Dogterom, M., and B. Yurke. 1997. Measurement of the force-velocity relation for growing microtubules. *Science* 278: 856–60. (*447*)

Drake, V. A., and A. G. Gatehouse. 1995. *Insect Migration: Tracking Resources Through Space and Time*. Cambridge, UK: Cambridge University Press. (*137*)

Drucker, E. G., and G. V. Lauder. 2002. Experimental hydrodynamics of fish locomotion: functional insights from wake visualization. *Integr. Comp. Biol.* 42: 243–57. (*141*)

DuBois, A. B., G. A. Cavagna, and R. Fox. 1974. Pressure distribution on the body surface of swimming fish. *Biol. Bull.* 60: 581–91. (*151*)

Dudley, R. 2000. *The Biomechanics of Insect Flight: Form, Function, Evolution*. Princeton, NJ: Princeton University Press. (*253, 467*)

Dudley, R., and P. DeVries. 1990. Tropical rain forest structure and the geographical distribution of gliding vertebrates. *Biotropica* 22: 432–34. (*262*)

Dudley, R., and C. P. Ellington. 1990. Mechanics of forward flight in bumblebees. II. Quasi-steady lift and power requirements. *J. Exp. Biol.* 148: 53–88. (*252*)

Dusenbery, D. B. 1996. *Life at Small Scale: The Behavior of Microbes*. New York: Scientific American Library. (*229, 234, 237, 449, 450*)

Eckman, J. E., A.R.M. Nowell, and P. A. Jumars. 1981. Sediment destabilization by animal tubes. *J. Mar. Res.* 39: 361–74. (*135*)

Edelin, C., and C. Atger. 1994. Stem and root tree architecture: questions for plant biomechanics. *Biomimetics* 2: 253–66. (*433*)

Ekholm, G. F. 1946. Wheeled toys in Mexico. *Am. Antiq.* 11: 222–28. (*512*)

Ellem, G. K., J. E. Furst, and K. D. Zimmerman. 2002. Shell clamping behavior in the limpet, *Cellara tramoserica*. *J. Exp. Biol.* 205: 539–47. (*429*)

Ellerby, D. J., and A. R. Ennos. 1998. Resistances to fluid flow of model xylem vessels with simple and scalariform perforation plates. *J. Exp. Bot.* 49: 979–85. (*197*)

Ellers, O., and M. Telford. 1992. Causes and consequences of fluctuating coelomic pressure in sea urchins. *Biol. Bull.* 182: 424–34. (*388*)

Ellington, C. P. 1984. The aerodynamics of hovering insect flight. *Philos. Trans. R. Soc. Lond. B* 305: 1–181. (*270*)

Ellington, C. P., K. E. Machin, and T. M. Casey. 1990. Oxygen consumption of bumblebees in forward flight. *Nature* 347: 472–73. (*459*)

Elliott, C.J.H. 1981. The expansion of *Schistocerca gregaria* at the imaginal ecdysis: the mechanical properties of the cuticle and the internal pressure. *J. Insect Physiol.* 27: 695–704. (*421*)

Ellis, C. H. 1944. The mechanism of extension of the legs of spiders. *Biol. Bull.* 86: 41–50. (*421*)

Emerson, S. B. 1978. Allometry and jumping in frogs: helping the twain to meet. *Evolution* 32: 551–64. (*474*)

Emerson, S. B., and D. Diehl. 1980. Toe pad morphology and mechanisms of sticking in frogs. *Biol. J. Linn. Soc.* 13: 199–216. (*429*)

Emerson, S. B., and M. A. R. Koehl. 1990. The interaction of behavioral and morphological change in the evolution of a novel locomotor type—flying frogs. *Evolution* 44: 1931–46. (*262*)
Emlet, R. B. 1982. Echinoderm calcite: a mechanical analysis from larval spicules. *Biol. Bull.* 163: 264–75. (*393*)
Ennos, A. R. 1987. A comparative study of the flight mechanism of Diptera. *J. Exp. Biol.* 127: 355–72. (*273*)
———. 1988. The importance of torsion in the design of insect wings. *J. Exp. Biol.* 140: 137–60. (*382*)
———. 1989. Inertial and aerodynamic torques on the wings of some Diptera. *J. Exp. Biol.* 142: 87–95. (*276*)
———. 1993a. The mechanics of the flower stem of the sedge *Carex acutiformis*. *Ann. Bot.* 72: 123–27. (*382*)
———. 1993b. The function and formation of buttresses. *Trends Ecol. Evol.* 8: 350–51. (*433*)
———. 2000. *Trees*. Washington, DC: Smithsonian Institution Press. (*431*)
Ennos, A. R., and A. H. Fitter. 1992. Comparative functional morphology of the anchorage systems of annual dicots. *Funct. Ecol.* 6: 71–78. (*431*)
Ennos, A. R., H. C. Spatz, and T. Speck. 2000. The functional morphology of the petioles of the banana, *Musa textilis*. *J. Exp. Bot.* 51: 2085–93. (*382, 384, 385*)
Eshbach, O. E., ed. 1952. *Handbook of Engineering Fundamentals*, 2d ed. New York: John Wiley & Sons. (*314, 321, 322, 380*)
Etnier, S. A. 2001. Flexural and torsional stiffness in multi-jointed biological beams. *Biol. Bull.* 200: 1–8. (*403*)
Etnier, S. A., and S. Vogel. 2000. Reorientation of daffodil (*Narcissus*: Amaryllidaceae) flowers in wind: drag reduction and torsional flexibility. *Am. J. Bot.* 87: 29–32. (*382, 385, 502*)
Evans, M.E.G. 1973. The jump of the click beetle (Coleoptera, Elateridae)—energetics and mechanics. *J. Zool. (Lond.)* 169: 181–94. (*475, 476*)
Eylers, J. P. 1976. Aspects of skeletal mechanics of the starfish *Asterias forbesi*. *J. Morphol.* 149: 353–67. (*398*)
Fairchild, L. 1981. Mate selection and behavioral thermoregulation in Fowler's toads. *Science* 212: 950–51. (*55*)
Farley, C. T., and C. R. Taylor. 1991. A mechanical trigger for the trot-gallop transition in horses. *Science* 253: 306–8. (*483*)
Federle, W., K. Rohrseitz, and B. Hölldobler. 2000. Attachment forces of ants measured with a centrifuge: better 'wax-runners' have a poorer attachment to a smooth surface. *J. Exp. Biol.* 203: 505–12. (*430*)
Feynman, R. P., R. B. Leighton, and M. Sands. 1963. *The Feynman Lectures on Physics*, vol. 1. Reading, MA: Addison-Wesley. (*13*)
Fish, F. E. 1982. Aerobic energetics of surface swimming in the muskrat *Ondatra zibethicus*. *Physiol. Zool.* 55: 180–89. (*292*)
———. 1990. Wing design and scaling of flying fish with regard to flight performance. *J. Zool. (Lond.)* 221: 391–403. (*262*)
———. 1994. Energy conservation by formation swimming: metabolic evidence from ducklings. Pp. 193–206 in L. Maddock, Q. Bone, and J. M. V. Rayner, eds. *Mechanics and Physiology of Animal Swimming*. Cambridge, UK: Cambridge University Press. (*292*)
———. 1996. Transitions from drag-based to lift-based propulsion in mammalian swimming. *Am. Zool.* 36: 628–41. (*143, 283*)
———. 1998. Imaginative solutions by marine organisms for drag reduction. Pp. 443–50 in J.C.S. Meng, ed., *Proceedings of the International Symposium on Seawater Drag Reduction*. Newport, RI. (*163*)
Fish, F. E., and R. V. Baudinette. 1999. Energetics of locomotion by the Australian water rat (*Hydromys chrysogaster*): a comparison of swimming and running in a semiaquatic mammal. *J. Exp. Biol.* 202: 353–63. (*292*)

Fish, F. E., and C. Hui. 1991. Dolphin swimming—a review. *Mammal Rev.* 21: 181–95. (*163, 294*)

Foelix, R. F. 1996. *Biology of Spiders*, 2d ed. New York: Oxford University Press. (*100*)

Ford, M. D., and J. H. Costello. 2000. Kinematic comparison of bell contraction by four species of hydromedusae. *Sci. Mar.* 64(Suppl. 1): 47–53. (*143*)

Foreman, T. K., and K. Linge. 1989. The importance of heel compression in the measurement of diurnal stature variation. *Appl. Ergon.* 20: 299–300. (*417*)

Forest Products Laboratory. 1999. Wood handbook: wood as an engineering material. FPL General Technical Report 113. Madison, WI: USDA Forest Service. (*305, 343*)

Frazzetta, T. H. 1966. Studies on the morphology and function of the skull in the Boidae (Serpentes). Part II: morphology and function of the jaw apparatus in *Python sebae* and *Python molurus*. *J. Morphol.* 118: 217–95. (*400*)

French, M. 1994. *Invention and Evolution: Design in Nature and Engineering*. Cambridge, UK: Cambridge University Press. (*458*)

Full, R. J., K. Earls, M. A. Wong, and R. L. Caldwell. 1993. Locomotion like a wheel? *Nature* 365: 495. (*490*)

Full, R. J., D. S. Stokes, A. N. Ahn, and R. K. Josephson. 1998. Energy absorption during running by leg muscles in a cockroach. *J. Exp. Biol.* 201: 997–1012. (*472*)

Full, R. J., and A. Tullis. 1990. Energetics of ascent: insects on inclines. *J. Exp. Biol.* 149: 307–17. (*490*)

Fuller, R. B. 1962. *Tensile-Integrity Structures*, no. 3,063,521. Washington, DC: U.S. Patent Office. (*398*)

Fulton, T. W. 1904. The rate of growth of fishes. *Annu. Rep. Fish. Board Scotland* 22(III): 141–241. (*70*)

Fung, Y. C. 1997. *Biomechanics: Circulation*, 2d ed. New York: Springer-Verlag. (*223*)

Gad-el-Hak, M. 1996. Compliant coatings: a decade of progress. *Appl. Mech. Rev.* 49: S147–57. (*163*)

Gafiychuk, V. V., and I. A. Lubashevsky. 2001. On the principles of the vascular network branching. *J. Theor. Biol.* 212: 1–9. (*211*)

Galileo G. 1637. *Dialogues Concerning Two New Sciences*. Trans. 1914, H. Crew and A. DeSalvio. New York: Macmillan. (*390*)

Gans, C. 1974. *Biomechanics: An Approach to Vertebrate Biology*. Philadelphia, PA: J. B. Lippincott Co. (*401, 470, 489*)

Gans, R. F. 1991. *Analytical Kinematics: Analysis and Synthesis of Planar Mechanisms*. Stoneham, MA: Butterworth-Heinemann. (*400*)

Gatehouse, A. G. 1997. Behavioral and ecological genetics of wind-borne migration by insects. *Annu. Rev. Entomol.* 42: 475–502. (*137*)

Gaylord, B., B. B. Hale, and M. W. Denny. 2001. Consequences of transient fluid forces for compliant benthic organisms. *J. Exp. Biol.* 204: 1347–60. (*295*)

Geiger, R., R. H. Aron, and P. Todhunter. 1995. *The Climate Near the Ground*, 5th ed. Braunschweig, Germany: Friedrich Vieweg & Sohn. (*175*)

Gibbs, D. A., E. W. Merrill, K. A. Smith, and E. A. Balazs. 1968. Rheology of hyaluronic acid. *Biopolymers* 6: 777–91. (*363*)

Gibbs, P. E., and G. W. Bryan. 1980. Copper—the major metal component of glycerid polychaete jaws. *J. Mar. Biol. Assoc. U.K.* 60: 205–14. (*333*)

Gillis, G. B. 1997. Anguilliform locomotion in an elongate salamander, *Siren intermedia*: effects of speed on axial undulatory movements. *J. Exp. Biol.* 200: 767–84. (*143*)

Glaser, A. E., and J.F.M. Vincent. 1979. The autonomous inflation of insect wings. *J. Insect Physiol.* 25: 315–18. (*421*)

Glasheen, J. W., and T. A. McMahon. 1995. Arms are different from legs: mechanics and energetics of human hand-running. *J. Appl. Physiol.* 78: 1280–87. (*486*)

———. 1996a. Size-dependence of water-running ability in basilisk lizards (*Basiliscus basiliscus*). *J. Exp. Biol.* 199: 2611–18. (*296*)

———. 1996b. Vertical water entry of disks at low Froude numbers. *Phys. Fluids* 8: 2078–83. (*296*)

Glemain, P., J. P. Cordonnier, L. LeNormand, and J. M. Buzelin. 1990. Urodynamic consequences of urethral stenosis: hydrodynamic study with a theoretical model. *J. d'Urol.* 96: 271–77. (*203*)

Goldman, J. A. 2001. *Effects of the Free Water Surface on Animals that Jump out of the Water*. Ph.D. dissertation, Duke University, Durham, NC. (*294*)

Goldstein, S. 1938. *Modern Developments in Fluid Dynamics*. Oxford, UK: Clarendon Press. Reprint, New York: Dover Publications, Inc., 1965. (*76*)

Goodship, A. E., L. E. Lanyon, and H. McFie. 1979. Functional adaptation of bone to increased stress. *J. Bone Joint Surg. Am.* 61: 539–46. (*506*)

Gordon, A. M., A. F. Huxley, and F. J. Julian. 1966. The variation in isometric tension with sarcomere length in vertebrate muscle fibers. *J. Physiol.* 184: 170–92. (*454*)

Gordon, J. E. 1976. *The New Science of Strong Materials*. Harmondsworth, UK: Penguin Books. Reprint, Princeton, NJ: Princeton University Press, 1984. (*302, 333*)

———. 1978. *Structures; or, Why Things Don't Fall Down*. New York: Plenum Press. (*52, 302, 312, 313, 316, 317, 319, 322, 333, 346, 370, 382, 425, 497*)

———. 1988. *The Science of Structures and Materials*. New York: Scientific American Library. (*302*)

Gosline, J. M. 1971. Connective tissue mechanics of *Metridium senile*. II. Visco-elastic properties and macromolecular model. *J. Exp. Biol.* 55: 775–95. (*341, 361, 416*)

Gosline, J. M., M. E. DeMont, and M. W. Denny. 1986. The structure and properties of spider silk. *Endeavour N. S.* 10: 37–43. (*310, 312, 316, 328, 344*)

Gosline, J. M., P. A. Guerette, C. S. Ortlepp, and K. N. Savage. 1999. The mechanical design of spider silks: from fibroin sequence to mechanical function. *J. Exp. Biol.* 202: 3295–303. (*310, 312, 314*)

Gosline, J. M., and R. E. Shadwick. 1996. The mechanical properties of fin whale arteries are explained by novel connective tissue designs. *J. Exp. Biol.* 199: 985–97. (*225*)

Gould, S. J. 1980. *The Panda's Thumb*. New York: W. W. Norton. (*15*)

———. 1981. Kingdoms without wheels. *Nat. Hist.* 90(4): 42–48. (*511*)

Gould, S. J., and R. C. Lewontin. 1979. The spandrels of San Marco and the panglossian paradigm: a critique of the adaptationist programme. *Proc. R. Soc. Lond. B* 205: 581–98. (*15*)

Graham, J. B., J. H. Gee, P. Motta, and I. Rubinoff. 1987. Subsurface buoyancy regulation by the sea snake *Pelamis platurus*. *Physiol. Zool.* 60: 251–61. (*98*)

Gray, J. 1936. Studies in animal locomotion. VI. The propulsive powers of the dolphin. *J. Exp. Biol.* 13: 192–99. (*162*)

———. 1953. *How Animals Move*. Cambridge, UK: Cambridge University Press. (*485*)

———. 1968. *Animal Locomotion*. London: Weidenfeld and Nicolson. (*485, 489*)

Green, D. M. 1981. Adhesion and the toe-pads of tree frogs. *Copeia* 1981: 790–96. (*429*)

Greene, D. F., and E. A. Johnson. 1990. The dispersal of winged fruits and seeds differing in autorotative behaviour. *Can. J. Bot.* 68: 2693–97. (*248*)

Greenewalt, C. H. 1962. Dimensional relationships for flying animals. *Smithson. Misc. Collect.* 144(2): 1–46. (*270*)

Griffin, T. M., and R. Kram. 2000. Penguin waddling is not wasteful. *Nature* 408: 929. (*487*)

Griffin, T. M., N. A. Tolani, and R. Kram. 1999. Walking in simulated reduced gravity: mechanical energy fluctuations and exchange. *J. Appl. Physiol.* 86: 383–90. (*480*)

Hainsworth, F. R. 1989. Wing movements and positioning for aerodynamic benefit by Canada geese flying in formation. *Can. J. Zool.* 67: 585–89. (*250*)

Haldane, J.B.S. 1928. On being the right size. Pp. 20–28 in *Possible Worlds*. New York: Harper. (*13, 44*)

Halfen, L. N. 1979. Gliding movements. Pp. 250–67 in W. Haupt and M. E. Feinleib, eds., *Physiology of Movements*. New York: Springer-Verlag. (*450*)

Hansen, B., and P. Tiselius. 1992. Flow through the feeding structures of suspension feeding zooplankton: a physical model approach. *J. Plankton Res.* 14: 821–34. (*242*)

Happel, J., and H. Brenner. 1983. *Low Reynolds Number Hydrodynamics*, 2d ed. Englewood Cliffs, NJ: Prentice-Hall. (*123, 196, 231, 233*)

Harris, J. E., and H. D. Crofton. 1957. Structure and function in the nematodes: internal pressure and cuticular structure in *Ascaris*. *J. Exp. Biol.* 34: 116–30. (*413*)

Harvey, E. N., D. K. Barnes, W. D. McElroy, A. H. Whiteley, D. C. Pease, and K. W. Cooper. 1944. Bubble formation in animals. I. Physical factors. *J. Cell. Comp. Physiol.* 24: 1–22. (*110*)

Harvey, P. H., and M. D. Pagel. 1991. *The Comparative Method in Evolutionary Biology*. Oxford, UK: Oxford University Press. (*15, 59*)

Hayward, A.T.J. 1971. Negative pressure in liquids: can it be harnessed to serve man? *Am. Sci.* 59: 434–43. (*102*)

Hebets, E. H., and R. F. Chapman. 2000. Surviving the flood: plastron respiration in the non-tracheate arthropod *Phrynus marginemaculatus* (Amblypygi: Arachnida). *J. Insect Physiol.* 46: 13–19. (*114*)

Hedenström, A., and F. Liechti. 2001. Field estimates of body drag coefficients on the basis of dives in passerine birds. *J. Exp. Biol.* 204: 1167–75. (*161*)

Heglund, N., and C. R. Taylor. 1988. Speed, stride frequency and energy cost per stride: how do they change with body size and gait? *J. Exp. Biol.* 138: 301–18. (*482*)

Heine, C. 1992. *Mechanics of Flapping Fin Locomotion in the Cownose Ray*, Rhinoptera bonasus *(Elasmobranchii: Myliobatidae)*. Ph.D. dissertation, Duke University, Durham, NC. (*271*)

Herreid, C. F., II, and R. J. Full. 1986. Energetics of hermit crabs during locomotion: the cost of carrying a shell. *J. Exp. Biol.* 120: 297–308. (*492*)

Hertel, H. 1963. *Structure-Form-Movement*. Trans. 1966, M. S. Katz. New York: Reinhold Publ. Co. (*293*)

Hettiaratchi, D.R.P., and J. R. O'Callaghan. 1978. Structural mechanics of plant cells. *J. Theor. Biol.* 74: 235–57. (*413*)

Hicks, J. W., and H. S. Badeer. 1992. Gravity and the circulation: "open" vs. "closed" systems. *Am. J. Physiol.* 262: R725–32. (*219*)

Higgins, R. P., and H. Thiel, eds. 1988. *Introduction to the Study of Meiofauna*. Washington, DC: Smithsonian Institution Press. (*242*)

Hill, A. V. 1950. The dimensions of animals and their muscular dynamics. *Sci. Prog. Oxf.* 150: 209–30. (*453*)

Hinton, H. E. 1976. Plastron respiration in bugs and beetles. *J. Insect Physiol.* 22: 1529–50. (*114*)

Hoerner, S. F. 1965. *Fluid-Dynamic Drag*. Vancouver, WA: Hoerner Fluid Dynamics. (*157, 294*)

Holbrook, M. N., M. W. Denny, and M.A.R. Koehl. 1991. Intertidal "trees": consequences of aggregation on the mechanical and photosynthetic properties of sea-palms *Postelsia palmaeformis* Ruprecht. *J. Exp. Mar. Biol. Ecol.* 146: 39–68. (*434, 435*)

Holbrook, M. N., and M. A. Zwieniecki. 2000. Embolism repair and xylem tension: do we need a miracle? *Plant Physiol.* 120: 7–10. (*111*)

Holmes, K. C., and M. A. Greeves. 2000. The structural basis of muscle contraction. *Philos. Trans. R. Soc. Lond. B* 355: 419–31. (*451*)

Holwill, M.E.J. 1975. The role of body oscillation in the propulsion of micro-organisms. Pp. 133–41 in T.Y.-T. Wu, C. J. Brokaw, and C. Brennen, eds., *Swimming and Flying in Nature*, vol 1. New York: Plenum Press. (*237*)

———. 1977. Low Reynolds number undulatory propulsion in organisms of different sizes. Pp. 233–42 in T. J. Pedley, ed., *Scale Effects in Animal Locomotion*. London: Academic Press. (*237*)

Houston, A. I. 1986. The optimal flight velocity for a bird exploiting patches of food. *J. Theor. Biol.* 119: 345–62. (*275*)

Hoyt, D. F., and C. R. Taylor. 1981. Gait and the energetics of locomotion in horses. *Nature* 292: 239–40. (*482*)

Huey, R. B., and P. E. Hertz. 1984. Effects of body size and slope on acceleration of a lizard (*Stellio stellio*). *J. Exp. Biol.* 110: 113–23. (*475*)

Hui, C. A. 1987a. Power and speed of swimming dolphins. *J. Mammal.* 68: 126–32. (*163*)

———. 1987b. The porpoising of penguins: an energy-conserving behavior for respiratory ventilation? *Can. J. Zool.* 65: 209–11. (*294*)

———. 1988a. Penguin swimming. I. Hydrodynamics. *Physiol. Zool.* 61: 333–43. (*271*)

———. 1988b. Penguin swimming. II. Energetics and behavior. *Physiol. Zool.* 61: 344–50. (*292*)

Hukins, D.W.L., and J. R. Meakin. 2000. Relationship between structure and mechanical function of the tissues of the intervertebral joint. *Am. Zool.* 40: 42–52. (*417*)

Hulbary, R. L. 1944. The influence of air spaces on the three-dimensional shapes of cells in *Elodea* stems, and a comparison with pith cells of *Ailanthus*. *Am. J. Bot.* 31: 561–80. (*413*)

Hunt, K. H. 1978. *Kinematic Geometry of Mechanisms*. Oxford, UK: Clarendon Press. (*396*)

Hunter, T., and S. Vogel. 1986. Spinning embryos enhance diffusion through gelatinous egg masses. *J. Exp. Mar. Biol. Ecol.* 96: 303–8. (*180*)

Huxley, A. F. 2000. Mechanics and models of the myosin motor. *Philos. Trans. R. Soc. Lond. B* 355: 433–40. (*451*)

Huxley, H. E. 2000. Past, present and future experiments on muscle. *Philos. Trans. R. Soc. Lond. B* 355: 539–43. (*451*)

Hyams, J. S., and C. W. Lloyd. 1994. *Microtubules*. New York: Wiley-Liss. (*447*)

Ingold, C. T. 1953. *Dispersal in Fungi*. Oxford, UK: Clarendon Press. (*234*)

Ingold, C. T., and S. A. Hadland. 1959. The ballistics of *Sordaria*. *New Phytol.* 58: 46–57. (*174*)

Isaacs, J. D., A. C. Vine, H. Bradner, and G. E. Bachus. 1966. Satellite elongation into a true "sky-hook." *Science* 151: 682–83. (*391*)

Jackson, A. P., J.F.V. Vincent, and R. M. Turner. 1988. The mechanical design of nacre. *Proc. R. Soc. Lond. B* 234: 415–40. (*319*)

Jacobs, M. R. 1954. The effect of wind sway on the form and development of *Pinus radiata* D. Don. *Aust. J. Bot.* 2: 35–51. (*505*)

Jaffe, M. J. 1973. Thigmomorphogenesis: the response of plant growth and development to mechanical stress. *Planta* 114: 143–57. (*504*)

Jenkins, F. A., Jr., K. P. Dial, and G. E. Goslow, Jr. 1988. A cineradiographic analysis of bird flight: the wishbone in starlings is a spring. *Science* 241: 1495–98. (*472*)

Jenkins, F. A. Jr., and N. H. Shubin. 1998. *Prosalirus bitis* and the anuran caudopelvic mechanism. *J. Vert. Paleontol.* 18: 495–510. (*471*)

Jensen, A. 1986. Why the best technology for escaping from a submarine is no technology. *Am. Heritage Invent. Technol.* 2(1): 44–49. (*95*)

Jensen, M. 1956. Biology and physics of locust flight. III. The aerodynamics of locust flight. *Philos. Trans. R. Soc. Lond. B* 239: 511–52. (*252*)

Jeronimidis, G. 1980. Wood, one of nature's challenging composites. Pp. 169–82 in J.F.V. Vincent and J. D. Currey, eds., *The Mechanical Properties of Biological Materials: Symposium of the Society for Experimental Biology* 34. Cambridge, UK: Cambridge University Press. (*344*)

Johansson, L. C., and U. M. Lindhe Norberg. 2001. Lift-based paddling in diving grebe. *J. Exp. Biol.* 204: 1687–96. (*280*)

Johnson, A. S. 1993. Sag-mediated modulated tension in terrebellid tentacles exposed to flow. *Biol. Bull.* 185: 10–19. (*328, 329*)

Johnson, A. S., and M.A.R. Koehl. 1994. Maintenance of dynamic strain similarity and environmental stress factor in different flow habitats: thallus allometry and material properties of a giant kelp. *J. Exp. Biol.* 195: 381–410. (*508*)

Jones, D. R., and D. J. Randall. 1978. The respiratory and circulatory systems during exercise. Pp. 425–501 in W. S. Hoar and D. J. Randall, eds., *Fish Physiology*, vol. 7. New York: Academic Press. (*153*)

Jordan, C. E. 1992. A model of rapid-start swimming at intermediate Reynolds number: undulatory locomotion in the chaetognath *Sagitta elegans*. *J. Exp. Biol.* 163: 119–37. (*280*)

Josephson, R. K. 1985. Mechanical power output from striated muscle during cyclic contraction. *J. Exp. Biol.* 114: 493–512. (*459*)

Josephson, R. K., J. G. Malamud, and D. R. Stokes. 2000. Asynchronous muscle: a primer. *J. Exp. Biol.* 203: 2713–22. (*459, 462*)

Jumars, P. A., and A.R.M. Nowell. 1984. Fluid and sediment dynamic effects on marine benthic community structure. *Am. Zool.* 24: 45–55. (*135*)

Kaiser, D. 2000. How do pili pull? *Curr. Biol.* 10: R777–80. (*450*)

Karassik, I. J., W. C. Krutzsch, W. H. Frazer, and J. P. Messina. 1976. *Pump Handbook*. New York: McGraw-Hill. (*199*)

Kassab, G. S., and Y.-C.B. Fung. 1995. The pattern of coronary arteriolar bifurcations and the uniform stress hypothesis. *Ann. Biomed. Eng.* 23: 13–20. (*214*)

Kelly, D. A. 1997. Axial orthogonal fiber reinforcement in the penis of the nine-banded armadillo (*Dasypus novemcinctus*). *J. Morphol.* 233: 249–55. (*416*)

———. 2002. The functional morphology of penile erection: tissue designs for increasing and maintaining stiffness. *Integr. Comp. Biol.* 42: 216–21. (*416*)

Ker, R. F. 1999. The design of soft collagenous load-bearing tissues. *J. Exp. Biol.* 202: 3315–24. (*417*)

Khanin, M. A., and I. B. Bukharov. 1991. Optimal arterial blood pressure. *J. Theor. Biol.* 148: 285–94. (*218*)

Kier, W. M. 1982. The functional morphology of the musculature of squid (Loliginidae) arms and tentacles. *J. Morphol.* 172: 179–92. (*418*)

———. 1985. The musculature of squid arms and tentacles: ultrastructural evidence for functional differences. *J. Morphol.* 185: 223–39. (*456, 462*)

———. 1992. Hydrostatic skeletons and muscular hydrostats. Pp. 205–231 in A. A. Biewener, ed., *Biomechanics—Structures and Systems: A Practical Approach*. Oxford, UK: Oxford University Press. (*420*)

Kier, W. M., and K. K. Smith. 1985. Tongues, tentacles and trunks: the biomechanics of movement in muscular-hydrostats. *Zool. J. Linn. Soc. Lond.* 83: 307–24. (*408, 418, 441*)

Kirkpatrick, S. J. 1994. Scale effects on the stresses and safety factors in the wing bones of birds and bats. *J. Exp. Biol.* 190: 195–215. (*500*)

Kitchener, A. C. 2000. Fighting and the mechanical design of horns and antlers. pp. 291–314 in P. Domenici and R. W. Blake, eds., *Biomechanics in Animal Behaviour*. Oxford, UK: BIOS Scientific Publishers. (*310*)

Klages, N.T.W., and J. Cooper. 1992. Bill morphology and diet of a filter-feeding seabird: the broad-billed prion *Pachyptila vittata* at South Atlantic Gough Island. *J. Zool. (Lond.)* 227: 385–404. (*264*)

Kleiber, M. 1961. *The Fire of Life. An Introduction to Animal Energetics*. New York: John Wiley & Sons. (*45*)

Kling, G. W., M. A. Clark, H. R. Compton, J. D. Devine, W. C. Evans, A. M. Humphrey, E. J. Koenigsberg, J. P. Lockwood, M. L. Tuttle, and G. N. Wagner. 1987. The 1986 Lake Nyos Gas Disaster in Cameroon, West Africa. *Science* 236: 169–75. (*111*)

Koehl, M.A.R. 1977a. Mechanical diversity of connective tissue of the body wall of sea anemones. *J. Exp. Biol.* 69: 107–25. (*361*)

———. 1977b. Mechanical organization of cantilever-like sessile organisms: sea anemones. *J. Exp. Biol.* 69: 127–42. (*374, 376, 503*)

———. 1982. Mechanical design of spicule-reinforced connective tissue: stiffness. *J. Exp. Biol.* 98: 239–67. (*399*)

———. 1993. Hairy little legs: feeding, smelling, and swimming at low Reynolds numbers. *Contemp. Math.* 141: 33–64. (*124*)

Koehl, M.A.R., K. J. Quillin, and C. Pell. 1995. Mechanical consequences of fiber orientation in the walls of hydraulic skeletons. *Am. Zool.* 35: 53a. (*409*)

Koehl, M.A.R., and J. R. Strickler. 1981. Copepod feeding currents: food capture at low Reynolds number. *Limnol. Oceanogr.* 26: 1062–73. (*239*)

Kram, R. 1996. Inexpensive load carrying by rhinoceros beetles. *J. Exp. Biol.* 199: 609–12. (*492*)

Kram, R., A. Domingo, and D. P. Ferris. 1997. Effect of reduced gravity on the preferred walk-run transition speed. *J. Exp. Biol.* 200: 821–26. (*480*)

Kramer, M. O. 1960. The dolphin's secret. *New Sci.* 7: 1118–20. (*163*)

Krogh, A. 1920. Studien über Tracheenrespiration. 2. Über Gasdiffusion in den Tracheen. *Pfluegers Arch. Gesamte Physiol. Menschen Tiere* 179: 95–112. (*214*)

———. 1941. *The Comparative Physiology of Respiratory Mechanisms*. Philadelphia, PA: University of Pennsylvania Press. (*207*)

LaBarbera, M. 1983. Why the wheels won't go. *Am. Nat.* 121: 395–408. (*512*)

———. 1989. Analyzing body size as a factor in ecology and evolution. *Annu. Rev. Ecol. Syst.* 20: 97–117. (*85*)

———. 1990. Principles of design of fluid transport systems in zoology. *Science* 249: 992–1000. (*213*)

LaBarbera, M., and S. Vogel. 1982. The design of fluid transport systems in organisms. *Am. Sci.* 70: 54–60. (*211*)

Lacoursière, J. O. 1992 A laboratory study of fluid flow and microhabitat selection by larvae of *Simulium vittatum* (Diptera: Simuliidae). *Can. J. Zool.* 70: 582–96. (*172*)

Lambert, Y., and J.-D. Dutil. 1997. Can simple condition indices be used to monitor and quantify seasonal changes in the energy reserves of Atlantic cod (*Gadus morhua*)? *Can. J. Fish. Aquat. Sci.* 54(Suppl): 104–12. (*71*)

Lande, R. 1980. Microevolution in relation to macroevolution. *Paleontology* 6: 235–38. (*510*)

Lane, D.J.W., A. R. Beaumont, and J. R. Hunter. 1985. Byssus drifting and the drifting threads of the young post-larval mussel, *Mytilus edulis*. *Mar. Biol.* 84: 301–8. (*137*)

Lang, T. G., and K. Pryor. 1966. Hydrodynamic performance of porpoises (*Stenella attenuata*). *Science* 152: 531–33. (*163*)

Langhaar, H. L. 1951. *Dimensional Analysis and Theory of Models*. New York: John Wiley & Sons. (*65*)

Lanyon, L. E. 1990. The relationship between functional loading and bone architecture. Pp. 269–84 in C. J. DeRousseau, ed., *Primate Life History and Evolution*. New York: Wiley-Liss. (*507*)

Lapennas, G. N., and K. Schmidt-Nielsen. 1977. Swimbladder permeability to oxygen. *J. Exp. Biol.* 67: 175–96. (*99*)

Lauder, G. V. 1984. Pressure and water flow patterns in the respiratory tract of the bass (*Micropterus salmoides*). *J. Exp. Biol.* 113: 84: 33–55. (*203*)

———. 2000. Function of the caudal fin during locomotion in fishes: kinematics, flow visualization, and evolutionary patterns. *Am. Zool.* 40: 101–22. (*281*)

Lawrence, J. M., and J. B. McClintock. 1988. Allocation of organic material and energy to the holdfast, stipe, and fronds in *Postelsia palmaeformis* (Phaeophyta: Laminariales) on the California coast, U.S.A. *Mar. Biol.* 99: 151–55. (*435*)

Lee, C. R., and C. T. Farley. 1998. Determinants of the center of mass trajectory in human walking. *J. Exp. Biol.* 201: 2935–44. (*82, 481*)

Lewin, R. A. 1995. Bioconvection. *Arch. Hydrobiol. Suppl.* 108: 67–73. (*138*)

Lewis, A. M. 1992. Measuring the hydraulic diameter of a pore or conduit. *Am. J. Bot.* 79: 1158–61. (*189*)

Lichtenegger, H. C., T. Schöbert, M. H. Bartl, H. Waite, and G. D. Stucky. 2002.

High abrasion resistance with sparse mineralization: copper biomineral in worm jaws. *Science* 298: 389–92. (*334*)

Lieberman, D. E., and A. W. Crompton. 1998. Responses of bone to stress: constraints on symmorphosis. Pp. 78–86 in E. R. Weibel, C. R. Taylor, and L. Bolis, eds., *Principles of Animal Design*. Cambridge, UK: Cambridge University Press. (*507*)

Lighthill, J. 1995. The role of the lateral line in active drag reduction by clupeoid fishes. Pp. 35–48 in C. P. Ellington and T. Pedley, eds., *Biological Fluid Dynamics: Symposium of the Society for Experimental Biology* 49. Cambridge, UK: Company of Biologists, Ltd. (*163*)

Lighton, J.R.B., G. A. Bartholomew, and D. H. Feener, Jr. 1987. Energetics of locomotion and load carriage and a model of the energy cost of foraging in the leaf-cutting ant *Atta colombica* Guer. *Physiol. Zool.* 60: 524–37. (*492*)

Lillywhite, H. B. 1993. Orthostatic intolerance of viperid snakes. *Physiol. Zool.* 66: 1000–1014. (*219*)

Lindstedt, S. L., P. C. LaStayo, and T. E. Reich. 2001. When active muscles lengthen: properties and consequences of eccentric contractions. *News Physiol. Sci.* 16: 256–61. (*461*)

Liu, H., R. Wassersug, and K. Kawachi. 1997. The three-dimensional hydrodynamics of tadpole locomotion. *J. Exp. Biol.* 200: 2807–19. (*143*)

Lodge, A. S. 1960. Rheological properties of concentrated polymer solutions. Pp. 5–26 in A. L. Copley and G. Stainsby, eds., *Flow Properties of Blood and Other Biological Systems*. New York: Pergamon Press. (*353*)

Lothian, N. V. 1921–22. The load carried by the soldier. *J. R. Army Med. Corps* 37: 241–63, 342–51, 448–58; 38: 9–24. (*491*)

Loudon, C., B. A. Best, and M.A.R. Koehl. 1994. When does motion relative to neighboring surfaces alter the flow through arrays of hairs? J. *Exp. Biol.* 193: 233–54. (*124, 242*)

Loudon, C., and M.A.R. Koehl. 2000. Sniffing by a silkworm moth: wing fanning enhances air penetration through and pheromone interception by antennae. *J. Exp. Biol.* 203: 2977–90. (*242*)

Loudon, C., and K. McCulloh. 1999. Application of the Hagen-Poiseuille equation to fluid feeding through short tubes. *Ann. Entomol. Soc. Am.* 99: 153–58. (*191*)

Loudon, C., and A. Tordesillas. 1998. The use of the dimensionless Womersley number to characterize the unsteady nature of internal flow. *J. Theor. Biol.* 191: 63–78. (*224*)

Lovvorn, J. R. 2001. Upstroke thrust, drag effects, and stroke-glide cycles in wing-propelled swimming by birds. *Am. Zool.* 41: 154–65. (*281*)

Lovvorn, J. R., and D. R. Jones. 1991. Effects of body size, body fat, and changes in pressure with depth on buoyancy and cost of diving in ducks (*Aythya* spp.). *Can. J. Zool.* 69: 2879–87. (*98*)

Lowell, R. B. 1985. Selection for increased safety factors of biological structures as environmental unpredictability increases. *Science* 228: 1009–11. (*498*)

Lucas, J. R. 1982. The biophysics of pit construction by antlion larvae (*Myrmeleon*, Neuroptera). *Anim. Behav.* 30: 651–64. (*364*)

Lugt, H. J. 1983. *Vortex Flow in Nature and Technology*. New York: Wiley-Interscience. (*132*)

Macdougal, D. T. 1925. Reversible variations in volume, pressure, and movement of sap in trees. *Carnegie Inst. Wash. Publ.* 365: 1–90. (*102*)

Maderson, P.F.A. 1964. Keratinized epidermal derivatives as an aid to climbing in gekkonid lizards. *Nature* 203: 780–81. (*429*)

———. 1970. Lizard glands and lizard hands: models for evolutionary study. *Forma Functio* 3: 179–204. (*429*)

Maerkel, K., and U. Roeser. 1992. Functional anatomy of the valves in the ambulacral system of sea urchins (Echinodermata: Echinoidea). *Zoomorphology* 111: 179–92. (*414*)

Maloiy, G.M.O., N. C. Heglund, L. M. Prager, G. A. Cavagna, and C. R. Taylor. 1986. Energetic cost of carrying loads: have African women discovered an economic way? *Nature* 319: 668–69. (*492*)

Mandelbrot, B. B. 1983. *The Fractal Geometry of Nature*. New York: W. H. Freeman. (*84*)

Marcinek, D., and M. LaBarbera. 1994. Quantitative branching geometry of the vascular system of the blue crab, *Callinectes sapidus* (Arthropoda, Crustacea): a test of Murray's law in an open circulatory system. *Biol. Bull.* 186: 124–33. (*213*)

Margaria, R. 1938. Sulla fisiologia e specialmente sul consumo energetico della marcia e della corsa a varie velocità ed inclinazioni del terreno. *Accad. Naz. Lincei Mem. Ser. VI* 7: 299–368. (*461, 480*)

Margulis, L., and K. V. Schwartz. 1998. *Five Kingdoms*, 3d ed. New York: W. H. Freeman. (*450*)

Marquis, D. 1927. *archy and mehitabel*. Garden City, NY: Doubleday and Co. (*373*)

Marsden, E. W. 1969. *Greek and Roman Artillery: Historical Development*. Oxford, UK: Clarendon Press. (*346*)

———. 1971. *Greek and Roman Artillery: Technical Treatises*. Oxford, UK: Clarendon Press. (*346*)

Marshall, C. D., L. A. Clark, and R. L. Reep. 1998. The muscular hydrostat of the Florida manatee (*Trichechus manatus latirostris*): a functional morphological model of perioral bristle use. *Mar. Mamm. Sci.* 14: 290–303. (*418*)

Martin, A. W., F. M. Harrison, M. J. Huston, and D. M. Stewart. 1958. The blood volume of some representative molluscs. *J. Exp. Biol.* 35: 260–79. (*209*)

Martinez, M. M., R. J. Full, and M.A.R. Koehl. 1998. Underwater punting by an intertidal crab: a novel gait revealed by the kinematics of pedestrian locomotion in air *versus* water. *J. Exp. Biol.* 201: 2609–23. (*487*)

Massare, J. A. 1988. Swimming capabilities of Mesozoic marine reptiles: implications for method of predation. *Paleobiology* 14: 187–205. (*283*)

Mastro, A. M., and A. D. Keith. 1984. Diffusion in the aqueous compartment. *J. Cell Biol.* 99: 180s–87s. (*181*)

Matlack, G. R. 1987. Diaspore size, shape, and fall behavior in wind-dispersed plant species. *Am. J. Bot.* 74: 1150–60. (*261*)

Mattheck, C. 1991. *Trees: The Mechanical Design*. Berlin: Springer-Verlag. (*431, 507*)

———. 1998. *Design in Nature: Learning from Trees*. Berlin: Springer-Verlag. (*373, 507, 508*)

Mattheck, C., and K. Bethge. 1990. Wind breakage of trees initiated by root delamination. *Trees Struct. Funct.* 4: 225–27. (*433*)

McComas, A. J. 1996. *Skeletal Muscle: Form and Function*. Champaign, IL: Human Kinetics. (*452*)

McCurley, R. S., and W. M. Kier. 1995. The functional biology of starfish tube feet: the role of a crossed-fiber helical array in movement. *Biol. Bull.* 188: 197–209. (*414*)

McCutchen, C. W. 1977. The spinning rotation of ash and tulip tree samaras. *Science* 197: 691–92. (*248*)

McGowan, C. 1999. *A Practical Guide to Vertebrate Mechanics*. Cambridge, UK: Cambridge University Press. (*402, 405, 465*)

McHenry, M. J., C. A. Pell, and J. H. Long. 1995. Mechanical control of swimming speed: stiffness and axial wave form in undulating fish models. *J. Exp. Biol.* 198: 2293–2305. (*143*)

McMahon, T. A. 1984. *Muscles, Reflexes, and Locomotion*. Princeton, NJ: Princeton University Press. (*454, 461*)

McMahon, T. A., and J. T. Bonner. 1983. *On Size and Life*. New York: Scientific American Books. (*59, 68, 259*)

Meinhardt, H. 1995. *The Algorithmic Beauty of Sea Shells*. New York: Springer-Verlag. (*89*)

Melkonian, M. 1992. *Algal Cell Motility*. New York: Chapman and Hall. (*450*)

Merz, A. J., and K. T. Forest. 2002. Bacterial surface motility: slime trails, grappling hooks, and nozzles. *Curr. Biol.* 12: R297–303. (*450*)
Metzger, C. 1893. Der Wind als massgebender Faktor für das Wachsthum der Bäume. *Mundener Forst.* 3: 35–86. (*505*)
Meyhöfer, E. 1985. Comparative pumping rates in suspension-feeding bivalves. *Mar. Biol.* 85: 137–42. (*203, 218*)
Middleman, S. 1972. *Transport Phenomena in the Cardiovascular System.* New York: John Wiley. (*194*)
Miller, P. L. 1966. The supply of oxygen to the active flight muscles of some large beetles. *J. Exp. Biol.* 45: 285–304. (*149*)
Millman, B. S. 1967. Mechanisms of contraction in molluscan muscle. *Am. Zool.* 7: 583–91. (*462*)
Minetti, A. E. 1995. Optimum gradient of mountain paths. *J. Appl. Physiol.* 79: 1698–1703. (*460, 491*)
———. 1998. The biomechanics of skipping gaits: a third locomotor paradigm? *Proc. R. Soc. Lond. B* 265: 1227–35. (*486*)
———. 2001. Walking on other planets. *Nature* 409: 467–69. (*486*)
Minetti, A. E., L. P. Ardigo, and F. Saibene. 1993. Mechanical determinants of gradient walking energetics in man. *J. Physiol.* 472: 725–35. (*490, 491*)
———. 1994. Mechanical determinants of the minimum energy cost of running in humans. *J. Exp. Biol.* 195: 211–25. (*491*)
Money, N. P. 1997. Wishful thinking of turgor revisited: the mechanics of fungal growth. *Fungal Genet. Biol.* 21: 173–87. (*413*)
Moore, J., and P. Willmer. 1997. Convergent evolution in invertebrates. *Biol. Rev.* 72: 1–60. (*509, 510*)
Morgan, J., and M.G.R. Cannell. 1994. Shape of tree stems: a reexamination of the uniform stress hypothesis. *Tree Physiol.* 14: 49–62. (*506*)
Morowitz, H. 1970. *Entropy for Biologists: An Introduction to Thermodynamics.* New York: Academic Press. (*206*)
Morris, M. J., G. Gust, and J. J. Torres. 1985. Propulsion efficiency and cost of transport for copepods: a hydromechanical model of crustacean swimming. *Mar. Biol.* 86: 283–96. (*143*)
Morrison, P., and P. Morrison. 1982. *Powers of Ten.* New York: Scientific American Books. (*43*)
Motokawa, T. 1984. Connective tissue catch in echinoderms. *Biol. Rev.* 59: 255–70. (*438*)
Muneoka, Y., and B. M. Twarog. 1983. Neuromuscular transmission and excitation-contraction coupling in molluscan muscle. Pp. 35–76 in A. S. M. Saleuddin and K. M. Wilbur, eds., *The Mollusca*, vol. 4. New York: Academic Press. (*29*)
Munk, W. H., and G. A. Riley. 1952. Absorption of nutrients by aquatic plants. *J. Mar. Res.* 11: 215–40. (*185*)
Murray, C. D. 1926. The physiological principle of minimum work. I. The vascular system and the cost of blood volume. *Proc. Natl. Acad. Sci. U.S.A.* 12: 207–14. (*211*)
Nachtigall, W. 1974. *Biological Mechanisms of Attachment.* New York: Springer-Verlag. (*424*)
Nathan, R., G. G. Katul, H. S. Horn, S. M. Thomas, R. Oren, R. Avissar, S. W. Pacala, and S. A. Levin. 2002. Mechanisms of long-distance dispersal of seeds by wind. *Nature* 418: 409–13. (*258*)
Nathan, R., and H. C. Muller-Landau. 2000. Spatial patterns of seed dispersal, their determinants and consequences for recruitment. *TREE* 15: 278–85. (*258*)
Nicklas, R. B. 1984. A quantitative comparison of cellular motile systems. *Cell Motil.* 4: 1–5. (*458*)
Nijhout, H. F. 1990. Metaphors and the role of genes in development. *Bioessays* 12: 1–6. (*503*)
Nijhout, H. F., and H. G. Sheffield. 1979. Antennal hair erection in male mosquitoes: a new mechanical effector in insects. *Science* 206: 595–96. (*444*)

Niklas, K. J. 1984. The motion of windborne pollen grains around conifer ovulate cones: implications on wind pollination. *Am. J. Bot.* 71: 356–74. (*234*)

———. 1992. *Plant Biomechanics: An Engineering Approach to Plant Form and Function*. Chicago, IL: University of Chicago Press. (*197, 310, 311, 321, 322, 344, 413, 422, 439, 499, 500, 505*)

———. 1994. *Plant Allometry: The Scaling of Form and Process*. Chicago, IL: University of Chicago Press. (*59*)

———. 1997. *The Evolutionary Biology of Plants*. Chicago, IL: University of Chicago Press. (*434*)

Niklas, K. J., and H.-C. Spatz. 2000. Wind-induced stresses in cherry trees: evidence against the hypothesis of constant stress level. *Trees Struct. Funct.* 14: 230–37. (*506*)

Niklas, K. J., J. Varne, and D. M. Berglund. 1999. Non-parametric statistical formulas for factors of safety in plant stems. *J. Theor. Biol.* 197: 135–47. (*500*)

Nobel, P. 1999. *Physicochemical and Environmental Plant Physiology*, 2d ed. New York: W. H. Freeman and Co. (*103, 113, 184*)

Norberg, U. M. 1990. *Vertebrate Flight: Mechanics, Physiology, Morphology, Ecology and Evolution*. Berlin: Springer-Verlag. (*253*)

Oberg, E., F. D. Jones, and H. L. Horton. 1984. *Machinery's Handbook*, 22nd ed. New York: Industrial Press. (*313, 373*)

Obst, B. S., W. M. Hamner, P. P. Hamner, E. Wolanski, M. Rubega, and B. Littlehales. 1996. Kinematics of phalarope spinning. *Nature* 384: 121. (*135*)

O'Dor, R. K. and J. A. Hoar. 2000. Does geometry limit squid growth? *ICES J. Mar. Sci.* 57: 8–14. (*86*)

Oliver, J. A. 1951. "Gliding" in amphibians and reptiles, with a remark on arboreal adaptation in the lizard *Anolis carolinensis* Voight. *Am. Nat.* 85: 171–76. (*262*)

O'Neill, P. L. 1978. Hydrodynamic analysis of feeding in sand dollars. *Oecologia* 34: 157–74. (*254*)

Osler, G.H.R., P. W. West, and G. M. Downes. 1996. Effects of bending stress on taper and growth of young *Eucalyptus regnans* trees. *Trees Struct. Funct.* 10: 239–46. (*505*)

Page, R. M. 1964. Sporangium discharge in *Pilobolus: a photographic study. Science* 146: 925–27. (*475*)

Paine, V. 1926. Adhesion of the tube feet in starfish. *J. Exp. Zool.* 45: 361–66. (*429*)

Pais, A. 1982. *'Subtle is the Lord . . .' The Science and the Life of Albert Einstein.* Oxford, UK: Oxford University Press. (*176*)

Pappenheimer, J. R. 1984. Contributions to microvascular research of Jean Léonard Marie Poiseuille. Chapter 1, pp. 1–10 in E. M. Renkin and C. C. Michel, eds., *Handbook of Physiology: The Cardiovascular System IV*. Betheda, MD: American Physiological Society. (*13*)

Parkes, E. W. 1965. *Braced Frameworks, An Introduction to the Theory of Structures*. Oxford, UK: Pergamon Press. (*394*)

Parry, D. A. 1949. The swimming of whales and a discussion of Gray's Paradox. *J. Exp. Biol.* 26: 24–34. (*163*)

Parry, D. A., and R.H.J. Brown. 1959. The hydraulic mechanism of the spider leg. *J. Exp. Biol.* 36: 423–33. (*421, 475*)

Parsons, P. E., and C. R. Taylor. 1977. Energetics of brachiation versus walking: a comparison of a suspended and an inverted pendulum mechanism. *Physiol. Zool.* 50: 182–89. (*486*)

Payne, K. 1998. *Silent Thunder: In the Presence of Elephants*. New York: Simon and Schuster. (*55*)

Pedley, T. J. 1983. Calculation of unstirred layer thickness in membrane transport experiments: a survey. *Q. Rev. Biophysics* 16: 115–50. (*184*)

Pedley, T. J., B. S. Brook, and R. S. Seymour. 1996. Blood pressure and flow-rate in the giraffe jugular vein. *Philos. Trans. R. Soc. Lond. B* 351: 855–66. (*219*)

Pennisi, E. 2002. Biology reveals new ways to hold on tight. *Science* 296: 250–51. (*430*)

Pennycuick, C. J. 1975. On the running of the gnu and other animals. *J. Exp. Biol.* 63: 775–99. (*453*)
———. 1982. The flight of petrels and albatrosses (Procellariiformes), observed in South Georgia and its vicinity. *Philos. Trans. R. Soc. Lond. B* 300: 73–106. (*265*)
———. 1990. Predicting wingbeat frequency and wavelength of birds. *J. Exp. Biol.* 150: 171–86. (*271*)
———. 1992. *Newton Rules Biology: A Physical Approach to Biological Problems*. Oxford, UK: Oxford University Press. (*59, 68, 85*)
———. 2002. Gust soaring as a basis for the flight of petrels and albatrosses (Procellariiformes). *Avian Sci.* 2: 1–12. (*295*)
Pennycuick, C. J., M. Klaassen, A. Kvist, and A. Lindstrom. (1996) Wingbeat frequency and the body drag anomaly: wind tunnel observations on a thrush nightingale (*Luscinia luscinia*) and a teal (*Anas crecca*). *J. Exp. Biol.* 199: 2757–65. (*161*)
Perry, T. O. 1982. The ecology of tree roots and the practical significance thereof. *J. Arboric.* 8: 197–211. (*433*)
Peters, R. H. 1983. *The Ecological Implications of Body Size*. Cambridge, UK: Cambridge University Press. (*59*)
Peterson, I. 1988. *The Mathematical Tourist: Snapshots of Modern Mathematics*. New York: W. H. Freeman. (*84*)
Peterson, J. A., J. A. Benson, M. Ngai, J. Morin, and C. Ow. 1982. Scaling in tensile "skeletons": structures with scale-independent length dimensions. *Science* 217: 1267–70. (*437*)
Petroski, H. 1985. *To Engineer is Human*. New York: St. Martin's Press. (*497*)
Pohl, P., S. M. Saparov, and Y. N. Antonenko. 1998. The size of the unstirred layer as a function of the solute diffusion coefficient. *Biophys J.* 75: 1403–09. (*184*)
Prandtl, L., and O. G. Tietjens. 1934. *Applied Hydro- and Aeromechanics*. Reprint, New York: Dover Publications, Inc., 1957. (*166*)
Prange, H. D., and K. Schmidt-Nielsen. 1970. The metabolic cost of swimming in ducks. *J. Exp. Biol.* 53: 763–77. (*292*)
Prosser, C. L., and F. A. Brown. 1961. *Comparative Animal Physiology*, 2d ed. Philadelphia, PA: W. B. Saunders Co. (*209*)
Prusinkiewicz, P. 1990. *The Algorithmic Beauty of Plants*. New York: Springer-Verlag. (*89*)
Pugh, L.G.C.E. 1971. The influence of wind resistance in running and walking and the mechanical efficiency of work against horizontal or vertical forces. *J. Physiol.* 213: 255–76. (*484*)
Purcell, E. M. 1977. Life at low Reynolds number. *Am. J. Phys.* 45: 3–11. (*181, 229*)
———. 1997. The efficiency of propulsion by a rotating flagellum. *Proc. Natl. Acad. Sci. U.S.A.* 94: 11307–11. (*143*)
Purohit, R. C., and S. D. Beckett. 1976. Penile pressures and muscle activity associated with erection and ejaculation in the dog. *Am. J. Physiol.* 231: 1343–48. (*416*)
Putz, F. E., P. D. Coley, K. Lu, A. Montalvo, and A. Aiello. 1983. Uprooting and snapping of trees: structural determinants and ecological consequences. *Can. J. For. Res.* 13: 1011–20. (*431*)
Quillin, K. J. 2000. Ontogenetic scaling of burrowing forces in the earthworm, *Lumbricus terrestris*. *J. Exp. Biol.* 203: 2757–70. (*411*)
Raffel, M., C. E. Willert, and J. Kompenhans. 1998. *Particle Image Velocimetry: A Practical Guide*. Heidelberg, Germany: Springer-Verlag. (*141*)
Randall, D. J., and C. Daxboeck. 1984. Oxygen and carbon dioxide transfer across fish gills. Pp. A263–314 in W. S. Hoar and D. J. Randall, eds., *Fish Physiology*, vol. 10. New York: Academic Press. (*152*)
Raschke, K. 1979. Movements of stomata. Pp. 383–441 in W. Haupt and M. E. Feinleib, eds., *Physiology of Movements*. New York: Springer-Verlag. (*422*)
Raup, D. M. 1966. Geometric analysis of shell coiling: general problems. *J. Paleontol.* 40: 1178–90. (*89*)

Rayner, J.M.V. 1999. Estimating the power curves of flying vertebrates. *J. Exp. Biol.* 202: 3449–61. (*275*)

Reynolds, O. 1883. An experimental investigation of the circumstances which determine whether the motion of water shall be direct or sinuous, and the law of resistance in parallel channels. *Philos. Trans. R. Soc. Lond.* 174: 935–82. (*121*)

Riedl, R. J. 1971. How much water passes through sandy beaches? *Int. Rev. Geol. Hydrobiol.* 56: 923–46. (*242*)

Riedl, R. J., and R. Machan. 1972. Hydrodynamic patterns of lotic sands and their bioclimatological implications. *Mar. Biol.* 13: 179–209. (*242*)

Riisgård, H. U. 1989 Properties and energy cost of the muscular piston pump in the suspension feeding polychaete *Chaetopterus variopedatus*. *Mar. Ecol. Prog.* Ser. 56: 157–68. (*203*)

Riisgård, H. U., and P. S. Larsen. 1995. Filter feeding in marine macro-invertebrates: pump characteristics, modelling and energy cost. *Biol. Rev.* 70: 67–106. (*200, 202, 203*)

Riskin, D. K., and M. B. Fenton. 2001. Sticking ability in Spix's disk-winged bat, *Thyroptera tricolor* (Microchiroptera: Thyropteridae). *Can. J. Zool.* 79: 2261–67. (*427*)

Roberts, T. J., R. L. Marsh, P. G. Weyand, and C. R. Taylor. 1997. Muscular force in running turkeys: the economy of minimizing work. *Science* 275: 1113–15. (*472*)

Rohrer, F. 1921. Der Index der Korperfulle als Mass des Ernahrungszustandes. *Muench. Med. Wochensch.* 68: 580–82. (*70*)

Rome, L. C. 1998. Matching muscle performance to changing demand. Pp. 103–13 in E. R. Weibel, C. R. Taylor, and L. Bolis, eds., *Principles of Animal Design*. Cambridge, UK: Cambridge University Press. (*456*)

Rose, C. W. 1966. *Agricultural Physics*. Oxford, UK: Pergamon Press. (*113*)

Roth, V. L., and J. M. Mercer. 2000. Morphometrics in development and evolution. *Am. Zool.* 40: 801–10. (*84*)

Rubega, M. A. 1997. Surface tension transport in shorebirds: how widespread is it? *Ibis* 139: 488–93. (*107*)

Rubega, M. A., and B. S. Obst. 1993. Surface-tension feeding in phalaropes: discovery of a novel feeding mechanism. *Auk* 110: 169–78. (*107*)

Rubenstein, D. I., and M.A.R. Koehl. 1977. The mechanisms of filter feeding: some theoretical considerations. *Am. Nat.* 111: 981–94. (*239*)

Rubin, C. T., and L. E. Lanyon. 1982. Limb mechanics as a function of speed and gait: a study of functional strains in the radius and tibia of horse and dog. *J. Exp. Biol.* 101: 187–211. (*507*)

Runham, N. W., P. R. Thornton, D. A. Shaw, and R. C. Wayte. 1969. Mineralization and hardness of the radular teeth of the limpet *Patella vulgata*. *Z. Zellforsch.* 99: 608–26. (*333*)

Ruppert, E. E., and K. J. Carle. 1983. Morphology of metazoan circulatory systems. *Zoomorphology* 103: 193–208. (*221*)

Salisbury, F. B., and C. Ross. 1969. *Plant Physiology*. Belmont, CA: Wadsworth Publishing Co. (*444*)

Salvadori, M. 1971. *Statics and Strength of Structures*. Englewood Cliffs, NJ: Prentice-Hall, Inc. (*387*)

Satter, R. L. 1979. Leaf movements and tendril curling. Pp. 442–484 in W. Haupt and M. E. Feinleib, eds., *Physiology of Movements*. New York: Springer-Verlag. (*422*)

Satterlie, R. A., M. LaBarbera, and A. N. Spencer. 1985. Swimming in the pteropod mollusc, *Clione limacina*. I. Behaviour and morphology. *J. Exp. Biol.* 116: 189–204. (*279*)

Schlessinger, W. H., J. T. Gray, D. S. Gill, and B. E. Mahall. 1982. *Ceanothus megacarpus* chaparral: a synthesis of ecosystem processes during development and annual growth. *Bot. Rev.* 48: 71–117. (*113*)

Schmidt-Nielsen, K. 1972. *How Animals Work*. Cambridge, UK: Cambridge University Press. (*95, 193*)

———. 1984. *Scaling: Why is Animal Size so Important?* Cambridge, UK: Cambridge University Press. (*59, 60, 87, 475*)
———. 1997. *Animal Physiology: Adaptation and Environment*, 5th ed. Cambridge, UK: Cambridge University Press. (*97, 181, 182*)
Scholander, P. F. 1954. Secretion of gases against high pressures in the swimbladder of deep sea fishes. 2. The rete mirabile. *Biol. Bull.* 107: 260–77. (*100*)
Scholander, P. F., H. T. Hammel, E. D. Bradstreet, and E. A. Hemmingsen. 1965. Sap pressure in vascular plants. *Science* 148: 339–46. (*103*)
Scholey, K. 1986. The climbing and gliding locomotion of the giant red flying squirrel, *Petaurista petaurista* (Sciuridae). Pp. 1–12 in W. Nachtigall, ed., *Biona Report 5: Bat Flight*. Heidelberg, Germany: Fischer. (*262*)
Schrodinger, E. 1944. *What is Life?* Cambridge, UK: Cambridge University Press. (*43, 176*)
Schulgasser, K., and A. Witztum. 1992. On the strength, stiffness, and stability of tubular plant stems and leaves. *J. Theor. Biol.* 155: 497–515. (*377*)
Seymour, R. S., A. R. Hargens, and T. J. Pedley. 1993. The heart works against gravity. *Am. J. Physiol.* 265(4): R715–20. (*219*)
Shadwick, R. E. 1992. Soft composites. Pp. 133–64 in J.F.V. Vincent, ed., *Biomechanics—Materials: A Practical Approach*. Oxford, UK: Oxford University Press. (*357*)
———. 1994. Mechanical organization of the mantle and circulatory system of cephalopods. *Mar. Behav. Physiol.* 25: 69–85. (*8, 218*)
Shadwick, R. E., and J. M. Gosline. 1985. Mechanical properties of the octopus aorta. *J. Exp. Biol.* 114: 259–84. (*218*)
———. 1994. Arterial mechanics in the fin whale suggest a unique hemodynamic design. *Am. J. Physiol.* 267: R805–18. (*225*)
Shadwick, R. E., C. M. Pollock, and S. A. Stricker. 1990. Structure and biomechanical properties of crustacean blood vessels. *Physiol. Zool.* 63: 90–101. (*218*)
Sheehan, S. 1986. *A Missing Plane*. New York: Putnam. (*70*)
Sherman, T. F. 1980. The cost of departure from optimal radii in microvascular networks. *J. Theor. Biol.* 136: 245–65. (*211*)
Shimeta, J., and P. A. Jumars. 1991. Physical mechanisms and rates of particle capture by suspension feeders. *Oceanogr. Mar. Biol. Annu. Rev.* 29: 191–257. (*239*)
Silvester, N. R., and M. A. Sleigh. 1985. The forces on microorganisms at surfaces in flowing water. *Freshwater Biol.* 15: 433–48. (*167, 170*)
Simmons, M. P., H. Ochoterena, and J. V. Freudenstein. 2002. Conflict between amino acid and nucleotide characteristics. *Cladistics* 18: 200–206. (*510*)
Slatyer, R. O. 1967. *Plant-Water Relationships*. London: Academic Press. (*184*)
Sleigh, M. A., and D. I. Barlow. 1976. Collection of food by *Vorticella*. *Trans. Am. Micros. Soc.* 95: 482–86. (*239*)
Sleigh, M. A., and J. R. Blake. 1977. Methods of ciliary propulsion and their size limitations. Pp. 243–56 in T. J. Pedley, ed., *Scale Effects in Animal Locomotion*. London: Academic Press. (*237*)
Smayda, T. J. 1970. The suspension and sinking of phytoplankton in the sea. *Annu. Rev. Oceanogr. Mar. Biol.* 8: 354–414. (*86*)
Smith, A. M. 1991. Negative pressure generated by octopus suckers: a study of the tensile strength of water in nature. *J. Exp. Biol.* 157: 259–72. (*426, 429*)
———. 1992. The role of suction in the adhesion of limpets. *J. Exp. Biol.* 161: 151–70. (*429*)
———. 1996. Cephalopod sucker design and the physical limits to negative pressure. *J. Exp. Biol.* 199: 949–58. (*103*)
Smith, A. M., W. M. Kier, and S. Johnsen. 1993. The effect of depth on the attachment force of limpets. *Biol. Bull.* 184: 338–41. (*427*)
Smith, L. S. 1963. Circulatory anatomy of the octopus arm. *J. Morphol.* 113: 261–66. (*210*)
Socha, J. J. 2002. Kinematics: gliding flight in the paradise tree snake. *Nature* 418: 603–4. (*262*)

Soluk, D. A., and D. A. Craig. 1990. Digging with a vortex: flow manipulation facilitates prey capture by a predatory stream mayfly. *Limnol. Oceanogr.* 35: 1201–6. (*136*)

Spedding, G. R. 1987. The wake of a kestrel (*Corvus monedula*) in slow flight. *J. Exp. Biol.* 125: 287–307. (*141, 277*)

Statzner, B., and T. F. Holm. 1982. Morphological adaptations of benthic invertebrates to stream flow—an old question studied by means of a new technique (laser-Doppler anemometry). *Oecologia* 53: 290–92. (*171*)

Steffensen, J. F. 1985. The transition between branchial pumping and ram ventilation in fishes: energetic consequences and dependence on water oxygen tension. *J. Exp. Biol.* 114: 141–150. (*152, 163*)

Stephenson, R., J. R. Lovvorn, M.R.A. Heieis, D. R. Jones, and R. W. Blake. 1989. Hydromechanical estimate of the power requirements of diving and surface swimming in lesser scaup, *Aythya affinis*. *J. Exp. Biol.* 147: 507–18. (*98*)

Stevens, E. D., and E. N. Lightfoot. 1986. Hydrodynamics of water flow in front of and through the gills of skipjack tuna. *Comp. Biochem. Physiol. A* 83: 255–59. (*191, 203*)

Steward, F. C. 1968. *Comp. Physiol. Growth and Organization in Plants*. Reading, MA: Addison-Wesley Publishing Co. (*449*)

Stone, H. A., and A.D.T. Samuel. 1996. Propulsion of microorganisms by surface distortions. *Phys. Rev. Lett.* 77: 4102–4. (*450*)

Stork, N. E. 1980. Experimental analysis of adhesion of *Chrysolina polita* (Chrysomelidae: Coleoptera) on a variety of surfaces. *J. Exp. Biol.* 88: 91–107. (*426, 429, 430*)

Strickler, J. R. 1975. Intra- and interspecific information flow among planktonic copepods: receptors. *Verh. Int. Ver. Limnol.* 19: 2951–58. (*229*)

———. 1977. Observation of swimming performances of planktonic copepods. *Limnol. Oceanogr.* 22: 165–70. (*230*)

———. 1985. Feeding currents in calanoid copepods: two new hypotheses. *Symp. Soc. Exp. Biol.* 39: 459–85. (*239*)

Stride, G. O. 1955. On the respiration of an aquatic African beetle, *Potamodytes tuberosus* Hinton. *Ann. Entomol. Soc. Am.* 48: 344–51. (*100, 153, 290*)

Suter, R. B., O. Rosenberg, S. Loeb, H. Wildman, and J. H. Long. 1997. Locomotion on the water surface: propulsive mechanisms of the fisher spider *Dolomedes triton*. *J. Exp. Biol.* 200: 2523–38. (*108*)

Swaine, M. D., and T. Beer. 1977. Explosive seed dispersal in *Hura crepitans*, L. (Euphorbiaceae). *New Phytol.* 78: 695–708. (*174*)

Swanson, C. P. 1973. *The Natural History of Man*. Englewood Cliffs, NJ: Prentice-Hall. (*15*)

Swartz, S. M. 1991. Strain analysis as a tool for functional morphology. *Am. Zool.* 31: 655–69. (*500*)

Swartz, S. M., M. B. Bennett, and D. R. Carrier. (1992) Wing bone stresses in free flying bats and the evolution of skeletal design for flight. *Nature* 359: 726–29. (*500*)

Swedmark, B. 1964. The interstitial fauna of marine sands. *Biol. Rev.* 39: 1–42. (*243*)

Tabor, D. 1979. *Gases, Liquids and Solids*, 2d ed. Cambridge, UK: Cambridge University Press. (*40, 182*)

Taghon, G. L., A.R.M. Nowell, and P. A. Jumars. 1980. Induction of suspension feeding in spionid polychaetes by high particulate fluxes. *Science* 210: 262–64. (*239*)

Tanaka, Y., and M. Hisada. 1980. The hydraulic mechanism of the predatory strike in dragonfly larvae. *J. Exp. Biol.* 88: 1–19. (*421*)

Tarkov, J. 1986. Engineering the Erie Canal. *Am. Heritage Invent. Technol.* 2(1): 50–57. (*35*)

Taylor, C. R. 1986. Carrying loads: the cost of generating muscular force. *News Physiol. Sci.* 1: 153–55. (*492*)

Taylor, C. R., S. L. Caldwell, and V. J. Rowntree. 1972. Running up and down hills: some consequences of size. *Science* 178: 1096–97. (*490*)

Taylor, C. R., G.M.O. Maloiy, E. R. Weibel, V. A. Langman, J.M.Z. Kamau, H. J. Seeherman, and N. C. Heglund. 1981. Design of the mammalian respiratory system. III. Scaling maximum aerobic capacity to body mass: wild and domestic mammals. *Respir. Physiol.* 44: 25–37. (*62*)

Taylor, G. M. 2000. Maximum force production: why are crabs so strong? *Proc. R. Soc. Lond. B* 267: 1475–80. (*456*)

Teale, E. W. 1949. *The Insect World of J. Henri Fabre*. New York: Dodd, Mead & Co. (*124*)

Telewski, F. W. 1995. Wind-induced physiological and developmental responses in trees. Pp. 237–63 in M. P. Coutts and J. Grace, eds., *Wind and Trees*. Cambridge, UK: Cambridge University Press. (*503*)

Telewski, F. W., and M. J. Jaffe. 1986. Thigmomorphogenesis: field and laboratory studies of *Abies fraseri* in response to wind or mechanical perturbation. *Physiol. Plant.* 66: 211–18. (*159, 503*)

Telford, M. 1983. An experimental analysis of lunule function in the sand dollar *Mellita quinquiesperforata*. *Mar. Biol.* 76: 125–34. (*254*)

———. 1985. Domes, arches and urchins: the skeletal architecture of echinoids (Echinodermata). *Zoomorphology* 105: 114–24. (*388*)

Tennekes, H. 1996. *The Simple Science of Flight: From Insects to Jumbo Jets*. Cambridge, MA: MIT Press. (*253, 259, 272, 278*)

Thom, A., and P. Swart. 1940. The forces on an aerofoil at very low speeds. *J. R. Aero. Soc.* 44: 761–70. (*252*)

Thomas, L. A., and C. O. Hermans. 1985. Adhesive interactions between the tube feet of a starfish, *Lepasterias hexactis*, and substrata. *Biol. Bull.* 169: 675–88. (*429*)

Thompson, D'Arcy W. 1942. *On Growth and Form*, 2d ed. Cambridge, UK: Cambridge University Press. (*16, 58, 69, 389, 393, 394, 412*)

Thompson, D'Arcy W. 1942. *On Growth and Form*. Abridged 1961, J. T. Bonner, ed. Cambridge, UK: Cambridge University Press. (*58*)

Thorington, R. W., Jr., and L. R. Heaney. 1981. Body proportions and gliding adaptations of flying squirrels (Petauristinae). *J. Mammal.* 62: 101–14. (*262*)

Thorp, W. H., and D. J. Crisp. 1947. Studies on plastron respiration. I. The biology of *Aphelocheirus* [Hemiptera, Aphelocheiridae (Naucoridae)] and the mechanism of plastron retention. *J. Exp. Biol.* 24: 227–69. (*113*)

Tidball, J. G., and T. L. Daniel. 1986. Elastic energy storage in rigored skeletal muscle cells under physiological loading conditions. *Am. J. Physiol.* 250: R56–64. (*473*)

Timoshenko, S., and S. Woinowsky-Krieger. 1959. *Theory of Plates and Shells*, 2d ed. New York: McGraw-Hill. (*387*)

Trager, G. C., J.-S. Hwang, and J. R. Strickler. 1990. Barnacle suspension feeding in variable flow. *Mar. Biol.* 105: 117–27. (*238*)

Trefil, J. S. 1984. *A Scientist at the Seashore*. New York: Scribners. (*285*)

Treloar, L.R.G. 1975. *The Physics of Rubber Elasticity*, 3d ed. Oxford, UK: Oxford University Press. (*328*)

Triantafyllou, M. S., G. S. Triantafyllou, and D.K.P. Yue. 2000. Hydrodynamics of fish-like swimming. *Annu. Rev. Fluid Mech.* 32: 33–53. (*282*)

Tricker, R.A.R., and B.J.K. Tricker. 1967. *The Science of Movement*. New York: American Elsevier. (*465*)

Tskhovrebova, L, J. Trinick, J. Sleep, and R. Simmons. 1997. Elasticity and unfolding of single molecules of the giant protein titin. *Nature* 387: 308–12. (*473*)

Tu, M. S., and M. H. Dickinson. 1996. The control of wing kinematics by two steering muscles of the blowfly (*Calliphora vicina*). *J. Comp. Physiol. A* 178: 813–30. (*472*)

Tucker, V. A. 1968. Respiratory exchange and evaporative water loss in the flying budgerigar. *J. Exp. Biol.* 75: 223–29. (*277*)

———. 1969. Wave-making by whirligig beetles (Gyrinidae). *Science* 166: 897–99. (*288, 293*)

———. 1975. The energetic cost of moving about. *Am. Sci.* 63: 413–19. (*282, 483*)

———. 2000. Gliding flight: drag and torque of a hawk and a falcon with straight and turned heads, and a lower value for the parasite drag coefficient. *J. Exp. Biol.* 203: 3733–44. (*161*)

Tucker, V. A., and G. C. Parrot. 1970. Aerodynamics of gliding flight in a falcon and other birds. *J. Exp. Biol.* 52: 345–67. (*257*)

Turner, J. S. 2000. *The Extended Organism: The Physiology of Animal-Built Structures.* Cambridge, MA: Harvard University Press. (*149*)

Tyree, M. T., and F. W. Ewers. 1991. The hydraulic architecture of trees and other woody plants. *New Phytol.* 119: 345–60. (*103*)

Tyree, M. T., and J. S. Sperry. 1988. Do woody plants operate near the point of catastrophic xylem disfunction caused by dynamic water stress? Answers from a model. *Plant Physiol.* 88: 574–80. (*103*)

Vallee, R. B. 1998. *Molecular Motors and the Cytoskeleton.* San Diego, CA: Academic Press. (*447*)

van der Wal, P., J. J. Videler, P. Havinga, and R. Pel. 1989. Architecture and chemical composition of the magnetite-bearing layer in the radula teeth of *Chiton olivaceus* (Polyplacophora). Pp. 153–66 in R. E. Crick, ed., *Origin, Evolution, and Modern Aspects of Biomineralization in Plants and Animals.* New York: Plenum Press. (*333*)

Van Dyke, M. 1982. *An Album of Fluid Motion.* Stanford, CA: Parabolic Press. (*117*)

Van Tuinen, M., D. B. Butvill, J.A.W. Kirsch, and S. B. Hedges. 2001. Convergence and divergence in the evolution of wading birds. *Proc. R. Soc. Lond. B* 268: 1345–50. (*510*)

Videler, J. J., and F. Hess. 1984. Fast continuous swimming of two pelagic predators, saith (*Pollachius virens*) and mackerel (*Scomber scombrus*): a kinematic analysis. *J. Exp. Biol.* 109: 209–28. (*143*)

Vincent, J.F.V. 1975. Locust oviposition: stress softening of the extensible intersegmental membranes. *Proc. R. Soc. Lond. B* 188: 189–201. (*314*)

———. 1982. The mechanical design of grass. *J. Mater. Sci.* 17: 856–60. (*340*)

———. 1990. *Structural Biomaterials*, rev. ed. Princeton, NJ: Princeton University Press. (*305, 307, 310, 313, 314, 317, 318, 319, 321, 328, 333, 340, 341, 354, 360, 361*)

Vincent, J.F.V., and G. Jeronimidis. 1991. The mechanical design of fossil plants. Pp. 21–36 in J.M.V. Rayner and R. J. Wooton, eds., *Biomechanics and Evolution.* Cambridge, UK: Cambridge University Press. (*413*)

Vincent, J.F.V., and P. Owers. 1986. Mechanical design of hedgehog spines and porcupine quills. *J. Zool. Lond. A* 210: 55–75. (*44, 77*)

Vogel, S. 1967. Flight in *Drosophila.* III. Aerodynamic characteristics of fly wings and wing models. *J. Exp. Biol.* 46: 431–43. (*252*)

———. 1970. Convective cooling at low speeds and the shape of broad leaves. *J. Exp. Bot.* 21: 91–101. (*84*)

———. 1977. Current-induced flow through sponges *in situ. Proc. Natl. Acad. Sci. U.S.A.* 74: 2069–71. (*149*)

———. 1978a. Organisms that capture currents. *Sci. Am.* 239(2): 128–39. (*149*)

———. 1978b. Evidence for one-way valves in the water-flow system of sponges. *J. Exp. Biol.* 76: 137–48. (*203*)

———. 1981a. Behavior and the physical world of an animal. Pp. 179–98 in P.P.G. Bateson and P. H. Klopfer, eds., *Perspectives in Ethology*, vol. 4. New York: Plenum Press. (*22*)

———. 1981b. *Life in Moving Fluids.* Boston: Willard Grant Press; Princeton, NJ: Princeton University Press. (*117*)

———. 1983. How much air flows through a silkmoth's antenna? *J. Insect Physiol.* 29: 597–602. (*125, 502*)

———. 1984. Drag and flexibility in sessile organisms. *Am. Zool.* 24: 37–44. (*159, 506*)

———. 1987. Flow-induced mantle cavity refilling in jetting squid. *Biol. Bull.* 172: 61–68. (*126, 153*)

———. 1989. Drag and reconfiguration of broad leaves in high winds. *J. Exp. Bot.* 40: 941–48. (*9, 382, 385, 506*)

———. 1994a. *Life in Moving Fluids*, 2d ed. Princeton, NJ: Princeton University Press. (*108, 159, 164, 168, 242, 271*)

———. 1994b. Dealing honestly with diffusion. *Am. Biol. Teacher* 56: 405–7. (*182*)

———. 1995a. Pressure *versus* flow in biological pumps. Pp. 297–304 in C. P. Ellington, and T. J. Pedley, eds., *Biological Fluid Dynamics: Symposium of the Society for Experimental Biology 49*.Cambridge, UK: The Company of Biologists, Ltd. (*201, 218*)

———. 1995b. Twist-to-bend ratios of woody structures. *J. Exp. Bot.* 46: 981–85. (*310, 321, 323, 354, 382, 385*)

———. 1998a. *Cats' Paws and Catapults: Mechanical Worlds of Nature and People*. New York: W. W. Norton. (*11, 158, 424, 458*)

———. 1998b. Exposing life's limits with dimensionless numbers. *Phys. Today* 51(11): 22–27. (*72, 290, 482*)

———. 2001. *Prime Mover: A Natural History of Muscle*. New York: W. W. Norton. (*347, 451*)

Vogel, S., C. P. Ellington, Jr., and D. C. Kilgore, Jr. 1973. Wind-induced ventilation of the burrow of the prairie dog *Cynomys ludovicianus*. *J. Comp. Physiol.* 85: 1–14. (*126, 149*)

Vogel, S., and C. Loudon. 1985. Fluid mechanics of the thallus of an intertidal red alga, *Halosaccion glandiforme*. *Biol. Bull.* 168: 161–74. (*158*)

Vogel, S., and M. N. Papanicolaou. 1982. A constant-stress creep testing machine. *J. Biomech.* 16: 153–56. (*356*)

Vogel, S., and S.A. Wainwright. 1969. *A Functional Bestiary*. Reading, MA: Addison-Wesley Publishing Co. (*53*)

Vollrath, F., and D. P. Knight. 2001. Liquid crystalline spinning of spider silk. *Nature* 410: 541–48. (*328, 344*)

Voltzow, J. 1985. Morphology of the pedal circulatory system of the marine gastropod *Busycon contrarium* and its role in locomotion (Gastropoda, Buccinacea). *Zoomorphology* 105: 395–400. (*221*)

———. 1990. The functional morphology of the pedal musculature of the marine gastropods *Busycon contrarium* and *Haliotis kamtschatkana*. *Veliger* 33: 1–19. (*420*)

Von Kármán, T. 1954. *Aerodynamics*. Ithaca, NY: Cornell University Press. (*158, 166*)

Wainwright, S. A. 1988. *Axis and Circumference*. Cambridge, MA: Harvard University Press. (*440*)

Wainwright, S. A., W. D. Biggs, J. D. Currey, and J. M. Gosline. 1976. *Mechanical Design in Organisms*. London, UK: Edward Arnold Reprint, Princeton, NJ: Princeton University Press, 1982. (*302, 305, 310, 312, 313, 314, 316, 317, 321, 328, 341, 344, 354, 378, 395, 403, 410, 413, 416*)

Wainwright, S. A., F. Vosburgh, and J. H. Hebrank. 1978. Shark skin: function in locomotion. *Science* 202: 747–49. (*415*)

Walker, J. A., and M. W. Westneat. 2000. Mechanical performance of aquatic rowing and flying. *Proc. R. Soc. Lond. B* 267: 1875–81. (*283*)

Wallace, J. B., and R. W. Merritt. 1980. Filter-feeding ecology of aquatic insects. *Annu. Rev. Entomol.* 25: 103–32. (*147*)

Walsby, A. E., and A. Xypolyta. 1977. The form resistance of the chitan fibers attached to the cells of *Thalassiosira fluviatilis* Hustedt. *Br. Phycol. J.* 12: 215–23. (*234*)

Ward, D. V., and S. A. Wainwright. 1972. Locomotory aspects of squid mantle structure. *J. Zool. (Lond.)* 167: 437–49. (*415*)

Washburn, J. O., and L. Washburn. 1984. Active aerial dispersal of minute wingless arthropods: exploitation of boundary layer velocity gradients. *Science* 223: 1088–89. (*234*)

Wassersug, R. J., and K. v.S. Hoff. 1985. The kinematics of swimming in anuran larvae. *J. Exp. Biol.* 119: 1–30. (*143*)

Watabe, S., and D. J. Hartshorne. 1990. Paramyosin and the catch mechanism. *Comp. Biochem. Physiol. B* 96: 639–46. (*462*)

Waterbury, J. B., J. M. Willey, D. G. Franks, F. W. Valois, and S. W. Watson. 1985. A cyanobacterium capable of swimming motility. *Science* 230: 74–75. (*450*)

Waters, N. E. 1980. Some mechanical and physical properties of teeth. Pp. 99–135 in J.F.V. Vincent and J. D. Currey, eds., *The Mechanical Properties of Biological Materials: Symposium of the Society for Experimental Biology* 34. Cambridge, UK: Cambridge University Press. (*310, 332, 340*)

Wauthy, G., M. Leponce, N. Banaï, G. Sylin, and J.-C. Lions. 1998. The backward jump of a box moss mite. *Proc. R. Soc. Lond. B* 265: 2235–42. (*247*)

Webb, P. W. 1974. Efficiency of pectoral fin propulsion of *Cymatogaster aggregata*. Pp. 573–84 in T.-Y. Wu, C. J. Brokaw, and C. Brennen, eds., *Swimming and Flying in Nature*, vol. 2. New York: Plenum Press. (*143, 281*)

———. 1975. Hydrodynamics and energetics of fish propulsion. *Bull. Fish. Res. Board. Can.* 190: 1–158. (*278*)

———. 1976. The effect of size on the fast-start performance of rainbow trout *Salmo gairdneri* and a consideration of piscivorous predator-prey interactions. *J. Exp. Biol.* 65: 157–77. (*475*)

———. 1979. Mechanics of escape response in crayfish (*Orconectes virilis*). *J. Exp. Biol.* 79: 245–63. (*283*)

———. 1988. Steady swimming kinematics of tiger musky, an esociform accelerator, and rainbow trout, a generalist cruiser. *J. Exp. Biol.* 138: 51–70. (*143*)

Webb, P. W., and D. Weihs. 1994. Hydrostatic stability of fish with swimbladders: not all fish are unstable. *Can J. Zool.* 72: 1149–54. (*115*)

Webber, D. M., and R. K. O'Dor. 1986. Monitoring the metabolic rate and activity of free-swimming squid with telemetered jet pressure. *J. Exp. Biol.* 126: 205–24. (*143*)

Weber, W. E., and E. Weber. 1825. *Mechanik der Menschlichen Gehwerkzeuge*. Trans. 1992, P. G. Maquet and R. Furlong. Berlin: Springer-Verlag. (*3*)

Weibel, E. R. 2000. *Symmorphosis: On Form and Function in Shaping Life*. Cambridge, MA: Harvard University Press. (*211, 222*)

Weibel, E. R., and D. M. Gomez. 1962. Architecture of the human lung. *Science* 137: 577–85. (*213*)

Weibel, E. R., C. R. Taylor, and L. Bolis, eds. 1998. *Principles of Animal Design*. Cambridge, UK: Cambridge University Press. (*501*)

Weihs, D. 1977. Periodic jet propulsion of aquatic creatures. *Fortschr. Zool.* 24(2–3): 171–75. (*279*)

Weihs, D., and G. Katzir. 1994. Bill sweeping in the spoonbill, *Platalea leucordia: evidence for a hydrodynamic function. Anim. Behav.* 47: 649–54. (*251*)

Weir, J. S. 1973. Air flow, evaporation and mineral accumulation in mounds of *Macrotermes subhyalinus* (Rambur). *J. Anim. Ecol.* 42: 509–20. (*149*)

Weis-Fogh, T. 1960. A rubber-like protein in insect cuticle. *J. Exp. Biol.* 37: 889–907. (*80*)

———. 1973. Quick estimates of flight fitness in hovering animals, including novel mechanisms for lift production. *J. Exp. Biol.* 59: 169–230. (*276*)

Weis-Fogh, T., and R. M. Alexander. 1977. The sustained power output from striated muscle. Pp. 511–25 in T. J. Pedley, ed., *Scale Effects in Animal Locomotion*. London: Academic Press. (*467*)

Weis-Fogh, T., and W. B. Amos. 1972. Evidence for a new mechanism of cell motility. *Nature* 236: 301–4. (*230*)

Wells, M. J. 1983. Circulation in cephalopods. Pp. 239–90 in K. M. Wilbur, ed., *The Mollusca*, vol. 5. New York: Academic Press. (*203*)

Went, F. W. 1968. The size of man. *Am. Sci.* 56: 400–13. (*12, 56*)

West, L. J., C. M. Pierce, and W. D. Thomas. 1962. Lysergic acid diethylamide: its effects on a male Asiatic elephant. *Science* 138: 1100–1103. (*63*)

Westneat, M. W. 1990. Feeding mechanics of teleost fishes (Labridae; Perciformes): a test of four-bar linkage models. *J. Morphol.* 205: 269–95. (*63*)

Westneat, M. W. 1991. Linkage biomechanics and evolution of the unique feeding mechanism of *Epibulus insidiator* (Labridae:Teleostei). *J. Exp. Biol.* 159: 165–84. (*401*)

Westneat, M. W., W. Hoese, C. A. Pell, and S. A. Wainwright. 1993. The horizontal septum: mechanisms of force transfer in locomotion of scombrid fishes (Scombridae, Perciformes). *J. Morphol.* 217: 183–204. (*470*)

Westneat, M. W., and P. C. Wainwright. (1989) Feeding mechanism of *Epibulus insidiator* (Labridae; Teleostei): evolution of a novel functional system. *J. Morphol.* 202: 129–150. (*401*)

Wetmore, K. L. 1987. Correlations between test strength, morphology and habitat in some benthic foraminifera from the coast of Washington. *J. Foraminferal Res.* 17: 1–13. (*498*)

White, C. M. 1946. The drag of cylinders in fluids at low speeds. *Proc. R. Soc. Lond. A* 186: 472–79. (*229*)

Whitt, F. R., and D. G. Wilson. 1982. *Bicycling Science*, 2d ed. Cambridge, MA: MIT Press. (*483*)

Wiesner, K. B. 1988. The role of damping systems. Pp. 789–802 in L. S. Beedle, ed., *Second Century of the Skyscraper*. New York: Van Nostrand Reinhold Co. (*382*)

Wigglesworth, V. B. 1987. How does a fly cling to the undersurface of a glass sheet? *J. Exp. Biol.* 129: 373–76. (*427*)

Wilcox, R. S. 1979. Sex discrimination in *Gerris remigis: role of a surface wave signal. Science* 206: 1325–27. (*288*)

Wildish, D., and D. Kristmanson. 1997. *Benthic Suspension Feeders and Flow*. Cambridge, UK: Cambridge University Press. (*238*)

Wilkie, D. R. 1950. The relation between force and velocity in human muscle. *J. Physiol.* 110: 249–80. (*456*)

———. 1960. Man as a source of mechanical power. *Ergonomics* 3: 1–8. (*457*)

Wilkie, I. C. 2002. Is muscle involved in the mechanical adaptability of echinoderm mutable connective tissue? *J. Exp. Biol.* 205: 159–65. (*438*)

Williams, T. A. 1991. A model of rowing propulsion and the ontogeny of locomotion in *Artemia* larvae. *Am. Zool.* 31: 60A. (*229*)

Williams, T. A., W. A. Friedl, M. L. Fong, R. M. Yamada, P. Sedivy, and J. E. Haun. 1992. Travel at low energetic cost by swimming and wave-riding bottlenose dolphins. *Nature* 355: 821–23. (*294*)

Wilson, A. M., M. P. McGuigan, A. Su, and A. J. van den Bogert. 2001. Horses damp the spring in their step. *Nature* 414: 895–99. (*473*)

Wilson, R. P., K. Hustler, P. G. Ryan, A. E. Burger, and E. C. Nöldeke. 1992. Diving birds in cold water: do Archimedes and Boyle determine energetic costs? *Am. Nat.* 140: 179–200. (*98*)

Wilt, F. 1968. Training for competitive running. Pp. 395–414 in H. B. Falls, ed., *Exercise Physiology*. New York: Academic Press. (*458*)

Withers, P. C. 1979. Aerodynamics and hydrodynamics of the 'hovering' flight of Wilson's storm petrel. *J. Exp. Biol.* 80: 83–91. (*264*)

Witztum, A., and K. Schulgasser. 1995. The mechanics of seed expulsion in Acanthaceae. *J. Theor. Biol.* 176: 531–42. (*444*)

Worsley, F. A. 1977. *The Great Antarctic Rescue: Shackleton's Boat Journey*. New York: W. W. Norton. (*265*)

Wu, T.Y.-T. 1977. Hydrodynamics of swimming at low Reynolds numbers. *Fortschr. Zool.* 24(2–3): 149–69. (*229*)

Young, C. M., and L. F. Braithwaite. 1980. Orientation and current-induced flow in the stalked ascidian *Styela montereyensis*. *Biol. Bull.* 159: 428–40. (*147, 403*)

Young, T. 1809. On the functions of the heart and arteries. *Philos. Trans. R. Soc. Lond.* 1809: 1–31. (*211*)

Zimmermann, M. H. 1971. Transport in the xylem. Pp. 169–220 in M. H. Zimmermann and C. L. Brown, eds., *Trees: Structure and Function.* New York: Springer-Verlag. (*203*)

SUBJECT INDEX

Entries in **boldface** refer to data; entries in *italics* refer to figures.